3rd International Symposium on Mechanics, Structures and Materials Science (MSMS 2022)

MATEC Web of Conferences Volume 358 (2022)

Tianjin, China
23 - 24 April 2022

Editors:

Jun Xiao
Xinwang Wu

ISBN: 978-1-7138-5350-3

Printed from e-media with permission by:

Curran Associates, Inc.
57 Morehouse Lane
Red Hook, NY 12571

proceedings.com

Some format issues inherent in the e-media version may also appear in this print version.

Printed with permission by Curran Associates, Inc. (2022)

For additional information, please contact EDP Sciences – Web of Conferences at the address below.

EDP Sciences – Web of Conferences
17, Avenue du Hoggar
Parc d'Activité de Courtabœuf
BP 112
F-91944 Les Ulis Cedex A
France

Phone: +33 (0) 1 69 18 75 75
Fax: +33 (0) 1 69 28 84 91

contact-edps@webofconferences.org

Additional copies of this publication are available from:

Curran Associates, Inc.
57 Morehouse Lane
Red Hook, NY 12571 USA
Phone: 845-758-0400
Fax: 845-758-2633
Email: curran@proceedings.com
Web: www.proceedings.com

TABLE OF CONTENTS

Preface: 3rd International Symposium on Mechanics, Structures and Materials Science (MSMS 2022)

MSMS 2022 includes original and peer-reviewed research papers from 3rd International Symposium on Mechanics, Structures and Materials Science (MSMS 2022), which held on April 23-24, 2022 in Tianjin City, China.

The mechanics, structures and materials science are experiencing considerable growth a few years ago, which involve the development of new technology requiring multidisciplinary simulation. The primary objective of this conference is to exchange knowledge among researchers, practicing engineers, technologists about the state of art in mechanics, structures, materials science and their applications. MSMS stimulates this growth by emphasizing fundamental advances that are relevant in dealing with problems of all length scales.

MSMS 2022 received 107manuscripts. And 68 submissions had been accepted by our reviewers. By submitting a paper to MSMS 2022, the authors agreed to the review process and understood that papers undergo a peer-review process. Manuscripts were reviewed by appropriately qualified experts in the field selected by the Conference Committee, who gave detailed comments and-if the submission was accepted the authors would submit a revised version that took into account this feedback. All papers were reviewed using a double-blind review process: authors declared their names and affiliations in the manuscript for the reviewers to see, but reviewers did not know each other's identities, nor did the authors receive information about who had reviewed their manuscript. The Committees of MSMS 2022 invested great efforts in reviewing the papers submitted to the conference and organizing the sessions to enable the participants to gain maximum benefit.

With our warmest regards,

Jun Xiao

Conference Organizing Chair
Tianjin, China

Statement of Peer review

In submitting conference proceedings to *Web of Conferences*, the editors of the proceedings certify to the Publisher that

1. They adhere to its **Policy on Publishing Integrity** in order to safeguard good scientific practice in publishing.
2. All articles have been subjected to peer review administered by the proceedings editors.
3. Reviews have been conducted by expert referees, who have been requested to provide unbiased and constructive comments aimed, whenever possible, at improving the work.
4. Proceedings editors have taken all reasonable steps to ensure the quality of the materials they publish and their decision to accept or reject a paper for publication has been based only on the merits of the work and the relevance to the journal.

Title, date and place of the conference

3rd International Symposium on Mechanics, Structures and Materials Science (MSMS 2022)
April 23-24, 2022
Tianjin, China

Proceedings editor(s):

Jun Xiao
Xinwang Wu

Date and editor's signature May 12, 2022

The Effect of Nb Content on Microstructure Evolution of GH4169 superalloy During Hot Deformation

Wen Chen [1], Peng Zhang [1], Jing Xie [1], Yang Liu [1] and Wenxia Zhao [2,*]

[1]China National Erzhong Group Deyang Wanhang Die Forging Co., LTD., Deyang 618000, Sichuan, China
[2] AECC Beijing Institute of Aeronautical Materials, Beijing 100095, China

Abstract: This research conducted a thermal compression deformation test on two GH4169 alloy samples with 5.40wt% and 5.21wt% Nb contents respectively, under deformation temperatures ranging from 900°C to 1030°C and the strain rates of 0.04s-1, 0.08s-1. Data from the deformation test were used to study the effects of deformation temperature and strain rate on the microstructure, δ phase morphology, and δ phase evolution of both alloy samples, so as to reveal the structural evolution mechanism. The results show that the long needle-like δ phase in the alloy microstructure with high Nb content undergoes deformation fracture and decomposition fracture during deformation and compression under high temperature, with the fracture decomposition temperature being 990°C. Moreover, microstructures of alloys with low Nb content were dominated by granular δ phases; as the deformation temperature increases or the strain rate decreases, the content of the δ phase gradually decreases, causing the alloy's dynamic recrystallization grain size and dynamic recrystallization volume fraction to gradually increase from deformation twins at low temperatures to complete recrystallization at high temperatures.

Keywords: GH4169 superalloy, Nb content, δ phase, Hot deformation, dynamic recrystallization

1. Introduction

GH4169 is a nickel-iron-based superalloy used at medium and high temperatures. It is an important material for aero-engine turbine discs[1-3]. Since turbine disks work under high temperatures and in high-stress environments, it is essential to ensure high-temperature strength and high resistance to low cycle fatigue to obtain a uniform and fine microstructure[4-8]. Generally, turbine disks are manufactured through multi-stage thermal deformation processes, and it is difficult to obtain a fine microstructure through satisfactory recrystallization and less grain growth. In GH4169 alloy, the body-centered tetragonal γ" phases (Ni3Nb) are the main strengthening phases, supplemented by the face-centered cubic structure γ' (Ni3AlTi), and δ phases of the orthogonal structure (Ni3Nb). The structure and properties of GH4169 alloy are very sensitive to the hot working process. Improper process control will cause coarse and mixed crystals, which will affect the long-lasting notch sensitivity, the toughness of the forging, etc.

The mechanical properties of GH4169 are very sensitive to the changes in microstructure, while the microstructure of the alloy is directly affected by the chemical composition[9-12]. The addition of Nb not only affects the GH4169 alloy's γ" phases, i.e., the strengthening phases, but also affects the δ phases. According to existing research, given that the Nb element content is 4.80wt%, and the static complete resolubilization temperature of δ phase is 990°C, but when the Nb element content is close to 5.50wt%, the static complete resolubilization temperature of the δ phase exceeds 1,030°C. Because the δ phases in GH4169 alloy can hinder the growth of grains by pinning the grain boundaries, a suitable temperature needs to be selected as the pretreatment temperature for large forgings. Maintaining a good microstructure state of the forging billet is the basis for preparing a turbine disk with excellent performance, and the selection of the temperature is mainly based on the patterns of dissolution and re-dissolution of the δ phases.

At present, research has been conducted on the dissolution behavior of δ phases during high-temperature deformation of GH4169 alloy, but there is no in-depth discussion on the evolution mechanism of δ phases during deformation and the effect of δ phase evolution on the high-temperature deformation mechanism of alloy[13,14]. In particular, the number of research conducted on the effect of Nb element content on the changes of alloy microstructure during deformation remains scarce[15,16]. Systematically studying the evolution process and structural characteristics of the δ phases in alloys with different Nb element contents is not only important for offering formulation guidance in the hot working process of GH4169 alloys, but also for providing a theoretical basis for the precise control of deformation structure and performance.

* Corresponding author: wenxia.zhao@outlook.com

2. Experimental Procedures

2.1 Materials and Methods

In this study, the material used was an alloy forging bar made from Fushun special steel with a diameter of 220 mm. Two contents of the Nb element were selected, and the chemical compositions (wt%) of the alloys are as shown in Table 1. A thermal simulation test sample was cut along the circumferential direction (equal radius) and machined into ϕ10mm×15mm.

Table 1 Chemical composition of test sample (wt%)

Element	C	Ni	Cr	Nb	Mo	Ti	Al	Mn	Si	B	Fe
HNB	0.030	54.20	18.32	5.41	2.95	0.96	0.52	0.080	0.060	0.005	Bal
LNB	0.030	54.18	18.13	5.20	2.95	0.98	0.54	0.080	0.070	0.005	Bal

The thermal compression test was carried out on the computer-controlled Gleeble-1500 thermal simulation test machine. Both ends of the specimen were lubricated with high-temperature glass lubricant. The deformation temperatures selected were 900°C, 930°C, 960°C, 990°C, 1020°C, the strain rates were 0.04 s-1, 0.08 s-1, and the strain exerted was 50%. After the sample was compressed to the status of the specified amount of strain, it was quickly water quenched to retain its high-temperature deformation structure. The compressed sample was cut along the axial direction using wire EDM, and then corroded with a mixed solution of 50mL H2O + 40mL HCl + 10mL HF after performing mechanical polishing. The microstructure was observed using OLYMPUS GX51 optical microscope and NanoNova450 scanning electron microscope.

2.2 The Effect of Deformation Temperature on Microstructure

The cross section of the deformed sample is shown in Figure 1, as can be seen that the deformation and temperature of core area is higher than edge aera, and microstructure of the core aera can reflect the real state under high temperature.

At 0.04 s-1 strain rate, 50% deformation, the volume fractions of the δ phases at the heart of the high-Nb-content alloy (HNB) were about 17.8%, 16.6%, 14.3%, 7.6%, and 0.67% respectively; the volume fractions of the δ phases at the heart of the low Nb content alloy (LNB) were about 3.8%, 1.9%, 1.6%, 0.2%, 0%, respectively. Microstructure of specimen during hot deformation at different temperature were shown in Fig. 2 and Fig. 3.

Zhang and Wang et al. [17,18] found that during high-temperature thermal compression and deformation, since the δ phases and the matrix γ phases are in a non-coherent relationship, it is not easy for moving dislocations to cut the δ phases, and the cutter tends to bend and bypass the δ phases. Also, the δ phases in the test alloy nucleate at the grain boundary and appear mainly in the form of short rods/long strips. During deformation, for the short rod-shaped δ phases, the movement dislocation occurs mainly via the bypass mechanism, but for clusters/long strip δ phases, it is difficult for moving dislocations to bypass and advance, causing an instance of piling up to occur near the δ phases, resulting in stress concentration. When the stress reaches the fracture limit of the δ phase, the cluster/long strip δ phases decomposes and fractures. Cai et al. [19] found that long strip δ phases show dissolution fracture when studying the static dissolution behavior of the δ phases. The reason is that under high-temperature conditions, a relatively large tension is generated on the interface between the phase and the matrix, which destroys the original interface stress balance, and creates grooves at the interface. Compared with the plane, the curvature radius of the curved surfaces on both sides of the groove is smaller, while the phases at the groove curved surface can be preferentially dissolved, which destroys the interfacial tension balance. To maintain balance, the groove needs to be further deepened, and this process needs to be repeated until the δ phases dissolve and break [20]. It can be seen that during compression deformation, the strain applied from the outside increases the subgrain boundary or high-density dislocation area inside the long strip δ phases, which in turn leads to the deformation and fracture of the δ phases. meanwhile, a large amount of moving dislocations accumulate near the long needle-shaped δ phases, and under the effect of the dislocation stress field, a fast path is provided for the diffusion of niobium atoms in the δ phases, while the accelerated diffusion of niobium promotes the dissolution and fracture of the long strip δ phases during the deformation process. It is precisely due to the combined effect of deformation fracture and dissolution fracture that the long needle-like δ phases spheroidize and decompose during high-temperature compression and deformation, transforming themselves into granular δ phases.

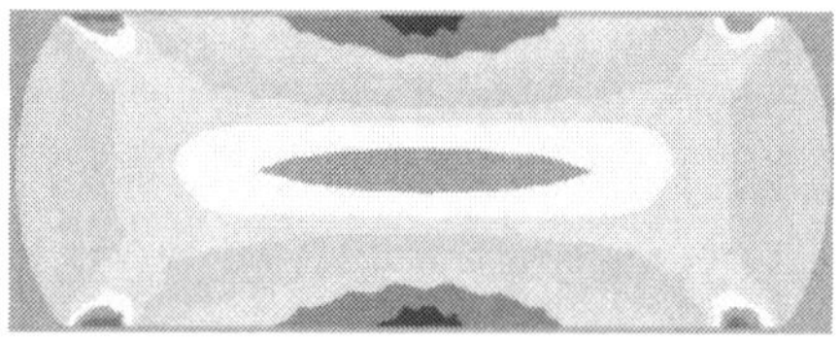

Figure 1. Cross-sectional schematic diagram

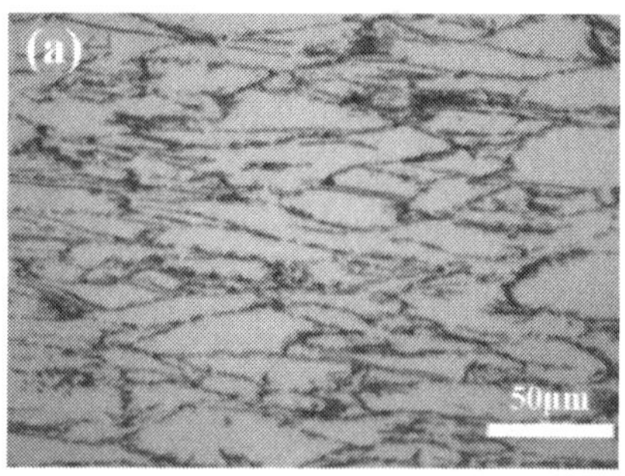

https://doi.org/10.1051/matecconf/202235801001

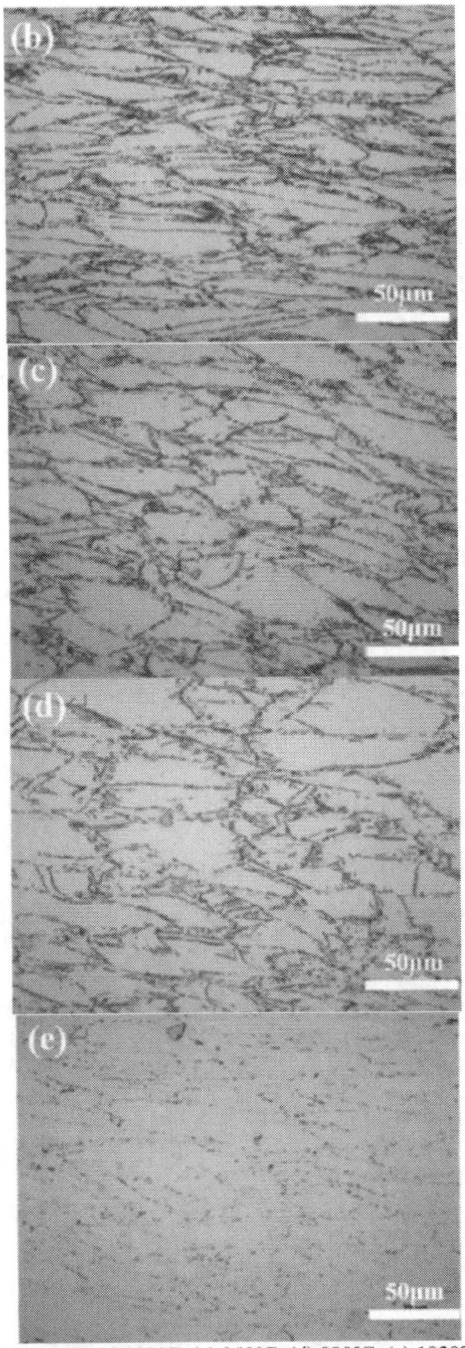

(a) 900℃; (b) 930℃; (c) 960℃; (d) 990℃; (e) 1020℃

Figure 2. Fiber structure of HNB samples at different temperatures under the deformation rate of $0.04s^{-1}$

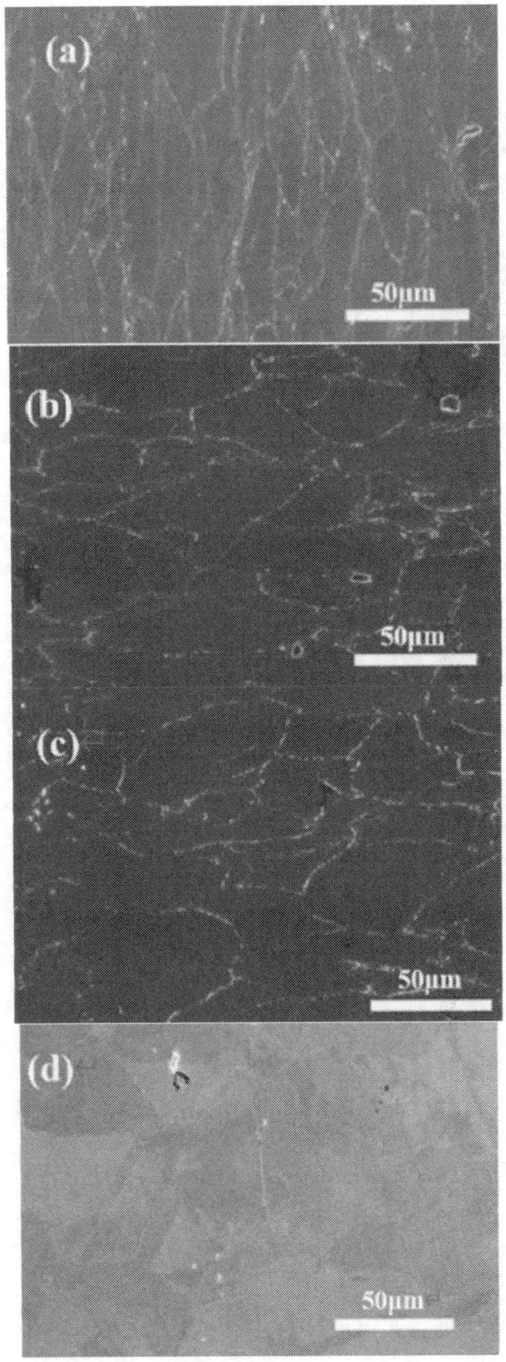

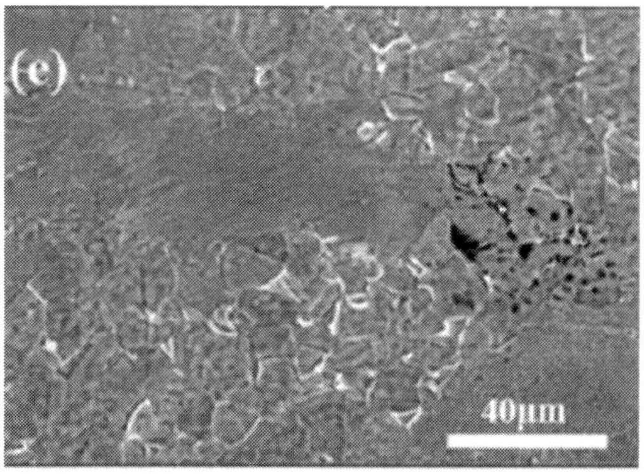

(a) 900°C; (b) 930°C; (c) 960°C; (d) 990°C; (e) 1020°C

Figure 3. Fiber structure of LNB samples at different temperatures under the deformation rate of $0.04s^{-1}$

2.3 The Effect of Strain Rate on Microstructure

After deformation at 990°C and under different strain rates, the microstructures of GH4169 alloy are similar were shown in Fig.4 and Fig.5, which are showing equiaxed grains formed by recrystallization, most of the long strip δ phases are obviously bent, twisted or broken, deformed and fractured, as well as dissolved and fractured, in addition to the obvious appearance of strip-like structures along the direction of applied force. However, in the microstructure deformed at the low strain rate, the content of the unbroken long strip δ phases is lesser than that at the high strain rate, the content of granular δ phases is larger, and the structure is more uniform, indicating that a low strain rate is conducive to the formation of a good deformed structure. The reason is that under deformation at high temperature, a low strain rate can prolong the deformation process, so that the dislocation cross slip and diffusion climbing process can be fully carried out, which is conducive to the progress of dynamic recrystallization and the decomposition as well as spheroidization of the long strip δ phases, thereby forming a uniform deformed structure.

The content of δ phases in GH4169 alloy with high Nb content gradually decreases, but the δ phases are not completely re-dissolved until 1020°C and are distributed as discontinuous particles on the grain boundary, increasing the dynamic recrystallization grain size of the alloy from ASTM 7.5 to ASTM 7.0. the δ phases in the low Nb content GH4169 alloy are generally re-dissolved at 990°C, and the dynamic recrystallization grain size of the alloy grows from ASTM7.5 to ASTM6.0.

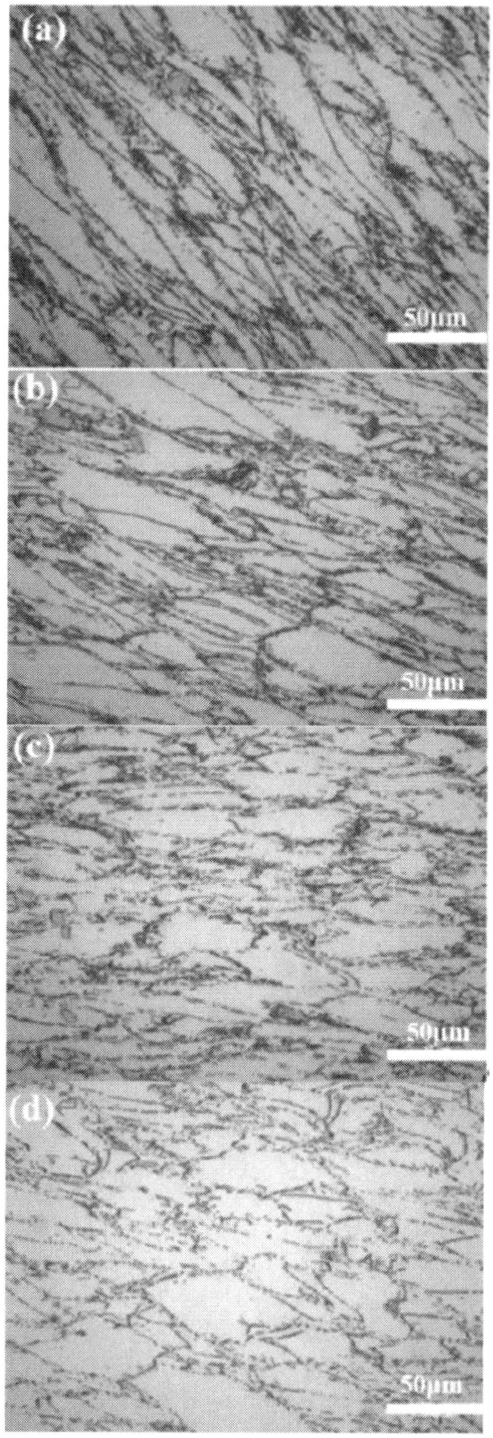

 https://doi.org/10.1051/matecconf/202235801001

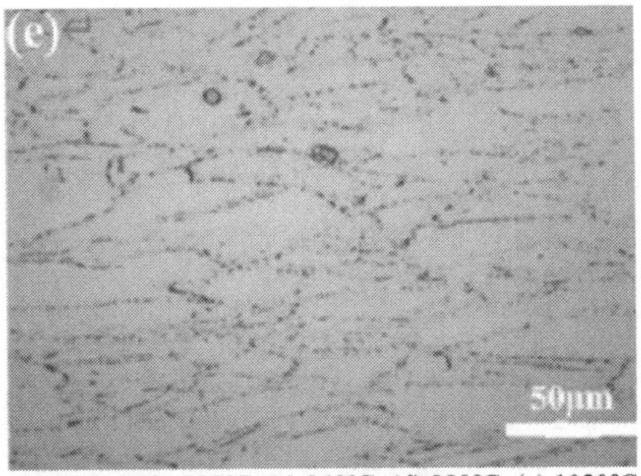

(a) 900°C; (b) 930°C; (c) 960°C; (d) 990°C; (e) 1020°C

Figure 4. Fiber structure of HNB samples at different temperatures under the deformation rate of $0.08s^{-1}$

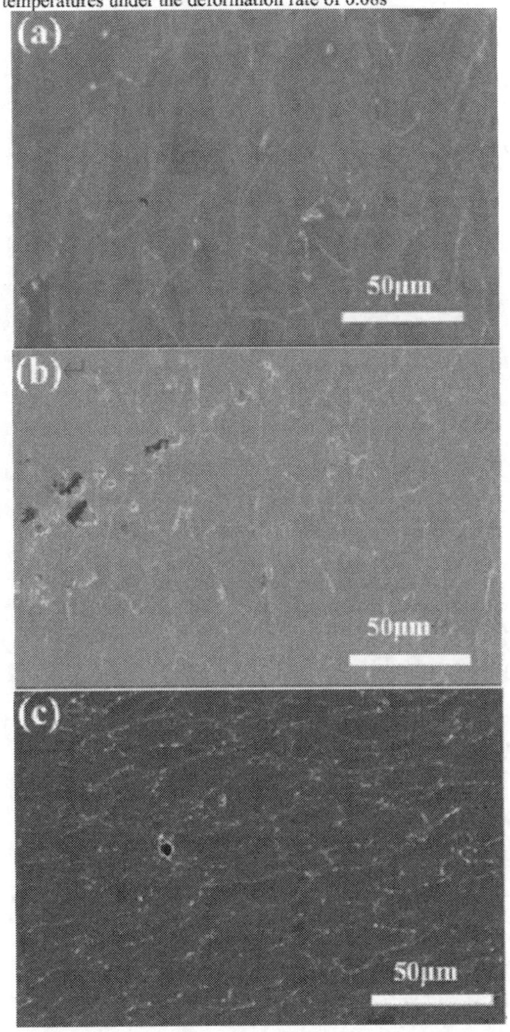

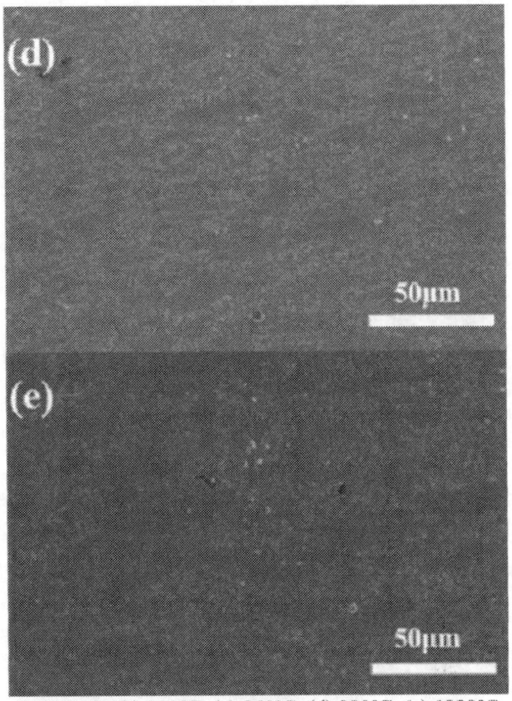

(a) 900°C; (b) 930°C; (c) 960°C; (d) 990°C; (e) 1020°C

Figure 5. Fiber structure of LNB samples at different temperatures under the deformation rate of $0.08s^{-1}$

3. Conclusion

(1) There are a large number of clusters/long strip δ phases distributed in the microstructure of the high Nb content GH4169 alloy. Due to the combined effects of deformation fracture and dissolution fracture during high-temperature compression deformation, the long strip δ phases are bent, twisted, and folded until spheroidization.
(2) The δ phases in the microstructure of the low Nb content GH4169 alloy are mainly forged rod-like/granules. With the increase of temperature, the short rod-like δ phases spheroidize and decompose, while the granular δ phases are re-dissolved, and the δ phases are completely re-dissolved at 1,020°C.
(3) With the increase of deformation temperature or decrease of strain rate, the content of δ phases in GH4169 alloy with high Nb content gradually decreases, but the δ phases are not completely re-dissolved until 1020°C and are distributed as discontinuous particles on the grain boundary, increasing the dynamic recrystallization grain size of the alloy from ASTM 7.5 to ASTM 7.0.
(4) With the increase of the deformation temperature or the decrease of the strain rate, the δ phases in the low Nb content GH4169 alloy are generally re-dissolved at 990°C, and the dynamic recrystallization grain size of the alloy grows from ASTM7.5 to ASTM6.0.

References

1. Kennedyrl. Allvac® 718plus TM , superalloy for the next forty years[C]. Superalloys 2005. Warrendale , PA: TMS, 2005:1–14 .
2. Sundararaman M, Mukhopadhyay P, Banerjee S. Precipation of the δ-Ni3Nb phase in two nickel base superalloys[J]. Metall Trans, A 1988;19A:453–65.
3. Collier JP, Wong SH, Phillips JC, Tien JK. The effect of varying Al, Ti, and Nb content on the phase stability of Inconel 718[J]. Metall Trans, A 1988;19A:1657–65.
4. Dix AW, Hyzak JM, Singh RP. Application of ultra fine grain alloy 718 forging billet[C]. Superalloys 1992. Warrendale, PA: TMS; 1992. p. 23–32.
5. Ruiz C, Obabueki A, Gillespie K. Evaluation of the microstructure and mechanical properties of Delta processed alloy 718[C]. Superalloys 1992. Warrendale, PA: TMS; 1992. p. 33–42.
6. Zhang Yun, Huang Xuebing, Wang Yong, et.al. Delta phase and deformation fracture behavior of Inconel 718 alloy[C]. Superalloys 718, 625, 706 and Various Derivatives Edited by E.A. Loria. Warrendale , PA: TMS ,1997:229–236.
7. Saied Azadian, Liu-Ying Wei, Richard Warren. Delta phase precipitation in Inconel 718[J]. Materials Characterization, 2004; 53:7–16.
8. Y. Desvallées, M. Bouzidi, F. bois, et al. Delta phase in Inconel 718: Mechanical properties and forging process requirements[J]. Superalloys 718,625,706 and Various Derivatives Edited by B.A. Loria. The Minerals, Metals&Materials Society, 1994.
9. S. Mahadevan, S. Nalawade, J. B. Singh, et al. Evolution of δ phase microstructure in alloy 718[J]. 7th International Symposium on superalloy 718 and derivatives. Edited by: E.A. Ott, J. R. Groh, A Banik, I, Dempster, T. P. Gabb, R, Helmink. TMS-2010.
10. H. Yuan, W. C. Liu. Effect of the δ phase on the hot deformation behavior of Inconel 718[J]. Materials Science and Engineering A 408 (2005) 281–289.
11. Hongjun Zhang, Chong Li, Qianying Guo. Delta precipitation in wrought Inconel 718 alloy; the role of dynamic recrystallization [J]. Materials Characterization, 133 (2017) 138–145.
12. Cai DY, Zhang WH, Nie PX, Liu WC, Yao M. Dissolution kinetics and behavior of phase in Inconel 718[J]. Trans Nonferrous Met Soc China 2003; 13(6):1338–1341.
13. Schafrik RE, Ward DD, Groh JR. Application of alloy 718 in GE aircraft engines: past, present and next five years[C]. Superalloys 718, 625, 706 and various derivatives. Warrendale, PA: TMS; 2001. p. 1–11.
14. Wang Y, Zhen L, Shao WZ, Yang L, Zhang XM. Hot working characteristics and dynamic recrystallization of delta-processed superalloy 718[J]. J Alloy Compd 2009;44:341–346.
15. K. Wang, M.Q. Li, J. Luo, et al. Effect of the δ phase on the deformation behavior in isothermal compression of superalloy GH4169[J]. Materials Science and Engineering A 528 (2011) 4723–4731.
16. Thomas A, El-Wahabi M, Cabrera JM, Prado JM. High temperature deformation of Inconel 718[J]. J Mater Process Technol 2006;177:469–72.
17. Zhang. H. Y, Zhang H, Cheng. M,et al. Deformation characteristics of δ phase in the delta-processedInconel 718 alloy [J]. Materials Characterization, 61 (2010) 49-53 .
18. Cone FP. Observations on the development of delta phase in IN718 alloy[C]. Superalloys 718, 625, 706 and various derivatives. PA: TMS; 2001. p. 323–332.
19. Cai DY, Zhang WH, Liu WC, Yao M. Dissolution behavior of δ phase in Inconel 718[J]. Journal of Iron and Steel Research, 14; 2002. p. 61–64. (in Chinese).
20. Si Jia-Yong, Chen Long, Liao Xiao-hang, et al. Microstructure Evolution of Delta-Processed GH4169 Alloy during Hot Deformation[J]. Materials For Mechanical Engineering, 2017 , 41 (4): 6–10. (in Chinese).

Preparation of High-barrier Polyethylene Terephthalate bottle by Microwave Plasma Enhanced Chemical Vapor Deposition

Lianhua Yin

Shanghai Publishing and Printing College, Shanghai, China

Abstract. In this work, we reported the investigation of diamond-like carbon (DLC) coating on the inner surface of polyethylene terephthalate (PET) bottles by Microwave Plasma Enhanced Chemical Vapor Deposition to improve their barrier properties. We used 2.45 GHz microwave to generate plasma, acetylene (C2H2) and argon (Ar) were used as monomer gas and diluted gas. The effects of plasma process parameters on deposition uniformity and barrier properties of thin films were studied in detail. In addition, we designed a new antenna shape to improve the uniformity of film deposition. We analyzed the microstructure and morphology of the DLC coating by atom force microscope (AFM). Fourier transform infrared spectroscope (FTIR) and Raman spectroscope were employed to investigate chemical composition and bonding structure. The deposition rate was obtained by Step Profiler. For the barrier properties, we measured the oxygen transmission rate (OTR) of the PET bottle coated with DLC film. It is noted that the uniformity of film deposition is affected by the process parameters and antenna shape, after optimizing the process parameters and antenna shape, the barrier property of PET bottles is improved obviously.

1. Introduction

A good packaging material must be able to simultaneously meet many requirements, such as convenient production, novel design, low cost, long shelf-life, retrievability and so on. Plastic is a popular packaging material, due to high cost performance, light quality, good mechanical strength, and most of the plastic materials are recyclable, especially polyethylene terephthalate (PET) material is used widely. However, although these materials have been used to protect the contents from external contamination, for certain products such as food or drink, people still have to try to prevent the infiltration of oxygen and ensure long-term storage and preservation. Therefore, the barrier improvement of PET packaging materials has become a very important research direction.

Diamond-like carbon (DLC) film is an amorphous carbon, its properties are very similar to diamond films, such as good optical properties, high hardness, high resistivity, unique tribological characteristic and so on [1-3]. Besides, the DLC structure is adjustable, as the different sp^2 and sp^3 hybridized orbital content demonstrate a different properties [4]. Compared to the diamond film, DLC film is easy to prepared by Chemical Vapor Deposition(CVD) or Plasma Enhanced Chemical Vapor Deposition (PECVD), such as ion beam deposition, magnetron sputtering, RF plasma-activated chemical vapor deposition, plasma source ion implantation and so on [5]. As a very efficient deposition method, microwave plasma was widely studied due to a high plasma density but no external the magnetic field which simplifies the device [6]. In practical applications, it is necessary to deposit film fast and evenly, we experimentally studied uniformity of DLC films deposition. As known the properties of the DLC film in mechanical, optical , barrier property and electrical are relevant to microstructure and morphology of the films, so in this work we analyzed the carbon bond characterization by Fourier transform infrared spectroscope (FTIR) and the microstructure and morphology of the DLC coating by atom force microscope (AFM). In the experiment, we explored how the quartz antenna length influenced the deposition distribution, some plasma parameters were also experimented like microwave power, duty cycle, monomer radio and work pressure. Finally, the barrier property of PET coated DLC film as application of was measured in this work.

This study aims to improve the barrier performance of PET bottles by DLC film coating. At the same time, the improvement of barrier performance is affected by the uniformity of film deposition, so the research will also focus on the uniformity of film deposition.

2. Experiments

2.1 Experimental set-up

We use microwave plasma chemical vapor deposition unit to deposit DLC film on internal surface of PET bottle, as shown schematic diagram of experimental set-up in Figure 1. The deposition system consists of plasma generation system, vacuum system, gas path system, cooling system. The operating frequency of the microwave source of the device is 2.45 GHz. The microwave energy was launched from the magnetron then transmitted by the rectangular waveguide, and coupled to the resonant cavity by antenna (the antenna was made up of two quartz tube and a copper pipe). The microwave surface wave propagation was realized on the plasma interface and the surface of the quartz medium, and then the high density plasma was generated. The discharge gas and the reaction monomer entered the reaction chamber through the internal pipe of the quartz antenna.

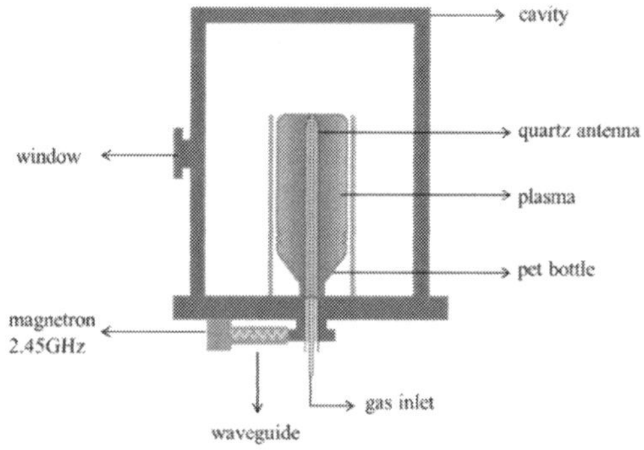

Figure 1. Schematic diagram of experimental set-up

2.2 Deposition

During deposition, we used acetylene C_2H_2 as carbon source, Argon (Ar) as diluted gas to deposit DLC film. DLC film was deposited on the inner wall of PET bottle (500 ml) to improve its barrier performance. In addition, other materials will also be used as substrates to assist in the analysis of the structure and performance of the film, including glass slides, monocrystalline silicon, polyethylene terephthalate (PET) web, and KBr tablets. Acetone, ethanol, deionized water were used as cleaning detergents respectively in ultrasonic condition, before samples were put into the chamber.

Prior to the deposition of DLC films, we pretreated substrates by Ar plasma, for the purpose of cleaning the substrate and activating the surface, which will benefit to the film nucleation and growth .The parameters of Ar discharge pretreatment were listed in table 1.

Table 1. The parameters of Ar plasma pretreatment

Gas	Ar
Microwave power	400W
Time	10s
Discharge mode	continuous

For coating deposition, we used pulse plasma. The pulse microwave power was 1500W, the duty cycle was 4:40 ms (t_{on}:t_{off}=4:40ms) , the gas pressure was 30 Pa, The flow ratio of argon and acetylene was 1:6, and exposure time was 3min. To improve coating uniformity, we adjusted the process parameters and used quartz antennas of different shapes, as described in detail later in this article.

Table2. The parameters of DLC film deposition

Base pressure	2 Pa
Discharge gas	Ar, C_2H_2
Microwave power	1500W
Discharge mode	Pulse
Duty cycle(t_{on} : t_{off})	4:40 ms
Gas pressure	30Pa
Gas flow ratio(C_2H_2:Ar)	6:1

3. Results and discussion

3.1 FTIR spectroscopy

Figure 2 is the FTIR spectroscopy of the deposited DLC film. The film deposition parameters were described in Table2 of this paper. The wave number around 1100cm^{-1} corresponds to the stretching vibration of the straight chain C-C bond and $(CH_3)_2CHR$. The hydrocarbon stretching vibration in the range of 2800-3100 cm^{-1} was appeared obviously, including sp^3 carbon-hydrogen bond (2800-3100 cm^{-1}) stretching vibration and sp^2 carbon-hydrogen bond (3000-3100cm^{-1}) stretching vibration.

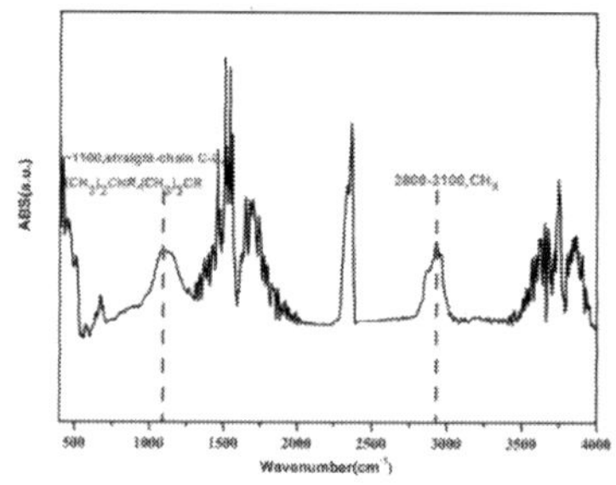

Figure 2. FTIR spectroscopy of the deposited DLC film

We analyzed the structure and composition of the DLC film in detail through Gaussian peak-differentiating and imitating the FTIR spectra in the wavenumber range of 2800-3100cm^{-1} as shown in Figure 3 The absorption peak is in the range of 2800~3100cm^{-1} mainly correspond to the component of CH_x $(0<x<4)$ group in the film, the absorption peaks around 2870, 2925 and 2960 cm^{-1} are corresponded to sp^3-CH_2 symmetric, sp^3-CH_2 asymmetric and sp^3-CH_3 asymmetric stretching respectively, the absorption band ranging from 3000 to 3100 cm^{-1} is sp^2-bonded carbon. The carbon in films is mainly presented in sp^3 hybridization, and a small amount of sp^2 hybridization. The results show that the main components of the deposited films are hydrogenated amorphous carbon.

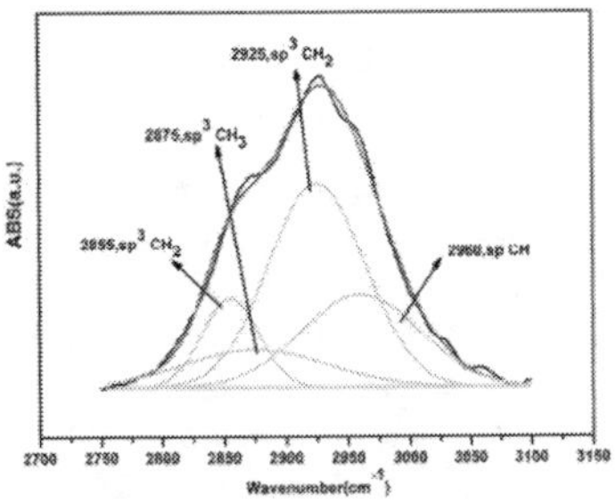

Figure 3. Gaussian peak-differentiating and imitating in the wavenumber range of 2800-3100cm^{-1}

3.2 Microstructure and morphology

Figure 4 shows the AFM morphology of DLC film and scan area range was 5×5 μm, the film deposition parameters were described in Table2 of this paper. The thickness of the deposited film measured by the step instrument was about 510nm, and the surface roughness (Ra) was 11.6nm after scanning, imaging, analyzing by atomic force microscope, which showed the surface of the film was significantly smooth. From AFM surface morphology image, we can examine the topography features and observe the surface smoothness, coating surface was dense and smooth with small particle size, and the image indicates that the film is compact and no obvious flaws and pores on the surface, which is very important for gas barrier properties.

Figure 4. AFM morphology of DLC film

3.3 Uniformity

In practical applications, the uniformity of the film deposited in large area is important, especially in the preparation of high barrier bottles by coating method. However, when the film is deposited by microwave PECVD, the microwave energy dissipates along the axis of the antenna, resulting in the uneven distribution of the film deposition rate along the antenna direction. In this paper, we studied the factors affecting the uniformity of film deposition in detail.

3.3.1 Plasma parameters.

Figure 5 shows the distribution of deposition rate along the axial direction of the antenna on different process parameters. Figure 5(a) shows the deposition trend with different microwave power, as seen, when the power was 1500W, the deposition rate of upper half part was lower than that of lower part, but as power increased to 2500W, the trend exchanged and the uniformity of film growth became better, the average deposition rate achieved around 100nm/min. Figure 5(b) is on various duty cycle t_{on}:t_{off} from 2:40 ms to 8:40 ms, as shown in the figure, when duty cycle value was small equal to 2:40 ms, the deposition rate was very low, although film deposition looked uniformly. The deposition attained above 100nm/min when t_{on}:t_{off}=8:40 ms, and the deposition rate during upper area around antenna was higher. As for the deposition uniformity, 6:40 ms was the best. When changing proportion of monomers, deposition rate distribution along axial direction along antenna is shown in Figure 5(c), from the figure, the uniformity seemed to become worse as the monomer ratio (C_2H_2:Ar in flow rate) increased, the film deposition uniformity was best as C_2H_2:Ar=1:1, while film grew more quickly in lower part than the upper as C_2H_2:Ar=8:1. Figure 5 (d) was on various pressure from 20 Pa to 40 Pa, it was observed that the deposition rate enhanced with the work pressure increased, when work pressure was 20 Pa, obtained the most uniform deposition, the middle part shown fastest rate on 40 Pa.

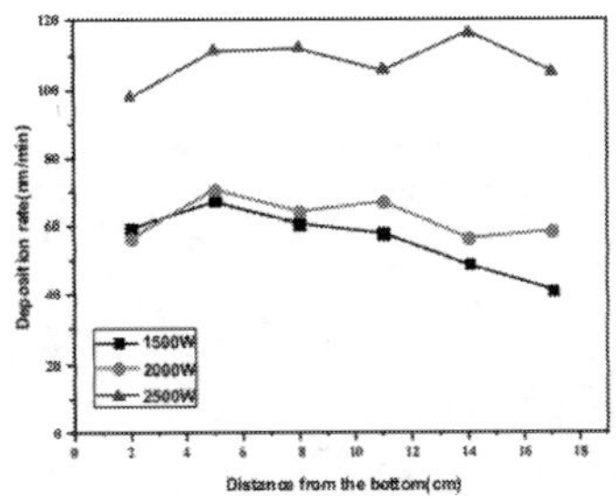

a

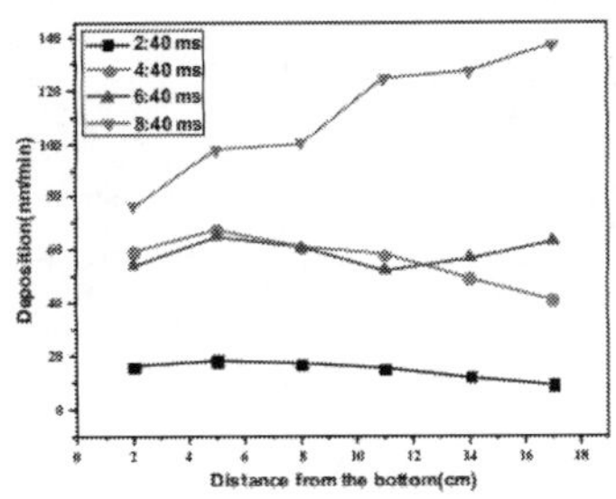

b

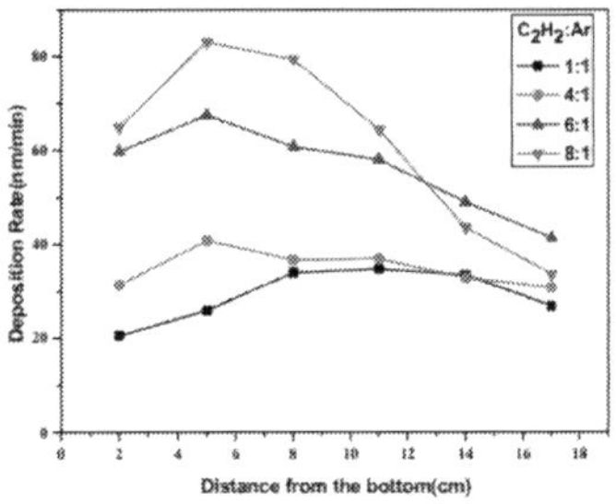

c

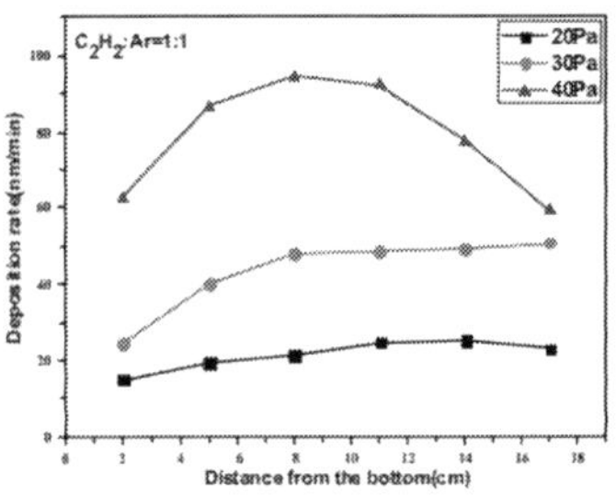

d

Figure 5. Deposition rate along the axial direction of the antenna (a) - on different microwave power; (b) - on duty cycle; (c) - with different monomer radio; (d) - with different work pressure

3.3.2 Antenna length.

In Figure 6 when quartz antenna length was shorter(about 20cm), the uniform was worse in the range of 0 cm to 17 cm from bottom, and the deposition rate was faster than that of longer antenna below 8 cm position, then reduced rapidly, in comparison, 25 cm antenna can achieve a uniform film deposition.

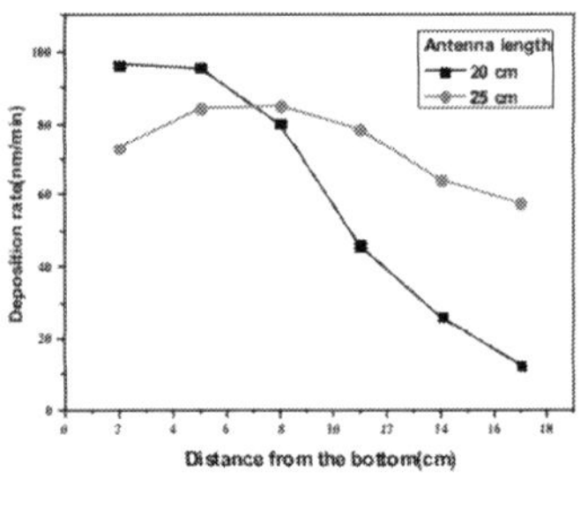

Figure 6. Deposition rate distribution along the axial direction of different antenna lengths

3.3.3 Antenna shape.

In this study, the quartz tube antenna, as the energy transmission medium, consists of an internal quartz tube (intake-tube), a copper tube in the middle (microwave energy transmission) and an external quartz tube (surface wave transmission medium). When the energy was transmitted along the quartz medium surface of conventional antenna (marked as Antenna-1, cylindrical structure with equal outer diameter), we deposited DLC film on the inner wall of the PET bottle, the film couldn't cover the bottom of the bottle.

Therefore, we try to optimize the structure of the antenna by adjusting the diameter of the upper part of the quartz antenna and changing the spatial distribution of the electromagnetic wave propagating along the dielectric surface, so as to improve coating coverage on the bottom of PET bottles. When the diameter of the upper end of the antenna (marked as Antenna-2) is enlarged, it is found that the antenna with this structure can improve the DLC deposition situation of pet bottle inner wall (The deposition rate distribution of new and former antennas is shown on Figure 7).

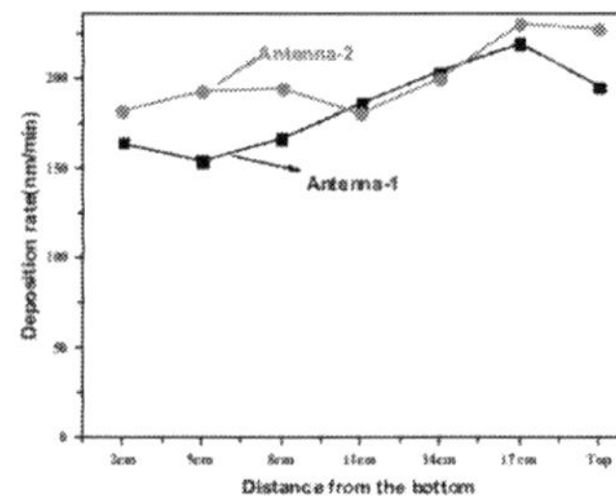

Figure 7. Deposition rate distribution of new and former antennas

3.4 The barrier property

We deposited DLC film on the inner wall of 500ml PET bottle by microwave PECVD technology to prepare high barrier bottle. It has been shown earlier in this article that the antenna structure has a significant effect on the distribution of film deposition. Therefore, we compared the effect of conventional antenna and modified antenna on improving the barrier property of PET bottles. Except for different antenna shapes, other parameters remain the same as shown in Table 3 (No.1 corresponds to conventional antenna, No.2 corresponds to improved antenna). The results show that the barrier properties of PET bottles can be improved obviously after coating DLC films by microwave PECVD technology. When using a conventional quartz antenna, the OTR of a 500ml PET bottle can be reduced from 0.058 cc/(pack-day) to 0.0038cc/(pack-day). More importantly, after adjusting the antenna shape, the OTR of PET bottle coated DLC film can as low as 0.0018 cc/(pack-day).

Table 3. Preparation process parameters and the corresponding OTR

Sample	Pulse microwave power(w)	Duty cycle ((t_{on}:t_{off}(ms))	Flow ratio (C_2H_2:Ar (sccm))	Pressure (Pa)	Time (min)	OTR (cc/(pack-day))
No.1 No.2	1500	4:40	6:1	30	15	0.0038 0.0018
Uncoated PET bottle						0.058

4. Conclusions

In this paper we produced High-barrier Polyethylene Terephthalate bottle deposited DLC films by Microwave Plasma Enhanced Chemical Vapor Deposition and investigated the composition and microstructure of the DLC film. In order to optimize the preparation method of High-barrier PET bottles, we also studied the uniformity of film deposition specifically. We mainly studied the influence of plasma parameters, antenna length and shape on DLC film deposition distribution. Results shows that, the antenna length and shape can influence the deposition uniformity obviously, and the antenna shape developed in this research can effectively improve the distribution of DLC film coating, so as to achieve the purpose of improving the barrier properties of pet bottles. In addition, the plasma parameters like microwave power, duty cycle, the proportion of monomers and inert gas and work pressure can also influence the growth rate and uniformity of DLC film.

In conclusion, this study found that the DLC film coating on the inner wall of pet bottles by Microwave Plasma Enhanced Chemical Vapor Deposition can greatly improve their barrier properties. Especially when optimized antenna shape is used for coating modification, the OTR of the PET bottle can be as low as 0.0018 cc/(pack-day), approximately 32 times lower and that will be very useful in package field.

Acknowledgments

This work was financially supported by "Chen Guang" Program of Shanghai Municipal Education Commission and Shanghai Education Development Foundation (19CGB11)

References

1. Geoffrey Dearnaley, James H. Arps, Biomedical applications of diamond-like carbon (DLC) coatings: A review, Surface and Coatings Technology, 200(2005). pp. 2518-2524.
2. N.D.Malleswararao.K, I.N.Niranjan Kumar, Investigation of tribological behaviour of DLC coating on hyper-eutectic Al-Si alloys, a review, Materials Today: Proceedings, 18(2019), pp. 2581-2589.
3. F. Wang, Z. Lu, L. Wang, G. Zhang, Q. Xue, Effect of tribochemistry on friction behavior of fluorinated amorphous carbon films against aluminum, Surface and Coatings Technology, 304(2016), pp. 150-159
4. .L.L. Liu, X.K. An, Z.Y. Ma, Z.Z. Wu, W. Tang, H. Lin, R.K.Y. Fu, X.B. Tian, P.K. Chu, F. Pan, Hard and adherent a-C:H gradient coatings by stress engineering, J. Alloys Compd., 765(2018), pp. 921-926.
5. J. Vetter, 60years of DLC coatings: Historical highlights and technical review of cathodic arc processes to synthesize various DLC types, and their evolution for industrial applications, Surface and Coatings Technology, 257(2014), pp. 213-240.
6. Hiroyuki KOUSAKA, Noritsugu UMEHARA, Kouichi ONO and Junqi XU, Microwave-Excited High-Density Plasma Column Sustained along Metal Rod at Negative Voltage, Japanese Journal of Applied Physics , 44(2005) , pp. 1154–1157.

Inspection and treatment of four pipe supports and hangers of 350MW unit

Tangqing Yuan [1,2,*], Liangliang Jian [3], Jidong Li [1,2], Yang Chen [1,2], Mingjian Jian [1], Huakai Zhang [1]

[1]Electric Power Research Institute of State Grid Shandong electric power company, No.2000 Wangyue Road, Shizhong District, Jinan , 250002, PR China
[2]Shandong Electric Power Industrial Boiler and Pressure Vessel Inspection Center Co., Ltd,No.2000 Wangyue Road, Shizhong District, Jinan , 250002, PR China
[3]Shandong Zhongshi Yitong Group Corporation Co.,Ltd, Ltd,No.2000 Wangyue Road, Shizhong District, Jinan , 250002, PR China

Abstract. Based on the on-site cold and hot state inspection of the four pipelines of a 350MW unit in Shandong, combined with the review and calculation of stress, design and installation data, operation and maintenance records and other data, this paper formulates a practical adjustment plan and construction scheme, implements comprehensive adjustment and treatment of the four pipeline supports and hangers, basically restores the normal working state of the supports and hangers and pipelines, and ensures the safe operation of the unit.

1 Introduction

After long-term operation, a large number of abnormalities, damages and failures often occur in the supports and hangers of steam and water pipes of utility boilers. The main causes are as follows:(1) the creep and stress relaxation of materials under long-term high-temperature operation make the stress of pipe system redistributed; (2) the installation of supports and hangers deviates from the original design or the original design is incomplete, resulting in significant changes in the load borne by supports and hangers; (3) deformation and corrosion of the external surface of supports and hangers during operation; (4) large impact load or violent vibration; (5) decline of spring performance after long-term operation, etc.[1-2]The above factors will change the stress of the pipe system and the bearing distribution of supports and hangers, increase the local deformation of the pipe section, increase the local stress of the pipe system and even exceed the basic allowable stress of the pipe. In addition, it will also cause a significant increase in the thrust and torque at the equipment interface connected with the pipe, resulting in damage and failure at the interface between the pipe and the equipment, so as to shorten the remaining service life of the pipe and seriously affect the safe and economic operation of the power plant. Therefore, it is very necessary to regularly check, check and adjust the supports and hangers of steam and water pipes of units in operation.

Taking a 350MW supercritical unit of a thermal power plant in Shandong as the engineering background, this paper carries out the inspection, calculation and adjustment of the supports and hangers of steam and water pipelines in operation, so as to realize the full cycle maintenance of the supports and hangers of steam and water pipelines after they are put into operation. The unit was put into operation in January 2019 and has operated for about 18000 hours so far. According to the working state requirements of supports and hangers, comprehensively inspect the supports and hangers under hot and cold conditions respectively; Recheck and calculate the stress of the four pipelines, and determine the supports and hangers to be adjusted through comparative analysis; Formulate and implement specific adjustment plans according to the requirements of relevant standards and regulations.

2 Function and working state requirements of supports and hangers

2.1 Constant spring hanger

It is used to bear the dead weight load of the pipeline, and its bearing capacity does not change with the change of the vertical displacement of the pipeline at the lifting point, that is, the load remains basically constant [3]. It is usually used where the vertical displacement of the pipeline is large or the transfer load needs to be limited. Working condition requirements: the hanger assembly is normal; The load displacement pointer is within the allowable range; No jamming, damage or abnormality of moving parts; The included angle between the hanger rod and the vertical line does not exceed the standard; The root bearing structure and auxiliary steel structure have no obvious deformation, and the main bearing weld has no macro cracks; The suspender and connecting accessories are not damaged or abnormal.

* Corresponding author: 13256774709@163.com

2.2 Variable force spring support and hanger

It is used to bear the dead weight load of the pipeline, and its bearing capacity changes with the change of the vertical displacement of the pipeline at the pipeline support and lifting point [4]. It is usually used where the vertical displacement of the pipeline is not very large, and plays an important role in reducing the thrust or torque borne by the equipment. Working condition requirements: the hanger assembly is normal; The hanger spring has no excessive compression or load loss; The load displacement pointer is within the allowable range; The included angle between the hanger rod and the vertical line does not exceed the standard; The root bearing structure and auxiliary steel structure have no obvious deformation, and the main bearing weld has no macro cracks; The suspender and connecting accessories are not damaged or abnormal.

2.3 Rigid hanger

It is used to bear the dead weight load of the pipeline and restrict the vertical downward displacement of the pipe system at the lifting point, which plays an important role in reducing the flexibility of the small pipe system, enhancing the stability of the pipeline and controlling the vibration of the pipeline[5].Working condition requirements: the hanger has no loss of load; The suspender and connecting accessories are not damaged or abnormal; The included angle between the hanger rod and the vertical line does not exceed the standard; The root bearing structure and auxiliary steel structure have no obvious deformation, and the main bearing welds have no macro cracks.

2.4 Sliding support

It supports the pipeline on the sliding base plate to bear the gravity load of the pipeline and restrict the vertical downward displacement of the pipeline fulcrum without limiting the horizontal displacement. Working condition requirements: the working surface shall be flat without jamming and void; The sliding bottom plate of the pipe part does not exceed the limit; For the sliding support with PTFE plate, the sliding bottom plate of the pipe part shall be covered with PTFE plate under cold and hot conditions.

2.5 Limit device

A device used to restrict or partially limit the displacement in one or more directions at the pipe support and lifting point. It plays an important role in restraining the unreasonable displacement of pipeline and controlling the vibration of pipeline. Working condition requirements: the assembly is intact; The structure shall be free of obvious deformation and the weld shall be free of cracks; No damage or missing parts; It can limit the displacement in the predetermined direction.

2.6 Fixed support

It belongs to limit support device. The fixed support can limit the linear displacement and angular displacement of the pipeline support point in three directions, and can bear the dead weight load of the pipeline. It plays an important role in ensuring pipeline stability and preventing pipeline vibration. It is often used where no displacement is allowed on the pipeline. Working condition requirements: the fixed support shall be firm and reliable without looseness, the structural parts shall be free of obvious deformation, the welds shall be free of cracking, and the concrete buttress shall be free of damage.

2.7 Damper

It is used to protect the pipeline from accidental dynamic conditions. When the pipeline is under normal conditions, it can adapt to the slow movement of the pipeline caused by thermal expansion and cold contraction, and has little restrictive effect on the pipeline [5]. It is mainly used to bear steam hammer, water hammer, pipeline exhaust reaction and seismic load. It has no obvious control effect on low-frequency high amplitude or high-frequency low amplitude vibration. Working condition requirements: the damper shall be free of oil leakage and oil leakage; No missing parts; No abnormal deformation of bearing elements including nearby steel structures; The welds of pin seat assembly and pipe part shall be free of cracks; The position of damper piston shall meet the requirements.

3 Inspection of pipe supports and hangers

Check the hot state of supports and hangers before the unit is overhauled and shut down, confirm the support and hangers to be replaced, put forward the purchase list in time, and test the cold state of supports and hangers after the unit's outage pipes are cooled to normal temperature. By means of telescope, camera, vertical line and ruler, the supports and hangers of the four pipes of Unit 3 are inspected in detail in hot and cold states. The inspection results are as follows:

- The layout of each pipeline system is reasonable,
- The piping system has no obvious vibration,
- There are rigid supports and hangers in each pipeline system to stabilize the pipeline or devices to prevent the remote transfer of load,
- the hot state and cold state inspection of the four pipe supports and hangers are shown in Table 1.

Table 1. Statistical table of supports and hangers with problems in four pipelines (unit: Group).

Serial number	Pipe name	Total number of hangers	Problem	Number of questions / total
1	Main steam pipeline	24	12	50.0%
2	Reheat hot section pipeline	22	6	27.3%
3	Reheat cold section pipeline	22	5	22.7%
4	Main water supply pipeline (high temperature section)	21	1	4.8%
5	Total of four pipelines	89	24	27.0%

4 Recheck calculation of pipeline stress

The recheck calculation of pipeline stress is based on DL / T 5366 (6), DL / T 5054 (7) and ASME B31 1. According to the requirements of the specification and the design parameters of the pipeline, the layout of the pipe system and supports and hangers, the special pipe stress analysis software CAESAR II 2016 is used to verify the correctness of the configuration, load and displacement of the supports and hangers, as well as the deformation, stress distribution of the pipeline and the restraint reaction of the supporting structure for the three-dimensional thermal displacement of the supports and hangers, the thrust at the connecting end of its equipment and the maximum stress value of the pipe system under the design state, Determine the weakest link or key part of the pipe and evaluate the safe operation state of the support and hanger system.

The strength failure of pipeline is mainly caused by the fracture failure caused by primary stress and fatigue fracture failure caused by secondary stress. The stress caused by the internal pressure and continuous load borne by the pipeline, and the stress caused by seismic load, water impact and continuous action impact of safety valve belong to primary stress. The stress caused by thermal expansion, additional displacement, internal and external wall temperature difference and other displacement constraints of the pipeline belongs to secondary stress.

The primary stress shall meet the requirements of the following formula:

$$\sigma_L = \frac{PD_i^2}{D_0^2 - D_i^2} + 0.75\frac{iM_A}{W} \leq 1.0[\sigma]^T$$

In equation: σ_L -- primary stress (MPA); P - design pressure (MPA); D_0 - Pipe outer diameter (mm); D_i - Tube inner diameter (mm); M_A - due to the combined moment of self-weight and other continuous external loads acting on the cross section of the tube, N.mm; W - Pipe Bending Section Factor (mm^3); $[\sigma]^T$ - Allowable stress (MPa) of steel at design temperature; i - Stress increase factor, and 0.75i should not be less than 1.

Secondary stresses shall meet the requirements of the following equation:

$$\sigma_E = \frac{iM_c}{W} \leq f\left[1.2[\sigma]^{20} + 0.2[\sigma]^T + \left([\sigma]^T - \sigma_L\right)\right]$$

In equation: σ_E - Secondary stress (MPa); $[\sigma]^{20}$ - Allowable stress (MPa) of pipe base material at 20℃; M_C - The synthetic moment (N.mm) caused by thermal expansion calculated from the full compensation value and the elastic modulus of steel at 20℃; f - Reduction factor of stress range.

According to the above theoretical basis, the CAESAR software is used to model and calculate the four pipelines according to the actual situation. The calculation model is shown in Figure 1-Figure 4. After cold and hot calculation, the maximum values of primary and secondary stress calculation and the number of nodes are extracted. As shown in Table 2, the results show that the cold and hot calculation of the four pipelines meet the allowable stress requirements specified in the standard.

Table 2. Calculation results of pipe stress.

Pipe name	Stress classification	Calculated value (MPa)	Permissible value (MPa)	Location, node
Main steam pipeline	Primary stress	47.24	82.20	Upper elbow of spring hanger No. 13, 250
	Secondary stress	231.85	251.84	Right elbow of spring hanger No. 28, 530
Reheat hot section pipeline	Primary stress	52.17	83.49	No. 470 Constant Hanger Front Tee, 640
	Secondary stress	190.84	250.41	No. 355 Constant Hanger Lower Elbow, 420
Reheat cold section pipeline	Primary stress	69.21	119.83	No. 617 Constant Hanger Right Tee, 170
	Secondary stress	92.53	207.70	No. 607 Constant Hanger Right Elbow, 130
Main water supply pipeline (high temperature section)	Primary stress	83.40	120.66	Rigid Hanger No. 1024, 110
	Secondary stress	79.94	157.82	No.1022 Constant Hanger Upper Side Elbow, 260

https://doi.org/10.1051/matecconf/202235801003

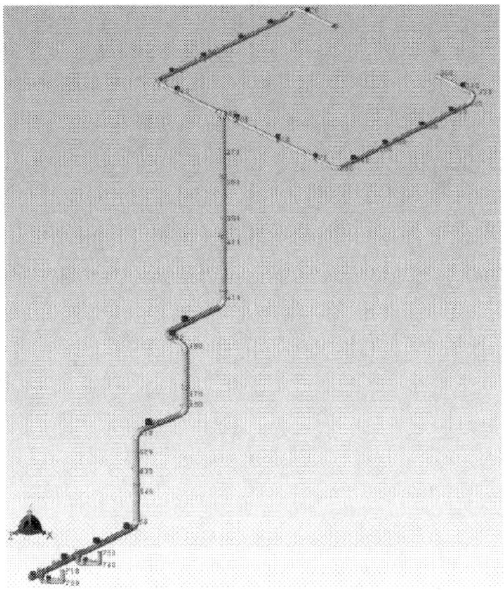

Fig. 1. Main steam pipeline.

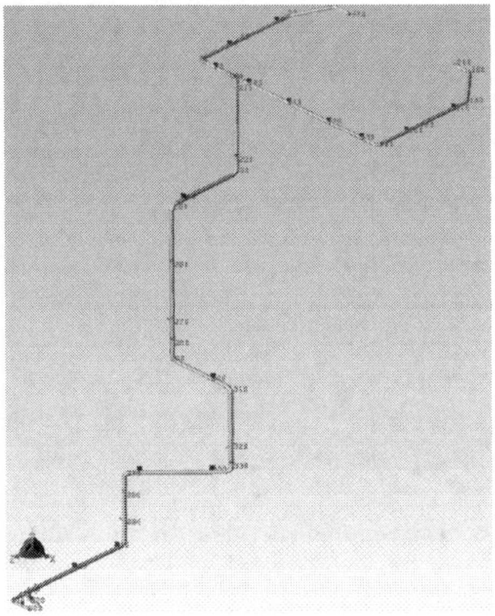

Fig. 2. Reheat hot section pipeline.

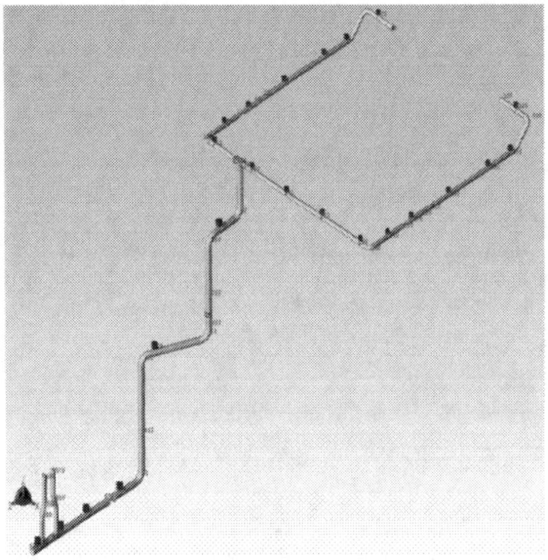

Fig. 3. Reheat cold section pipeline.

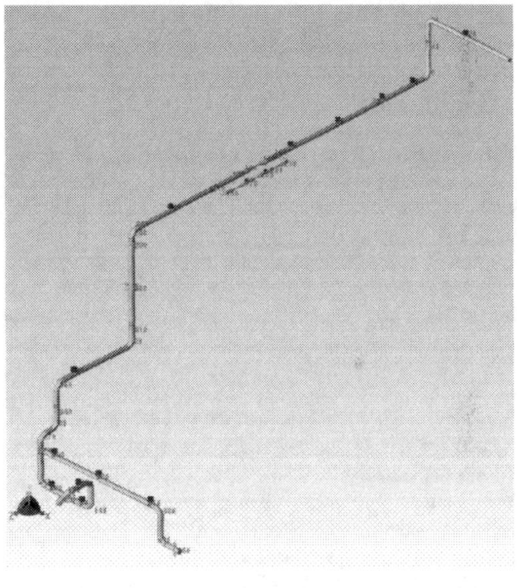

Fig. 4. Main water supply pipeline (high temperature section).

5 Adjustment and treatment of pipe supports and hangers

According to DL/T 616-2006 Maintenance & adjusting guide for thermal power plant Steam-water piping and support-hangers, design and installation drawings of four major pipeline supports and hangers, hot and cold inspection results, pipeline stress distribution and practical experience in adjustment and treatment of supports and hangers, practical adjustment measures are formulated on the basis of careful analysis of the causes of support and hanger faults and defects Adjust the functional parts and connectors of the damping device to

 https://doi.org/10.1051/matecconf/202235801003

eliminate the existing hidden dangers and ensure the safe operation of the pipeline.

Adjustment and treatment sequence of supports and hangers: for single line pipeline, it shall be carried out from the furnace top down in sequence; For multi line pipelines, they shall also be carried out in parallel and in sequence. After the pipe supports and hangers are adjusted, the load will be transferred and redistributed. Therefore, the supports and hangers at adjacent lifting points shall be retested and adjusted if necessary to ensure that the adjacent supports and hangers are still in normal working condition. After the operation of the unit, the working state under hot working conditions shall be checked, and the existing problems shall be analyzed and adjusted to make the whole support and hanger system in normal working state under cold and hot working conditions. When preparing the technical report, the report shall include the standard basis for inspection and adjustment, cold and hot inspection data of supports and hangers, stress analysis data, displacement and load verification data, support and hanger adjustment measures, construction scheme, working status before and after adjustment, distribution diagram of pipeline supports and hangers, conclusions and suggestions, etc.

6 Conclusion

Through the on-site inspection, stress recheck calculation, adjustment and treatment of the four pipelines of a 350MW unit in Shandong, this paper realizes the full cycle maintenance technology of steam water pipeline supports and hangers, eliminates the hidden dangers of supports and hangers, improves its operation state and improves the health level of pipelines. It is suggested that the technicians of the power plant should conduct a thermal visual inspection on the supports and hangers of the main steam and water pipelines of the unit once a year according to the requirements of relevant standards and regulations, and record the inspection results in the file. During each overhaul, all pipe parts, roots, connectors and functional parts of pipe supports and hangers shall be inspected and maintained, and records shall be made, and problems found shall be handled in time.

References

1. Yan Jialing. Inspection and adjustment of steam water pipe supports and hangers of ultra supercritical units [J]. Engineering construction, 2017,49 (03): 97-99
2. Guo Yanjun, Chen Yueguang. Support and hanger failure and safety analysis of steam water pipeline in power plant [J]. Power equipment, 2007, 8 (4): 62-64
3. NB/T 47038-2013 Constant spring support and hanger
4. NB/T 47039-2013 Variable spring hangers
5. DL/T 1113-2009 The inspection code of pipe hangers and supports for power plants
6. DL/T 5366-2014 Technical code for stress calculating of steam/water piping in power plant
7. DL/T 5054-2016 Code for design of stream/water piping of fossil-fired power plant

Thermal Decomposition of Epoxy Resin Under SF6 Atmosphere

Ren Yang[1], Hongyu Yang[2], Yuhan Wu[2], Dezhi Lin[2], Jun Xue[3], Xiao Zhang[2], Weiqiang Zhang[2, *]

[1]State Grid Shaanxi Electric Power Research Institute, Xi'an 710054, China
[2]Xi'an Key Laboratory of Organometallic Material Chemistry, School of Chemistry and Chemical Engineering, Shaanxi Normal University, Xi'an 710119, China
[3]State Grid Shaanxi Electric Power Company, Xi'an 710054, China

Abstract. The latent fault diagnosis of GIS equipment relies on the electrothermal corrosion information of the decomposition products of SF_6. The thermal decomposition of epoxy resin under atmosphere (SF_6-Epoxy) is essential to reveal the relationship between carbon-based characteristic decomposition components and the degradation of insulating materials. The thermal decomposition process of SF_6-Epoxy was measured using synchronous thermogravimetry(TG)/differential scanning calorimetry(DSC), which unveiled the three pyrolysis stages of temperature range and heat release. The online gas chromatography identified nine decomposition components containing carbon, sulfur and oxygen elements. According to the panoramic pyrolysis reaction mechanism of the SF_6-epoxy system, CO_2, CH_4, SOF_2, H_2S and H_2 were proposed as the characteristic decomposition components for the thermal deterioration of SF_6-Epoxy insulating system.

1. Introduction

With the continuous improvement of the capacity level of the national grid, GIS circuit breakers because of their prominent advantages, such as small footprint, safe and reliable operation, and long access cycles, are widely used in the field of high-voltage and ultrahigh voltage power delivery, which ensures the safe deployment and precise control of power delivery [1]. During long-term charged operation, sulfur hexafluoride (SF_6) and solid insulation materials (epoxy resin epoxy, polytetrafluoroethylene PTFE) within GIS circuit breakers undergo electrothermal decomposition, which leads to overall degradation of the circuit breaker and forms a latent failure hazard [2-4]. The detection of SF_6 oxygen and sulfur decomposition products SO_2, SOF_2 and H_2S can initially diagnose the latent faults of GIS circuit breakers [5-6]. Abnormal working conditions such as arc, abnormal discharge and local overheating in the circuit breaker cause the decomposition of organic solid insulating materials, forming a large number of carbon-based characteristic decomposition products CO, CO_2, CS_2, COS, CH_4, CF_4 [7]. Considering the element balance, the formation of these carbon-based decomposition products must be related to the degradation of organic solid insulating materials in the circuit breaker. However, the information on the degradation elements of the insulating materials contained in these carbon-based decomposition products has not been effectively used, which limits the engineering application of the characteristic decomposition product analysis method in the detection of latent faults. Deeply mining of the electrothermal corrosion information of insulating materials carried by carbon-based decomposition products has important scientific significance and engineering practical value for online monitoring and fault diagnosis of GIS equipment. Epoxy resins are a class of commonly used organic insulating materials, which are widely used in the production and manufacture of GIS high-voltage electrical equipment such as pot insulators, solid insulating switch cabinets, side sleeve of insulated exchange valve [8]. When these devices suffer from local overheating failure, the epoxy type of materials undergo thermal deterioration. Studying the thermal decomposition reaction of sulfur hexafluoride and epoxy resin system (SF_6-epoxy) to elucidate the relationship between the decomposition products of SF_6 and the degradation of insulating materials is a fundamental way to solve the contradiction between the unitary nature of characteristic components and the diversity of sources. At present, SF_6 epoxy thermal decomposition reaction is poorly studied, and many key questions need to be deeply investigated regarding whether SF_6 is directly involved in the thermal decomposition of epoxy [9] or its secondary decomposition reaction is coupled with the thermal decomposition reaction of epoxy, especially the formation mechanism of carbon-based characteristic decomposition products. *Tang et al.*, first studied the thermal decomposition process of epoxy resin under an atmosphere of SF_6 and detected CF_4, CO_2 and a small amount of CS_2, and proposed CF_4 as a characteristic decomposition product to judge the failure of local overheating on the surface of epoxy resin [10]. While similar pyrolysis experiments only found trace amounts of CF_4 (< 10 ppm) [11]. Considering that arcing, partial discharge, or flash discharge also caused the

* Corresponding author: zwq@snnu.edu

 https://doi.org/10.1051/mateconf/202235801004

decomposition of SF_6 to produce CF_4, *Zhang et al.*, suggested that CF_4 cannot be used as a judgment for early potential overheating failure. The accurate correspondence between the characteristic decomposition products and the thermal decomposition reaction of insulating materials is the main challenge in judging the latent failure by using the carbon based characteristic decomposition components. In this paper, a synchronous TG / DSC technique was applied to systematically investigate the thermal and mass changes of the thermal decomposition process of SF_6 epoxy. The gas-solid decomposition product composition of the thermal decomposition of bisphenol A-type epoxy resin under an atmosphere of SF_6 was comprehensively analyzed by combining on-line gas chromatography and ambient SEM. The system aligned the thermal decomposition panoramic reaction course of SF_6 epoxy at each weight-loss stage to escape gas and combed out the characteristic decomposition product generation rules, which would be useful for monitoring thermal corrosion of epoxy insulation and provided the experimental basis for diagnosis of GIS latent fault.

2. Result and Discussion

The thermal decomposition of any epoxy resin material starts from the dehydration of the secondary hydroxyl groups on the polymer chain, and then undergo a series of carbon-oxygen bonds, carbon-nitrogen bonds, and carbon-carbon bonds to break, and finally form outgassing products [12]. The entire thermal decomposition process of epoxy resin involves the coupling of multiple chemical transformations, which is a typical complex non-equilibrium multiphase thermodynamic system. In the atmosphere of nitrogen, the pyrolysis of epoxy resin mainly occurs in the range of 400°C, showing a single weight loss; in the air, the oxidation of oxygen accelerates the decomposition of epoxy resin. The main pyrolysis stage of weight loss is advanced to the range of 350°C. The pyrolysis of epoxy resin further coupled with heterogeneous oxidation and decomposition, resulting in a secondary weight loss stage at 521°C [13]. In order to clarify the thermal decomposition reaction process of the SF_6-Epoxy system, We firstly uses the temperature-programming technique to investigate the coupling relationship between the SF_6-Epoxy thermal decomposition heating rate, holding time and the reaction of each decomposition stage, and accurately measures the activation energy of the relevant characteristic decomposition reaction.

2.1 The thermal decomposition process of SF6 - Epoxy system

Fig. 1 presents the synchronous thermal weight loss differential scanning calorimetry (TG-DSC) curves of the SF6 epoxy thermal decomposition process. TG experiment results showed that the first stage dehydration process occurred in the range of 258-342°C for epoxy resin under SF6 atmosphere, with a weight loss of 18%. The second stage thermal decomposition occurred in the range of 342-481 °C with a weight loss of 11%. The final decomposition occurred in the range of 481-629°C, with a weight loss of 46%. Compared to the main thermal weight loss range (around 400 °C) of N2 epoxy resin,[14] SF6 accelerated the thermal decomposition process of epoxy materials obviously. The decomposition temperature range of SF6 increased by nearly 150 °C, which indicated that SF6 molecules have chemical corrosion effect on polymer chains. The synchronous DSC curves show that the corresponding characteristic decomposition peak temperatures of these three decomposition stages are 296 °C, 408 °C, and 551 °C. By integrating DSC curve heat flow area, we evaluated the heat release in the primary decomposition stage Δ H = 0.911 kJ / g, the heat release in the secondary decomposition stage Δ H = 0.598 kJ / g, and the ultimate decomposition exotherm Δ H = 9.963 kJ / g.

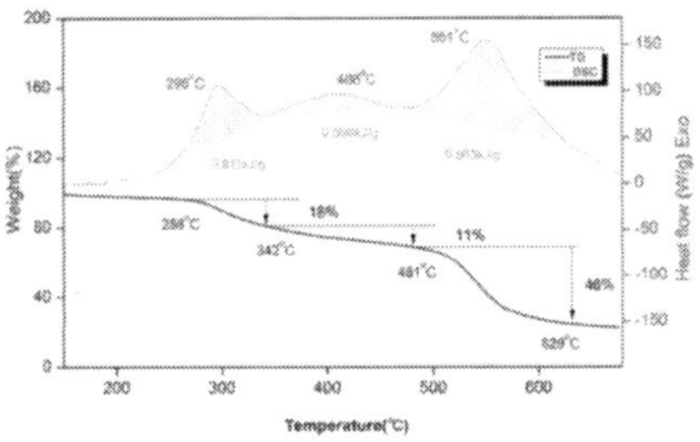

Fig. 1 TG-DSC of thermal decomposition of SF_6-Epoxy (heating rate 20 ℃/min)

The activation energies of the related decomposition processes are calculated from the sample thermal weight loss changes under different heating rates [15]. Table 1 lists the SF_6 epoxy system thermal decomposition parameters under five heating rates with a heating rate of 5-25 °C · min^{-1}. The modification of the heating rate did not change the coupling relationship of the three decomposition stages, and increasing the heating rate only accelerated the total weight loss process of the samples.The Kissinger and Flynn wall ozwa methods are commonly used methods to calculate the activation energies for thermal decomposition of polymers [15]. The ramp rates of 15, 20, and 25 °C/ min and the peak temperatures corresponding to the DTG curves were chosen to calculate apparent activation energy of primary decomposition (E_p) and final decomposition (E_t). By the Kissinger method, E_p was 61.3 kJ/mol and E_t was 256.24 kJ/mol,while E_p was 55.03 kJ/mol and E_t was 255.98 kJ/mol by the Flynn Wall Ozwa method.

Table 1 The influence of heating rate on SF6-Epoxy system

Itemβ/(°C·min^{-1})	5	10	15	20	25
T_i [a](°C)	233	252	249	258	256
T_{exo1} [b] (°C)	267	288	283	296	306
$ΔH_1$ [c](kJ/g)	1.527	1.317	1.003	0.911	0.780
T_1-T_2 (°C)	233-318	252-340	249-344	258-342	256-362
W_1 [d](%)	23	22	16	18	24
T_{exo2} (°C)	366	376	367	408	398
T_{exo3} (°C)	--	--	411	-	-
$ΔH_2$(kJ/g)	2.929	0.728	0.695	0.598	0.504
T_2-T_3 (°C)	318-457	340-462	344-462	342-481	362-487
W_2 (%)	29	15	19	11	10
T_{exo4}(°C)	495	511	538	551	556
$ΔH_3$(kJ/g)	31.49	13.52	11.49	9.963	8.722
T_3-T_4 (°C)	457-535	462-630	462-610	481-629	487-550
W_3 (%)	36	32	33	46	27
W_{end} [e](%)	88	69	68	75	61

 https://doi.org/10.1051/matecconf/202235801004

[a] The temperature at which weight loss begins; [b] The temperature corresponding to the exothermic peak; [c] The heat flow area of DSC curve to calculate the heat release in the primary decomposition stage; [d] The weight loss; [e] The total weight loss.

It can be concluded that the initial stage of thermal decomposition reaction in SF_6 epoxy system may mainly involve the dehydration process with small apparent activation energy, while the terminal thermal decomposition may involve the formation of C-F with large apparent activation energy of corresponding reaction. The secondary decomposition process involves complex chemical reactions and the whole system mass changes less since the characteristic decomposition peak of DTG curve is not obvious.

2.2 Thermal decomposition products of SF6 - Epoxy system

Reaction force field simulations revealed that the thermal decomposition of epoxy resin mainly produced escaped gas molecules such as CH_2O, H_2O, CO_2, CH_4 and H_2 [16]. In order to comprehensively investigate the thermal decomposition process of the SF_6 epxoy system, three characteristic decomposition stages, the primary pyrolysis stage (below 300°C), the secondary pyrolysis stage (300°C-480°C) as well as the terminal pyrolysis stage (above 480°C) were investigated. We analyzed the products composition of the escaped gas using online gas chromatography. SEM was used to study the morphology and elemental composition of the solid products.

Table 2 lists the three characteristic thermal decomposition stages from 270-517°C with the escape gas composition generated by the SF_6 epoxy system. During the primary decomposition stage below 300°C, the pyrolysis escape gas products are mainly CO, SO_2 as well as a small amount of hydrogen. This stage of thermal decomposition reaction is mainly the dehydration process of the epoxy resin bulk with that of SF_6. As the pyrolysis temperature increases, the decarbonylation process of the epoxy resin intensifies, with increasing CO concentration. SF_6 epoxy pyrolysis consisted of a series of complex chemical transformation producing a total of nine escape gases CO, CO_2, COS, CH_4, CS_2, SOF_2, SO_2, H_2S and H_2 during the secondary pyrolysis stage at 300-482°C. The secondary reactions involved include, the oxidation of CO by SO_2 to produce COS, which is further hydrolyzed to produce H_2S and CS_2. The water-gas shift of CO and H_2O produce H_2 to CO_2. The epoxy chain contains an active methyl groups in the bipartite position and increasing temperature leads to demethylation. When the temperature increased to 394°C, the concentration of CH_4 increased significantly compared to that at 352°C or 323 °C. The concomitant increase in CO, CO_2 and decrease in H_2 concentration within the final pyrolysis stage above 450 °C, indicating that the dehydration of the epoxy resin complete. The significant decrease of CH_4 content indicated that the demethylation mainly occurred at the secondary thermal decomposition. The decarbonization was through the entire pyrolysis process. The concentration of CO increased, same as the corresponding SOF_2, COS. H_2S is the hydrolysis products of COS. At 517 °C for 44 min, H_2 and CH_4 were no longer contained in the pyrolysis gas. The concentrations of CO and CO_2 decreased sharply. When the holding time is extended to 67 min, the concentrations of CO and CO_2 further decrease, indicating that the final pyrolysis is the decarbonylation and decarboxylation.

To further verify the generation pathway of SF_6 epoxy gas products, Fig.4-6 aligned the micromorphologies of the epoxy bulk, secondary pyrolysed solid products with secondary pyrolysed solid products. The elemental contents were listed in Table 3-5. After undergoing ultimate pyrolysis, the epoxy was completely pyrolyzed with no solid product remaining. Microtopographic analysis revealed the irregular bulk solid packing at the micrometer scale for the bulk of the epoxy. Both the primary pyrolyzed solid product and the secondary pyrolyzed solid product showed irregular bulksolid packing were observed. From the above figure, it is

Table 2 Characteristic gaseous products of thermal decomposition of SF6-Epxoy system

	T/ °C	Concentration of Characteristic escaping gaseous product (uL / L)								
		CO	CO_2	COS	CS_2	CH_4	H_2	H_2S	SO_2	SOF_2
Primary thermolysis	270	2.364	-	-	-	-	4.296	-	46.06	-
	288	7.43	-	-	-	-	13.63	1.022	66	-
Secondary thermolysis	300	23.84	87.66	0.7215	1.426	3.716	78.3	5.299	62.43	2.616
	305	40.92	85.66	4.232	6.162	12.57	28.8	193.8	155.8	2.385
	358	110.9	259.1	9.168	2.373	17.88	19.35	572.1	117.1	1.563
	451	144.4	388.4	15.53	-	31.99	22.98	205.5	107	2.744
	482	488.8	806	38.1	9.403	27.76	26.87	202.7	184.3	10.46
Final thermolysis	517	302.4	2443	35.56	-	3.84	2.274	20.04	887.5	291.5
	517[a]	189	1502	13.95	-	-	-	15.35	538.2	99.44
	517[b]	101.2	987.3	8.886	-	-	-	11.75	-	52.95

[a] Heat preservation for 44 min; [b] Heat preservation 67 min.

Figure 2 ESEM image of epoxy resin

Figure 3 ESEM image of primary pyrolysed solid product

Figure 4 ESEM image of econdary pyrolysed solid product

Table 3 Epoxy element concentration

Epoxy resin		
Element	Wt%	At%
CK	74.15	78.34
OK	15.64	12.41
NK	10.21	9.25

Table 4 Primary pyrolysed solid element concentration

Primary pyrolysed solid product		
Element	Wt%	At%
CK	70.50	75.42
OK	21.63	17.37
NK	7.87	7.22

Table 5 Secondary pyrolysed solid element concentration

Secondary pyrolysed solid product		
Element	Wt%	At%
CK	70.50	75.42
OK	21.63	17.37
NK	7.87	7.22

found that the raw epoxy primary carbonized solid product and the secondary carbonized solid product. These solid decomposition products have less fine agglomerated morphologies relative to the overall morphology of the epoxy resin. The size of the solid becomes larger as the pyrolysis deepening. EDS spectroscopy analysis showed that the elemental composition of the raw epoxy with the solid decomposition products mainly contained carbon, oxygen, and nitrogen, while sulfur and fluorine elements were absent. These results show that during the thermal decomposition process, SF_6 does not directly produce multiphase thermal erosion reaction on epoxy resin. The thermal decomposition of SF_6 epoxy mainly includes the rapid pyrolysis of epoxy resin, and the secondary reaction of SF_6 and gaseous product of epoxy resin. The carbon to oxygen element ratios of the solid samples were 6.3:1, 4.3:1;3.6:1, respectively, indicating that the decarbonylation process of the epoxy resin was significantly faster than the dehydration process, which was consistent with the changes of concentration of gaseous products CO and H_2O related product (H_2S, SOF_2) .

2.3 Summary

According to the synchronous TG/DSC/GC analysis, we propose a panoramic reaction course for thermal decomposition of SF_6-epoxy system (Fig. **5**). The decomposition process starts with the dehydration of epoxy resin. Sequentially, the hydrolysis reaction of SF_6 mutually coupled with the decarbonylation reaction of epoxide. The secondary reactions occur between H_2O, CO, SO_2 which converted to CO_2, COS, CS_2, H_2, and H_2S. These secondary reactions occur mainly in the secondary decomposition stage, in which CO reacts with SO_2 to produce COS, CS_2. From the correspondence consideration of thermal attack of insulating materials, CO is a singular source with SO_2, but its secondary reaction is complicated and not an ideal thermal failure marker characteristic decomposition products. CO_2, although a more complex source, formed as a result of the hydrolysis of CO and COS, with a dramatic increase in its content and a predominant epoxide thermal decarboxylation process at the terminal decomposition stage. Therefore, CO_2 can be used as a characteristic decomposition component for determining the deterioration of epoxy resin hot corrosion at high temperatures (> 500°C). As a singular source characteristic decomposition component, CH_4 can directly correspond to the thermal deterioration of epoxy insulator at 300-500°C. SOF_2, H_2S and H_2 are all produced upon hydrolysis, and a combination of these characteristic components can determine the ultra-low concentration of water of SF_6 epoxy system.

3. Conclusion

In this paper, the material and energy changes of thermal decomposition process of SF_6-Epoxy system were systematically studied, and the temperature range and exothermic heat that characterized the three characteristic pyrolyses were determined. Nine carbon based and sulfoxy decomposition products derived from the pyrolysis of SF_6 epoxy are recognized by on-line GC. After comprehensively analyzing the composition of the decomposition products of gas solids, in this paper, the panoramic reaction course of SF_6 epoxy pyrolysis is resolved, and this information provides an experimental basis for the detection of thermal decomposition failures of epoxy. CO_2 is the characteristic decomposition component in determining the deterioration of the thermal corrosion of epoxy resins at high temperatures (> 500°C), CH4 directly corresponds to the deterioration of the thermal degradation of the epoxy insulation that occurs at 300-500°C, while SOF_2, H_2S, and H_2 are important evidence to determine the micro water content of SF_6-Epoxy system. Because of the complex secondary reactions, CO and SO_2 are not suitable as thermal failure marker characteristic decomposition components.

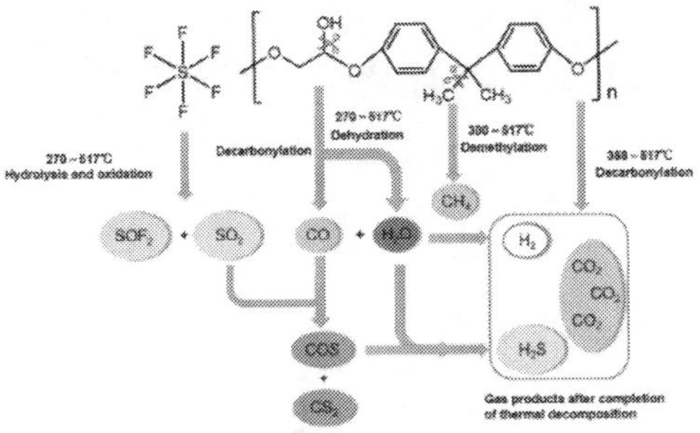

Figure 5. Panoramic analysis of pyrolysis reactions of the SF6 -Epoxy insulating system

Acknowledgements

Project Supported by State Grid Shaanxi Electric Power Company Science and Technology Project (No. 5226ky19004D).

References

1. X. Y. Qin, X. H. Lv, Y. Chen., High Volt. Appar., 57, 72 (2021)
2. X.L. Yan, C.Y.Wang, G. Song, High Volt. Appar., 49, 1 (2013)
3. S. C. Ji, L. P. Zhong, K. Liu, Proc. CSEE, 35, 2318 (2015)
4. J. Tang, D. Yang, F. P. Zeng, Tran. CES, 31, 41 (2016)
5. X. L. Yand, G. Song, C. Y. Wang, Elec. Power Auto. Eq., 34, 83 (2014)
6. Q. Yao, Y. Wang, Y. Liu, High Volt. Appar., 46, 53 (2010)
7. X. L. Yan, Y. S. Ji, C. Y. Wang, Smart Grid, 4, 653 (2016)
8. S. V. Levchik, E. D. Weil, Polymer Intl., 53, 1901 (2010)
9. T. C. Zhao, X. She, Q. Li, High Volt. Appar., 56, 103 (2020)
10. L. Cheng, J. Tang, X. J. Huang, High Volt. Tech., 41, 453 (2015)
11. H. Wen, X. X. Zhang, IEEE Tran. Dielec. Elec. Insl., 26, 1411 (2019)
12. F. Mustata, N. Tudorachi., J. Therm. Anal. Calorim., 125, 97(2016)
13. C. Ma, D. Sánchez-Rodríguez, T.Kamo, J. Hazard. Mater., 412, 125329 (2021)
14. K. Q. Zhou, R. Gao, X. D. Qian, J. Hazard.Mater., 338, 343 (2017)
15. P. Peng, P. Ren, Q. M. Li, High Volt. Tech., 46, 3622 (2020)
16. H. Jin, R. H. Li, X. B. Meng, High Volt. Appar., 54, 44 (2018)
17. H. Wen, X. X. Zhang, R. Xia, Mater., 12, 75 (2019)

Synthesis of silver coatings by brush plating with cyanide-free solution

Baoshuai Du*, Zhongwen Zhang, Shuai Suo, Xinmei Li

State Grid Shandong Electric Power Research Institute, Jinan 250003, China

Abstract. Silver coating can be used to improve the surface properties of components made of copper in terms of corrosion resistance, wear resistance and electrical contact conductivity. A cyanide-free solution based on $Na_2S_2O_3$ is proposed for the brush plating of silver coating on copper substrate. XRD, SEM and microhardness tester were used to characterize the microstructure and properties of the coatings. It is found that this cyanide-free solution can deposit silver coating with high efficiency. Coatings consist of nano-sized grains which present as aggregated nodular morphology. The brush plated coatings show microhardness of 122.47 Hv, which is promising in the engineering application of electrical contact components.

Keywords: Cyanide-free solution, silver coating, brush plating, microstructure

1. Introduction

For electrical contact components made of copper, in order to increase their corrosion resistance and improve their electrical conductivity, a protective coating is normally applied to the surface of the components. It has been proved that silver coating is effective in terms of enhancing the surface properties beyond the capability of the substrate[1,2]. Electrodeposition is often used to perform the fabrication of the silver metallic coating. Moreover, driven by the safety and environment concern, there have been a lot of investigations devoted to the development of cyanide-free electroplating bath[3].

Brush plating, or selective plating, uses an anodic brush to deliver the electroplating solution to the cathodic workpiece. Direct contact between the anode and cathode is prevented by the electrically insulating cathode which covers the insoluble brush. The brushing action disturbs the hydrodynamic boundary layer which results in fast solution movement. Thus, although the brush plating is characterized by the same principle with electroplating, it possesses its own features, such as portability and high efficiency[4,5]. It has been proved that brush plating is promising in the on-site repair of copper electrical contact components with silver coating.

In this study, a cyanide-free solution is developed for the brush plating of silver coating. Phase constituents and microstructure of the coating was characterized. It is expected that the developed bush plating technique can be used for the repair and remanufacturing of copper components used in electric power industry.

2. Experimental material and methods

Copper plates with dimensions of 60 mm×30 mm×2 mm were used for the bush plating process. Cathodic clean with alkaline solution was conducted before the bush plating. This is followed by anodic etch in a weak acidic electrolyte solution in order to improve the coating adhesion. The brush plating solution is composed of $Na_2S_2O_3$(200-220 g/L), K2S2O5(40-60 g/L), AgNO3(30-50 g/L) and additives (1g/L). Solution temperature was kept in the range of 30-40°C. Electrical brush plating was performed manually at a working voltage of 1.7-2.5 V.

Surface morphology was examined with JSM-7800F scanning electron microscopy (SEM) and micro-zone chemical was conducted using the attached X-ray energy disperse spectroscopy (EDS). Phase constituents were identified using DX 2700 diffractometer with Cu k_α source operated at 40 mA and 35 kV. The step size and scan speed were set at 0.02° and 1°/min respectively and the range of scan angle is from 20° to 100°. Microhardness of the brush plated coating is evaluated using a microhardness tester (Tukon 1102) with loading of 50 g and duration time of 12 s.

* Corresponding author: dubaoshuai@163.com

3. Results and discussion

3.1 Appearance and coating thickness

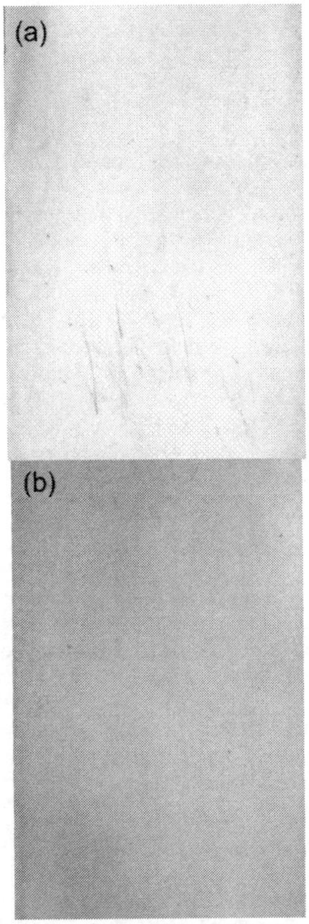

Fig. 1 Typical coating appearance: (a) 1.7 V; (b) 2.5V

The brush plated coatings show smooth and uniform visual appearance with white color, which is different from other cyanide-free electrodeposited silver coatings characterized by a shining mirror bright deposit[3]. Nevertheless, the coating is coherent and defect-free, indicating its potential to be used as protective coating on the copper substrate. Besides, it can be found that coating deposited with higher voltage exhibit a greyer color, which may be induced by the high energy input during the depositing process.

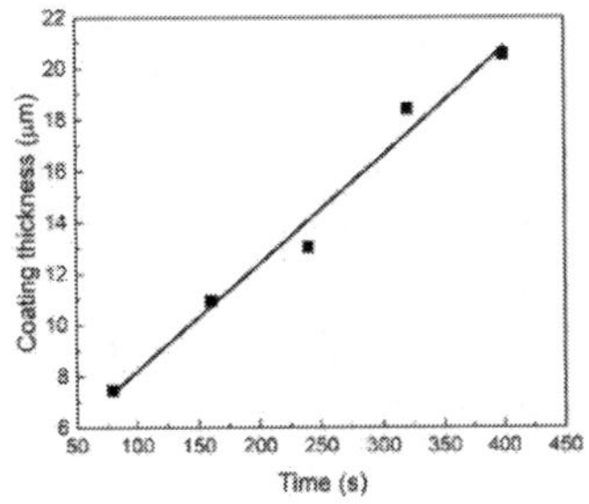

Fig. 2 Coating thickness versus brushing time

The variation of coating thickness with brush plating time is shown Fig. 2. It is apparent that the increase of brush plating time causes a linear increase of coating thickness. The linear fitting function between the brushing time and coating thickness is given by:

$$CT = 3.9836 + 0.04206 \times t \quad (1)$$

where CT is the coating thickness, and t is the brushing time. It ca be found that final coating thickness is more than 20 μm with brushing time within 400 s. Compared with traditional electroplating process, brush plating shows a higher efficiency.

3.2 Phase identification and grain size

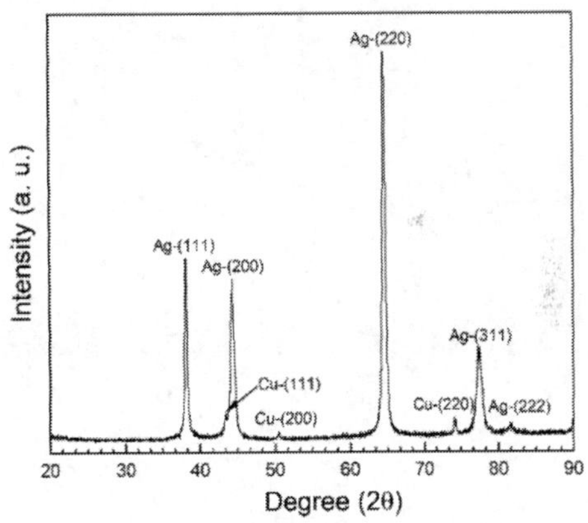

Fig. 3 XRD pattern of the brush plated silver coating

Phase identification is performed using the XRD and diffraction peaks are presented in Fig. 3. XRD pattern of the brush plated coating shows the diffraction peaks of silver and copper. It is evident that the synthesized coating is composed of pure silver with no other phases. Diffraction peaks of copper is originated from the substrate.

Grain size of the brush plated coating was quantified by applying the Scherrer equation of the (111) peaks. The Scherrer equation is known as:

$$D = \frac{K\lambda}{\beta \cos\theta} \quad (2)$$

where K is dimensionless shape factor, λ is the wave length of the X ray, β is the line broadening at half the maximum intensity, and θ is the Bragg angle. According to equation (2), the calculated crystallite grain size is 26 nm, which is in the nano-size range. Thus, it shows that with proper solution bush plating can be used to fabricate silver coating with very fine crystal structure.

3.3 Microstructure and microhardness

Fig. 4 SEM images of the brush plated coating: (a) ×1000; (b) ×5000

SEM images of the brush plated coating with voltage of 1.7 V is shown in Fig. 4. It can be found the coating surface is homogeneous and covered with aggregated grains. The well-formed silver crystallites are apparent on the surface. The crystallites present as nodular morphology. This kind of appearance is typical for the electrodeposited coatings. For brush plating, the repetitive movement of the brush can effectively interrupt the grain growth of the deposit and motivate the crystal nucleation, thus resulting in finer grain size.

Microhardness measurement shows that the average microhardness is 122.47 Hv and 105.39 for the coating bush plated with 1.7 V and 2.5 V respectively. Compared with other electroplating Ag coatings, this is a relatively higher value[6,7]. This is may be due to the fine crystal size presented in the plated coating. For components used as the electrical contact, adhesive wear resistance is often considered. And it is generally accepted that adhesive wear resistance is governed by the Archard law. According to this law, higher hardness indicates better wear resistance. Thus, conclusion can be made that this brush plated coating shows promising as wear resistant coating for electrical contact components.

4. Conclusion

The brush plated silver metallic coating has been developed for engineering application, which uses a cyanide-free solution. Phase identification shows that the coating consists of pure silver with crystal size of 26nm. The brush plated coating consists of fine aggregations of nodular silver crystallites. The highest microhardness of the brush plated coating is Hv 122.47, which can be used as protective coating electrical contact components.

Acknowledgments

This work was financially supported by Science and Technology Project of State Grid Shandong Electric Power Company (No. 520620210020, Research on Fabrication and Application of Electric Brush Plated High Performance Silver Alloy Coatings).

References

1. Dadvand, Mina, Nazila. Pulse Electrodeposition of Nanostructured Silver-Tungsten-Cobalt Oxide Composite from a Non-Cyanide Plating Bath[J]. Journal of the Electrochemical Society, 2014.
2. Jiao Z B , Schuh C A . Nanocrystalline Ag-W alloys lose stability upon solute desegregation from grain boundaries[J]. Acta Materialia, 2018, 161:194-206.
3. Liu A M , Ren X F , Wang C , et al. DMH and NA-based cyanide-free silver electroplating bath: a promising alternative to cyanide ones in microelectronics[J]. Ionics, 2021, 27(1):417-422.
4. Zhong Z , Clouser S J . Nickel–tungsten alloy brush plating for engineering applications[J]. Surface and Coatings Technology, 2014, 240:380-386.

5. Behera P , Rajagopalan S K , Brahimi S , et al. Effect of brush plating process variables on the microstructures of Cd and ZnNi coatings and hydrogen embrittlement[J]. Surface and Coatings Technology, 2021, 417(3):127181.
6. Bs A, Sj B, Sd A, et al. A comparative study of electrodeposition routes for obtaining silver coatings from a novel and environment-friendly thiosulphate-based cyanide-free electroplating bath[J]. Surface and Coatings Technology, 2021.
7. Wang Y, Tay S L, Zhou X, et al. Influence of Bi addition on the property of Ag-Bi nano-composite coatings[J]. Surface and Coatings Technology, 2017, 349.

Cause and mechanism analysis of cracking and leakage of water wall tube

Jie Zhao [1,*], Fangyuan Song [2]

[1] Inner Mongolia Power (Group) Co.,Ltd, Hohhot 010020, China
[2] Inner Mongolia Electric Power Training Center, Huhhot 010020, China

Abstract: In this paper, the cracking and leakage cause of the water wall tube was studied by means of macro morphology analysis, chemical composition analysis, microstructure detection, mechanical properties test and energy dispersive X-ray spectroscopy (EDS) analysis. The results indicated that the cracking of water wall reducer is mainly related to high temperature fatigue and sulfur corrosion. During operation, under the combined action of thermal expansion and contraction caused by frequent boiler load changes and axial alternating stress caused by pipe wall vibration, thermal fatigue cracks are gradually formed on the outer wall of the steel pipe toward the fire side. Under the combined action of the alternating stress and S corrosion, the cracks are gradually spread from outer to inner wall, and finally penetrate the tube wall, causing leakage. Finally, the corresponding preventive measures of water wall leakage are put forward to improve the safety and economy of power plant boiler.

Keywords: water wall reducer; thermal fatigue; sulfur corrosion; leakage.

1. Introduction

As an important energy industry, thermal power generation industry is closely related to people's daily necessities and is also an important original driving force for the development of national economy. In the current power industry, thermal power generation has always been in the absolute dominant position of China's power industry. Therefore, the thermal power industry has an important impact on the modernization development. The continuous optimization and efficient development of thermal power generation is an important guarantee for our country to complete the socialist modernization drive. In thermal power generating units, utility boiler is one of the unstable equipment, the safety condition of which limits the safety of the whole unit. However, the operation status of the water wall determines the safety of the boiler to a great extent. Because the water wall of utility boiler is not only affected by the high temperature and high pressure inside the pipe, but also by the wear and corrosion on the both sides of flue gas and working medium, thus the working environment the water wall is very bad.

In recent years, with the rapid development of thermal power in China, more and more supercritical and ultra supercritical units have been put into use. With the extension of operation time, the problem of explosion and leakage caused by early failure and cracking of water wall gradually appears and becomes more and more prominent. Some power plants even shut down frequently and unplanned, which has become an important factor affecting the safe and stable operation of ultra supercritical units [1-2]. Water wall is the heating surface distributed on the inner wall of boiler furnace and cooled by water. It is also the main heating surface in the furnace. Its reliability is directly related to the safe operation of power plant units [3-6]. With the development of boiler to large capacity and high parameters, the temperature of boiler water wall increases accordingly. In May 2021, a water wall tube leakage failure occurred in a power plant, and the specification of the water wall tube in question was Φ6×7 mm, 20G. The boiler described in this paper is a single drum natural circulation, ultra-high pressure parameter and one-time intermediate reheat solid slag discharge pulverized coal boiler manufactured by Harbin Boiler Factory with the model of HG-670/13.7-YM13. The high-temperature reheater is of W-shaped structure and arranged in the horizontal flue downstream, with a total of 114 rows. The design working pressure is 2.4 MPa, the design working temperature is 540 °C and the boiler was put into operation in 2002. In order to find out the causes of water wall tube leakage and avoid the recurrence of similar accidents, this paper makes a comprehensive inspection and Analysis on the leaking tube section and its adjacent tube section of water wall tube.

* Corresponding author: 158062831@qq.com

2. Experimental results and analysis

2.1 Macroscopic morphology observation and analysis

It can be seen in Fig. 1 (a) that there is one leakage point on the tube wall on the fire side of the water-cooled wall tube. The size of the leakage point is small, and there is no obvious expansion of the pipe diameter near the explosion opening, while a large number of transverse cracks can be seen on the outer wall surface of the steel pipe on the opposite side of the explosion opening. In addition, there is a certain thickness of oxide skin on the tube wall of the water wall, and there are no signs of wear and other mechanical damage, as shown in Fig. 1 (b).

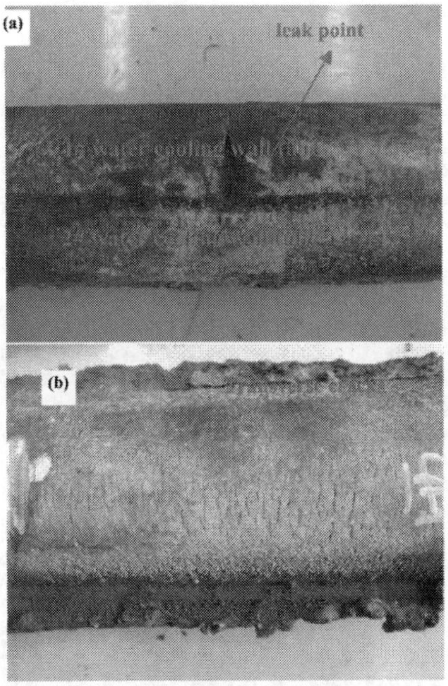

(a) leak point (b) transverse cracks on the opposite side of leakage point

Fig.1 Macroscopic morphology of different parts of water wall reducer

2.2 Chemical composition detection and analysis

According to the requirements of GB/T 4336-2016[7], the SPECZROMAXx desktop direct reading spectrometer is used to analyze the element chemical composition of water wall tubes, and the results are shown in Table 1. It can be seen that the content of each element in the water wall tube meets the requirements of the standard.

Table 1 Chemical composition test results of the water wall reducer

detection element	C	Si	Mn	P	S
measured value	0.19	0.26	0.52	0.008	0.001
GB/T 5310-2017[8]	0.17～0.23	0.17～0.37	0.35～0.65	≤0.025	≤0.015

2.3 Microstructure detection and analysis

According to the requirements of DL/T 884-2019 [9], the microstructure of water wall tube was detected by Axio Observer.Alm type metallographic microscope. It can be seen from the metallographic structure observation that the water wall tube has a small leakage point, and the cut decreases from the outer wall to the inner wall, presenting funnel-shaped. The matrix structure near the explosion opening is consisted of ferrites and pearlites, and no deformation and elongation grain are found the tip. The matrix structure near the blast hole is consisted of ferrites and pearlites, the carbides in the pearlite area are begun to disperse, while the pearlites morphology is still obvious, and the spheroidization level of which is rated as level 2, as shown in Fig. 2(a)~Fig. 2(b). In addition, a lot of transverse cracks are found near the burst, and the cracks are initiated from the outer wall and expanded to the inner wall. The shape of the crack is nearly wedge-shaped with a sharp tail end, and the interior is full of corrosions. The matrix structure of the water wall tube is mainly consisted of ferrites and pearlites, and the carbides in pearlite area are begun to disperse, while the pearlite morphology is still obvious, the spheroidization level of whicu is rated to level 2, as shown in Fig. 2(c) and Fig. 2(d).

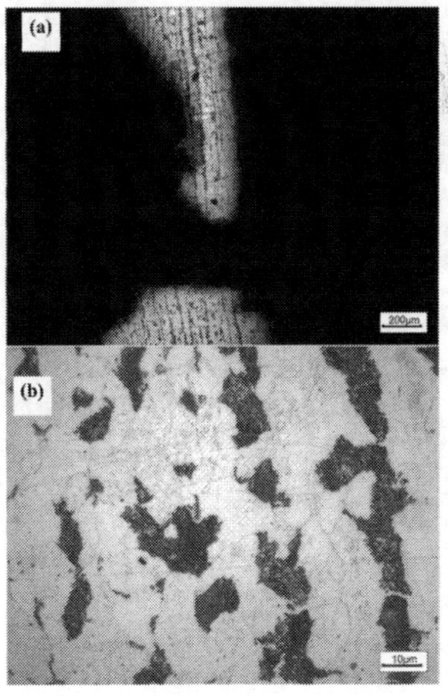

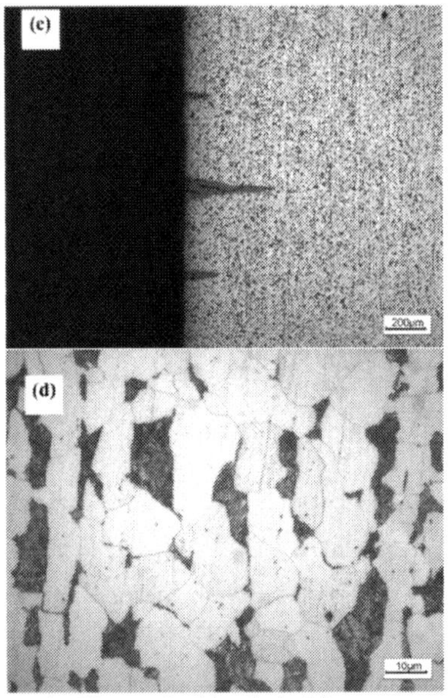

(a) crater tip (b) near the blast hole (c) transverse cracks (d) matrix

Fig.2 Microstructure morphology of different parts of water wall reducer

2.4 SEM observation of micro morphology and EDS analysis

It can be seen form the observation results of SEM micro morphology that the transverse crack of water wall tube wall is mainly transgranular, and the crack is filled with corrosion products, as shown in Figure 3. The chemical elements of corrosion products are then analyzed by EDS energy spectrometer, and the results are shown in Table 2. It can be seen from the table that the corrosion products rare mainly consisted of Fe and O, a small amount of S elements are also detected[10-12].

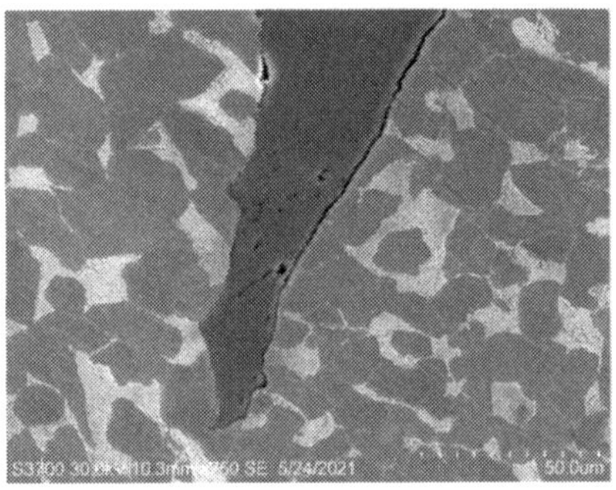

Fig.3 SEM micrograph of crack tip

Table 2 Energy spectrum analysis results of main chemical elements in local area of crack tip

main element composition of the crack tip	Fe	0	S
proportion of each element/%	54.60	44.51	0.89

2.5 Testing and analysis of mechanical properties

Take samples from the water wall tube for tensile performance test, and the results are shown in Table 3. It can be seen that the mechanical properties of the exploded water wall tube are all met the requirements of standard values.

Table 3 Test results of mechanical properties for water wall tubes

Sample	σ_b/MPa	σ_s/MPa	δ/%
water wall tube	307	453	28
Standard value[8]	245	410 ~ 450	≥24

2.6 Discussion and analysis

The macroscopic morphology observation results shown that the water wall tube has one leakage point, and the diameter near the leakage point is not expanded. Multiple transverse cracks can be seen on the surface of water wall. The chemical composition detection results revealed that the material of water wall tube meets the standard requirements. The results of the microstructure observation indicate that there are no elongated grains formed at the burst tip of the water wall tube. In addition, there are multiple transverse cracks extending from the outer wall to the inner wallwwater wall are found far from the burst, and the cracks are filled with amount of corrosion products containing a small amount of S element, indicating that the formation of cracks is related to high-temperature sulfur corrosion. The test results of mechanical properties show that the mechanical properties of the exploded water wall tube meet the standard requirements. To sum up, the cracking and leakage of water wall pipe are mainly related to high temperature fatigue and S corrosion.

3. Conclusions and suggestions

The cracking and leakage of water wall tube are mainly related to thermal fatigue and S corrosion. During the operation of the boiler, transverse fatigue cracks are gradually formed at the outer wall of the water wall tube under the long-term action of axial alternating thermal stresses. Then the cracks are gradually expand from the outer wall to the inner wall under the combined action of S element corrosion and alternating stresses, and finally penetrate the tube wall to cause leakage. At last, it is suggested that the nearby water wall pipes should be inspected first, and the water wall tubes with similar cracks should be replaced in time. In addition, regularly clean the scale layer on the fire side of the water wall to reduce the high-temperature S corrosion. Finally, the metal technical supervision of heating surfaces at all levels should be strengthened to effectively avoid the recurrence of the same type of failure happening again.

Reference

1. M. Nurbanasari. Investigation of leakage on water wall tube in a 660 MW supercritical boiler[J]. Journal of Failure Analysis & Prevention, 2014, 14(5): 657-661.
2. X.C. Song, X.J.Wu, Y.H. Kang. The method and apparatus of magnetic nondestructive testing for boiler waterwall tube[J]. Key Engineering Materials, 2004, 270-273: 642-646.
3. L.B. Chen. An analysis and investigation on the leakage in water-wall tube of profiled flue of the coal-burning boiler[J]. Journal of Shandong University (Engineering Science), 2005, 35(2): 14-17.
4. F. Ren, J. Xu, J. Si. Failure analysis of a water wall tube[C]. Journal of Physics Conference Series, 2020, 1635: 012072.
5. P. Sun, C.C. Zhao, N. Wu et al. Water Wall Tube Leakage of 600 MW Power Plant Boiler[J]. Foundry Technology, 2017, 38(11): 2664-2666.
6. Y.T. Yan, L.B. Niu, Y.X. Long, et al. Leakage cause analysis on water wall tube of a supercritical 600 MW e unit boiler[J]. PTCA (PARTA: PHYS.TEST.), 2019, 55(11): 779-803.
7. National Technical Committee for Steel Standardization. Carbon and low-alloy steel-determination of multi-element contents-spark discharge atomic emission spectrometric method (routine method): GB/T 4336-2016[S]. Beijing: China Standards Press, 2016.
8. National Technical Committee for Steel Standardization. Seamless steel tubes and pipes for high pressure boiler: GB/T 5310-2017[S]. Beijing: China Standards Press, 2017.
9. Technical Committee for metal material standards of power station in power industry. Power plant metallographic inspection and assessment guideline: DL/T 884-2019[S]. Beijing: China Electric Power Press, 2019.
10. Y.B. He, W.J. Wang, C.M. Li, et al. Causes and prevention of high temperatures sulfur corrosion of water wall tubes of supercritical concurrent boiler[J]. Electric Power Construction, 2010, 31(10): 61-66.
11. Z. Yin, H. Cai, Z.T. Hou, et al. Study on typical transverse crack of water wall tube of 1000 MW domestic ultra-supercritical units[J]. Journal of Science and Engineering, 2015, 33(03): 442-447.
12. H. Xu, X.H. Xiong, X. Liu, et al. Study on high temperature corrosion products composition of water wall tube in a utility boiler and assessment of reconstruction[J]. Clean Coal Technology, 2019, 25(02): 144-148.

https://doi.org/10.1051/matecconf/202235801007

Dynamic balance design and simulation verification of reed system of loom beating up mechanism

Kang Wang*, Sai Guo

School of Mechanical Engineering, Tiangong University, Tianjin, China

Abstract. In order to reduce the vibration caused by the eccentricity of the reed system in the beating up mechanism of the loom, the centrifugal force generated by the eccentric reed system is decomposed into two parallel base planes, and a dynamic balance design scheme based on the principle of counterweight is established. Based on structural analysis and theoretical calculation, the reasonable balance weight mass, structure and assembly position are determined, and the reed system simulation model based on Adams/view platform is constructed. The correctness of the dynamic balance design scheme is verified by rigid body dynamics simulation and error analysis. The analysis results show that the centrifugal force simulation value F=347.008N, the design value F1=347.222N, and the error is 0.06%, which meets the requirements of mechanical accuracy design. It provides an important technical reference for dynamic balance design, vibration reduction and noise reduction of fixed axis rotation system.

1. Introduction

The high-speed running mechanism must be accompanied by greater inertial force and moment of inertia, which leads to greater vibration and noise of the machine. This problem has become a restrictive factor to ensure the stable operation of looms and improve product quality.
Kuchar and Maciej studied the influence of reed system vibration on fabric [1] and the simulation experiment is also done [2]. Chung Feng Jeffrey Kuo,Studied and analyzed the dynamic model of beating up mechanism [3]. Celik, H. I discussed the kinematic law of beating up mechanism [4]. Katunskis.J respectively carried out theoretical analysis and experimental verification on the beating up stage [5].
This paper takes the reed system as the research object, and carries out a series of research on it, such as structural analysis, theoretical calculation, CAD modeling, dynamic simulation and error analysis, so as to provide theoretical basis and technical reference for the vibration and noise reduction and high-speed research and development of loom beating up mechanism.

2. Reed system structure and its kinematic law

2.1 Reed system structure

The reed system is the core component of the loom, as shown in figure 1. Its components include: reed, reed base, reed base foot and rotating shaft. The reed is supported in the reed base. The reed, reed base, reed base foot and rotating shaft are fixedly connected in turn. The whole reed system can be regarded as a rigid body. The rotating shaft is installed on the box through the bearing, and the transmission mechanism drives its operation, so as to realize the weft beating operation process.

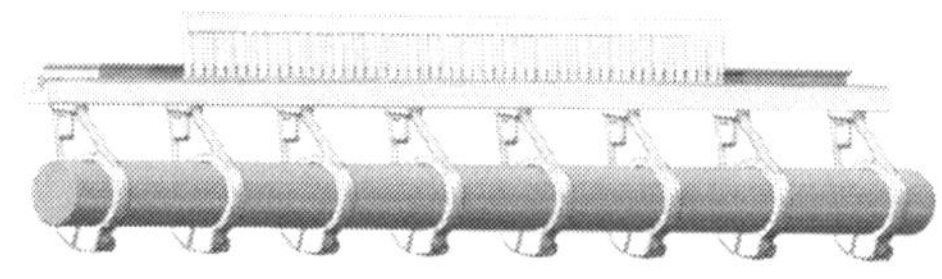

Figure 1. Structural diagram of reed system.

2.2 Kinematic law

In this paper, the four-bar linkage is used as the transmission mechanism, and its schematic diagram is shown in figure 2. The rectangular coordinate system is established with the crank rotation center O_1 as the origin, the limit position of the front dead center of the beating up mechanism $O_1A_1B_1C_1$ as the initial position, the crank rotates counterclockwise, O_2 is the rotating shaft of the reed system, L_1 is the crank, L_2 is the connecting rod, L_3 is the rocker and the swing angle of the rocker is ψ.In this paper, the complex vector method is used to solve the motion law of reed system with MATLAB, and the crank speed is set to 1000 r/min.

* Corresponding author: experience202109@163.com

https://doi.org/10.1051/matecconf/202235801007

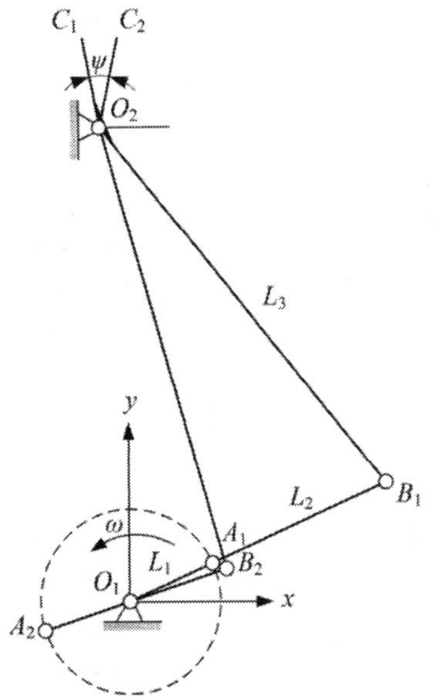

Figure 2. Four link beating up mechanism.

The eccentric reed system will inevitably produce centrifugal force during operation, which acts on the frame through the bearing, resulting in severe vibration and noise of the machine. It can be seen from figure 4 that this motion is variable speed fixed axis swing, the maximum angular velocity of return is 25.459 rad/s, the angular velocity is the largest, and the centrifugal force is also the largest.

3. Principle and method of dynamic balance

3.1 Principle analysis

In the dynamic balance design of the rotor, firstly, the eccentric mass in each rotation plane shall be determined through structural analysis, and then the number, size and position of counterweight required to make the rotor achieve dynamic balance shall be calculated according to the distribution of eccentric mass.
The reed system is approximately equivalent to a particle m, and figure 3 shows the design principle of dynamic balance.

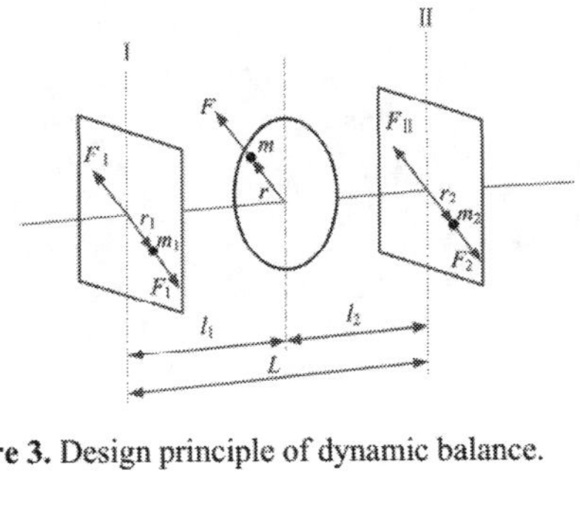

Figure 3. Design principle of dynamic balance.

According to theoretical mechanics, the force can be decomposed into two components parallel to it. According to the dynamic balance design principle shown in figure 3, two balance base planes I and II are selected as the installation position of counterweight, and the centrifugal force F is decomposed into planes I and II. Obviously, in order to make the reed system reach the dynamic balance state, it is only necessary to apply an appropriate balance mass m_1 and m_2 in planes I and II respectively, so as to generate two centrifugal inertial forces F_1 and F_2 in the opposite direction, and finally make the sum of the centrifugal inertial forces in the two planes equal to zero. In figure 3, F is the centrifugal force generated by the eccentric mass m, where L_1=204mm, L_2=204mm, r=26.509mm, r_1=40.56mm, r_2=40.56mm.

3.2 Counterweight calculation

According to the static equilibrium conditions, the following equilibrium equations exist in the equilibrium bases I and II:

$$F_{\rm I} + F_1 = 0 \tag{1}$$

$$F_{\rm I} + F_{\rm II} = 0 \tag{2}$$

Where: F_1 and F_2 are the centrifugal forces generated by the balance weights m_1 and m_2.
According to the rotor dynamics theory and figure 3, F_1 and F_2 can be expressed as:

$$F_1 = F\frac{l_1}{L} = mr\omega^2\frac{l_1}{L} \tag{3}$$

$$F_2 = F\frac{l_2}{L} = mr\omega^2\frac{l_2}{L} \tag{4}$$

Replace equation (3) into static equilibrium equation (2) and eliminate it ω^2 available:

$$mr\frac{l_1}{L} + m_1 r_1 = 0 \tag{5}$$

The mass of the reed system measured in the CAD environment is 40.416kg, and the counterweights m_1 and m_2 in the balance base surfaces I and II are 13.2 kg calculated by equation (5).

3.3 Balance weight structure design

According to the center of mass position and counterweight mass, a sector balance weight is designed in the balance base plane to offset the decomposition forces $F_{\rm I}$ and $F_{\rm II}$. The design parameters in the two balance base planes are the same, so the structure of the balance weight is also the same. The section of the sector balance weight designed in this paper is shown in figure 6, R_0=60mm, R_1=40mm, R=152mm, α=120°. The balance weight is made of steel with mass density ρ=8000 kg/m^3, the thickness of the balance weight h=80.8mm is calculated from formula (5). According to the geometric parameters of the balance weight, its three-dimensional solid model is established in Solidworks environment.
The installation position and initial phase of the balance weight are determined according to the dynamic balance design, and the balance weight is assembled on the rotating shaft. The reed system model is shown in figure 5. The mass center of the model is measured in

Solidworks. The overall mass center of the assembled model is basically consistent with the rotation center of the rotating shaft, which verifies the correctness of the structural design of the balance weight.

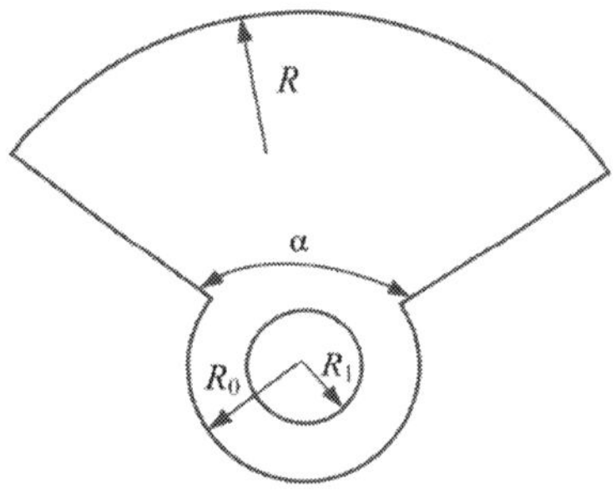

Figure 4. Schematic diagram of balance weight.

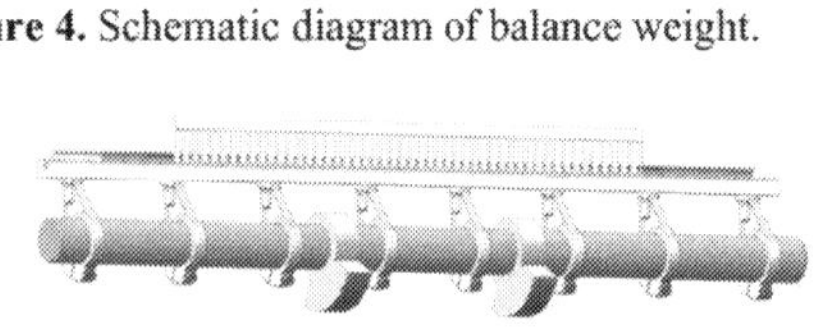

Figure 5. 3D CAD model of balance weight.

4. Simulation Verification

4.1 Model construction

The three-dimensional model is imported into Adams/view environment through the interface program, and the material properties, constraints and connecting pairs between components are defined. The simulation is set to two swing cycles. Because the cubic spline interpolation method is intuitive, accurate derivation and smooth curve, the cubic spline interpolation method is adopted.

4.2 Dynamic simulation

The mass center displacement curve of the balance weight is obtained through data post-processing, as shown in.

It can be seen from that when the reed system moves with the specified law, the displacement amplitude of the mass center of the balance weight is 40.5mm, which is consistent with the design parameters of the rotation radius of the mass center of the balance weight. The mass center displacement law changes periodically with time, and the curve is smooth without obvious fluctuation, which meets the requirements of rotor dynamics design and speed law.

shows the mass center velocity curve of the balance weight. It can be seen from the curve in figure 8 that the curve is the zero point, that is, the reed system reverses, the return speed amplitude is v=1.0326m/s. The expression of centrifugal force is:

$$F = m_1 \frac{v_1^2}{r_1} \tag{6}$$

Substitute the parameters to calculate the centrifugal force of the balance weight, and the simulation result is F=347.008N; Because the two balance weights are centrosymmetric, the centroid velocity and centrifugal force are the same.

4.3 Error analysis

According to the dynamic balance theory, the design parameters m, r,ω,l_1, and L are substituted into equations (3) and (4) respectively to calculate the design value of centrifugal inertia force of balance weight: F_1=347.222N. In order to verify the accuracy of dynamic balance design, the error between centrifugal force design value F_1 and simulation value F is analyzed.

$$\Delta_1 = \frac{F_1 - F}{F_1} \times 100\% \tag{7}$$

Calculate the balance weight design error Δ_1 as 0.06% according to equation (7). According to engineering design specifications and experience, mechanical design errors are unavoidable in practice. According to the error calculation results obtained in this paper, the error between the dynamic balance theoretical design value and the dynamic simulation value meets the design requirements, which verifies the correctness of the design of the rotating shaft reed system, and shows that the dynamic balance theoretical calculation result is accurate and the design scheme is reasonable and feasible.

5. Conclusion

In this paper, the shape, mass and assembly position of the balance weight are determined through the dynamic balance design and calculation of the reed system, so that the center of mass of the whole reed system is consistent with the rotation center, The centrifugal force simulation value of the balance weight is F=347.008N, the design value is F_1=347.222N, and the error is 0.06%, which meets the design requirements of mechanical accuracy, effectively offsets the centrifugal inertia force generated in the swing process, so as to reduce the vibration and noise of the machine, The dynamic simulation results verify the correctness of the dynamic balance design scheme.

It can be concluded that the dynamic balance method established in this paper is practical and effective, and is easy to realize in engineering. It provides an important technical reference in the dynamic balance design, vibration and noise reduction of fixed axis rotation system.

References

1. Kuchar.M, The impact of the frequency of reed vibrations on improving the conditions in thickening dense technical fabrics [J], TEKSTIL VE KONFEKSIYON, 26 (2016) 380-384.
2. Kuchar.M, Vibratory Thickening of Weft Threads in a Weaving Loom–Simulation Tests [J], FIBRES & TEXTILES in Eastern Europe, 2013, pp. 59-64.
3. Chung-Feng Jeffrey Kuo,Te-Li Su,Chia-Hong Chen. Dynamic Modeling and Control of a Beat-Up

 https://doi.org/10.1051/matecconf/202235801007

Mechanism [J], Polymer-Plastics Technology and Engineering, 57 (2008) 367-375.

4. Celik.H.I, Kinematic analysis and synthesis of the beat-up mechanism for handmade carpet looms [J], JOURNAL OF THE TEXTILE INSTITUTE, 101 (2010) 882-889.
5. Katunskis.J, Theoretical and experimental beat-up investigation [J], FIBRES & TEXTILES IN EASTERN EUROPE, 12 (2004) 24-28.

Detection method of heterotropic fiber based on improved YOLOv5

Hengli Zuo [1,2], Yuhong Du [1,2,*]

[1] School of Mechanical Engineering, Tiangong University, Tianjin, China
[2] Tianjin Key Laboratory of Modern Electromechanical Equipment Technology, Tianjin, China

Abstract. Aiming at the problems of inaccuracy and poor real-time detection of heterosexual fiber in cotton cleaning process, a target detection model of heterosexual fiber based on YOLOv5 network was proposed to realize fast and accurate identification and location of heterosexual fiber in cotton. YOLOv5 was selected as the basic target detection model, and depth separable convolution was introduced to reduce the number of parameters of the detection model and improve the detection speed. Combined with SE module of channel attention mechanism, it can reduce irrelevant information interference and enhance feature expression ability. Comparative ablation tests were performed on YOLOv5 model before and after modification. The experimental results show that the improved YOLOv5 model has a mAP of 91.6% and a frame rate of 83 frames /s. The improved YOLOv5 model can not only improve the detection accuracy but also improve the detection speed, which can better meet the requirements of accuracy and real-time detection of cotton foreign fiber.

1. Introduction

Heterogenous fiber is one of the main factors affecting cotton quality. With the rapid development of domestic textile industry, cotton cleaning industry has been vigorously developed, cotton cleaning has become an important factor affecting the development of textile industry. In the industrial environment, it is difficult to quickly and accurately detect the opposite sex fibers similar to cotton, which will affect the quality and production efficiency of cotton textiles. The rapid and accurate detection of foreign fiber mixed in raw cotton is the focus of cotton spinning industry and the key problem to be solved.

In recent years, the research on foreign fiber detection of cotton has made some progress. Fortier C A[1] and Cintron M S[2] both realized the detection of foreign fibers in cotton by Fourier transform infrared spectroscopy. Du Yuhong[3] et al. adopted the method of clustering statistical RGB image information to realize foreign fiber identification and positioning. Zhang Hang, Qiao Xi[4] et al. proposed a hyperspectral image segmentation method of plastic film to achieve foreign fiber detection and recognition by combining spectral analysis and image processing technology. The research on the detection of heterosexual fiber is mainly based on the color, shape, texture and other characteristics of the segmentation, but the white heterosexual fiber similar to cotton is difficult to carry out image segmentation, so it is difficult to be used in the industrial detection of heterosexual fiber. At present, researchers have applied deep learning target recognition technology to agriculture, which can better achieve target detection of substances. Du Yuhong[5] et al. realized the detection and recognition of foreign fibers based on the Faster R-CNN network. Zhao DeAn[6] et al. realized underwater crab recognition based on deep convolutional neural network YOLOv3 and optimized Retinex algorithm to enhance image details. Wu Xing[7] et al. proposed a lightweight model based on YOLOv3 network to achieve rapid and accurate detection of Apple. Now commonly used networks have their own characteristics, the detection model needs to be modified according to the characteristics of foreign fiber and the detection environment.

To solve the problem of the identification of foreign matter detecting under industrial environment, to promote deep learning in the textile industry the application of target detection, in this paper, polypropylene fiber silk, cloth, feather, hemp rope, plastic film, chemical fiber, seven kinds of common straight hair fiber as the research object, based on the single phase detection algorithm YOLOv5 as the infrastructure of the detection algorithm, Combined with SE module of channel attention mechanism and depth devoluble convolution, DSCE-Yolov5 method of heterotropy detection was established to achieve rapid and accurate recognition and classification of all kinds of heterotropy, and to provide theoretical basis for the later heterotropy elimination.

* Corresponding author: DYH202@163.com

https://doi.org/10.1051/matecconf/202235801008

2. Improved YOLOv5 foreign fiber detection model

YOLOv5 is the latest algorithm in YOLO series, which has made great improvement in model size and detection speed. But YOLOv5 the original model is in the public training data sets, for the building of the opposite sex in our study the polypropylene fiber data set silk, chemical fiber and transparent plastic film and white cotton under certain lighting conditions the characteristics of color close, hard to detect, need to improvement and optimization of the model, in order to more quickly identify and detect its objectives. Since the detection of heterotropic fibers is used in the actual industry, the detection model needs smaller volume and faster detection speed. In this paper, standard convolution in YOLOv5 feature extraction network is replaced by deeply separable convolution, and a Pooling Block Attention Module (DSCE) is proposed combining the SE Module of Attention mechanism. This module is added to YOLOv5 network to solve the problem of no attention preference in detection. DSCE module uses the method of first compression and then expansion to enhance the learning of key points sampled from shallow features.

The improved YOLOv5 network structure as shown in figure 1, not only the depth of separable convolution can be used to reduce the pressure calculation of attention mechanism module, and convolution can be calculated more deep shallow characteristic graph extraction for important information characteristic figure, further to extract the foreign key information, improve the whole detection algorithm for foreign effect.

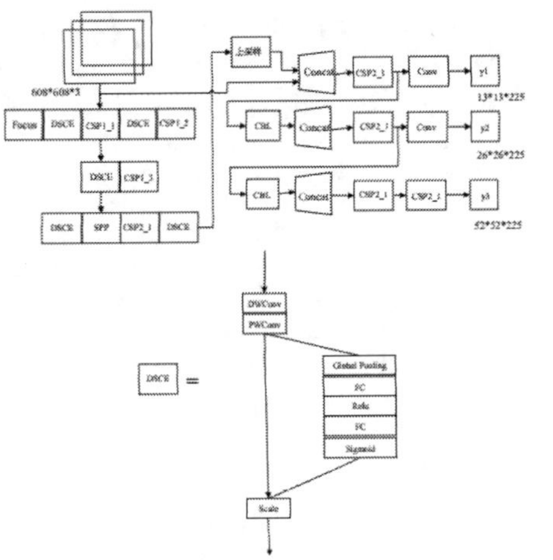

Figure 1 Structure of DSCE-YOLOV5 network model

3. Experiment and result analysis

3.1 Introduction of experimental environment and data set

The hardware platform used for this model training and test is the training environment based on Intel I5-9300HCPU, the graphics card is Nvidia GTX1660Ti and 8GB memory. The software platform is Win10+ Pytorch1.8.0 +Python3.7.10+ CUDA10.0+ VS2017+ OpencV2.0.

The experimental materials are mainly heterosexual fibers removed from cotton in the actual production of 10 cotton spinning enterprises such as Gucheng Silver Spinning Company in Hubei Province. The raw materials for the experiment were PP yarn, cloth, feather, hemp rope, plastic film, chemical fiber and hair. This paper studied the detection and identification of these 7 kinds of heterosexual fibers. The image examples of 7 kinds of cotton foreign fibers are shown in Figure 2. In order to improve sample diversity, the light intensity was changed during image collection, and the type, quantity, size and position of heterosexual fibers in each image were randomly selected and placed. A total of 1000 images were collected, and then the data set was enlarged to 5000 images through translation, rotation and cropping.

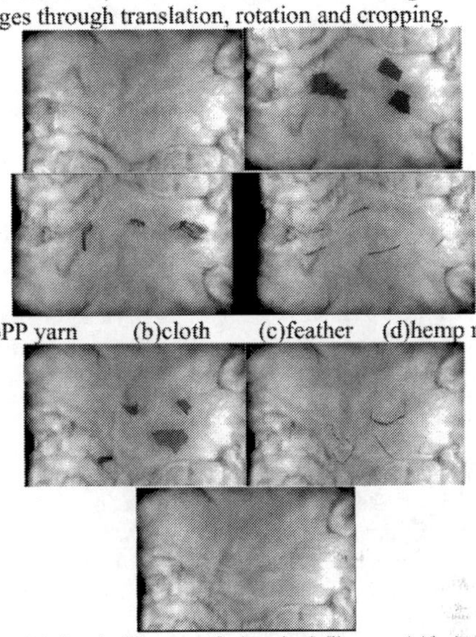

(a)PP yarn (b)cloth (c)feather (d)hemp rope

(e)plastic film (f)chemical fiber (g)hair

Figure 2. Example of cotton foreign fiber image

3.2 Data Preparation

3.2.1 Data set markup

When labeling the data, Labelimg tool was used for box-selected labeling, which was divided into 7 types of heterotropic fibers. The labeling example was shown in FIG. 3, and 3750 labeled images were finally sorted out as training data sets. Another 1,250 images were used as a test data set.

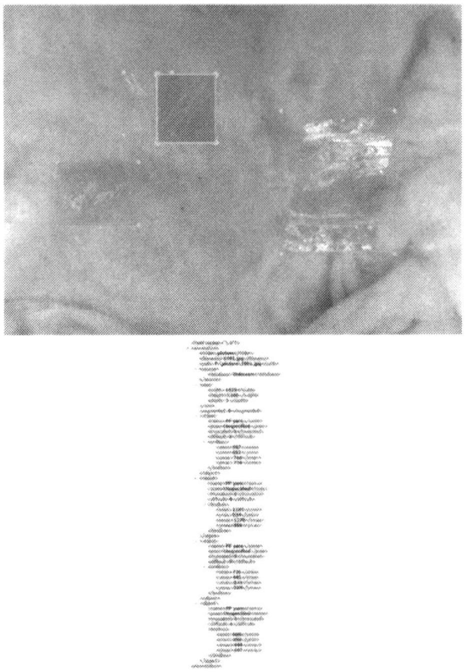

Figure 3. Example markup of an image

3.2.2 The setting of the anchor

In the prediction network of YOLOv3, each prediction feature layer has three pre-set prior frames. During the training process, appropriate prior frames are selected according to the size of the target size in the training picture for frame regression output prediction frame. K-means clustering method is added to YOLOv5 network. All pre-calibrated training data of real frames are clustered to appropriate prior frames, and then data training is performed according to the clustered prior frames, so that the results of prediction frames are closer to labeled real frames.

In this paper, an adaptive prior frame method is proposed. A part of data is trained by pre-set prior frame. During the training process, the difference of prediction results is calculated by constantly comparing the prediction box of training results with the marked real box, and the size of prior frame is adjusted according to the difference result to make prediction. The prediction frame size is close to the real frame size by iterative update, and the small target can be recognized more accurately. Thus, the anchor value is (15,75)、(19,27)、(29,105)、(30,50)、(41,26)、(43,72)、(51,106)、(58,47)、(67,75)。

3.3 Experiment and result analysis

3.3.1 Evaluation index

In this paper, the performance of heterociliary measurement method was evaluated by the following indicators[8]:

(1)The accuracy and recall rates are calculated as follows：

$$P=\frac{TP}{TP+FP} \tag{1}$$

$$R=\frac{FP}{FP+TN} \tag{2}$$

Where, TP represents the number of correctly divided into positive samples, FP represents the number of incorrectly divided into positive samples, and FN represents the number of incorrectly divided into negative samples.

(2)Average accuracy and average accuracy are calculated as follows：

$$AP=\int_0^1 P(\mathrm{R})\ \mathrm{dR}\,(10) \tag{3}$$

$$mAP=\frac{\sum_{i=0}^{N} AP_i}{N} \tag{4}$$

Where, N represents the number of all target categories.

(3)Detection speed：

The detection speed refers to the number of images (frames) that the target detection network can detect per second. It is expressed as FPS (frames per second).

3.3.2 Ablation experiments on a foreign fiber dataset

The change of foreign fiber detection performance caused by the change of network structure was gradually verified through the detection ablation experiment of foreign fiber. Four network models including YOLOv5s, YOLOv5s combined with depth-detachable convolution, SE-YOLOv5s and DSCE-YOLOv5S were trained in the four experiments respectively. The name can be used to correspond to the structural changes mentioned above. The experimental process and experimental test results are shown in Figure 4 and Table 1, where the curve rises successively from below: YOLOv5s combined with depth-detachable convolution, YOLOv5s, DSCE-YOLOv5s and SE-YOLOv5s.

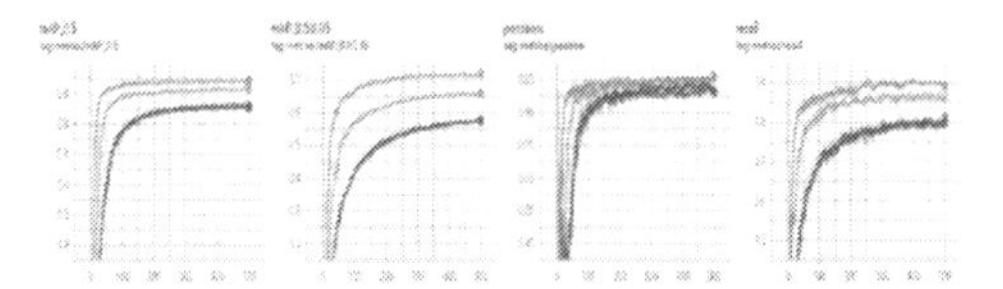

Figure4 Experimental data of each model on the heterogenous fiber data set

Table 1 Ablation experiment based on foreign fiber data set

method	Depth separable convolution	SE module	FPS/(frame*s-1)	mAP@0.5/%	AP@0.5% PP yarn hair
YOLOv5s			67	86.4	85.8 74.8
YOLOv5s +depth-separable convolution	√		92	85.6	83.7 73.4
YOLOv5s +SE		√	58	94.3	93.6 91.9
The algorithm in this paper	√	√	83	91.6	90 87.3

Figure 4 and Table 1 are the experimental results of foreign fiber detection ablation experiment. It can be found that compared with the original YOLOv5s model, the detection accuracy is reduced by 0.93% and the detection speed is increased by 37.3% by using the depth-separable convolution to replace the ordinary convolution in the trunk network by reducing the number of parameters. Compared with the original YOLOv5s model, use the SE module can improve the detection accuracy by 9.1% and reduce the detection speed by 13.4% by reducing the loss of each channel. Compared with the original YOLOv5s, the accuracy of the foreign fiber detection model is improved by 6% and the detection speed is improved by 23.9% using the depth-separable convolution and SE module. It indicates that the combination of the depth separable convolution proposed in this paper instead of ordinary convolution and the SE module of attention mechanism can improve the detection accuracy of foreign fiber detection model and improve the detection speed.

3.3.3 Analysis of foreign fiber detection results

The partial foreign fiber detection results before and after the improvement of YOLOv5s network are shown in Figure 5, where (a) is the original input image, (b) is the detection effect picture of YOLOv5s, and (c) is the effect picture of the improved algorithm in this paper. Due to problems such as illumination brightness and small target, it is difficult to directly detect the polypropylene filament, transparent plastic film and small feather with similar color to cotton in Figure 5 (a) in the original picture. It can be seen from Figure 5 (a) and (b) that the original YOLOv5s mis-detected polypropylene fiber and missed detection of feathers, and the improved DCE-YOLOv5s can accurately detect foreign fibers when polypropylene fiber and transparent plastic film are similar to cotton, as shown in Figure 5 (c). However, it is difficult to completely detect the whole plastic film with transparent part. Feathers are still detected even when some foreign fiber features are lost due to the small size of feathers. This is because each detection scale of DCE-YOLOv5s integrates the feature information of other scales. By redistributing the weight of feature channel, the characteristic information of foreign fiber can be effectively enhanced, the interference of foreign fiber surrounding can be suppressed, and the detection of foreign fiber can be enhanced. Therefore, compared with the original YOLOv5s, DSCE-YOLOv5s can more accurately detect the opposite fiber in the image, and the improved network can reduce the loss of image information to obtain more information, improve the integrity of effective information, and effectively improve the problem of missed detection and overdetection.

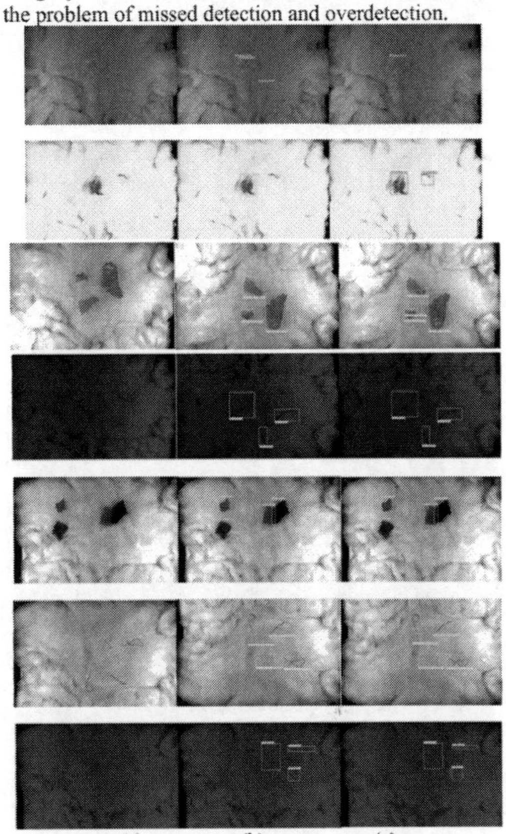

(a) (b) (c)

Figure 5 Comparison of foreign fiber detection results(a)master map, (b)YOLOv5s, (c)DSCE-YOLOv5s

4. Conclusion

In order to solve the problem of foreign fiber recognition in the industrial environment, this paper proposed an improved YOLOv5s algorithm to identify the foreign fiber in cotton, and the main conclusions are as follows:

(1) An improved YOLOv5s network model for heterofiber detection is proposed, which is suitable for industrial environment. The data processing method of image data enhancement is adopted, and the accuracy is up to 91.6%, and the frame rate is up to 83F /s.

(2) According to the characteristics of the heterotropic fiber data set, the size of the target frame of YOLOv5 model was adjusted and updated, and an improved feature extraction network based on the depth separable convolution and attention mechanism SE module was

introduced to propose a heterotropic fiber recognition method based on the improved YOLOv5s model.

(3) Experiments on the test set show that the detection model in this paper can accurately detect the different fibers with similar color to cotton. Compared with YOLOv5s model, the recognition accuracy of this model is 6 percentage points higher, and the detection speed is 2.4 percentage points higher. Compared with the existing models with the same or similar objects, the DCE-YOLOv5s model in this paper has higher recognition accuracy and faster recognition speed.

References

1. Fortier C A, James E R, Michael S C. Identification of cotton and cotton trash components by Fourier transform near-infrared spectroscopy[J]. Textile Research Journal, 2011,81(3):230-238.
2. Michael S C, James E R. Identification of common cotton contaminants using an FTIR microscope with a focal plane array detector[J]. AATCC Journal of Research,2017,4(6):12-17.
3. Du Yuhong, Wang Jiafu, Jiang Xiuming, et al. Graphic Detection Algorithm of Cotton Foreign Fiber Based on Clustering Statistical Analysis [J]. Journal of Textile Science and Technology,2015, 36(03):135-139.
4. Zhang Hang, Qiao Xi, Li Zhenbo. Hyperspectral Image Segmentation method of mulching film in lint [J]. Transactions of the Chinese Society of Agricultural Engineering,2016, 32(13):161-167.
5. Du Yuhong, Dong Chaoqun, Zhao Di, et al. Application of improved Faster RCNN model in identification of cotton heterogenies [J]. Advances in laser and optoelectronics,2020, 57(12):132-141.
6. Zhao De 'an, Liu Xiaoyang, Sun Yueping, et al. Underwater Crab Recognition Method based on machine vision [J]. Transactions of the Chinese Society for Agricultural Machinery, 2019, 50(03):151-158.
7. Wu XING, Qi Zeyu, Wang Longjun, et al. Apple Detection Method Based on Lightweight YOLOv3 convolutional Neural Network [J]. Transactions of the Chinese Society for Agricultural Machinery,2020, 51(08): 17-25.
8. BELL S, ZITNICK C L, BALA K, et al. Inside-outside net: detecting objects in context with skip pooling and recurrent neural networks [C]//Proceedings of 2016 IEEE Conference on Computer Vision and Pattern Recognition. Washington D.C.,USA:IEEE Press,2016:2874-2883.

Experimental Study on Tribological Properties of Polymer-based Composite Nano-additives Suitable for Armored Vehicle Engine Lubricating Oil

Bingli Fan [1, 2, 3], Zhengjie Li [1, 2], Annan Sun [1, 2], Yiwei Guo [1, 2], Xiaowen Qi [1, 2, 3], Changxin Liu [1, 2, *]

[1] School of Mechanical Engineering, Yanshan University, Qinhuangdao 066004, China
[2] Key Laboratory of Self-Lubricating Spherical Plain Bearing Technology of Hebei Province, Yanshan University, Qinhuangdao 066004, China
[3] Aviation Key Laboratory of Science and Technology on Generic Technology of Self-Lubricating Spherical Plain Bearing, Yanshan University, Qinhuangdao 066004, China.

Abstract. Lubricating oil is known as "the blood to maintain the normal operation of machinery", and additives are an indispensable and important part of lubricating oil. In this paper, the Polymer-based Composite Nano-additive (PCNA) suitable for armored vehicle engine lubricating oil was developed, the physical and chemical properties of lubricating oil containing additives were carried out. The results showed that adding an appropriate amount (3 vol.%) of additives could improve the low-temperature fluidity of lubricating oil and improve the kinematic viscosity of lubricating oil without affecting the viscosity-temperature characteristics of lubricating oil. The four-ball testing machine, high temperature end face testing machine and engine bench are used to measure the anti-friction lubrication effect and anti-wear ability improvement effect of the additive on lubricating oil. The experimental results show that the additive can effectively improve the antifriction, lubrication and anti-wear performance of lubricating oil. In addition, the addition of the additive can effectively improve the running state of the engine, and the engine noise is obviously reduced. Finally, the additives developed in this paper are compared with three common brand additives in the market by using the high temperature end face testing machine. The results show that the additive developed in this paper has certain advantages in improving the antifriction and lubrication effect of lubricating oil.

Key words: Polymer-based Composite Nano-additives; physical and chemical properties; Anti-friction lubrication; Anti-wear properties.

1. Introduction

Lubricating oil is known as "the blood to maintain the normal operation of machinery" [1], and the quality of lubricating oil has a huge impact on the service performance and life of mechanical equipment. Engine lubricating oil consists of base oil and additives [2,3]. The quality of base oil and the performance of additives determine the performance of lubricating oil. Due to the large amount of work and wide involvement in improving the quality of base oil, it is difficult to improve the quality of base oil. In contrast, the development of additives is much easier. Therefore, the research on improving the performance of lubricating oil mainly focuses on the research on additives [4].

Since 1930s, Exxon Company has carried out research on additives, and successively developed a series of lubricating oil additives with different functions, such as pour point depressant, antioxidant and anti-corrosion agent, detergent, ashless antioxidant, friction improver and ashless additive. By 1990s, the global pattern dominated by Lubrizol, Infineum, Chevron and Afton was basically formed, and the above four additive companies occupied 85% of the global market share of lubricating oil additives [5].

Pour point depressant, antioxidant and anti-corrosion agent, viscosity improver and ashless antioxidant in additives are closely related to the physical and chemical properties of base oil. Therefore, the research on the above additives is mainly led by lubricating oil companies and large additive companies. However, the academic circles focus on the research of friction modifiers, extreme pressure additives and other additives. In recent years, the research on additives such as friction modifiers and extrusion additives has mainly focused on carbon-based additives represented by graphite [6], graphene [7], carbon nanotubes [8], nano-diamonds, and layered inorganic additives represented by molybdenum disulfide,

* Corresponding author: cxliu@ysu.edu.cn

tungsten disulfide [9-10], and soft metal powders represented by nano-copper [11], nano-nickel [12], and inorganic represented by cesium trifluoride, cerium borate [13], as well as ionic liquids [6] (such as choline ionic liquids) and so on. Among them, nano-diamond has unique advantages because of its high hardness, good chemical stability and thermal stability. For example, Zhang Chuanan et al. [14] developed a lubricating oil anti-wear additive NGAW containing nano-diamond. When the addition amount of the additive is 5%, under the test condition of 392N test load, the friction and wear reduction rates of liquid paraffin wax, No.20 machine oil, CD30 oil and SF30 oil are 57%, 39%, 25% and 15% respectively. Moreover, the increase rates of unit load are 39.2%, 15.3%, 70% and 40%, respectively, which indicates that NGAW additive has good extreme pressure and anti-wear performance for different lubricating materials. At the same time, XPS analysis of the chemical composition of the friction surface shows that there are two substances on the friction surface: carbon in adsorbed organic compounds and carbon in diamond, which indicates that there is a surface film containing nano-diamond on the friction surface. Hu Xiaoli [15] used a four-ball friction and wear tester to study the friction and wear properties of two chemically modified nano-diamond particles in water-based solution. The results showed that when 0.1% nano-diamond sulfonic acid derivative was added to the water-based solution, its bearing capacity could be increased by about 10% and the diameter of wear spots could be reduced by 4.4%. In addition, metal additives such as nano-copper and nano-chromium and mineral additives such as molybdenum disulfide with layered structure are also widely used. Wang Xiaoli et al. [16] compared and studied the repair effect of copper additive with particle size of 20 nm on friction pairs under different friction test conditions, The results show that the nano-copper additive can form a protective soft film composed of clusters with a particle size of 0.1 m and low shear strength on the worn surface during the friction process, which can reduce the adhesion wear and abrasive wear, so that the nano-copper additive exhibits good anti-wear and antifriction performance. Zhao Jinzhen et al. [17] first obtained three kinds of nano-copper lubricants with 100SN, 150SN, 500SN, etc. as base oils through in-situ preparation method, and then carried out tribological and extreme pressure performance experiments on the above three kinds of lubricants. The results show that the nano-copper lubricating oil prepared in situ in 100SN base oil has higher bearing capacity and good antifriction and anti-wear properties, which can increase the maximum non-jamming load of the base oil by 27%, and at 392N and 1450r/min, the friction coefficient and wear spot diameter of the base oil can be reduced by 3.8% and 20%, respectively. However, nano-copper prepared in situ in 150SN and 500SN base oils has no obvious effect on the bearing capacity of lubricating oil. MoS_2 itself has a lamellar structure, and the binding force between layers is van der Waals force, which is easy to slide in the friction process, so it has excellent antifriction effect; For example, nanospherical MoS_2 prepared by Huo Yingjie et al. [18] as a lubricating oil additive significantly improves the extreme pressure performance of lubricating oil; Wohengzhou et al. [19] also found that compared with MoS_2 particles, nano-MoS_2 is more prone to chemical reactions and forms a surface film containing MoO_3 on the worn surface of steel balls. The extreme pressure, anti-wear and anti-friction properties of nano-MoS_2 additive are better than those of ordinary MoS_2. However, with the continuous progress of gasoline engine technology in recent years and the increasingly stringent environmental protection requirements, it is required to reduce the sulfur and phosphorus elements contained in lubricating oil as much as possible without reducing the anti-wear, anti-friction and high-temperature stability of engine oil [20]. At present, Molyvan855 (molybdenum content 10 wt.%) produced by R.T. Vanderbilt Company and Sakura-lube 700 (molybdenum content 4.4 wt.%) produced by Asahi Kasei Co., Ltd. In the market represent the advanced level of non-sulfur and phosphorus oil-soluble organic molybdenum additives [21,22]. Li Xiaolei and others have carried out research on the application of inactive oil-soluble organic molybdenum additives in low viscosity lubricating oils [22,23]. However, the above additives are mineral additives or oil-soluble additives, and the research on polymer-based composite Nano-additives (PCNA) is less and insufficient.

In this paper, the PCNA for engines of armored vehicle engine and automobiles was developed, and its physical and chemical properties were tested. The antifriction and lubricating effect of the additive on lubricating oil was investigated by tribological test and bench drag test. This study will provide reference for the subsequent research and development of such additives.

2. Experimental part

2.1 Preparation and Addition of PCNA

The polyisobutylene succinimide used to prepare composite Nano-additives in this study is of industrial grade and purchased from Shanghai McLean Biochemical Technology Co., Ltd.. Polyisobutylene (molecular weight 1300) was provided by Aladdin Reagent (Shanghai) Co., Ltd.. Nano boron nitride particles and nano aluminum nitride particles were purchased from Shanghai China Metallurgical New Materials Co., Ltd. With a particle size ≤ 30 nm. PA025 of Tiancheng Meijia Lubricating Oil (Beijing) Co., Ltd. is used as base oil without further treatment. A preparation method of the Polyisobutylene-based Composite Nano-additives (PCNA) is as follow:

(1) After grinding the polyisobutylene and polyisobutylene-based succinimide at high speed for 60-90 minutes, they were sprayed for homogenization, spraying pressure of 30-70MPa, and spraying duration of 60-120 minutes; Then dispersed by ultrasound for 120 minutes and extracted for later use;

(2) After mixing and shearing nano boron nitride, the nano aluminum nitride and the PA025 for 30-90 minute, mixed with mixed with polyisobutylene-based succinimide, they were sprayed for homogenization, spraying pressure of 90-130MPa, and spraying duration of 30-150 minutes;

Then dispersed by ultrasound for 90 minutes and extracted for standby;
(3) After adding the materials obtained in the steps 1 and 2 into a reaction kettle, heating to 55 DEG C, fully stirring for 70 minutes, cooling and filtering, the PCNA was obtained and shown in Fig. 1.

Fig. 1 PCNA

The experimental base oil are Kunlun brand general lubricating oil for heavy load power transmission (10W-40, CF-4 grade) and Great Wall Golden Star lubricating oil (10W-40), hereinafter referred to as K lubricating oil and C lubricating oil respectively. The 100ml lubricating oil and sufficient additives were added into a 500ml beaker in turn, then put the beaker under normal temperature and stir magnetically for 5min, and collect it for later use after stirring.

2.2 Physical and chemical performance test

The effects of additives on the physical and chemical properties of lubricating oil were evaluated by using kinematic viscosity and pour point as indexes. According to the standard GB/T265, the kinematic viscosity of samples at 40 °C and 100 °C was measured by Spike U-VISC 210 Automatic Viscometer (Fig. 2a). The Viscosity Tester is produced by Spitzer Technology (Beijing) and fully meets the requirements of ASTM D445 and D446 standards.

According to GB/T3535, a Multi-purpose Pour Point Tester (Fig. 2b) is used to measure the minimum temperature at which a sample can flow when cooled under specified conditions. The Pour Point Tester is produced by Hunan Jinshi Petrochemical Instrument Co. Ltd, which is suitable for measuring the freezing point and pour point of petroleum products. The equipment meets the standards of GB/T510 "Determination of Solidification Point of Petroleum Products" and GB/T3535 "Determination of Pour Point of Petroleum Products". The temperature control range is -70 ~ 0°C.

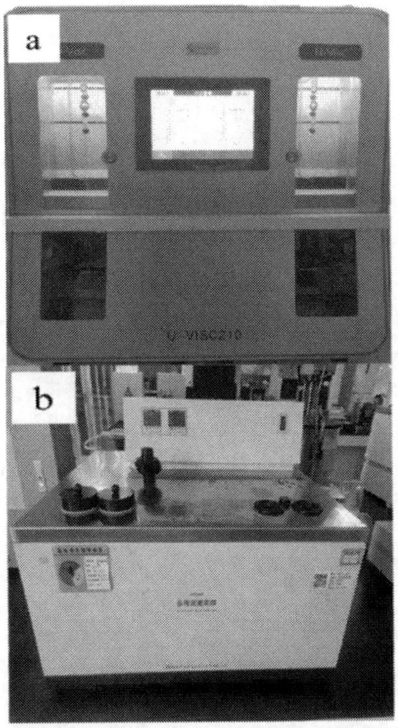

Fig. 2 Equipment for characterization of physical and chemical properties, a) Spike U-VISC 210 Automatic Viscometer and b) Multi-purpose Pour Point Tester.

2.3 Tribological performance test

According to the regulations of GB/T 3142-2019, the effects of different content additives on the anti-wear performance of lubricating oil were investigated by using a four-ball friction tester and taking the comprehensive wear value (ZMZ) as the evaluation index. The four-ball friction tester used is lever type, manufactured by Jinan Shunmao Test Instrument Co. Ltd., model MRS-10G, which meets the requirements of ASTM D2783 and SH/T 0189-92.

The MDW-5G screen-display high-temperature end-face friction and wear testing machine (Jinan Chenda Co., Ltd., China) was used to evaluate the effect of additives on the anti-friction performance of lubricating oil. The schematic diagram of the movement of the specimen is shown in Fig. 3. The material of the friction upper and lower samples refers to the piston ring and cylinder wall of a certain type of armored vehicle diesel engine. The material of the upper sample is 65Mn spring steel, and the material of the lower sample is RuT350 vermicular graphite cast iron. The upper sample is a ring with an inner diameter of 34 mm and an outer diameter of 40 mm, and the lower sample is a disk with a diameter of 43 mm. The setting parameters of the testing machine: the spindle speed is 1000r/min, the heating temperature is 80°C, and the applied load is 330N. The above test conditions were obtained by investigating an engine manufacturer located in Shijiazhuang, Hebei, China.

 https://doi.org/10.1051/matecconf/202235801009

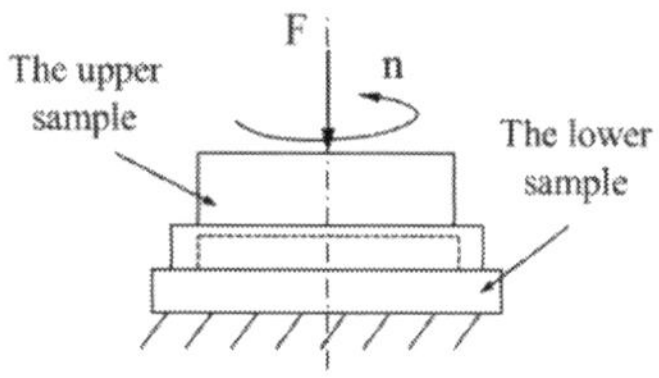

Fig. 3 Specimen motion diagram

2.4 Verification of bench reverse drag test

The 250kW Engine AC Power Dynamometer Test Bench (Fig. 4) was selected to carry out the evaluation experiment of the improvement effect of additives on the antifriction and lubrication effect of engine lubricating oil. The bench is produced by China Nantong Xindanuo Measurement and Control Technology Co., Ltd. And conforms to the provisions of GB/T 18297-2001 Performance Test Methods for Automobile Engines and GB/T 19055-2003 Reliability Test Methods for Automobile Engines. The engine used in the test is GW2.8TC diesel engine of Great Wall, which was purchased from Baoding Great Wall Scrap Car Recycling and Dismantling Co, Ltd. The bench reverse towing test is carried out according to GB/T 18297-2001 Automobile Engine Performance Test Method.

Fig. 4 250KW Engine AC Power Dynamometer Test Bench

3. Results and Discussion

3.1 Physical and chemical performance test

The kinematic viscosity and viscosity index of lubricating oil with additive content of 0%, 3% and 6% at 40°C and 100°C are shown in Fig. 5a and Fig. 5b, respectively, and the experimental base oil is K lubricating oil. The viscosity index is obtained by calculation [24]. It can be seen that with the increase of additive content, the kinematic viscosity of lubricating oil at 40°C and 100°C gradually decreases (Fig. 5a), while the viscosity index of lubricating oil remains unchanged first and then decreases (Fig. 5b). It is worth noting that the kinematic viscosity of lubricating oil decreased more obviously at low temperature (40°C) than at high temperature (100°C).

The standard of lubricating oil with good kinematic viscosity is that it has good fluidity at low temperature and can form a good oil film at high temperature [25]. The viscosity index refers to the degree of change of viscosity with temperature. The higher the viscosity index of lubricating oil is, the better its viscosity-temperature performance is, whereas the worse it is [26]. In addition, viscosity index is also related to equipment wear, and lubricating oil with higher viscosity index can greatly reduce the possibility of equipment wear [27]. To sum up, combined with the results of Fig. 5a and Fig. 5b, it can be seen that a proper amount (3%) of the additive can improve the kinematic viscosity of the lubricating oil without weakening the viscosity-temperature characteristics of the lubricating oil, thereby improving the starting performance of the engine and reducing the starting wear of the engine.

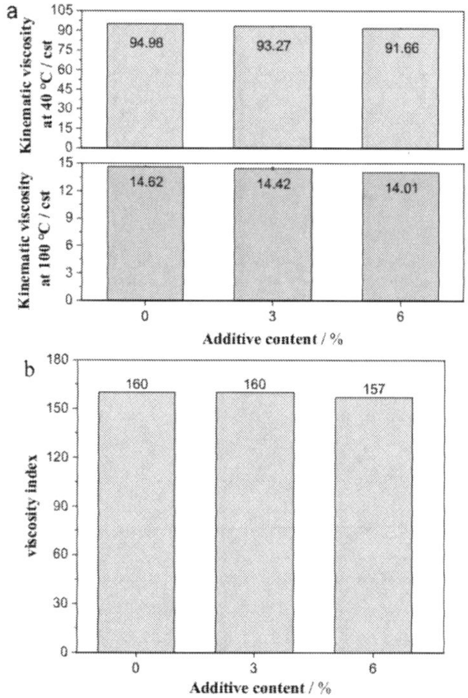

Fig. 5 Kinematic visibility a) and visibility index b) as function of additional content

Table 1 shows the pour point measurement results of lubricating oils with additive content of 0%, 3%, and 6%, and the base oil used in the experiment is K lubricating oil. It can be seen that with the increase of additive content, the pour point of lubricating oil decreases gradually. Pour point refers to the lowest temperature at which the oil sample to be tested keeps flowing under the corresponding specified conditions [28]. The lower the pour point, the better the low temperature fluidity of the lubricating oil [29]. The results shows that the additive can improve the low temperature fluidity of lubricating oil.

Table 1 Oil pour point measurement results

Additive content (%)	Oil pour point (°C)
0	-36
3	-39
6	-42

https://doi.org/10.1051/matecconf/202235801009

1.1 Tribological performance test

The combined wear value (ZMZ) test of lubricating oil with additive content of 0%, 3% and 6% was carried out by using a four-ball testing machine. The results are shown in Fig. 6, and the base oil used in the test is K lubricating oil. It can be seen that the combined wear value (ZMZ) of lubricating oil can be improved by adding different amounts of lubricating oil additives. The combined wear value is an index of the extreme pressure resistance of the lubricant. The greater the combined wear value, the higher the wear resistance of the lubricant [30]. It indicates that the additive can improve the wear resistance of lubricating oil, and when the additive content is 3%, the wear resistance improvement effect is the best.

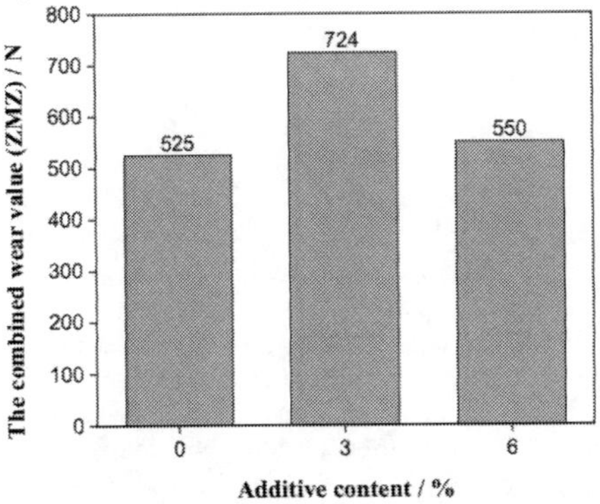

Fig. 6 The combined wear value of the oil

Fig. 7 shows the high-temperature end-face friction coefficient values of lubricating oils with additive content of 0%, 3%, and 6%. The base oil used in the experiment is K lubricating oil. It can be seen that after adding different contents of additives, the friction coefficient in the stable stage is much lower than that of pure lubricating oil, and the friction coefficient decreases from 0.15 to 0.03~0.05.

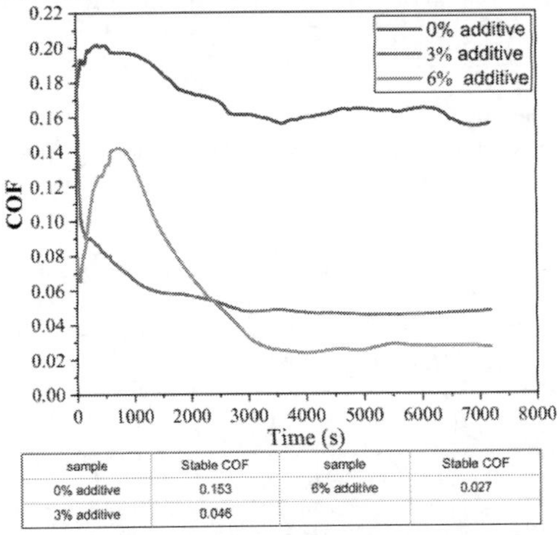

sample	Stable COF	sample	Stable COF
0% additive	0.153	6% additive	0.027
3% additive	0.046		

Fig. 7 The effect of additives on coefficient of friction

Based on the results of four-ball test and high temperature end face friction test, it can be seen that when the additive content is 3%, the lubricating oil can obtain good antifriction and lubrication effect and increase certain wear resistance.

3.2 Bench drag test

In order to further confirm the antifriction and lubricating effect of additives on lubricating oil, after completing the material-level test, this paper uses the engine bench to carry out the engine reverse drag experiment with the engine friction torque, friction work, noise and other evaluation indicators. The base oil used in the experiment is K lubricating oil. The experimental results are shown in Fig. 8. Adding different amounts of additives to the lubricating oil will reduce the friction torque value of the engine (Fig. 8a). Based on the friction torque results of 1500R and 2000R, adding 3% additive to lubricating oil has the best effect on reducing the friction torque of the engine using armored vehicle lubricating oil as lubricating medium, for example, at 1500R, the friction torque is reduced by 3.91%; At 2000R, the friction torque is reduced by 3.6%.

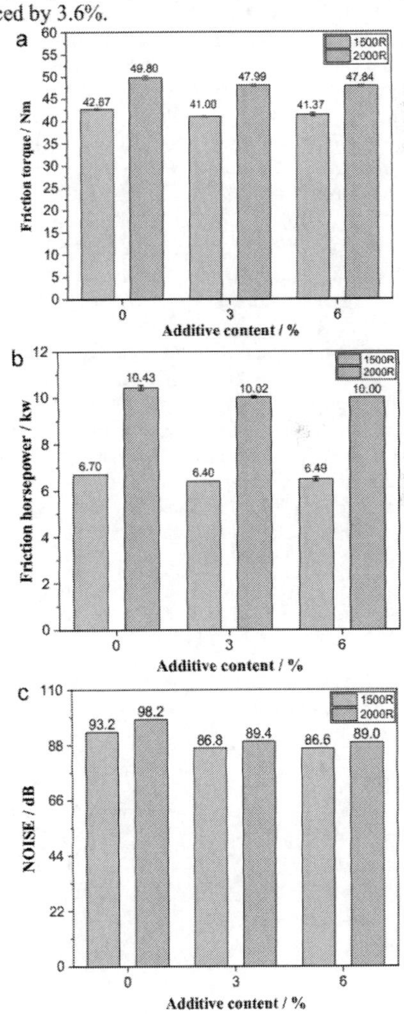

Fig. 8. Experimental results of engine reverse drag. a) Friction torque, b) Friction power, c) Noise.

The measurement results of engine friction power with lubricating oil containing different additive amounts as lubricating medium are shown in Fig. 8b. Friction power is the power consumed when the dynamometer drives the engine to rotate. It can be seen that the variation law of friction power with additive content is consistent with that of friction torque with additive content.

The experiment results of engine noise with pure lubricating oil and armored vehicle lubricating oil containing different amounts of additives as lubricating medium are shown in Fig. 8c. It can be seen that the addition of additives in different proportions reduces the engine noise. This shows that the addition of additives reduces the engine noise and improves the engine running state.

3.3 Comparative Experiment of Different Brand Additives

After completing the above experiment, in order to explore the difference between the additives developed in this paper and the common additives in the market in improving the antifriction and lubrication effect of lubricating oil, considering that most of the common additives in the market are lubricating oil additives for gasoline engines, three brand additives commonly found in the market that focus on improving the anti-friction lubrication effect of lubricating oil are selected in this paper. Taking Class C lubricating oil as the base oil, adding 3% of the above additives to the lubricating oil, the high temperature end face friction experiment was carried out. The obtained results were compared with the additives developed in this paper. Experimental conditions: spindle speed, 1000r/min, applied load, 330N, heating temperature, 80°C, experimental time, 45min. The experimental results are shown in Fig. 9. It can be seen that compared with the other three brands of additives, the additives developed in this paper have better antifriction and lubricating effects on lubricating oil, and the friction coefficient fluctuates less and the friction process is more stable.

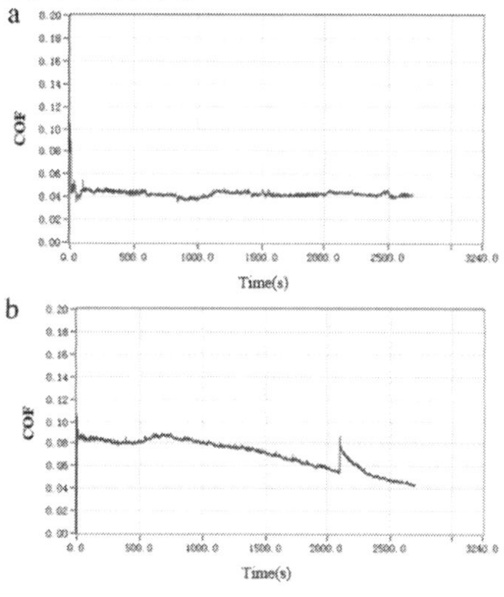

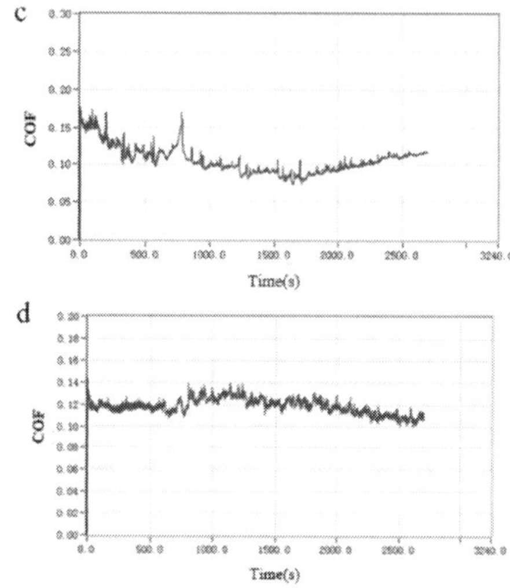

Fig. 9. Comparison experience of lustration performance of variable brand oil additives, a) PCNA, b) A brand, c) 3M brand, d) 1B brand.

The results obtained from frictional experiment of different brand additives indicate that the additives developed in this paper have certain advantages in improving the antifriction and lubrication effect of lubricating oil.

4. Conclusion

In this paper, the PCNA suitable for armored vehicle engine lubricating oil was developed. Firstly, the additive was added to lubricating oil, and its effects on the physical and chemical properties of lubricating oil were investigated. Secondly, using four-ball testing machine, high-temperature end testing machine and engine bench, the improvement effect of composite Nano-additives on anti-wear, anti-friction and lubrication performance of lubricating oil was explored. Finally, the additives developed in this paper are compared with three common brand additives in the market by using the high temperature end face testing machine. Through the above experiments, the following conclusions can be drawn:

(1) The additive developed in this paper can improve the low-temperature fluidity of lubricating oil, improve the kinematic viscosity of lubricating oil, improve the starting performance of engine and reduce the starting wear of engine without affecting the viscosity-temperature characteristics of lubricating oil.

(2) The additive can effectively improve the antifriction and lubricating effect of lubricating oil, and can also improve the wear resistance of lubricating oil when added in an appropriate amount (3%).

(3) The additive can effectively improve the running state of the engine and reduce the noise of the engine.

(4) Compared with the three common additives in the market, the additive has certain advantages in improving the antifriction and lubrication effect of lubricating oil.

Acknowledge

This study was supported by the National Defense Basic Scientific Research Project(Grant No. JCKY2019407D002),Hebei Innovation Capability Improvement Plan Project (Grant No. 19244006D), Aviation Scientific Fund Project(Grant No. 20184599001), Aviation Scientific Fund Project (Grant No. 20200045099001), Collaborative Education Project of Industry University Cooperation of the Ministry of Education (Grant No. 202102009003), and Doctoral Fund Project of Yanshan University (Grant No. BL18057).

References

1. Song Zeng-hong, Yan Yu-cai, Qiao Dan, et al. Research progress of Lubricant additives. LUBRICARING OIL. Vol. 34 (2019) No. 05, p. 16-22.
2. ZHANG Jing-yun, Zhai Qing-ge, SHI Shun-xiang, et al. Development status of lubricating base oil technology and additives. Yunnan Chemical Technology. Vol. 46 (2019) No. 08, p. 130-132.
3. GUO Wen-juan. Research Progress of Lubricant Additives for Engine. Synthetic Materials Aging and Application. Vol. 49 (2020) No. 03, p. 121-124.
4. ZHENG Kun, YANG Hong. Application status and technical development of lubricating oil additives. Petrochemical Standards and Quality of China. Vol. 40 (2020) No. 21, p. 141-143.
5. Fu Xisheng, Pan Yuanqing. Market status, technical progress and development trend of oil additives. Petroleum Commercial Technology. Vol. 34 (2016) No. 03, p. 4-15.
6. Li Chunfeng, Luo Xinmin, Hou Bin. Surface modification and Tribological Performance of Expanded Graphite (EG) as Lubricating Oil Additives. LUBRICATION ENGINEERING. Vol. 32 (2007) No. 07, p. 111-113 +121.
7. ZHANG Li-xiu, ZHAO Yue, WEI Xiao-yi, et al. Effect of Ionic Liquids on Dispersion and Lubricity of Graphene Lubricating Oil [J]. LUBRICATION ENGINEERING. Vol. 45 (2020) No. 10, p. 28-35.
8. GUO Xiao-yan, PENG Yi-tian, HU Yuan-zhong, et al. Tribological Behavior and Mechanism of Carbon Nanotubes as Oil Additives. LUBRICATION ENGINEERING. Vol. 32 (2007) No. 11, p. 95-97.
9. LU Yan-jun, KANG Jian-xiong, ZHANG Yong-fang, et al. Research progress of anti-friction and anti-wear of piston-cylinder liner system in internal combustion engine. Journal of Traffic and Transportation Engineering. Vol.20 (2020) No. 04, p. 21-34.
10. Shi Chen, Mao Da-heng, Feng Hao. Preparation of tungsten disulfide motor oil and its tribological characteristics. Journal of Central South University of Technology. Vol. 14 (2007) No. 5, p. 673-678.
11. Wang Jing, Guo Xiaochuan, Mi Hongying, et al. Effect of Nano-coppers with Different Dimensions on Tribological Properties of Lithium Grease. China Petroleum Processing & Petrochemical Technology. Vol. 22 (2021). No. 01. p. 118-126.
12. XIA Yan-qiu, JIN Shou-ri, SUN Wei-ming, et al. Effect of Nano Grade Copper Powder on wear and Friction Properties of Lubricant. LUBRICATION ENGINEERING. (1999) No. 03, p. 33-34.
13. Lingtong Kong, Hua Hu, Tianyou Wang, et al. Synthesis and Surface Modification of the Nanoscale Cerium Borate as Lubricant Additive. Journal of Rare Earths. Vol. 29 (2011) No. 11, p. 1095-1099.
14. ZHANG Chuan-an, QIAO Yu-lin, CHI Jun-cheng, et al. Tribological properties of nano-diamond additive for lubricating oil. China Surface Engineering. Vol. 55 (2002) No. 02, p. 29-32.
15. Hu Xiaoli, Zhang Xiaoqing, Lei Hong. Research on the Tribological Properties of Water-based Lubricant with the Ultra-fined Diamond Particles. JOURNAL OF SHANGHAI DIANJI UNIVERSITY. Vol.11 (2008) No. 01, p. 24-27.
16. Wang Xiaoli, Xu Binshi, Xu Yi, et al. Study on Friction and Wear Behavior and Mechanism of nano-Cu Additive in Lubrication Oils. TRIBOLOGY. Vol. 27 (2007) No. 03, p. 235-240.
17. ZHAO Jin-zhen, HUANG Wei. In-situ Direct Preparation of Nano-copper Lubricant Oil and study on Tribological Properties. LUBRICATION ENGINEERING. Vol. 36 (2011) No. 10, p. 37-39.
18. Huo Yingjie, Hou Suoxia, Zhang Haoqiang, et al. Controllable Hydrothermal Method Preparation of Nano Molybdenum Disulfide and Its Extreme Pressure Property. LUBRICATION ENGINEERING. Vol. 42 (2017) No. 09, p. 107-110.
19. WO H Z, HU K H, HU X G. Tribological Properties of MoS2 Nanoparticles as Additive in a Machine Oil. TRIBOLOGY. Vol. 24 (2004) No. 01, p. 33-37.
20. Huang Wenxuan.Application Guide of Lubricant Additives. CHINA PETROCHEMICAL PRESS. 2003, p. 262-265.
21. WANG Guo-dong, LIU Yan, LI Lai-ping, et al. DEVELOPMENT OF THE ORGANIC MOLYBDENUM [J]. CHINA MOLYBDENUM INDUSTRY. Vol. 36 (2012) No. 6, p. 1-5.
22. Zou Yang, Zhang Zitong, Li Xiaolei, et al. Influence of Organic Molybdenum on the Tribological Property of Low-viscosity Oil [J/OL]. Surface Technology.
23. Dai Yuanjing, Li Xiaolei, Ling Yanli, et al. Tribological Properties of Non-thiophosphorous Organic Molybdenum. Surface Technology. Vol. 49 (2020) No. 09, p. 35-44 +71.
24. Information on: http://www.chinalubricant.com/tools/mixture.html.
25. Feng Yujuan.Importance of kinematic viscosity detection of lubricating oil. Metallurgical Equipment. Vol. 239 (2017) No. 02, p. 285-287.

26. Song Yuefeng. Study on the modification of plant oil as environmentally friendly lubricants base stocks (Master, Hefei University of Technology, China 2014). p. 3-4.
27. Xing Suobin. The Study on influence of lubricating oil on the life of gear reducer (Master, North China Electric Power University, China 2017). p. 18-19.
28. ZHANG Hao, ZHANG Zheng-dong, ZHANG Bo, et al. Effect of Test Conditions on Oil Pour Point Test Results. OIL AND GAS TREATING AND PROCESSING. Vol. 32 (2014) No. 05, p. 33-36 +9.
29. SUN Li-peng, WANG Zhi-hong. Study on Production Process and Product Performance of High-pressure Hydrogenated Lubricating Oil Base Oil. Chemical Management. (2020) No. 26, p. 54-55.
30. Qiu Jianwei, Li Jianming, Xue Weiguo, et al. Study on the Synthesis and Tribological Performances of M ethoxyethane D ia lky l d ith iocarbam ate as Lubricant Additive. LUBRICATION ENGINEERING. Vol. 34 (2009) No. 10, p. 69-71 +83.

Study on Critical Concentration of Dynamic Gelation of New Chromium Crosslinker

Ruibo Cao*, Guochao Liu, Wenlin Chen, Liangfeng Hu, Wei Yan, Haibo Liu

Research Institute of Exploration and Development of Daqing Oilfield Company Ltd., Daqing, China

Abstract: A chrome gel system with 13 million molecular weight and 19 million molecular weight polymer and DQ-8 cross-linking agent is dynamically gelled under different conditions of 50-mesh and 100-mesh screens under different shear rates. Concentration limit. The experiment shows that the critical concentration of 13 million molecular weight gelation is 500-700mg/L, and the critical concentration of new cross-linking agent in the system is 80mg/L-90mg/L. The system of new cross-linker is cut by steel mesh. The post-gelation time is prolonged. Between 6h-10h, after 12 days of testing, the viscosity of the chrome gel system after gelation can still remain above 1000mPa·s, and the viscosity of the ungelatinized solution drops below 10mPa·s.

1. Introduction

In daqing oil field, the new chrome crosslinking agent and polymer configuration of chromium gel system achieved good development effect, become the important component of the polymer flooding, in order to further optimize the new crosslinking agent in oil field application, according to the reservoir property and fluid characteristics of test area, the new crosslinking agent for polymer configuration parameters optimization.The new crosslinking agent is more targeted in well selection and layer selection in the application process, and the parameters in the system are quantified to better guide the field application, so as to achieve the purpose of reducing cost and increasing efficiency[1-3].

2. Experimental part

2.1 Experimental materials

Polymer: molecular weight 13 million, 19 million, effective content 90%

New crosslinking agent: Cr^{3+} crosslinking agent independently developed by Daqing Oilfield, the model is DQ-8, and the effective content is 3%.

Water used for the experiment: water from the South Third Injection Station of Daqing Oilfield.

Experimental equipment: Brinell viscometer, thermostat, high temperature and high pressure core gripper, steel fake core, etc.

Experimental temperature: 45℃.

2.2 Experimental steps:

(1) Take polymers with different molecular weights, use clean water to configure mother liquor, and age for more than 12 hours.

(2) When preparing the target solution, clear water was used, and a new crosslinking agent was added into the target solution of polymers with different molecular weights to prepare the chromium gel system.

(3) the configured target liquid is loaded into the intermediate container, and the 50 mesh and 100 mesh screen are loaded into the steel dummy core, so that the gel system configured with different concentrations of polymers and new crosslinking agents is shears by different mesh screens.Simulate different shear rates.

(4) The initial viscosity of the chromium gel system was measured by connecting the sample at the exit end, and the viscosity of the system was measured again at the sample points at different times.

2.3 Experimental scheme

The concentration of the two molecular weight polymers was increased by 100mg/L, and the concentration of the new crosslinking agent was increased by 10mg/L. The experimental schemes of 50 mesh and 100 mesh were shown in the table below:

Table 1. Experimental protocol

Polymer type	Polymer concentration	Chromium crosslinker concentration	Polymer type	Polymer concentration	Chromium crosslinker concentration
13 million	500 mg/L	70 mg/L	19 million	400 mg/L	70 mg/L
	600 mg/L	80 mg/L		500 mg/L	80 mg/L
	700 mg/L	90 mg/L		600 mg/L	90 mg/L

* Corresponding author: caoruibo@cnpc.com.cn

3. Experimental results and analysis

3.1 The critical concentration of 50 mesh chromium gel system

3.1.1 Critical gelling concentration of 13 million polymer - new crosslinker system.

The prepared chromium gel system was prepared by using 13 million molecular weight and new chromium crosslinking agent. The configured chromium gel system was shearing through the steel core equipped with 50-mesh screen. The initial viscosity of the solution was measured at the exit section and the viscosity of the system was detected at different monitoring points[4].

As can be seen from Figure 1, when the concentration of the new crosslinking agent is 70mg/L and 80mg/L, the molecular weight of the polymer is 13 million, and the concentration varies from 500mg/L to 700mg/L, the viscosity of the chromium gel system shows a sharp decline after rising, but the viscosity fails to remain stable. With the increase of the coagulation time, the viscosity of the system drops sharply after dehydrating and breaking the glue.It eventually dropped to less than 10mPa·s.When the polymer concentration reaches 700mg/L and the cross-linking agent concentration reaches 90mg/L, the viscosity of the chromium gel system rises rapidly. When the peak value reaches 3000mPa·s, after a long time of viscosity detection, it can still reach more than 1500mPa·s.Therefore, for the 13 million molecular weight polymer, after the shear of 50 mesh screen, the critical gelling concentration of the system is HPAM700mg/L+Cr90mg/L[5-7].

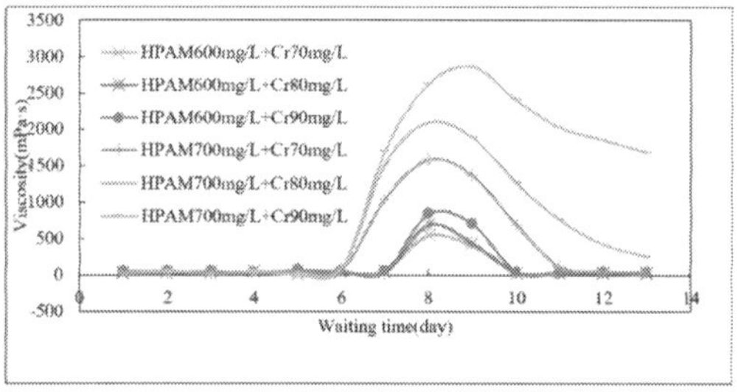

Figure 1. 50 mesh screen 13 million molecular weight different formula concentration curve

3.1.2 Critical gelling concentration of 19 million polymer-new crosslinker system.

The chromium gel system was prepared by using 19 million molecular weight and the new chromium cross-linking agent. The configured chromium gel system was shearing through the steel core equipped with a 50-mesh screen. The initial viscosity of the solution was measured at the exit section, and the viscosity of the system was detected at different monitoring points.

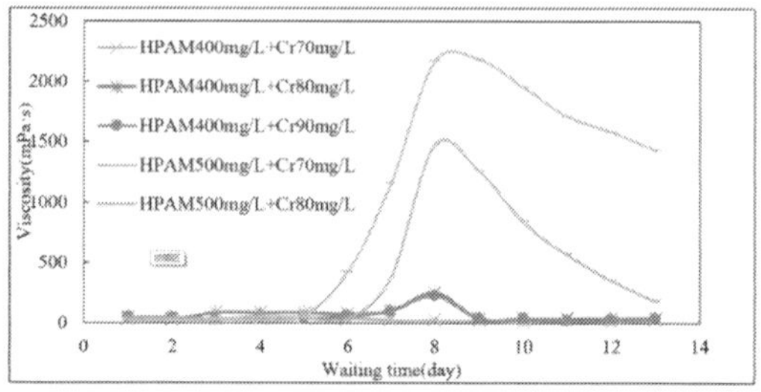

Figure 2. 50 mesh screen 19 million molecular weight different formula concentration curve

The figure 2 shows that when the chromium gel system also after 50 mesh sieve of shear, due to the increase of polymer molecular weight, so new crosslinking agent and polymer gel critical concentration decreases, when the polymer concentration below 500 mg/L, although crosslinking agent concentration of 90 mg/L, but the system did not gel, drops rapidly after the peak viscosity, Finally, the viscosity of the system was lower than 10mPa·s. Only when the concentration of the polymer reached 500mg/L and the concentration of the new chromium cross-linking agent reached 80mg/L, the system formed gel. When the viscosity of the system was measured at different detection points, the viscosity was still higher than 1000mPa·s. The critical gelling concentration of chromium gel system is HPAM500mg/L+Cr80mg/L.

Compared with 13 million and 19 million molecular weight polymers under the shear condition of 50 mesh screen, when the polymer concentration is the same, the larger the molecular weight is, the closer the three-dimensional network structure between the molecules is, and the lower the critical concentration of gumming. Under the same polymer molecular weight and the same concentration conditions, the viscosity of the system gradually increased with the increase of the concentration of chromium crosslinker, but the effect of increasing the concentration of crosslinker on the system concentration was less than that of increasing the concentration of polymer and the molecular weight of polymer. Therefore, in the field application process, we can reasonably plan the field injection by analyzing the weight of the influence of polymer molecular weight, polymer concentration and crosslinking agent concentration on the chromium gel system concentration[8].

3.2 The critical concentration of 100 mesh chrome gel system

3.2.1 Critical gelling concentration of 13 million polymer - new crosslinker system.

The chromium gel system was prepared by using 13 million molecular weight and the new chromium crosslinker. The configured chromium gel system was shears through the steel core equipped with a 100-mesh screen. The initial viscosity of the solution was measured at the exit section, and the viscosity of the system was detected at different monitoring points.

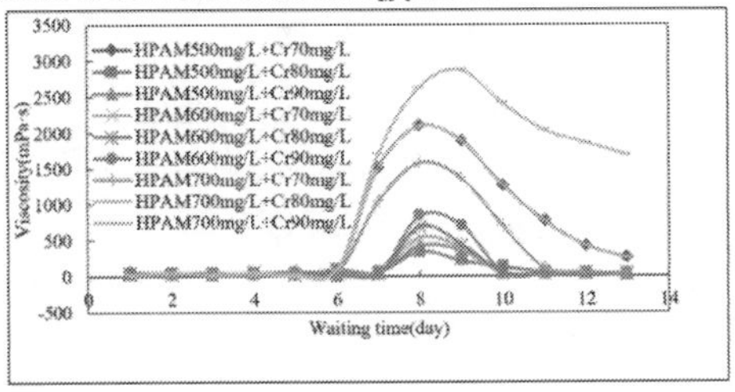

Figure 3. 100 mesh screen 13 million molecular weight different formula concentration curve

As shown in the figure above, when the concentration of the polymer and the new crosslinking agent was the same as that of the 50-mesh screen, the shear rate increased with the increase of the mesh number, so the concentration of the polymer increased by 100mg/L under the condition that the concentration of the crosslinking agent and the molecular weight of the polymer were the same. Although at the concentration of HPAM700mg/L+Cr80mg/L,The viscosity of the system is maintained for a long time, and the peak viscosity is above 2000mPa·s. However, in the end, the chromium gel system is still dehydrated and broken, and the viscosity drops sharply, and finally drops to less than 10mPa·s. Therefore, under the condition of 100 mesh screen and 13 million molecular weight,The critical gelling concentration of chromium gel system is HPAM700mg/L+Cr90mg/L.

3.2.2 Critical gelling concentration of 19 million polymer-new crosslinker system.

The chromium gel system was prepared by using 19 million molecular weight and new chromium crosslinking agent. The configured chromium gel system was passed through the steel core equipped with a 100-mesh screen for shearing. The initial viscosity of the solution was measured at the exit section, and the viscosity of the system was detected at different monitoring points.

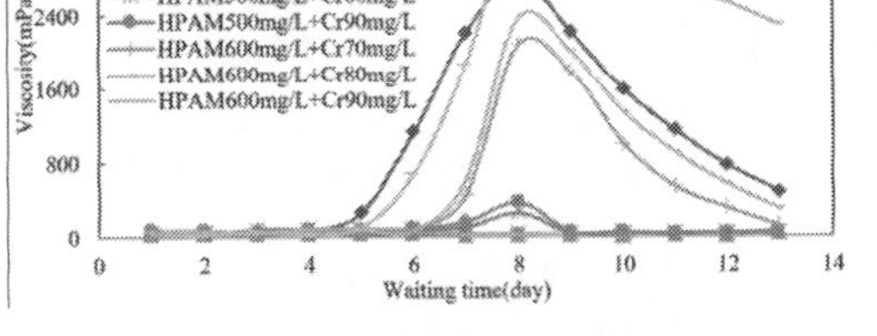

Figure 4. 100 mesh screen 19 million molecular weight different formula concentration curve

As shown in Fig. 4, under the condition of the same mesh mesh, the molecular weight increased to 19 million, so under the condition of the same crosslinking agent, the polymer concentration decreased to a certain extent. However, since the molecular weight difference of the polymer was not large, the system could only form gel when the concentration of the crosslinking agent reached 90mg/L.However, under the condition of 19 million molecular weight, the peak viscosity reached more than 3500mPa ·s, which was higher than about 2800mPa·s of 13 million molecular weight. Although the chromium gel system with the concentration of polymer at 600mg/L and the concentration of cross-linking agent below 90mg/L also had the phenomenon of temporary gelation, the final viscosity still could not remain stable.Therefore, under the condition of 100 mesh screen and 19 million molecular weight, the critical gel-forming concentration of the chromium gel system is HPAM600mg/L+Cr90mg/L.

Under the condition of 100 mesh sieve, compared to 13 million and 19 million molecular weight allocation of chromium gel system compared with 50 mesh sieve conditions, two kinds of molecular weight under the condition of critical gel concentration are less than 50 mesh sieve, due to the increase of shear rate, serious damage to between polymer molecular structure, decrease caused by molecular interaction, show the system viscosity reduced[9].

4. Conclusion

(1) The critical gelling concentration of chromium gel system with different mesh mesh and different molecular weight is shown in the table below:

Table 2. Critical concentration of gelation of chrome gel system under different mesh screens and different molecular weight conditions

Screen mesh	Polymer molecular weight	Polymer concentration	Concentration of the new crosslinker
50 mesh	13 million	600mg/L	90 mg/L
	19 million	500mg/L	80 mg/L
100 mesh	13 million	700 mg/L	90 mg/L
	19 million	600 mg/L	90 mg/L

(2) The gelation time of the new chromium crosslinker is generally 4-6 days. After shearing, the gelation process of the system becomes slow, which can provide reference for the injection time in the field.

(3) When the mesh number of the screen increases, the shear rate increases, and the critical gelling concentration of both the polymer and the new crosslinking agent increases. Under the condition of the same crosslinking agent concentration, the critical gelling concentration of the 19 million molecular weight polymer is less than 13 million polymer.

References

1. Song Chengzhe. Study on Mechanical Properties of Salt Resistant Frozen Polymer Repair Mortar for Bridge Repair in Seasonal Frozen Area [J].Northern Communications,2018(08):6-9.

https://doi.org/10.1051/mateccconf/202235801010

2. Tang Zewei, Mu Lijun, Zhou Zhiping.The feasibility experiment of nanofiltration concentrated hydrogel profile control system [J]. Environmental Protection of Oil and Gas Field, 2018, 28(04):30-33+61.
3. Hao Jinsheng. Adaptability Evaluation of Low Content Salt Resistant Polymer under High Salinity Water Condition [J]. Chemical Engineering and Equipment, 2018(08):103-105.
4. Fu Guoqiang, Wang Keliang.Effect of alternate injection of gel and surfactant on oil displacement after water flooding [J]. Modern chemical industry,2016,45(03):495-497+500.
5. Yu Kai. Study and determination of the type of control and flooding agent in Naiman oilfield [J]. Chemical management,2016(05):164.
6. Fu Guoqiang. Influencing factors of polyacrylamide /Cr(Ⅲ) gel system and profile control effect [D]. Northeast Petroleum University,2016.
7. Chen Tengwei, Yang Huan, Zhang Huanghe.Effect of water-soluble ions on the properties of polymer-organic chromium gel system [J]. Journal of Yangtze University (Natural Science Edition), 2015, 12(19): 11-15+3.
8. Zhou Wanfu, Wang Xin, Zhou Quan.Self-healing properties of betaine sulfonate type polyacrylamide/chromium gel [J]. Journal of Northeast Petroleum University, 2015,39(04):88-94+11.
9. Li Yu-fei. Study on the synthesis and retarding crosslinking performance of organic chromium [D]. Northeast Petroleum University, 2015.

Study of quantitative structure-property relationship for density of ionic liquids based on Monte Carlo optimization

Xingang Jia *, Wenzhen Wang, Bo Yang, Chunbao Du

Chemistry and Chemical Engineering College, Xi'an Shiyou University, Xi'an, China

Abstract. Ionic liquids (ILs) have attracted increasing interests and applications due to its unique physiochemical properties. Density is a vital physical property of ILs. In this work, a comprehensive collection of density data is conducted on 184 variable ILs. The study of quantitative structure-property relationship (QSPR) is carried out for the selected density data of ILs using simplified molecular input line entry specification (SMILES) as the representation of the molecular structure of ILs by means of CORAL software. QSPR relationships were constructed with the balance of correlations (BC) and the classic scheme. Results from three random splits displayed desirable models for predicting the external test set with the correlation coefficient (R2) and cross validated correlation coefficient (Q2) in ranges of 0.8234–0.9770 and 0.7599–0.9745, respectively. The best predictions obtained by the balance of correlations along with the global SMILES descriptors are included in the modeling process. The average statistical characteristics of the external test set are the following: n =36, R2 =0.9770, Q2= 0.9745, standard error of estimation (s)=0.023, mean absolute error (MAE) =0.018 and Fischer F-ratio (F)=1443

1. Introduction

Ionic liquids (ILs) are composed of organic cation and organic/inorganic anion with melting point less than 100°C. Because of their unique physical and chemical properties, such as negligible vapor pressure[1-2], non-flammability[3-5], high ionic conductivity[6-8], large liquid range, and good ability to dissolve organic and organic materials[9-10], ILs have been intensively researched and been successfully applied in many fields during the past decades [11-18]. While, the wide range of combinations of anion and cation can lead to a great number of ILs which are also regarded as task specific materials. That is the reason that the development of these ILs is increasingly being pursued by many researchers around the world.

Density of ILs is a fundamental and important physical property which is required in many design problems of chemical engineering process calculations, analysis and material science. For example, the appropriate value of density is desired for process piping, transfer rates, equipment sizing and other unit process in various chemical, oil, clay and food industry. However, due to the unbelievably high number of possible ILs, Experimental data for density of ILs are still scare and always only available for a few classes of common researched ILs. More density data and deep understanding of the property are demanded for designing new processes or developing task-specific ionic liquids.

For this purpose, the methods of predicting the density of ILs are desirable. Quantitative structure-property relationship (QSPR) is a robust and useful computational tool used to predict the properties of homologous series [19]. Some various QSPR models have been proposed for the prediction of density of ILs. For example, Palomar et al [20-21] predicted the density of imidazolium-based ionic liquids by combining molecular descriptor with artificial neural network (ANN) algorithm. Trohalaki et al. [22] employed electrostatic, quantum mechanical and thermodynamic descriptors obtained from CODESSA software to derive QSPR model to predict the density of triazolium bromides. Keshavarz et al [23] introduced two simple correlations to predict the density of a wide range of ionic liquids based on the size, elemental composition and type of cations and anions. Valderrama et al. [24] advised ANN and group contribution method (GCM) to estimate the density of 103 of ionic liquids. Shen et al. [25] utilized GCM combined with Patel–Teja equation of state to estimate the density of ionic liquids at ambient temperature and atmospheric pressure. El-Harbawi et al [26] developed a QSPR model based on the combination of multiple linear regression (MLR) and polynomial equation using the same descriptors reported by Shen et al. Among the QSPR models mentioned above, some complex molecular descriptors and algorithm were used to develop correlation models for the estimation of the density of ILs. In this work, a new model is proposed based on the simplified molecular input line entry specification (SMILES) descriptor.

* Corresponding author: jiaxingang76@xsyu.edu.cn

2. Materials and methods

2.1 Density data of ILs

The experimental density data at 298.15K and 100Kpa for 184 pure ILs were collected from IL ThermoDatabase [27] The studied ILs covers a wide range of cations including imidazolium, pyridinium, pyrrolidinium, piperidinium, guanidinium, phosphonium, ammonium, and different anions such as and different anions such as hexafluorophosphate, tetrafluoroborate, halide, bis (trifluo-
romethylsulfonyl) mide, bis(perfluoroethylsulfonyl)imide, carboxylic acid, alkyl sulfate, dialkylphosphate, trifluoro-methyl sulfonate, trifluoroacetate, alkoxy-alkylsulfates, dicyanamide, tricyanomethanide, tris(trifluoromethy lsulfonyl) methide, aminoacid, etc.

2.2 Method

SMILES has been employed as an alternative descriptor in the QSPR models [28-29]. The SMILES of 184 ionic liquids were obtained from the free software ChemSketch [30].The CORAL software generated the SMILES-based optimal descriptors, which are described as follows:

$$\begin{aligned} {}^{SMILES}DCW(\text{Threshold}, N_{epoch}) = {} & \alpha\sum CW(S_k) + \beta\sum CW(SS_k) \\ & +\gamma\sum CW(SSS_k) + x\cdot CW(NOSP) \\ & +y\cdot CW(\text{HALO}) + z\cdot CW(\text{BOND}) \\ & +w\cdot CW(\text{PAIR}) \end{aligned} \quad (1)$$

Where DCW is for correlation weight of a molecular descriptor; The Sk is one character (e.g., C, N, O, etc.) or two characters which cannot be separated (e.g., Cl, Fe, Br, etc.); The SSk is a combination of two SMILES characters; The SSSk is a combination of three SMILES characters. α, β, γ, x, y, z and w can be 1 or 0 [31]. NOSP stands for the chemical elements: nitrogen, oxygen, sulfur and phosphorus; HALO stands for halogen elements except for iodine element. BOND symbolizes a mathematical function for the presence or absence of double, triple or stereo chemical bonds. CW means the correlation weight of a molecular descriptor. The numerical value of CW(Sk), CW(SSk) and CW(SSSk) are obtained from Monte Carlo method. Nepoch and Threshold are parameters of Monte Carlo optimization. Combinations of their values produce possibility of generating various results of the SMILES-based optimal descriptor. The best model was obtained with the best combination of the threshold and the Nepoch for the correlation coefficient of the external validation set. The correlation coefficient between density and DCW(Threshold, Nepoch) is a mathematical function of the correlation weights of the SMILES attributes. The predictive model was built up by the following equation.

$$\rho = C_0 + C_1 \times {}^{SMILES}DCW(Threshold, N_{epoch}) \quad (2)$$

There are two basic modeling methods in CORAL software. The first one is the mode of classic scheme based on the split of all collected ILs in the training set and validation set. The second one is the balance of correlations based on the split into three sets: training, calibration and validation sets. The training set is the core part of the model. The task of the training set is to calculate correlation weights which can give as large as possible correlation between the experimental and predicted density for the training set. The calibration set can detect the beginning of over-fitting. The validation set is the final estimator of the predicted QSPR model. In this paper, the calculation for 184 density data of ILs were performed in two versions for three random splits. Version 1 was based on the balance of correlation and version 2 was based on the classic scheme.

Table 1. Comparison of statistical characteristics of the models between the classic scheme and the balance of correlation

T	N_{act}	Probe	Training set				Calibration set				Validation set				
			N_t	R^2_t	S_t	F_t	N_c	R^2_c	S_c	F_c	N_v	R^2_v	S_v	F_v	R^2_m
Balance of correlation															
Split 1															
1	274	1	66	0.9930	0.014	9083	65	0.9929	0.016	8846	53	0.8832	0.054	386	0.8259
1	274	2	66	0.9930	0.014	9096	65	0.9930	0.016	8945	53	0.8822	0.054	382	0.8286
1	274	3	66	0.9931	0.014	9153	65	0.9930	0.016	8895	53	0.8777	0.056	366	0.7992
Average				0.9930	0.014	9111		0.9930	0.016	8895		0.8810	0.055	378	0.8179
2	199	1	66	0.9923	0.015	8270	65	0.9920	0.017	7830	53	0.8904	0.052	414	0.8312
2	199	2	66	0.9920	0.015	7907	65	0.9901	0.018	6326	53	0.8922	0.051	422	0.8764
2	199	3	66	0.9924	0.015	8352	65	0.9921	0.017	7944	53	0.8851	0.053	393	0.8390
Average				0.9922	0.015	8177		0.9914	0.017	7367		0.8892	0.052	410	0.8489
3	190	1	66	0.9920	0.015	7890	65	0.9918	0.017	7660	53	0.8998	0.049	458	0.8538
3	190	2	66	0.9920	0.015	7963	65	0.9920	0.017	7830	53	0.8998	0.049	458	0.8686
3	190	3	66	0.9923	0.015	8222	65	0.9923	0.017	8117	53	**0.9003**	0.049	461	0.8558
Average				0.9921	0.015	8025		0.9921	0.017	7869		0.9000	0.049	459	0.8594
Split 2															
1	288	1	74	0.9724	0.031	2541	74	0.9909	0.015	7837	36	0.9499	0.033	645	0.9488
1	288	2	74	0.9726	0.031	2560	74	0.9913	0.015	8198	36	0.9705	0.025	1117	0.9616
1	288	3	74	0.9719	0.031	2493	74	0.9920	0.014	8889	36	0.9622	0.028	866	0.9610
Average				0.9723	0.031	2531		0.9914	0.015	8308		0.9609	0.029	876	0.9571
2	239	1	74	0.9723	0.031	2527	74	0.9906	0.016	7607	36	0.9502	0.032	649	0.9465
2	239	2	74	0.9717	0.031	2470	74	0.9921	0.014	9023	36	0.9582	0.030	780	0.9511
2	239	3	74	0.9715	0.031	2451	74	0.9918	0.015	8690	36	0.9616	0.028	852	0.9584
Average				0.9718	0.031	2483		0.9915	0.015	8440		0.9567	0.030	760	0.9520
3	200	1	74	0.9716	0.031	2461	74	0.9903	0.016	7348	36	**0.9740**	0.024	1276	0.9579
3	200	2	74	0.9712	0.031	2432	74	0.9909	0.015	7803	36	0.9642	0.027	917	0.9619
3	200	3	74	0.9715	0.031	2456	74	0.9907	0.015	7653	36	0.9718	0.024	1171	0.9697
Average				0.9714	0.031	2450		0.9906	0.016	7601		0.9700	0.025	1121	0.9632
Split 3															
1	299	1	82	0.9895	0.017	7571	74	0.9676	0.031	2148	28	0.8116	0.073	112	0.7347
1	299	2	82	0.9894	0.017	7467	74	0.9679	0.031	2171	28	0.8004	0.076	104	0.7124
1	299	3	82	0.9890	0.018	7160	74	0.9679	0.030	2172	28	0.8211	0.070	119	0.7457
Average				0.9893	0.017	7400		0.9678	0.031	2164		0.8111	0.073	112	0.7309
2	231	1	82	0.9892	0.017	7300	74	0.9675	0.031	2146	28	**0.8839**	0.056	198	0.8607
2	231	2	82	0.9886	0.018	6965	74	0.9675	0.031	2146	28	0.8746	0.058	181	0.8454
2	231	3	82	0.9889	0.018	7116	74	0.9675	0.031	2143	28	0.8480	0.065	145	0.7918
Average				0.9889	0.018	7127		0.9675	0.031	2145		0.8688	0.060	175	0.8326
3	215	1	82	0.9879	0.018	6507	74	0.9666	0.031	2085	28	0.8308	0.068	128	0.7794
3	215	2	82	0.9872	0.019	6146	74	0.9674	0.031	2136	28	0.7867	0.077	96	0.7058
3	215	3	82	0.9875	0.019	6307	74	0.9671	0.031	2118	28	0.7813	0.078	93	0.6880
Average				0.9875	0.019	6320		0.9670	0.031	2113		0.7996	0.075	105	0.7244
Classic scheme															
Split 1															
1	321	1	127	0.9962	0.010	32555					57	0.9135	0.051	581	0.9061
1	321	2	127	0.9960	0.010	31372					57	0.9072	0.053	538	0.8848
1	321	3	127	0.9953	0.011	26594					57	0.9020	0.054	506	0.8845
Average				0.9958	0.011	30173						0.9076	0.053	542	0.8918
2	249	1	127	0.9956	0.011	28028					57	0.9063	0.053	3	0.8833
2	249	2	127	0.9960	0.010	31274					57	0.9122	0.052	572	0.8708
2	249	3	127	0.9960	0.010	31474					57	0.9081	0.053	543	0.8949
Average				0.9959	0.011	30259						0.9089	0.053	549	0.8830
3	225	1	127	0.9960	0.010	31083					57	**0.9141**	0.051	585	0.9012
3	225	2	127	0.9958	0.011	29814					57	0.9141	0.051	585	0.9093
3	225	3	127	0.9957	0.011	28886					57	0.9020	0.054	506	0.8746
Average				0.9958	0.011	29928						0.9100	0.052	559	0.8951
Split 2															
1	310	1	146	0.9793	0.025	6809					38	0.8711	0.059	243	0.7882
1	310	2	146	0.9790	0.025	6712					38	0.8955	0.054	308	0.7936
1	310	3	146	0.9792	0.025	6764					38	0.8986	0.052	319	0.8180
Average				0.9791	0.025	6762						0.8884	0.055	290	0.7999
2	259	1	146	0.9787	0.025	6620					38	0.8456	0.066	197	0.7231
2	259	2	146	0.9789	0.025	6682					38	0.8883	0.056	286	0.7908
2	259	3	146	0.9786	0.025	6588					38	0.8742	0.060	250	0.7570
Average				0.9787	0.025	6630						0.8693	0.061	244	0.7569
3	240	1	146	0.9790	0.025	6713					38	**0.9001**	0.052	324	0.8170
3	240	2	146	0.9781	0.025	6420					38	0.8251	0.070	170	0.7009
3	240	3	146	0.9784	0.025	6512					38	0.8855	0.055	278	0.7988
Average				0.9785	0.025	6548						0.8702	0.059	258	0.7722
Split 3															
1	313	1	155	0.9799	0.024	7501					29	0.8942	0.053	220	0.8701
1	313	2	155	0.9802	0.024	7606					29	0.8518	0.064	149	0.7861
1	313	3	155	0.9802	0.024	7605					29	0.8733	0.058	179	0.8344
Average				0.9801	0.024	7570						0.8731	0.058	183	0.8302
2	259	1	155	0.9793	0.024	7270					29	**0.9185**	0.047	293	0.9072
2	259	2	155	0.9794	0.024	7321					29	0.8938	0.053	219	0.8738
2	259	3	156	0.9795	0.024	7373					29	0.8423	0.065	139	0.7907
Average				0.9794	0.024	7321						0.8849	0.055	217	0.8572
3	234	1	155	0.9793	0.024	7289					29	0.8808	0.056	192	0.8506
3	234	2	155	0.9789	0.025	7146					29	0.8858	0.055	202	0.8657
3	234	3	155	0.9791	0.024	7210					29	0.8754	0.058	183	0.8423
Average				0.9791	0.024	7215						0.8807	0.056	192	0.8529

The Nact is the number of SMILES attributes classified as not rare; N is the number of ILs in set; R2 is Correlation coefficient; S is standard error of estimation; F is Fischer F-ratio; Subscript t is for training set; Subscript c is for calibration set; Subscript v is validation set; R2m metric should be N >0.5 [32].

3. Results and discussion

Optimal descriptors were calculated with SIMLES. The balance of correlations (version 1) and the classic scheme (version 2) have been applied in the Monte Carlo method optimization. The statistical characters of models of two versions were shown in Table 1. The values of statistical characters were obtained in 3 probes of the Monte Carlo optimization for all splits in both versions. It was seen from Table 1 that the prediction with CORAL is acceptable because of its good statistical quality for all splits. The best model was a model with a threshold of 3 for split 2 in version 1. Figure 1 graphically displayed best models for all splits. The figure showed that there was an excellent agreement between experimental density and predicted density using the best models.

The values of the best T and the Nepoch for all splits in both versions presented the following QSPR models for the calculated density of ILs according to equation 2.

QSPR mathematical model of density for Split 1 in version 1:

$$\rho=1.2263243(\pm0.0002445)+0.0101727(\pm0.0000102)\times DCW(3,19) \quad (3)$$

(training set: n=66, R^2=0.9923, Q^2=0.9919, s=0.015, MAE=0.010, F=8253, calibration set: n=65, R^2=0.9921, Q^2=0.9916,s=0.017,MAE=0.012, F=7892, validation test: n=53, R2=0.8949, Q2=0.8880,s=0.050,MAE=0.028, F=434)

QSPR mathematical model of density for Split 2 in version 1:

$$\rho=1.1785844(\pm0.0005943)+0.0103934(\pm0.0000188)\times DCW(3,7) \quad (4)$$

(training set: n=74 R^2=0.9666, Q^2=0.9652, s=0.034, MAE=0.018, F=2084, calibration set: n=74, R^2=0.9830, Q^2=0.9820, s=0.021, MAE=0.015, F=4169, validation test: n=36, R2=0.9770, Q2=0.9745, s=0.023,MAE=0.018, F=1443)

QSPR mathematical model of density for Split 3 in version 1:

$$\rho=1.1412170(\pm0.0004769)+0.0071543(\pm0.0000178)\times DCW(2,4) \quad (5)$$

(training set: n=82 R^2=0.9753, Q^2=0.9737, s=0.026, MAE=0.020, F=3458, calibration set: n=74, R^2=0.9566, Q^2=0.9547, s=0.036, MAE=0.021, F=1587, validation test: n=28, R^2=0.8971, Q^2=0.8854,s=0.052,MAE=0.030, F=227)

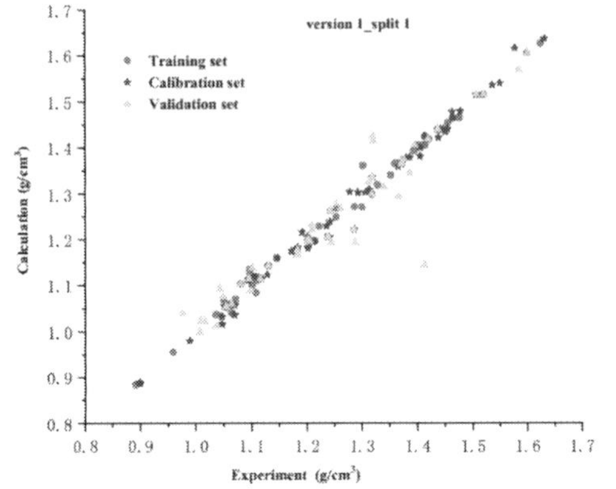

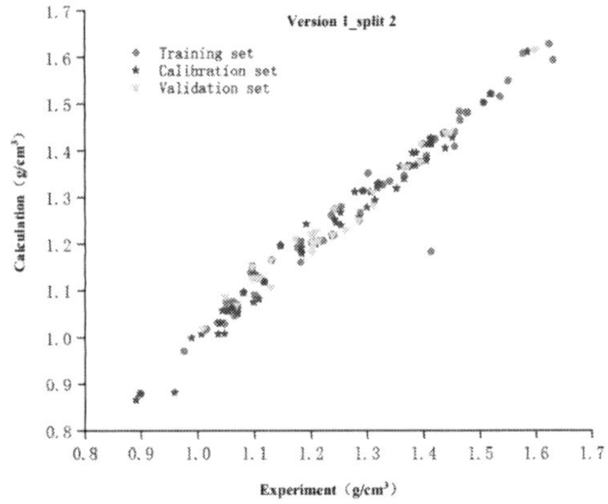

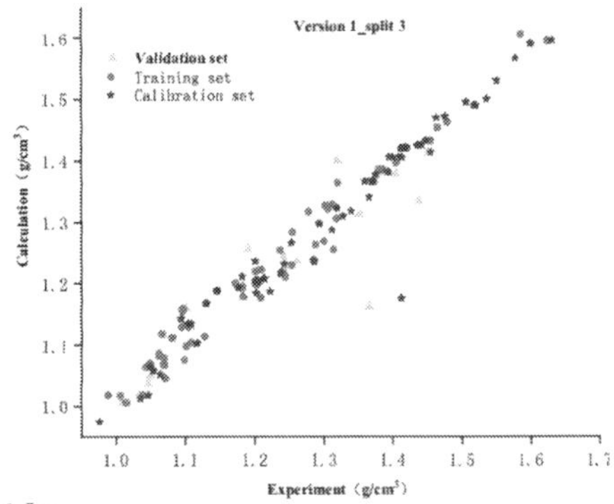

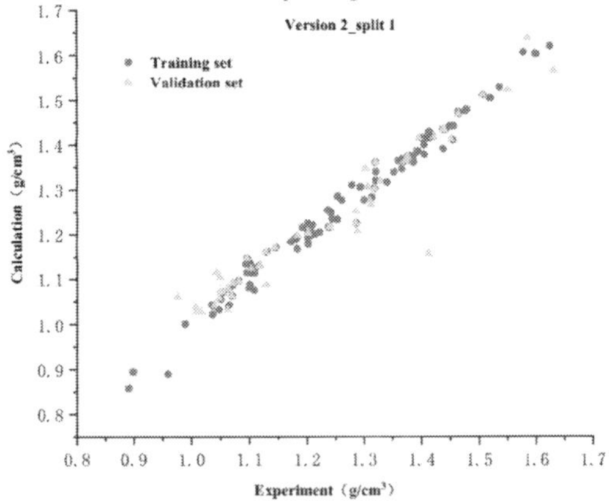

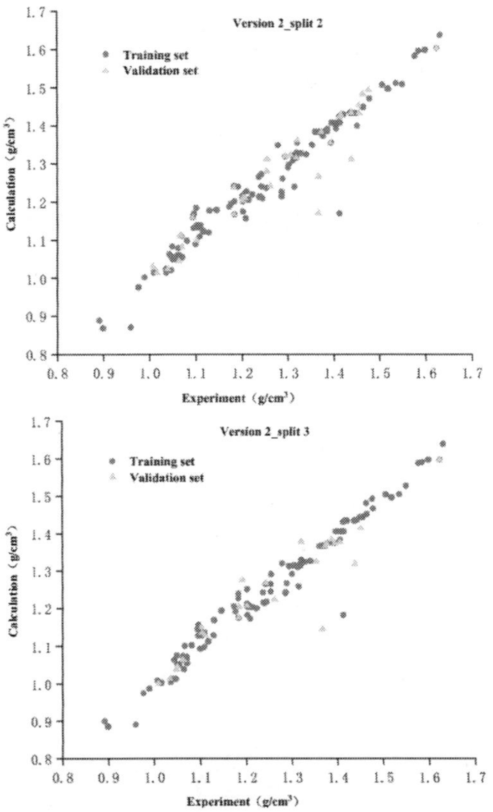

Figure 1 Graphical exhibition of QSPR models for all applied splits of two versions for density of ILs QSPR mathematical model of density for Split 1 in version 2:

$$\rho=1.1524604(\pm0.0002331)+0.0170517(\pm0.0000184)\times DCW(3,5) \quad (6)$$

(training set: n=127 R^2=0.9856, Q^2=0.9851, s=0.020, MAE=0.015, F=8584, validation test: n=57, R^2=0.9282, Q^2=0.9235, s=0.047,MAE=0.028, F=711)

QSPR mathematical model of density for Split 2 in version 2:

$$\rho=1.0997165(\pm0.0003689)+0.0071090(\pm0.0000084)\times DCW(1,3) \quad (7)$$

(training set: n=146 R^2=0.9589, Q^2=0.9579, s=0.035, MAE=0.022, F=3358, validation test: n=38, R^2=0.9038, Q^2=0.8961, s=0.049,MAE=0.029, F=338)

QSPR mathematical model of density for Split 3 in version 2:

$$\rho=1.1195132(\pm0.000178)+0.0179471(\pm0.000013)\times DCW(3,3) \quad (8)$$

(training set: n=155 R^2=0.9847, Q^2=0.9843, s=0.021, MAE=0.014, F=9849, validation test: n=29, R^2=0.8234, Q^2=0.7599, s=0.080, MAE=0.043, F=126)

Table 2. Criteria of predictability of best QSPR models

	Balance of correlations			Classic scheme		
	Split 1	Split 2	Split 3	Split 1	Split 2	Split 3
$r^2_i(x,y)$						
n	53	36	28	57	38	29
r^2	0.8949	0.9770	0.8971	0.9282	0.8442	0.8837
r^2_0	0.8897	0.9744	0.8799	0.9158	0.8429	0.8694
rr^2_0	0.8936	0.9763	0.8967	0.9271	0.8281	0.8836
$(r^2-r^2_0)/r^2$	0.0057	0.0026	0.0191	0.0134	0.0016	0.0162
$(r^2-rr^2_0)/r^2$	0.0014	0.0007	0.0005	0.0012	0.0191	0.0000
k	0.9949	1.0032	0.9921	0.9976	0.9955	0.9902
kk	1.0036	0.9965	1.0063	1.0010	1.0021	1.0080
$r^2_i(y,x)$						
r^2	0.8949	0.9770	0.8971	0.9282	0.8442	0.8837
r^2_0	0.8936	0.9763	0.8967	0.9271	0.8281	0.8836
rr^2_0	0.8897	0.9744	0.8799	0.9158	0.8429	0.8694
$(r^2-r^2_0)/r^2$	0.0014	0.0007	0.0005	0.0012	0.0191	0.0000
$(r^2-rr^2_0)/r^2$	0.0057	0.0026	0.0191	0.0134	0.0016	0.0162
k	1.0036	0.9965	1.0063	1.0010	1.0021	1.0080
kk	0.9949	1.0032	0.9921	0.9976	0.9955	0.9902
R^{*2}_m (test)	0.8626	0.9516	0.8788	0.8969	0.7371	0.8811
R^{*2}_m (average)	0.8467	0.9397	0.8291	0.8608	0.7754	0.8296
ΔR^2_m	0.0318	0.0239	0.0992	0.0721	0.0766	0.1030

From equation 4 to equation 8, R^2 is the correlation coefficient, Q^2 is the leave-one-out cross-validated correlation coefficient; s is the standard error of estimation; MAE is a mean absolute error and F is the Fisher ration. From the statistical point of view, Q^2 can be used to determine the predictive ability of the model.Higher value of Q^2 means better model prediction. Table 2 displays statistical criteria of the predictability of the best models for all splits in two versions. It is concluded that the predictability for best models of all splits is good and robust according to the criterion [32-33]. The 184 ILs experimental and calculated density values for best model, namely split 2 in version 1 are showed in Table S1.

4. Conclusion

The density of an ionic liquids is an essential physical property which is important in industrial process design involving ionic liquids. In this paper, CORAL software was first applied for building up QSPR models for predicting the density of ionic liquids. Density data set of 184 pure ILs at ambient temperature and pressure were used for developing and validating the model by two different versions built in CORAL software. The results show that the proposed QSPR models are robust and have good predictive ability by using the global SMILES descriptors. The version of the balance of correlations displays higher accuracy than the version of the classic scheme. The average statistical characteristics of the external test set for the best QSPR model are the following: n=36, R^2=0.9770, Q^2=0.9745, standard error of estimation (s)=0.023, mean absolute error (MAE) =0.018 and Fischer F-ratio (F)=1443.

This work was supported by the by the National Natural Science Foundation of China (No.211720002) and Shaanxi Provincial Education Department (No. 20JK0839).

References

1. R.L. Gardas, J.A.P. Coutinho, Fluid Phase Equilibration. 266 (1-2), 195(2008).
2. J.D. Holbrey, K.R. Seddon, Clean Prododuction. Process. 1, 223(1999).
3. Yu G, Chen X, Asumana C, et al. AIChE journal, 57, 507(2011)
4. Lei Z, Chen B, Koo Y M, et al. Chemical Reviews,117, 6633(2017)
5. Diallo A O, Len C, Morgan A B, et al. Separation and purification technology, 97, 228(2012)
6. Yoshizawa M, Narita A, Ohno H. Australian journal of chemistry, 57, 139(2004)
7. MacFarlane D R, Forsyth M, Izgorodina E I, et al. Physical Chemistry Chemical Physics, 11, 4962(2009)
8. Kunze M, Montanino M, Appetecchi G B, et al. The Journal of Physical Chemistry A, 114, 1776(2010)
9. Earle M J, Seddon K R. Pure and applied chemistry, 72, 1391(2000)
10. Das R N, Roy K. Toxicology Research, 1,186(2012)
11. Abbott A P, McKenzie K J. Physical Chemistry Chemical Physics, 8, 4265(2006)
12. Koel M. Critical Reviews in Analytical Chemistry, 35, 177(2005)
13. Opallo M, Lesniewski A. Journal of Electroanalytical Chemistry, 656, 2(2011)
14. Fontanals N, Borrull F, Marcé R M. TrAC Trends in Analytical Chemistry, 41, 15(2012)
15. Sasikumar B, Arthanareeswaran G, Ismail A F. Journal of Molecular Liquids, 266,330(2018)
16. Rykowska I, Ziemblińska J, Nowak I. Journal of molecular liquids, 259, 319(2018)
17. Kim B K, Lee E J, Kang Y, et al. Journal of industrial and engineering chemistry, 61, 388(2018)
18. Ho T D, Canestraro A J, Anderson J L. Analytica Chimica Acta, 695, 18(2011)
19. Katritzky A R, Karelson M, Lobanov V S. Pure and Applied Chemistry, 69, 245(1997)
20. Palomar J, Ferro V R, Torrecilla J S, et al. Industrial & engineering chemistry research, 46, 6041(2007)
21. Palomar J, Torrecilla J S, Ferro V R, et al. Industrial & engineering chemistry research, 47, 4523(2008)
22. Trohalaki S, Pachter R, Drake G W, et al. Energy & Fuels, 19, 279(2005)
23. Keshavarz M H, Pouretedal H R, Saberi E. Journal of Molecular Liquids, 216,732(2016)
24. J.O. Valderrama, A. Reátegui, R.E. Rojas, Industrial & engineering chemistry research, 48, 3254(2009)
25. Shen C, Li C, Li X, et al. Chemical engineering science, 66, 2690(2011)
26. El-Harbawi M, Samir B B, Babaa M R, et al. Arabian Journal for Science and Engineering, 39, 6767(2014)
27. https://ilthermo.boulder.nist.gov/ accessed on 2020-3-16
28. Toropov A A, Toropova A P, Benfenati E, et al. Molecular diversity, 14, 821(2010)
29. Toropov A A, Toropova A P, Martyanov S E, et al. Chemometrics and Intelligent Laboratory Systems, 109, 94(2011)
30. ACD/ChemSketch, version 2020.2.0, Advanced Chemistry Development, Inc., Toronto, ON, Canada, www.acdlabs.com, 2021.
31. Toropova A P, Toropov A A, Benfenati E, et al. Journal of computational chemistry, 32, 2727(2011)
32. Golbraikh A, Tropsha A. Journal of molecular graphics and modelling, 20, 269(2002)
33. Ojha P K, Mitra I, Das R N, et al. Chemometrics and Intelligent Laboratory Systems, 107(2011)

Research on rock triaxial cyclic loading and unloading characteristics based on PFC3D

Ziyuan Qi [1,2], Jiancheng Xiao [1,2,*], Jingjing Lu [1,2], Hui Zhou [1,2], Huaguo Wang [3]

[1] State Key Laboratory of Geomechanics and Geotechnical Engineering, Institute of Rock and Soil Mechanics, Chinese Academy of Sciences, Wuhan, 430071, China
[2] University of Chinese Academy of Sciences, Beijing, 100049, China
[3] College of Energy and Mining Engineering, Shandong University of Science and Technology, Qingdao, 266590, China

Abstract. Numerical methods provide a new way for solving complex geotechnical engineering problems. Firstly, we review the emergence and development of numerical methods, and introduce the basic principles of particle flow discrete element and its application in geotechnical engineering. After that, a method of servo combination of cylinder and loading plate is used, the numerical simulation of the three-dimensional rock in triaxial loading and unloading test is carried out by PFC3D software, and the loading and unloading mechanical characteristics of rock mass under the linear elastic contact constitutive model and parallel bonded contact constitutive model are studied respectively, and we obtain the triaxial loading and unloading stress-strain curves of rock under the elastic constitutive model and elastoplastic constitutive model. The residual strains and hysteresis phenomena of the rock under the two constitutive models are analysed, and the feasibility of the particle flow discrete element method in studying the mechanical characteristics of rocks is demonstrated.

1. Introduction

With the continuous improvement of computer performance, the role of numerical methods in solving geotechnical engineering problems has gradually increased. Rock mass is a kind of geological body with discontinuity, heterogeneity and anisotropy, its characteristics make it difficult to describe the constitutive relationship, and the boundary conditions are often very complicated when mechanical analysis of rock mass is carried out, which causes many difficulties to engineering analysis. The emergence of numerical methods provides a fast and accurate analysis method for rapid analysis of geotechnical engineering problems.

The more commonly used methods to solve geotechnical problems are the finite element method (FEM) and the discrete element method (DEM). The theoretical basis of FEM is the principle of minimum potential energy, which treats the research object as a continuum, that is, each unit displacement maintains a single-valued continuum. The theoretical basis of DEM is Newton's second law, which treats rock masses divided by joints and fissures as rigid bodies. For the discontinuous rock mass existing in nature, the use of DEM for simulation has greater advantages, making DEM an important means of simulating rock mass.

An important discrete method in DEM is particle flow discrete element. One of its representative software is PFC software released by ITASCA company. In recent years, a series of researches have been carried out based on PFC software at home and abroad.

2. Application of particle flow discrete element in geotechnical engineering

2.1 Development of the discrete element method

The discrete element method was proposed by Cundall P.A. in the early 1970s and was initially used to study the mechanical characteristics of discontinuous media such as rock masses. Later, Cundall P.A. and Strack proposed a two-dimensional discrete element simulation method.

Currently, experts and scholars from many countries are working to promote the development of discrete elements. The earliest discrete element systems were limited to dealing with discrete blocks of stiffness, after which the methods were expanded. In the 1980 year, ITASCA developed the discrete element UDEC to be put on the market. Later, Lorig and Brady developed a program to deal with discrete element-boundary element coupling, and Cundall P.A. developed 3DEC, a 3D numerical simulation software to simulate rock masses with joints in the 1988 year [1]. China has also conducted a series of researches on discrete element software, and in the China Rock 2019 conference, several experts and scholars gave lectures on discrete element to explain, Liu Chun et al. of Nanjing University developed the high-performance discrete element software MatDEM, which greatly improved the efficiency of the operation.

* Corresponding author: xiaojiancheng19@mails.ucas.ac.cn

2.2 Application state of particle flow discrete element in simulating rock mechanics test

The mechanism of rock masses failure is of great importance to mechanics theory, underground engineering construction and geological disaster management. The discontinuous media theory divides rock-like material into computational models consisting of rigid particles or blocks, the discrete element method using numerical test methods has become an important breakthrough point in solving the failure mechanism of rock-like media from a mesoscopic perspective. [2] Scholars at home and abroad have carried out many researches and explorations on this issue. The research result of Xu et al. [3] show that the materials mechanical characteristics of complete specimens under compression can be simulated more accurately by using PFC. Liu et al. [4] use PFC to research the time-expansion effect and brittle-ductile-plastic transition character of marble. Huang et al. [5] use particle flow to simulate the macro and meso-mechanical behaviours of red sandstone containing two pre-existing non-coplanar fissures and analysed its strength and failure characteristics.

Based on this, a servo loading method combining loading plate and cylinder is adopted in this paper to study the simulation process of conventional triaxial loading and unloading of rock mass under elastic constructive model and elastoplastic constructive model respectively, and obtain stress-strain curves, and analyse the mechanical behaviour and mechanism of rock mass during the loading and unloading process, and study the mechanical characteristics of rock mass under cyclic loading and unloading conditions, and then verify the feasibility of particle flow discrete element in simulating triaxial test.

3. Basic principle of particle flow discrete element

3.1 Particle flow discrete element

After the development of the UDEC program by Cundall P.A., the in-depth research based on the theory of discrete elements continues with the development of the BALL program, which is used to calculate the mechanical characteristics of discrete discontinuous media bodies into particles. The PFC (Particle Flow Code) program, which currently succeeds PFC3D and PFC2D, is based on the discrete element method (DEM) framework and consists of computational engine and graphical operation interface.

PFC is mainly used to simulate the movement and interaction of particles of limited size. The particles in the PFC model are rigid bodies with mass, which can be translational and rotational. The interaction between the particles has internal inertial force, moment or paired contact force.

The advantages of particle flow discrete element method are that it can get rid of the limitation of deformation, so it is easy to deal with discontinuous media problems, and can better deal with the problems related to multi-field coupling. For the non-unique deformation problems of rock cracking, separation, etc., the principles and results can be fed back by showing the dynamic process. However, the particle flow discrete element method also has its own shortcomings. Firstly, the software is similar to FLAC3D, the pre-processing of modelling is difficult, and it is not easy for beginners to get started, especially it is very difficult to establish complex models, and secondly, the calibration of software parameters is difficult, and the particle flow method lacks the test of engineering applications.

3.2 Mechanics of the discrete element method for particle flow

The movement pattern of the particles in the particle flow discrete element method is basically the same as the principles of kinematics in the DEM, that is, Newton's second law is used to analyse the movement state of a single particle.

$$F_i - \beta_g v_i = m\frac{\Delta v}{\Delta t} \tag{1}$$

$$M_i - \beta_g \omega_i = I\frac{\Delta \omega_i}{\Delta t} \tag{2}$$

In the above equation, F and M are the unbalance force and unbalance moment, and β is the overall damping of the system.

Particle flow contact model is the core principle of PFC, and it is important to simulate the contact mode between two particles in PFC. With the development of PFC, the contact model between particles has been diversified, such as rigid body contact, surface tension generated by liquid bridge and viscous force due to the fluid's own sticky. The following is a brief introduction of the parallel bonding model among the common contact models of the particle flow discrete element method. [6]

First of all, we know that rock-like materials are friction-like materials, while metals belong to artificially formed crystalline materials, their material characteristics are different, mainly in: ① The shear strength of rocks is related to the cohesion of rock particle materials and the angle of internal friction, while metal materials do not have this characteristic. ② Rock materials are multiphase materials, with the characteristics of solid, water and gas, so they are affected by hydrostatic pressure. ③ Rocks have double strength characteristics, which in turn determines that rock materials have the characteristics of hardening and softening. The difference in material characteristics determines that rock-like materials have many different mechanical characteristics from metallic materials, mainly in the compression hardness, isobaric yield characteristic, dilatancy, hardening and softening characteristics of rock-like materials, etc.

It can be seen that the mechanical characteristics of rock-like materials are very different from those of metal-like materials. The characteristics of rock-like materials determine that the shear yield and failure of rocks must take into account the friction angle of the particles inside the rock, so the yield conditions and failure criteria of rock materials are different from those of metal materials.

The Mohr-Coulomb criterion is widely used for rock mass as the yield condition, which reflects the strength characteristics of rocks greatly and is suitable for building the constructive model of geotechnical materials, and is widely used in the geotechnical engineering industry now.

$$|\tau| = c + \sigma \tan\varphi \tag{3}$$

In the above equation, τ is the shear stress (shear strength) on the shear surface; σ is the normal stress on the shear surface; c is the bonding force; and φ is the internal friction angle.

In PFC, when the parallel bonding model is used, as shown in Fig. 1, the stiffness of the material with the parallel bonding model consists of both contact stiffness and bonding stiffness, and the parallel bonding component works together with the linear component to provide elastic interaction between the contact objects, once the contact stress between the particles exceeds the strength limit, the bonding breaks and the bonding is lost, and the model between particles degenerates into a linear contact model. The contact bond model can only transfer force, while the parallel bond model can transfer force and moment, so the parallel bond model can reflect the constructive relationship between the rock particles more accurately when simulating the contact between the rock particles.

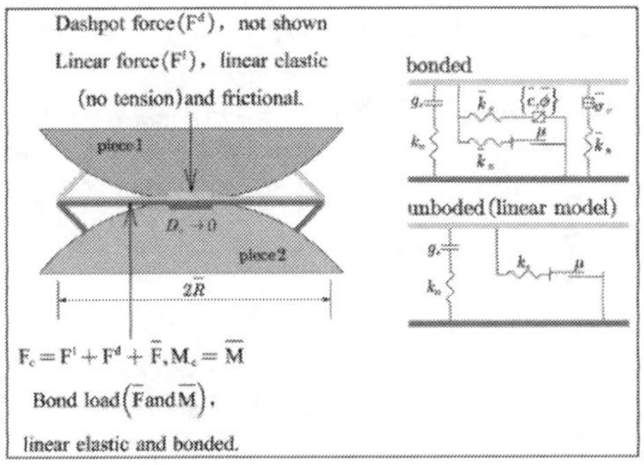

Fig. 1. Schematic diagram of the parallel bonding model between particles

4. Numerical test of particle flow discrete element triaxial loading and unloading

4.1 Principles of mechanics and computation

For the PFC software, the wall can only be given displacement and velocity, and there is no way to give pressure to the wall, so the stress from the triaxial test is converted to the displacement and velocity of the wall.

In this model, there exists a cylindrical wall in contact with the edge of the spherical particles, since particles are spheres, then the contact mode between the particles and the wall is point contact, and the lateral displacement of the spherical particles occurs due to the application of axial pressure, and the extrusion of the surrounding wall occurs, and the lines between the contact points of the spherical particles and the wall and the centres of the spherical particles are always horizontal, so the force of the spherical particle on the wall is also horizontal, and let the force of each particle on the wall force is F_i, then the sum of the force of each particle is the combined force on the particle. [7]

$$F = \sum F_i \tag{4}$$

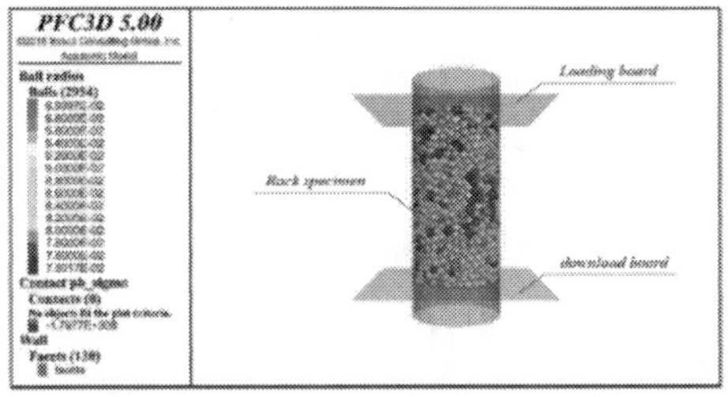

Fig. 2. Schematic diagram of loading mechanism

Then the stresses generated at each boundary can be calculated by calculating the average stress of the particles on the wall according to Newton's third law, by dividing the combined force acting on the wall by the area where the interaction between the wall and the particles.

$$\sigma = \frac{F}{A_{eff}} \tag{5}$$

The strain can be calculated by the following equation.

$$\varepsilon = \frac{L - L_0}{\frac{1}{2}(L_0 + L)} \tag{6}$$

In the above equation, L is the length of the specimen at a moment in the corresponding direction and L_o is the original length of the specimen. For the cylindrical specimens in this paper, the following equations can be used to calculate the axial and normal strains.

$$\varepsilon_H = \frac{H - H_0}{\frac{1}{2}(H_0 + H)} \tag{7}$$

$$\varepsilon_R = \frac{R - R_0}{\frac{1}{2}(R_0 + R)} \tag{8}$$

The kinematic parameters of the wall are controlled to ensure that the axial and lateral stresses remain constant throughout the loading process of the specimen, which is performed by calling the get_gain and servo functions of the FISH language. A cycle is set and the sum of the stresses of each spherical particle on the lateral wall is collected once in each cycle to adjust the wall kinematic parameters, and minimize the difference between measured stress and demanded stress. According to the related literatures, the following algorithm can be used for the servo mechanism.

$$\dot{u}^{(\omega)} = G\left(\sigma'_{measured} - \sigma_{measured}\right) = G\Delta\sigma \qquad (9)$$

G in the above equation is the incremental function obtained by using the get_gain function, the maximum increment that causes the wall boundary force within a certain Δt is:

$$\Delta F^{(\omega)} = k_n^{(\omega)} N_c \dot{u}^{(\omega)} \Delta t \qquad (10)$$

In the above equation, N_c is the number of particles in contact with the wall, and $\Delta F^{(\omega)}$ is the average stiffness of each particle contact surface, the average restraint stress is:

$$\Delta\sigma^{(\omega)} = \frac{k_n^{(\omega)} N_c \dot{u}^{(\omega)} \Delta t}{A} \qquad (11)$$

In the above equation, A is the effective area where the cylindrical wall and the spherical particles are in contact and there is a force. At this point, the difference between the measured value of the restraint stress and the demand value needs to be checked. In practice, it is necessary to use the amplification factor method to derive the following expression.

$$\left|\Delta\sigma^{(\omega)}\right| < \alpha\left|\Delta\sigma\right| \qquad (12)$$

Substituting equation (11) into equation (12), we can get:

$$\left|\frac{k_n^{(\omega)} N_c \dot{u}^{(\omega)} \Delta t}{A}\right| < \alpha\left|\Delta\sigma\right| \qquad (13)$$

Introducing the incremental function G above, increment can be obtained by derivation.

$$G = \frac{\alpha A}{k_n^{(\omega)} N_c \Delta t} \qquad (14)$$

According to the related literatures, we learn that in numerical simulations, α=0.5.

4.2 Establishment of particle flow discrete element model

According to the principles introduced in the 4.1, the model can be built. First of all, the domain is created to determine the limits of the area where the model is to be built, and then three walls are set up, respectively the loading and download boards of infinite size and the side confining plate. After that, the particles are generated. First, a region is created using geometry, and then spherical particles are generated in the geometry region by using the ball command. At this point, it will be found that the locations of some particles are beyond the range of geometry and wall, in this case the run is likely to appear that the particles are extruded from the wall, so we can reduce the size of the sphere in the X, Y, and Z directions to make the particles shrink, so as to avoid the bad effect of this situation on the simulation. The model diagram is shown in Fig. 3.

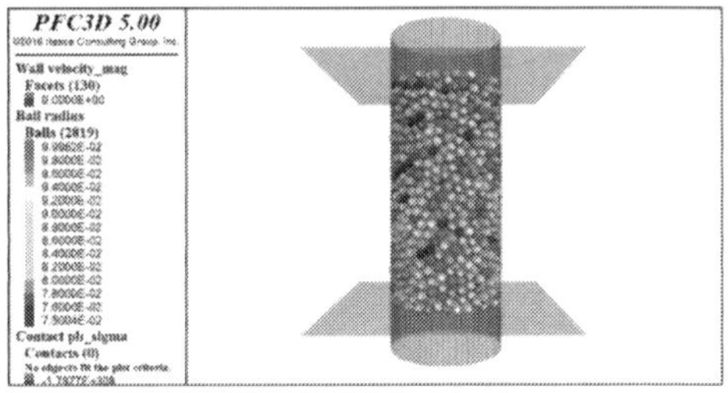

Fig. 3. Schematic diagram of the establishment of PFC model

In this model, the height of the model is 40 mm, and the width is 20 mm. In generating the spherical particles, the radius has a size between 0.075 mm and 0.1 mm. 10001 random numbers are used. Fig. 3 shows that 2819 particles were generated, and the line contact model was used between the particles.

After the particles were established, the principles command flow introduced in the 4.1 will be executed and the constrained stress state of the wall can be read, and the loading of the specimen will be performed.

4.3 Mechanical response and analysis of specimen under triaxial loading and unloading

4.3.1 Numerical simulation of elastic material specimen under triaxial loading and unloading

First, multiple plots need to be established to record the state curve of the specimen and the loading plates under the hist. After that, we use the command flow to load the specimen. First of all, the axial pressure is applied, the loading and download boards need to gradually reach the required loading pressure, so it is necessary to use the principles described in the 4.1 to use FISH language to form a circular structure, set the displacement within each step, and feedback the force of the particles on the wall at this time. When the demand pressure is not reached in one step, increasing the average velocity, and gradually approaching the demanded axial pressure. When the program gets the feedback that the axial pressure reaches the demand pressure, exiting the cycle and loading at this speed, thus obtaining a uniform axial loading speed, as shown in Fig. 4, the loading principle of the lateral confining pressure is the same as that of the axial pressure, that is, the cyclic structure is used to achieve the required test pressure, and unloading after loading is completed.

Making the specimen an elastic material, setting the specimen material mechanical parameters are as follows: tensile strength is 1×10^5 Pa, shear strength is 1×10^5 Pa, friction coefficient is 0.5, axial pressure is 2.3 MPa,

 https://doi.org/10.1051/matecconf/202235801012

confining pressure is 1 MPa, then performing triaxial loading and unloading.

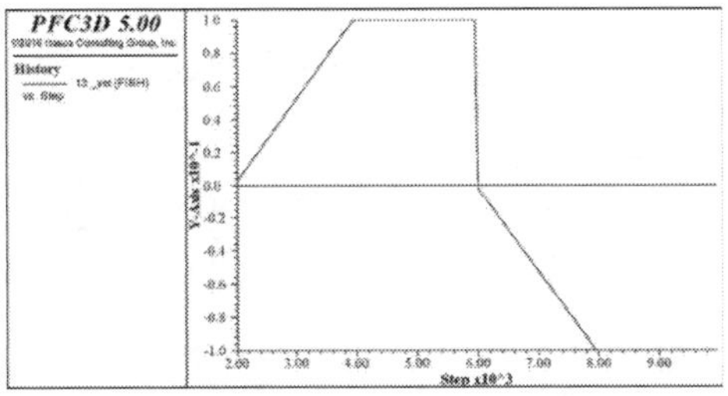

Fig. 4. Velocity image of the wall under loading

After the command flow is finished, running the program to load the specimen. During the loading process, the displacement and contact force cloud charts of the particles can be viewed, as shown in Fig. 5,

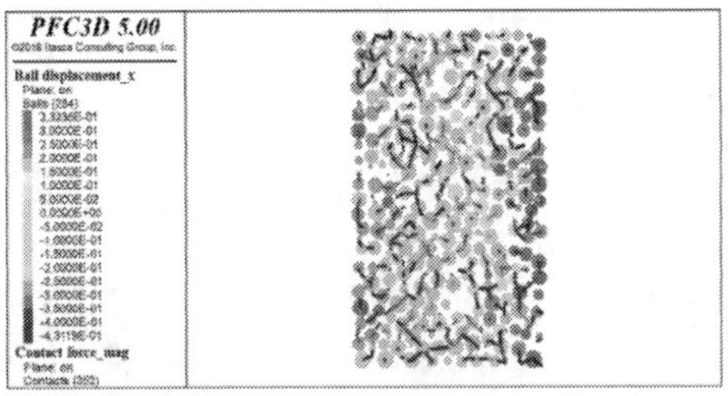

Fig. 5. Cloud chart of displacement profile at a certain moment during loading

After the PFC program is calculated, we can view the process curves during loading and unloading, as shown in Fig. 6, Fig. 7. The full stress-strain curve of the triaxial compression loading and unloading test of the elastic material specimen is shown in Fig. 6. Starting loading, the axial stress increases to the point D, and the axial displacement is 0, indicating that the rock is compact. Unloading is performed after loading to the point E, then the curve goes back to the point F. Unlike the conventional full stress-strain curve of elastic material specimen, the two curves of loading and unloading of the specimen shown in Fig. 6. do not overlap, because the microfractures and open structural planes between the constituent particles of the rock are gradually compacted during the loading process, the feature better reflects the compaction characteristics of the rock material.

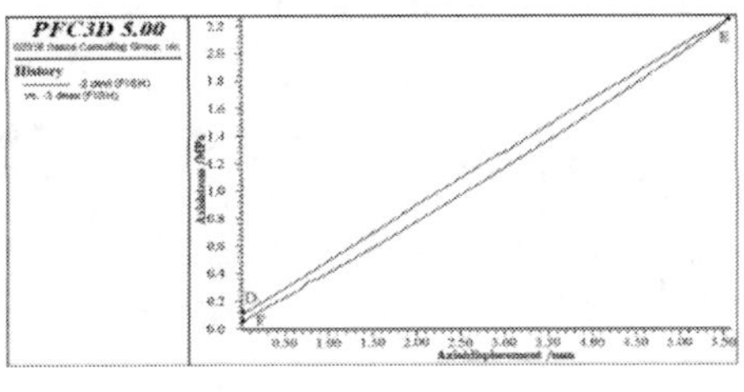

Fig. 6. Full stress-strain curve of elastic material under triaxial loading and unloading

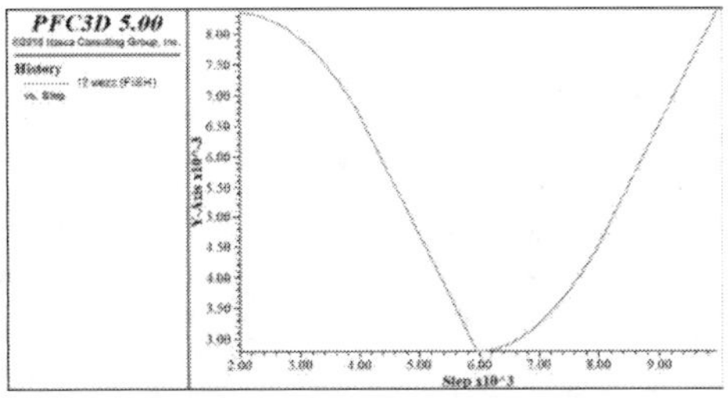

Fig. 7. Axial strain curve of elastic material under triaxial loading and unloading

4.3.2 Numerical simulation of elastoplastic material specimen under triaxial loading and unloading

Using the above principles, we can create elastoplastic specimen by changing material parameters, setting the specimen material mechanical parameters are as follows: tensile strength is 0.5×10^5 Pa, shear strength is 0.5×10^5 Pa, friction coefficient is 0.5, axial pressure is 3 MPa, confining pressure is 1 MPa, then performing triaxial loading and unloading.

Using the above principle to bring the force to the demand restraint stress, as shown in Fig. 8.

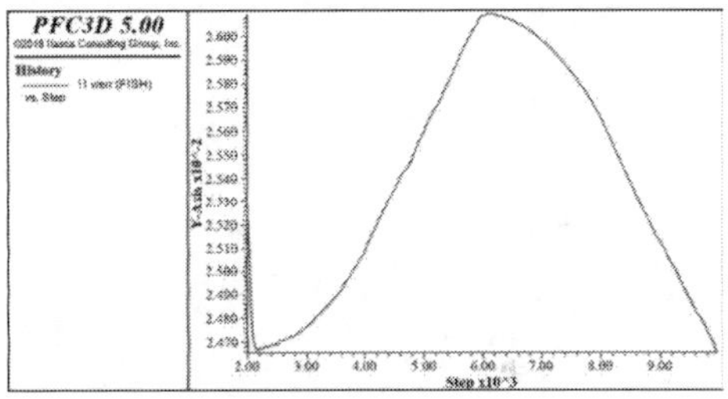

Fig. 8. Lateral strain curve of elastoplastic material under triaxial loading and unloading

After the loading is completed, velocity image of elastoplastic material under triaxial loading, full stress-strain curve of elastoplastic material under triaxial loading and unloading and axial strain curve of elastoplastic material under triaxial loading and unloading can be obtained, as shown in Fig. 9, Fig. 10, Fig. 11.

The full stress-strain curve often triaxial compression loading and unloading test of the elastic-plastic material specimen is shown in Fig. 10. In the early stage of loading, the stress-strain curve is convex upward. In the elastic phase, the stress and strain of rock is basically linear. In the post-peak phase, unloading starts at the point A, it can be seen that the unloading curve does not coincide with the loading curve, and the strain is not returns to 0, i.e., the point B. In this case, the recoverable deformation is called elastic deformation, expressed as ε_e, which is the length indicated by BC. The unrecoverable deformation is called plastic deformation or residual deformation, expressed as ε_p, i.e., the length from the origin to the point B. The plastic deformation of the specimen is generated and loading to the point A again, it can be seen that the loading and unloading curves of the rock cannot be completely overlapped, and the unloading curve always lags behind the loading curve, forming a closed plastic hysteresis loop with a sharp leaf shape. The area enclosed by the plastic hysteresis loop is the energy absorbed and consumed by the inelastic deformation part of the rock structural planes, which is called plastic work ΔW. The energy is used for adjustment of rock structure, structural planes compaction, or relative slip and movement of the structure. Continuing to load, it can be seen that the full stress-strain curve will have multiple steep falls after the peak, indicating that the rock produces multiple local brittle fractures.

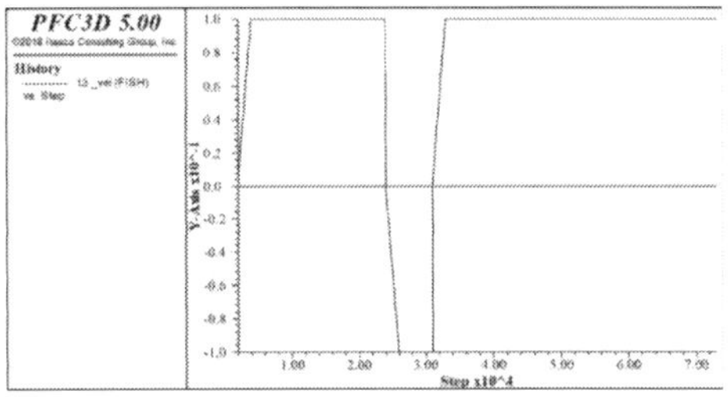

Fig. 9. Velocity image of elastoplastic material under triaxial loading

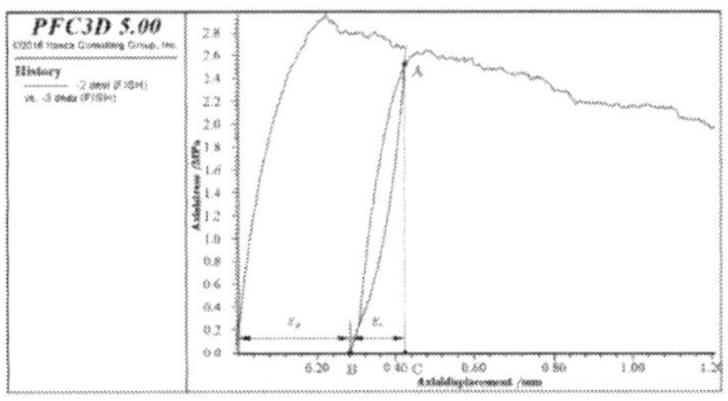

Fig. 10. Full stress-strain curve of elastoplastic material under triaxial loading and unloading

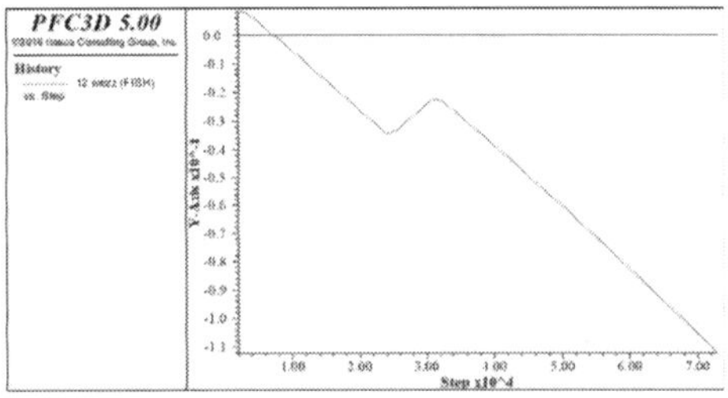

Fig. 11. Axial strain curve of elastoplastic material under triaxial loading and unloading

4.4 Analysis of triaxial loading and unloading test results

Based on the above numerical simulation tests as well as the analysis results, the following conclusions can be drawn.

(1) The triaxial loading and unloading program can simulate the triaxial test well, and good conclusions can be drawn for elastic and elastoplastic materials, the loading and unloading principles in the 4.1 have good usability, and can be extended to obtain triaxial loading and unloading tests for multiple stress paths and multiple specimens.

(2) For the elastic material, it can be seen from the stress-strain curve that the residual stress after the loading and unloading process is 0. For the elastoplastic material, a large residual deformation is generated during the post-peak loading and unloading process, which is the same phenomenon as the triaxial test results, indicating that the test results are accurate.

(3) During the loading process, it can be found that the lateral displacements of particles at the periphery are larger and the particles closer to the axis have less lateral displacement. For the elastoplastic material, it is difficult to increase axial pressure when it is added to 3.56 MPa, because the material has already undergone plastic failure and the support capacity has been reduced.

5. Conclusion

The following conclusions can be drawn from the analysis of this paper.

(1) Numerical simulation is one of the important engineering aids under the increasingly complex engineering conditions, and its highly efficient computational capability provides an important reference for the prevention and control of engineering disasters.

(2) Particle Flow Code (PFC), as one of the important branches of discrete element method (DEM), is gradually being widely used in geotechnical engineering because of its ability to provide multiple contact relationships between particles and its powerful computational capability.

(3) Particle flow contact model is the core principle in PFC, with the development of PFC, the contact between particles has been diversified, such as rigid body contact, surface tension generated by liquid bridge and viscous force due to the fluid's own sticky etc.

(4) The triaxial test can be simulated by using PFC. By the algorithm of converting the force into the indenter velocity, the specimen consisting of particle flow is triaxial loaded, by setting the mechanical parameters of the material, the material is set as an elastic material and an elastic-plastic material respectively, and the stress-strain curves of the two materials are obtained by loading respectively. It can be seen from the stress-strain curves of these two materials that the elastic material residual strain is 0, while the elastoplastic material has a larger residual strain after loading and unloading repeatedly. It can be seen that the PFC can better simulate the triaxial test from the numerical simulation results.

(5) Using this triaxial test model, it can be extended to realize numerical simulation of various tests by changing the shape of the specimen, such as hollow thin-walled cylinder, thick-walled cylinder, cube, etc., using these principles, some other loading and unloading processes can also be realized.

Regarding the calibration of some parameters, the author will further discuss in the future study. ←

References

1. W.H.Wang, X.B.Li, Discrete element method and its application in geotechnical engineering, *Rock and Soil Engineering Technology*, 19, 177-188 (2005)
2. H.C.Zhu, PFC and application cast of caving study, *Chinese Journal of Rock Mechanics and Engineering*, 25, 1927-1931 (2006)
3. J.M.Xu, Z.L.Xie, H.T.Jia, Simulation of mesomechanical properties of limestone using particle flow code, *Rock and Soil Mechanics*, 31(S2), 390-395 (2010)
4. N.Liu, C.S.Zhang, W.J.Chu, Simulating time-dependent failure of deep marble with particle flow code, *Chinese Journal of Rock Mechanics and Engineering*, 30, 1989-1996 (2011)
5. Y.H.Huang, S.Q.Yang, Particle flow simulation of macro and meso-mechanical behavior of red sandstone containing two pre-existing non-coplanar fissures, *Chinese Journal of Rock Mechanics and Engineering*, 33, 1644-1653 (2014)
6. Y.Yang, S.G.Tang, Simulation of particle flow using discrete element method, *China Powder Science and* Technology, 05, 38-43 (2006)
7. H.T.Liu, X.H.Cheng, Size effect analysis of coarse-grained soil by discrete element method, *Rock and Soil Mechanics*, 30 (Plus 1), 287-292, (2009)

Cause analysis of surface micro-cracks on 12Mn pipeline steel

Yushi Tian [1, 2, *], Shaoyang Chu [2], Lijun Xu [2], Jun Wei [2], Shengtao Qiu [1, 2], Rong Zhu [1]

[1] School of Metallurgical and Ecological Engineering, University of Science and Technology Beijing, China
[2] National Engineering Research Center of Continuous Casting Technology, Central Iron and Steel Research Institute, Beijing, China

Abstract. In order to find out the causes of defects in 12Mn pipeline steel produced by a Chinese steel pipe plant, the characteristics of the defects are analyzed and discussed by means of metallographic microscope, scanning electron microscope, ASPEX scanning electron microscope and transmission electron microscope and so on. The physical and chemical tests results show that there are several typical micro-cracks after slight grinding the pipe outer skin in which a few large particles of brittle alumina reside, which possess the macro-characteristics of continuously curving and extending along the radial direction as well as many clear and smooth inner edges and few secondary branches, so that the cracks are not experienced a long and high-temperature oxidation process. The industrial tests indicate that most of the defects have been proved occurred in the quenching and tempering process of the steel pipe production. The main cause of steel pipe cracking is the stress concentration arose from some large brittle inclusions in the quenching process. When the critical strength of the steel pipe is exceeded by quenching stress, a micro-crack will generate and propagate along the grain boundary. These defects of steel pipes could be eliminated by means of taking steelmaking process optimization measures.

1. Introduction

12Mn pipeline steel is a kind of high-quality tempered pearlite seamless pipeline steel developed by a steelmaking company in China, which has already widely applied in natural gas, oil and other long-distance pipeline projects for the characteristics of high strength, toughness, good weldability and low cost. With the purpose of meeting requirements of microstructure and properties, 12Mn hot-rolled pipe is usually needed to be quenched and tempered [1-3]. However, a large number of steel pipes are found a typical defect which is subcutaneous micro-crack on the pipe outer surface after an ultrasonic flaw detection. They can be delivered normally after surface grinding at the expense of increasing the cost of labor and reducing the efficiency of production.

There has been much literature about the surface defects of steel pipes. The main causes are improper quenching process [4-5], unreasonable piercing process [6] and defects of billet itself, such as surface cracks [7-8], surface scratches [9] and large nonmetallic inclusions [10].To find out the cause concerning defects of 12Mn steel pipe, the characteristics and formation mechanism of defects on 12Mn steel pipe were analyzed through a series of tests, for example, metallography, scanning electron microscope, transmission electron microscope and so on. Furthermore, industrial tests were carried out to verify the analysis results. On this basis, the improved process measures are put forward to eliminate the defects on the outer surface of the finished steel pipe and to improve the quality of product.

2. Experiment scheme

2.1 Experimental material

The experimental material was the hot rolled seamless pipe produced by a steelmaking company in China. The main chemical composition is shown in Table 1. The dimension of finished pipe was Φ 168mm × 10.97mm × 12m. The production process was round billet → inspection → annular heating furnace → piercing process→ hot rolling → magnetic flux leakage testing → quenching and tempering → ultrasonic testing → finishing → warehousing.

Table 1. Main chemical composition (wt. %) of test steel

C	Si	Mn	P	S	N	Alt
0.09~0.14	0.25~0.40	1.15~1.35	≤0.018	≤0.008	≤0.008	0.02~0.045

2.2 Physical testing and chemical analysis

The defects bearing samples which were cut from the steel pipe with flaw detection alarm were machined into a size of 10 mm × 10 mm × 15 mm by wire cutting and then rubbed crack surfaces with sandpaper before polishing. In order to obtain the microstructures near cracks as well as morphology and composition of inclusions and so on, optical microscope, several detection methods such as ASPEX scanning electron microscope and energy dispersive spectrometer were used to analyze the

* Corresponding author: yushi_tian@tom.com

processed samples etched by 4% nitric acid alcohol solution at room temperature for about 2s ~ 5s. After that, several thin layers samples with thickness of about 0.6 mm were cut off from the defects bearing samples before rubbing and polishing to 30 μm by fine grit sandpapers, and then examined by transmission electron microscope.

2.3 Industrial tests

In order to determine the process of cracks causing, the steel pipes were subject to ultrasonic flaw detection immediately after hot rolling, and then quenched by water to room temperature where holding at 850℃ for 50 min and tempered by air cooling where holding at 550 ℃ for 75 min before ultrasonic flaw detection once again, as shown in Figure 1. At the same time, it was necessary to carry out the quality tracking and statistic works.

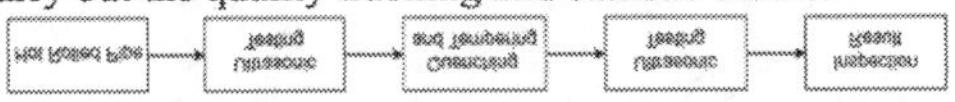

Figure 1. Flowchart of industrial tests

3. Experimental results and discussion

3.1 Macro-characteristics of cracks

Generally, most of the defects on steel surface are not easy to be found by naked eyes after ultrasonic flaw detection. It is observed that there are several typical micro-cracks after slight grinding the pipe outer skin, as shown in Fig. 2 (a). These micro-cracks are mainly linearly distributed under the outer skin of the steel pipes, most of which are in the shape of slender single line, with the length of about 5 ~ 30cm as well as the width of about 30 ~ 100 μm and the depth of generally less than 1mm. Fig. 2 (b) shows the morphology of the polished cross section of a crack bearing sample, which possesses macro-characteristics of continuously curving and extending along the radial direction as well as many clear and smooth inner edges and few secondary branches.

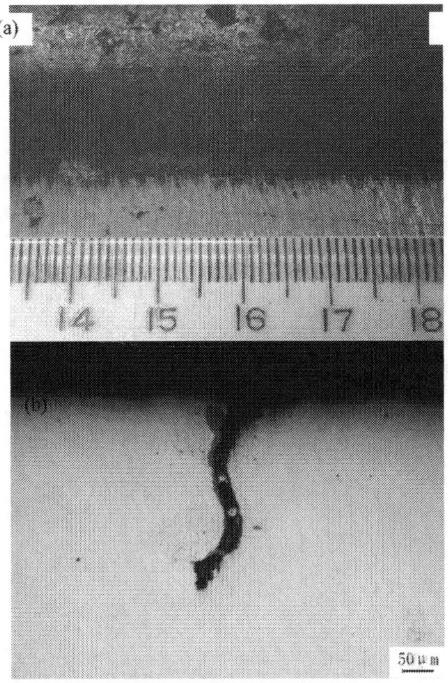

Figure 2. Morphology of typical surface micro cracks (a) after grinding; (b) crack depth

3.2 Micro-characteristics of cracks

3.2.1 Metallographic structure

The defects bearing samples were etched with 4% nitric acid alcohol solution after being polished, the microstructures near a crack are shown in Fig. 3. The ferrite grains are slightly coarse near the crack end region, while relatively uniform near the crack tip region, as shown in Fig. 3 (a), which indicates that the region where a crack is located can bear a strong thermal stress during quenching. There is no obvious decarburization phenomenon on both sides of the crack, which shows that the crack is not experienced a long and high-temperature oxidation process. The extension line of the crack tip could be clearly visible in the area where the crack extends along the radial direction of the steel pipe, as shown in Fig. 3 (b), and the matrixes around the crack tip are considered as tempered pearlites which distribute evenly.

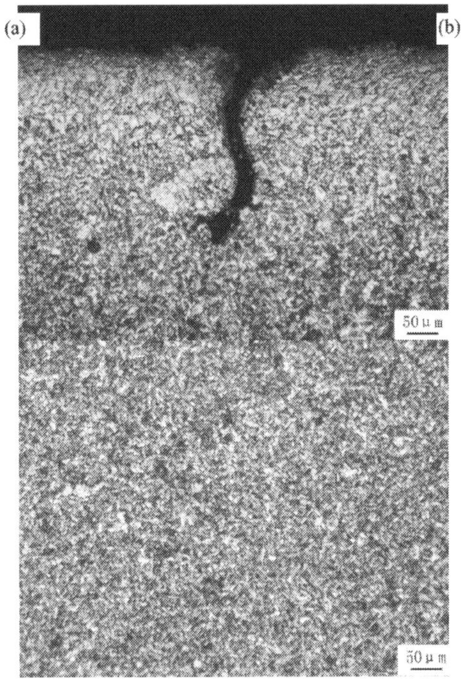

Figure 3. Metallographic structure of surface crack of quenched sample
(a) crack body; (b) crack tip

3.2.2 Morphology and composition of inclusions

Fig. 4 and Tab. 2 describe the morphology and chemical compositions of inclusions in the crack. The inclusions which are in shape of oval and distributed in the middle and tip of the crack are mainly identified as single-phase alumina with the size of 20 ~ 50 μm. The chemical compositions of the inclusions are almost the same, and there has been found a small amount of titanium element in the crack. Generally, alumina inclusions whose size are less than 100 μm are mostly deoxidized products. In addition, they are difficult to float up and to remove by liquid slag in molten steel, and eventually remain in solidified billet. Alumina inclusions, whose sizes are hardly changed during hot deformation, are characterized by brittleness. If the large size of brittle inclusions gathers in steel pipe, stress concentration point can originate from hot deformation or heat treatment process, which is the most important reason for steel pipe cracking.

In the quenching stage of steel pipe production, the surface layer of steel pipe bears the strong dual effects of thermal stress and transformation stress, which are easy to form a large stress concentration around the large-size alumina inclusions. Once the instantaneous stress exceeds the fracture strength of the steel matrix, micro-cracks will form and extend along the low strength grain boundary.

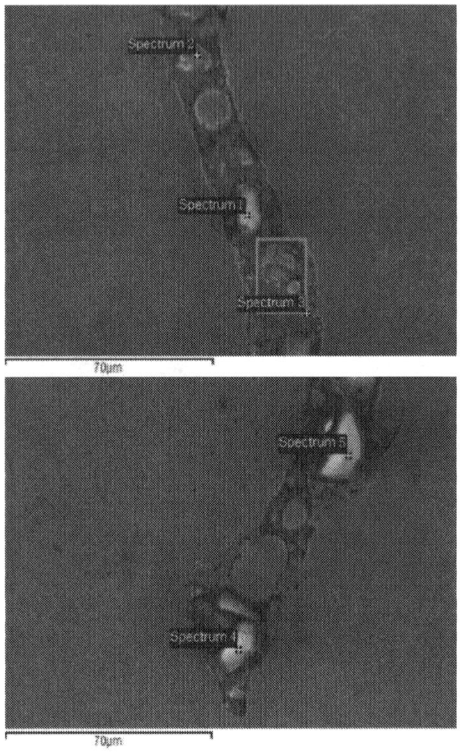

Figure 4. SEM morphology of the inner surface crack of the sample

Table 2. Chemical compositions of inclusions (wt. %)

Spectrum	C	O	Al	Fe
1	/	52.52	46.16	0.88
2	0.54	21.09	/	76.72
3	3.99	27.01	/	64.37
4	0.43	52.37	46.24	0.68
5	0.33	51.64	45.87	0.86

3.2.3 The type and distribution of inclusions.

ASPEX scanning electron microscope was used to detect inclusions on the across section with an area of 24.8 mm2 near the crack, and the size of inclusions is characterized by average diameter. The results show that a total of 1249 inclusions are detected. Fig. 5 (a) shows the main types of inclusions analyzed, in which oxide inclusions account for the highest proportion, followed by sulfide and alumina. Fig. 5 (b) shows the size distributions of the main types of inclusions, it can be seen that most of the inclusions are within 10 μm in diameter, especially that within 2 ~ 4 μm, but there are still a small number of large-size inclusions over 20 μm.

ASPEX can record the coordinate position of inclusions in the detection area, where Y represents the radial direction of steel pipe and X represents the tangent direction of steel pipe circumference. It can be concluded that the distribution of inclusions on cross section of the sample is not uniform, as shown in Fig. 6 and Fig. 7. Many inclusions gather together in an area of the cross section in which there are a few large inclusions scattered, whose size are larger than 30 μm, which contributes to favorable condition for stress concentration during steel pipe quenching.

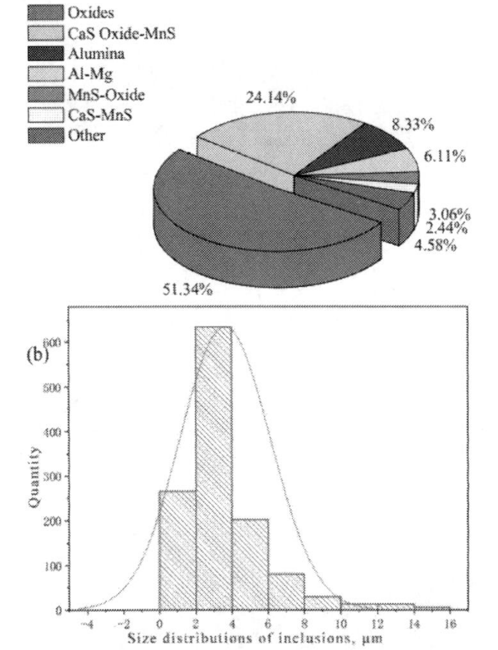

Figure 5. Main types and size distributions of inclusions (a) main types; (b) the histogram of size distributions

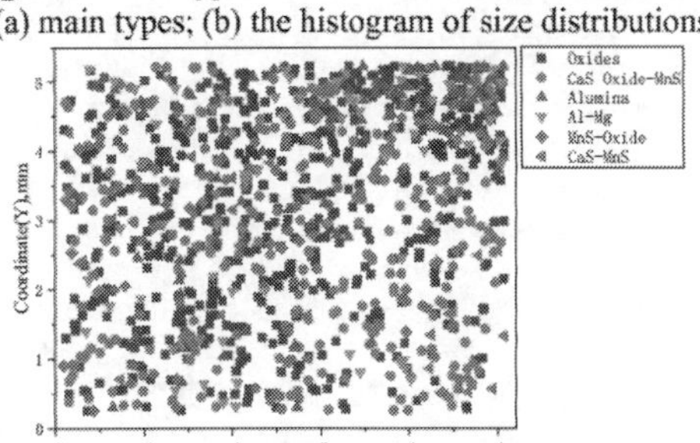

Figure 6. Location distributions of main types of inclusions

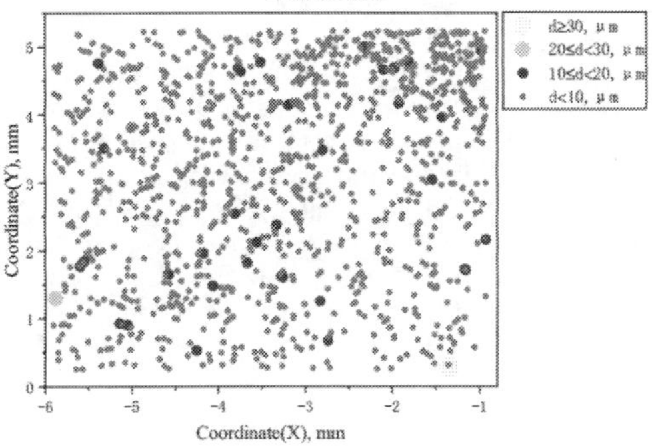

Figure 7. Location distributions of main types of inclusions

3.2.4 Morphology of precipitates

Fig. 8 shows a transmission electron microscope photograph of the matrix near a crack, where Fig. 8(a) and Fig. 8(b) show the typical morphology and location of precipitates, most of which are in the shape of spherical with the size of about 100 ~ 300nm, and distributed along the grain boundary. The diffraction pattern of Fig. 8 (d) shows that these precipitates are granular cementite, and that no other types of precipitates are found.

Fig. 9(c) shows the morphologies of ferrite grain distributed near the crack, which have the characteristic of strip shape. This indicates that the recover behaviors of most ferrite grains only occur at 550 ℃ tempering temperature, whose dislocation cells and dislocation lines gradually disappear and dislocation density decreases as well as multilateralization occurs, so that sub-grains are formed. The mechanical mixtures of the recovered ferrites and granular cementite form tempered pearlite, which effectively eliminate the residual internal stress in the steel pipe and obtain good comprehensive mechanical properties. All the properties of tempered steel pipes without cracks meet the requirements of API 5L.

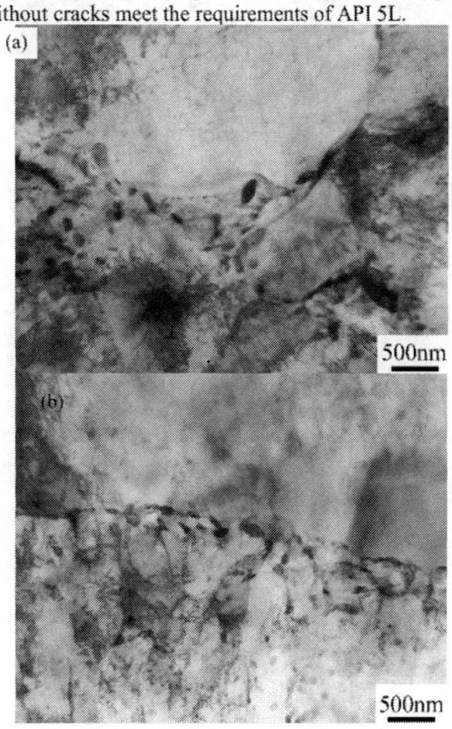

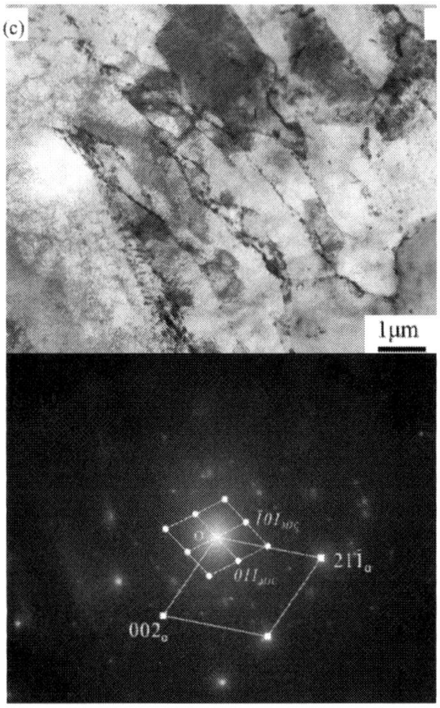

Figure 8. Precipitates near the crack
(a) and (b) the location of the precipitates; (c) grain shapes; (d) diffraction pattern

3.3 Ultrasonic Testing

In order to determine the source process of steel pipes cracking, there were 158 hot rolled pipes selected from the same casting scheme for two industrial tests, where 54 steel pipes were tested in the first time and 104 steel pipes were tested in the second time. Fig. 9 shows the flaw detection results of the industrial tests and indicates that there is 1 unqualified steel pipe occurred after hot rolling process in the first test while 0 unqualified steel pipe in the second test, and 2 unqualified steel pipes occurred after quenching and tempering process in the first test while 11 unqualified steel pipes in the second test.

To sum up, there is a total of 1 unqualified steel pipe occurred after hot rolling process with the defect rate of 0.63% and a total of 13 steel pipes occurred after quenching and tempering process with the defect rate of 8.23% in all 158 steel pipes tested. Therefore, it's almost certain that most of the defects have been proved occurred in the quenching and tempering process, and that almost all of them are considered as micro-cracks under the outer skin of the steel pipes.

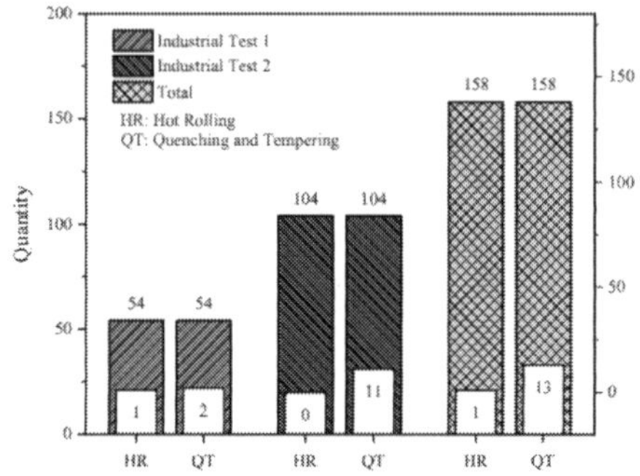

Figure 9. Industrial test results

3.4 Discussion

All the defects which are not easy to be directly observed by naked eyes and almost considered as tiny and slender micro-cracks under surface skin occur under the outer skin of the steel pipes. Microstructures of the cross section along the radial extension direction near the crack are distributed nonuniformly, where ferrite grains near the crack end area grow coarser than that of crack tip area and a few large alumina brittle inclusions reside in the crack. There is no significant difference between microstructures of both sides of the crack and that of the tip area. With few secondary branch cracks and no traces of high temperature oxidation, the inner surface edge of the crack is generally smooth and has no obvious decarburization phenomenon which is the typical characteristic of quenching crack [11-12], so that the micro-pores around the inclusions are not directly exposed to air during the process of deforming and rolling. The industrial flaw detection test results directly show that most cracks bearing steel pipes are found in the quenching and tempering stage of the production process, which proves that the micro-cracks are produced in the steel pipe quenching process. Due to the existence of brittle inclusions as well as the absence of oxidation and decarburization phenomenon in the cracks, the main cause of steel pipe cracking is the stress concentration arose from some large brittle inclusions in the quenching process. When the critical strength of the steel pipe is exceeded by quenching stress, a micro-crack will generate and propagate along the grain boundary.

In order that increasing cleanliness of steel billets should be effective it must be planned systematically. The total oxygen content in steel can be more stringent controls for decreasing large inclusions on optimizing the steelmaking process, such as adjusting soft blowing time, improving protective casting, avoiding slag entrapment, using a new type tundish covering agent and so on. The subcutaneous micro-cracks of steel pipes have been eliminated since taking steelmaking process optimization measures, so that the product quality and productivity of steel pipes are significantly improved.

4. Conclusion

(1) There are several typical micro-cracks after slight grinding the pipe outer skin, which possess macro-characteristics of continuously curving and extending along the radial direction as well as many clear and smooth inner edges and few secondary branches and are mainly linearly distributed under the outer skin of the steel pipes, most of which are in the shape of slender single line, with the length of about 5 ~ 30cm as well as the width of about 30 ~ 100 μm and the depth of generally less than 1mm.

(2) The region where a crack is located can bear a strong thermal stress during quenching. The crack is not experienced a long and high-temperature oxidation process. The extension line of the crack tip could be clearly visible in the area where the crack extends along the radial direction of the steel pipe, and the matrixes around the crack tip are considered as tempered pearlites which distribute evenly.

(3) The inclusions which are in shape of oval and distributed in the middle and tip of the crack are mainly identified as single-phase alumina with the size of 20 ~ 50 μm. Many inclusions gather together in an area of the cross section in which there are a few large inclusions scattered, whose size are larger than 30 μm, which contributes to favorable condition for stress concentration during steel pipe quenching. Most of precipitates are in the shape of spherical with the size of about 100 ~ 300nm, and distributed along the grain boundary. The diffraction pattern shows that these precipitates are granular cementite, and that no other types of precipitates are found.

(4) There is a total of 1 unqualified steel pipe occurred after hot rolling process with the defect rate of 0.63% and a total of 13 steel pipes occurred after quenching and tempering process with the defect rate of 8.23% in all 158 steel pipes tested. Therefore, it's almost certain that most of the defects have been proved occurred in the quenching and tempering process, and that almost all of them are considered as micro-cracks under the outer skin of the steel pipes.

(5) The main cause of steel pipe cracking is the stress concentration arose from some large brittle inclusions in the quenching process. When the critical strength of the steel pipe is exceeded by the quenching stress, a micro-crack will generate and propagate along the grain boundary. These defects of steel pipes could be eliminated by means of taking steelmaking process optimization measures

References

1. L. Morsdorf, E. Emelina, B. Gault, et al, Carbon redistribution in quenched and tempered lath martensite, Acta Materialia 205(2021) 1-53.
2. M. Liu, C. H. Wang, Y. C. Dai, et al, Effect of quenching and tempering process on sulfide stress cracking susceptibility in API-5CT-C110 casing steel, Materials Science and Engineering A 688(2017) 378-387.
3. S. Bordbar, M. Alizadeh, S. H. Hashemi, Effects of microstructure alteration on corrosion behavior of welded joint in API X70 pipeline steel, Materials & Design 45(2013) 597-604.
4. X. H. Wang, M. H. Tang, J. M. Zhao, Cause Analysis and Countermeasure of Quenching Crack of 25MnV Steel Oil Casing with Large Diameter, Hot Working Technology 46(2017) 256-259.
5. X. Y. Kong, R. Q. Ma, H. Y. Yu, et al, Analysis on quenching crack of 40CrMnMo steel center tube and its counter measures, Heat Treatment of Metals 35(2010) 91-93.
6. J. J. Cao, W. W. Zhao, Y. Li, et al, Failure analysis on inner surface crack of 37Mn5 hot-rolled steel tube, Heat Treatment of Metals 41(2016) 228-231.
7. K. Zhang, Y. Y. Wu, D. C. Dong, Causes and solutions of outer fold on surface of 20 steel pipes, China Measurement & Test 39(2013) 28-30.
8. J. Wei, H. P. Hong, X. J. Liao, et al, Analysis on formation of outer fold of 37Mn5 steel tube and process control, Special Steel 31(2010) 30-32.
9. Y. Huang, B. F. He, A. R. Wang, et al, Analysis of reason for surface enfoldment defect of 20# seamless steel tube, Industrial Heating 45(2016) 5-7.
10. Y. M. Lang, X. H. Nie, G. F. Wang, et al, Cracking reason analysis of 20 steel seamless pipe, Hot Working Technology 45(2016) 257-260.
11. S. Šolić, B. Podgornik, V, Leskovšek, The occurrence of quenching cracks in high-carbon tool steel depending on the austenitizing temperature, Engineering Failure Analysis, 92(2018) 140-148.
12. C. J. Wang, Crack cause analysis of 42CrMo steel thick wall pipe, Heat Treatment of Metals, 39(2014) 164-166.

Research on a mechanical model of coal and rock around well based on variable temperature

Sheng Yu[1,2], Jie Zhang[1,2,*], Xin Li[1,2], Cuinan Li[3] , Rongxin Li[1,2]

[1]Southwest Petroleum University State Key Laboratory of "Oil and Gas Reservoir Geology and Development Engineering," Chengdu, China
[2]School of Petroleum and Natural Gas Engineering, Southwest Petroleum University, Chengdu, China
[3]Engineering Technology Research Institute, PetroChina Southwest Oil and Gas Field Company, Guanghan, China

Abstract. To analyze the mechanical properties of coal and rock under variable temperature, the mechanical model of coal and rock around the well under variable temperature is studied based on elastic-plastic mechanics and thermodynamic theory. The control equations of coal and rock deformation, seepage field, and temperature field around the well considering temperature and clay mineral content are derived. On this basis, the finite element numerical simulation is carried out by using the relevant parameters obtained from coal and rock mechanics experiments. The numerical simulation results show that the coal rock strength, peak stress, and peak strain show a decreasing trend when the temperature increases and the simulation results of the coal samples are consistent with the experimental results, which verifies the reasonableness of the mechanical model.

1. Introduction

During the CBM extraction process, the temperature of the coal seam changes dramatically due to geological disturbances and heat exchange of drilling fluids [1]. The change in temperature field will cause changes in the seepage and stress fields of the coal seam, resulting in changes in the rock mechanics of the coal seam rock, which will cause the collapse of the coal bed methane well wall [2].

Changes in temperature can cause changes in the mechanical parameters of the rock, including compressive strength, Poisson's ratio, modulus of elasticity, and so on [3,4]. Some previous models of coal rock mechanics have been developed, mainly considering the effects of ground stress and pore pressure on rock mechanical parameters, while less research has been conducted on the effects of temperature. Lu et al.[5]unified the characterization of the intrinsic and structural stresses of coal rocks through the equivalent pore pressure coefficient, modified the KarlTerzaghi equation, and revealed the variation of mechanical properties of gas-bearing coal with effective stress; Wu Shiyue et al.[6]derived the relationship between adsorption properties, swelling stress and surface pressure of coal rocks and the relationship between effective stress and swelling stress, and analyzed the mechanical properties of coal rocks under the action of adsorption and swelling stress. The results of the model solution were compared with the experimental results, and the model's accuracy was verified. At present, the research on the mechanical properties of coal rocks under different temperature conditions is mainly carried out through mechanical experiments, and the research on the dynamic changes of the temperature, seepage and stress fields is not deep enough, while the research on the mechanical model of coal rocks considering the temperature effect is less or imperfect. Therefore, this paper establishes a mechanical model of coal rocks around the well considering temperature and clay mineral content, analyses the influence law of variable temperature effect on the mechanical properties of coal rocks, and compares the experimental results of coal rock mechanics with the numerical simulation results to verify the accuracy of the mechanical model.

2. Control equation of coal and rock around the well under variable temperature

2.1 Control equations for coal rock deformation around wells

The deformation of the coal rock around the well is influenced by three main factors, namely temperature, coal seam pressure, and ground stress. These three factors act together to cause complex deformation of the coal seam rock.

Assuming that the coal seam rock mass is an isotropic elastic medium, the strain equation caused by coal seam temperature, pressure, and in-situ stress follows.

$$\varepsilon_T = \frac{\beta_T}{3}\Delta T; \varepsilon_{PY} = -\frac{K_Y}{3}\Delta P; \varepsilon_{PX} = \frac{2\rho RTaK_Y}{9V_m} In(1+bP); \varepsilon_\sigma = \varepsilon_n^E + \varepsilon_m^E \quad (1)$$

where β_T is the coefficient of thermal expansion of the coal bed rock, 1/K; ΔT is the coal bed temperature change,

* Corresponding author: 392509026@qq.com

K; K_Y is the volume compression factor, MPa^{-1}; ΔP is the coal bed pressure change, MPa; ρ is the apparent density of the coal bed rock, t/m^3; R is the Pratt's gas constant, where $R = 8.3143 J/(mol \cdot K)$; a is the ratio of limiting coal rock adsorption to mass, m^3 /t; V_Q is the molar volume of the gas, where $V_Q = 22.4 \times 22^{-3} m^3 / mol$; b is the adsorption equilibrium constant, MPa^{-1}; P is the coalbed methane pressure; Where ε_n^E and ε_m^E are the strain due to elastic deformation of the clay mineral and the strain due to elastic deformation of the rock skeleton respectively, %.

The occurrence of elastic strain in coal seam rocks should follow Hooke's Law.

$$\begin{cases} \varepsilon_\sigma = \dfrac{1}{2}\left(\dfrac{1}{G_n} + \dfrac{1}{G_m}\right)\left(\sigma^g - \dfrac{\upsilon}{1+\upsilon}\sigma^g\right) \\ \sigma^g = \sigma_1 + \sigma_2 + \sigma_3 \end{cases} \tag{2}$$

Where G_n, Gm are the coefficients of lamé of clay minerals and rock masses, MPa; υ is the Poisson's ratio; σ^G is the effective stress, MPa; σ_1 , σ_2 , σ_3 are the maximum principal stress, intermediate stress and minimum principal stress, MPa.

The deformation of the coal seam rock is affected by all three of these factors, so by superimposing equations (1) and (2), we can obtain the equation for controlling the deformation of the coal rock around the well.

$$\begin{aligned} \varepsilon &= \varepsilon_T + \varepsilon_{px} + \varepsilon_{pY} + \varepsilon_e \\ &= \frac{\beta_Y}{3}\Delta T + \frac{2\rho RTaK_Y}{9V_m} In(1+bP) - \frac{K_Y}{3}\Delta P + \frac{1}{2}\left(\frac{1}{G_n} + \frac{1}{G_m}\right)\left(\sigma^g - \frac{\upsilon}{1+\upsilon}\sigma^g\right) \end{aligned} \tag{3}$$

2.2 Governing equation of coal and rock seepage field around the well

Assuming that the flow of CBM molecules in the pores of the rock is a one-dimensional Darcy seepage, the molecular flow equation, the adsorption equation and the equation of state are as follows.

$$v = \frac{\kappa}{\mu}\nabla P; Q = \left(\frac{abcP}{1+bP} + \varphi_T \frac{P}{P_m}\right)\rho_m; \rho_g = \frac{\rho_m P}{\xi P_m} \tag{4}$$

where v is the percolation velocity of CBM molecules, m/s; k is the permeability of the coal bed rock, mD; ∇P is the molecular pressure of CBM per unit burial depth, Pa/m; μ is the dynamic viscosity of CBM molecules, Pa·s; c is the correction factor; φ_T is the coal rock porosity, %; ξ is the CBM compression factor, related to the temperature T; ρ is the density of CBM at standard conditions, kg /m^3; P_n is the pressure of CBM at standard conditions, the value is taken as 1 atm.

The process of molecular flow of coal bed methane satisfies the law of conservation of mass with the following equation.

$$\frac{\partial Q}{\partial t} + \nabla \cdot (\rho_g v) = I \tag{5}$$

Taking equations (4) into equation (5) gives the following equation.

$$\left[2\varphi_T + \frac{2abcP_m}{(1+bp)^2} + \frac{2abcP_m}{1+bP}\right]\frac{\partial P}{\partial t} + 2P\frac{\partial \varphi_T}{\partial t} - \nabla \cdot \left(\frac{\kappa}{\mu}\nabla p^2\right) = I \tag{6}$$

Of which:

$$\frac{\partial \varphi_T}{\partial t} = \left(1 - \frac{K'}{K_g}\right)\frac{\partial \varepsilon_V}{\partial t} + \frac{1-\varphi_T}{K_g}\frac{\partial P}{\partial t}; \alpha = 1 - \frac{K'}{K_g}; K' = \frac{2(\omega G_n + (1-\omega)G_m)(1+\upsilon)}{3(1-2\upsilon)} \tag{7}$$

Where K' is the bulk modulus of the coal rock mass, MPa; K_g is the skeletal bulk modulus, MPa; and ω is the clay mineral content, %.

Putting formulas (7) into formula (6), the governing equation of the seepage field of coal and rock around the well can be obtained.

$$I = \left[2\varphi_T + \frac{2abcP_m}{(1+bP)^2} + \frac{2abcP_m}{1+bP}\right]\frac{\partial P}{\partial t} + 2P\left[\left(1 - \frac{K'}{K_g}\right)\frac{\partial \varepsilon_V}{\partial t} + \frac{1-\varphi_T}{K_g}\frac{\partial P}{\partial t}\right] - \nabla \cdot \left(\frac{\kappa}{\mu}\nabla P^2\right) \tag{8}$$

2.3 Controlling equations for the temperature field of the coal rock around the well

The variation of the coal seam rock's temperature field follows the first and second laws of thermodynamics so that the following equation can be obtained.

$$dK + dU = \delta Q_d + \delta W; \delta W = \sigma_{ij} d\varepsilon_{ij}; ds = \frac{\delta Q_d}{T} \tag{9}$$

Where dK and dU are the increment of kinetic energy and internal energy per unit volume of rock per unit time, respectively, J; δ_W and δ_{Qd} are the work done and the heat obtained per unit volume of rock per unit time, J.

According to the references[7,8], it follows that

$$\begin{aligned} dU &= \rho C_V dT + \sigma_{ij}^d \varepsilon_{ij} + \theta T_0 d\varepsilon_V \\ &+ \theta_{PX} T_0 \left[T_0 In(1+bP)\frac{\partial a}{\partial T} + T_0 \frac{aP}{1+bP}\frac{\partial b}{\partial T} + aIn(1+bP)\right] d\varepsilon_V \end{aligned} \tag{10}$$

Of which:

$$\begin{cases} \chi = \dfrac{E\upsilon}{(1+\upsilon)(1-2\upsilon)} = \dfrac{2G\upsilon}{1-2\upsilon}; G = \dfrac{E}{2(1+\upsilon)}; \theta_t = \dfrac{(3\chi + 2G)\beta}{3} \\ \theta_{pY} = \dfrac{(3\chi - 2G)K_Y}{3}; \theta_{px} = \dfrac{(2\rho RK_Y)(3\chi + 2G)}{9V_m}; G = (\omega G_n + (1-\omega)G_m) \end{cases} \tag{11}$$

Combining equations (9-11), the control equation of the temperature field of coal and rock around the well can be obtained:

$$\begin{aligned} \delta Q_d &= dK + dU - \delta W = \rho C_V dT + \sigma_{ij} d\varepsilon_{ij} + \theta_T T_0 d\varepsilon_V \\ &+ \theta_{PX} T_0 \left[T_0 In(1+bP)\frac{\partial a}{\partial T} + T_0 \frac{aP}{1+bP}\frac{\partial b}{\partial T} + aIn(1+bP)\right] d\varepsilon_V + dK \end{aligned} \tag{12}$$

where C_V is the constant volume-specific heat of coal seam rock mass, J/(kg·K).

3. Verification of mechanical model of coal and rock around the well under variable temperature

To verify the mechanical model established in the previous paper, triaxial compression experiments were carried out on coal rock specimens using the RTR-1000 high-temperature and high-pressure triaxial rock mechanics testing system. Three temperature conditions (40.0°C, 70.0°C, and 100.0°C) were set to obtain the specimens' rock mechanics parameters and stress-strain curves under different temperature conditions to provide the basic parameters required for the numerical model. Finally, the experimentally obtained coal rock mechanical parameters and the derived control equations were input into the numerical model for numerical simulation studies, and the reasonableness of the mechanical model was verified by comparing and analyzing the numerical simulation results with the experimental results. The experimental results of coal rock mechanics can be seen in Table 2 and Figure 2.

3.1 Establishment of geometric model and parameter setting of coal and rock around the well

In this paper, a three-dimensional geometric model of the coal seam rock was established, the model size was consistent with the mechanical test, and the size of Φ25mm×50mm was set. The geometric model and the meshing results are shown in Fig 1.

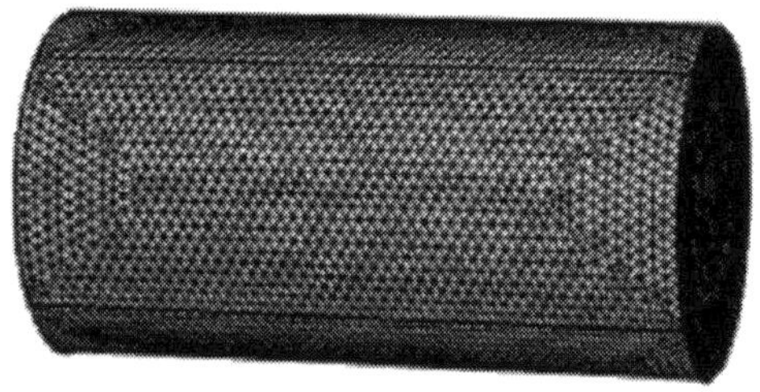

Figure 1. Geometric model and meshing of the coal sample.

Based on the rock mechanics parameters obtained from the previous coal rock mechanics experiments and the results of the related literature[11], the various model parameters required for the numerical simulations were input into the COMSOL software for the simulations, and some of the model parameter settings are shown in Table 2.

Table 1. Table of parameters for the numerical coal rock model.

Name of coal sample parameters	Unit	Parameter values
Modulus of elasticity (E_m)	GPa	6.36
Poisson's ratio (υ)	/	0.33
Cohesion (c)	MPa	16.31
Internal friction angle (Φ)	°	29.8
Test temperature (T)	°C	40.0～100.0
The density of the coal sample (ρ_s)	Kg/m³	1590
The density of coal bed methane (ρ_g)	Kg/m³	0.714
Coal bed methane compression factor (K_Y)	Kg/m³/Pa	0.9982e-5
Initial penetration rate	m²	0.6e-13
Initial porosity	/	0.09
Coal bed methane viscosity	Pa*S	1.34e-5
Reference pressure adsorption constant (a)	m³/Kg	14.5
Adsorption equilibrium constant (b)	m³/Kg	0.73
Lamé constant (G)	GPa	1.387
Coefficient of expansion of coal rock mass (β)	1/K	1.16×10^{-4}
Clay content (ω)	/	0.5%
Modulus of elasticity of clay (E_n)	GPa	1.27

3.2 Numerical simulation results and analysis of coal and rock around the well

Figures 2, 3, and 4 show the relationship curves between the main stress difference and axial strain in the coal seam rock at temperatures of 40°C, 70°C, and 100°C. The two curves in the figure represent the experimental and simulated values.

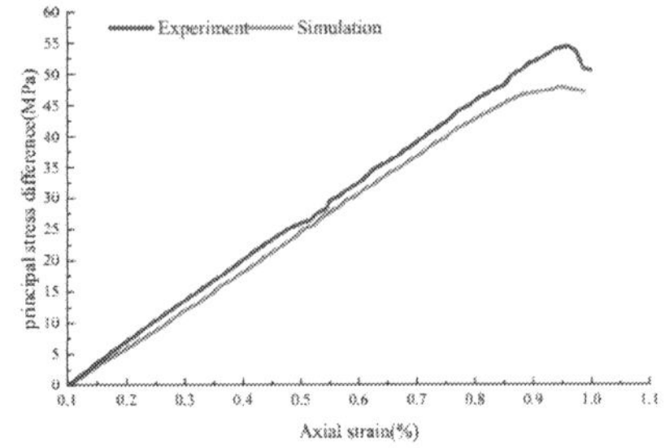

(a)Temperature is 40°C

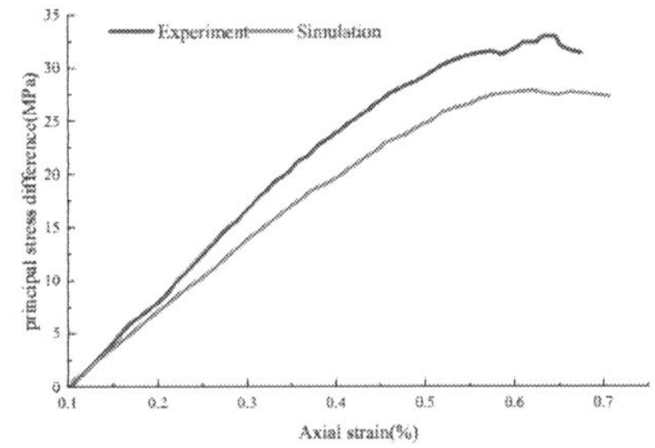

(b)Temperature is 70°C

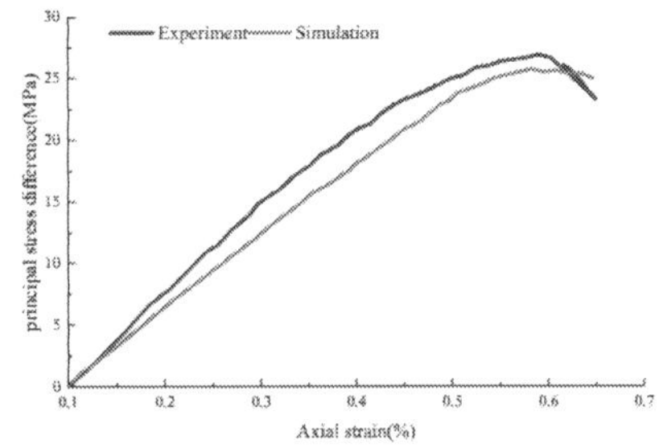

(c)Temperature is 100°C

Figure 2. Principal stress difference-axial strain curves for coal rocks at different temperatures.

From Figures 2, 3, and 4, it can be seen that the peak principal stress difference of the coal rock specimens gradually decreases as the temperature increases from 40°C to 100°C, and the axial strain when the peak stress is reached shows a decreasing trend. It can also be seen that the stress-strain curves of the experimental and simulated specimens are in better agreement, and the trend of change is consistent.

Table 2. Comparison of experimental and numerical simulation results for coal rock specimens.

Temperature/°C	Compressive strength/MPa		Relative error /%	Peak strain/%		Relative error /%
	Experimental	Simulation		Experimental	Simulation	
40	84.5	78.9	6.63	0.9584	0.9460	1.29
70	65.5	57.8	11.8	0.6324	0.6194	2.06
100	58.0	55.8	3.8	0.5900	0.5822	1.32

As can be seen from Table 2, The relative errors between the experimental and simulated values for the compressive strength of the coal rock specimens were 3.8-11.8%, while the relative errors between the experimental and simulated values for the peak axial strain were 1.29-2.06%, both of which were within a reasonable margin of error. Therefore, the results obtained from the mechanical experiments of the coal rock specimens are in good agreement with the numerical simulation results, which indicates that the established mechanical model of the perimeter coal rock is reasonable and can provide a theoretical basis for the study of the mechanical properties of the perimeter coal rock under different temperature conditions.

4. Conclusion

(1) By deriving the deformation control equation of coal and rock around the well, the seepage field control equation of coal and rock around the well and the temperature field control equation of coal and rock around the well, the mechanical model of coal and rock around the well under the action of variable temperature is established.

(2) As the temperature rises from 40°C to 100°C, the coal rock strength weakens, the peak principal stress difference decreases, and the peak strain decreases.

(3) The results of the numerical simulation of the perimeter coal rock specimens agree with the mechanical test results, the relative error is within a reasonable range, and the mechanical model established is reasonable and can be used for the study of the mechanical properties of the perimeter coal rock under different temperature conditions.

References

1. N. Zhang, "Coupled thermo-fluid-solid analysis of coal rock damage evolution and gas seepage," Ph.D. thesis, China University of Mining and Technology, 2017.
2. G. B. Stracher., T. P. Taylor, "Coal fires burning out of control around the world: thermodynamic recipe for environmental catastrophe," International Journal of Coal Geology, vol. No. 1-2(2004), pp. 7-17.
3. B. Zhou., J. Xu., S. J. Peng, et al, "Test system for the visualization of dynamic disasters and its application to coal and gas outburst," International Journal of Rock Mechanics & Mining Sciences, vol. (2019), pp. 104083.
4. S. F. Liang., S. Z. Fang., G. H. Wei, et al, "Mechanical properties of siliceous sandstone after high-temperature action," Journal of Zhengzhou University, vol. 42, No. 3(2021), pp. 87-92.
5. P. Lu., Z. W. Shen., G. W. Zhu, et al, "Effective stress and mechanical deformation damage characteristics of gas-bearing coal," Journal of the University of Science and Technology of China, vol. No. 6(2001), pp. 687-693.
6. S. Y. Wu., W. Zhao, "Effective stress analysis of adsorbed CBM coal," Journal of Rock Mechanics and Engineering, vol. No. 10(2005), pp. 1674-1678.
7. A. A. Borisenko, "Effect of gas pressure on stresses in coal strata," Soviet Mining Science, vol. 21, No. 1(1985), pp. 88-92.
8. M. H. Li, "Mechanical properties of coal rocks and spatial and temporal evolution of gas transport undermining conditions," Ph.D. thesis, Chongqing University, 2013.
9. B. Hu, "Research on the Mechanics of Well Wall Instability in Coal Seam in Honghe Oilfield," Ph.D. thesis, Southwest Petroleum University, 2015.

MATEC Web of Conferences **358**, 01015 (2022) https://doi.org/10.1051/matecconf/202235801015
MSMS 2022

Infrared Spectroscopy Method for Detecting the Content of Antioxidant Irganox 1076 in Polypropylene

Qiukai Zheng [1], Jingjing Fan [2,*]

[1] Chemistry & Chemical and Environmental Engineering College, Weifang University, Weifang, Shandong 26106 China
[2] Biological and Agricultural Engineering College, Weifang University, Weifang, Shandong 261061 China

Abstract. A method for rapid determination of antioxidant Irganox 1076 in polyolefin by Fourier transform infrared spectroscopy was established. By comparing the infrared spectra of Irganox 1076 and polypropylene, the absorption peak of Irganox 1076 at 1738 cm^{-1} was selected as the detection peak. Within the concentration range involved, the standard curve equation of Irganox 1076 in PP is $y = 0.00593 + 0.12681x$, and the linear regression coefficient is 0.98742. The standard curve was used to determine the content of Irganox 1076 in the actual sample, and the maximum relative error was 6.85%. The method is accurate, rapid and sensitive, which can be used for on-line monitoring of processing and using polyolefin products.

1. Introduction

Polyolefin is a kind of polymer material with the largest amount at present. In view of its excellent performance, low price, rich raw materials and easy processing, polyolefin is widely used. Food packaging materials are called "special food additives". It is the last process of modern food industry. Food packaging materials industry is developing rapidly with a rapid momentum. Plastic packaging is the main material of food packaging because of its convenient carrying and processing. On the other hand, polyolefins are easily affected by various environmental factors such as high temperature, ultraviolet, high load and microorganisms during processing, storage and use. The main reason for aging is that the structure or components have weaknesses that are easy to cause aging, such as unsaturated double bonds, branched chains, carbonyls, hydroxyl groups at the end, and so on. The molecular chain will break and form free radicals. If this free radical is not eliminated in time, it will further cause the fracture of plastic molecules, and finally make the relative molecular weight of plastics become very low and lose value. The macroscopic performance is the color changes and mechanical properties decline, that is, the degradation or aging of polymer materials [1].

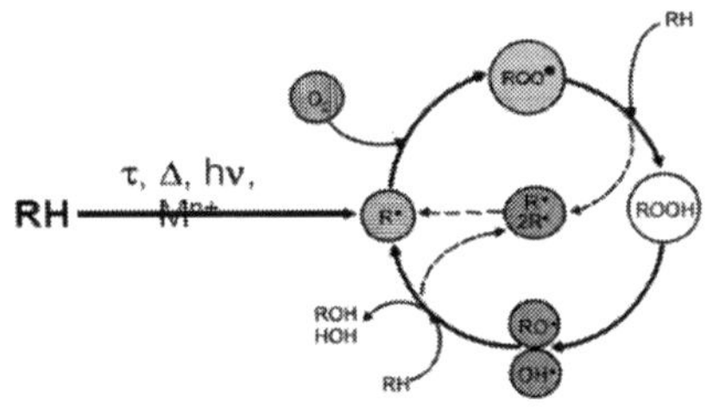

Figure 1. The Automatic oxidation of polymers

Without changing the original processing technology, adding antioxidant is the most effective means to slow down the aging of polyolefins [2~5]. The most common is phenolic antioxidants, which mainly capture free radicals by releasing hydrogen atoms. Common phenolic antioxidants include BHT, BHA, TBHQ, Irganox 1076 and Irganox 1010 [6]. However, almost all antioxidants are toxic. In particular, antioxidants used in food packaging materials will gradually penetrate into food and damage health if they contact with food for a long time in the preservation process [7~9]. As a kind of environmental estrogen, phenolic antioxidants can accumulate in human body. It has been studied that high-dose BHA can cause cancer [10], BHT can inhibit the activity of human biotransferase and affect the metabolism of organisms [11], TBHQ has adverse effects on human liver, spleen and stomach, and has acute toxicity and chronic carcinogenic effects [12]. Therefore, the addition of such antioxidants in food packaging materials has attracted more and more attention and has been strictly controlled. EU regulations Commission Directive 2002/72/EC, Commission

* Corresponding author: zqiukai@163.com

Directive 2004/19/EC and China's GB 9685-2008 hygienic standard for the use of additives for food containers and packaging materials stipulate that the specific migration limits (SML) of BHA, BHT and irganox1076 are 30.0, 3.0 and 6.0 mg/kg respectively. Whether it is the quality control of polymer products or the safety detection of food packaging materials, the rapid detection of antioxidant content in polyolefins is of great significance. At present, the commonly used detection methods of antioxidant content in polymer materials are chromatography [13~17], mass spectrometry [18,19], spectroscopy [20], but most of these methods have the problems of long time-consuming, cumbersome operation, destructive pretreatment of samples and so on. Fourier transform infrared spectroscopy (FTIR) overcomes the above problems and has the advantages of easy sample preparation and rapid analysis results. In this study, the representative polyolefin material polypropylene (PP) was selected, the content of antioxidant Irganox 1076 was determined by Fourier transform infrared spectroscopy, the standard curve was established, and the precision and accuracy of the results of this method were verified.

2. Materials and Methods

FTIR-650 Fourier transform infrared spectrometer, Tianjin Gangdong Technology Development Co., Ltd; PP-045 powder, Maoming Shihua Dongcheng Chemical Co., Ltd; Antioxidant Irganox 1076, BASF Ag; Toluene, analytically pure.

Antioxidant Irganox 1076 was held in toluene to prepare a solution with a mass fraction of 1%. The solutions of different quality were well-mixed with PP. The samples were processed on a hot bench at 180 ℃ for 3 min before being pressed into thickness of 1mm transparent sheets to be tested. In this process, toluene volatilizes completely. The testing results showed the contents of Irganox 1076 at different positions of the samples were uniform. The mass fractions of Irganox 1076 in the prepared samples were 0, 1.0%, 2.0%, 3.0%, 4.0% and 5.0%, respectively.

3. Results and Discussion

3.1 Characteristic peak of Irganox 1076

Fourier transform infrared spectrometer was used to detect the spectra of antioxidant Irganox 1076, PP samples with and without Irganox 1076. The obtained spectra are shown in Figure 1. Comparing the spectra of three samples, the antioxidant Irganox 1076 has an obvious absorption peak at 1738cm^{-1}, which is the characteristic absorption peak of carbonyl structure in antioxidant, while PP sample has no obvious absorption peak near this wavenumber.

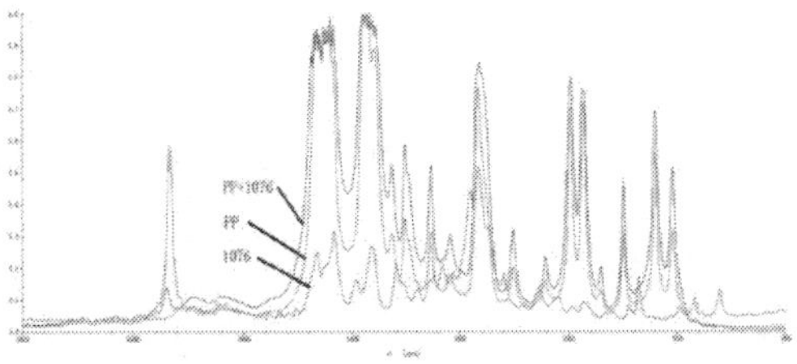

Figure 2. The FTIR curves of Irganox 1076, non-stabilized and stabilized PP with Irganox 1076

3.2 Standard curve of antioxidant Irganox 1076 in PP

Fourier transform infrared spectrometer was used to detect PP samples containing Irganox 1076 with mass fractions of 0, 0.2%, 0.4%, 0.6%, 0.8% and 1.0% respectively. The absorption peaks at 1738cm^{-1} in the obtained spectrum are shown in Figure 2.

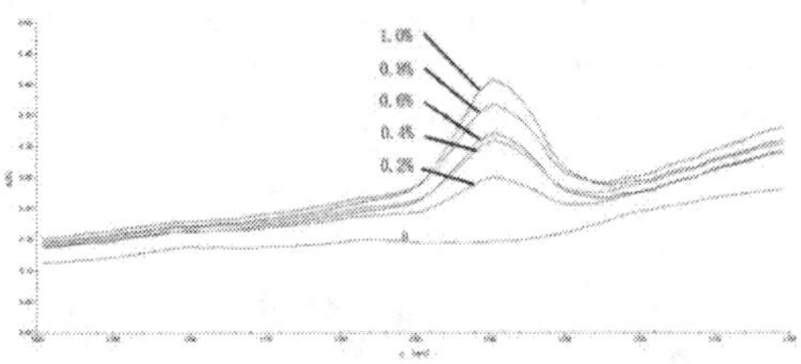

Figure 3. The FTIR curves of PP stabilized with Irganox 1076 in various concentrations at 1738cm^{-1}

Although all the samples were made in the same mold, there were still unavoidable nuances in the thickness of different samples. In order to eliminate the influence of the difference of sample thickness on the intensity of characteristic absorption peak of Irganox 1076, the characteristic absorption peak of PP sample needs to be selected as a reference. Comparative analysis of the spectra of the three samples in Figure 1 showed that PP has an obvious absorption peak at 841cm^{-1}, while the antioxidant Irganox 1076 has no absorption peak near this wavenumber. This absorption peak was selected as the reference. The mass fraction of antioxidant Irganox 1076 in PP sample was taken as the abscissa, and the ratio of the characteristic absorption peak area A_1 of antioxidant Irganox 1076 at 1738cm^{-1} to the reference absorption peak area A_2 of PP at 841cm^{-1} was drawn as the ordinate to establish the standard curve, as shown in Figure 3.

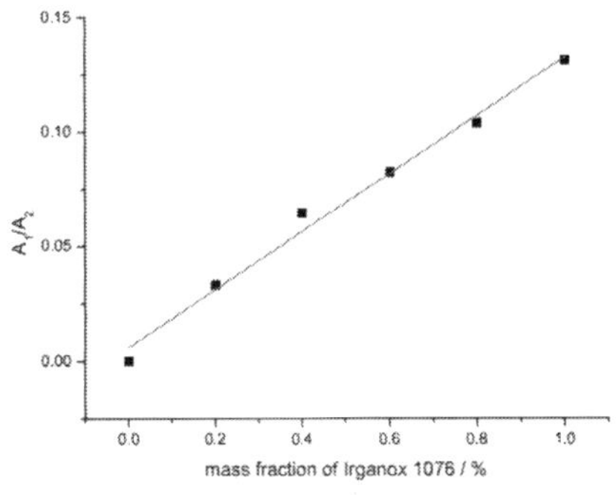

Figure 4. The standard curve of Irganox 1076 contents in PP

As shown in the figure above, the ratio A_1/A_2 of the characteristic absorption peak of antioxidant Irganox 1076 to the reference absorption peak area of PP had a linear relationship with the addition amount of antioxidant Irganox 1076 in PP. The linear regression equation is y = 0.00593 + 0.12681x, the linear regression coefficient is 0.98742.

3.3 Precision and accuracy verification

PP samples with Irganox 1076 content of 0.25% were prepared and tested for 5 times by Fourier transform infrared spectrometer. The determination results were calculated according to the standard curve. It can be shown that the minimum relative error of the determination result of Irganox 1076 content in PP sample is 1.58% and the maximum is 6.85% in Table 1.

Table 1. Precision of test results

Sample	mass fraction/%		Relative Error/%
	True value	Measured value	
1	0.25	0.246	1.58
2	0.25	0.258	3.33
3	0.25	0.245	1.98
4	0.25	0.256	2.57
5	0.25	0.233	6.85

PP samples with Irganox 1076 content of 0.05%, 0.25%, 0.45%, 0.65% and 0.85% were prepared and detected by Fourier transform infrared spectrometer. The determination results were calculated according to the standard curve according to the above method. It can be shown that the minimum relative error of the determination result of Irganox 1076 content in PP sample is 0.25% and the maximum is 4.32% in Table 2.

Table 2. Accuracy of test results

Sample	mass fraction/%		Relative Error/%
	True value	Measured value	
1	0.05	0.048	4.32
2	0.25	0.256	2.35
3	0.45	0.436	3.16
4	0.65	0.648	0.25
5	0.85	0.837	1.58

4. Conclusion

A method for rapid detection of antioxidant Irganox 1076 in PP by Fourier infrared spectroscopy was established. The ratio of the characteristic absorption peak area of Irganox 1076 to PP was used as the response parameter to eliminate the influence of sample thickness on the test results. The detection method is simple and fast, and is suitable for the rapid detection of various polyolefin products.

Acknowledgments

This work was financially supported by Shandong Provincial Natural Science Foundation ZR2020MC174.

References

1. Celina M C. Polymer Degradation and Stability, 2013, 98(12): 2 419-2 429.
2. Rabek J F. Photostabilization of Polymers [M]. New York: Elsevier Pub, 1990.
3. Lau W W Y, Pan J Q. Journal of the IES. 1995, 35(3): 31-38.
4. Lau W W Y, Pan J Q. Polymer Light Stabilizers and Antioxidants, in desk Reference of Functional Polymer [M]. New York: ACS Press, 1996.
5. Pospisil J. Polymer Degradation and Stability, 1995, 49(1): 99-110.
6. Gómez-Estaca J, López-de-Dicastillo C, Hernández-Muñoz P, et al. Trends in Food Science & Technology. 2014, 35(1): 42-51.
7. Lin Q B, Li X M, Song H, et al. Journal of Analytical Science2010, 26(06): 631-635.
8. Cheung K L, Yu S, Pan Z, et al. Chemical Research in Toxicology, 2011, 24(5): 670-676.
9. Liu Y F, Li Z H, Li J L, et al. Food & Machinery, 2012, 28(06): 251-255.
10. Soto-Vaca A, Gutierrez A, Losso J N, et al. Journal of Agricultural and Food Chemistry. 2012, 60(27): 6 658-6 677.
11. Holaas E, Bohne V B, Hamre K, et al. Journal of Agricultural and Food Chemistry. 2008, 56(23): 11 540-11 549.
12. Cheung K L, Yu S, Pan Z, et al. Chemical Research in Toxicology. 2011, 24(5): 670-676.

13. Dopico Garcia M S, López Va J M, Bouzab R, et al. Analytica Chimica Acta, 2004, 521(2): 179-188.
14. Wang B, Gao L H, Lu X Y, et al. Chinese Journal of Analysis Laboratory, 2017, 36(07): 831-836.
15. ZOU Z X, LI Y P, LIN Y, et al. Chinese Journal of Analysis Laboratory, 2017, 36(09): 1067-1070.
16. Xiong Z Q, Wang L B, et al. Chinese Journal of Chromatography, 2011, 29(3): 273-276.
17. Chen L W, Wu C S, Wang Y, et al. Journal of Instrumental Analysis, 2016, 35(02): 206-212.
18. Alin J, Hakkarainen M. Journal of Agricultural and Food Chemistry, 2011, 59(10): 5 418-5 427.
19. Zhang X L, Liu Y, Gong Z G, et al. Chinese Journal of Chromatography, 2014, 32(08): 827-831.
20. Bian L Q, Gu H P, Ding D X, et al. Petrochemical Technology, 2005(06): 587-590.

Preparation of nano α-Fe_2O_3 and γ-Fe_2O_3 and used for elemental mercury removal

Zengqiang Tan [1,*], Fanhai Kong [2]

[1] Huaneng Changjiang Environmental & Technology Co. Ltd., Beijing 100031, China
[2] Xi'an Thermal Power Research Institute Co. Ltd., Xi'an, Shannxi 710054, China

Abstract: The rod-shaped nano α-Fe_2O_3 and γ-Fe_2O_3 were prepared by a hydrothermal method. And the crystal phase structure, particle size, morphology and surface area of the synthesized nano α-Fe_2O_3 and γ-Fe_2O_3 were characterized by XRD and BET analysis. The removal of gaseous elemental mercury by nano α-Fe_2O_3 and γ-Fe_2O_3 was carried out on a fixed bed reactor. The effects of oxygen, bed temperature, particle size and acid flue gas components have been discussed. The results show that elemental mercury could be oxidized by nano α-Fe_2O_3 and γ-Fe_2O_3 with the presence of oxygen.

1. Introduction

The elemental mercury (Hg^0) emitted from coal combustion is a serious environmental problem, since its high toxicity and high volatility. Mercury released into the environment can be transformed into methyl-Hg, which could bio-accumulate in fish. The removal of Hg^0 is a key point for mercury removal [1].

The noble metals (such as gold and palladium) have good performance for the oxidation of Hg^0, but they are too expensive. The manganese oxides loaded on some carriers (e.g Al_2O_3) displayed significant oxidation capacity for Hg^0 at higher temperature [2]. The denitration device should be installed upstream of particulate control devices to obtain the optimum reaction temperature, which results in catalyst deactivation for exposure to fly ash. So developing catalysts operating at lower temperatures has been of great interest. For example, an adsorbent composed of active nano-layer on mesoporous silica substrate was developed by Abu-Daabes and Pinto [3], which was used to remove mercuric chloride under low-temperature (<135 °C).

Particles can catalyze some reactions that are not feasible in particle-free systems [4]. Recently, the nano-sized metal oxides have attracted more attention for excellent catalytic decomposition performance of pollutants. Especially, the nano-sized metal oxides have densed surface unsaturated sites and smaller particle size, which could enhance catalytic capability of catalyst [5]. Among the nano-sized metal oxides, nano iron oxide has advantages of catalytic activity, toxicity resistance, and good performance in removal of carbon monoxide and sulfur dioxide [6].

The catalytic removal performance of Hg^0 by different crystal structure of iron oxide (α-Fe_2O_3 and γ-Fe_2O_3) has been controversial. Some studies concluded that γ-Fe_2O_3 is the main iron phase in ash, and γ-Fe_2O_3 played the dominant role in the catalytic removal of Hg^0 [7,8]. But Grant et al [9, 10] reported that the two crystalline structure of iron oxide had no significant removal effects for the Hg^0. Therefore, it is necessary to conduct detailed study of α-Fe_2O_3 and γ-Fe_2O_3 used for mercury removal. More importantly, the potential of the material in the nanoparticle form has not been explored.

In this study, α-Fe_2O_3 and γ-Fe_2O_3 nanoparticles were synthesized to remove Hg^0 and the mechanism was also studied.

2. Experimental

2.1 Synthesis of α-Fe_2O_3 and γ-Fe_2O_3 nanoparticles

The hydrothermal method was used to synthesize nano α-Fe_2O_3 catalyst. A starting material was $Fe(OH)_3$ prepared by adding a stoichiometric mixture of $FeCl_3 \cdot 6H_2O$ and NaOH. After vigorously stirred for 40 minutes at room temperature and pH of 12, the particle was separated and put into a high-pressure hydrothermal apparatus at 180°C for 2 hour, the yellow product was centrifuged, rinsed with distilled water, and calcined at 500~600°C for 4 hour. The obtained deep red α-Fe_2O_3 was collected for the following experiments and characterization. Polymer PEG was added to the reaction system as oxidant, surfactant and dispersing agent. The nano-α-Fe_2O_3 and PEG-400 were mixed and stirred fully for some time, then calcined at 400 °C to obtain pure crystalline γ-Fe_2O_3.

* Corresponding author: tanzq2008@126.com

2.2 Characterization of nanoparticles

The morphological observation of synthesized samples was conducted with a scanning electron microscope (Holand Quanta 200). The nitrogen adsorption apparatus (Micromeritics ASAP 2020) was used to determine the pore size distribution and Brunauer-Emmett-Teller (BET) surface area. The synthesized α-Fe_2O_3 and γ-Fe_2O_3 nanoparticles were investigated using X-ray powder diffraction (XRD) analysis (Holand) to understand the crystal phase structure. Using Scherrer equation, the crystal sizes of γ-Fe_2O_3 and α-Fe_2O_3 were calculated using formula (1).

$$D = \frac{K\lambda}{(\beta - \beta_1)\cos\theta} \quad (1)$$

Where, θ is diffraction angle, β_1 is half-intensity width from instrumental broadening, β is half-intensity width of the relevant diffraction peak, λ is X-ray wave length, K is half width, D is grain diameter.

2.3 Experimental method

The reactor consists of a quartz reactor insert tube (20 mm I.D.), the catalyst was put in the inner tube. The thickness of the sample bed was maintained around 3 mm. The simulated flue gas was N_2 stream containing O_2 (10%) and CO_2 (12%). The Hg^0 concentration was controlled by adjusting the water bath temperature and the N_2 flow rate passing through the mercury saturator. A "semi-continuous" mercury analyzer (VM3000, German) was used to monitor the Hg^0 concentration. 0.25 g of catalyst was used in each test, the flue gas rate was held at 2 L/min, the inlet Hg^0 concentration was 40 μg/m^3.

The Hg^0 removal performance is expressed by removal efficiency η according to formula (2). Wherein C_{outlet} and C_{inlet} represent the outlet and inlet Hg^0 concentration (μg/m^3).

$$\eta = (1 - C_{outlet} / C_{inlet}) \times 100\% \quad (2)$$

3. Results and discussion

3.1 Characterization

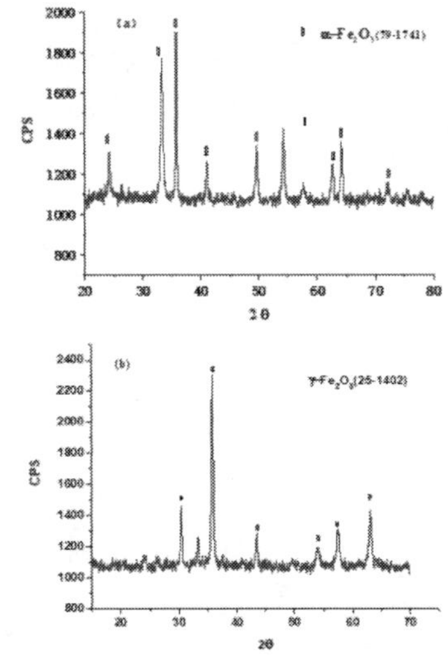

Fig. 1 X-Ray diffraction pattern of the synthesized nano-particle (a: α-Fe_2O_3 b: γ-Fe_2O_3).

The synthesized catalyst's XRD patterns of are shown in Fig.1. The narrow and high intensity peak at 33.142°and 35.654°could be observed from Fig.1(a). The single-phase formation of α-Fe_2O_3 structure could be confirmed corresponding to the JCPDS no. 174179. There are two obvious crystal phases around the peak at 2θ=36.5 and 2θ=63°(Fig.1(b)), which can be assigned to γ-Fe_2O_3 (JCPDS 25-1402). These above results suggested that the preparation of γ-Fe_2O_3 from α-Fe_2O_3 was successful.

Table 1. Properties of catalysts

Sample	BET Surface Area (m^2/g)	Total Pore Volume (cm^3/g)	Average pore size(nm)	Particle size (nm)
Nano α-Fe_2O_3	43.76	0.11	16.61	18.13
Nano γ-Fe_2O_3	36.54	0.09	15.12	27.24

Table 1 shows the properties of pure samples, the Fe_2O_3 nanoparticles with different size were prepared with different concentration of reactants and pH value. The BET surface area of samples was found to be in the range of 36.5~43.7 m^2/g, and pore size of nano samples was in the range of 15.1~16.6 nm.

3.2 Effect of reaction temperature

The reaction temperature plays an important role in the heterogeneous removal of Hg^0 by γ-Fe_2O_3 and α-Fe_2O_3 under O_2+N_2+CO_2 atmosphere (BL atmosphere). The effects of temperature on Hg^0 removal by nano Fe_2O_3 were evaluated in the range of 80-150°C.The oxidation efficiency of Hg^0 by γ-Fe_2O_3 and α-Fe_2O_3 increased with the temperature increase as shown in Fig.2.

The enhancement performance of Hg^0 removal by γ-Fe_2O_3 and α-Fe_2O_3 with increasing temperature could attribute to the active sites increase for Hg^0 [11]. The chemical adsorption of Hg^0 could be enhanced by temperature increase. The surface functional groups and reaction temperature are significant in both adsorption and surface reactions.

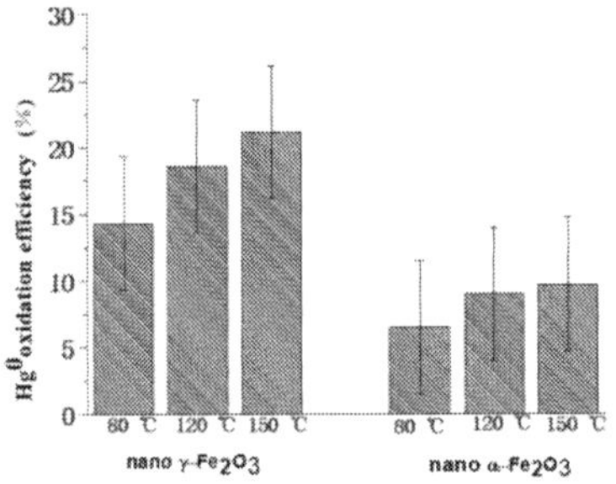

Fig.2 Hg^0 removal efficiency under different bed temperature.

Furthermore, the Hg^0 removal efficiency of γ-Fe_2O_3 is greater than α-Fe_2O_3 under the same operation conditions. The γ-Fe_2O_3 is spinel crystal structure with cation vacancy, the vacancies are distributed uniformly. But the α-Fe_2O_3 belongs to trigonal corundum-typed structure, and Fe^{3+} is surrounded by six equidistant oxygen. The metal-oxygen bond energy of α-Fe_2O_3 is -196.4 kcal/mol, smaller than that of γ-Fe_2O_3 (-191.7 kcal/mol). Moreover, the oxygen dissociation pressure of α-Fe_2O_3 is $10^{-13.8}$ torr, and that of γ-Fe_2O_3 is $10^{-8.8}$ torr. This implies that the oxygen of γ-Fe_2O_3 has higher thermodynamic chemical potential than that of the α-Fe_2O_3, companied with more favorable catalytic activity. This theory was confirmed by some researchers, who studied the catalytic reactivity of α-Fe_2O_3 and γ-Fe_2O_3 to butene, which showed that lattice oxygen of γ-Fe_2O_3 has higher activity than that of α-Fe_2O_3 [12,13].

3.3 Effect of oxygen

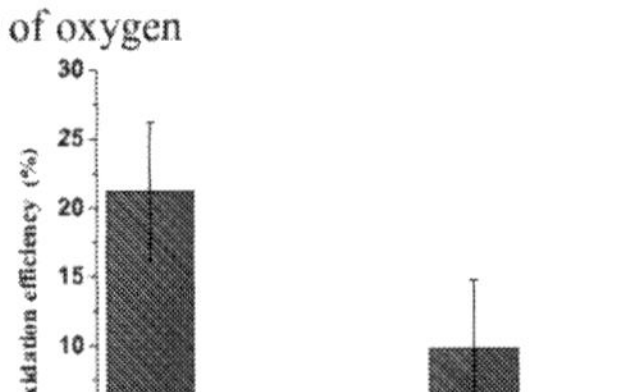

Fig.3 Hg^0 removal efficiency at 150°C with different oxygen concentration.

To identify the influence of oxygen on Hg^0 removal, experiments were conducted with varied concentration of oxygen (0% and 6%) at 150°C. The Hg^0 removal efficiency was found to depend on oxygen (Fig.3). With the presence of 6% oxygen, the Hg^0 removal efficiency by γ-Fe_2O_3 and α-Fe_2O_3 increased from 0% to 21% and 10%, respectively. Li [5] found that CO could be oxidized by nano Fe_2O_3 through complicated route even without O_2. However, the experimental results proved that nano Fe_2O_3's Hg^0 removal ability was negligible without O_2. This observation is consistent with another study [14], and no obvious Hg^0 adsorption was observed when a gas stream of Hg^0/N_2 passed through the V_2O_5 catalyst.

The promotion effect of oxygen on Hg^0 removal could be explained by the Mars-Maessen mechanism [15]. The process of Hg^0 oxidation involves electron transfer. The adsorbed Hg^0 could react with lattice oxygen provided by the metallic oxides (M_xO_y), and the M_xO_y is Fe_2O_3, and the intermediate product M_xO_{y-1} could be Fe_3O_4 or FeO.

3.3 Effect of particle size

To study the effect of particle size on Hg^0 removal, the nano iron oxide catalyst and common iron oxide were tested under the BL atmosphere. The properties of iron oxide and mercury removal performance were summarized in Table 2. The common iron oxide is purchased iron oxide sieved with 200 mesh.

Table 2. The effect of property of iron oxide to mercury removal

	Nano α-Fe_2O_3-1	Nano α-Fe_2O_3-2	Common α-Fe_2O_3
BET surface area(m^2/g)	43.76	31.25	17.98
Total pore volume (cm^3/g)	0.11	0.091	0.02
Average pore width(nm)	16.61	14.86	6.10
XRD size (nm)	18.1	42.3	<7.0μm
Inlet concentration(μg/m^3)	40	40	40
90% breakthrough time(min)	45	35	<1

As can be seen from Table 2, the common α-Fe_2O_3 penetrated quickly under BL atmosphere, indicating that normal-sized α-Fe_2O_3 had almost no mercury removal ability. But the 90% breakthrough time of the nano α-Fe_2O_3 could last for more than 30 minutes. This observation implied that the unique catalytic property of nano α-Fe_2O_3 to the Hg^0. It possibly resulted from quantum size effect of nanomaterials and larger surface area, and larger surface area could provide more surface active sites for Hg^0 removal.

Fig.4 showed the Hg^0 removal efficiency by three catalysts with the presence HCl. The results were similar to that of BL atmosphere. The Hg^0 removal efficiency of ordinary-sized α-Fe_2O_3 was quite low, but the two nano-sized Fe_2O_3 showed a higher mercury removal efficiency. Furthermore, with the increased surface area and decreased particle size, the mercury removal efficiency had an increasing trend.

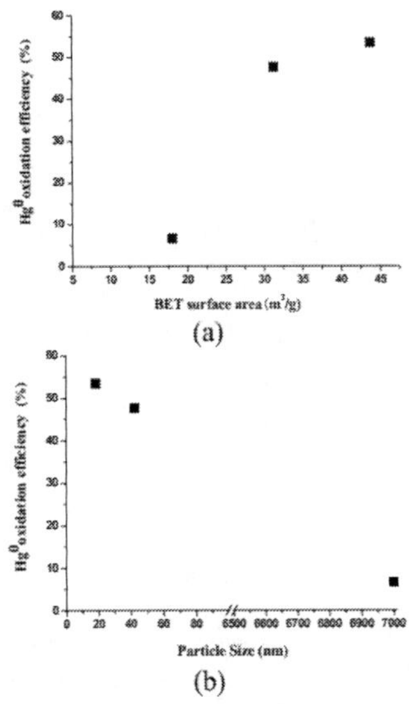

Fig.4 The effect of surface area(a) and particle size(b) to Hg^0 removal

4. Conclusions

In this study, the nano α-Fe_2O_3 and nano γ-Fe_2O_3 were synthesized successfully by hydrothermal method. The removal performance of Hg^0 by nano α-Fe_2O_3 and nano γ-Fe_2O_3 was studied with a fixed-bed reactor.

The normal-sized iron oxides had negligible mercury adsorption capacity under the BL atmosphere, but the nano α-Fe_2O_3 and nano γ-Fe_2O_3 showed obvious capacity for Hg^0. Furthermore, the nano γ-Fe_2O_3 showed greater oxidative capacity under the BL atmosphere. The increase of bed temperature and oxygen content were found to promote the mercury removal.

Acknowledgments

Project Supported by National Natural Science Foundation of China (51976072), Tsinghua-Huaneng Basic Energy Joint Research Institute (U20YYJC01), Scientific Research Project of Xi'an West Boiler Environmental Protection Engineering Co., Ltd. (GA-21-TZK01).

References

1. Ying Li, Liu YX, Yang W, Liu L, Pan JF. Adsorption of elemental mercury in flue gas using biomass porous carbons modified by microwave/hydrogen peroxide [J]. Fuel, 2021, 291, 120152.
2. Yao T, Duan Y, Zhu C. Investigation of mercury adsorption and cyclic mercury retention over MnOx/γ-Al_2O_3 sorbent [J]. Chemosphere, 2018, 202:358-365.
3. Abu-Daabes MA, Pinto NG. Synthesis and characterization of a nanostructured sorbent for the direct removal of mercury vapor from flue gases by chelation [J], Chem. Eng. Sci. 2005, 60, 1901–1910.
4. Hu J, Geng X, Duan Y. Effect of mechanical-chemical modification process on mercury removal of bromine modified fly ash [J]. Energy & Fuels, 2020, 34(8):9829-9839.
5. Li P, Donald E, Miser SR. The removal of carbon monoxide by iron oxide nanoparticles [J]. Appl. Catal. B: Environ. 2003, 43,151-162.
6. Reddy BV, Rasouli F, Hajaligol MR. Novel mechanism for removal of CO by Fe2O3 clusters. Fuel 2004, 83(11-12), 1537-1541.
7. Kurien U, Hu Z, Lee H. Radiation enhanced uptake of Hg0(g) on iron (oxyhydr) oxide nanoparticles[J]. Rsc Advances, 2017, 7(71):45010-45021.
8. Zhang Y, Mei D, Wang T, Wang J, Gu Y, Zhang Z, et al. In-Situ Capture of Mercury in Coal-Fired Power Plants Using High Surface Energy Fly Ash [J]. Environmental Science and Technology. 2019; 53(13):7913-20.
9. Grant ED, Raymond AD, Constance LS. Fixed- bed sdudies of the interactions between mercury and coal combustion fly ash [J]. Fuel Process Technol 2003, 82, 197-123.
10. Galbreath K C, Zygarlicke C J, Tibbetts J E. Effects of NOx, α-Fe_2O_3, γ-Fe_2O_3, and HCl on mercury transformations in a 7-kW coal combustion system [J]. Fuel Process Technol 2004, 86, 429-448.
11. Khedr MH, Abdel KS, Nasr MI. Effect of temperature on the catalytic oxidation of CO over nano-sized iron oxide [J]. Materials Science and Engineering A 2006, 430(1-2),40-45.
12. Antua-Nieto C, Rodríguez E, López-Antón M, García R, Martínez-Tarazona M. Carbon materials loaded with maghemite as regenerable sorbents for gaseous Hg0 removal [J]. Chemical Engineering Journal. 2020, 387:124151.
13. Trobajo J R , Antuna-Nieto C , Rodriguez E , et al. Carbon-based sorbents impregnated with iron oxides for removing mercury in energy generation processes [J]. Energy, 2018, 159(9):648-655.
14. He S, Zhou JS, Zhu Y, Luo ZY, Ni M, Cen KF. Mercury oxidation over a vanadia-based selective catalytic reduction catalyst [J]. Energy Fuels. 2008, 23,253–259.
15. Granite E, Pennline H, Hargis R. Novel sorbents for mercury removal from flue gas [J]. Ind Eng Chem Res 2000, 39, 1020-1029.

Immobilizing metallocene by solid polymethylaluminoxane "sMAO" for slurry-phase ethylene-1-hexene copolymerization

Chaofan Wu [1], Tengjie Ge [2], Zeyu Jiang [2], Yanfeng Jiang [2], Qian Zhang [1] and Jian Li [1,*]

[1] College of Chemical Engineering and Material Science, Tianjin University of Science and Technology, Tianjin 300457, P. R. China
[2] Daqing Petrochemical Research Center, Petrochemical Research Institute of CNPC, Daqing 163714, P. R. China

Abstract. In the field of olefin catalytic polymerization, metallocene catalysts show higher activity than traditional Ziegler-Natta catalysts. In this study, benzoic acid catalyzed the controlled hydrolysis of trimethylaluminum (TMA) to methylaluminoxane (MAO). After pyrolysis, an insoluble form of solid polymethylaluminoxane (sMAO) was synthesized. The influence of synthesis conditions on the particle size and morphy was systematically explored. The synthetic sMAO is used as a carrier to support the metallocene catalyst, and the catalyst is applied to the ethylene/1-hexene copolymerization. The experimental results show that the activity can be as high as 12149.8 gPE·(gcat)-1 when the slurry reaction is catalyzed in a 5L kettle.

1. Introduction

Polyolefins account for nearly half of the world plastic production.[1] Thanks to their low cost and versatility, they are used in packaging, automotive, construction, pharmaceutical/medical, and electronics/electrical applications. Low pressure olefin polymerization processes that use supported coordination catalysts (Ziegler-Natta, Phillips, and metallocene) are the main route for making polyolefins.

The discovery of metallocene-based catalysts for the polymerization of a-olefins has opened the possibility of synthesizing new materials. The single site metallocene catalysts offer several advantages with respect to the traditional Ziegler–Natta catalysts, as narrow molar mass distribution, control of comonomer incorporation, high activity, and low residual metal content.[2]

Methylaluminoxane (MAO) is the most commonly used activator and co-catalyst for transition metal containing, single-site complexes in olefin polymerisation.[3] The combination of MAO, an inert inorganic carrier material (most commonly silica) and a precatalyst complex have been described by Severn as the ''Holy Trinity'' of supported single-site catalysts.[4] However, there is also growing interest in single-site catalysts in particle forming polymerisation processes that are free from an inert inorganic carrier. These require an activating support material, which can be combined with the precatalyst complex in a single synthetic step. This process has several advantages from an industrial viewpoint, as it reduces the time and energy-intensive drying steps and hence can lower manufacturing costs. Furthermore, these 'self-supported' catalyst systems have an advantage over silica-supported systems as the complex loading can be increased significantly, which leads to correspondingly higher polymerisation activities. [5-6]

In 2013, Tosoh Finechem Corporation reported, in the patent literature, an insoluble form of solid methylaluminoxane (sMAO) formed via controlled hydrolysis of trimethylaluminium with benzoic acid.[7] We have recently reported the laboratory scale synthesis and detailed characterisation of sMAO, and demonstrated its function as a solid-phase support, scavenger and activator in slurry-phase ethylene polymerisation.[8-10]

2. Experimental

2.1 Materials

All manipulations were carried out using standard Schlenk techniques under N_2, orin an glovebox under N_2. All glassware was dried at 160 °C overnight prior to use. Hexane and toluene were dried and degassed using an MBraun SPS-800 solvent purification system. Dried solvents were collected, degassed and stored over N_2 in K mirrored ampoules.Trimethylaluminum, Benzoic acid were purchased from Sigma-Aldrich.

2.1.1 Preparation of sMAO:

In a typical procedure an ampoule containing a 30 mL solution of TMA (2mol/L) in toluene was cooled to 15 °C with rapid stirring, and benzoic acid (3.019 g, 2.478 mmol) was added under a flush of N2 over a period of 30 minutes. Effervescence (presumably methane gas) was observed and the reaction mixture appeared as a white suspension, which was allowed to warm to room temperature. After 30 minutes the mixture appeared as a colorless solution and was heated at 70 °C for 32 h (a stir rate of 500 rpm was used throughout these experiments). The mixture obtained was a colorless solution free of gelatinous material. Above mixtrue was concentrated to 16mL by

* Corresponding author: lijian@tust.edu.cn

vocumm and subsequently heated at 100 °C for 14 h. The reaction mixture was cooled to room temperature and hexane (150 mL) added, resulting in the precipitation of a white solid which was isolated by filtration, washed with hexane (2 x 100 mL) and dried in vacuo for 3 h. Total yield = 3.478 g (86% based on 40.3 wt% Al).

2.1.2 Immobilized catalysts:

To a Schlenk flask charged with sMAO (300 mg, 4.58 mmolAl) and rac-(EBI)ZrCl2 (9.6 mg, 0.0229 mmol) was added toluene (40 mL), and the resulting orange dispersion was heated at 80 °C for 2 h with regular swirling. The mixture was allowed to cool to room temperature and a pink-orange solid settled below a colorless supernatant solution. The supernatant was removed by decantation and the remaining slurry was dried in vacuo overnight, to afford a free-flowing orange solid. Total yield = 249 mg (86% based on 41.3 wt% Al).

2.2 Polymerization

The immobilized catalyst (nearly 40.0 mg), triisobutylaluminum scavenger (150 mg), and hexane (2 L) were added to a high-pressure reactor. Ethylene gas was continuously fed into the ampoule at 1.0MPa overpressure at 70 °C. After 30 minutes, the reaction was stopped by removing the ampoule from the oil bath, and degassing in vacuum. The polymer was isolated on a frit and vacuum dried at room temperature for 3 hours. Each polymerization experiment was conducted at least twice to ensure the reproducibility of the corresponding outcome, and mean activities are quoted in units of gPE/gcat.

2.3 Characterization

Diameter of particles were measured by Mastersizer 2000 Laser Diffraction Particle Size Analyzer (Malvern Instruments, Britain). The microspheres size distribution was characterized as SPAN value, defined as follows (Eq. (1)): where D90, D50 and D10 are the particle sizes at which 90%, 50% and 10% of the distribution fall below.

$$SPAN = \frac{D90 - D10}{D50} \quad (1)$$

Scanning electron microscopy (SEM) measurements were conducted using Philips XL30 ESEM and Hitachi S-4800 instrument at an operation voltage of 20.0 keV and 0.7 keV. Elemental analysis was performed by energy dispersive spectroscopy (EDS) attached to the SEM.
Thermograms were obtained using a PerkinElmer Diamond differential scanning calorimeter (DSC) system. PE productions were heated from 25 to 250 °C at a heating rate of 10 °C /min under constant purging of N_2 at 20 mL/min.

3. Results and discussion

(a)

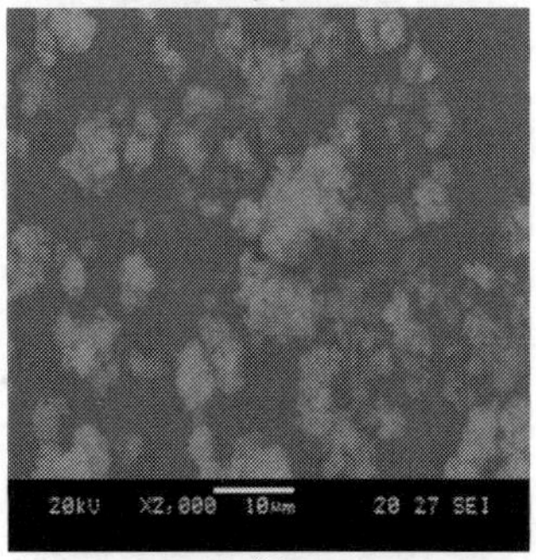

(b)

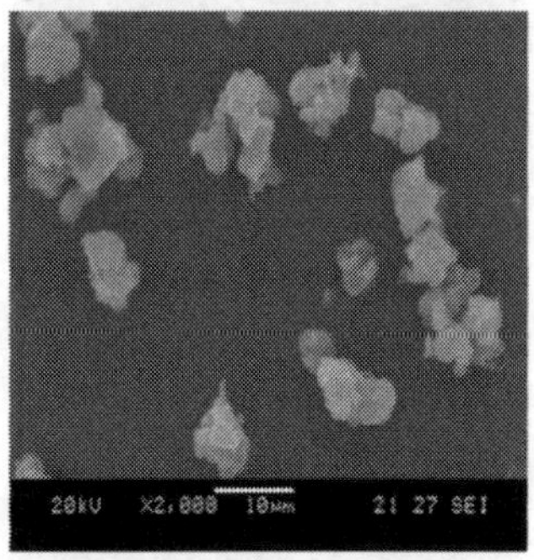

(c)

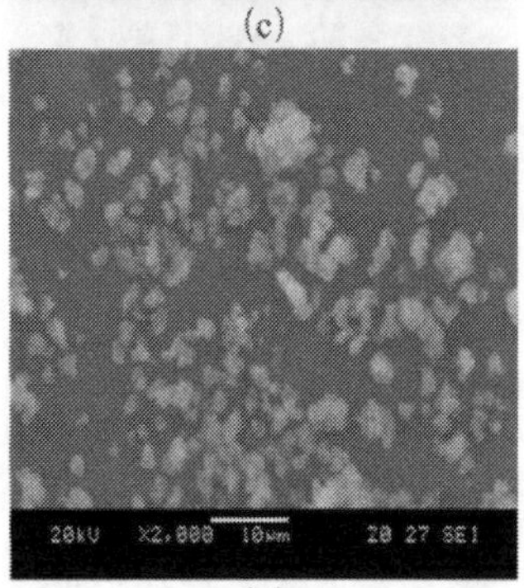

(d)

Figure 1. SEM images of the catalyst with different toluene/hexane volume ratio(a) 1:3; (b) 1:4; (c) 1:5; (d) 1:6

Changing the volume ratio of toluene and hexane may influence the morphology and diameter of the sMAO supports (Figure 2). At lower toluene/hexane molar ratios, the sMAO particles exhibited nonspherical morphology (Figure 1a), while individually adding hexane led to an increase in the dispersity of the catalyst particles (Figure 1b), because the solubility of MAO in hexane is lower than it in toluene. Although the mean particle diameter was obtained in the 0.5-1 μm range with a narrow distribution as well as spherical shape, the aggregation was obvious. To eliminate the aggregation, the toluene /hexane volume ratio was increased to 1:4. The diameter of sMAO ranged from 3-5 μm, and distributed homogenously (Figure 1c). Upon increasing the volume ratio to 1:6, a spherical morphology with an approximately 1-2 μm particle diameter was observed (Figure 1d).

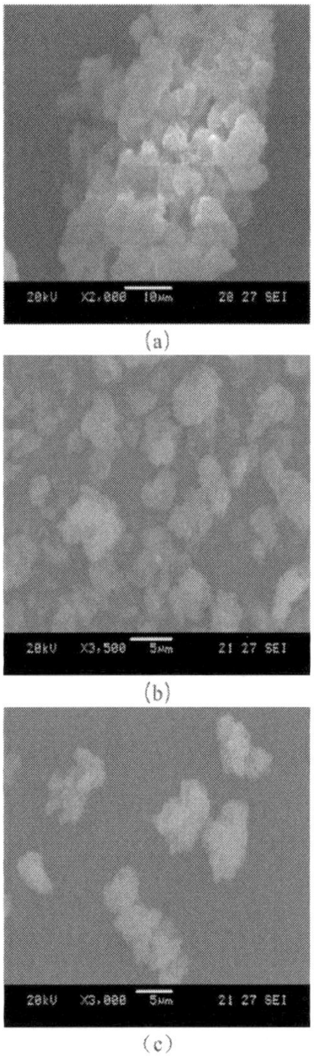

Figure 2. SEM images of the catalyst with different hexane-addting temperatures 50°C; (b) 35°C; (c) 20°C

Figure 2 shows the effect of the hexane-addting temperature on the diameter and morphology of the sMAO. The particles diameter exhibits a relation with the hexane-addting temperature and the D50 diameter of the catalyst increases from 27.8 μm to 6.75 μm as the temperature was changed from 50 °C to 20 °C. The surface morphology of the sMAO particles was also related to the temperature, with the particles obtained at higher temperatures exhibiting a broader diameter distribution. The small particles display regular spherical morphology while the larger particles collapsed (Figure 2b) or even aggregated (Figure 2a). The lower temperature (20 °C) is the optimum temperature that helps to maintain the smooth surface and uniform diameter of the particles (Figure 2c).

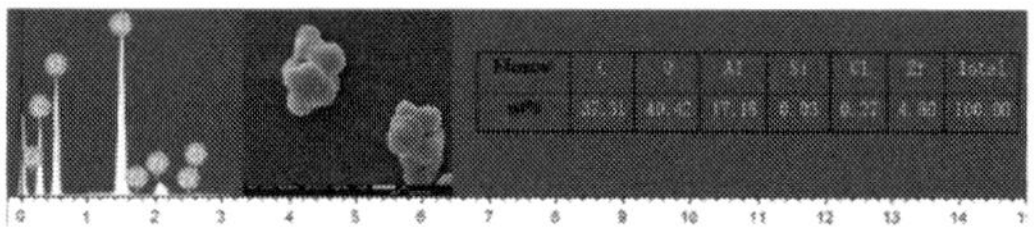

Figure 3 Energy dispersive X-ray spectrometry results for the catalyst

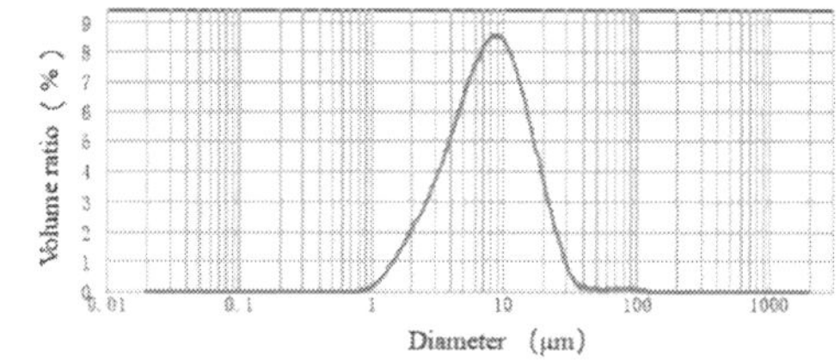

Figure 4 Particle size distribution of the catalyst

EDS measurements of the catalyst show that the C, O, Al, Cl and Zr elements were distributed homogeneously in the catalyst system (Figure 3). Figure9 shows the size distribution of the catalyst particles. The results show that the catalyst particle distribution is concentrated at approximately 7μm with a narrow particle size distribution and with less fine content (Figure 4). Since the polymer particle morphology is mainly determined by the morphology of the catalysts through the replication phenomenon, the low fine content of the catalysts results in a low content of polyethylene fine powder, avoiding equipment blockage.

Table 1. Effect of 1-hexene addition on polymerization activity

No.	Amount of 1-hexene		Amount of catalyst (mg)	Amount of PE (g)	Activity (gPE·(gcat)$^{-1}$)
	Volume(mL)	v/v(%)			
1	0	0	39.7	243	6120.9
2	10	0.5	41.6	378	9086.5
3	20	1.0	42.1	398	9453.7
4	30	1.5	41.1	405	9854.0
5	40	2.0	41.4	503	12149.8
6	60	3.0	41.8	413	9880.4
7	80	4.0	41.8	291	6961.7

Conditions:, ethylene(1MPa), 2h, cocatalyst((i-Bu)$_3$Al),Al/Zr (3500), solvent(2L), 80°C.

Obviously enhancement in activity was observed for ethylene/1-hexene copolymerisation compared to ethylene homopolymerisation (6120.9 gPE/gcat). When the ratio of 1-hexene increased from 0%-2.0%, A large increase in ethylene copolymerization was presented (from 6120.9 to 12149.8 gPE/gcat).Many theories have been proposed for the positive comonomer effect, including fracturing of catalyst particles exposing new sites, the formation of new active species by coordination of a-oleffins, and activation of dormant active sites as well as improved diffusion of ethylene close to the catalytic site, which improves polymerisation activity. However, further increase the amount of 1-hexene (from 2.0%-4.0%) lead to a decrease of activity, indicating that the negative comonomer effects outweigh the positive effects. The negative effects of comonomer addition are proposed to be due to competitive binding between ethylene and a-oleffins and, if the rate of migratory insertion of the aoleffin is slower than that of ethylene, the rate of chain propagation will decrease leading to a decrease in polymerization activity. The negative effects of comonomers on ethylene polymerisation activity may also be due to slower rates of insertion; the increased steric bulk of a-oleffin comonomers in the polymer chain can lead to reduced rates of ethylene insertion.

4. Conclusion

In this report we have developed an optimized laboratory-scale synthesis for solid polymethylaluminoxane (sMAO). This amorphous solid has been extensively characterized using SEM imaging and BET physisorption provide insight into the surface structure and porosity of sMAO. sMAO is shown to be a very promising bifunctional support and activator for single site metallocene-based pre-catalysts in slurry-phase ethylene polymerization reactions.

Acknowledgments

The authors thank the financial support from the Project of Petrochemical Research Institute of CNPC (2020B-2060, 2020B-2061)

References

1. Markus Klapper,* Daejune Joe, Sven Nietzel, Joseph W. Krumpfer, and Klaus Müllen*. Olefin Polymerization with Supported Catalysts as an Exercise in Nanotechnology. Chemistry of Materials, 2013, 26(1): 802-819
2. Madelyn M. Stalzer, Tobin J. Marks, Massimiliano Delferro, Supported Single-Site Organometallic Catalysts for the Synthesis of High-Performance Polyolefins. Catalysis Letters, 2015, 145(1):3-14
3. Harmen S. Zijlstra[a,b] and Sjoerd Harder*[b,c]. Methylalumoxane – History, Production, Properties, and Applications. European Journal of Inorganic Chemistry, 2015, 1: 19-43
4. Aline C. dos Ouros, Michèle O. de Souzab and Heloise O. Pastore*,Metallocene Supported on Inorganic Solid Supports: an Unfinished History. Journal of the Brazilian Chemical Society, 2014, 25(12): 2164-2185
5. Alexander F R Kilpatrick, Jean-Charles Buffet, Peter Nørby, Nicholas H Rees, Nicholas P Funnell, Saovalak Sripothongnak, and Dermot O'Hare. Synthesis and characterization of solid polymethylaluminoxane "sMAO"; a bifunctional activator and support for slurry-phase ethylene polymerization. Chemistry of Materials, 2016, 28(20): 7444-7450.
6. Alexander F. R. Kilpatrick, a Harry S. Geddes, Jean-Charles Buffet, b Zoë R. Turner, a a Andrew L. Goodwin *b and Dermot O'Hare *a. Polymethylaluminoxane organic frameworks (sMAOF) – highly active supports for slurry phase ethylene polymerization. 2021, 11(16):5472-5483
7. Kaji, E.; Yoshioka, E. Solid Polymethylaluminoxane Composition and Method for Manufacturing Same. US Patent. 8,404,880 B2 2013.
8. Thomas J. Williams, a Jessica V. Lamb,a Jean-Charles Buffet, a Tossapol Khamnaenb and Dermot O'Hare *a. Synthesis of ultra-high molecular weight poly(ethylene)-co-(1-hexene) copolymers through high-throughput catalyst screening. RSC Advances, 2021, 11, 5644-5650
9. T. J. Williams, J.-C. Buffet, Z. R. Turner and D. O'Hare, Group 4 permethylindenyl constrained geometry complexes for ethylene polymerisation catalysis, Catal. Sci. Technol., 2018, 8, 5454–5461.
10. J. V. Lamb, J.-C. Buffet, Z. R. Turner and D. O'Hare, Group 4 permethylindenyl complexes for slurry phase polymerisation of ethylene, Polym. Chem., 2019, 10, 1386–1398.

A Verticality Adjustment Technology for AP1000 Refueling Machine Mast

Changdi Zuo, Jianlei Dong, Haidong Zhao, Jun Xie, Di Mu

Shandong Nuclear Power Co. LTD., Yantai, 265116 China

Abstract: The refueling machine is used to handle the nuclear fuel assembly in the nuclear island, and is used to transport fuel between the fuel upender basket and reactor vessel core. During Unit 1 of a nuclear power plant initial fuel loading, it was found that the inner mast operated with large friction, which caused frequent triggering of overload protection during fuel operation, and the refueling machine was difficult to be operated normally. According to the unique design characteristics of the inner mast of the AP1000 refueling machine, the verticality adjustment method of the mast of the AP1000 refueling machine is developed. By adjusting the verticality of the outer mast, the concentricity of the inner and the outer mast, the levelness of the equilibrium arm of the inner mast, and the gap between the guide roller of the outer mast and the inner mast, the problem of high friction in the operation of the inner mast is solved. At the same time, the verticality of the inner mast during operation is ensured. This method has been successfully applied in the first overhaul of a power plant. It can be used to guide the installation and adjustment of the refueling machine mast with similar structure.

Key words: Refueling machine; Mast; Verticality; Total station

1. Foreword

The AP1000 core consists of 157 fuel assemblies[1], and the theoretical minimum gap between adjacent fuel assemblies is only 1.02mm. This requires that the inner mast of the refueling machine has a very high running verticality when the fuel assembly is moved up and down. Otherwise, when the refueling machine is loaded and refueled in the on index position, it is easy to cause scratches between the operated fuel assembly and the adjacent ones, which will maybe damages the fuel assembly intermediate grid.

In order to prevent damage during the fuel assembly up and down movement, the AP1000 refueling machine[2][3] is designed with overload and underload protection control logic. If the parameters such as the verticality of the outer mast, the concentricity of the inner and outer mast, the levelness of the balance arm of the inner mast, the clearance between the guide roller of the outer mast and the inner mast are not adjusted properly, sometimes it can meet the verticality requirements of the inner mast. But the running friction of the inner mast will still be too large, which will affect the fuel loading and unloading efficiency and the life of the guide roller.

During initial fuel loading of a nuclear power plant, it was found that the friction of the inner mast reached about 30kg at an operation altitude of 2800mm-3500mm, which was very close to the protection value of 68kg when the fuel assembly was loaded. The refueling machine can only run at reduced speed, seriously affect the loading speed.

2. Idea of adjusting the mast of AP1000 refueling machine

The AP1000 refueling machine does not rely on guide rollers to ensure vertical movement of the inner mast. The AP1000 refueling machine is designed to ensure that the inner mast maintains a vertical posture in a free hanging state. The guide roller is only used to limit the movement of the inner mast in the horizontal direction. Therefore, as long as the verticality of the outer mast is ensured, the guide roller is installed on a vertical line, and the gap between the guide roller and the inner mast meets the requirements of the drawing, the running verticality and friction of the inner mast will naturally meet the requirements.

Therefore, the first step should be to adjust the verticality of the outer mast, the second step to adjust the horizontality of the balance arm of the inner mast, the third step to adjust the concentricity of the inner and outer masts, and the fourth step to adjust the gap between the inner mast and the guide roller.

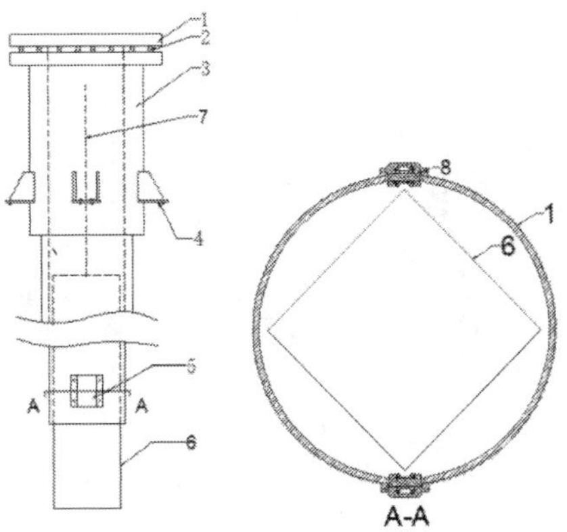

Fig. 1 Diagram of Mast Installation
1-Outer mast; 2-Spherical bearing; 3-Stationary mast; 4-Gasket used for outer mast verticality adjustment; 5-Guide roller assembly; 6-Inner mast; 7-Steel wire rope; 8-Gasket

As shown in Figure 1, the AP1000 mast is supported by the part 3 installed on the trolley, and the part 1 is installed on the part 3 via the part 2. Due to the existence of the part 2, the part 1 can rotate together with the part 6 for a certain amount angle. On the outer surface of the part 1, seven pairs of parts 5 are mounted from top to bottom.
The verticality of part 1 can be adjusted by changing the thickness of part 4. The gap between parts 5 and 6 can be adjusted by changing the thickness of part 8.

2.1 Adjustment of verticality of outer mast

The machining of the outer mast adopts one-time clamping, one-time positioning, and one-time machining to complete all mechanical features, with high machining accuracy. Therefore, using certain machined features on the outer mast can reflect the verticality of the mast installation.
According to this idea, a part 3 with a thread at one end and a cone at the other end is produced, the part 3 is screwed into the threaded hole of the part 1, and the conical point of the part 3 reflects the center of the threaded hole of the part 1. Screw the part 3 into the threaded holes at the uppermost and lowermost positions on the same side of the part 1 and the part 2, and hang the plumb in the upper part 3, as shown in Figure 2. The difference in the distance between the plumb line and the machined flange surfaces of parts 1 and 2 reflects the verticality in this direction, and the distance between the plumb line and the lowermost tooling reflects the verticality in another orthogonal direction. According to the measurement results, adjust the thickness of part 4 in Figure 1 until the verticality of the outer mast reaches ±4.76mm.

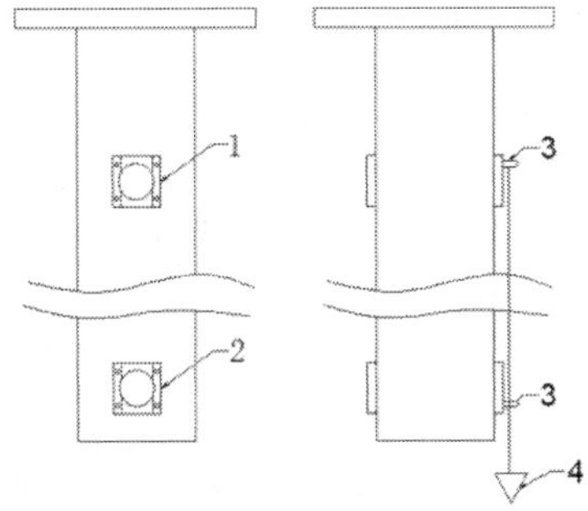

Fig. 2 Diagram of Verticality Adjustment for Outer Mast
1-The first pair of guide roller assembly mount; 2- The seventh pair of guide roller assembly mount; 3- Outer mast verticality adjustment tool; 4-plumb

2.2 Adjustment of inner mast balance arm

By adjusting the lengths of part 1 and part 3 as shown in Figure 3 to ensure that part 2 is kept horizontal, so as to ensure that part 4 remains concentric with the outer mast during the up and down movement, which can reduce the chance of the inner mast and the guide roller rubbing against each other, thereby reducing friction.

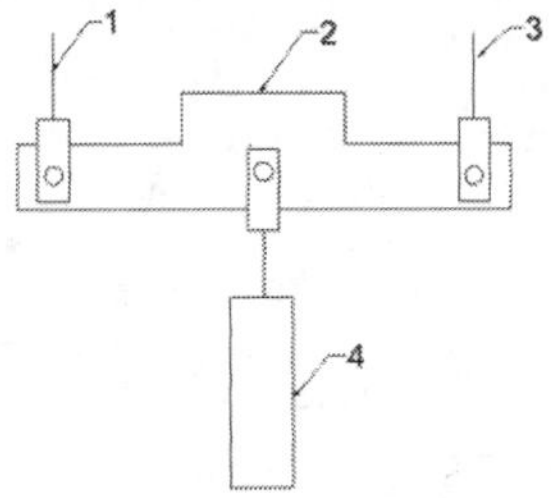

Fig. 3 Diagram of Levelness Adjustment for Equilibrium Arm
1-The first steel wire rope; 2-Balance arm; 3- The second steel wire rope; 4-Inner mast

2.3 Concentricity adjustment between inner and outer masts

Adjust the position of the motor wire rope drum assembly so that the part 2 in Figure 4 is in the center of the part 1. Whether the part 2 is in the center of the part 1 can be characterized by adjusting the distance and angle between the four right-angled sides of the part 2 and the reserved machining line of the part 1. The purpose is to ensure that the inner and outer masts are concentric and reduce the running friction of the inner mast.

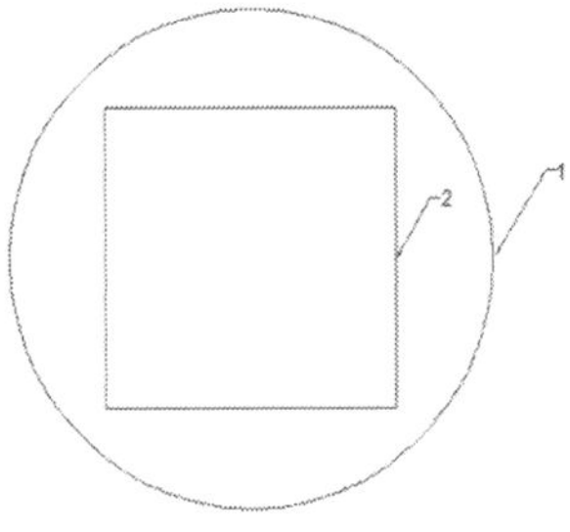

Fig. 4 Diagram of the Concentricity Adjustment of the Inner and the Outer Mast
1-Upper flange of outer mast; 2-Inner mast;

2.4 Adjustment of guide roller clearance

According to the design drawings, the theoretical gap between the inner mast and the guide roller is 1.6mm, which is converted into a gap of 1.13mm between the side of the inner sleeve and the side of the guide wheel.

Lower the inner mast until the bottom of it is slightly lower than the position of the seventh pair of guide roller seats. Without gaskets, install the guide roller on one side, push the inner mast to make it stick to the guide roller, and plug the other side between the guide roller and the inner mast with a 2.3mm thick feeler gauge, measure the clearance between guide roller and its seat. Half of the measured value is the theoretical thickness of the shims added to the guide rollers on the north and south sides. Since the outer mast is not absolutely vertical, in order to ensure the uniform clearance between the inner mast and the guide roller, the thickness of the gaskets added on both sides should be adjusted based on the theoretical thickness and the verticality of the outer mast measured after adjustment per section 2.1 of this article, withdraw part of the gasket from one side of the outer mast and add it to the other side.

Using the tooling as shown in Figure 5, take the seventh pair of guide wheels as the benchmark to ensure that the distance between the outer surface of the guide roller seat and the vertical line is equal, and adjust the thickness of the gasket between the guide roller and its seat to make the seven guide rollers at the same side are installed on a vertical line, and the other side is installed in the same way.

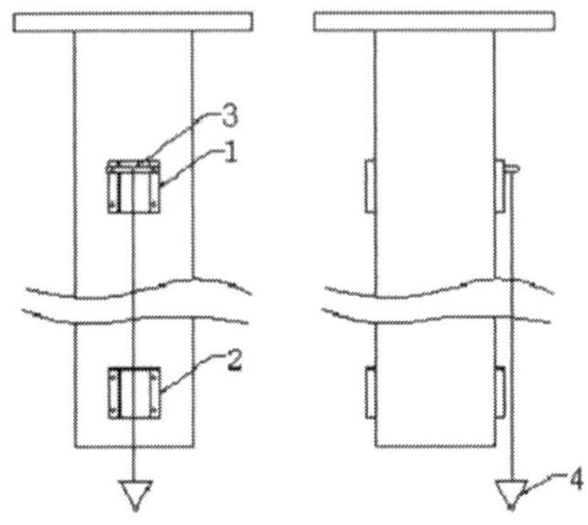

Fig. 5 Diagram of Gap Adjustment between the Guide Roller of the Outer Mast and the Inner Mast
1-The first pair of guide roller assembly; 2- The seventh pair of guide roller assembly; 3-Guide roller clearance adjustment tool; 4-Plumb

3. Check and accept

After adjustment, stick the target reflector at the position of the inner mast close to the fuel assembly gripper, and use the total station to measure the running verticality of the inner mast. Since the positioning accuracy of the refueling machine is ±3.18mm, the inner mast running verticality shall not be greater than 3.18mm. It should be noted that, the coil spring of the reel wrapped with fuel gripper cylinder compressed air supply pipe, the underwater camera cable reel coil spring, and the gripper proximity switch cable reel coil spring apply eccentric force on the top of the inner mast. When the refueling machine is not loaded, the eccentric force has a great influence on the posture of the inner mast, which will cause the measured verticality of the inner mast to be large. It is necessary to untie the connection between the air supply pipe, the cable and the top of the inner mast to eliminate the influence of the eccentric force before measuring.

During commissioning, simulate the process of inserting fuel assemblies into the core with the inner mast loaded with dummy fuel assembly and measure the running verticality of the inner mast, the measured value is required to be within 1mm.

The running friction of the inner mast is also verified. It is divided into no-load acceptance test and load acceptance test. It is required that the mast can run smoothly at the highest speed without jamming, overload and underload. Friction is generally controlled within 10kg.

4. Conclusion

The inner mast maintains a vertical posture in the free hanging state, and the guide roller is only used to limit the movement range of the inner mast in the horizontal direction, which is the unique design concept of the AP1000 refueling machine. According to this, a method

 https://doi.org/10.1051/matecconf/202235801018

for adjusting the verticality of the mast of the refueling machine was developed. During the first overhaul of a nuclear power plant, this method was applied to solve the problem of large running friction of the inner mast of the refueling machine, and at the same time ensured the inner mast verticality during operation.

References

1. LIU Zhan, RONG Jian, ZHANG Li, WANG Hai-yang, SHEN Li-feng. Design Analysis of Large-Scale Advanced Generation III Pressurized Water Reactor at Home and Abroad. China Nuclear Power, Volume 13, No. 3, June 2020, Pages 403-407
2. GUO Ke-ke. Characteristics Analysis of AP1000 Fuel Handling Equipments. Progress Report on China Nuclear Science & Technology, Volume 2, Oct 2011, Pages 118-121
3. ZHANG Yi, ZHANG Wei-dong, LIAN Shu-sheng. Structural Analysis of AP1000 Refueling Machine Prototype. Lifting and Transporting Machinery, No.11, 2014, Pages 65-69

MATEC Web of Conferences **358**, 01019 (2022) https://doi.org/10.1051/matecconf/202235801019
MSMS 2022

Stress Analysis of Floating Photovoltaic Power Plants under Wind Conditions in 8x8 Arrangement

Bingqing Yuan [1,2], Yaoping Bei [1,2], Mingyu Zhang [3], Xia Deng [3,*]

[1] China three gorges renewables (group) Co., Ltd, Beijing, China
[2] Three gorges renewable energy huainan solar power generation Co., Ltd., Huainan, China
[3] Share power Co., Ltd., Nanjing, China

Abstract. This paper adopts Sharepower solar floating photovoltaic power station unit. The structure is simulated and analyzed, the strength of a single solar structure support is analyzed, the photovoltaic array structure is analyzed, and the rectangular and square structures with different wind directions are analyzed.

1. Introduction

In this paper, the numerical simulation analysis of the floating photovoltaic power station under the given wind field conditions is carried out, and the force of the photovoltaic panel components and the floating body is analyzed, and the load is provided for the anchoring calculation. The numerical simulation of the floating power station under the simultaneous action of wind and waves is carried out, and the stress of the floating body is analyzed [1].

2. Geometric modelling

In this project, the incoming wind speed is extended along the positive direction of the x-axis. Figure 1 is the windward direction model, and Figure 2 is the leeward direction model.

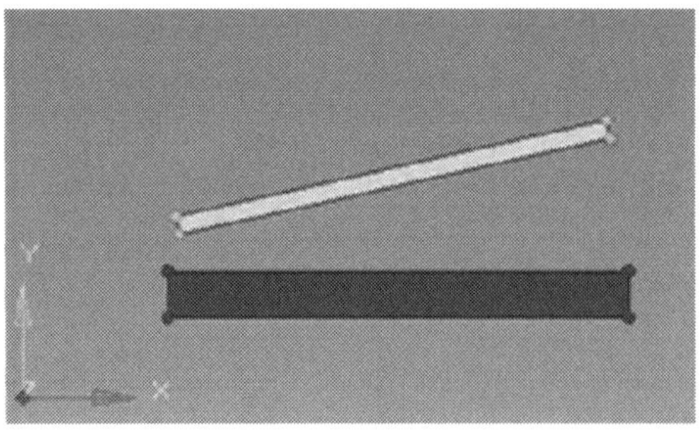

Figure 1 Windward direction model

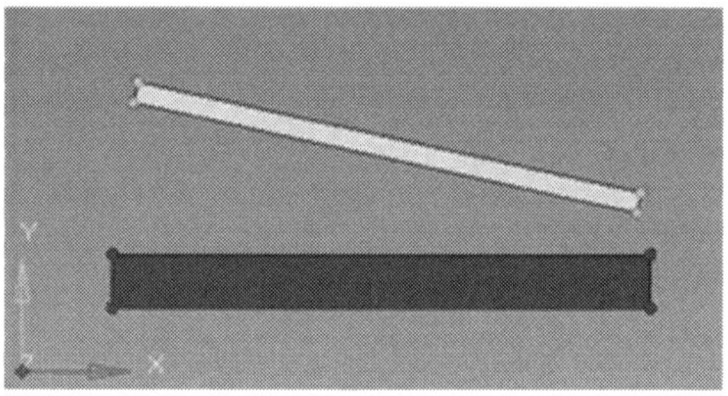

Figure 2 Leeward direction model

Table 1 Surface mesh size

Parameter value	Value
Basic size	1000mm
The total thickness of the boundary layer	8mm
Boundary Layer	8
Boundary layer growth rate	1.4
Block1	80mm
Block2	160mm
Block3	320mm

Import the surface mesh in bdf format into star-ccm+, divide the volume mesh according to the parameter values in Table 1, and refine locally around the floating power station, as shown in Figure 3.

* Corresponding author: dengxia@sharepower.com.cn

https://doi.org/10.1051/matecconf/202235801019

2.1 Boundary conditions

Under the action of wind, the inlet is the velocity inlet, the outlet is the pressure outlet, and the incoming flow velocity is 30m/s;

Under the combined action of wind and waves, the wave height is set to 0.46m, and the water velocity is set to 1m/s.

2.2 Calculation parameters

Three-dimensional unsteady incompressible turbulent flow
K-e turbulence model
No slip wall, regardless of roughness
Time step 0.01s
Physical calculation time 10s
Monitor the force in the x and y directions

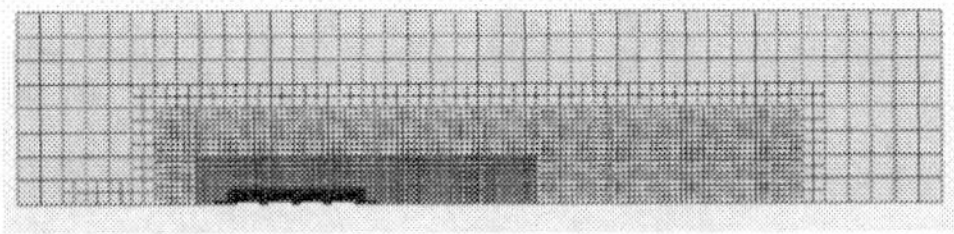

Figure 3 Face mesh

The floating power station consists of two parts: a floating body and a photovoltaic panel assembly. Simplify some parts to create 3D geometric models based on float and assembly dimensions and float spacing. Add a wind field outside the model to simulate the flow field of the floating power station under a certain wind speed.

3. The operating conditions analysis

The 8*8 square matrix model is affected by wind and waves at the same time, the wind speed is 30m/s, and the wave height is 0.46m.

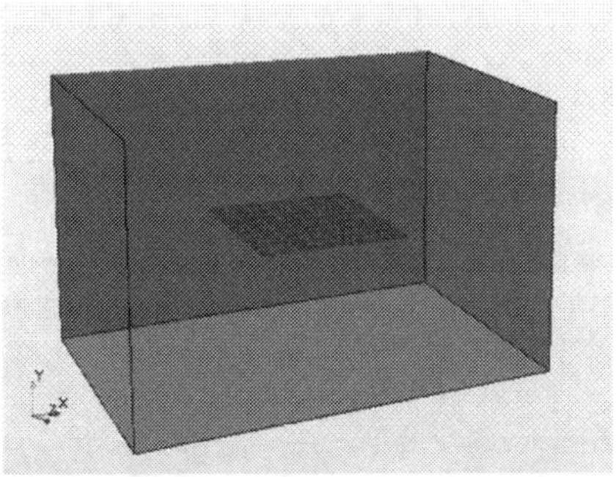

Figure 4 8*8 square matrix model

The force of the floating body also shows a periodic law with the periodic motion of the wave, and the law of the same row of floating bodies is similar, as shown in Figure 4.

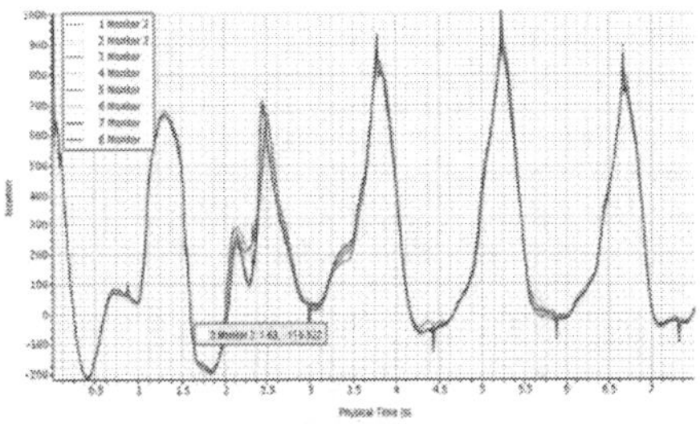

X-axis force of the first row of floating bodies

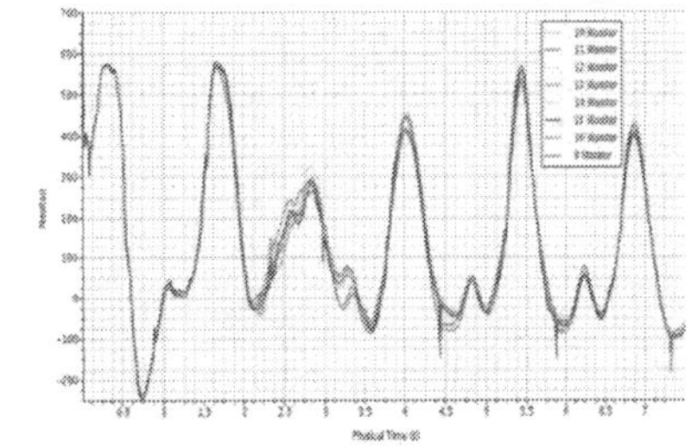

X-axis force of the second row of floating bodies

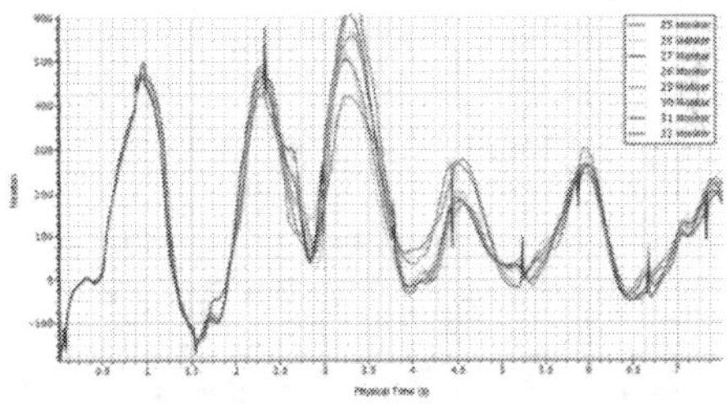

X-axis force of the fourth row of floating bodies

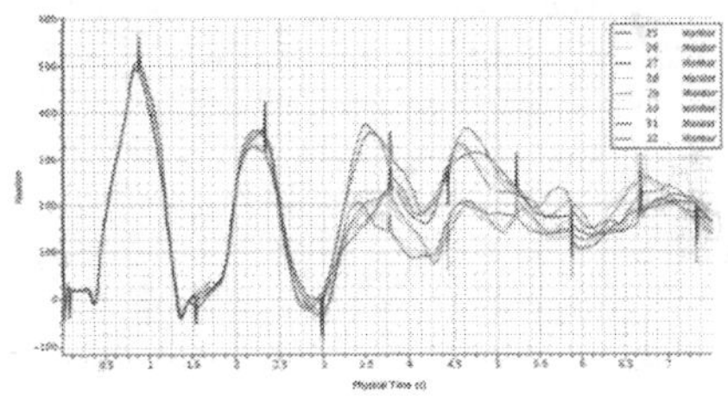

X-axis force of the eighth row of floating bodies

Figure 5 Periodic law of force on floating bodies in rows 1, 2, 4, and 8

Due to the similarity of the periodicity of the force on the same row of floating bodies, a certain column of data can be taken as the reference value in the analysis. In this project, the fourth column of data is taken as the reference value, see Table 2.

Table 2 Average horizontal force of floating bodies in the fourth row of the 8*8 square matrix

4th row of floats	1	2	3	4	5	6	7	8
Average horizontal force (N)	239.717	112.293	60.5771	84.809	107.015	141.954	131.785	182.751

Table 3 The total force of 4 kinds of phalanx floating bodies

Square	The number of east-west direction	The number of north-south direction	The total force of the floating body (N)
Square 1	96	56	590112
Square 2	96	112	1159968
Square 3	144	80	1251504
Square 4	192	104	2157120

Because the wave model in star-ccm+ software has certain limitations, the wave will not attenuate when encountering obstacles. Referring to the actual situation, the wave inside the square matrix has almost no force on the floating body. If the total force of the first three rows, the last two rows and the two rows on both sides of the square matrix is calculated, the results are shown in Table 4.

Table 4 The total force of the floating body around the phalanx

Square array	East-west number	North-south number	Horizontal force of the first row (N)	Horizontal force of the second row (N)	Horizontal force of the third row (N)	Horizontal force of the middle row (N)	Horizontal force of the second row Force (N)	The horizontal force of the next row (N)	The total force of the floating body (N)
	A	B	F1	F2	F3	F4	F5	F6	(F1+F2+F3+F5+F6)*A+F4*(B-5)*4
Square 1	96	56	240	112	60	111	132	183	92436
Square 2	96	112	240	112	60	111	132	183	117300
Square 3	144	80	240	112	60	111	132	183	137988
Square 4	192	104	240	112	60	111	132	183	183540

4. Conclusion

In this paper, numerical simulation analysis is carried out for the 8*8 arrangement, and the dual effects of wind and waves are analyzed in a targeted manner. Through the analysis of the force situation, it can be found that this arrangement can meet the strength requirements.

Acknowledgments

This work was financially supported by the scientific research project of China Three Gorger Corporation. (Contract No.: 202003043)

References

1. Z. Hameed, Y.S. Hong, Y.M. Cho, S.H. Ahn, C.K. Song. Condition monitoring and fault detection of wind turbines and related algorithms: A review. Renewable & Sustainable Energy Reviews, 13, 1-39 (2009).
2. R. Hemmati. Optimal design and operation of energy storage systems and generators in the network installed with wind turbines considering practical characteristics of storage units as design variable. Journal of Cleaner Production, 185, 680-693 (2018).

3. H. Li, Z. Chen. Overview of different wind generator systems and their comparisons. Iet Renewable Power Generation, 2, 123-138 (2008).
4. Z.X. Li, Y. Jiang, Q. Guo, C. Hu, Z.X. Peng. Multi-dimensional variational mode decomposition for bearing-crack detection in wind turbines with large driving-speed variations. Renew. Energy, 116, 55-73 (2018).
5. M. Liserre, R. Teodorescu, F. Blaabjerg. Stability of photovoltaic and wind turbine grid-connected inverters for a large set of grid impedance values. Ieee Transactions on Power Electronics, 21, 263-272 (200

https://doi.org/10.1051/matecconf/202235801020

Simulation analysis of single solar floating photovoltaic panel structure based on wind direction change

Yaoping Bei [1,2], Bingqing Yuan [1,2], Fei Feng [3], Xia Deng [3,*]

[1] China three gorges renewables (group) Co., Ltd, Beijing, China
[2] Three gorges renewable energy huainan solar power generation Co., Ltd., Huainan, China
[3] Share power Co., Ltd., Nanjing, China

Abstract. This paper adopts Sharepower solar floating photovoltaic power station unit. The structure is simulated and analysed, the strength of a single solar structure support is analysed, the photovoltaic array structure is analysed, and the rectangular and square structures with different wind directions are analysed.

1. Introduction

Water photovoltaic is a novel photovoltaic layout. In this paper, the vibration photovoltaic panel support is numerically simulated, and various working conditions are selected for stress value and isoline analysis to analyze the strength of the support [1-4]. The supporting structure of floating photovoltaic power station needs to be portable. Firstly, the stress of the supporting structure is analyzed by numerical simulation. The of the support under different conditions is calculated and analyzed, and the distribution law of the maximum stress point is obtained The rationality of the support structure is verified by combining the 3D calculation results of CFD under typical working conditions

2. Analysis modelling

The overall wind load and flow load are numerically studied. The numerical investigation of CFD calculation verifies the feasibility of using the simplified floating body model. The calculated results are in good agreement with the wind tunnel test results; Through real-scale modeling and 3D simple calculation of the square array, the overall wind load at different wind directions is obtained. The law of change under; Through the analysis of the load distribution law under the north wind condition of the maximum load wind direction, a 2.5D calculation strategy is proposed, and the numerical test is obtained. The number of cells is 995767, as shown in Figure 1:

Figure 1. Model schematic

3. Calculate the result

As shown in Figure 2 to Figure 5, the stress nephogram and maximum stress distribution position of all supports on the right and left are shown. Under this working condition, the whole support is in a tensile state.

The overall stress nephogram of the left support is shown in Figure 2:

* Corresponding author: dengxia@sharepower.com.cn

 https://doi.org/10.1051/matecconf/202235801020

Figure 2. Overall stress nephogram of left support

From the stress nephogram, it can be seen that the stiffness of position 1 is small, and the stress at the arc is the largest, which is 27.777mpa, as shown in Figure 3: however, it can still meet the system stress requirements and has good stability.

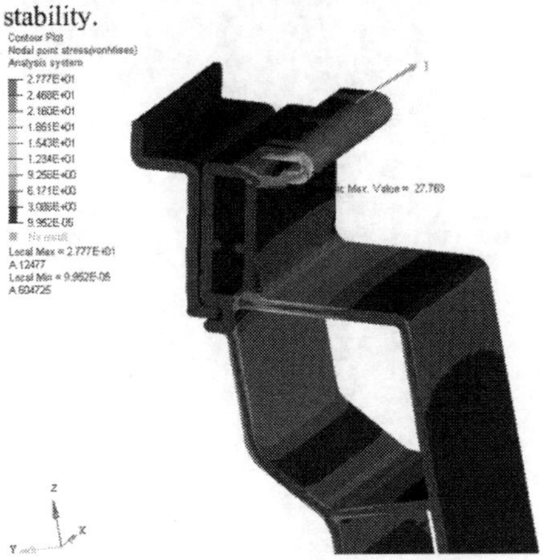

Figure 3. Local stress nephogram of left support

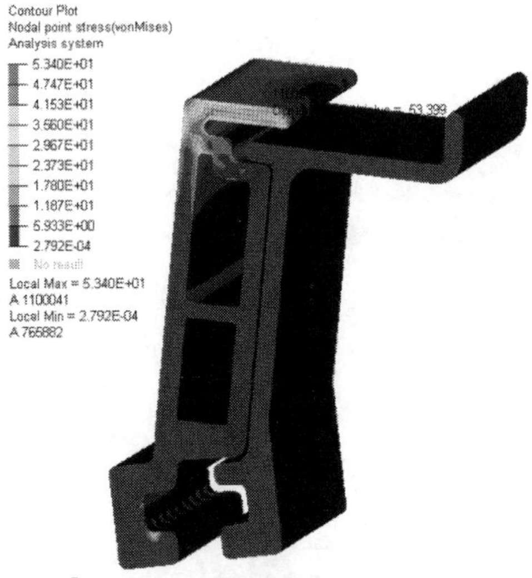

Figure 4. Cloud diagram of overall stress of right support

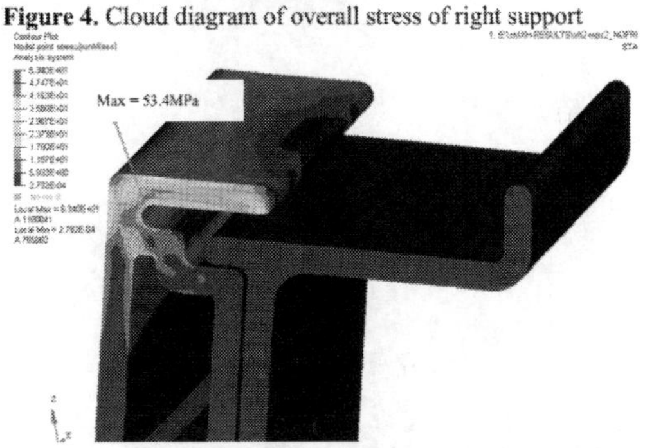

Figure 5. Cloud diagram of local overall stress of right support

According to the calculation, the maximum stress of the left support and the right support are 27.777mpa and 53.4mpa respectively, which is far lower than the yield limit of aluminum alloy material 200MPa, so it meets the strength design requirements and has a very high safety factor.

3.1 Operating conditions 1- North-South wind analysis results

In order to find the maximum stress distribution position of the left and right supports, this simulation is carried out. Under the test conditions, the support is in traction state. After calculation, the maximum stresses of the left support and the right support are 7.577mpa and 8.494mpa respectively, See Figure 6-7 for details.

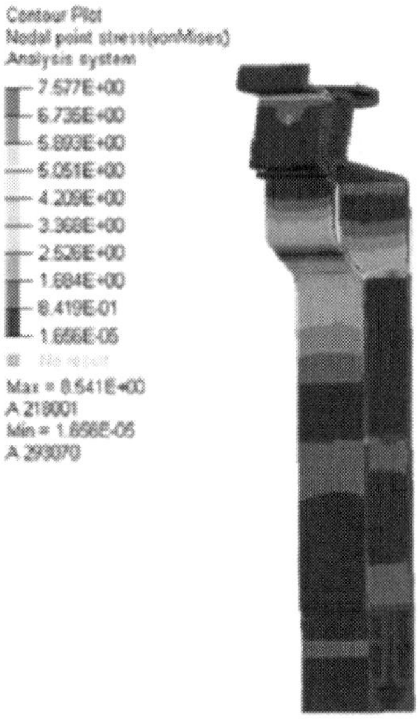

Figure 6. Left Bracket Local Stress Cloud Map

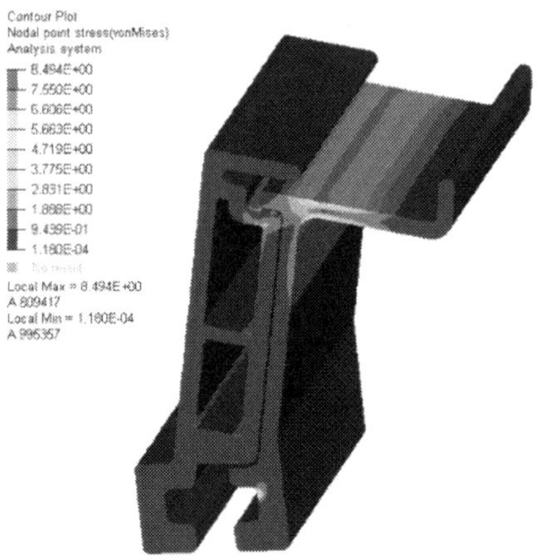

Figure 7. Left Bracket Local Stress Cloud Map

3.2 Operating conditions 2- North-South leeward analysis results

According to the calculation, the maximum stress of the left support and the right support are 27.777mpa and 53.4mpa respectively, which is far lower than the yield limit of aluminum alloy material 200MPa, so it meets the strength design requirements and has a very high safety factor, as shown in Figure 8-9.

Figure 8. Left Bracket Local Stress Cloud Map

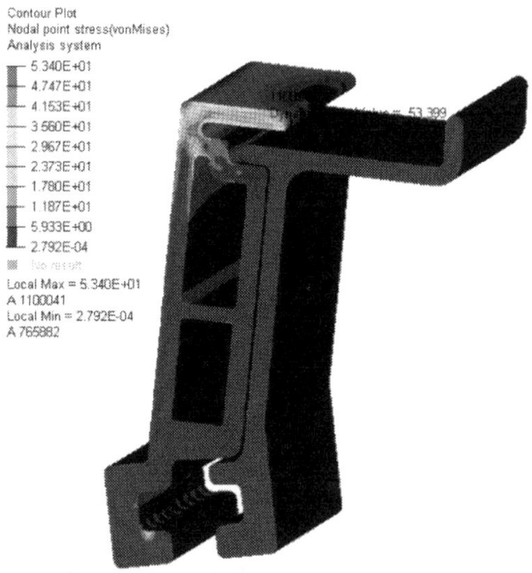

Figure 9. Right Bracket Local Stress Cloud Map

3.3 Operating conditions 3-East-West wind analysis results

In order to find the maximum stress distribution position of the left and right supports, this simulation is carried out. Under the test conditions, the support is in traction state. After calculation, the maximum stresses of the left support and the right support are 20.53mpa and 18.65mpa respectively, See Figure 10-11 for details.

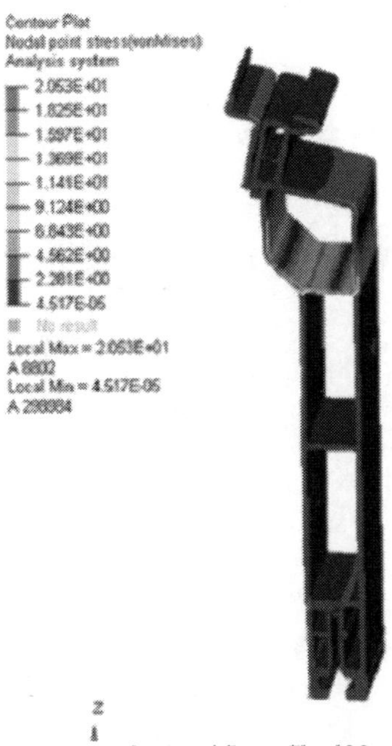

Figure 10. Left Bracket Local Stress Cloud Map

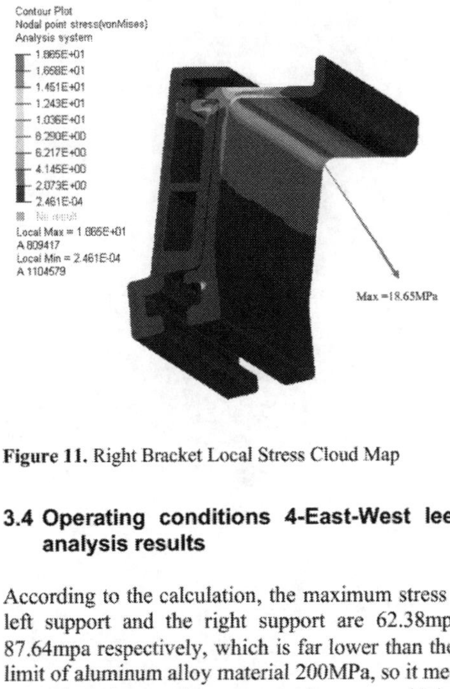

Figure 11. Right Bracket Local Stress Cloud Map

3.4 Operating conditions 4-East-West leeward analysis results

According to the calculation, the maximum stress of the left support and the right support are 62.38mpa and 87.64mpa respectively, which is far lower than the yield limit of aluminum alloy material 200MPa, so it meets the strength design requirements and has a very high safety factor, as shown in Figure 12-13.

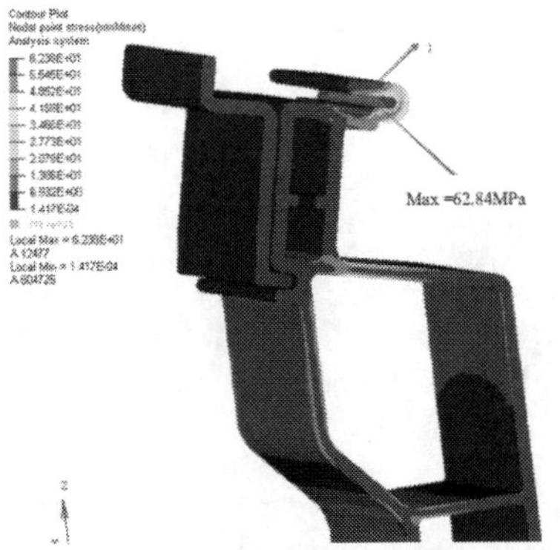

Figure 12. Left Bracket Local Stress Cloud Map

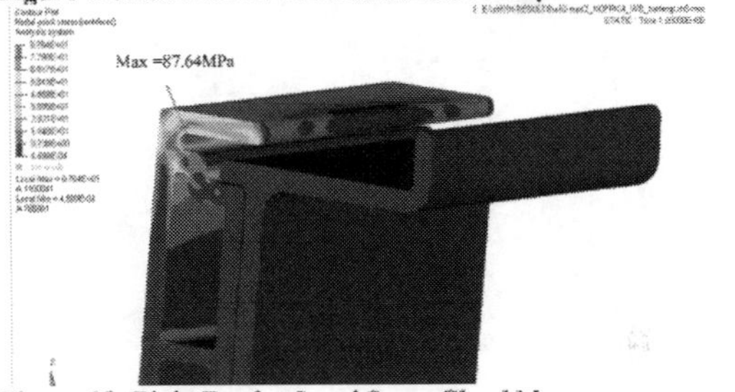

Figure 13. Right Bracket Local Stress Cloud Map

4. Conclusion

Floating photovoltaic power station has the characteristics of many modules and large scale, and there is no mature method to evaluate its flow load Firstly, the stress of supporting structure is analyzed by numerical simulation. The of the support under different conditions is calculated and analyzed, and the distribution law of the maximum stress point is obtained Combined with the 3D calculation results of CFD under typical working conditions, the rationality of the support structure is verified All designs are far lower than the yield strength of aluminum alloy and meet the design requirements. In the next step, the structure can be optimized to obtain the material saving scheme.

Acknowledgments

This work was financially supported by the scientific research project of China Three Gorges Corporation. (Contract No.: 202003043)

References

1. D. Chen. New opportunities, developments and challenges of Chinese photovoltaic power plants. Electronic Engineering & Product World for Engineering Managers & Designers, 2017(5): 3-5.
2. K. Branker. A review of solar photovoltaic levelized cost of electricity. Renewable and Sustainable Energy Reviews, 2011, 15(9): 4470- 4482.
3. M. Darren. Photovoltaic Technologies[J].Energy policy.2008,36(12).
4. P. Parker. Residential solar photovoltaic market stimulation: Japanese and Australian lessons for Canada. Renewable & sustainable energy reviews.2008,12(7).

Study on the Synthesis of Water-dispersible Acrylic Resin Containing Hydroxyl Group`

Ma Jing [1, 2], Mu Rong [1, 2], Daowei Lai [1, 2], Huaming Li [3], Yong Tao [3], Xiao Fei [3], Chunyu Zhou [1, 3], Chufeng Sun [1, 2, *]

[1] School of Chemical Engineering, Northwest Minzu University, Lanzhou 730030, China
[2] Key Laboratory of Environment-Friendly Composite Materials of the State Ethnic Affairs Commission, Gansu Provincial Biomass Function Composites Engineering Research Center, Lanzhou 730030, China
[3] Northwest Yongxin Paint & Coatings Co., Ltd., Lanzhou 730030, China

Abstract: Hydroxy acrylic resin is a long-chain polymer material formed by cleavage and recombination of double bonds of various acrylates and methacrylates under the action of initiators. Among them, MMA and BA provide the rigidity of the molecular chain and For flexibility, HEMA provides the hydroxyl groups required for cross-linking and curing, and MAA gives the resin water solubility. By rationally designing different monomer dosages, resin products with high gloss and high hardness are prepared, and through paint making experiments, the paint film Performance is tested.

Keywords: Water-borne resin; Acrylic; Hydroxy resin

1. Introduction

With the rapid development of economy and society, the problem of environmental pollution has become a problem that must be solved in the process of economic development[1-2]. Eco-friendly coatings have attracted the attention of researchers because of their use of water or eco-friendly solvents as a dispersive medium, which greatly reduces the use of volatile organic solvents (VOCs) in their products[3]. The film-forming substances used in coating products can be divided into: resin, emulsion, dispersion and so on. Because the vast majority of emulsion products in the high-speed mixing process is prone to the phenomenon of milk breaking, resulting in product failure. Although the dispersion has the advantages of emulsion and resin, its high[4].

Acrylic resin has high hardness, high gloss characteristics, is widely used in vehicle protection, equipment protection, infrastructure construction and other fields[5-6]. However, acrylic resin in poor water resistance, low solid content, high viscosity and other deficiencies, in order to reduce the viscosity of the resin and improve the solid content, the introduction of the silicon carbonate shrinkgies (E10-p) to modify the resin research, E10-p as a reactive solvent[7-8], can reduce the VOC content of the resin, and because E10-p has epoxy group, At high temperature[9], the hydronization of the key forms a hydroxyl, and the esterification reaction with organic acids, forming an ester base with a large spatial bit resistance effect, which can effectively reduce the viscosity of the resin[10-11].

2. Experiment

2.1 Materials

Methyl methacrylate (MMA), butyl acrylate (BA), methacrylic acid (MAA), hydroxyethyl methacrylate (HEMA),glycidyl tert-carbonate (E10-p) ,propylene glycol butyl ether (PnB), N, N-dimethylethanolamine (DMEA), tert-butyl peroxybenzoate (TBPB) , 2,4-diphenyl-4-methyl-1-pentene (AMSD) are all industrial grade. Deionized water is homemade in the laboratory.

2.2 Instruments

BT100-1F peristaltic pump, ZNHW intelligent constant temperature electric heating jacket, 78-1 magnetic heating stirrer, RW20 stirrer, FS-1 experimental disperser, DZF vacuum drying box, NETZSCH DSC-214 differential scanning calorimeter, QFZ-II adhesion tester.

2.3 Synthesis of water-dispersible hydroxy acrylic resin

Add a certain amount of PnB, 50% of E10-pto a four-necked flask equipped with a stirring device, a thermometer, and a condenser. Warm to 140°C and hold the reaction for 30 minutes. The remaining monomer and 90% TBPB and AMSD Mix evenly, slowly add to the flask within 3h, hold for 1h, add 5% of TBPB and AMSD to react for 1h, add the remaining TBPB and AMSD to warm to 150 °C for 1h, cool to 80 °C, add a certain amount of DMEA adjusts the pH, quickly stirs for 30 minutes,

* Corresponding author: chufengsun@xbmu.edu.cn

adds a certain amount of deionized water under high-speed stirring, and mixes for 30 minutes at high speed to obtain a water-dispersible hydroxy acrylic resin. The water-dispersible hydroxy acrylic resin formulation is shown in Table 1.

Table 1 Water-dispersed hydroxyl acrylic resin formula

name	Dosage /%	effect
MMA	52-60	Provide hardness
HEMA	14-18	Provides hydroxyl
BA	2-7	Provides flexibility
MAA	4-6	Provide water solubility
E10-p	4-6	Reduce viscosity
TBPB	3-5	Initiator
AMSD	1-2	Chain transfer agent

2.4 Preparation of coatings

Take a certain amount of resin, add water-based defoamer, substrate wetting agent, film-forming aid, deionized water under high-speed stirring, and mix evenly at high speed to obtain an aqueous hydroxy-containing acrylic varnish. The curing agent is DNW-5500 isocyanate curing agent. The acrylic varnish formulation is shown in Table 2.

Table 2 Acrylic varnish formulations

name	Dosage /%
Resin	60.0
Defoamer	0.1
Substrate wetting agent	0.1
Film-forming aid	0.4
water	39.4

2.5 Test

2.5.1 Calculation of theoretical hydroxyl content

In the resin synthesis process, the calculation formula of the theoretical hydroxyl content is shown in Formula 1. The theoretical hydroxyl content of the resin can be calculated by the molecular weight and mass fraction of hydroxyethyl methacrylate, and the theoretical hydroxyl value = theoretical hydroxyl content×33. However, since -OH reacts with -COOH during the reaction, it is necessary to perform a titration measurement on the actual acid value and subtract the difference between the theoretical acid value and the actual acid value to obtain an approximate actual hydroxyl value. The calculation formula of the theoretical acid value is shown in Equation 2.

$$\text{Hydroxyl content} = \frac{m\,(OH)}{M\,(OH)} \times m \times 17$$

m (OH): Mass fraction of hydroxyethyl methacrylate
M (OH): Molecular weight of hydroxyethyl methacrylate
m: Amount of synthetic resin

Formula 1 Calculation formula of theoretical hydroxyl content

$$\text{Theoretical acid number} = \frac{m\,(COOH) \times 56100}{86.1}$$

m (COOH): Methacrylic acid mass fraction

Formula 2 Calculation formula of theoretical acid value

2.5.2 Acid value determination

Take about 1g of neutralized resin, accurate to 0.001g, take 50g of 1: 1 mixed hydrogenated toluene and ethanol solution to completely dissolve the resin, add 3-4 drops of phenolphthalein indicator, use 0.1mol/L potassium hydroxide The standard solution was titrated, and the end point was that the solution was initially reddish and did not change color within 30s. Measure the amount of potassium hydroxide standard solution and calculate the actual acid value of the resin.

2.5.3 DSC analysis

The dried resin was measured for the glass transition temperature (Tg) using a differential scanning calorimeter. The test was performed under the conditions that the protective gas was nitrogen, the temperature range was -10-100°C, and the heating rate was 10K / min.

2.5.4 Solid content test

Take a certain amount of resin, weigh the resin (accurate to 0.01), and dry it at 140°C for 2h. After taking out and cooling sufficiently, measure the remaining weight. Solid content = weight after drying / weight before drying × 100%.

2.5.5 Varnish performance test

According to n (-NCO): n (-OH) = 1.5: 1, mix the curing agent and acrylic resin uniformly, add appropriate amount of substrate wetting agent and defoaming agent, dilute to the construction viscosity, spray on the iron plate after grinding After the surface is dried, it is dried in an oven at 80°C. for 30 minutes, and cured at room temperature for 7 days to test the performance of the paint film.

Adhesion test: Spray the paint on the surface of the iron plate with a thickness of 25-35 μm, bake it at 80°C for 30 minutes, and maintain it at room temperature for 7 days, then use an adhesion tester to test it.

Impact resistance test: The coating was sprayed on the surface of the iron plate with a thickness of 50±5 μm, dried at 80°C. for 30 minutes, and maintained at room temperature for 7 days, and then tested using an impact tester.

Gloss Test:The colored paint was applied to the surface of the iron plate with a thickness of 50±5 μm, dried at 80°C. for 30 minutes, and maintained at room temperature for 7 days, and then measured using a 60 ° angle gloss meter.

Thermal Reserve Performance Test: Take a certain amount of paint and store it in an oven at 50±2°C, and observe the change of paint state.

Freeze-thaw performance test: Take a certain amount of paint and place it in the environment of -18±2 °C for 16h, take it out and place it in the environment of 23±2 °C for 8h, repeat it several times, and observe the change of paint state.

3. Results and discussion

3.1 Acid value and water solubility

By changing the amount of MAA introduced, the resin was prepared with a neutralization degree of 100%. The water dispersibility of the prepared resin was studied. The relationship between the acid value and water solubility of the resin is shown in Table 3.

Table 3 Relationship between acid value and water solubility

MAA introduction	Theoretical acid value / mgKOH / g	Actual acid value / mgKOH / g	Water soluble	Resin state
2.0%	13.72	7.24	Insoluble / 3h stratification	Milky opaque
2.5%	17.15	9.55	Insoluble / 3h stratification	Milky opaque
3.0%	20.58	10.88	Soluble / 48h stratification	Milky translucent
3.5%	24.01	13.96	Soluble / 72h stratified	translucent
4.0%	27.43	17.39	Soluble / 30d without delamination	translucent
4.5%	30.86	21.08	Soluble / 30d without delamination	Transparent
5.0%	34.29	23.86	Soluble / 30d without delamination	Transparent

There is an antagonistic relationship between the water solubility and water resistance of the resin. In order to ensure that the resin has good water solubility and stability and also has good water resistance, the amount of MAA introduced should be controlled between 4.5% and 5%. The acid value is between 30.86-34.29 mgKOH / g, and the actual acid value is between 21.08-23.86 mgKOH / g.

3.2 Paint film performance test

By adjusting the ratio of soft and hard monomers and functional monomers, acrylic resin dispersions with different glass transition temperatures and hydroxyl content were prepared. Samples of different resins were prepared and the properties of the samples were tested. Table 4 shows.

Table 4 Paint film performance test table

project	CY-1	CY-2	CY-3	CY-4	CY-5	CY-6
Tg / °C	20	30	30	35	40	40
Hydroxyl content /%	2.5	2.5	3.3	3.3	3.5	4.0
Solid content /%	42.4	43.1	42.8	42.7	42.5	42.8
Surface drying time / min	60	45	30	25	20	15
Adhesion / level	1	1	1	1	2	2
Impact resistance / level	1	1	1	1	2	2
Gloss (60 °)	86	87	88	89	88	86
Thermal storage stability ≧	30d	30d	30d	30d	30d	30d
Freeze-thaw stability ≧	5 times	5 times	5 times	5 times	5 times	5 times

According to the testing of paint film properties, when the glass transition temperature is below 30°C, the paint film has a low hardness and a long surface drying time, and the phenomenon of re-adhesion has occurred in a long-term wet state. When the glass transition temperature is above 40 °C, the hardness of the paint film is high, but the adhesion and impact resistance are poor. It was finally determined that under this scheme, the optimal theoretical glass transition temperature should be controlled between 30-35°C.

The increase of hydroxyl content is beneficial to the crosslinking of the paint film, and at the same time can improve the water solubility of resin and the hardness of the paint film, but the increase of hydroxyl content will lead to increased resin viscosity, the scheme will control the theoretical hydroxyl content at 3.3%, the theoretical hydroxyl value in 108.95 mgKOH /g.

3.3 Actual parameters

By theoretical calculation of the resin, the theoretical hydroxyl content of the resin was determined to be 3.30%, the theoretical hydroxyl value was 108.95 mgKOH /g, and the theoretical acid value was 34.29 mgKOH/g. Due to some side reactions in the reaction process, the actual values of the parameters of each part need to be measured. After titration testing of the resin, it was finally determined that the actual acid value was about 21.5-24 mgKOH/g, the actual hydroxyl content was 2.95-3.05%, and the actual hydroxyl value was 97.35-100.65 mgKOH/g. The resin parameters are shown in Table 5.

Table 5 Determination of resin parameters

parameter	Theoretical value	Actual value
Acid value	30.86-34.29 mgKOH /g	21.0-24.0 mgKOH /g
Hydroxyl content	3.30%	2.95-3.05%
Hydroxyl value	108.95 mgKOH /g	97.35-100.65 mgKOH /g

3.4 DSC analysis

The DSC-214 differential scanning calorimeter was used to detect the samples. The test temperature range is -10-100 °C, the heating rate is 10 K/min, and the sample mass is 5-10 mg. The weight loss curve is recorded, and its thermal stability and phase transition temperature are analyzed by the curve. It can be seen from Figure 1 that the glass transition temperature of this resin is 29.72 °C, the specific heat change is 1.158 J/(g*K), and there is no exothermic peak of crystal melting, so it can be inferred that the synthesized product did not Crystallization. The design glass transition temperature of this product is 30 °C, and the actual test glass transition temperature is 29.72 °C, which is basically consistent with the design value.

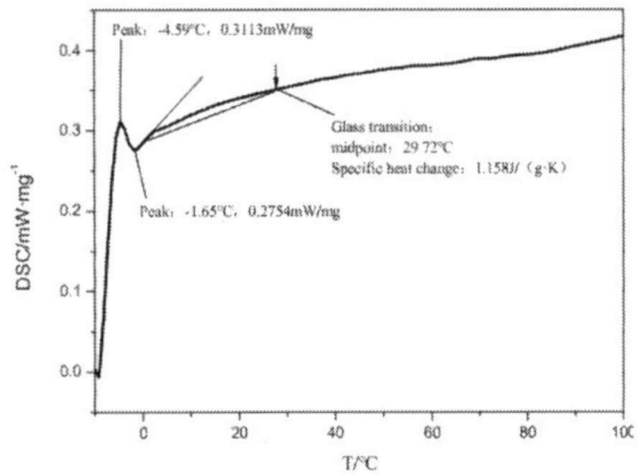

Figure 1 Resin DSC

4. Conclusions

By synthesizing the water-dispersed hydroxy acrylic base resin to determine the synthesis parameters of the base resin, it is finally determined that the amount of MAA introduced in the base resin should be controlled between 4.5%-5%, and the theoretical acid value is 30.86-34.29 mgKOH/g , The actual acid value determined by titration is between 21.08-23.86 mgKOH/g; the theoretical hydroxyl content is 3.30%, the theoretical hydroxyl value is 108.95 mgKOH/g, the actual hydroxyl value is 97.35-100.65mgKOH/g; Tg should be controlled at 30°C about. The prepared base resin has good properties in terms of adhesion, flexibility, hardness, gloss, storage properties, etc. The 60° gloss test result is around 90°.

Acknowledgments

This work is supported by the National Natural Science Foundation of China (Grant No. 51975561) and the Key Research and Development Project of Gansu Province (Grant No. 21YF5WA064).

References

1. Wu M, Ge S, Jiao C, et al. Improving electrical, mechanical, thermal and hydrophobic properties of waterborne acrylic resin-glycidyl methacrylate (GMA) by adding multi-walled carbon nanotubes[J]. Polymer, 2020, 200: 122547.
2. Guo X, Ge S, Wang J, et al. Waterborne acrylic resin modified with glycidyl methacrylate (GMA): formula optimization and property analysis[J]. Polymer, 2018, 143: 155-163.
3. Kazemi M E, Shanmugam L, Li Z, et al. Low-velocity impact behaviors of a fully thermoplastic composite laminate fabricated with an innovative acrylic resin[J]. Composite Structures, 2020, 250: 112604.
4. Kazemi M E, Shanmugam L, Chen S, et al. Novel thermoplastic fiber metal laminates manufactured with an innovative acrylic resin at room temperature[J]. Composites Part A: Applied Science and Manufacturing, 2020, 138: 106043.
5. Shahabi M, Movahedi Fazel S, Rangrazi A. Incorporation of chitosan nanoparticles into a cold-cure ortho-dontic acrylic resin: Effects on mechanical properties[J]. Biomimetics, 2021, 6(1): 7.
6. Ajay R, Suma K, Ali S A. Monomer modifications of denture base acrylic resin: a systematic review and meta-analysis[J]. Journal of pharmacy & bioallied sciences, 2019, 11(Suppl 2): S112.
7. Zoccolotti J O, Tasso C O, Arbeláez M I A, et al. Properties of an acrylic resin after immersion in antiseptic soaps: Low-cost, easy-access procedure for the prevention of denture stomatitis[J]. PloS one, 2018, 13(8): e0203187.
8. Jiang W, Jin X, Li H, et al. Modification of nano-hybrid silicon acrylic resin with anticorrosion and hydrophobic properties[J]. Polymer Testing, 2020, 82: 106287.
9. Badaró M M, Bueno F L, Makrakis L R, et al. Action of disinfectant solutions on adaptive capacity and virulence factors of the Candida spp. biofilms formed on acrylic resin[J]. Journal of Applied Oral Science, 2021, 29.
10. Procópio A L F, Da Silva R A, Maciel J G, et al. Antimicrobial and cytotoxic effects of denture base acrylic resin impregnated with cleaning agents after long-term immersion[J]. Toxicology in Vitro, 2018, 52: 8-13.
11. Sun C, Dai J, Zhang H, et al. Preparation and characterization of spherical titania and their influence on self-cleaning and anticorrosion properties of acrylic resin[J]. Progress in Organic Coatings, 2019, 128: 21-31.

https://doi.org/10.1051/matecconf/202235801022

Study on mechanical properties of traffic pavement materials under fatigue loading

Guotong Feng [1,*], Wenqing Wang [2], Liuting Ma [2] and Fengjun Liu [2]

[1] Hangzhou Urban-Rural Construction & Development Research Institute, Hangzhou, China
[2] Zhejiang University Urban-Rural Planning & Design Institute, Hangzhou, China

Abstract. The durability of pavement subgrade materials is the most important indicator for judging their merits as well as their mechanical properties. The durability of pavement subgrade material is mainly reflected in its water stability, because in the actual pavement environment, the subgrade is subjected to the infiltration of groundwater, immersion and the washout of pressure water under the concrete slab. The subgrade is subjected to a series of damage caused by freeze-thaw cycles and wet-dry cycles due to changes in temperature. In view of this, this chapter has targeted. In view of this, this chapter focuses on the frost resistance, water stability, scour resistance, dry shrinkage and sulfate erosion resistance of iron tailings as pavement base material. The study is carried out to simulate the real situation as much as possible to provide more detailed and accurate guidance for the actual engineering application. The study will provide more detailed and accurate guidance for the actual engineering application.

1. Introduction

The compressive rebound modulus specimen is a cylindrical specimen made by the static compression molding method [1]. The pavement subgrade material is not an ideal elastomer, and the stress-strain relationship is non-linear. Moreover, since the stress distribution in each structural layer of the pavement is not uniform, the modulus is mostly cut-line modulus. The cut line modulus is the slope of the line connecting two points on the stress-strain curve, reflecting the average condition of the stress-strain relationship in a certain working stress range. In pavement materials, these two points are generally chosen as the origin on the stress-strain curve and the stress is half of the ultimate strength or at a specified strai[3, 4]n.

In addition, when the strain in the modulus value is the total strain including the rebound and residual strain components, it becomes the deformation modulus; if the strain only includes the rebound strain after unloading, it is called the rebound modulus. Since the elastic analysis theory is used to analyze the stresses and strains in the pavement structure, and the value of the rebound strain is related to the fatigue damage of the pavement, the modulus of resilience is usually used to characterize the deformability of the material in pavement design[5, 6].

The curve shown in the figure is the modified rebound modulus curve after excluding the significant deviation points. It can be seen from the figure that the incorporation of fiber significantly reduces the rebound modulus value of the specimen, and when the amount of polypropylene fiber is incorporated, the rebound modulus of the specimen decreases compared with that of the specimen without polypropylene fiber[7, 8]. The main reason for this phenomenon is that the initial properties of polypropylene fibers are better and the low temperature brittleness is obvious, compared with the traditional brittle cement stabilized gravel material, the tensile and compressive deformation capacity of the composite material is greatly improved after the addition of polypropylene fibers, and the bending fatigue strength is excellent. The increase in flexibility ensures that the aggregate material can withstand large shrinkage deformation without causing damage[9, 10].

The durability performance of pavement base material is the most important index to judge its merit as well as the mechanical performance. The durability of pavement subgrade material is mainly reflected in its water stability, because in the actual pavement environment, the subgrade is subjected to the infiltration of groundwater, immersion and the washout of pressure water under the concrete slab when the vehicle and other loads pass through, humidity and temperature changes caused by freeze-thaw cycles and wet-dry cycles and a series of damage[11, 12].

This results in a large temperature difference between day and night in winter. Therefore, the freeze-thaw cycle test standard is proposed for the first time in the newly introduced Experimental Procedures for Inorganic Bond Stabilized Materials for Highway Engineering[13, 14]. The specimens of freeze-thaw cycle are cylindrical specimens made by static compression molding method. The specimens were immersed in water for hours and then taken out, frozen at for hours, and then thawed in water tank for hours, so that the cycle times to measure the lateral limitless compressive strength. The ratio of the lateral limit compressive strength of the two groups of specimens was obtained from their freezing coefficients,

* Corresponding author: 13692428@qq.com

and the results are shown in the table. It can be seen from the graph that the freezing coefficient of all the ratios are above, and the change of the freezing coefficient of the specimens with different polypropylene fiber doping is not obvious, which means that the freezing stability of each ratio is better and the amount of polypropylene fiber doping has little effect on the freezing stability of the specimens[15, 16].

2. Mechanical performance analysis

2.1 Compaction experiments

Compaction experiments were carried out on cement stabilized iron tailings, lime stabilized iron tailings, cement-lime stabilized iron tailings and cement and bio-coagulase stabilized iron tailings respectively, in order to plot the relationship curve of water content and dry density of stabilized soil, so as to determine its optimum water content and maximum dry density and provide compaction standards for engineering design and field construction. The ratio of cement: gravel: iron tailing mass is used as an example to explain in detail the determination method of the best water content and maximum dry density of this ratio. The maximum particle size of the experimental material should be controlled as follows.

Preset a gradient for the water content and calculate the total amount of water to be added to the specimen based on the formula. Spray the water evenly on the test material and mix it well, close and moisten it with a plastic pocket, then add the cement to the test material and mix it to a homogeneous state. When the cement is added to the wonk, the compaction experiment must be carried out within. The test material will be loaded into the compaction cylinder in five times, each compaction before leveling its surface and slightly compacted, and then heavy hammer compaction times. After the last compaction experiment is completed, the test piece should not exceed the top of the barrel, beyond the test piece should be scrapped. After the compaction is completed, the specimen is taken out and weighed with a stripper. After weighing and breaking, the representative sample is weighed in quarters of about grams, and its mass is recorded and put into the oven with the temperature set to dry, and when the sample is weighed twice before and after, the mass of the sample no longer changes, and its water content is measured at this time. From the formula, the dry density of the sample is calculated. With the water content as the horizontal coordinate and the dry density as the vertical coordinate, the relationship curve between dry density and water content is drawn. If the experimental data fails to form a continuous and complete convex curve, additional experiments should be done at this time until the curve.

2.2 Effect of cement dose on unconfined compressive strength

The effect of cement dose on the unconfined compressive strength of the specimens was the greatest, so the effect of cement dose on the unconfined compressive strength of the specimens was carried out first. Since the iron tailings are very fine grained, the cement dose experiment was carried out with reference to the standard of cement stabilized fine-grained soil. The national standard only specifies the requirement of day unconfined compressive strength, but in order to better reflect the development pattern of specimen strength, this experiment was conducted simultaneously with day and day specimens of unconfined compressive strength experiment. The experimental results are shown in Fig. With the increase of cement dose, the unconfined compressive strength of the day and sky specimens increased significantly. However, when the cement dose is day lateral limit compressive strength is only, and does not meet the strength requirements of inorganic bond stabilization materials for highway engineering, which is used as the base material for high-speed and first-class roads. And according to engineering experience, the amount of cement to do pavement base material generally does not exceed, too high cement dose is easy to cause cracks in the pavement and thus affect the normal use of the pavement. Therefore it is not desirable to increase the cement dosage only to increase the lateral limitless compressive strength. It is necessary to explore other suitable admixtures or binder materials to improve the unconfined compressive strength. Considering that the unconfined compressive strength of the specimen is close to this strength index when the cement dose is, the cement dose is set as a comprehensive technical and economic factor.

2.3 Effect of lime admixture on unconfined compressive strength

Lime is also a common inorganic binder. However, lime alone is used to excite iron tailings specimens that disintegrate after immersion in water, and the specimens have no strength. Therefore, lime and cement were chosen simultaneously as excitation agents for stabilizing iron tailings. The dose of cement was fixed as appropriate and the dose of lime was in turn appropriate. But many cracks appeared. The unconfined compressive strengths measured after days of immersion were and could not meet the requirements of the national standard for the unconfined compressive strength of the specimens in days.The mechanical properties are shown in Table 1.

Table 1. Mechanical properties.

Numb le	Materi al 1	Materi al 2	Materi al 3	Materi al 4	Materi al 5
1	0.836	0.826	0.893	0.803	0.873
2	0.909	0.913	0.934	0.924	0.914
3	0.953	0.964	0.989	0.999	0.989

From the above experiments, it was found that the addition of lime did not have the expected results. The reason may be that the plasticity index of iron tailings is low, and the content of clay particles is small, so it is difficult for lime to play an effective role, but affects the strength of the specimen. The effect of crushed stone admixture on the unconfined compressive strength was to improve the gradation by adding crushed stone as aggregate. It can be seen from the figure that the increase of crushed stone dosing helps to improve the unconfined compressive strength, but when the crushed stone dosing exceeds, the unconfined compressive strength decreases. Therefore, the optimum amount of crushed stone is determined. Although the crushed stone is mixed with the amount of unconfined compressive strength of the day to meet the strength requirements of the highway and primary highway base, but because the mining road needs to drive tons of heavy trucks, the strength requirements as close as possible to even reach the upper limit of national regulations. Therefore, at this stage, the strength still can not meet the actual requirements, still need to further explore.

3. Determination of aggregate mix ratio

The cement dosage has a great influence on improving the lateral limit compressive strength of the specimens, and ranks second among the factors affecting the lateral limit compressive strength of the specimens. However, in practical application, it is found that although the increase of cement dosage can improve the lateral limit compressive strength of specimens, too high cement dosage will lead to cracks in inorganic aggregates after a period of time, which will affect the effectiveness and life of inorganic materials. Moreover, according to the requirements in the national experimental regulations on inorganic aggregate materials. The mechanical properties of the material after optimization are shown in Table 2.

Table 2. Mechanical properties.

Numble	Material 6	Material 7	Material 8	Material 9	Material 10
1	0.996	0.986	0.998	0.980	0.989
2	0.989	0.996	0.998	0.989	0.998
3	0.993	0.994	0.999	0.999	0.999

Therefore, although the increase of cement dose can significantly improve the lateral limitless compressive strength of the specimen, the cement dose is set as considering the other properties of the material. Crushed stone is mainly added to inorganic aggregates as an aggregate to improve the gradation of inorganic aggregates. Appropriate addition of some crushed stone aggregate can not only improve the gradation, but also increase the lateral limitless compressive strength of the specimens. However, when the amount of crushed stone is increased to, the unconfined compressive strength of the aggregate starts to decrease compared with the ratio of crushed stone. The excessive amount of crushed stone leads to a significant increase of coarse particles in the aggregate and an increase of pore space, which affects the unconfined compressive strength of the specimens. Therefore, the optimum amount of crushed stone is set as the total mass fraction of the aggregate.

In order to improve the service life of pavement materials, on the premise of meeting the compressive strength of road base materials, this chapter focuses on experimental research on the mechanical properties of pavement materials in terms of the short board a strain tensile strength, flexural tensile strength, and compressive rebound modulus to try to achieve a reasonable use of materials.

4. Split tensile strength test

The tensile strength is the ultimate strength or maximum stress when the material is subjected to axial tension. The tensile strength of a pavement subgrade material is mainly provided by the internal cohesion of the material. The tensile strength of the pavement material is related to the composition of the material, especially the interfacial bonding properties of the bonding material and the bonded material components. The incorporation of alkene fibers does not significantly increase the labor-cracking tensile strength of the aggregates. When the fiber dosage exceeded, the tensile strength of the specimens increased significantly, and the tensile strength of the specimens with the fiber dosage was higher than that of the specimens without polypropylene fiber, which was quite obvious. After that, when the amount of fiber is used, the tensile strength begins to decrease.

The measured day, day labor cracking tensile strength results and the analysis of the results are shown in the table, table. From the data in the above table, it can be seen that when the fiber dosage is, the specimen's, flexural tensile strength reaches, respectively, after the compressive strength starts to decrease with the increase of fiber dosage. Corresponding to the same ratio of specimens, no lateral compressive strength were, that is, the ratio of flexural tensile strength and no lateral compressive strength were, all meet the Sha Aimin in the "road base engineering" proposed the most suitable ratio of flexural tensile strength and no lateral compressive strength should be greater than the conditions. In addition, the ratio of flexural tensile strength to unconfined compressive strength of the specimen is greater than the corresponding ratio of flexural tensile strength to unconfined compressive strength, which indicates that with the growth of time, the growth of flexural tensile strength of the specimen is more significant than the growth of unconfined compressive strength.

5. Compressive resilient modulus test

The stress-strain relationship of a material under load is customarily reflected by the modulus of stress to strain ratio. The modulus of resilience is one of the key factors affecting the thickness of each layer in the pavement composition, and a small change in the resilience modulus of the soil base can have a large impact on the thickness of the pavement structure, especially in asphalt pavements. And pavement design indicators and other indicators of pavement performance, such as top surface compressive strain, top surface bending and sinking of the soil base and other major parameters are inseparable from the modulus of resilience, are determined by the state of the soil base, so a reasonable and accurate evaluation and selection of the modulus of resilience of the soil base is of great importance.

The loss of strength of the specimens after freeze-thaw cycles is mainly due to the following two reasons. The first is the liquid expansion pressure. Generally speaking, the base material has more pores and is a porous material. When water freezes, it has to expand in volume. When the water content in the pores exceeds a certain critical value, a large pressure will be generated due to the volume expansion effect. When it is freeze-thaw cycle action experiment, it is the internal pore water freezing and swelling effect, the additional internal stress generated will cause extrusion and damage to the pore wall of the pavement material. The second main reason is the penetration pressure. As the water in the capillary of the test piece is a solution containing several soluble salts. When the salt concentration increases force news, the freezing temperature decreases. Therefore, partial freezing of the solution in the micro-pores of the base material will result in osmotic pressure caused by water freezing.

Many domestic scholars' research concluded that the strength and stability of the roadbed are closely related to the moisture content of the roadbed and affect the strength of the pavement structure to a certain extent. Therefore, it is very important to study the water stability of iron tailings as pavement subgrade material. Cylindrical specimens made by static compression molding method for dry and wet cycles. The fixed specimen ratio is cement: gravel: iron tailings: bio-coagulase; fiber admixture were, specimens were divided into two groups after maintenance, one group for dry and wet cycle experiments, the other group as a reference, maintenance of the same time. The test specimens of the experimental group were immersed in water for hours, and then taken out of the air dry hours, so that the cycle times to measure the lateral limit compressive strength.

6. Conclusion

The addition of polypropylene fiber helps to improve the splitting tensile strength of the aggregate, and the splitting tensile strength of the specimens reaches the highest value when the fiber dosage is, which is higher than the splitting tensile strength of the specimens without the addition of polypropylene fiber.

When the fiber dosage is, the splitting tensile strength of the specimen reaches the highest value, and the compressive strength starts to decrease with the increase of fiber dosage. Corresponding to the same proportion of specimens, no lateral compressive strength, respectively.

When the fiber dosage is,, the ratio of flexural tensile strength to unconfined compressive strength of the specimen days is greater than, which meets the requirement that the most suitable ratio of flexural tensile strength to unconfined compressive strength should be greater than that derived by domestic scholars.

With the growth of time, the growth of flexural tensile strength of all specimens is more significant than the growth of unconfined compressive strength. The incorporation of polypropylene fiber significantly reduced the rebound modulus value of the specimens, and the larger the amount of polypropylene fiber, the more obvious the decrease of the rebound modulus value of the specimens. When the amount of polypropylene fiber was added, the rebound modulus of the specimens decreased compared with that of the specimens without polypropylene fiber.

References

1. Ma Q, Guo R, Zhao Z, et al. Mechanical properties of concrete at high temperature—A review[J]. Construction and Building Materials, 2015, 93: 371-383.
2. Ming N C, Siaw N L, Putra Jaya R, et al. Image Analysis and Mechanical Properties of Asphalt Mixture with Waste Plastic[C]//Key Engineering Materials. Trans Tech Publications Ltd, 2022, 912: 135-152.
3. Fonseca M, Capitão S, Almeida A, et al. Influence of Plastic Waste on the Workability and Mechanical Behaviour of Asphalt Concrete[J]. Applied Sciences, 2022, 12(4): 2146.
4. Yu X, Zadshir M, Yan J R, et al. Morphological, Thermal, and Mechanical Properties of Asphalt Binders Modified by Graphene and Carbon Nanotube[J]. Journal of Materials in Civil Engineering, 2022, 34(5): 04022047.
5. Eisa M S, Mohamady A, Basiouny M E, et al. Mechanical properties of asphalt concrete modified with carbon nanotubes (CNTs)[J]. Case Studies in Construction Materials, 2022, 16: e00930.
6. Xuan D, Zhan B, Poon C S. Assessment of mechanical properties of concrete incorporating carbonated recycled concrete aggregates[J]. Cement and Concrete Composites, 2016, 65: 67-74.
7. Novak J, Kohoutkova A. Mechanical properties of concrete composites subject to elevated temperature[J]. Fire Safety Journal, 2018, 95: 66-76.
8. Pacheco J, De Brito J, Chastre C, et al. Experimental investigation on the variability of the main mechanical properties of concrete produced with coarse recycled concrete aggregates[J]. Construction and Building Materials, 2019, 201: 110-120.
9. Bieliatynskyi A, Yang S, Pershakov V, et al. Peculiarities of the use of the cold recycling method for the restoration of asphalt concrete pavements[J]. Case Studies in Construction Materials, 2022: e00872.
10. Yan K, Li Y, Long Z, et al. Mechanical behaviors of asphalt mixtures modified with European rock bitumen and waste cooking oil[J]. Construction and Building Materials, 2022, 319: 125909.
11. Saadeh S, Al-Zubi Y, Katawal P, et al. Biochar effects on the performance of conventional and rubberized HMA[J]. Road Materials and Pavement Design, 2021: 1-17.
12. Mohammed A, Aliyu I, Sulaiman T A, et al. Mechanical Properties of Asphalt Concrete with Crumb Rubber as Partial Replacement for Coarse

Aggregate[J]. Nigerian Journal of Engineering, 2021, 28(3).

13. Abbad El Andaloussi N, Zaoui A. Enhancing Mechanical and Self-Healing Properties of Asphalt Binder Through the Incorporation of Nano-silica[J]. International Journal of Pavement Research and Technology, 2021: 1-14.
14. Yang S, Sun S, Qin L, et al. Microstructure and Meso-Mechanical Properties of Asphalt Mixture Modified by Rubber Powder under a Multi-Scale Effect[J]. Coatings, 2021, 11(11): 1321.
15. Kadhim Y N, Hussain W A M, Abdulrasool A T. The effect of animal bone ash on the mechanical properties of asphalt concrete[J]. Civ. Eng. J, 2021, 7: 1741-1752.
16. Valentin J, Trejbal J, Nežerka V, et al. A comprehensive study on adhesion between modified bituminous binders and mineral aggregates[J]. Construction and Building Materials, 2021, 305: 124686.

https://doi.org/10.1051/matecconf/202235801023

Effect of annealing temperature on the photocatalytic performance of ZnO@MoS_2 composite materials

Weijia Yang [1, 2, a, *], Lihua Liu [1], Jingjing Zhao [1], Zeyi Guan [1], Bingshang Lu [1], Fengming Wang [1], Xin He [1, 2], Yuechun Fu [3]

[1] School of Applied Physics and Materials, Wuyi University, Jiangmen 529020, China
[2] Jiangmen flexible electrode materials and their application engineering and technology research center, Wuyi University, Jiangmen, Guangdong 529020, China
[3] School of Resources, Environment and Materials, Guangxi University, Nanning 530004, China

Abstract. In this work, we report the ZnO@MoS2 composite materials prepared by a hydrothermal method, and the effect of annealing temperature on the photocatalytic performance of ZnO@MoS2 composite materials. It is found that annealing temperature plays an important role in the formation of MoS2 surface morphology, resulting in the dense films and a reduction of nanostructure nummber and active sites by enhancing temperature. Furthermore, ZnO@MoS2 composite materials annealed at 50 oC possess a higher photocatalytic degradation efficiency of 62.88% on methylene blue solution compared with ZnO materials or other samples.

Keywords: Hydrothermal method; ZnO@MoS2 composite materials; annealing temperature; photocatalytic performance.

1. Introduction

ZnO is a typical third-generation semiconductor material. ZnO has many advantages such as good photoelectric performance [1], abundant source materials, good biological and environmental compatibility, and good photocatalytic performance [2]. ZnO has been widely used in light-emitting diodes [1], photodetectors [3, 4], biosensors [5], sewage treatment [2, 6], etc. However, ZnO materials are prone to hydrolysis, which severely limits the application of ZnO. Especially in sewage treatment, ZnO needs to be soaked in sewage for a long time, and hydrolysis will drastically reduce its photocatalytic performance. Therefore, the surface of ZnO needs to be modified with a protective layer. While avoiding the hydrolysis of ZnO, it is necessary to further improve its photocatalytic performance [6]. MoS2 has stable chemical properties, does not hydrolyze, and can form a heterojunction with ZnO, which significantly improves the photocatalytic performance [6]. This work will hydrothermally grow the MoS2 nanosheet film layer on a film composed of ZnO micron/nanopillars to obtain a ZnO@MoS2 composite material. At the same time, this work also studies the effect of annealing process on the photocatalytic performance of ZnO@MoS2 composite materials.

2. Experimental details

Firstly, ZnO micro/nano pillars on ITO substrates were prepared by a hydrothermal method. The precursor solution was prepared with zinc acetate as the zinc source, hexamethyltetramine as the catalyst, polyethylene glycol as the buffer, and deionization as the solvent. The required reagents are added to deionized water, and magnetically stirred at 60 °C for 1 hour to obtain a clear and uniform precursor solution of 0.02 mol/L. The prepared precursor solution and the cleaned indium tin oxide (ITO) substrate were put into the autoclave, and the reaction solution was kept at 85 °C for 5 minutes to obtain the ZnO micron/nano column sample. And 2 minutes clean of deionized soak was used to remove residual reagents.

Then, the MoS2 shell layer was prepared on the ZnO micron/nano column sample using the hydrothermal method again. In order to avoiding the corrosion of the ZnO micron/nano column samples caused by the second hydrothermal, Mo2O3, high-purity S powder, octylamine and absolute ethanol were taken as the reaction system for the synthesis of MoS2, and the reaction time is 6 h. After cleaning and drying, the samples were annealed at 50 oC, 100 oC and 150 oC to improve the crystalline quality of MoS2 and the combination with ZnO, in order to obtain better photocatalytic performance.

* Corresponding author: yangweijia5377@126.com

After the preparation, X-ray diffractometer (XRD, X'Pert Pro MFD), micro-focus Raman spectrometer (LabRAM HR UV-NIR), and high-resolution field emission scanning electron microscope (SEM, ZEISS Sigma 500) were used to detect the prepared samples. A photochemical reaction instrument (OCRS-K) was applied to test the photocatalytic performance of the photocatalytic performance on methylene blue solution. And an ultraviolet (UV) spectrophotometer (Shimadzu UV-2550) was employed to detect the photocatalytic effect of the prepared samples.

3. Results and discussion

Fig. 1 shows the XRD patterns of ZnO@MoS2 composite materials without and with different annealing temperature. The star marks are corresponding to the peaks of ITO. The other peaks are assigned to ZnO peaks. There is no single peak belonged to MoS2. It seems that MoS2 does not have grown on ZnO. In fact, based on the PDF car of MoS2 in the XRD database, the peaks' locations of MoS2 are the same with those of ITO or ZnO. Besides, according to the reference of [6], MoS2 should be prepared on ZnO. Therefore, it is necessary to employ other techniques to identify the presence of MoS2.

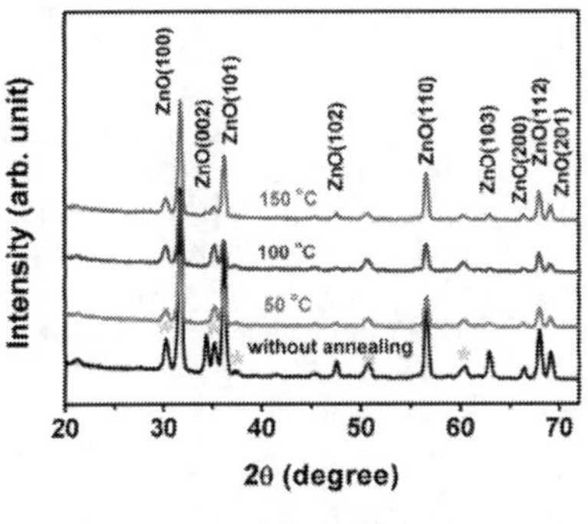

Fig.1 XRD patterns of $ZnO@MoS_2$ composite materials without and with different annealing temperature

SEM is taken to study the surface morphology of the prepared samples, as shown in Fig. 2(a) to 2(c). Before the growth of MoS_2, the ZnO surfaces are very smooth [7]. Significantly, dense nanosheets have been coated on the surfaces of ZnO materials after the preparation of MoS2. Normally, such nanosheets should be MoS_2. That is to say, ZnO@MoS2 composite materials have been successfully prepared on ITO substrates. Furthermore, as the annealing temperature is increased, the MoS_2 film consisted of nanosheets become dense, and the number of nanosheets or nanostructure is decreased. This indicates that annealing temperature plays an important role in the formation of MoS2 film surface.

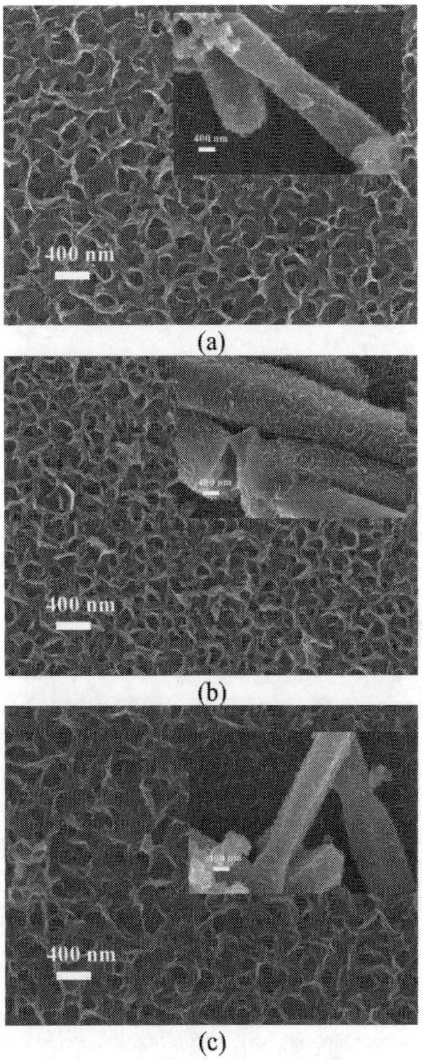

Fig.2 $ZnO@MoS_2$ composite materials with different annealing temperature: (a) 50 oC, (b) 100 oC, and (c) 150 oC

Raman spectra are carried out to confirm the MoS2, too. As shown in Fig. 3, seven sharp peaks can be found in the typical ZnO material. When it turns to the ZnO@MoS2 composite materials, most peaks of ZnO have been gone, and a wide peak is located in 328.5 cm-1, which should be composed of few peaks. It has been reported that MoS2 have two strong peaks located in about 405 cm-1 and 355 cm-1 in the Raman spectra [8]. Besides, after annealing, due to oxidation, new peaks are located in around 280 cm-1 and 335 cm-1, which is assigned to molybdenum oxide [9]. Indication, ZnO materials also have a peak in 332.5 cm-1 as shown in Fig.3. Based on what has been mentioned above, the wide peak located in 328.5 cm-1 may be mainly composed of the response of ZnO@MoS2 and a small part of molybdenum oxide. This result clearly means that MoS2 has been prepared on ZnO material. Moreover, during the annealing process, MoS2 has been slightly oxidized into molybdenum oxide.

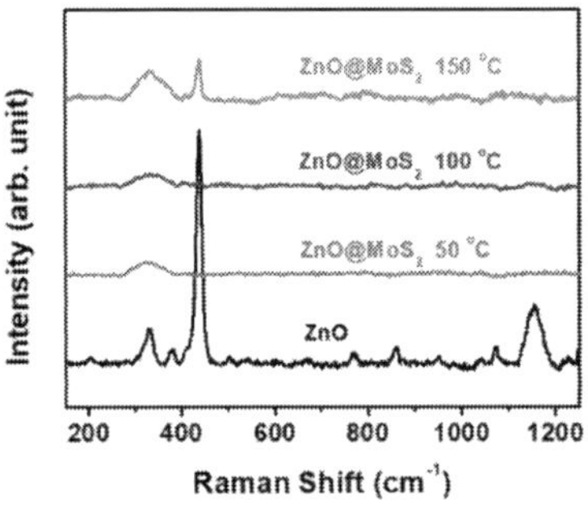

Fig.3 Raman spectra of ZnO and ZnO@MoS$_2$ composite materials

Fig. 4(a) is the absorption spectra of methylene blue solution treated with ZnO and ZnO@MoS2 composite materials. It can be seen clearly that ZnO@MoS2 composite materials has the best photocatalytic degradation efficiency. And unfortunately, further enhancing the annealing temperature leads to a decrease of photocatalytic degradation efficiency for the ZnO@MoS2 composite materials. Normally, photocatalytic degradation efficiency (E) can be calculated with the below formula [10]:

$$E = \frac{C_0 - C}{C_0} \times 100\% \quad (1)$$

Herein, C0 is the original concentration of methylene blue solution, C is the concentration of methylene blue solution after the photocatalytic degradation reaction. According to this formula and the data from Fig.4(a), photocatalytic degradation efficiency for the ZnO materials and ZnO@MoS2 composite materials annealing at 50 oC, 100 oC and 150 oC are 58.13%, 62.88%, 40.73% and 30.16% respectively (in Fig. 4(b)). Compared to the ZnO materials, ZnO@MoS2 composite materials annealing at 50 oC own more nanostures and active sites, as illustrated in Fig. 2(a). Accordingly, more active sites result in higher photocatalytic degradation efficiency. Unfortunately, in a higher annealing temperature, active sites (Uusually point defects) on the MoS2 surfaces will be repaired through oxidation. 62.88% photocatalytic degradation efficiency in this word is compared to that of MoS2/MoO3 composite materials [10].

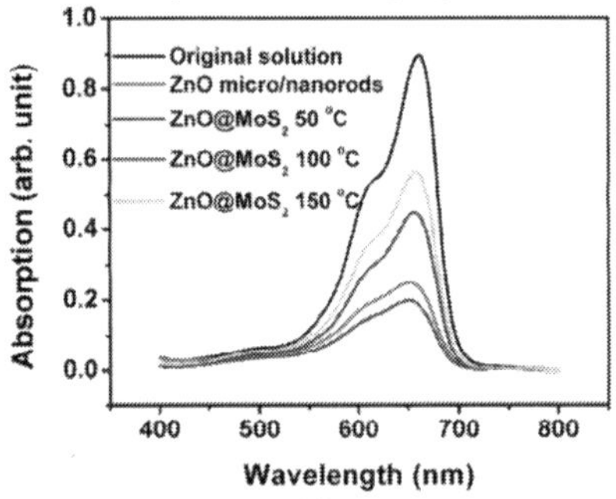

(a)

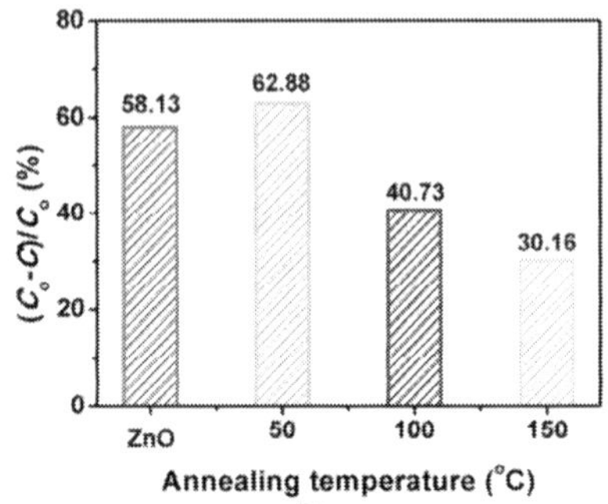

(b)

Fig.4 (a) Absorption spectra of methylene blue solution treated with ZnO and ZnO@MoS$_2$ composite materials; (b) Photocatalytic degradation efficiency of ZnO and ZnO@MoS$_2$ composite materials on methylene blue solution

4. Conclusion

ZnO@MoS2 composite materials have been prepared by a hydrothermal method. XRD, SEM, Raman spectra and UV spectrophotometer are deployed to analyze the structural, surficial, and optical properties of ZnO@MoS$_2$ composite materials with annealing. It is found that annealing temperature significantly affects the surface morphology of ZnO@MoS$_2$ composite materials, and leads to a reduction of active sites as the annealing temperature is over 50 oC. It is also found that ZnO@MoS$_2$ composite materials annealing at 50 oC reveal the best photocatalytic degradation efficiency of 62.88% on methylene blue solution. Higher annealing temperature reduces the photocatalytic degradation efficiency of ZnO@MoS$_2$ composite materials due to the reduction of active sites caused by oxidation. This work indicates that the annealing process should be improved to ensure the active sites.

Acknowledgements

This project is supported by open foundation of Guangxi Key Laboratory of Processing for Non-ferrous Metals and Featured Materials, Guangxi University (Grant No. 2020GXYSOF03); the Science Foundation for Young Teachers Projects of Wuyi University (2018td03); the University Student Innovation and Entrepreneurship Training Program Project of Guangdong Province (S202011349077); the Cooperative Education Platform of Guangdong Province ([2016]31); the Innovative Leading Talents of Jiangmen (Jiangmen(2019)7); the Key Laboratory of Optoelectronic materials and Applications in Guangdong Higher Education (2017KSYS011); the Science and Technology Projects of Jiangmen City (Jiangke [2018]359, [2018]352, 2018JC01006); the College Students Innovation and Entrepreneurship Training Project of Wuyi University (18KWL01).

References

1. S.D. Baek, P. Biswas, J.W. Kim, et al. Low-Temperature Facile Synthesis of Sb-Doped p-Type ZnO Nanodisks and Its Application in Homojunction Light-Emitting Diode, ACS applied materials & interfaces, 8 (2016) 13018-13026.
2. C. Chen, W. Mei, W. Yu, et al. Enhanced sunlight-driven photocatalytic property of Mg-doped ZnO nanocomposites with three-dimensional graphene oxide/MoS2 nanosheet composites, RSC Advances, 8 (2018) 17399-17409.
3. K. Hu, F. Teng, L. Zheng, et al. Binary response Se/ZnO p-n heterojunction UV photodetector with high on/off ratio and fast speed, Laser & Photonics Reviews, 11 (2017) 1600257.
4. Y. Ning, Z. Zhang, F. Teng, et al. Novel Transparent and Self-Powered UV Photodetector Based on Crossed ZnO Nanofiber Array Homojunction, Small, 14 (2018) e1703754.
5. J.E. Eixenberger, C.B. Anders, K. Wada, et al. Defect Engineering of ZnO Nanoparticles for Bioimaging Applications, ACS applied materials & interfaces, 11 (2019) 24933-24944.
6. S.E. Islam, D.R. Hang, C.H. Chen, et al. Facile and Cost-Efficient Synthesis of Quasi-0D/2D ZnO/MoS2 Nanocomposites for Highly Enhanced Visible-Light-Driven Photocatalytic Degradation of Organic Pollutants and Antibiotics, Chemistry, 24 (2018) 9305-9315.
7. J. Liu, M. Liu, Y. Liu, et al. Rapid Hydrothermal Growth of ZnO Nanorods on a Magnetron Sputtered Thick ZnO Seed Layer, Key Engineering Materials, 815 (2019) 9-14.
8. Y. Sun, Synthesis, characterization and performance of MoS2 nanoflowers, Master Thesis, Fujian Normal University, 2009(In Chinese).
9. C. Cai, F. Wu, Y. Fang, et al. Hydrothermal Synthesis of Flower-Like MoS2 Nanoparticle, Hans Journal of Nanotechnology 3(2013) 19-23(In Chinese).
10. R. Song, Q. Xi, X. Shi, et al. Preparation and Photocatalytic Property of MoO3@MoS2 Composite, Hans Material Sciences, 8 (2018) 535-541(In Chinese).

Self-assembly of ZnO nanodots on glass substrates by a facile sol-gel method

Weijia Yang [1, 2, a, *], Jingjing Zhao [1], Lihua Liu [1], Zeyi Guan [1], Xin He [1, 2]

[1] School of Applied Physics and Materials, Wuyi University, Jiangmen 529020, China
[2] Jiangmen flexible electrode materials and their application engineering and technology research center, Wuyi University, Jiangmen, Guangdong 529020, China

Abstract. In this work, we report the self-assembly of ZnO nanodots on the glass substrates by a sol-gel method. Firstly, the precursor solution consisted of zinc acetate and the solvent of ethylene glycol or propylene glycol is employed to fabricate precursor thin film by spin coating. Secondly, the as-grown precursor thin film is annealed at a low temperature to form the self-assembled ZnO nanodots. The effects of solvent type, precursor solution concentration and annealing time on the morphologies of the self-assembled ZnO nanodots were investigated. In addition, the self-assembly evolution of the ZnO nanodots is proposed in order to helping people to understand the formation of the nanodots by mean of sol-gel method.

Keywords: Self-assembly; ZnO nanodots; sol-gel method; precursor solution concentration; annealing time

1. Introduction

In recent years, ZnO nandots arrays have been widely utilized as high-performance electrodes for lithium-ion batteries [1], fluorescent paper sensor [2], gas sensor [3, 4], humidity sensor [5], quantum reinforcing material [6], template for synthesis of nanorods/nanowires [7] and so on [8-11], owing to their excellent physical and chemical properties [8, 9, 12, 13]. In order to achieving and applying the ZnO nanodots, many methods have been created to synthesize the ZnO nanodots[3, 14, 15]. Gruzintsev et al. reported that two-dimensional ordered array of the ZnO quantum nanodots can be realized by means of the synthetic opal with a diameter of 279 nm [16]. It has been demonstrated by Choi et al. that highly-ordered ZnO nanodot arrays can be obtained by electrosynthesis with colloidal polystyrene spheres and electrodeposited polypyrrole templates (double template) [17]. Metal-organic chemical vapor deposition (MOCVD) also can be used to selectively grow ZnO nanodots on focused-ion beam nanopatterned substrates [18]. Interestingly, by mean of the step edges on R-face sapphire, ZnO nanodot arrays can be formed by MOCVD [19]. Similarly like MOCVD, pulsed laser deposition can be deployed to prepare ZnO naondots, too [20]. Amazingly, it was reported by Kaupužs et al. that Zn and ZnO nanoparticles can be formed by laser radiation on ZnO single crystals [21]. Erol demonstrated that ZnO nanoparticles synthesized by sol-gel process reveal outstanding humidity sensing properties [5]. Similarly like sol-gel process, Riedel et al. employ spray methods to fabricate ZnO nanodots [7], which are subsequently applied for synthesizing ZnO nanorods array. Nowadays, environmental protection is getting more and more people's attention. As a result, green synthesis of ZnO nanodots is necessary. It has been demonstrated by Jacob et al. that stable, ultra-small ZnO quantum dots (nanodots) can be achieved by a facile, green, solution approach, which have radius less than 2 nm [6]. Based on the concept of green synthesis, it is superior to prepare ZnO nanodots by sol-gel method with environmentally friendly materials.

Here, we focus on the ZnO nanodots for synthesizing ZnO nanorods. Uniform size and uniform distribution ZnO nanodots are considered as the best template for growing ZnO naonrods. According previous studies, sol-gel method would be a better approach for preparing ZnO nanodots due to its mild reaction system and low growth temperature that is usually lower than 100 oC [5, 10, 12, 22]. It is believed that both solvent and annealing temperature play an important role in sol-gel method [5, 10]. However, systematic research on this aspect is still relatively rare. Furthermore, the formation evolution of ZnO nanodots is often neglected.

In this work, we report the effect of solvent type, solution concentration and annealing temperature on the morphological properties of ZnO nanodots. Furthermore, the formation evolution of the ZnO nanodots is also proposed.

* Corresponding author: yangweijia5377@126.com

2. Experimental

ZnO nanodots were prepared on glass substrates by a sol-gel method. Firstly, zinc acetate (Zn(Ac)2, purity of 99.5%) put into the solvent (ethylene glycol or propylene glycol) to prepare 0.15 mol/L mixture precursor solution by mean of magnetic stirring for 60 min at room temperature. Subsequently, 0.75 mL precursor solution was put on the middle position of the clean glass substrate, which was set in the spin coater. Afterwards, the glass substrates were rotated for 30 s at 500 rpm and then rotated for 120 s at 1500 rpm to form the precursor thin films. The as-grown precursor thin films were transferred into the annealing furnace, following an annealing process at 70 oC for 3.0 h. During the annealing process, ZnO nanodots were gradually formed.

In the second part, precursor solutions with solution concentrations of 0.01, 0.015, 0.02 and 0.025 mol/L were prepared respectively. And then ZnO nanodots were grown in the same process but annealing at 90 oC for 30 min.

In the third parts, similarly, precursor solutions with solution concentration of 0.01 mol/L were utilized to synthesize ZnO nanodots by annealing at 90 oC for 0.5, 1.0, 2.0 and 3.0 h, respectively.

The morphology of ZnO nanodots samples was measured by mean of scanning electron microscope (SEM, ZEISS Sigma 500).

3. Results and discussion

Fig. 1(a)-(d) represent the SEM images of ZnO nanodots self-aggregating on glass substrates, and the solvent is varied, and the annealing treatment of 70 °C for 3 h remains unchanged. As expected, when ethylene glycol is chose as the solvent, ZnO nanodots are self-aggregated on the glass substrates, which are confirmed by Fig. 1(a). ZnO nanodots, the white points in Fig. 1(a), can be seen clearly. Further study by the high magnification SEM image shows that ZnO nanodots with a half-spheric morphology are almost fully covered on the glass substrate, the size of ZnO nanodots is not uniform and most of them land in the range from 7.7 to 18.5 nm, as shown in Fig. 1(b). Furthermore, there is a significant height difference between the large size ZnO nanodots and the small size ones. On the contrary, when it turns to the solvent of propylene glycol, uniform sized ZnO nanodots can be observed in Fig. 1(c). High magnification SEM image further confirms that ZnO nanodots are more uniform distribution in the size with a diameter of 12.8-21.5 nm (Fig. 1(d)), and no significant height difference can be found in Fig. 1(d). Distinctly, there is a large or small space between the ZnO nanodots. The different result caused by the solvent type may be due to the difference in the volatility and molecular weight between thylene glycol and propylene glycol. Compared with ethylene glycol, propylene glycol has larger viscosity and molecular weight value, which makes it suitable for creating space between ZnO nanodots, possibly resulting in uniform size of the ZnO nanodots. Therefore, the ZnO nanodots made from the solvent of propylene glycol are better match with the demands for the subsequent growth of ZnO nanorods/nanowires[7].

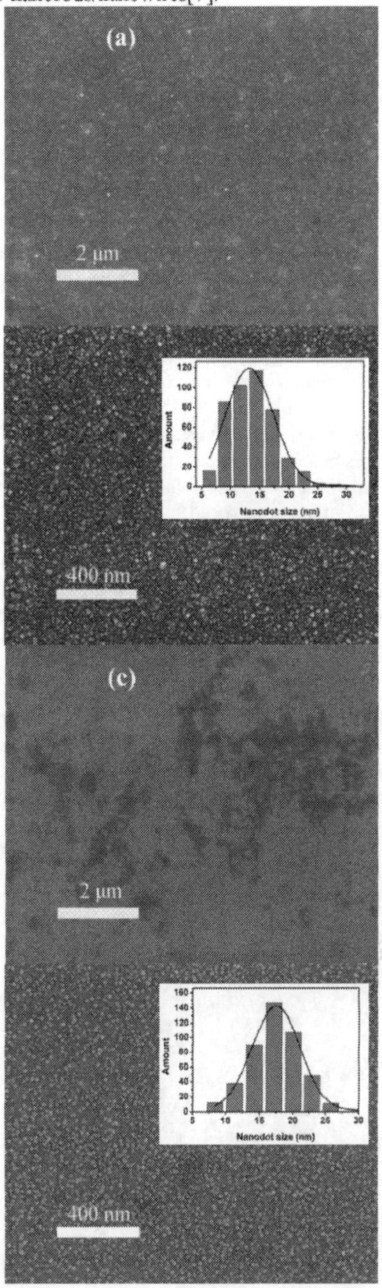

Fig. 1. Self-assembly of ZnO nanodots on glass substrates with different solvent: (a) ethylene glycol and (c) propylene glycol. (b) and (d) are their corresponding high magnification SEM images, respectively. The inserts are their corresponding size distribution maps, respectively.

In this part, we focus on the effect of the precursor solution concentrations on the morphology of ZnO nanodots. The Zn(AC)2 in the propylene glycol are of 0.01, 0.015, 0.02 and 0.025 mol/L respectively, and the formation process is described as above, but the annealing time is reduced to 0.5 h. Fig. 2(a) displays the SEM image of ZnO nanodots that made from the 0.01 mol/L precursor solution. The coated precursor film initiates its self-aggregation to form a long strip during the annealing process. Due to uneven film thickness precursor', quite a lot of large size nanosheets also can be seen in Fig. 2(a). In contrast, with the increase of the precursor solution concentration, it is found that the morphology of the self-assembled ZnO nanodots on the glass substrates is significantly different. By increasing the precursor solution concentration from 0.01 to 0.015 mol/L, the coherent aggregation of ZnO nanodots with flat tops can be observed in Fig. 2(b). When further increasing the precursor solution concentration to 0.02 mol/L, the size of ZnO nanodots with a half-spheric morphology becomes apparently smaller than those made from the small precursor solution concentration, whereas the density of ZnO nanodots greatly increase, as shown in Fig. 2(c). If the precursor solution concentration is increased to the value of 0.025 mol/L, the glass substrate is covered by ZnO thin film (nanofilm) and some ZnO nanodots (white points) are found on the ZnO thin film' surface (Fig. 2(d)). Based on above results, it can be concluded that the suitable precursor solution concentration range for self-assembling ZnO nanodots with the solvent of propylene glycol is from 0.01 to 0.02 mol/L, which is mainly due to the increased number of crystallized core during annealing process. It is easily believed that the number of crystallized core is increased with the precursor solution concentration. Once it is over than a certain value, taking the case of 0.025 mol/L precursor solution as an example, the ZnO nanodots will become into ZnO nanofilm. In addition, compared with the annealing time of 3 h, it is also found that shortening annealing time can induce ZnO nanodots too. Therefore, it is necessary to optimize the annealing time for self-assembling ZnO nanodots with a better morphology. This will be studied in the next part.

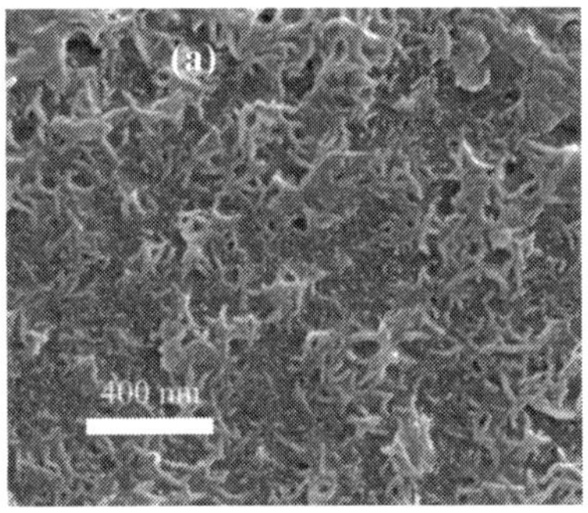

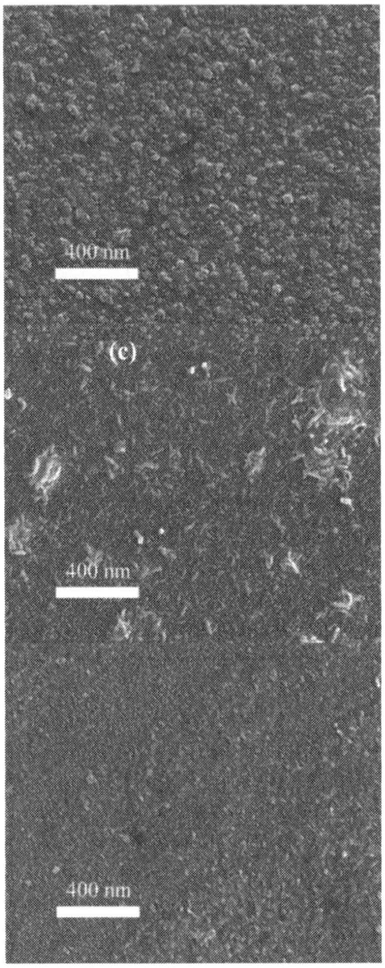

Fig. 2 Self-assembly of ZnO nanodots on glass substrates with different propylene glycol solvent concentration: (a) 0.01 mol/L, (b) 0.015 mol/L, (c) 0.02 mol/L and (d) 0.025 mol/L

In fact, the size and distribution of the self-aggregated ZnO nanodots on the glass substrates are mainly depended on the precursor solution concentration, the annealing process, and the thickness of the precursor thin film. It has been demonstrated that annealing temperature plays a critical role in the size and uniformity of the self-aggregated nanodots. In details, higher annealing temperature is able to result in a reduced uniformity in the size distribution of the nanodots. On the contrary, lower annealing temperature certainly takes a longer time for self-assembling nanodots. However, lower annealing temperature means more safety and higher controllability in the morphology of nondots. For the self-aggregated ZnO nanodots, 90 oC is considered as a superior value of annealing temperature based on our previous experiments. By comparing Fig. 1(d) and Fig. 2(b), it can be concluded that the morphology of the self-aggregated ZnO nanodots is significantly affected by the annealing time. In order to finding the suitable annealing time and studying formation evolution for self-assembling ZnO nanodots, annealing time values of 0.5, 1, 2 and 3 h are set to synthesize ZnO nanodots with a precursor solution concentration of 0.01 mol/L, as shown in Fig. 3(a)-(d). When the annealing time is shorter than 0.5 h, precursor thin film is still retained and no ZnO nanodots can be found. By increasing the annealing time to 0.5 h, the numbers of the self-assembled ZnO nanodots are apparently enhanced, and the nanostrips are still retained as shown in Fig. 4(a). It should be noted that these nanostrips are made of the nanodots, which means that the nanostrips would eventually break into the self-assembled ZnO by increasing the annealing time. The nanostrips disappear completely and only self-assembled ZnO nanodots have been obtained on the glass substrates when using an annealing time of 1 h (Fig. 2(b)). The average size and density of the self-assembled ZnO nanodots are 15 nm and about 6.6×1010 cm-2, respectively, as illustrated in Fig. 3(b). These small ZnO nanodots would gradually merge with large one, which lead to a reduction of density accordingly. For 2 h annealing, a few self-assembled ZnO nanodots become into large square or triangular prism nanodots from the half-spheric nanodots while others change into smaller ones (Fig. 3(c)). The largest square ZnO nanodot has a size of ~141 nm. In addition, the density of the self-assembled ZnO nanodots is decreased obviously. If the annealing time is further lengthened to 3 h, the size of the large sized ZnO nanodots will keep enlarging, while small nanodot will get smaller and smaller until it disappears completely, and accordingly, the density of the self-assembled ZnO nandots will keep reducing, as shown in Fig. 3(d). The largest sized of ZnO nanodot is further increased to 306 nm that is about 2 time of the value annealed for 2 h.

In this work, our aim is to synthesize ZnO nanodots for growing ZnO nanorods, especially core-shell ZnO-based nanorods. Therefore, annealing time of 1 h would be an appropriate value for synthesizing ZnO nanodots that matches our demands. Interestingly, after 1 h annealing, if the annealing time is further lengthened, a few ZnO nanodots become much larger while other become smaller than those annealed for 1 h.

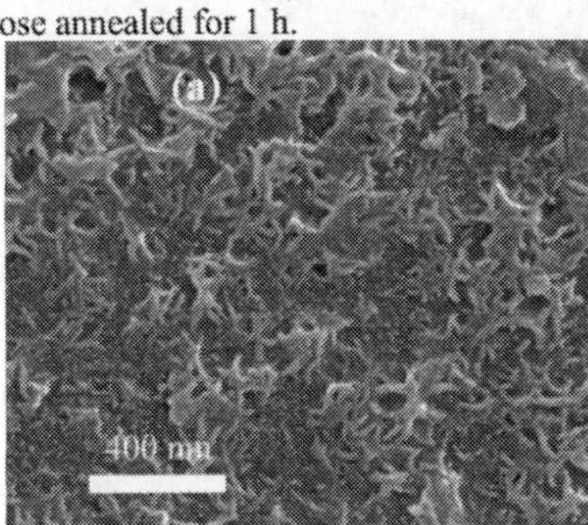

Fig. 3 SEM images of the self-aggregated ZnO nanodots on glass substrates with an annealing time of (a) 0.5, (b) 1, (c) 2 and (d) 3 h, respectively, by madding from 0.01 mol/L precursor solution concentration.

From what has been mentioned above, the self-assembly evolution of ZnO nanodots can be summarized as shown in Fig. 4. Firstly, the precursor solution is made into precursor thin film by mean of spin coating. Secondly, in the initial state of annealing at 90 oC, zinc acetate may hydrolyze accompanying the evaporation of alcohol. Taking ethanol solvent as an example, this reaction would be summarized as equation (1). Indeed, the Zn(OH) will be broke into ZnO and H2O, which could be displayed as equation (2).

$$Zn(AC)_2 + (CH_2OH)_2 \rightarrow Zn(OH)\downarrow + CH_2(AC)\text{-}CH_2(OH)\uparrow \quad (1)$$

$$Zn(OH)_2 \rightarrow ZnO\ (s) + H_2O\ (g) \quad (2)$$

When the equation (1) mainly occurs, it is possible to observe the nanostrips on the glass substrates, as shown in Fig. 2(a) and Fig. 3. Thirdly, as the annealing time is lengthened, the equation (2) turns up and results in the formation of self-assembled ZnO nanodots, and gradually it will be able to become the main reaction that produce more and more ZnO nanodots, as shown in Fig. 1(c), Fig. 2(b) and Fig. 3. However, after the self-assembled ZnO nanodots have been achieved completely, the size of some ZnO nanodots enlarges and others become smaller if the annealing time keeps lengthening, as show in Fig. 2(c) and (d).

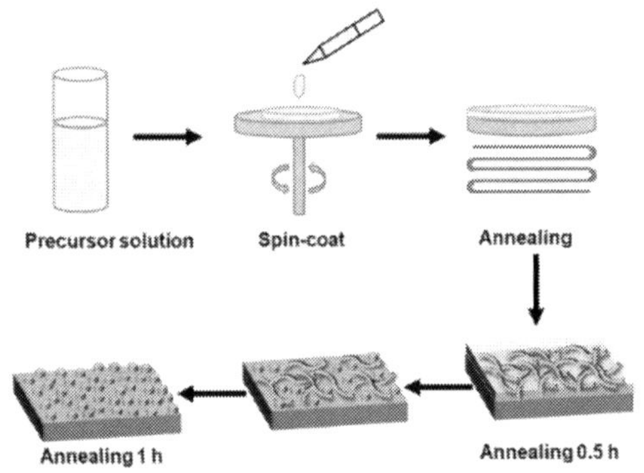

Fig. 4 Schematic diagram for the self-assembly formation of ZnO nanodots on glass substrates

4. Conclusion

By mean of sol-gel method, ZnO nanodots have been self-assembled on the glass substrates at a low annealing temperature. It is found that propylene glycol solvent is beneficial to forming uniform sized self-assembled ZnO nanodots with distinct space between nanodots when compared with ethanol solvent. The precursor solution concentration plays an important role in the formation of the assembled ZnO nanodots. 0.01-0.015 mol/L would be a superior value for self-aggregating ZnO nanodots that are good templates for synthesizing ZnO nanorods in a short annealing time. Furthermore, annealing time has demonstrated to be a key factor that affects the morphology, size and uniformity of the self-assembled ZnO nanodots. To match the aim for synthesizing core-shell ZnO-based nanorods, 1 h may be considered as a suitable annealing time value. In addition, the self-assembly evolution of the ZnO nanodots is proposed.

Acknowledgements

This project is supported by open foundation of Guangxi Key Laboratory of Processing for Non-ferrous Metals and Featured Materials, Guangxi University (Grant No. 2020GXYSOF03); the Science Foundation for Young Teachers Projects of Wuyi University (2018td03); the University Student Innovation and Entrepreneurship Training Program Project of Guangdong Province (S202011349077); the Cooperative Education Platform of Guangdong Province ([2016]31); the Innovative Leading Talents of Jiangmen (Jiangmen(2019)7); the Key Laboratory of Optoelectronic materials and Applications in Guangdong Higher Education (2017KSYS011); the Science and Technology Projects of Jiangmen City (Jiangke [2018]359, [2018]352, 2018JC01006); the College Students Innovation and Entrepreneurship Training Project of Wuyi University (18KWL01).

References

1. T. Yang, Y. Liu, Z. Huang, J. Liu, P. Bian, C.D. Ling, H. Liu, G. Wang, R. Zheng, In situ growth of ZnO nanodots on carbon hierarchical hollow spheres as high-performance electrodes for lithium-ion batteries, Journal of Alloys and Compounds, 735 (2018) 1079-1087.
2. X. Liu, Y. Yang, X. Xing, Y. Wang, Grey level replaces fluorescent intensity: Fluorescent paper sensor based on ZnO nanoparticles for quantitative detection of Cu2+ without photoluminescence spectrometer, Sensors and Actuators B: Chemical, 255 (2018) 2356-2366.
3. M.W. Maswanganye, K.E. Rammutla, T.E. Mosuang, B.W. Mwakikunga, The effect of Co and In combinational or individual doping on the structural, optical and selective sensing properties of ZnO nanoparticles, Sensors and Actuators B: Chemical, 247 (2017) 228-237.
4. T. Zhao, L. Zhang, L. Li, G. Zhang, Kankan, K. Shi, Synthesis, characterization and sensing properties of ZnO-modified BN–FeB49, Journal of Alloys and Compounds, 600 (2014) 130-136.
5. A. Erol, S. Okur, B. Comba, Ö. Mermer, M.Ç. Arıkan, Humidity sensing properties of ZnO nanoparticles synthesized by sol–gel process, Sensors and Actuators B: Chemical, 145 (2010) 174-180.
6. N. Mary Jacob, T. Thomas, Digestive ripening and green synthesis of ultra-small (r<2nm) stable ZnO quantum dots, Ceramics International, 40 (2014) 13945-13952.
7. W. Riedel, Y. Fu, Ü. Aksünger, J. Kavalakkatt, C.-H. Fischer, M.C. Lux-Steiner, S. Gledhill, ZnO and ZnS nanodots deposited by spray methods: A versatile tool for nucleation control in electrochemical ZnO nanorod array growth, Thin Solid Films, 589 (2015) 327-330.
8. Y. Yin, Y. Sun, M. Yu, X. Liu, B. Yang, D. Liu, S. Liu, W. Cao, M.N.R. Ashfold, Arrays of nanorods composed of ZnO nanodots exhibiting enhanced UV emission and stability, Nanoscale, 6 (2014) 10746-10751.
9. C. Zhang, F. Zhu, H. Xu, W. Liu, L. Yang, Z. Wang, J. Ma, Z. Kang, Y. Liu, Significant improvement of near-UV electroluminescence from ZnO quantum dot LEDs via coupling with carbon nanodot surface plasmons, Nanoscale, 9 (2017) 14592-14601.
10. Y. Xia, Z. Tang, Monodisperse inorganic supraparticles: formation mechanism, properties and applications, Chemical Communications, 48 (2012) 6320.
11. Q. Liu, Z. Sun, Y. Dou, J.H. Kim, S.X. Dou, Two-step self-assembly of hierarchically-ordered nanostructures, Journal of Materials Chemistry A, 3 (2015) 11688-11699.
12. X. Chen, X. Song, W. Qiao, X. Zhang, Y. Sun, X. Xu, W. Zhong, Y. Du, Solvent-directed and anion-

modulated self-assemblies of nanoparticles: a case of ZnO, CrystEngComm, 18 (2016) 9139-9151.

13. M. Ali, H. Rahaman, D.S. Rahman, S. Nath, S.K. Ghosh, Water/n-heptane interface as a viable platform for the self-assembly of ZnO nanospheres to nanorods, CrystEngComm, 16 (2014) 7696-7700.
14. A.N. Mallika, A. Ramachandra Reddy, K. Sowri Babu, C. Sujatha, K. Venugopal Reddy, Structural and photoluminescence properties of Mg substituted ZnO nanoparticles, Optical Materials, 36 (2014) 879-884.
15. H. Zhong, J. Wang, M. Pan, S. Wang, Z. Li, W. Xu, X. Chen, W. Lu, Preparation and photoluminescence of ZnO nanorods, Materials Chemistry and Physics, 97 (2006) 390-393.
16. A.N. Gruzintsev, V.T. Volkov, G.A. Emelchenko, I.A. Karpov, W.M. Maslov, G.M. Michailov, E.E. Yakimov, Luminescence of two-dimensional ordered array of the ZnO quantum nanodots, obtained by means of the synthetic opal, Thin Solid Films, 459 (2004) 111-114.
17. B.B. Choi, N. Myung, K. Rajeshwar, Double template electrosynthesis of ZnO nanodot array, Electrochemistry Communications, 9 (2007) 1592-1595.
18. S.-W. Kim, T. Kotani, M. Ueda, S. Fujita, S. Fujita, Selective growth of ZnO nanodots prepared by metalorganic chemical vapor deposition on focused-ion beam-nanopatterned substrates, Physica E: Low-dimensional Systems and Nanostructures, 21 (2004) 601-605.
19. K. Kametani, H. Imamoto, S. Fujita, Formation of ZnO nanodot arrays along the step edges on R-face sapphire by metalorganic chemical vapor deposition, Physica E: Low-dimensional Systems and Nanostructures, 32 (2006) 33-36.
20. R. O'Haire, A. Meaney, E. McGlynn, M.O. Henry, J.R. Duclère, J.P. Mosnier, Growth of crystalline ZnO nanostructures using pulsed laser deposition, Superlattices and Microstructures, 39 (2006) 153-161.
21. J. Kaupužs, A. Medvids, P. Onufrijevs, H. Mimura, Origin of n-type conductivity in ZnO crystal and formation of Zn and ZnO nanoparticles by laser radiation, Optics & Laser Technology, 111 (2019) 121-128.
22. A.u.H.S. Rana, J.Y. Lee, Y.-P. Hong, H.-S. Kim, Transient Current Response for ZnO Nanorod-Based Doubly Transparent UV Sensor Fabricated on Flexible Substrate, physica status solidi (RRL) - Rapid Research Letters, 12 (2018) 1800001.

https://doi.org/10.1051/matecconf/202235801025

Preparation and characterization of a novel emphatically charged strengthened chitosan composite nanofiltration membrane

Tao Mu [1,*], Qiang Wang [2], Liang Cao [3], Xiaohang Zhu [1], Zhiru Zhao [1]

[1] School of Materials Science and Engineering, Yingkou Institute of Technology, Yingkou, China.
[2] Avic Sino (Yingkou) high-tech Co., Ltd., Yingkou, China.
[3] Foreign Language Teaching and Research Department, Yingkou Institute of Technology, Yingkou, China.

Abstract. In this paper, chitosan was grafted with acrylic acid and positively charged elements, and a series of reinforced nanofiltration membranes were prepared by coating modified chitosan polymer with polysulfone ultrafiltration membrane as support. The structure of chitosan derivatives and polymers was characterized by infrared spectrum, and the membrane structure was characterized by scanning electron microscope and atomic force microscope. The performance of composite nanofiltration membrane was closely related to the structure of polymer and the electrical properties of positively charged elements. The rejection rate of 50% molar ratio composite nanofiltration membrane to CaCl2 reached 97.4%. The corresponding flux is 430.7 Lm-2h-1. The retention order was: CaCl2 > NaCl > Na2SO4. The tensile strength increased by 27.3%. Acid resistance (5%HCl)) was improved by 50.5%. Alkali resistance (5%NaOH) increased by 42.2%. The results showed that the positive charge enhanced chitosan composite nanofiltration membrane had excellent performance. The charge effect had no effect on NaCl and Na2SO4, but had significant effect on CaCl2. The membrane is a typical positive charged nanofiltration membrane suitable for separating high-valent cations from weak acids or weak bases.

1. Introduction

Nowadays, Nanofiltration (NF) membrane, as a new type of separation membrane between reverse osmosis (RO) membrane and ultrafiltration (UF) membrane, has been developed actively [1]. Thanks to its special features like high seperation efficiency and low energy consumption required in operation, NF has been connected solely or in combination with other division handle in numerous areas such as water softening treatment of squander water, oil industry, nourishment handling and so on[2,3].

In this paper, a composite NF membrane, based on multi-layer composite structure, was designed, with polysulfone UF membrane serving as the foundation to offer mechanical strength and the positively charged polymer of adjusted chitosan as the best layer to supply filtration work. In arrange to upgrade mechanical quality and electrical property of NF layer, crylic corrosive and emphatically charged compound were joined to chitosan through hydroxylation. At that point the polymer was gotten by free radical polymerization of adjusted chitosan. A arrangement of NF layer were arranged with this plan. The application prospect of the film will be overseas.

2. Experiment

2.1 Materials and approaches

A self-made device was utilized within the layer execution test. The conductivity of the arrangement was measured by DDS-307 conductivity meter (Shanghai Leici Instrument Production line). The chemical composition of the altered chitosan and polymer was measured by Range One infrared spectrometer (PerkinElmer). The morphology of NF film was watched by S4800 filtering electron microscopy (HITACHI) and Measurement Symbol transmission electron microscopy (Bruker company). LDS- 20KN ductile testing machine (Huikai Testing Instrument Co., Ltd.) was utilized for film malleable quality test.

Chitosan [Mw≥20000 Da, degree of deacetylation (DD≥90%)]; N-methyl pyrrolidone, azodiisobutyronitrile (AIBN), methanesulfonic acid, acrylic acid, $SOCl_2$ and polyvinylpyrrolidone, gained from Sinopharm Chemical Reagent Co. Ltd, were all expository review acquired from Sinopharm Chemical Reagent Co. Ltd; acidic corrosive, acetone, glutaraldehyde and polyvinyl liquor were gotten from Shenyang Xinxi reagent Manufacturing plant; polysulfone was obtained from Shanghai Shuguang chemical plastics mechanical enterprise.

* Corresponding author: mutao@yklg.edu.cn

2.2 Preparation of modified chitosan

The structure of the positively charged compound (M_1) was shown in Figure 1.

Chitosan was modified by acrylic acid and positively charged compound in different ratio. The schemes was shown in Figure 2.

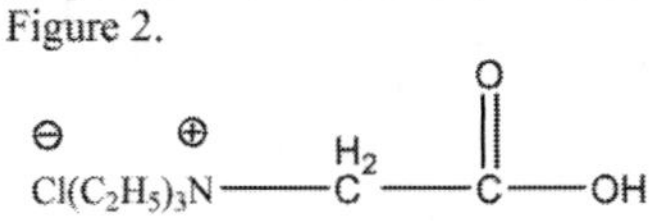

Figure 1. The structure of positively charged compound

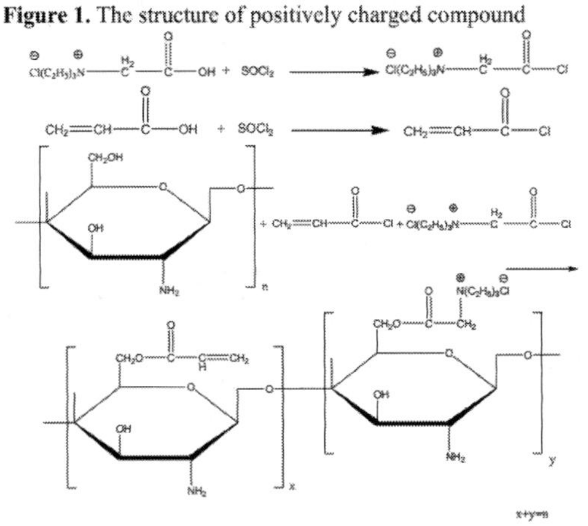

Figure 2. Scheme of modified chitosan synthesis Common forms to adjust chitosan preparation.

(1) Responding with $SOCl_2$ at 40°C for 3 hours, acrylic corrosive and emphatically charged compound chlorides were arranged. The items were decontaminated through distillation.

(2) Picked up from the final step, the acyl chlorides were broken down in chloroform, and is included to the chitosan methane sulfonic corrosive arrangement dropwise for more than 3.5 hours. Table 1 appears the proportion between emphatically charged compounds and acyl chloride. After the response, cool down the response blend to 4°C for 10 hours some time recently aceton precipitation treatment. Filt the items twice and dry beneath vaccuum condition.

Table 1. Esterification reaction feeding

Chitosan derivatives	n_{cts}/mol	$n_{丙烯酰氯}$ /mol	m_{M1}/ mol	A/%
$D_{1\text{-}0}$	0.005	0.005	0	0
$D_{1\text{-}1}$	0.005	0.0049	0.0001	2
$D_{1\text{-}2}$	0.005	0.00475	0.00025	5
$D_{1\text{-}3}$	0.005	0.0045	0.0005	10
$D_{1\text{-}4}$	0.005	0.004	0.001	20
$D_{1\text{-}5}$	0.005	0.0025	0.0025	50
$D_{1\text{-}6}$	0.005	0.0015	0.0035	70
$D_{1\text{-}7}$	0.005	0.0005	0.0045	90
$D_{1\text{-}8}$	0.005	0	0.005	100

Note:A: molar ratio of positively charged compound and acyl chlorides.

2.3 Preparation of polymer

The general process of polymer preparation is shown in Figure 3.

The adjusted chitosan was broken down in dry tetrahydrofuran and AIBN was included as initiator. The polymerization was carried out beneath the assurance of nitrogen at 60 °C for 24 hours. The response blend was accelerated with methanol some time recently filtration. The item was washed once more with hot methanol and dried beneath vacuum to get polymer.

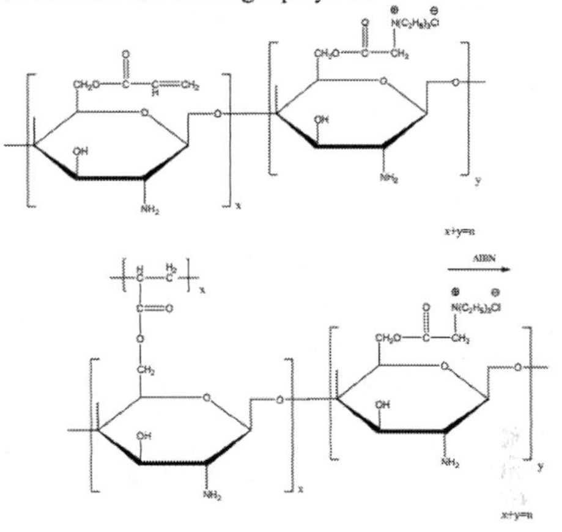

Figure 3. Scheme of polymer synthesis

2.4 Preparation of UF membranes

The polysulfone UF film was arranged by stage reversal, and the process was indicated as below[4]:

(1) 4.2g polysulfone was broken down in 25.7g N-methylpyrrolidone. 0.12g acetone and 0.075g polyvinylpyrrolidone (PVP) were included into the arrangement to create a casting arrangement, and the insoluble pollutions were evacuated by G2 sand channel. The arrangement was at that point permitted to stand for 10 hours.

(2) The casting arrangement was connected to a chunk of cloth (80 gap) dressing on the glass. To begin with, the dissolvable within the film was partially evaporated at room temperature for one diminutive, and after that the film was exchanged to a water shower for forming.

2.5 Preparation of composite NF membranes[5]

The polymer was broken down in 2.5ml 6% acidic corrosive arrangement with 0.06% polyvinyl liquor (porogen). The casting arrangement was gotten by deoxidizing the over solution.

The polysulfone ultrafiltration layer was settled on the glass. The pouring fluid was at that point coated on the ultrafiltration layer. The recently shaped layer was vaporized at room temperature for 60 seconds and after that crosslinked with 1% glutaraldehyde. The composite films were arranged after 16 hours at room temperature.

2.6 Permeation experiment

Flux and dismissal were calculated based on the conditions 1 and 2.

$$F = V / A t \quad (1)$$

Where: F is the flux; V is the volume of the penetrating liquid passing through the film; A is the compelling range of layer ($0.93cm^2$); t is the time for saturation.

$$R = (1-C_p/C_0) \times 100\% \quad (2)$$

Where: R is the dismissal; C_p and C_0 are the concentrations of the penetrated liquid and bolster individually.

The concentration was supplanted by the conductivity of salt arrangements since the 1000mg/L arrangements were exceptionally weaken arrangement in this research.

3. Results and discussions

3.1 Infrared analysis

3.1.1 Infrared examination of altered chitosan.

The assimilation band at 1750cm-1 was related to the extending vibration of ester carbonyl; the assimilation band at 1200cm-1 was related to the extending vibration of ether linkage; the assimilation band at 1700cm-1-1600cm-1 was related to the extending vibration of carbon-carbon twofold bond; and the assimilation band at 1039cm-1 was related to the extending vibration of C-N in quaternary ammonium gather.

3.1.2 Infrared analysis of polymer.

The comes about appeared comparing with the adjusted chitosan, twofold bond assimilation top vanished at 1600cm-1-1700cm-1. The polymerization of altered chitosan was fruitful.

3.2 Impact of molar proportion on the dismissal and flux of composite NF membrane

The test of composite nanofiltration layer was conducted after a pre-pressure at 0.4 Mpa for 0.5 hour. The connections between molar proportion and film execution were appeared in Figure 4 and Figure 5.The dismissal for $CaCl_2$ expanded as the expanding of molar proportion till 50%, which was from 96.2% to 97.4%. When molar proportion was higher than 50%, the dismissal dropped. Whereas the dismissal of NaCl and Na_2SO_4 diminished. The dismissal of Na_2SO_4 was littler than NaCl. The flux expanded all the time as appeared in Figure 5. The information shown the structure of polymer, the charged characteristic of M_1 and right rate was vital for the composite film execution. The acrylic corrosive made the polymer thickness expanded, which driven to thick structure in surface of NF film. So the dismissal of composite layer was tall for salt arrangement. The work of M_1 was Donnon exclusion [6].Since the dynamic layer had quaternary ammonium bunch dissemination and permitted a more grounded shock of Ca^{2+} than Na^+, the arrange of dismissal: $CaCl_2 > NaCl > Na_2SO_4$. The reason of $NaCl > Na_2SO_4$ was that the fascination was more grounded for SO_4^{2-} than Cl^-. The drift of $CaCl_2$ can be clarified as takes after: the electrostatic impact fortified with the increment of molar proportion, which driven to the expanding of the dismissal. The dismissal started to diminish when molar proportion surpassed 50%, which was the result that the polymer consistency diminished. It was ordinary emphatically charged nanofiltration layer. The nanofiltration membrane of 50% molar proportion was the most execellent.

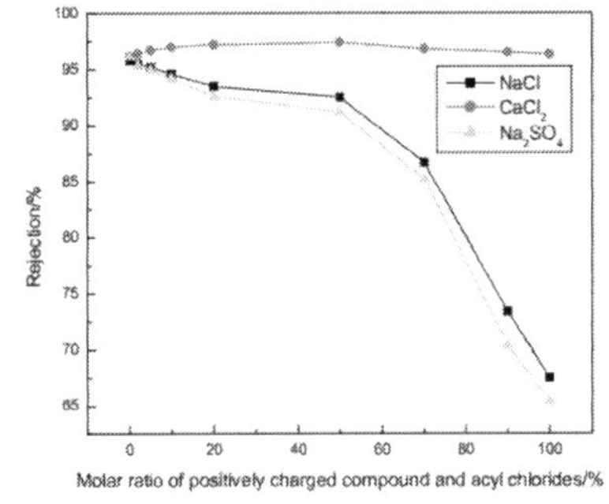

Figure 4. Effect of molar ratio on the rejection of the composite membrane

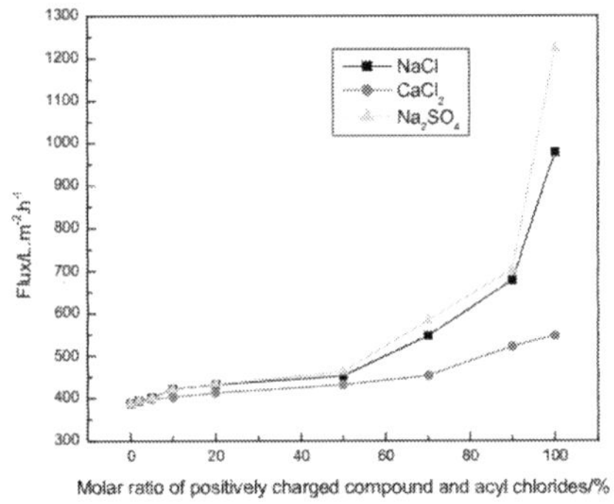

Figure 5. Effect of molar ratio on the flux of the composite membrane

3.3 Investigation of mechanical quality

The tensile strength test was carried out on the film with a width of 10mm. The tensile strength decreased from 9.2N to 6.6N with increasing molar ratio. The reason was that the viscosity of the polymer decreased with the increase of molar ratio. The tensile strength of 50% molar nanofiltration membrane was 27.3% higher than that of 100% molar nanofiltration membrane.

3.4 Analysis of acid resistance

The acid resistance test was carried out with hydrochloric acid (5%) at 0.4 MPa. When the molar ratio increased from 36.8 to 20.2 h, the acid resistance decreased with the increase of the molar ratio. The reason was that the viscosity of the polymer decreased with the increase of the molar ratio, resulting in the decrease of the polymer density. Acid resistance of 50% molar ratio nanofiltration membrane increased 50.5% comparing with 100% molar ratio nanofiltration membrane.

Analysis of alkali resistance

Under the condition of 0.4 MPa, the alkali resistance test was carried out with 5% sodium hydroxide (NaOH). From 31.8 h to 18.7 h, the alkali resistance decreased with the increase of the molar ratio. The reason was consistent with acid resistance analysis. The alkali resistance of 50% mole ratio nanofiltration membrane was 42.2% higher than that of 100% mole ratio nanofiltration membrane.

3.5 Structure characteristic of composite membrane

The 50% molar proportion cross area and surface were characterized by S4800 filtering electron magnifying instrument, as appeared in Fig. 6. The surface of the composite layer was amplified by 1000 times, as appeared in Fig. 6 (a), and (b) the cross segment of the layer was amplified by 600 times. The surface of the composite layer was thick and there were a few arched zones, demonstrating that the surface of the composite layer was not totally smooth, but there were fine gel particles. They shaped the dynamic layer of the composite film. The cross area of the composite film comprised of two layers. The upper layer was a thick crosslinked layer. The lower portion was a light polysulfone back layer. The combination of the best thick useful layer and the free supporting layer kept the fabulous execution of the composite layer.

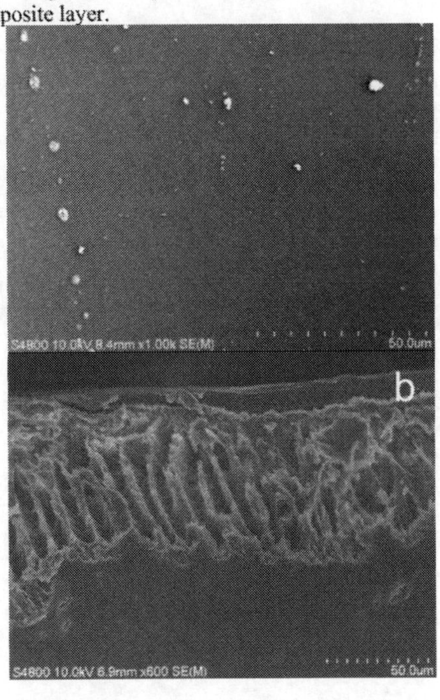

Figure 6. The surface and cross-section images of the composite membrane

4. Conclusion

The composite nanofiftration film detailed in this paper had two-layer structure: the upper one was the lean and thick cross-linked layer, which played a key part in division; the lower layer was the polysulfone bolster layer, with a sponge-like permeable surface. The upper layer was arranged with polymers of acrylic corrosive and emphatically charged compound adjusted chitosan. The dismissal rate of 50% composite nanofiltration layer for $CaCl_2$ was 97.4% for $CaCl_2$ with 50% molar proportion composite nanofiltration film, comparing flux was 430.7 $Lm^{-2}h^{-1}$. the arrange of dismissal: $CaCl_2$ > NaCl > Na_2SO_4.Tensile quality of film expanded 27.3%. The HCl(5%) resistance expanded 50.5%. The NaOH(5%) resistance expanded 42.2%. This amazing execution was the comes about of the layer structure and electric property adjustment through the presentation of acrylic corrosive and emphatically charged compound. The coming about composite NF film was appropriate for isolating the multivalent cation solutes from pitifully acidic arrangement or pitifully soluble arrangement. The film had a wide application prospect.

Acknowledgments

Financial support from Natural Science Foundation of Liaoning Province General Program (Grant No: 20180550492), The Excellent Talents Support Program of Yingkou Institute of Technology (ETSTYKIOT) and Doctoral enterprise project of China Yingkou are gratefully acknowledged.

References

1. J. Chen, W. X. Shi, L.H. Zhang and R.J. Zhang, A novel polyester composite nanofiltration membrane formed by interfacia polymerization of pentaerythritol (PE) and trimesoyl chloride (TMC), APPL SURF SCI. 416 (2017) 152-159.
2. Y. L. Chen, F. Liu, Y. Wang, H. B. Lin and L. Han, A tight nanofiltration membrane with multi-charged nanofilms for high rejection to concentrated saltsJ. Membr. Sci. 537 (2017) 407-415.
3. Y. Gao, A.M.S. de Jubera, B.J. Mariñas and J.S. Moore, Nanofiltration membranes with modified active layer using aromatic polyamide dendrimers, Adv. Funct. Mater. 23 (2013) 598-564.
4. M.B. Wu, Y. Lv, H.C. Yang, L.F. Liu, X. Zhang and Z.K. Xu, Thin film composite membranes combining carbon nanotube intermediate layer and microfiltration support for high nanofiltration performances, J. Membr. Sci. 515 (2016) 238-244.
5. Y. Lv, H.C. Yang, H.Q. Liang, L.S. Wan and Z.K. Xu, Nanofiltration membranes via co-deposition of polydopamine/polyethylenimine followed by cross-linking, J. Membr.Sci. 476 (2015) 50-58.
6. R.S. Harisha, K.M. Hosamani, R.S. Keri, S.K. Nataraj and T.M. Aminabhavi, Arsenic removal from drinking water using thin film composite nanofiltration membrane, Desalination 252 (2010) 75-80.

Study on preparation and performance of additive flame retardant DOPO-HQ modified epoxy composite

Chenghui Wang, Lele Wang, Yingqiang Zhang*

Shanghai Institute of Technology, Shanghai, China

Abstract. An additive flame retardant was prepared from DOPO-HQ. Fourier transform infrared spectroscopy (FT-IR) was used to analyze its chemical structure. It was added to the experimental formula to prepare EP composite material, and the limiting oxygen index experiment and thermogravimetric analysis experiment were carried out on the material, and the scanning electron microscope experiment was carried out on the residual carbon after combustion. The results show that the modified epoxy resin has a limiting oxygen index of 29.5%, and there is no droplet during combustion; thermogravimetric analysis experiments show that the temperature at which the thermal weight loss is 5%wt decreases, but the residual carbon rate increases. Scanning electron microscopy (SEM) test results show that carbon layer formed by the combustion products of epoxy resin is denser..

Keywords: Additive flame retardant, epoxy resin, limiting oxygen index.

1. Introduction

As we all know, epoxy resin has a very wide range of applications in coatings, adhesives, electronic packaging materials, aerospace materials, etc. due to its excellent machinable property, bonding properties, corrosion resistance and low curing shrinkage. However, as an organic polymer material containing carbon, hydrogen, and oxygen, epoxy resin is extremely flammable, seriously threatening human life and property safety, and greatly restricts its application[1]. Therefore, the flame retardant Research and development is very necessary.

Traditionally: Add-on halogen flame retardant has obvious flame retardant effect, but it will release carcinogen dioxin and toxic gas during the combustion process, and its use is restricted[2]. At present, based on people's requirements for an environmentally friendly and healthy living environment, a new type of high-efficiency, low-toxic, low-smoke, green and environmentally friendly organic flame retardant has become a research hotspot. Phosphorus-containing flame retardants catalyze and promote the formation of carbon from polymer materials during the thermal degradation process, which not only improves the flame retardant performance of the material, but also effectively reduces the emission of smoke and corrosive gases during the combustion of the material[3]. Among them, DOPO-HQ molecules have excellent thermal stability, heat resistance, flame retardant, and chemical stability have become a research trend. As a new member of the flame retardant family, silicone flame retardants have gradually attracted people's attention in the 1980s[4]. Because of their excellent flame retardant, good processing and mechanical properties, they are environmentally friendly and have broad development prospects. Phosphorus-silicon flame retardant is not only a green flame retardant product but also has a synergistic effect. In addition, organosilicon catalyzes the degradation of organophosphorus into acid during the degradation process, and promotes the dehydration of polymer into carbon. The formed carbon layer is polymerized with lower viscosity silicon oxide, which increases the viscosity, strength, and density of the carbon layer. , Wrap the debris or gas generated by polymer degradation to form a dense insulation layer that insulates oxygen and heat transfer, and suspends or relieves the burning of polymer materials[5].

In this study, flame retardant DOPO-HQ and silane coupling agent KH-550 were selected for synthesis to synthesize a new type of phosphorous-silicon flame-retardant monomer, and its structure was characterized by FT-IR. The epoxy resin was added with flame retardant and their properties were characterized by limiting oxygen index.

2. Experimental

2.1 Materials

Isophorone diisocyanate(IPDI), analytical grade, Shanghai Organic Pharmaceutical Chemical Production Co., Ltd.; dibutyltin dilaurate (DBTDL), analytical grade, Shanghai Organic Pharmaceutical Chemical Production Co., Ltd.; N,N-dimethyl formamide (DMF), analytical

* Corresponding author: sh_yqzhang@sina.com

grade, Shanghai Organic Pharmaceutical Chemical Production Co., Ltd.; flame retardant 10-(2,5-dihydroxyphenyl)-10-hydro-9-oxa-10-phosphenanthrene-10-oxide(DOPO-HQ), analytical grade, Aladdin Chemical Reagent Co., Ltd.; silane coupling agent KH-550 industrial grade, Shanghai Inco Industry Co., Ltd. ; Epoxy resin E-51, industrial grade, Guangzhou Fufei Chemical Technology Co., Ltd.; 4,4-diaminophenylsulfone, analytical grade(DDS), Jingmen Dongxin Biotechnology Co., Ltd.

2.2 Preparation of additive DOPO-HQ-IPDI-KH550

Put 20ml DMF, 6.28g DOPO-HQ into the four-necked flask, blowed in nitrogen, set the temperature to 110°C, added 4.44g IPDI and 0.1ml catalyst, reacted for 16 hours, then added 4.42g KH-550 to the mixture and reacted for 4 hours , To obtain an orange transparent liquid. After the liquid was cooled, added deionized water to precipitate the flame retardant, filtered with suction, washed several times, and put it in a 60°C blast drying oven for 4 hours to obtain 15 g of flame retardant.

2.3 Preparation of flame-retardant epoxy resin

Put 10 g of epoxy resin E-51 into a four-necked flask, heated it at 120°C for 15 minutes, added flame retardants with different proportions to it to be transparent, and then added 3.162 g of curing agent DDS to it to react 30min, poured into the preheated mold, vacuumed for 30min at 25°C, vacuumed for 30min at 120°C, and cured at 160°C for 2h to obtain the cured product of ring flame-retardant epoxy resin. The flame retardant formula is shown in Table 1:

Table 1. Experimental formula of flame retardant variable

Group	DOPO-HQ	Flame retardant
1	5%	0%
2	5%	5%
3	5%	10%
4	5%	15%
5	5%	20%

3. Results and discussion

3.1 FT-IR analysis of flame retardant

The FT-IR spectra of the chemical structure characterization of flame retardant are shown in Figure1. Flame retardant was obtained by the reaction of IPDI with KH-550 and flame retardant DOPO-HQ. The key lies in the reaction of -NCO group and -OH group. As can be seen from the figure 1, At a wavenumber of 3410cm-1, it is the phenolic hydroxyl group absorption peak, 2090cm-1 is the C=O double bond absorption peak, 1640cm-1 is the benzene ring absorption peak, 1460 cm-1 is the P=O double bond absorption peak, 1390 cm-1 It is the -CH3 absorption peak, 1140 cm-1 is the C-O-C absorption peak, 928 cm-1 is the Si-O absorption peak, 758 cm-1 is the -NH absorption peak, and there is no -NCO characteristic at 2205-2270 cm-1.The characteristic peak of -NCO group disappears, indicating that the compound has been successfully synthesized.

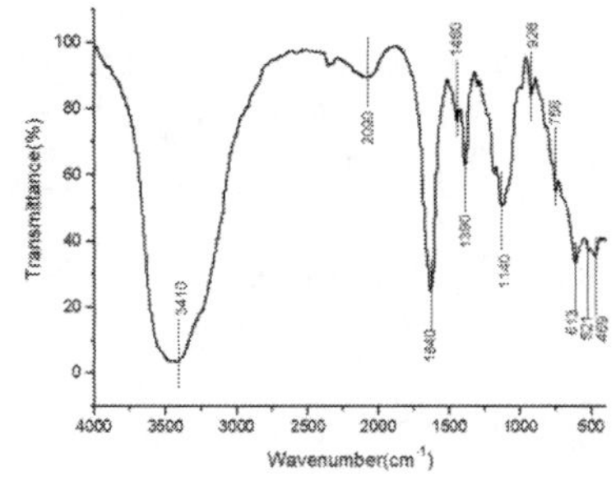

Figure 1. FT-IR spectra of DOPO-HQ-IPDI-KH550

3.2 Limiting oxygen index analysis of composite materials

The limiting oxygen index experiment was carried out on five groups of epoxy resin composites with different flame retardant formulations, and the results are shown in Figure 2: As the flame retardant content increases, the limiting oxygen index of epoxy resin increases. Compared with EP, the LOI of flame retardant/EP-5wt%,flame retardant/EP-10wt%, flame retardant /EP-15wt% and flame retardant /EP-20wt% samples increased to 25.5%, 28.5%, 29.5%, 28.5%. The epoxy cured product cannot be self-extinguished during the test, and there is a phenomenon of melting and dripping. For flame retardant modified epoxy cured products, the flame can be self-extinguished after they are evacuated from the igniter. With the increase of flame retardant content, it takes less time for the flame to self-extinguish. This may be because the organic phosphorus thermally degraded to produce phosphoric acid and other compounds to play a solidified phase flame retardant mechanism, promote the surrounding dehydration into char, and form a dense and heat insulation carbon layer, which effectively inhibits the spread of flammable gas and heat, and improves the material resistance flammability.

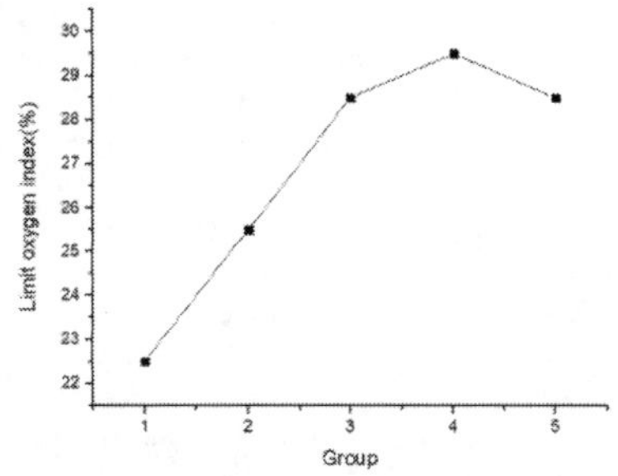

Figure 2. Limit oxygen index of epoxy resin composite

https://doi.org/10.1051/matecconf/202235801026

3.3 Thermogravimetric analysis test of composite materials

Thermogravimetric analysis tests were carried out on epoxy resin and epoxy resin with the highest limiting oxygen index, and the results are shown in Figure 3. The degradation of epoxy resin is one-time degradation, and the initial degradation temperature is 81.5°C. The temperature at which the thermal weight loss is 5%wt is 399.5°C. It decomposes rapidly in a narrow temperature range of 399.5°C-459.5°C, with a thermal weight loss of 76%. After the temperature rises to 484°C, the curve tends to be flat, the degradation is complete, and the carbon residue content of the compound is 16.11%. The epoxy resin with 15% flame retardant is degraded into a one-time degradation, and the initial degradation temperature is 77.5°C. The temperature at which the thermal weight loss is 5%wt is 283.36°C. It decomposes rapidly in a narrow temperature range of 346°C-400°C, and the thermal weight loss is 66%. After the temperature rises to 440°C, the curve tends to be flat, the degradation is complete, and the carbon residue content of the compound is 25.57%. This may be because the lower bond energy of organic silos in flame retardant degrades first, which promotes the degradation of organic phosphorus into acids, exerts the synergistic flame retardant effect of phosphorus - silicon interaction, forms a denser and more stable thermal insulation layer, delays or stops epoxy resin degradation.

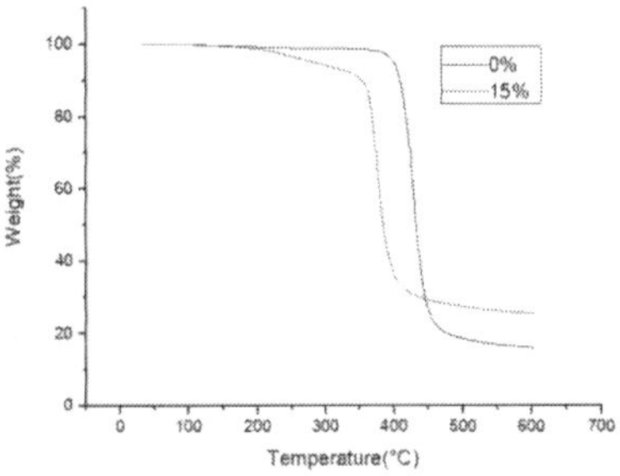

Figure 3. Thermogravimetric curve of epoxy resin composite

3.4 Scanning electron microscope analysis of residual carbon after combustion

The morphology and structure of the carbon layer formed when the material is burned is very important to the flame retardant of the material. Therefore, it is necessary to study the morphology and structure of the carbon layer formed after the epoxy curing material is burned in order to understand its flame-retardant mechanism. Figure 4 is a partial SEM picture of the residual carbon formed after the epoxy cured product is burned, where (a), (b), (c), (d) and (e) are 0%, 5%, 10%, 15%, and 20%flame retardant added. The residual carbon layer formed by 0%, 5% composite rubber has large and many pores, while the residual carbon layer formed by 10%, 15% and 20% composite rubber has almost no pores, especially the picture shown in Figure d, which has a denser structure. The ground insulates the oxygen and heat on the surface of the material, and at the same time verifies the results shown in the limiting oxygen index.

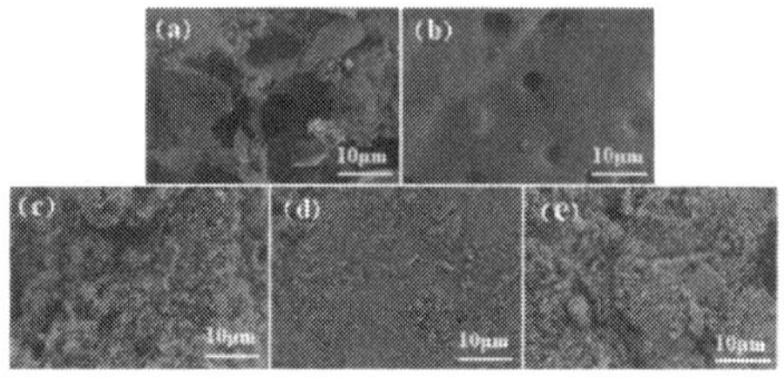

Figure 4. Scanning Electron Microscope for Residual Carbon of Epoxy Resin Composite

3.5 Tensile performance test of epoxy resin composite

Tensile performance tests were conducted on five groups of epoxy resin composites with different flame retardant content, and the results are shown in Figure 5: The tensile strength of the epoxy cured product is 19.6 MPa. When flame retardant is introduced into the cured product, its tensile strength is reduced, from 19.6 MPa of EP to 6.5 MPa. This may be because the addition of the bulky and rigid DOPO group in the flame retardant component limits the formation of macromolecular chains, which weakens the force between the three-dimensional cross-linked network structure of the cured product, which leads to the tensile strength of the cured product reduce.

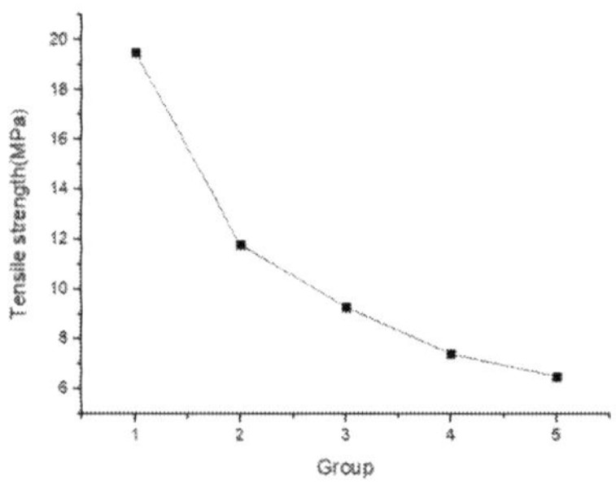

Figure 5. Tensile curve of epoxy resin composite

4. Conclusion

A phosphorus-containing additive type special agent was prepared, and a composite structural material was prepared. The structure of flame retardant was characterized by FT-IR. In the analysis of the limiting oxygen index of the material, the flame retardant has a good flame retardant effect. The composite material has no droplets, no toxic gas generation, low quality loss, and good stability. In this study, the composite material with 15% flame retardant has better flame retardant effect. It has a higher limiting oxygen index, less mass loss, and a denser carbon layer structure. In summary, it shows that flame retardant has a wide application potential in flame-retardant epoxy resin composite products.

Acknowledgments

This study was supported by the Association for the Promotion of Science / Education Foundation Program of Shanghai (LM201874, LM201909), the Science and Technology Innovation Action Plan Project of Shanghai (15520503400).

References

1. A.B Morgan, V Benin, D A Klosterman, et al. Organophosphorus-hydrazides as potential reactive flame retardants for epoxy, J. Journal of Fire Sciences, 2019, 38(1):28-52.
2. J Zhang, X Ming, S Chen, et al. A bio-based hyper-branched flame retardant for epoxy resins,J. Chemical Engineering Journal, 2019,381-391.
3. Q Zhang, J Wang, S Yang, et al. Synthesis of a P/N/S-based flame retardant and its flame retardant effect on epoxy resin, J. Fire Safety Journal, 2020, 113(19):102-111.
4. X D Qian, M Lei, et al. Organic/inorganic flame retardants containing phosphorus, nitrogen and silicon: Preparation and their performance on the flame retardant of epoxy resins as a novel intumescent flame retardant system, J. Materials Chemistry and Physics, 2014, 143(3):1243-1252.
5. Y Shuang, Y Hu, Q Zhang. Synthesis of a phosphorus–nitrogen-containing flame retardant and its application in epoxy resin, J. High Performance Polymers, 2018, 31(2):417-518.

Self-assembly of dumbbell-shaped supramolecules with simulated annealing and Monte Carlo simulation

Bo Du [1, 2, 3, *], Qing Wang [1], Congying Li [1] and Hong Li [3]

[1] China Institute of Building Standard Design & Research Co., Ltd., Beijing, China
[2] Department of Polymer Science and Engineering, Tianjin University, Tianjin, China
[3] Beijing National Technology Transfer Center, Chinese Academy of Sciences Beijing, China

Abstract. The self-assembly structures of dumbbell-shaped supramolecules are investigated using a simple sphere-rod-sphere (SRS) model. Through the simulated annealing Monte Carlo simulation under NVT condition, two variations of this model, i.e. the rigid/semi-rigid model, are compared by their order parameters. We found that, by varying the interactions of this SRS model, various nanostructures such as cross-linked networks, clusters phases, bundle-like nanowires and other order phases are discovered. The formation mechanism of these novel nanostructures is discussed.

1. Introduction

The self-assembly of supramolecules have received particular attentions in past decades, especially in recent advance in molecular nanostructure determination and nanotechnology. [1, 2] Early in the year 1923, Vorländer[3] had recognized that the anisotropic shape of the molecule is an important or even the dominant factor in determining the self-assembly behavior of liquid crystalline. Rigid supramolecule has the similar character of liquid crystal molecule which has non-spherical rigid shape and possesses anisotropic properties in thermal conductivity, optics and electromagnetics. These properties enable these liquid-crystal-like molecules applied in many important fields, such as the computer and electronic industry.

In the current study, we focus on a kind of dumbbell-shaped rigid supramolecule, which has similar molecular shape as linear liquid crystal but with two tunable terminal groups. This class of linear liquid crystal may have more complex phase structures and transitions which are still not well understood right now. In this study, we construct a simplified model to simulate the self-assembly structure of this kind of supramolecule, and the general properties of them may provide more potential applications and understandings.

2. Models

In this work, the self-assembly of dumbbell-shaped supramolecule is constructed by a simplified sphere-rod-sphere (SRS) model, see details in Fig. 1. Here, the interaction between spheres employs the Lennard Jones type potential, the interaction between ellipsoids uses Gay-Berne (GB) potential, and the interaction between ellipsoid and sphere employs a variant of GB potential. [4] There are two variations of this model, a semi-rigid model constrained with a harmonic potential and a completely rigid model.

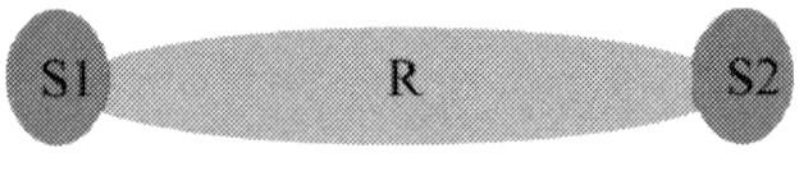

Figure 1. Schematic illustration of the sphere–rod–sphere model.

2.1 The Lennard-Jones potential

The Lennard-Jones potential (also referred to as the LJ-6-12 potential) is a mathematically simple model that approximates the interaction between a pair of neutral atoms or molecules.

$$U^{LJ}(r_{ij}) = 4\varepsilon[(\frac{\sigma}{r_{ij}})^{12} - (\frac{\sigma}{r_{ij}})^{6}] \quad (1)$$

where ε is the depth of the potential well, σ is the finite distance at which the inter particle potential is zero. In our simulation, $\sigma = 0.2$ and $\varepsilon = 1$, the energy is in units of ε.

* Corresponding author: dubo2008@tju.edu.cn

2.2 The Gay-Berne potential

The Gay-Berne (GB) potential can be regarded as an anisotropic and shifted version of the Lennard-Jones (LJ) interaction suitable for uniaxial molecules, where the strength ε and the range parameter σ depend on the orientations of the two particles and their intermolecular vector. The interaction between ellipsoids uses Gay-Berne [5] (GB) potential. The Gay–Berne potential was first introduced to model the site-site potential for a pair of anisotropic but cylindrically symmetric molecules. The functional form is based on the Gaussian overlap model proposed by Berne, et al.[6] The model is thus defined by the following expressions:

$$U(\overline{u_i},\overline{u_j},\hat{r}_{ij}) = 4\varepsilon(\overline{u_i},\overline{u_j},\hat{r}_{ij})\{[\frac{\sigma_0}{r_{ij}-\sigma(\overline{u_i},\overline{u_j},\hat{r}_{ij})+\sigma_0}]^{12} - [\frac{\sigma_0}{r_{ij}-\sigma(\overline{u_i},\overline{u_j},\hat{r}_{ij})+\sigma_0}]^{6}\} \quad (2)$$

where $\overline{u_i}$ and $\overline{u_j}$ are unit vectors describing the orientations of the two molecules and $\hat{r}_{ij}$ is the intermolecular vector. The strength parameter has the form:

$$\varepsilon(\overline{u_i},\overline{u_j},\hat{r}_{ij}) = \varepsilon_0 \varepsilon'^{\mu}(\overline{u_i},\overline{u_j},\hat{r}_{ij}) \varepsilon^{\nu}(\overline{u_i},\overline{u_j}) \quad (3)$$

$$\varepsilon(\overline{u_i},\overline{u_j}) = [1-\chi^2(\overline{u_i}\cdot\overline{u_j})^2]^{-1/2} \quad (4)$$

$$\varepsilon'(\overline{u_i},\overline{u_j},\hat{r}_{ij}) = 1-\frac{\chi'}{2}\{\frac{[\hat{r}_{ij}\cdot(\overline{u_i}+\overline{u_j})]^2}{1+\chi'(\overline{u_i}\cdot\overline{u_j})} + \frac{[\hat{r}_{ij}\cdot(\overline{u_i}-\overline{u_j})]^2}{1-\chi'(\overline{u_i}\cdot\overline{u_j})}\} \quad (5)$$

The range parameter is defined by the following equation:

$$\sigma(\overline{u_i},\overline{u_j},\hat{r}_{ij}) = \sigma_0\{1-\frac{\chi}{2}[\frac{(\overline{u_i}\cdot\hat{r}_{ij}+\overline{u_j}\cdot\hat{r}_{ij})^2}{1+\chi(\overline{u_i}\cdot\overline{u_j})} + \frac{(\overline{u_i}\cdot\hat{r}_{ij}-\overline{u_j}\cdot\hat{r}_{ij})^2}{1-\chi(\overline{u_i}\cdot\overline{u_j})}]\}^{-1/2} \quad (6)$$

The first parameters we considered for the GB potential are chosen to fit the interaction curves of two molecules arranged in a linear fashion for side-by-side and end-to-end orientations. The fitted parameters were: $\mu = 2$, $\nu = 1$, $\sigma_e/\sigma_s = 5$, $\varepsilon_e/\varepsilon_s = 5$, $\sigma_0 = 0.2$.

2.3 The variant Gay-Berne potential

The interaction between ellipsoid and sphere employs a variant of GB potential.

$$\sigma(\hat{r}_{ij},\overline{u_j}) = \sigma_0[1-\chi(\hat{r}_{ij}\cdot\overline{u_j})^2]^{-1/2} \quad (7)$$

$$\varepsilon(\hat{r}_{ij},\overline{u_j}) = \varepsilon_{RS}\left[1-\left(1-\left(\frac{\varepsilon_E}{\varepsilon_S}\right)^{1/\mu}(\hat{r}_{ij}\cdot\vec{u}_{ij})\right)\right]^{\mu} \quad (8)$$

In our simulation, $\varepsilon_E/\varepsilon_S = 5$ and $\varepsilon_{RS} = 1.5$.

3. Results and discussion

We have performed off-lattice Monte Carlo (MC) simulations of a system of dumbbell shaped mesogens described above in the canonical (*NVT*) ensemble simulating samples of N =648 , and 1024 particles modeled with the coarse grained pair potential described before. The phase diagram has been explored by performing the simulated annealing MC runs. With different parameters, they exhibit isotropic (*I*) phases, various cross-linked (*CL*) networks, clusters (*C*) phases, bundle-like nanowires (*C*/*CL*) and other order phases. After the simulated annealing calculation, every final configuration is stabilized with a long time equilibrium run. The total energy E of the system, as a function of the Monte Carlo step, decreases rapidly and reaches equilibrium, as can be seen in Fig.2. (a).

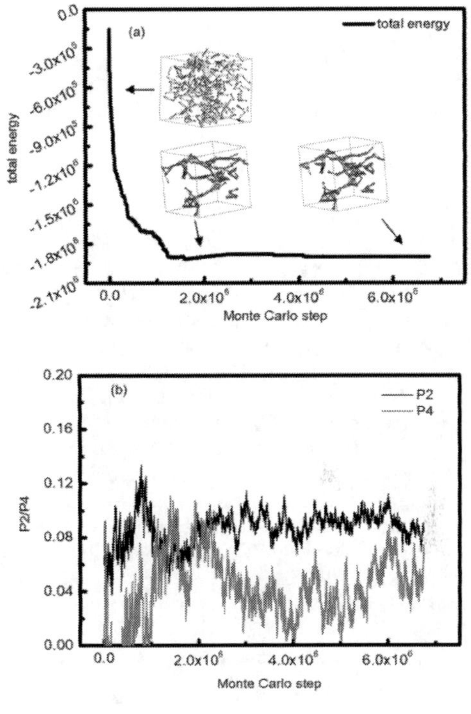

Figure 2. Equilibrium for the simulated system ε0 =1, (a) The total energy of the system as a function of the Monte Carlo step;(b) The P2 and P4 of the system as a function of Monte Carlo step.

In order to characterize the phase behavior of the system the second-rank orientational order parameter [7, 8] and the fourth-rank orientational order parameter[7, 9] are computed via:

$$S_{\alpha\beta} = \frac{1}{N}\sum_{i=1}^{N}\frac{1}{2}(3e_{i\alpha}e_{i\beta} - \delta_{\alpha\beta}) \quad (9)$$

where α, β = x, y, z are the indices referring to the laboratory frame, $\delta_{\alpha\beta}$ the Kronecker symbol, and <...> the ensemble average. P_2 the largest eigenvalue of the order parameter $S_{\alpha\beta}$.

$$< P_4 >= \frac{35}{8} < \cos^4\beta > - \frac{30}{8} < \cos^2\beta > + \frac{3}{8} \quad (10)$$

where $\cos\beta = \vec{u}\cdot\vec{d}$, the director $\vec{d}$ is the eigenvector corresponding to the P_2.

The order parameter is a crucial factor in relating to experiment, since it determines the observed anisotropy in second/fourth rank tensor properties. Fig. 2(b) shows the evolution of P_2 and P_4 with the Monte Carlo step. As one can see from the plot, both P_2 and P_4 are small at high temperatures indicating the isotropic (*I*) phase. Ideally, P_2 and P_4 equal zero in the isotropic (*I*) phase but in practice attain small positive values of about 0.1. However, a non-vanishing value of *S* in the isotropic (*I*) phase is also due to a small but significant finite-size effect [8]. Focusing on the lower T^* first, P_2 increases with temperature down to 26 (Monte Carlo step = 9×10^5) indicating the new phase, that is *CL* phase, exists. When the depth of the potential well between rods ε_0 = 25, the resulting difference becomes obvious in Fig. 3(b). The clusters (*C*) phases is found at T^* = 10 (Monte Carlo step = 1.5×10^6). In contrast, the comparison of Fig. 2(b) and Fig. 3(b) at low temperature shows that the latter one has a lower P_2 value. This tendency for translational order along the molecular axis grows stronger as the system is cooled below the *I-C* phase boundary. This can also be seen in the evolution of the total energy, as plotted in Fig. 3(a).

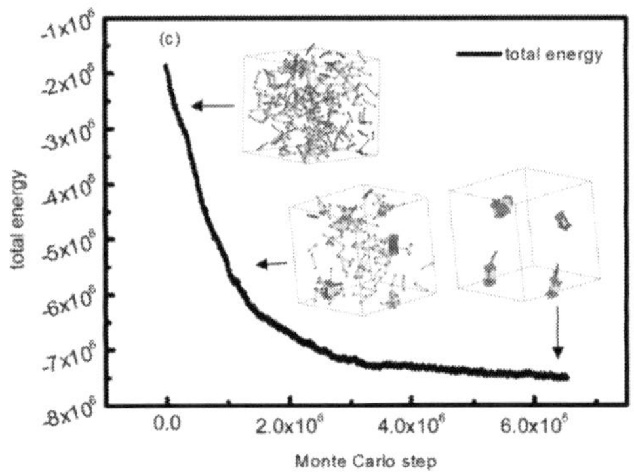

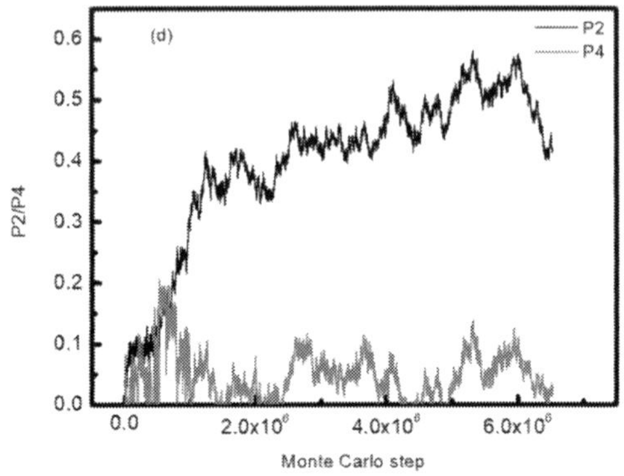

Figure 3. Equilibrium for the simulated system ε0 =25, (a) The total energy of the system as a function of the Monte Carlo step;(b) The P2 and P4 of the system as a function of Monte Carlo step.

To further reduce the computational load, we further apply the rigid model to check the value of P_2 with ε_0 changing from 1 to 25 in Fig. 4. The results indicate that many phase structures can be obtained by tuning the value of ε_0 and the regular changing with ε_0 may be utilized to design novel nano materials.

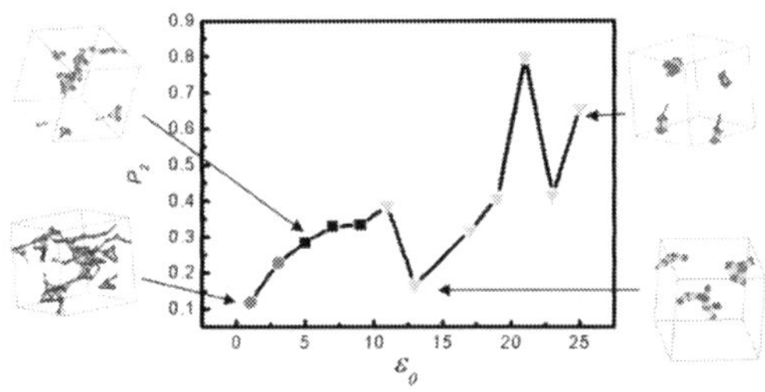

Figure 4. Plots of P2 as function of ε0. Different phases are denoted with different symbols: Cross Linked networks (CL ●), Clusters (C ▼), C/CL(■).

4. Conclusion

The phase structures of a simplified dumbbell-shaped supramolecule model are systematically studied with simulated annealing MC simulation. The design of the model with detailed interaction formulism is described. By varying the interaction parameters, various self-assembly structures such as isotropic, cross-linked networks, clusters phases, bundle-like nanowires and other order phases are obtained. This flexible model may serve as a generic model to study the phase structures and transitions of a large class of symmetric or asymmetric dumbbell-shaped supramolecules. Related studies are on the way.

References

1. Fejer S N, Chakrabarti D, Wales D J. Soft Matter. 2011, 7(7): 3553-3564.
2. Whitesides G M, Grzybowski B. Science. 2002, 295(5564): 2418-2421.
3. Bates M A, Luckhurst G R. Journal of Chemical Physics. 1996, 104(17): 6696-6709.
4. Michel D J, Cleaver D J. Journal of Chemical Physics. 2007, 126(0345063).
5. Gay J G, Berne B J. Journal of Chemical Physics. 1981, 74(6): 3316-3319.
6. Bernet B J, Pechukast P. Journal of Chemical Physics. 1972, 56(8): 4213-4216.
7. Chakrabarti D, Wales D J. Physical Review Letters. 2008, 100(12): 127801.
8. Mazza M G, Greschek M, Valiullin R, et al. Physical Review E. 2011, 83(5): 51704.
9. Zannoni C. Journal of Materials Chemistry. 2001, 11(11): 2637-2646.

https://doi.org/10.1051/matecconf/202235801028

Property and Structure Characterization of High Molecular Weight Polyacrylonitrile Polymers Initiated by 2,2'-Azobis (2-methyl propionamide) Dihydrochloride Using Aqueous Deposited Polymerization Technique

Yaqi Zhao*, Banglei Liang, Zhenxin Zhao and Qiao Feng

School of Materials and Chemical Engineering, Henan University of Urban Construction, Pingdingshan, China

Abstract. Powdery high molecular weight polyacrylonitrile (HMW-PAN) polymers were successfully synthesized by aqueous deposited polymerization (ADP) method with 2,2'-azobis (2-methyl propionamide) dihydrochloride (AIBA) as the initiator and itaconic acid (IA) as comonomer. The utilization of IA into the ADP system was conducive to relax the proceeding of thermo-oxidation process. Differential scanning calorimeter (DSC) curves of obtained PAN copolymers showed the doublet peaks character under air atmosphere, attributing to the exothermic cyclization and oxidative reactions during heat-treatment process. With the utilization of IA in the reaction system, crystallinity and grain sizes of the PAN polymers decreased. The stretching vibration of nitrile group (C≡N) at 2240cm-1 was the strongest peak in both Fourier transform infrared spectra (FTIR). And the stretching vibration and bending of CH2 were the stronger peaks in FTIR. Amido group (NH2), carbonyl group (C=O) and C-O single bond existed in FTIR of PAN homopolymer and copolymers, indicating occurrence of C≡N hydrolysis during polymerization process except carboxyl group (COOH) introduced by IA.

1. Introduction

High molecular weight polyacrylonitrile (HMW-PAN) copolymer is one of the best precursors to prepare high quality carbon fibers [1-3]. During the preparation process of HMW-PAN copolymers, some vinyl comonomers containing flexible functional groups have been focused by many investigators, such as acylic acid (AA), itaconic acid (IA), methylacrylate (MA), vinyl acetate (VAc), and so on [4-10]. And their thermal properties and structural characters had a direct effect on PAN copolymer spinning dopes. Therefore, the introduction of a few comonomers greatly enhanced the internal mobility of PAN's segments, depressed the onset of cyclized temperature during oxidation, and improved the final spinnability of PAN dopes.

Compared with traditional homogeneous solution polymerization, HMW-PAN can be prepared by aqueous polymerization system, because of none chain transfer reactions and partially-solubility of acrylonitrile (AN) in water (H_2O) system. As one kind of heterogeneous solution polymerization, there have been some reports about aqueous deposited polymerization (ADP) of AN. But many complex initiating systems containing alkali metallic ions were usually employed to initiate their copolymerizations. This is not beneficial to manufacture high quality PAN-based precursors for carbon fibers [11-15]. So it is necessary to search an appropriate initiator for the production of HMW-PAN copolymers. By now, there have been some reports about the important water-soluble initiator ammonium persulphate with none alkali metallic ions [16-19]. But there have been few reports about the water-soluble azo initiator, which can also initiate the PAN's copolymerization through the water-soluble radical in water phase. And HMW-PAN can be also obtained using the ADP technique.

In this paper, as a single organic and water-soluble initiator without alkali metallic ions, 2,2'-azobis (2-methyl propionamide) dihydrochloride (AIBA) was used as the initiator, which is a relatively moderate initiator to control the polymerization process and thermal properties of PAN copolymers [20-22]. And a commonly used comonomer-itaconic acid (IA) was copolymerized with AN using the ADP technique. Futhermore, there are also few reports about detailed property and structure characterization of powdery HMW-PAN polymers directly obtained by this initiator. As a result, thermal properties of PAN polymers obtained were investigated by thermal analyzer. And their physical and chemical structures were discussed by wide-angle X-ray diffraction (WAXD) and Fourier transform infrared spectra (FTIR).

* Corresponding author: zhyq5891@163.com

2. Experimental SEction

2.1 Original Materials

AN (Zhengzhou Paini Chemical Reagent Factory, China) and IA (Tianjin Guangfu Fine Chemical Research Institute, China) were analytically purity used as the reaction monomers without further treatment. AIBA (Shanghai Kaiyang Biological Technology Co., Ltd, China) with 98% purity was directly used. And deionized H_2O was adopted as the reaction medium during the polymerization process.

2.2 ADP of PAN

Powdery PAN homopolymer and PAN copolymer with different weight fractions of IA (0~10wt%) in the feed, i.e. Homo-PAN and various P(AN-co-IA)s, were synthesized under nitrogen (N_2) atmosphere using a three-necked flask with deionized H_2O at 60°C. The synthesized samples were described as Homo-PAN, PI-1, PI-2, PI-3, and PI-4, corresponding to 0wt%, 1wt%, 2wt%, 4wt%, 10wt% IA in the feed, respectively. While the concentration of AIBA was 1wt% based on the total monomer concentration at 22wt%. Good agitation was used to ensure that the reaction heat could be thrown off in the polymerization process. The reaction time lasted 2h. And then the precipitated products were washed successively with water and methanol, and dried to a constant weight under vacuum at 50°C.

2.3 Measurements

Conversions of the polymerizations were tested using weight method. Viscosity average molecular weights (M_v) of Homo-PAN and P(AN-co-IA)s were determined by Ubbelolohde viscometer method at (30±0.2)°C using dimethylsulfoxide (DMSO) solvent. The intrinsic viscosity $[\eta]$ equation was calculated by one-point method as follows:

$$[\eta]=\frac{\sqrt{2(\eta_{sp}-\ln\eta_r)}}{c},[\eta]=2.865\times10^{-4}M_v^{0.768} \quad (1)$$

where the needed PAN weight was about 0.04~0.05g in 50mL DMSO.

TA Instruments Co. SDT Q600 (USA) thermal analyzer was used to get simultaneous thermal analysis (STA) curves of different PAN polymers, including differential scanning calorimeter (DSC) and thermal gravity analysis (TGA) curves. Their STA curves with the temperature range from room temperature to 450°C were tested at a 10°C/min heating rate under air atmosphere. Also the DSC curves under N_2 atmosphere were measured using PerkinElmer DSC 8500 (USA) thermal analyzer with the aluminum crucible under aluminum cover, whose temperature range was from room temperature to 400°C.

Wide-angle X-ray diffraction (WAXD) of Homo-PAN and P(AN-co-IA)s between 5~50° were measured at a 3.6°/min step rate under room temperature by PANalytical X'Pert Pro analyzer (Holland) All the WAXD measurements were well done under the same conditions. Crystallinity index (X_c) was calculated by the integral area method according the references [23-25]. And the crystalline planar spacing d and crystallite size (L_c) were determined by Scherrer equation and Bragg equation [25, 26].

FTIR of Homo-PAN and P(AN-co-IA)s were recorded on PerkinElmer Frontier (USA) spectrometer for 16 times. The transmission mode was used on finely pure powdery samples using direct compression method without KBr. The testing range was from 400cm^{-1} to 4000cm^{-1}.

3. Results and discussions

3.1 Acquirement of HMW-PAN polymers

As detailed in Table 1, the M_v values measured by the method above are approximately above 16×10^4. It is evidently investigated that the synthesized PAN polymers almost had the higher molecular weights than the results in the references [12,13,26]. The main reason is that H_2O was adopted as reaction medium in ADP process. Due to the presence of H_2O, chain transfer reactions were prevented, which could increase average molar weights of PAN polymers. While in the homogeneous solution polymerization process of AN and vinyl comonomers, chain transfer reactions to reaction solvents, such as DMSO, dimethylformamide (DMF) and so on, would lower molecular weights of the final PAN copolymers [12,13,26]. Furthermore, relative researches and monomer reactivity ratios of AN and IA have been reported by Zhao et al [16,17], which is deduced that the IA comonomer has the higher reactivity than AN monomer. So, the samples PI-1 and PI-2 with 1wt% and 2wt% IA in the feed had the higher M_v values than the Homo-PAN. When the IA amount reached 4wt%, the IA amount in the PAN main chains became much more enough to prevent the active chain growth. And the M_v value of sample PI-3 became lower. While the conversions of samples PI-2 and PI-3 had become higher than samples Homo-PAN and PI-1. However, only the 14.09% products were prepared with its M_v value at 16.18×10^4 about sample PI-4 copolymer with 10wt% IA. Thus, more IA comonomer in the feed was not conducive to produce HMW-PAN copolymer with high conversion.

Table 1. ADP results of AN/IA under different monomer ratios

Samples	AN/IA(w/w)	Conversion (%)	M_v ($\times10^4$)
Homo-PAN	100/0	76.39	40.60
PI-1	99/1	66.10	55.94
PI-2	98/2	82.91	52.93
PI-3	96/4	83.12	41.74
PI-4	90/10	14.09	16.18

3.2 STA Analysis of PAN Polymers

Different DSC curves of powdery Homo-PAN and different P(AN-co-IA)s under air and N_2 atmosphere are listed in Figure 1(a) and Figure 1(b) . The parameters obtained from the exotherms including the initiation temperature (T_i, the temperature when the exothermic reactions begin), the termination temperature, i.e., the final temperature (T_f, the temperature when the exothermic reactions end), their difference ($\Delta T = T_f - T_i$), the exothermic peak temperature (T_p, the evident exothermic peak in DSC curves), the evolved heat (ΔH) and the velocity of evolving heat ($\Delta H/\Delta T$) are described in Table 2.

DSC exotherms of different PAN polymers are due to the oxidative reactions and cyclization reactions under different atmospheres [1-3, 13]. The obtained exotherm parameters also reflect different reaction characters. The T_i values for all the P(AN-co-IA)s were much smaller than Homo-PAN, indicating that a greater release of the exothermic initiation process with IA in the feed. T_i values showed a change tendency to become lower with the increase of required IA amounts in the polymerization system. It can be concluded that IA played a significant role in initiating the exothermic reaction of different PAN copolymers. However, the T_p and T_f values showed few changes, indicating that T_p and T_f were not greatly affected by the increase of the IA comonomer [1-3, 13, 16, 25-27].

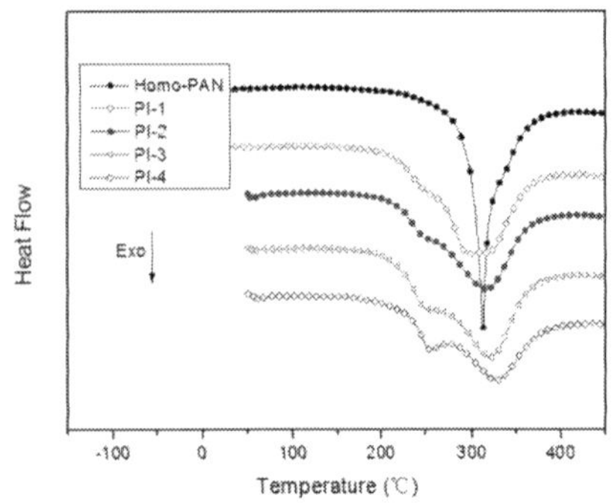

(a)

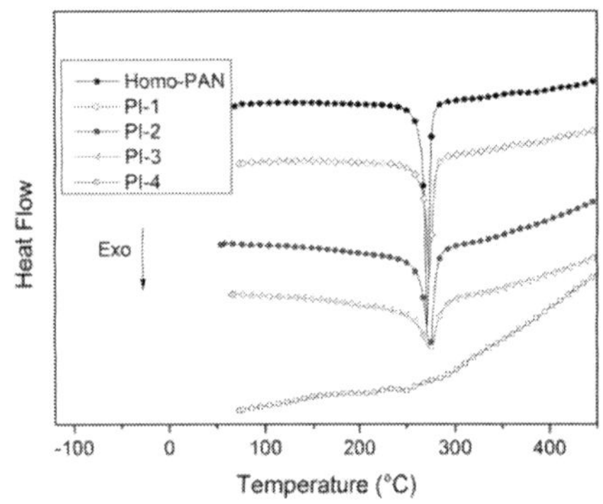

(b

Figure 1. DSC curves of Homo-PAN and different P(AN-co-IA)s under air atmosphere from STA results and DSC curves of Homo-PAN and different P(AN-co-IA)s under N_2 atmosphere

Table 2. Parameters of DSC curves of Homo-PAN and different P(AN-co-IA)s

samples	atmosphere	T_i (°C)	T_p (°C)	T_f (°C)	ΔT (°C)	ΔH (J·g^{-1})	$\Delta H/\Delta T$ (J·g^{-1}·°C^{-1})
Homo-PAN	Air	220	312	375	155	4194	27.06
PI-1		212	315	377	165	4303	26.08
PI-2		206	317	380	174	4968	28.55
PI-3		199	249,319	385	186	4920	26.45
PI-4		187	255,327	389	202	3917	19.39
Homo-PAN	N_2	267	271	275	8	445	55.63
PI-1		268	273	280	12	418	34.83
PI-2		264	274	281	17	353	20.76
PI-3		260	276	285	25	231	9.24
PI-4		237	251	310	83	220	2.65

Another phenomenon deduced from DSC curves was the doublet peaks character in different P(AN-co-IA)s under air atmosphere. While in case of Homo-PAN, the doublet character was not very clear. It could be relevant to the different reaction mechanism during heat-treatment process of different PAN polymers. Furthermore, with the increase of IA amounts, much more intricate changes were found in DSC exotherms of different P(AN-co-IA)s, which was probably due to the superposition of the two peaks. The doublet peaks might be due to the exothermic cyclization and oxidative reactions during heat-treatment process. And the reactions might happen more or less simultaneously. However, initiation of oxidation reaction occurred before the onset of cyclization. The lower-temperature exothermic peak was attributed to the primary oxidation (including dehydrogenation) and cyclization reactions. And the higher-temperature exothermic peak was attributed to the secondary oxidation reactions leading to the polymer chain scission. The evolution of HCN, CO_2 and CO also happened in this region. There is no essential difference to the experimental results by the other many workers [13, 28-30].

While in the N_2 atmosphere, the different PAN polymers had higher T_i values, lower T_f, T_p, ΔT and ΔH values than those in air atmosphere. The main reason was due to strong oxidation reactions under air atmosphere of different PAN polymers. In addition, the Homo-PAN had the minimum ΔT and the maximum $\Delta H/\Delta T$. At the same time, the broader (higher ΔT) and the smoother (lower $\Delta H/\Delta T$) exotherm were found in different PAN copolymer than in Homo-PAN. It is indicated that the evolving heat was concentrative and expeditious in Homo-PAN. The IA comonomer slowed the rate of exothermic reaction. It is obvious that the IA comonomer has role of relieving exothermic reactions of PAN polymers.

3.3 TGA Analysis of PAN Polymers

TGA curves of powdery Homo-PAN and different P(AN-co-IA)s are shown in Figure 2. Obviously, weight loss reactions of all PAN polymers included three reaction stages, named as micro-scale weight loss zone, intense weight loss zone and slow weight loss zone, which were related to small molecules loss and intense exothermic reactions. And the three stages were combinated into the preoxidation process of PAN-based precursors, whose difference reflected the different degree and rate of the preoxidation reaction. According to the TGA curves, the initiation decomposition temperature (T_{id}) and different weight loss values obtained from the TGA curves including w_1 (%), w_2 (%) and w_3 (%) corresponding to the three stages were listed in Table 3.

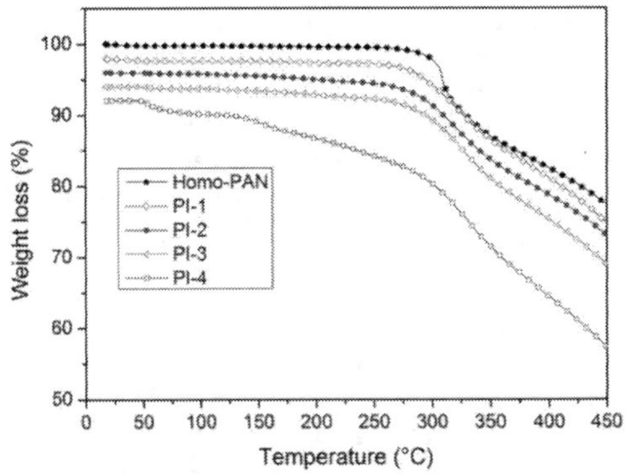

Figure 2. TGA curves of Homo-PAN and different P(AN-co-IA)s under air atmosphere from STA results

Table 3. Parameters for TGA curves of Homo-PAN and different P(AN-co-IA)s

Samples	T_{id} (°C)	w_1 (%)	w_2 (%)	w_3 (%)
Homo-PAN	303	0.5	12.9	22.3
PI-1	282	0.7	11.4	23.3
PI-2	277	1.1	11.5	23.0
PI-3	275	1.2	13.7	25.1
PI-4	132	1.7	10.3	34.5

It is investigated that the first stage occurred at the higher temperature to the Homo-PAN with none IA, with its largest weight loss slope seen from Figure 2. All P(AN-co-IA)s had ease exothermic reactions under air atmosphere, with its lower T_{id} values. When the IA amount was less than 4wt%, the w_1 and w_3 weight loss values had low rise trend. Due to the introduction of IA comonomer in the feed, the increase of w_1, w_2 and w_3 were attributed to higher IA content in the PAN main chains, which might promote the pyrolysis reaction. At the same time, when the IA amount reached 10wt%, the w_1 and w_3 reached the maximum values. That is to say, sample PI-4 had much more weight loss at the first and third stage. However, the sample PI-3 had the higher w_2 value and the sample PI-4 with the lowest w_2 value. Higher IA content in sample PI-3 was benefit to the exothermic reaction, which made the w_2 value become higher. While the T_{id} value of sample PI-4 was so lower that it made the exothermic reaction occurred much more early. And the sample PI-4 also had the lowest M_v value. Therefore, the w_2 value became a little smaller. It was probably attributed to interactive affect of higher IA contents in the final P(AN-co-IA) polymers and their M_v values [1-3, 13, 25-30].

3.4 WAXD Analysis of PAN Polymers

WAXD curves of Homo-PAN and different P(AN-co-IA)s measured under room temperature are depicted in Figure 3. It is shown that all the PAN polymers had the strongest diffraction peaks appearing at 2θ around 17°. They were corresponded to (100) crystalline plane of the pseudohexagonal cell or (200) reflections of the orthorhombic structure with their *d* values equaling about 0.53nm [31-33]. It is noticed that the centers of the strongest diffraction peaks hardly changed with the increase of required IA amounts. Therefore, *d* values of different PAN polymers basically remained unchanged. But the changes of peak intensity and peak area caused changes of crystallinity in different PAN polymers.

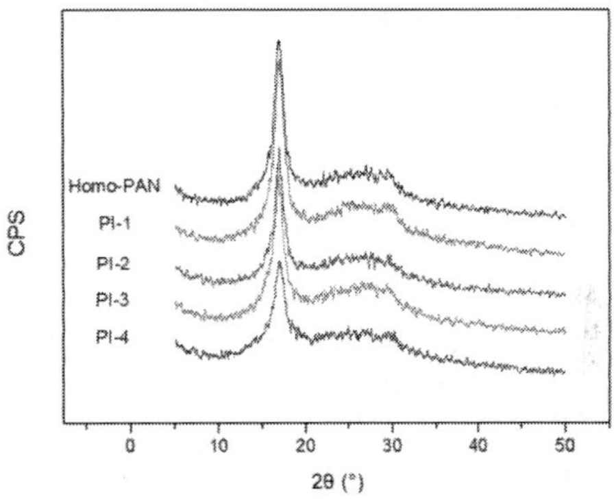

Figure 3. WAXD curves of Homo-PAN and different P(AN-co-IA)s

Then all the d, L_c and X_c values can be deduced and listed in Table 4. It is seen that the Homo-PAN with none IA in the feed had the best crystallinity than the other PAN copolymers under room temperature, which was relevant to destructive effect of introduced IA comonomer. And the destructive effect became larger with the increase of IA less than 4wt%. Fewer orderly arrangement zones were acquired in the P(AN-co-IA)s, which would decrease their crystallinity and grain sizes. Therefore, the sample PI-4 with the highest IA content had the lowest L_c and X_c values.

Table 4. Parameters for WAXD curves of Homo-PAN and different P(AN-co-IA)s

Samples	2θ (°)	d (nm)	FWHM (°)	L_c (nm)	X_c (%)
Homo-PAN	16.92	0.524	1.298	6.125	53.1
PI-1	17.10	0.519	1.272	6.251	48.9
PI-2	16.98	0.522	1.270	6.260	48.0
PI-3	16.95	0.523	1.273	6.245	47.5
PI-4	16.98	0.522	1.409	5.642	43.3

3.5 FTIR Analysis of PAN Polymers

FTIR of powdery Homo-PAN and P(AN-co-IA)s are described in Figure 4. It is obvious that the stretching vibration (ν) of nitrile group (C≡N) at 2240 cm^{-1} was the strongest peak in FTIR spectra. And the 2240 cm-1 band remained unchanged in the Homo-PAN and P(AN-co-IA)s. It is indicated that the AN units appeared the status of uninterrupted long sequences [34]. A shoulder-like absorption peak around 2190 cm^{-1} near the C≡N group might be due to the νC=NH formed during the ADP process [35]. Besides, the stretching vibration around 2940cm^{-1} and bending vibration (δ) around 1450cm^{-1} of methylene group (CH_2) were the stronger peak in the two spectra. During ADP of AN and IA, C≡N would be changed into amide group ($CONH_2$) or carboxyl group (COOH) owing to hydrolysis under water bath heating [28,36]. Thus, a double broad absorption peak at 3628cm^{-1} and 3546cm^{-1} belonging to stretching vibration of NH_2 and a weak absorption peak at 1621cm^{-1} belonging to bending vibration of NH_2 could be exhibited in FTIR of Homo-PAN and P(AN-co-IA)s. Simultaneously, a weak and broad band at 3248cm^{-1} might belong to double-frequency band of δNH_2 (DFB-δNH_2) at 1624cm^{-1}.

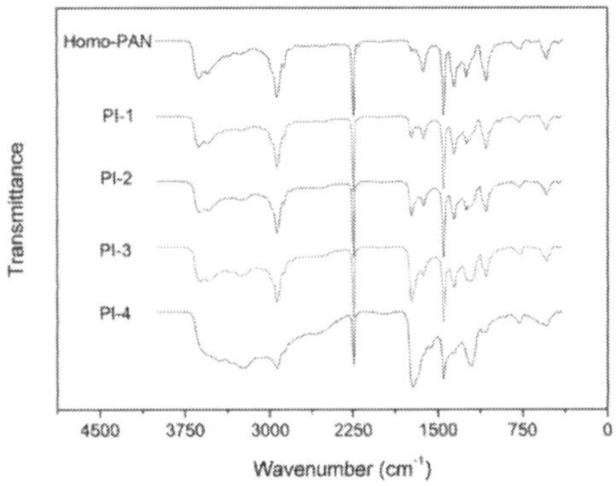

Figure 4. FTIR of Homo-PAN and different P(AN-co-IA)s

Furthermore, a weak absorption peak at 1735cm^{-1} presented in Homo-PAN was possibly attributed to carbonyl group (C=O) formed in hydrolysis process. Compared to Homo-PAN, the absorption band around 1735cm^{-1} in P(AN-co-IA) was stronger than Homo-PAN because IA is a carboxyl compound, which includes two carboxyl groups in one molecule. As a result of the two reasons above, the intensity of νC=O is stronger than that of δNH_2. The weak stretching vibration of hydroxyl group (OH) was superposed to DFB-δNH_2 at 3248cm^{-1}. So the absorption band at 3248cm^{-1} in P(AN-co-IA)s was a little stronger than that in Homo-PAN. The C-H vibrations of different modes were shown in the band regions at 1370~1350cm^{-1} and 1270~1220cm^{-1} [35]. It is reported by Minagawa et al that the stereospecific absorption bands at 1250 and 1230cm^{-1} might be assigned to the wagging mode of the C-H group coupled with rocking mode of the CH_2 group[37]. Due to excess hydrolysis of the C≡N group, an indistinct absorption band around 1180cm^{-1} might be assigned to νC-O bond.

4. Conclusion

Powdery HMW-PAN polymers were synthesized by ADP technique using AIBA as initiator. According to investigations of thermal properties and structure characters using TA, WAXD and FTIR, some conclusions can be obtained as follows:

The different PAN polymers synthesized had the higher molecular weights and higher conversions with the required IA amounts at 1wt%, 2wt% and 4wt%.

The use of IA into the polymerization system was available for relaxing the thermooxidation process of different PAN copolymers. Simultaneously, DSC curves of the PAN copolymers under air atmosphere revealed the doublet peaks phenomenon, which was not evident in Homo-PAN and in N_2 atmosphere. And TGA curves of the PAN polymers were divided into three stages. More IA used in the ADP system made the P(AN-co-IA)s have lower initiation decomposition temperature and larger loss weight value.

(3) IA in the feed caused destructive effect to the crystal structure of PAN polymers, which made the crystallinity and grain sizes of the PAN polymers decrease. The nitrile group (C≡N) stretching vibration at 2240cm^{-1} and stretching vibration and bending of CH_2 were very strong in both FTIR. In Homo-PAN, a weak absorption at 1730cm^{-1} was attributed to the C=O group stretching vibration because of hydrolysis of C≡N. However, it became stronger in different P(AN-co-IA)s with the increase of required IA dosages in the feed.

Acknowledgments

This work was financially supported by the fund of National Science Foundation of China (Grant No. 51803047), and the fund of Science and Technology Key Project of Henan Province in China (Grant No. 202102210035), and the fund of Young Key Teachers Cultivating Program of Henan Province Universities in China (Grant No. 2019GGJS227), and the fund of Academic Technology Leader Program of Henan University of Urban Construction (Grant No. YCJXSJSDTR201909).

References

1. A.K. Gupta, D.K. Paliwal, P. Bajaj, Acrylic precursors for carbon fibers, J. Macromol. Sci. R. M. C. 31 (1991) 1-89.
2. K. Sen, S.H. Bahrmi,
P. Bajaj, High performance acrylic fibers, J. Macromol. Sci. R. M. C. 36 (1996) 1-76.
3. P. Bajaj, A.K. Roopanwal, Thermal stabilization of acrylic precursors for the production of carbon fibers: an overview, J. Macromol. Sci. R. M. C. 37 (1996) 97-147.
4. G.H. Olive, S. Olive, The chemistry of carbon fiber formation from polyacrylonitrile, Adv. Polym. Sci. 51 (1983) 1-60.

5. W.J. Howard. The glass temperatures of polyacrylonitrile and acrylonitrile-vinyl acetate copolymers, J. Appl. Polym. Sci. 5 (1961) 303-307.
6. M.M. Colemanand, G.T. Sivy, Fourier transform IR studies of the degradation of polyacrylonitrile copolymers-I: Introduction and comparative rates of the degradation of three copolymers below 200°C and under reduced pressure, Carbon 19 (1981) 123-126.
7. G.T. Sivy, M.M. Coleman, Fourier transform IR studies of the degradation of polyacrylonitrile copolymers-II: Acrylonitrile/methacrylic acid copolymers, Carbon 19 (1981) 127-131.
8. M.M. Coleman, G.T. Sivy, Fourier transform IR studies of the degradation of polyacrylonitrile copolymers-III: Acrylonitrile/vinyl acetate copolymers, Carbon 19 (1981) 133-135.
9. G.T. Sivy, M.M. Coleman, Fourier transform IR studies of the degradation of polyacrylonitrile copolymers-IV: acrylonitrile/acrylamide copolymers, Carbon 19 (1981) 137-139.
10. J.S. Tsai, C.H. Lin, Effect of comonomer composition on the properties of polyacrylonitrile precursor and resulting carbon fiber, J. Appl. Polym. Sci. 43 (1991) 679-685.
11. A. Maity, M. Biswas, Kinetics and mechanism of the K2CrO4-NaAsO2 redox-initiated aqueous polymerization of acrylonitrile, J. Appl. Polym. Sci. 96 (2005) 276-280.
12. P. Bajaj, D.K. Paliwal, A.K. Gupta, Acrylonitrile-acrylic acids copolymers. 1. Synthesis and characterization, J. Appl. Polym. Sci. 49 (1993) 823-833.
13. A.K. Gupta, D.K. Paliwal, P. Bajaj, Effect of the nature and mole fraction of acidic comonomer on the stabilization of polyacrylonitrile, J. Appl. Polym. Sci. 59 (1996) 1819-1826.
14. J.R. Ebdon, T.N. Huckerby, T.C. Hunter, Free-radical aqueous slurry polymerizations of acrylonitrile: 1. End-groups and other minor structures in polyacrylonitriles initiated by ammonium persulfate/sodium metabisulfite, Polymer 35 (1994)250-256.
15. J.R. Ebdon, T.N. Huckerby, T.C. Hunter, Free-radical aqueous slurry polymerizations of acrylonitrile: 2. End-groups and other minor structures in polyacrylonitriles initiated by potassium persulfate/sodium bisulfite, Polymer 35 (1994) 4659-4664.
16. Y.Q. Zhao, C.G. Wang, Y.X. Wang, B. Zhu, Aqueous deposited copolymerization of acrylonitrile and itaconic acid, J. Appl. Polym. Sci. 111 (2009) 3163-3169.
17. Y.Q. Zhao, C.G. Wang, M.J. Yu, C.S. Cui, Q,F. Wang, B. Zhu, Study in monomer reactivity ratios of acrylonitrile/itaconic acid in aqueous deposited copolymerization system initiated by ammonium persulfate, J. Polym Res. 16 (2009) 437-442.
18. C.S. Cui, C.G. Wang, Y.Q. Zhao. Acrylonitrile/ammonium itaconate aqueous deposited copolymerization, J. Appl. Polym. Sci. 102 (2006) 904-908.
19. C.S. Cui, C.G. Wang, Y.Q. Zhao. Monomer reactivity ratios for acrylonitrile-ammonium itaconate during aqueous-deposited copolymerization initiated by ammonium persulfate, J. Appl. Polym. Sci. 100 (2005) 4645-4648.
20. J.P. Liu, X.M. Yang, X.F. Hu, L. Hu, Recent research progress on the synthesis and application of water soluble azo initiators, Chemical Reagents, 33 (2001) 317-323.
21. S.S. Wang, Y.S. Chen, Q. Ouyang, J. Huang, J.X. Yang, H. Liu, Y.F. Liu, Effect of azo initiators on structure and thermal stabilization behavior of PAN and its precursor, China Synthetic Fiber Industry, 40 (2017) 11-16.
22. Z.H. Xie, Y.S. Chen, Q. Ouyang, X.F. Wang, J.X. Yang, J. Huang, Study on acrylonitrile solution polymerization initiated by azobiscyanopentanoic acid, China Synthetic Fiber Industry, 38 (2015) 13-17.
23. J.P. Bell, J.H. Dumbleton, Changes in the structure of wet-spun acrylic fibers during processing, Text. Res. J. 41 (1971) 196-203.
24. G. Hinrichsen, Structural changes of drawn polyacrylonitrile during annealing, J. Polym. Sci. Polym. Symp. 38 (1972) 303-319.
25. A.K. Gupta, R.P. Singhal, Effect of copolymerization and heat treatment on the structure and X-ray diffraction of polyacrylonitrile, J. Polym. Sci. Polym. Phys. 21 (1983) 2243-2262.
26. P. Bajaj, K. Sen, S.H. Bahrmi, Solution polymerization of acrylonitrile with vinyl acids in dimethylformamide, J. Appl. Polym. Sci. 59 (1996) 1539-1550.
27. Y.Q. Zhao, C.G. Wang, Y.J. Bai, G.W. Chen, M. Jing, B. Zhu, Property changes of powdery polyacrylonitrile synthesized by aqueous suspension polymerization during heat-treatment process under air atmosphere, J. Colloid Interf. Sci. 329 (2009) 48-53.
28. A.K. Gupta, D.K. Paliwal, P. Bajaj, Effect of an acidic comonomer on thermooxidative stabilization of polyacrylonitrile, J. Appl. Polym. Sci. 58 (1995) 1161-1174.
29. W.N. Turner, F.C. Johnson, The pyrolysis of acrylic fiber in inert atmosphere. I. Reactions up to 400°C, J. Appl. Polym. Sci. 13 (1969) 2073-2084.
30. Y. Tsuchiya, K. Sumi, Thermal decomposition products of polyacrylonitrile, J. Appl. Polym. Sci. 21 (1977) 975-980.
31. O.P. Bahl, R.B. Mathur, K.D. Kundra, Structure of PAN fibres and its relationship to resulting carbon fibre properties, Fibre Sci. Technol. 15 (1981) 147-151.

32. T.H. Ko, P. Chiranairadul, H.Y. Ting, C.H. Lin, The effect of modification on structure and dynamic mechanical behavior during the processing of acrylic fiber to stabilized fiber, J. Appl. Polym. Sci. 37 (1989) 541-552.
33. M.J. Yu, Y.J. Bai, C.G. Wang, Y. Xu, P.Z. Guo, A new method for the evaluation of stabilization index of polyacrylonitrile fibers, Mater. Lett. 61 (2007) 2292-2294.
34. P. Bajaj, M. Padmanaban, Copolymerization of Acrylonitrile with 3-chloro, 2-hydroxy-propyl acrylate and methacrylate, J. Polym. Sci. Polym. Chem. 21 (1983) 2261-2270.
35. T. Usami, T. Itoh, H. Ohtani, S. Tsuge, Structural study of polyacrylonitrile fibers during oxidative thermal degradation by pyrolysis-gas chromatography, solid-state carbon-13 NMR, and Fourier-transform infrared spectroscopy, Macromolecules 23 (1990) 2460-2465.
36. S.B. Deng, R.B. Bai, J.P. Chen, Behaviors and mechanisms of copper adsorption on hydrolyzed polyacrylonitrile fibers, J. Colloid Interf. Sci. 260 (2003) 265-272.
37. M. Minagawa, K. Miyano, M. Takahashi, Infrared characteristic absorption bands of highly isotactic poly(acrylonitrile), Macromolecules 21 (1998) 2387-2391.

 https://doi.org/10.1051/matecconf/202235801029

Rapid Determination of Antioxidant Irganox 1076 in Polyethylene

Qiukai Zheng [1], Dan Hu [2], Jingjing Fan [3,*]

[1] Chemistry & Chemical and Environmental Engineering College, Weifang University, Weifang, Shandong 26106 China
[2] Qidong Animal Disease Prevention and Control Center, Qidong, Jiangsu 226200 China
[3] Biological and Agricultural Engineering College, Weifang University, Weifang, Shandong 261061 China

Abstract. A method was established to quickly determine the content of antioxidants in polyethylene by Fourier transform infrared spectroscopy. By analyzing the FTIR spectra of the antioxidant Irganox 1076 and PE, the characteristic peak of the antioxidant Irganox 1076 in PE was determined to be the absorption peak at 1738cm-1. According to the standard curve of the antioxidant, the content of the antioxidant had a good linear relationship with the absorption intensity of its characteristic peak, and the linear regression coefficient of the equation was 0.97363. The relative error was less than 7%. The method is accurate and reliable, allowing for rapid quantitative analysis of trace antioxidant content in polyolefin.

1. Introduction

Polyolefin is widely used in the fields of household appliances, packaging and building materials because of its rich source of raw materials, low production cost and good comprehensive performance. It has become an indispensable material in human life and production and occupies an important position in the national economy [1]. Polyethylene (PE) is a kind of polyolefin which has been widely used in many fields. However, polyolefin in the preparation, storage and application is inevitably affected by factors such as heat, light, oxygen, heavy metal ions and mechanical shear [2], resulting in discoloration and mechanical physical properties degradation [3]. Its products turn yellow, hard and brittle, and even completely lose their use value [4, 5]. In order to inhibit its oxidative degradation, antioxidants are generally added to polyolefins [6, 7]. Antioxidants can be divided into main antioxidants, auxiliary antioxidants and metal passivators according to different action mechanisms. The main antioxidants are divided into hindered phenol type and hindered amine type according to the chemical structure formula. Among them, hindered amine type antioxidants are heavily polluted and dark in color. They are mostly used in black, dark plastic or rubber products. Auxiliary antioxidants mainly include phosphorus and sulfur antioxidants [8]. When used together with main antioxidants, they play a synergistic role [9]. On the one hand, they decompose the generated peroxide into alcohol compounds [10]; On the other hand, reduce the main antioxidant inactivated by oxidation [11]. Metal passivators form extremely stable complexes with active metal ions to prevent oxidative degradation of polymers [12-14]. Controlling the amount of antioxidants is conducive to improving the quality of polyolefin products, reducing costs and improving economic benefits. Therefore, it is necessary to establish a simple, rapid and accurate analytical method for determining the content of trace antioxidants in PE. Due to the low antioxidant content in polyolefins, the pretreatment of samples is cumbersome. There are many methods for analyzing the additive content in polyolefins, such as ultraviolet spectrum [15], mass spectrometry [16], nuclear magnetic resonance [17] extraction [18], chromatography [19], thermal analysis [20], infrared spectroscopy [21], and so on. However, these methods have problems such as high cost, long test cycle, and difficult online analysis, which are hard to meet the needs of rapid analysis. Infrared spectroscopy is a simple, fast and accurate analytical method. In this paper, the qualitative and quantitative analysis of irganox 1076, a trace antioxidant in PE, was studied by infrared spectroscopy, which has important practical significance for analyzing the effect and modification of antioxidants in polyolefin.

2. Experimental Section

2.1 Raw materials and auxiliaries

High-density polyethylene (HDPE), GX-103, Hirst, Germany; Irganox 1076, commercial product; toluene, analytical purity.

2.2 Instruments and Equipment

Fourier transform infrared spectrometer (FTIR), Tianjin Gangdong Science and Technology Development Co., Ltd. The wave number of the spectrogram ranges from 4000 to 200 cm^{-1}, the resolution is 4 cm^{-1}, and the data is analyzed with software origin8.0.

* Corresponding author: zqiukai@163.com

2.3 Preparation of samples

The antioxidant Irganox 1076 was weighed with an electronic balance with an accuracy of 0.0001 g, dissolved in toluene, weighed different amounts of the solution and mixed well with PE, processed for 3 min on a hot table at 160°C, and pressed into a sample of a certain thickness with a mold.

3. Results and Discussion

3.1 Determination of characteristic absorption peaks

The antioxidant Irganox 1076, PE and PE samples with 0.8% Irganox 1076 were tested by infrared spectroscopy, and the appropriate absorption peaks were intercepted as shown in Figure 1.

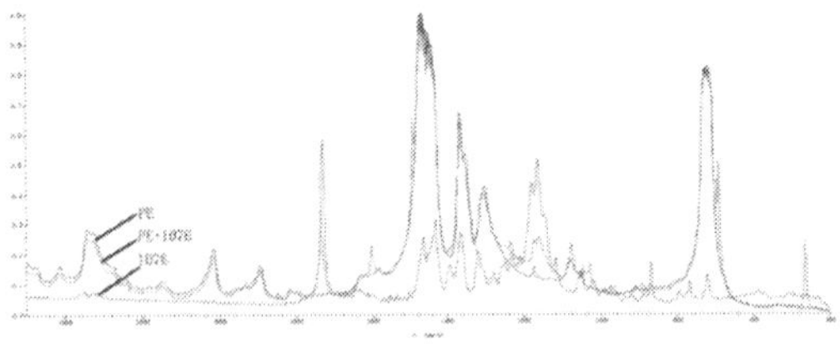

Figure 1. The FTIR curves of Irganox 1076, non-stabilized and stabilized PE with Irganox 1076

It can be seen from Figure 1 that Irganox 1076 has a strong absorption peak at 1738cm^{-1}, there is no strong absorption peak around, and PE has basically no absorption at this wave number. Therefore, the absorption peak at 1738cm^{-1} is selected as the characteristic absorption peak of antioxidant Irganox 1076 for quantitative analysis. PE samples containing 0.8% Irganox 1076 have a significant absorption peak at the wave number 1738cm^{-1} without being affected by the PE absorption peak, so it is desirable to take this absorption peak as a characteristic absorption peak of the antioxidant Irganox 1076.

3.2 The drafting of standard curves

The PE and Irganox 1076 solutions were mixed evenly and prepared into standard samples with antioxidant Irganox 1076 content of 0 to 1.0% (0, 0.2%, 0.4%, 0.6%, 0.8%, 1.0%, respectively). The specimens were pressed into samples with a thickness of 1 mm by a mold for infrared spectrum test and analysi. The characteristic absorption peaks of Irganox 1076 with different contents are shown in Figure 2.

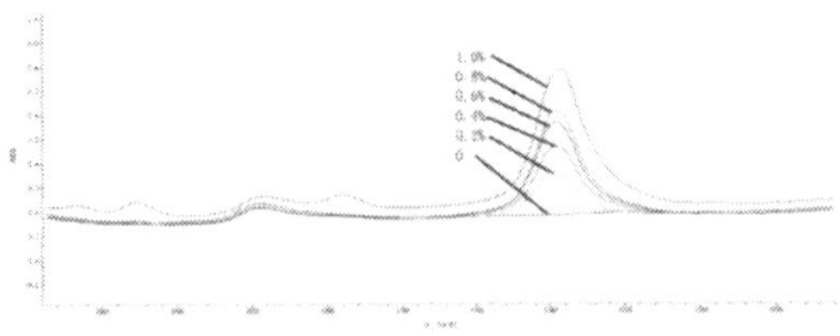

Figure 2. The FTIR curves of PE stabilized with Irganox 1076 in various concentrations at 1738cm^{-1}

As can be seen from Figure 2, the wave number 1738cm^{-1} absorption peak area increases with the increase of antioxidant content. After baseline correction, the characteristic absorption peak area (A_1) of the antioxidant was calculated using Fourier infrared spectroscopy software. In order to eliminate the effect of sample thickness differences, the characteristic absorption peak of the PE sample needs to be selected as a reference. Comparing the spectra of the three samples in Figure 1, PE has an absorption peak at 2017cm^{-1}, while the antioxidant Irganox 1076 has no absorption peak here, and the absorption peak can be selected here as a reference. After baseline correction, the characteristic absorption peak area of the antioxidant (A_2) is calculated using Fourier infrared spectroscopy software. The mathematical model was established by using Origin8.0 analysis software to obtain the standard curve of A_1/A_2 and its mass fraction, as shown in Figure 3.

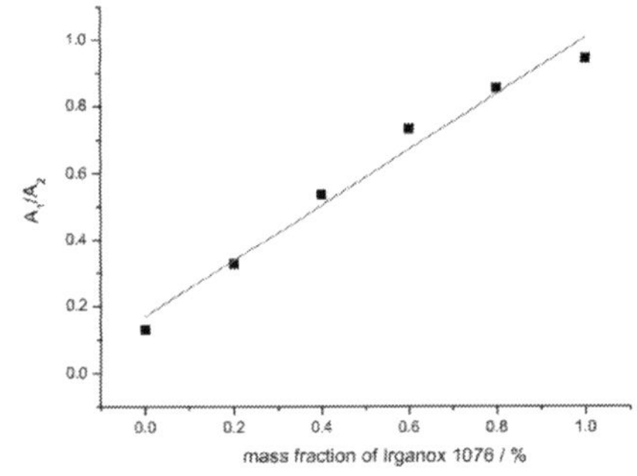

Figure 3. The standard curve of Irganox 1076 contents in PP

It can be seen from Figure 3 that the ratio of the characteristic absorption peak area of antioxidant Irganox 1076 to the reference absorption peak area of PE is positively correlated with the antioxidant content. The fitting equation is $y = 0.17007 + 0.83474x$, and the correlation coefficient of the equation is 0.97363. This indicates that the standard curve has a good correlation.

3.3 Accuracy verification

In order to verify the reliability of the fitting equation, five PE / antioxidant Irganox 1076 samples with different antioxidant contents were prepared for verification. The antioxidant Irganox 1076 characteristic peak absorption peak area and PE reference absorption peak area were determined, and the ratio was calculated. The antioxidant content was calculated from the fitted equation, and compared with the true value, as shown in Table 1.

Table 1. Accuracy of test results

Sample	Mass Fraction/%		Relative Error /%
	True Value	Measured Value	
PE-1	0.05	0.051	1.30
PE-2	0.25	0.256	2.26
PE-3	0.45	0.479	6.42
PE-4	0.65	0.616	5.20
PE-5	0.85	0.820	3.55

As can be seen from Table 1, the relative errors are within 7%, and the reliability of the fitted equation is high, which basically meets the needs of the actual detection.

4. Conclusion

(1) The quantitative standard curve of the antioxidant Irganox 1076 in the PE/antioxidant Irganox 1076 sample is y=0.17007+0.83474x, and the correlation coefficient of the equation is 0.97363;
(2) The relative errors between the measured value and the calculated value of the antioxidant Irganox 1076 in the PE/antioxidant Irganox 1076 samples are less than 7%, and the maximum is 6.42%, which basically meets the needs of actual detection.

Acknowledgments

This work was financially supported by Shandong Provincial Natural Science Foundation ZR2020MC174.

References

1. Mei G, Beccarini E, Caputo T, et al. Recent Technical Advances in Polypropylene [J]. Journal of Plastic Film and Sheeting, 2009, 25(2): 95 ~ 113.
2. In-Jin S, Won C, Jae-Gyu S, et al. Experimental and Analytical Study of the DC Breakdown Characteristics of Polypropylene Laminated Paper with a Butt Gap Condition Considering the Insulation Design of Superconducting Cable [J]. Japanese Journal of Applied Physics, 2014, 53(83): 4 ~ 8.
3. Wang X, Xing W, Tang G, et al. Synthesis of a Novel Sulfur-bearing Secondary Antioxidant with a High Molecular Weight and its Comparative Study on Antioxidant Behavior in Polypropylene with two Commercial Sulfurbearing Secondary Antioxidants Having Relatively Low Molecular Weight [J]. Polymer Degradation and Stability, 2013, 98(11): 2391 ~ 2398.
4. Xie J, Xu R, Chen X, et al. Influence of Heat-treatment Temperature on the Structure and Properties of Polypropylene Microporous Membrane [J]. Polymer Bulletin, 2015, 73(1): 265 ~ 277.
5. Maeda E A, Santos A F, Silva L G A, et al. Chemical, Physical, and Mechanical Properties Evolution in Electron Beam Irradiated Isotactic Polypropylene [J]. Materials Chemistry and Physics, 2016, 169: 55 ~ 61.
6. Umile G. Spizzirri G C, Nevio Picci and Francesca Iemma. Recent Development in the Synthesis of Eco - Friendly Polymeric Antioxidants [J]. Current Organic Chemistry, 2014, 18: 2912 ~ 2927.
7. Wang L, Li K, Zhu G, et al. Preparation and Properties of Highly Branched Sulfonated Poly(ether ether ketone)s Doped with Antioxidant 1010 as Proton Exchange Membranes [J]. Journal of Membrane Science, 2011, 379(1 ~ 2): 440 ~ 448.
8. Pénzes G, Domján A, Tátraaljai D, et al. High Temperature Reactions of an Aryl–alkyl Phosphine, an Exceptionally Efficient Melt Stabiliser for Polyethylene [J]. Polymer Degradation and Stability, 2010, 95 (9): 1627 ~ 1635.
9. Hassanpour S, Khoylou F. Synergistic Effect of Combination of Irganox 1010 and Zinc Stearate on Thermal Stabilization of Electron Beam Irradiated HDPE/EVA both in Hot Water and Oven [J]. Radiation Physics and Chemistry, 2007, 76(11-12): 1671 ~ 1675.
10. Kriston I, Pénzes G, Szijjártó G, et al. Study of the High Temperature Reactions of a Hindered Aryl Phosphite (Hostanox PAR 24) used as a Processing Stabiliser in Polyolefins [J]. Polymer Degradation and Stability, 2010, 95(9): 1883 ~ 1893.
11. Papanastasiou M, Mcmahon A W, Allen N S, et al. Atmospheric Pressure Photoionization Mass Spectrometry as a tool for the Investigation of the Hydrolysis Reaction Mechanisms of Phosphite Antioxidants [J]. International Journal of Mass Spectrometry, 2008, 275(1-3): 45 ~ 54.
12. Allen N S, Barcelona A, Edge M, et al. Aspects of the Thermal and Photostabilisation of High Styrene–butadiene Copolymer (SBC) [J]. Polymer Degradation and Stability, 2006, 91(6): 1395 ~ 1416.
13. Cogen J M, Hilmer A J. Crystal Polymorphism in Antioxidant- metal Deactivator 1,2-bis (3,5-di-tert-butyl-4-hydroxyhydrocinnamoyl)hydrazine[J]. Polymer Degradation and Stability, 2008, 93(12): 2193 ~ 2198.
14. Morlat-Therias S, Fanton E, Gardette J-L, et al. Photochemical Stabilization of Linear Low-density Polyethylene/clay Nanocomposites: Towards Durable Nanocomposites [J]. Polymer Degradation and Stability, 2008, 93 (10): 1776 ~ 1780.
15. Métois P, Scholler D, Bouquant J, Feigenbaum A. Alternative test methods to control the compliance of polyolefin food packaging materials with the European Union regulation: the case of aromatic

antioxidants and of bis(ethanolamine) antistatics based on ^{1}H-NMR and UV-visible spectrophotometry [J]. Food Addit Contam, 1998, 15(1): 100-111

16. Vit A I, Galbraith M N, Hodqkin J H, et al. Quantitative analysis of additives in polyethylene by MID mass spectrometry [J]. Polym Degrad Stab, 1993, 42(1): 69 ~ 73
17. Schilling F C, Kuck V J. Determination of stabilizer concentrations in Polyethylene [J]. Polym Degrad Stab, 1991, 31(2): 141 ~ 152
18. Pinto A M, Taylor L. Evolution of a method for quantitative supercritical fluid extraction 330 antioxidant from high density polyethylene [J]. J Chromatoga, A, 1998, 811(1 ~ 2): 163 ~ 170
19. Thilen M, Shishoo R. Optimization of experimental parameters for the quantification of ploymer additives using SFE/HPLC [J]. J Appl Polym Sci, 2000, 76(6): 938 ~ 946
20. Karlsson K, Assarqren C, Gedde U W. Thermal analysis for the assessment of antioxidant content in polyethylene [J]. Polym Testing, 1990, 6(9): 421 ~ 431
21. Karstang T V, Henriksen A. Infrared spectroscopy and multivariate calibration used in quantitative analysis of additives in high-density polyethylene. Chemometrics and Intelligent Laboratory Systems [J]. 1992, 14(1 ~ 3): 331 ~ 339

Methods and exploration for mechanical testing and microscopic testing of loess under freeze-thaw cycles

Qingbo Liu [1], Kai Zhao[1], Laishuan Zhao [1, 2], Linhao Wang [1, 2]

[1] China Railway No.3 Bureau Group Investment Co., Ltd. Taiyuan, Shanxi,030006 China
[2] Tianshui Sanyangchuan Tunnel Traffic Engineering Co., Ltd. Tianshui, Gansu,741030 China

Abstract: The stability of loess subgrade is affected by many factors. Freeze-thaw cycles are a common factor that causes the strength of loess subgrades to be damaged. In order to study the influence of freeze-thaw cycles on the stability of loess subgrades, reveal the mechanism of freeze-thaw cycles on the strength of loess subgrades. It is necessary to sample the loess soil samples in this area, and perform compaction test, freeze-thaw cycle test, unconfined compression test and electron microscope scanning test on the soil sample to analyze the influence of freeze-thaw cycle on the strength and microstructure of loess soil. . In this way, the potential relationship between the microstructure and the strength of the soil can be obtained. The results found that the freeze-thaw cycle of loess led to a trend of first increase and then decrease in soil strength. The microstructure showed a trend of smaller soil particles and increased soil voids. The potential relationship between the microstructure and the strength of the soil was analyzed. It explains the conclusion that the strength of the soil first increases and then decreases.

Keywords: Loess; The freeze-thaw cycle; Mechanical properties; Microstructure

1. Introduction

The loess is in the cyclical change of climate and seasons, and the soil is also in the cyclical state of freezing and thawing of cooling and warming at this time, and this kind of freezing and thawing has a great influence on the physical and mechanical properties of loess. Previous studies by scholars have shown that for loess that has undergone multiple freeze-thaw cycles, the freeze-thaw cycles will weaken the shear strength of the loess, the ability to resist deformation is weakened, and the structure of the loess will be destroyed. Therefore, the performance of the loess under the freeze-thaw cycle The research has an important impact on the construction of the loess area. Yang Gengshe et al. [1] used nuclear magnetic resonance technology to study the distribution of pores in undisturbed loess under the action of freeze-thaw, and believed that the content of micro and small pores increased, while the content of medium and large pores decreased under alternate freeze-thaw cycles. Liu Lei [2] conducted indoor triaxial tests on Taiyuan loess under different initial conditions, and believed that the shear strength of Taiyuan loess showed a decreasing trend with the increase of water content. When the water content was large, its deformation characteristics showed strain hardening. Shear shrinkage, when the water content is small, its deformation characteristics are shown as strain softening and dilatancy. Yuan Kangfeng [3] used conventional compression apparatus and triaxial apparatus to study the law between the shear strength and structural parameters of loess on a certain slope in southern Shanxi. When the moisture content is low, the structure is stronger. When the pressure is constant, the shear strength of loess decreases with the increase of water content, and when the water content is constant, the shear strength of loess increases with the increase of pressure. When the moisture content is constant, the shear strength increases with the increase of structural parameters. The greater the structure, the stronger the macroporous structure and bonding strength, and the greater the shear strength. Aoyama et al. [4] found through experiments that the cohesion of the soil body decreased under the action of freezing and thawing, but the internal friction angle hardly changed. Liu Leqing [5] found that with the decrease of freezing and thawing temperature, the unconfined compressive strength of unconfined loess and remolded loess decreased, but the unconfined compressive strength of remolded loess changed unevenly, indicating that temperature has an effect on remolded loess. The intensity of the impact is serious. CHAMBERLAIN EJ [6] and other studies have found that the freeze-thaw cycle will change the structure of the soil. When water freezes under the action of freezing, the frost heave force will be generated. Due to the action of ice, larger pores or cracks will be formed, and the nature of the soil It also matters a lot. Ye Wanjun et al. [7] analyzed that during the freezing and thawing process of soil, the significant increase of moisture in the frozen zone is an important reason for the attenuation of soil strength and the increase of pores, and the relationship between soil strength and pore area is inversely proportional. Edwin J et al. [8] observed the changes in the permeability and structure of four types of fine-grained soils with the freeze-thaw process, and believed that the freeze-thaw cycle can cause a decrease in porosity and an increase in vertical permeability. Although no exact relationship has been established, it can be It is determined that the soil with the highest plasticity index has the largest change in permeability. Van Bochove et al. [9] believed that the freeze-thaw cycle changed the physical and biological structure of silty clay. Maria HohmannPorebska [10] believes that the cohesion of geological parameters is greatly affected by freezing and thawing. Chamberlain EJ, Blouin SE et al. [11] performed three freeze-thaw cycles on four cohesive soils with a plasticity index ranging from 0 to 20 under a confining pressure of 1.7 to 140 kPa. analyze. Under the conditions of vertical unidirectional freezing and thawing and pressure and sealing, the consolidated sample will be strengthened due to the freezing and thawing cycle, and the void ratio will decrease, but the vertical permeability will increase; for fine-grained soils, vertical permeability will increase. The increase in performance is due to the formation of vertical polygonal cracks in the sample during the freeze-thaw cycle; for coarse-grained soil, it is mainly due to the flocculation of fine particles in the soil, which leads to a decrease in the volume of the coarse-grained soil. small. And it is proposed that the occurrence of vertical cracks is mainly due to the formation of negative pore water pressure and water migration during the freezing process. .Boonsinsuk P[12] and others through repeated freezing and thawing tests, found that as the number of freezing and thawing increases, the shear strength index damage degree of composite soil materials increases. Wang Jing [13] found that although the number of freeze-thaw cycles is different, the lower the negative temperature during freezing, the more significant the strength of the soil after melting. However, when the negative temperature is lower than -10°C during freezing, the attenuation of the strength will reach a steady state. . Ndèye Fatou Ngom [14] uses a generalized cylinder to model the soil microstructure on the basis of computer high-resolution three-dimensional tomography (CT) soil samples, and the pores in the soil can be extracted by a set threshold. Qin Hui [15] carried out the wave velocity test of compacted loess, analyzed the change law of shear wave velocity and compression wave velocity of compacted loess under different freeze-thaw cycle conditions, and found that with the increase of the number of freeze-thaw cycles and water content , The form of soil sample failure presents a trend of transition from brittle failure to plastic failure.

2. Experiment method

It is required to conduct compaction test, freeze-thaw cycle test, unconfined compression test and scanning electron microscope test on soil samples to analyze the influence of freeze-thaw cycle on the strength and microstructure of loess soil.

Compaction test According to the "Code for Design of Highway Subgrade" [20], a heavy-duty compactor is used to carry out compaction tests on the modified expansive soil with different lime content. The preparation method of dry compaction test specimens for lime-improved expansive soil: (1) After the soil sample is air-dried to a moisture content of less than 10%, it is passed through a 5 mm sieve, and the soil sample under the sieve is taken for later use. (2) Determine the moisture content of the air-dried soil sample and divide the soil sample into several parts (each soil sample is 2.5-3.0 kg). (3) After adding water to each soil sample to the set moisture content, seal it in a plastic bag for 24 hours so that the soil sample is fully infiltrated and the moisture content is uniform. (4) Perform compaction after mixing, and determine the maximum dry density and optimal moisture content of different samples according to the test method.

2.1 Preparation method of improved soil

The selected soil samples are dried, crushed, and sieved. Mix the admixture with the treated loess in proportion. Calculate the water quality according to the water content of the fly ash soil, add the weighed water to the mixture, stir it evenly, and let it stand in a humidifier for 24 hours. The moisturized mixture is placed in a mold for uniform compaction, and the mixture in the mold is compacted from both ends by three times of compaction. Use a demolding instrument to eject the test piece from the mold. In order to prevent moisture loss, seal the prepared sample with a plastic film and place it in a curing box (the temperature of the curing box is (20±1)°C, and the relative humidity is (96±2)%), in the humidity box Keep in good health for 7 days. Make the test piece with a degree of compaction greater than 95%, and then perform multiple freeze-thaw cycles on the soil sample. In the experiment, try to use the same batch of test pieces and perform different freeze-thaw cycles.

2.2 Freeze-thaw cycle test method

It is necessary to simulate or measure the main concentration range of the water content of loess within this depth based on the test results of the soil temperature at the borrow site and taking into account the depth of the borrow.

In order to facilitate the direct shear test directly after the freeze-thaw cycle, the direct shear ring knife can be directly used for sample preparation, and the soil samples with different moisture contents are prepared by the drip method according to the test requirements, and the soil samples are wrapped with plastic film to keep the moisture content. Change. Since the moisture content of the soil sample will migrate to the surface during the freezing process, in order to make the moisture content of the soil sample even after melting, it is melted outdoors and allowed to stand for 24 hours, and then a direct shear test is performed. [16]

The freeze-thaw cycle adopts a temperature-controlled freezer. The temperature setting of the temperature cycle process should be as close as possible to the ambient temperature of the test site, and the setting temperature should be as low as -10 degrees Celsius. Place the specimen in the freezer for 24 hours to simulate a frost heave During the process, the temperature of the freezer is adjusted to positive degrees Celsius, and the test piece is placed in the freezer for 24 hours to simulate a melting process, which is a complete freeze-thaw cycle process. If necessary, temperature sensing tools can be set in the soil to observe the soil temperature in real time.

2.3 Microstructure research methods

In the past, optical microscope, CT and other testing technologies, due to their low resolution, could not meet the requirements for studying soil images. The current high-resolution image testing technology available is mainly scanning electron microscopy. Try to use the same material and the same batch of specimens with the same manufacturing process to prepare soil samples through different freeze-thaw cycles. After the soil sample is successfully prepared, the soil sample is carefully broken apart to expose a fresh surface, and trimmed with a knife into small pieces that meet the irradiation requirements, and the microstructure changes are observed.

Dehydration of soil samples: The samples used in the scanning electron microscope need to be in a completely dry state, so the processed soil samples need to be dehydrated. The dehydration technology of soil samples is divided into three types: drying method, air-drying method and freeze-drying method. The air-drying method is to make the soil sample reach a testable standard in a natural state in a cool, dark and ventilated environment; the drying method is generally to place the soil in a drying box at 105°C and keep it for 12 hours to reach the soil. Completely dry; freeze-drying method uses liquid nitrogen freezing technology to instantaneously freeze the soil sample, and then depressurize it in a freeze dryer, so that the liquid phase of the soil is sublimated. Combining the three dehydration techniques, the drying method can dry the soil sample in a shorter time and more economically. It should be noted that the disturbance to the soil sample is minimized during the drying process.

Gold spraying of soil samples: As the two-ash modified loess is not a good conductor, it is necessary to spray gold technology to enhance the conductivity of the soil samples, so as to obtain a good microstructure image of the soil. Use ion sputtering instrument to spray gold on the soil sample, and then paste the soil sample on the disc with conductive tape for use.

When scanning, a typical position should be selected for scanning. The selected scanning point should be at a position where the particles of the lime-ash soil are evenly distributed and the pores are well distributed, avoiding large particles or large pores, so that the microstructure image of the improved loess is typical and representative. The magnification of the microscopic image determines whether the image can truly reflect the microstructure of the soil. Excessive magnification will cause the image to be unrepresentative, and too small a magnification will result in blurred imaging and large errors. It should be selected according to the actual situation.

3. Experimental results

In the freeze-thaw cycle test of loess, the strength of the soil is affected by water content, confining pressure, compaction, as well as the loess itself and the improved soil. When the water content is low, the changes in water content are very sensitive to changes in the structural properties of loess. As the confining pressure gradually increases, the change range of structural parameters becomes smaller. The increase in confining pressure makes the structural parameters of the soil show a trend of first becoming larger and then smaller. The larger the confining pressure, the more stable the structural parameters and the smaller the strain value. When the degree of compaction is different, the strain when it becomes stable is also different [17]. Multiple freeze-thaw cycles did not significantly affect the overall moisture content of the loess samples.

3.1 The influence of freeze-thaw cycles on the compressive strength of loess

After the freeze-thaw cycle of the soil sample with high moisture content, the surface of the soil sample appeared pitted, the content of large pores increased significantly, the surface became looser, and there were tiny cracks, and there were also small cracks in the contact with the ring knife. The surface changes of soil samples with low water content are not obvious.

When the number of freeze-thaw cycles is greater than a certain number of times, the compressive strength tends to be stable; the compressive strength of the modified loess with 2% cement dosage has the largest decrease rate, and the modified loess specimens break and fail after 9 freeze-thaw cycles. The 3% cement dosage improves the resistance of loess. The second rate of compressive strength decline is caused by the low cement dosage and the poor resistance to freezing and thawing of the structure formed by cement hydration products and soil particles. When the cement dosage or compaction coefficient is small, the strength of the modified loess will be greatly reduced under the effect of freezing and thawing, and the freezing resistance will be poor [18].

Under the same freezing and thawing temperature, the relationship between the unconfined compressive strength of remolded loess and the number of freezing and thawing cycles is basically the same as that of undisturbed loess, and both show strain softening. The first few freeze-thaw cycles will increase the strength of the reshaped loess. The freeze-thaw cycles destroy the skeleton of the reshaped loess particles. The large pores between the solid particles are filled with small particles, and the overall connection of the reshaped loess is improved. The results of repeated freezing and thawing cycles on the strength are consistent with the undisturbed loess, and all decrease with the decrease of freezing and thawing temperature. The overall trend is to increase first, then decrease and then increase [19].

3.2 The influence of freeze-thaw cycles on cohesion and internal friction angle

The freeze-thaw cycle has a significant influence on the cohesive force, and the cohesive force decreases exponentially with the increase of the number of freeze-thaw cycles under the same moisture content. For different moisture content, the greater the moisture content, the greater the reduction in cohesion after freezing and thawing cycles. After the same freeze-thaw cycle, as the freezing temperature decreases, the cohesive force does not change much after freezing and thawing.
The internal friction angle increases slightly with the increase of the number of freeze-thaw cycles, and the final increase is not large. The main reason is that the proportion of large pores in the soil sample decreases after the freeze-thaw cycle, which leads to an increase in the contact points between soil particles. Causes a slight increase in the angle of internal friction.

3.3 The effect of freeze-thaw cycles on microstructure

With the increase of the number of freeze-thaw cycles, the number of small particles of remodeled loess gradually increases, the large pores are transformed into small pores, the arrangement of the particles is loose, and the cementing capacity between the particles decreases. As the number of freeze-thaw cycles increases, the primary structure of loess is damaged more severely. Most of the large particles are decomposed into small particles and filled into the gaps between the particles, so that the particles are arranged tightly and the connection capacity is enhanced. Freezing and thawing will cause changes in the soil particle system. With the increase in the number of freeze-thaw cycles, the pore area ratio of the soil shows a trend of first increasing and then becoming stable, and the small pores in the soil continue to transform into medium and large pores. , The strength of the soil is negatively correlated with the pore area [7].

4. Conclusions

(1) Summarize the precautions in the freeze-thaw cycle experiment of loess, consider the actual situation of the project, and set the temperature in the freeze-thaw cycle of the test loess to simulate the actual situation. The addition of admixtures may enhance the water retention and heat preservation capacity of the loess.
(2) Sort out some loess strength tests under dry-wet cycle conditions, introduce the methods and procedures of compaction test, as well as the humidification and drying methods in dry-wet cycle experiments, and the precautions in electron microscope scanning experiments.
(3) Summarize the influence of freeze-thaw cycles on the strength of loess. With the increase of the number of freeze-thaw cycles, the strength of loess first increases and then decreases. Summarize the influence of freeze-thaw cycles on the microstructure of loess. With the increase in the number of freeze-thaw cycles, the experimental loess showed a trend of smaller soil particles and larger soil voids, and the original structure of the loess was seriously damaged.

References

1. Yang Gengshe, You Ziyu, Wu Di, Zhao Luqing. Experimental study on the relationship between the pore size distribution of undisturbed loess under freeze-thaw environment and its mechanical properties[J]. Coal Engineering, 2019, 51(03): 107-112.
2. Liu Lei. Research on the influence of the moisture content of Taiyuan loess on the parameters of Duncan-Chang model[J]. Roadbed Engineering, 2010(02):169-171.
3. Yuan Kangfeng, Yuan Panfeng. The correlation between the structural properties and strength indexes of loess[J]. Journal of Liaoning Technical University (Natural Science Edition), 2015, 34(03): 373-376.
4. Aoyama K, Ogawa S, Fukuda M. Temperature dependencies of mechanical properties of soils subjected to freezing and thawing[C]//Kinosita S, Fukuda M. Proceedings of the 4thinternational symposium on ground freezing, Sapporo, Japan. Rotterdam: A Balkema, 1985: 217-222.
5. Liu Leqing, Zhang Wuyu, Zhang Bingyin, Gu Yuxi, Xie Banglong. Experimental study on the unconfined compressive strength and microscopic laws of loess under freeze-thaw cycles[J]. Hydrogeology and Engineering Geology, 2021, 48(04): 109 -115.
6. CHAMBERLAIN E J, GOW ANTHONY J. Effect of freezing and thawing on thepermeability and structure of soils[J]. Engineering Geology, 1979, 13(1-4): 73-92.
7. Ye Wanjun, Chen Yiqian, Zhang Dengfeng, Bai Yang. Macro and micro experiment research on the influence of water migration on the strength of compacted loess under freeze-thaw action[J]. China Journal of Highway and Transport, 2021, 34(06): 27-37.
8. Chamberlain E J, Gow A J. Effect of Freezing and Thawing on the Permeability and Structure of Soils[J]. Engineering Geology, 1979, 13(1):73-92.
9. Bochove V, Eric, Prevost, et al. Effects of Freeze-Thaw and Soil Structure on Nitrous Oxide Produced in a Clay Soil. [J]. Soil Science Society of America Journal, 2000.
10. Hohmann-Porebska M. Microfabric effects in frozen clays in relation to geotechnical parameters[J]. Applied Clay Science, 2002, 21(1-2):77-87.
11. Chamberlain E J. Physical changes in clays due to frost action and their effect on engineering structures [J]. 1989.
12. Yong R N, BOOSINSUK P, YIN C W P. Alteration of soil behaviour after cyclic freezing and thawing[C]//International symposium on ground freezing. 4. 1985: 187-195.
13. Wang Jing. Research on mechanical properties and micro-mechanism analysis of subgrade soil in seasonal freezing area after freeze-thaw cycles[D]. Jilin University, 2012.

14. Ngom N F, Monga O, Mohamed M M O, et al. 3D shape extraction segmentation and representation of soil microstructures using generalized cylinders[J]. Computers & geosciences, 2012, 39: 50-63.
15. Qin Hui, Liu Xin, Lan Xingxing. Experimental study on the evaluation of the strength characteristics of compacted loess under the action of freeze-thaw cycles by wave velocity[J/OL]. Journal of Engineering Geology: 1-11[2021-08-16]. https ://doi.org/10.13544/j.cnki.jeg.2021-0176.
16. Zhang Hui. Research on the moisture migration and strength of loess under freezing and thawing[D]. Xi'an University of Architecture and Technology, 2014.
17. Wen Hao. Study on the changes of mechanical properties of loess soil under freeze-thaw cycles[J]. Shanxi Transportation Science and Technology, 2021(01): 39-41+53.
18. Jiang Yingjun, Wang Hanyue, Qiao Huaiyu, Yue Weimin, Dong Xin. Stability of cement-improved loess roadbed under the action of water, dry-wet and freeze-thaw cycles[J]. Science Technology and Engineering, 2020, 20(35): 14592-14599.
19. Liu Leqing, Zhang Wuyu, Zhang Bingyin, Gu Yuxi, Xie Banglong. Experimental study on unconfined compressive strength and microscopic law of loess under freeze-thaw cycles [J]. Hydrogeology and Engineering Geology, 2021, 48(04):109 -115.
20. Ministry of Communications of the People's Republic of China. Highway Subgrade Design Code: JTG D30—2004 [S]. Beijing: People's Communications Publishing House, 2004.

Exploration on mechanical test method of improved loess under dry-wet cycles

Laishuan Zhao [1, 2], Tiehai Wu [1], Shijie Yu [1, 2], Baogui Rong [1, 2]

[1] China Railway No.3 Bureau Group Investment Co., Ltd. Taiyuan, Shanxi,030006 China
[2] Tianshui Sanyangchuan Tunnel Traffic Engineering Co., Ltd. Tianshui, Gansu,741030 China

Abstract: The stability of loess subgrade is affected by many factors. Dry-wet cycle is a common factor that causes damage to the strength of loess subgrade. In order to study the influence of dry-wet cycle on the stability of loess subgrade, reveal the mechanism of dry-wet cycle damage to the strength of loess subgrade It is necessary to sample the loess soil samples in this area, and perform compaction test, dry-wet cycle test, unconfined compression test and scanning electron microscope test on the soil sample to analyze the influence of dry-wet cycle on the strength and microstructure of loess soil. . In this way, the potential relationship between the microstructure and the strength of the soil can be obtained. The results show that the loess dry-wet cycle causes the loess's tensile strength and compressive strength to decrease with the increase of the number of dry-wet cycles. The dry-wet cycle effect gradually reduces the shear structural strength of the undisturbed loess and eventually disappears. Scanning electron microscope tests found that the humidification process caused slight movement of the particles, while the drying process resulted in permanent deformation of small pores.

Keywords: Loess; Dry-wet cycle; Mechanical properties; Microstructure.

1. Introduction

The loess is in the cyclical change of climate and seasons, and the soil is also in the cyclical state of alternating dry and wet receiving rainfall and evaporation at this time, and such alternating dry and wet have a great impact on the physical and mechanical properties of loess. Previous studies by scholars have shown that the loess that has undergone multiple drying and wet cycles will weaken the shear strength of the loess, the ability to resist deformation, and the structure of the loess will be destroyed. Therefore, the performance of the loess under the dry-wet cycle The research has an important impact on the construction of the loess area.

Study on the strength of loess under dry and wet cycles. Hu Wenle et al. [1] conducted a large number of consolidated and undrained triaxial tests and concluded that the shear strength is closely related to the water content, compactness and confining pressure. Barzegar [2] conducted a large number of experiments on reshaping soil by simulating the indoor dry-wet cycle environment. The results showed that in the dry-wet cycle of increasing drying, the softening of soil by humidification is an important factor in reducing soil strength. The repeated occurrence of the wet cycle will increase the intensity changes caused by the softening of the soil. Wang Li et al.[3] found through the direct shear test of loess that no matter how much the clay content of the loess is, as the moisture content increases, the c value of the sample first increases and then decreases, while the Φ value shows It is a linear decline. Wang Wei et al. [4] found through the study of remolded loess that the peak strength and residual strength of remolded loess are similarly affected by moisture content and normal stress, and the moisture content can be used to estimate the residual strength. AL-Homoud AS [5] found that the water-soil characteristic curve of expansive soil under dry-wet cycle conditions changes with the change of its water content, and the shear strength changes of soil samples with the same suction are also different. Wu Junhua [6] conducted repeated dry-wet shrinkage tests on reshaped expansive soils, and finally concluded that the repeated dry-wet shrinkage effects will significantly reduce the shear strength of the soil, and the attenuation of cohesive force is higher than Angle of internal friction. Cheng Jiaming et al. [7] studied the influence of dry-wet cycles on the shear strength of solidified loess through direct shear tests, and showed that the strength of solidified loess decreased overall after drying-wet cycles, but it was much higher than that of plain loess. Zhao Jingang et al. [8] established a fill slope model in Ankang expansive soil area, conducted dry-wet cycle tests on expansive soil slopes, combined with theoretical analysis and experimental research, and used dry-wet cycles to determine the dynamics of physical quantities of expansive soil slopes. The influence law is studied, and the research results show that, among the factors that affect the moisture of expansive soil, sunshine and cloudy days are indirect factors, while stagnant water and rainfall are direct factors. Mu Huandong et al. [9] studied the influence of moisture content on the shear strength parameters and strength attenuation characteristics of loess under dry-wet cycles through direct shear tests. The shear strength is reduced, and the strength attenuation is most obvious when one dry-wet cycle is applied; the more the number of dry-wet cycles, the smaller the amplitude of strength attenuation. Yan Yajing et al. [10] studied the changes in the shear strength of unsaturated remolded loess and its parameters after being affected by soluble salt, and concluded that soluble salt has a significant non-linear weakening of the shear strength and cohesion of unsaturated loess.

For the study of the loess structure under the dry-wet cycle. Diel J et al. [11] believed that the dry-wet cycle not only affects the easily visible macroporous soil (> 8μm), but also affects the mesoporous soil. The lower the bulk density, the greater the change in macropores. Repeated wet and dry cycles did not cause random changes in the distance between particles and pores. Microstructural evolution of expansive clay during drying–wetting cycle [12] observed and analyzed and obtained the distribution of adsorbed water and capillary water, and measured the changes of adsorbed water and capillary water content with matrix suction during the drying-wetting cycle. The evolution of pores and macropores under suction is studied. It is observed that the micropores undergo permanent deformation during the dry-wet cycle. Ye Wanjun et al. [13] believed that with the progress of the dry-wet cycle, the contact form between paleosol particles gradually transitioned from surface-surface contact to edge-surface contact, and finally developed into edge-edge or point-surface contact. At the same time, the proportion of small pores gradually decreases, the content of large and medium pores gradually increases, the particle abundance value continues to increase, and the fractal dimension gradually decreases. Romero E et al. [14] found that during the dry-wet cycle, the soil microstructure changes significantly, accompanied by volume changes, which may strongly affect soil properties, such as compressibility, shear strength, and hydraulic conductivity. During the drying process, most of the shrinkage deformation occurs in the capillary zone, and the interfacial surface tension tightens the skeleton through capillary force. The discharge of pore water occurs sequentially from larger pores to smaller pores. The author divides water into three categories, capillary water, adsorbed water and strongly bound water. In the process of wet and dry cycles, capillary water is easier than adsorbed water. Discharge, and strongly bound water is always present. It also introduces the application status and development of two commonly used techniques for studying the microstructure of partially saturated soils, namely mercury intrusion method and environmental scanning electron microscopy. Research on the properties of lime-improved expansive soil under dry-wet cycles_Qiu Qinyuan et al. [15] conducted dry-wet cycle experiments on cement-improved loess, and found that after 8 dry-wet cycles, the edge cracks of the 3% sample gradually expanded. When the sample is separated from the ring knife, the shrinkage is obvious, and the soil at the edge of the sample peels off more seriously, and there are large penetration cracks; the sample with 5% content has obvious shrinkage, and the number of cracks at the edge of the sample increases. The phenomenon of block spalling is more serious than the previous few times, but it can ensure the integrity of the sample as a whole, and there is no obvious penetration crack; the sample with 7% content is complete with a little shrinkage, but there is still no obvious crack, nor is there any The phenomenon of soil block peeling occurs. Wang Fei et al. [16] used expansive soil and loess as the research objects. The shrinkage index line shrinkage rate and volume shrinkage rate of expansive soil decreased with the increase of the number of drying and wet conditions, and finally stabilized. Shi Gang et al. [17] selected samples from representative highway engineering test sites and applied SEM images to study the microstructure of loess. They found that the geological age changes from new to old, and the degree of weathering to soil has increased weakly. And the corresponding structure types mainly include four types of scaffolding macropores, embedded micropore microcemented structure, flocculent, clotted cemented structure, and embedded microporous semi-cemented structure.

2. Experiment method

2.1 Compaction test method

Compaction test According to the "Code for Design of Highway Subgrade" [18], a heavy-duty compactor is used to carry out compaction tests on the modified expansive soil with different lime content. The preparation method of the dry compaction test sample for lime-improved expansive soil:(1) After the soil sample is air-dried to a moisture content of less than 10%, it is passed through a 5 mm sieve, and the soil sample under the sieve is taken for later use. (2) Determine the moisture content of the air-dried soil sample and divide the soil sample into several parts (each soil sample is 2.5-3.0 kg).(3) After adding water to each soil sample to the set moisture content, seal it in a plastic bag for 24 hours so that the soil sample is fully infiltrated and the moisture content is uniform.(4) Perform compaction after mixing, and determine the maximum dry density and optimal moisture content of different samples according to the test method.

2.2 The method of improved soil preparation

The selected soil samples are dried, crushed, and sieved. Mix the admixture with the treated loess in proportion. Calculate the water quality according to the water content of the modified soil, add the weighed water to the mixture, stir it evenly, and let it stand in a humidifier for 24 hours. The moisturized mixture is placed in a mold for uniform compaction, and the mixture in the mold is compacted from both ends in three compactions. Use a demolding instrument to eject the test piece from the mold. In order to prevent moisture loss, seal the prepared sample with a plastic film and place it in a curing box (the temperature of the curing box is (20±1)°C, and the relative humidity is

(96±2)%), in the humidity box Keep in good health for 7 days. Make the test piece with a degree of compaction greater than 95%, and then perform multiple freeze-thaw cycles on the soil sample. In the experiment, try to use the same batch of test pieces and perform different freeze-thaw cycles.

2.3 Test method of dry and wet cycle

According to the test results of the soil moisture content of the soil at the borrow site, and taking the depth of the soil into consideration, the main concentration range of the water content of the loess within this depth is to be simulated or measured [19].

Humidification, for samples that require a dry-wet cycle, the samples need to be humidified. First measure the quality and moisture content of the sample, calculate the amount of water that needs to be added to reach the predetermined moisture content, and then use a syringe to titrate the sample to add water, and after the sample reaches the predetermined moisture content, put the sample in a moisturizing cylinder and let it stand for 24 hours Above, make the moisture in the sample uniform.

The dry-wet cycle is divided into two processes: drying and humidification. After drying and humidifying once, it is called a dry-wet cycle.

Dry, let the soil body stand for more than 24 hours, the sample with uniform moisture in the sample will be air-dried naturally, and the sample needs to be weighed at intervals. When the quality required by the lower limit moisture content is reached, stop air-drying and put it in a moisturizer Keep moisture in the cylinder for more than 24 hours to make the water in the sample evenly distributed.

After the sample is air-dried and humidified to the target moisture content, the soil sample is sealed and cured for 24 hours to ensure sufficient moisture transfer, and then the next dry-wet cycle of the sample is performed again [21]. For the long-term multi-period dry-wet cycle simulation, Huang D proposed a 30-day cycle and a 180-day dry-wet cycle simulation. Drying in a blast drying oven at 50°C for 24 hours achieves the purpose of dehumidification. Shake in 100 ml of distilled water for 5 minutes in a shaker, and let it stand for 24 hours to achieve the humidification effect.

3. Experimental results

3.1 The influence of dry-wet cycle on the strength of loess

The strength of loess is mainly divided into three parts: tensile strength, compressive strength, and shear strength.

(1) Tensile strength

The tensile stress-strain curve of loess is obviously different under the influence of moisture content and the number of wet and dry cycles, but they are all brittle failures, and the curve is basically a strain hardening type; the lower the soil moisture content, the stage of the stress-strain curve of loess The more obvious it is; the number of dry-wet cycles will not change the law of the stress-strain curve of loess under the same moisture content, but it will reduce the tensile stress and tensile strain of the loess, and as the moisture content becomes smaller, this decrease The more obvious the phenomenon.

The tensile strength of loess is obviously affected by the number of wet and dry cycles. With the increase of the number of wet and dry cycles, the attenuation amplitude of its tensile strength decreases. After one dry and wet cycle, the attenuation amplitude of the tensile strength of loess is the largest. It tends to be stable after the second wet and dry cycle.

(2) Compressive strength

Under the condition of uniaxial compression, the stress-strain curve of the loess sample increases nonlinearly, decreases rapidly after reaching the peak, and finally stabilizes to a certain value. The number of dry-wet cycles will not change the law of the compressive stress-strain curve of loess under the same moisture content, but it will reduce the ultimate compressive strain of loess, and the smaller the moisture content, the more obvious this reduction phenomenon.

The number of dry-wet cycles has a very obvious influence on the unconfined compressive strength of loess. The compressive strength decreases continuously with the increase of the number of dry-wet cycles. After one dry-wet cycle, the unconfined compressive strength of loess attenuates the most. However, with the increase of the number of wet and dry cycles, the attenuation amplitude gradually weakened, and it basically stabilized after several wet and dry cycles.

(3) Shear strength

The increase in the number of dry and wet cycles affects the shear strength index of loess. Under the same conditions, the increase in the number of dry and wet cycles and moisture content weakens the cementation of the loess particles, resulting in the shear strength, cohesion and internal friction angle of the soil. In loess with high moisture content, the reduction of cohesive force is much smaller than that of loess with low moisture content; the internal friction angle of undisturbed and remolded loess gradually increases with the increase in the number of dry and wet cycles Decrease, but not much. The attenuation value of the shear strength parameters of the undisturbed and reshaped loess after several cycles of drying and wetting is reflected, especially the attenuation value of the cohesive force is relatively large. The dry-wet cycle effect will gradually reduce the shear structural strength of the undisturbed loess, and eventually disappear, making its strength properties close to those of remolded loess.

3.2 Changes in microstructure

The study of soil microstructure has gradually changed from qualitative analysis to quantitative research. With the development of instruments, there are more and more testing methods. The combination of electron microscope and computer image analysis is a better method to study soil microstructure , You can get basically all the microstructure parameters of the soil.

Without the effect of dry-wet cycle, the particles are tightly arranged, and the initial structure strength is greater. After being humidified, the bound water film around the soil particles is thickened, reducing the sliding friction between the particles, causing slight movement between the particles; while in the drying stage, the continuous loss of moisture causes damage to the structure of the sample. Local pores increase, which leads to a decrease in occlusal friction between particles. After three dry-wet cycles, the internal structure of the soil sample has tended to a new stable equilibrium state, after which the cycle action has little effect on the strength of the soil. With the increase in the number of wet and dry cycles, the tiny pores inside the loess gradually become larger and the porosity increases [22].

During the dry-wet cycle, the soil microstructure changes significantly, accompanied by volume changes, which may strongly affect soil properties, such as compressibility, shear strength, and hydraulic conductivity. During the drying process, most of the shrinkage deformation occurs in the capillary zone, and the interfacial surface tension tightens the skeleton through capillary force. The discharge of pore water occurs sequentially from larger pores to smaller pores. The internal moisture of loess during the dry-wet cycle can be divided into three categories: capillary water, adsorbed water and strongly bound water. During the dry-wet cycle , Capillary water is easier to discharge than adsorbed water, and strongly bound water always exists. For saline loess, with the increase in the number of dry-wet cycles, the accumulated salt expansion continuously destroys the original structure of the soil, accompanied by the fragmentation of the particles, resulting in the gradual decrease of the occlusal friction of the soil, so the overall strength is continuously decreasing. Trend [23].

4. Conclusions

(1) The precautions in the loess dry-wet cycle experiment are summarized. Considering the actual situation of the project, the humidity of the test loess dry-wet cycle should be controlled to simulate the actual situation. The addition of admixtures may enhance the moisture retention capacity of the loess.

(2) Sort out some loess strength tests under dry-wet cycle conditions, introduce the methods and procedures of compaction test, as well as the humidification and drying methods in dry-wet cycle experiments, and the precautions in electron microscope scanning experiments.

(3) Summarized the changes in the microstructure of loess by the dry-wet cycle. The humidification process leads to slight movement of particles, while the drying process leads to permanent deformation of small pores. After several cycles of dry and wet, the internal structure of the soil sample has tended to a new stable equilibrium state, after which the cyclic effect has little effect on the strength of the soil.

References

1. Hu Wenle, He Pengli, Liu Hua. Analysis of the variation law and optimal mix ratio of basalt fiber loess shear strength[J]. Chinese Journal of Geological Hazard and Control, 2019, 30(4): 92-97.
2. BarzegarAR, Oades JM, Rengasamy P. Soil structure degradation and mellowing of compacted soils by saline-sodic solutions[J]. Soil Sci Soc Americ Journal 1996;60(2):583-8.
3. Wang Li, Li Xi'an, Hong Bo, et al. Test of the influence of clay content on the shear strength of loess [J]. Coal Geology and Prospecting, 2019, 47(3): 179-185.
4. Wang Wei, Luo Yasheng. Ring shear test study on the shear strength of remolded loess [J]. Bulletin of Soil and Water Conservation, 2017, 37(5): 110-114.
5. Al-Homoud A S, Basma A A, Malkawi A I H, et al. Cyclic Swelling Behavior of Clays[J]. Journal of Geotechnical Engineering. 1995,121(7):562-565.
6. Wu Junhua, Yang Song. Measurement of matrix suction of expansive soil under dry-wet cycles and experimental study of its influence on shear strength[J]. Rock and Soil Mechanics, 2017, 03:678-684.
7. Cheng Jiaming, Wang Yinmei, Miao Shichao, etc. Study on the dry-wet cycle characteristics of solidified loess[J]. Journal of Engineering Geology, 2014, 22(02): 226-232.
8. Zhao Jingang, Wang Jiading, Gu Tianfeng, et al. Deformation characteristics and mechanism of expansive soil fill slope under rainfall-evaporation cycle[J]. Journal of Natural Disasters, 2013, 06:116-124.
9. Mu Huandong, Deng Yahong, Li Rongjian. Study on the influence of dry-wet cycles on the shear strength of loess in the ground fissure zone[J]. Journal of Engineering Geology, 2018, 26(05): 1131-1138.
10. Yan Yajing, Wen Baoping, Huang Zhiquan. The influence of soluble salt on the shear strength of Lanzhou unsaturated remolded loess[J]. Rock and Soil Mechanics, 2017, 38(10): 2881-2887.
11. Diel J, Vogel H J, Schlüter S. Impact of wetting and drying cycles on soil structure dynamics[J]. Geoderma, 2019, 345: 63-71.
12. Ma T, Wei C, Yao C, et al. Microstructural evolution of expansive clay during drying–wetting cycle[J]. Acta Geotechnica, 2020, 15(8).
13. Ye Wanjun, Wu Yuntao, Yang Gengshe, Jing Hongjun, Chang Shuibin, Chen Ming. Study on the fine microstructure and macromechanical properties of ancient soils under the action of dry-wet cycles[J]. Chinese Journal of Rock Mechanics and Engineering, 2019 , 38(10): 2126-2137.
14. Romero E, Simms PH (2008) Microstructure investigation inunsaturated soils: a review with special attention to contribution of mercury intrusion porosimetry and environmental scanningelectron microscopy. Geotech Geol Eng 26(6):705–727. https://doi.org /10.1007/s10706-008-9204-5
15. Qiu Qinyuan. Research on the properties of lime-improved expansive soil under dry-wet cycle[J]. Roadbed Engineering, 2021(01): 79-83.

16. Wang Fei, Li Guoyu, Mu Yanhu, Zhang Peng, Wu Yahu, Fan Shanzhi. Experimental study on the deformation characteristics of compacted loess under dry-wet cycles[J]. Rock and Soil Mechanics, 2016, 37(08): 2306- 2312+2320.
17. Shi Gang, Zhi Xilan, Ma Jianqin, et al. Composition and microstructure characteristics of loess and foundation bearing capacity[J]. Highway and Transportation Science and Technology, 2005, 22(12): 46-49
18. Ministry of Communications of the People's Republic of China. Highway Roadbed Design Code: JTG D30—2004 [S]. Beijing: People's Communications Publishing House, 2004.
19. Song Yang, Liu Siyuan, Wang Chenjing. The influence of moisture content and dry-wet cycles on the deformation characteristics of undisturbed loess[J]. Journal of Liaoning Technical University (Natural Science Edition), 2021, 40(02): 148-155.
20. Yuan Zhihui. Study on the change mechanism of strength and microstructure of loess under dry-wet cycles[D]. Chang'an University, 2015.
21. Huang D, Niu D, Su L, et al. Chloride diffusion behavior of coral aggregate concrete under drying-wetting cycles[J]. Construction And Building Materials, 2021, 270: 121485.
22. Pan Zhenxing, Yang Gengshe, Ye Wanjun, etc. Study on the mechanical properties and meso-damage of undisturbed loess under dry-wet cycles[J]. Journal of Engineering Geology, 2020.
23. Ma Junze, Zhang Weibing, Zhang Xiao. Experimental study on the strength characteristics of sulfated saline soil under dry-wet cycles[J]. Chinese Journal of Underground Space and Engineering, 2020, 16(03): 704-713.

 https://doi.org/10.1051/matecconf/202235801032

Influence of preparation conditions on the physical structure and surface properties of enteromorpha clathrate bio-char

Lu Liu*, Yali Zheng, Peng Gong, Guangcai Shao, Xu Huang

School of Energy and Power Engineering, Jiangsu University, Zhenjiang, China

Abstract. Bio-char with rich pore structure was obtained by pyrolysis of enteromorpha clathrate (EC) and subsequent activation process. The effect of pyrolysis methods, KOH concentrations used in activation and heating rate was studied. A new method that combined pyrolysis and activation into one step was proposed. The bio-char obtained via slow pyrolysis at 450°C and 700°C had a certain pore structure. The bio-char obtained via fast pyrolysis had almost no effective pore structure and a large amount of organic matters still exist in the bio-char, while it had the largest specific surface area after activation by KOH at 800°C. Therefore, bio-char that is a kind of by-product in the process of making bio-oil by fast pyrolysis of EC is worthy of further exploration. As the KOH concentration of the impregnated solution increased, the specific surface area first increased and then decreased. The bio-char obtained via impregnation and activation with 3 mol/L KOH had the highest specific surface area (1128.85 m2/g) and pore volume (0.789 cm3/g). If the processes of pyrolysis and activation were combined into one step via mixing KOH and EC, the structure of cell tissue in the EC was reserved. The sample that prepared by mixing 2 g dried EC with 0.1 g KOH powder has the highest specific surface area (724.66 m2/g) and better pore structure. The best heating rate was 5°C/min for impregnation method and 2°C/min for one step method.

1. Introduction

Bio-char is a kind of carbon based material that has been attracting increasing attention because of its abundant raw materials and high adsorption capacity [1]. Among them, seaweed bio-char has a low price and contains active ions such as K and Na, which has great research values [2]. Enteromorpha clathrate (EC), as a kind of common marine green algae, is produced in the southeastern coastal areas of China. EC has a high efficiency in absorbing nutrients and a fast growth rate, which has a good auxiliary effect on the ecological restoration of the ocean. However, with the continuous marine pollution, the ocean eutrophication is becoming serious, leading to the increase of EC production. Meanwhile, with the developing of bio-oil production by pyrolysis of EC, a large amount of bio-char will be inevitably produced in this process [3]. Compared with activated carbon, this type of bio-char eliminates the carbonization process and then reduces costs in carbon production. If the bio-char from oil production is activated and applied to the field of industrial processes, it will be a very promising comprehensive utilization approach [4].

The operation conditions in the pyrolysis process have significant impacts on the characteristics of bio-char. Jian et al. [5] found that with the increase of pyrolysis temperature, the PH and ash content of bio-char tended to increase. Antal et al. [6] found that the bio-char yield and carbon content decreased as the heating rate increased. Bandosz et al. [7] found that the specific surface area and pore volume of the bio-char decreased if the pyrolysis time was extended from 30 min to 1 h at 950°C.

The activation process is an important step in the modification of bio-char. The activation methods can be divided into physical activation and chemical activation. Ahmad et al. [8] used CO_2 and H_2O as activating gas to treat bio-char from palm trees and the specific surface area of activated carbon prepared at 792-1079°C reached 1084 m^2/g, 85.9% of which belonged to the micropores. The characteristics of bio-char after physical activation are related to the activated gas, reaction conditions, and the type of bio-char. Ghouma et al. [9] found that different bio-char (pine wood chips, walnut shells, etc.) produced under different activation conditions (time, temperature, etc.) was different in the adsorption capacity. After optimizing the conditions, the best bio-char had an adsorption capacity of 131 mg/g for NO_2, while the worst was only 25.5 mg/g. Nabais et al. [10] used CO_2 and to activate the bio-char after pyrolysis and found that the use of water vapor as activating gas can make the bio-char have wider pore size distribution, while CO_2 can make the bio-char have larger pore volume. Rostamian et al. [11] used KOH as activator and the specific surface area of the activated bio-char reached 2201 m^2/g. Compared with KOH, NaOH had less corrosiveness and lower price. However, the surface area of the bio-char activated by NaOH was less than 1 m^2/g with less microporous. Zhao et al. [12] used $CaCl_3$ in activation to increase the number of basic groups on the surface, the degree of aromatization and the adsorption capacity. Sun et al. [13] used H_3NO_4

* Corresponding author: lliuenergy@ujs.edu.cn

as an activator and found that the prepared bio-char had a specific surface area of 1252 m^2/g with more micropores. In this paper, we use EC to produce bio-char and study the influence of pyrolysis methods, KOH concentrations and heating rate on the physical and chemical properties of bio-char. The feasibility of explore bio-char from bio-oil production by fast pyrolysis is discussed.

2. Experimental

2.1 Preparation of bio-char

EC bio-char is prepared by two methods: fast pyrolysis and slow pyrolysis.

Fast pyrolysis method: The fast pyrolysis device is a stainless steel reactor. First, the reactor was heated to 550°C at a heating rate of 15°C/min, and then N_2 gas passed over the reactor at a flow rate of 1 L/min for 5 minutes. The flow rate was set to 200 mL/min and the dry EC powder was put into the reactor and maintained 15 minutes. After the reaction, the bio-char was collected and denoted as EC-FP.

Slow pyrolysis method: 5 g of dry EC powder was placed in a tubular furnace and passed 400 mL/min N_2 as a protective gas. After draining the air in the furnace, temperature of the tubular furnace was first raised to 100°C at a heating rate of 10°C/min and kept for 30 minutes at 100°C, then raised to 450°C or 700°C at 1°C/min and kept for 30 minutes. After cooling to room temperature, the sample was taken out and denoted as EC-450 and EC-700.

2.2 Activation of EC bio-char

2 g of bio-char prepared in section 2.1 was placed in KOH solution for 24 h and then filtered and dried at 110°C for 12 h. Then the sample was put into the tubular furnace with a N_2 flow at 400 mL/min as a protective gas. The activation is from room temperature to 800°C at a heating rate of 2°C/min or 5°C/min and maintained 30 minutes at 800°C. The cooled sample was taken out, washed with 1 mol/L hydrochloric acid solution and deionized water to pH=7, filtered and dried at 110°C for 12 h. The sample was denoted as EC-A-BM-C, A represents the pyrolysis method in section 2.1, B represents the mole of the KOH solution, and C represents the heating rate.

2.3 Preparation of EC bio-char by one-step activation method

2 g of dry EC powder mixed with KOH powder and put into a tubular furnace with 400 mL/min N_2. The activation was from room temperature to 800°C at 1°C/min or 2°C/min and maintained 30 min at 800°C. The cooled sample was taken out and denoted as EC-2-D-C, D represents the quality of the KOH, and C represents the heating rate.

2.4 Characterization analysis

N_2 adsorption and desorption at liquid nitrogen temperature (77 K) was performed on the NOVA 3000e (Kangta). The pretreatment of the samples was at 200°C for 3 hours. Fourier transform infrared (FTIR) spectrometer test was carried out in the Nicolet iS50 (USA Thero Electron Corporation). The bio-char was mixed with KBr and pressed. The scanning range was 4000~400 cm^{-1}, and the resolution was 4 cm^{-1}. The scanning electron microscope (SEM) was obtained using the S-4800 (Hitachi).

3. Results and discussion

3.1 Influence of pyrolysis method

To study the structural characteristics of EC after different pyrolysis processes, the bio-char produced by fast pyrolysis and slow pyrolysis at 450°C and 700°C were analyzed. According to Table 1, the EC-FP obtained by the fast pyrolysis had no effective pore structure and EC-450 obtained by slow pyrolysis at 450°C had a low specific surface area, 84% of which was macropores. The specific surface area of EC-700 obtained by slow pyrolysis at 700°C was 43.48 m^2/g, and 81% was microporous. It can be seen that the pyrolysis rate and temperature have a significant effect on the pore structure of the bio-char.

After being impregnated and activated with KOH at a concentration of 2 mol/L, the specific surface area of the bio-char was greatly improved. Although EC-FP had almost no pore structure, after activation, the specific surface area of EC-FP-2M-5 was the largest (801.99 m^2/g), 79% of which was microporous. The specific surface area of EC-700-2M-5 and EC-450-2M-5 reached 560.42 m^2/g (72% microporous) and 480.49 m^2/g (88.2% microporous).

Table 1. Specific surface area (S_{BET}) and pore parameters of the EC bio-char.

Samples	$S_{BET}(m^2/g)$	$V(cm^3/g)$	$S_{mic}(m^2/g)$	$S_{ext}(m^2/g)$
EC-FP	~0	—	—	—
EC-450	1.96	0.003	0.32	1.64
EC-700	43.48	0.031	35.35	8.13
EC-FP-2M-5	801.99	0.605	631.18	170.82
EC-450-2M-5	480.49	0.311	424.03	56.46
EC-700-2M-5	560.42	0.434	400.94	159.47
EC-FP-1M-5	492.53	0.383	376.91	115.62
EC-FP-3M-5	1128.85	0.789	961.59	167.26
EC-FP-5M-5	1078.70	0.772	879.86	198.85
EC-FP-7M-5	912.43	0.641	740.68	171.75
EC-2-0.05-1	42.27	0.033	32.33	9.94
EC-2-0.1-1	594.24	0.451	433.67	160.58
EC-2-0.2-1	499.45	0.372	387.64	111.81
EC-FP-3M-2	883.98	0.647	709.74	174.24
EC-2-0.1-2	724.66	0.554	521.84	202.82
EC-2-0.1-5	589.17	0.469	417.56	171.60

Figure 1(a) shows the FTIR results of the bio-char prepared by different pyrolysis methods. It can be seen that the peak positions of EC-FP, EC-700 and EC-450 were same, indicating that the functional groups of the samples were same, including C–H, C–O, C=O, C=C, O–H, etc. The absorption intensity of EC-FP was the highest, followed by EC-450, and EC-700 was the weakest, indicating that EC-FP has a large number of surface functional groups because of incomplete pyrolysis. The peak at 700-1000 cm^{-1} corresponds to C–H bond [14]. The prominent peaks at 1095 cm^{-1} and 1013 cm^{-1} correspond to C–O bonds; the other prominent peaks at 1741 cm^{-1} and 1605 cm^{-1} correspond to C=O of carboxyl or carbonyl groups [15, 16]. The tiny absorption peak at 2371 cm^{-1}

corresponds to the C≡C bond; the absorption peaks at 3436 cm^{-1} and 3309 cm^{-1} correspond to the stretching vibration of the O–H bond [15, 17]. EC-FP and EC-450 had absorption peaks at 1487 cm^{-1} and 1416 cm^{-1} that correspond to aromatic bonds, while it was not observed for EC-700 [17].

After being impregnated and activated by KOH, the FTIR of bio-char tended to be consistent (Figure 1(b)), indicating that although the pre-pyrolysis methods were different, the types and proportions of surface functional groups after activation were similar. The peak corresponding to the C–O bond shifted to 1052 cm^{-1}. Compared with the peaks of EC-450 and EC-700, the peaks of EC-450-2M-5 and EC-700-2M-5 had increased intensities, indicating the number of C–O functional groups on the surface of the bio-char increased after activation. The absorption peaks corresponding to the C=O bond shifted to 1684 cm^{-1} and 1573 cm^{-1}. The functional group corresponding to the aromatic bond disappears.

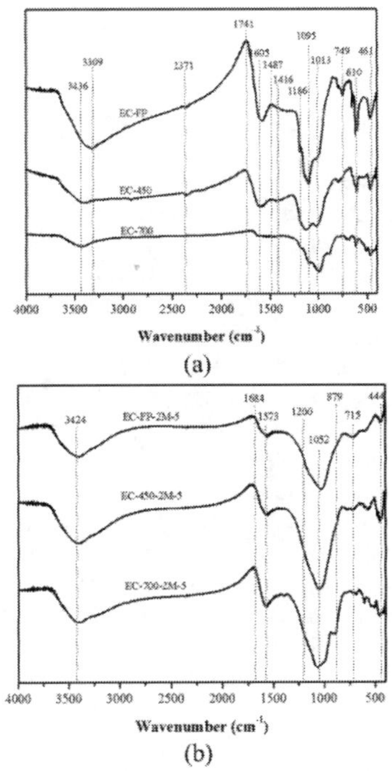

Figure 1. FTIR spectra of the bio-char (a) prepared by fast pyrolysis and slow pyrolysis at 450°C and 700°C, and (b) activated by 2 mol/L KOH after fast pyrolysis and slow pyrolysis.

3.2 Influence of KOH concentration in impregnation

In order to study the effect of impregnation concentration on the surface properties of bio-char, the fast pyrolyzed bio-char was impregnated with 1-7 mol/L KOH solution and activated at 800°C. As the solution concentration increased, the specific surface area first increased and then decreased, and the pore volume was positively correlated with the specific surface area (Table 1). This is because when the concentration of the KOH solution was low, there was less KOH adhesion on the surface of the bio-char, and it was not fully etched when activated; as the solution concentration increased, the activation became more fully; but when the solution concentration was too high, the KOH condensed into agglomerates on the surface of the bio-char, resulting in a poor etching effect. The specific surface area and pore volume of the bio-char prepared by impregnation and activation with 3 mol/L KOH solution were the highest, reaching 1128.85 m^2/g and 0.789 cm^3/g, respectively, and 85% were micropores. Figure 2(a) is the SEM images of the sample EC-FP-3M-5. It can be seen that the sample had a layered structure with well-developed pores distributed inside.

3.3 One-step preparation method

During the experiments, we found that by mixing dry EC powder with KOH powder at different quality and then pyrolyzing it, bio-char with a high specific surface area can also be obtained. The specific surface area and pore structure of bio-char prepared with different mixing ratios of EC and KOH are shown in Table 1. When 0.05g KOH powder was mixed with EC, the specific surface area was only 42.27 m^2/g, and the pore volume was only 0.033 cm^3/g. When 0.1 g KOH powder was mixed with EC, the specific surface area increased to 594.243 m^2/g with a pore volume of 0.451 cm^2/g. However, when 0.2 g KOH powder was mixed with EC, its specific surface area and pore volume were reduced.

Figure 2(b) is SEM images of sample EC-2-0.1-1. The cell tissue structure of this sample was obvious with lamellar layers, and the surface was distributed with more circular pores, whose pore size was smaller than that EC-FP-3M-5. Figure 2(c) shows the FTIR spectrum of EC-2-0.1-1. Compared with EC-FP-3M-5, the absorption peaks were shifted to the left. There were many chaotic peaks before 734 cm^{-1}, and these peaks are mainly C-H bonds of functional groups. In addition to the main absorption peaks of 1589 cm^{-1}, 1149 cm^{-1} and 1095 cm^{-1}, there are also strong absorption peaks at 993 cm^{-1} and 919 cm^{-1}, representing the C–O bond. Bio-char prepared by one-step method has more abundant surface functional groups.

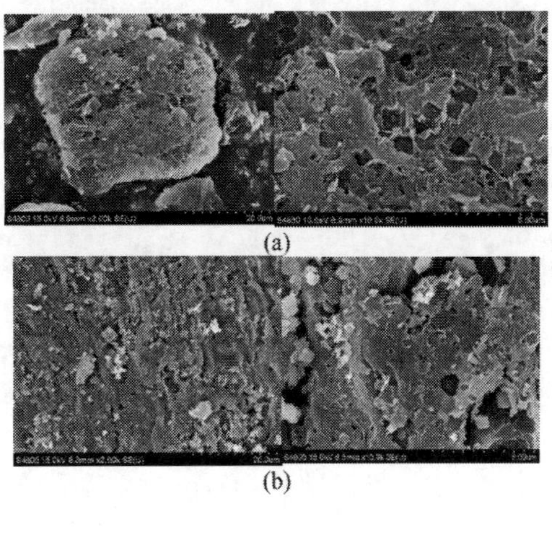

(a)

(b)

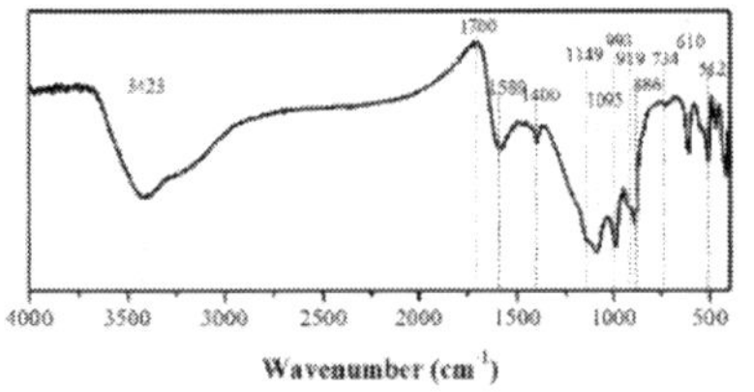

(c)

Figure 2. SEM images of (a) EC-FP-3M-5 and (b) EC-2-0.1-1; (c)FTIR spectra of EC-2-0.1-1.

3.4 The effect of heating rate

In order to study the effect of heating rates on EC bio-char, EC-FP-3M and EC-2-0.1 were prepared with different heating rates. It can be seen from Table 1, the pore characteristics of the sample prepared by impregnation with an activation heating rate of 5 °C/min was better than 2 °C/min. For the bio-char obtained by mixing and pyrolysis of EC powder and KOH powder, the sample prepared at a heating rate of 2 °C/min had the highest specific surface area, reaching 724.658 m^2/g.

4. Conclusion

EC bio-char with abundant pore structure was obtained and effects of the pyrolysis method, activation method, impregnation concentration and heating rate on the physical and chemical properties of the bio-char was studied. The bio-char prepared by fast pyrolysis before activation had the highest specific surface area. Therefore, as a kind of by-products in producing bio-oil by fast pyrolysis of EC, bio-char can be used as raw material to prepare active carbon.

Before activation, the bio-char obtained by slow pyrolysis had a certain pore structure, while the bio-char obtained by fast pyrolysis had almost no effective pore structure. After KOH activation, the specific surface area of the bio-char had a great increase. The specific surface area and pore volume of bio-char prepared with 3 mol/L KOH solution and 5°C/min heating rate were the highest, reaching 1128.85 m^2/g. Although the number and proportion of surface functional groups of bio-char obtained by different pyrolysis methods had a great difference, they got unanimity after activation.

EC bio-char can also be prepared by the one-step method, which has more abundant surface functional groups and obvious cell tissue structure. The bio-char prepared by mixing 2 g EC and 0.1 g KOH powder with 2°C/min heating rate had a higher specific surface area and a better pore structure.

Acknowledgments

This work was supported by the National Natural Science Foundation of China (51906090) and Jiangsu Government Scholarship for Overseas Studies (JS-2019-221).

References

1. S. Tao, C. Li, X. Fan, G. Zeng, L. Pei, Z. Xing, Q. Wen, W. Zhao, D. Luo, C. Fan, Activated coke impregnated with cerium chloride used for elemental mercury removal from simulated flue gas, Chem. Eng. J. 210 (2012) 547-556.
2. H. Lee, R.S. Park, H.W. Lee, Y. Hong, Y. Lee, S.H. Park, S.C. Jung, K.S. Yoo, J.K. Jeon, Y.K. Park, Adsorptive removal of atmospheric pollutants over pyropia tenera chars, Carbon. Lett. 19 (2016) 79-88.
3. W. Chen, W.M. Zhang, J. Meng, Z. Xu, Researches on biochar application technology, Eng. Sci. 13 (2011) 83-89.
4. A.T. Tag, G. Duman, S. Ucar, Effects of feedstock type and pyrolysis temperature on potential applications of biochar, J. Anal. Appl. Pyrol. 120 (2016) 200-206.
5. M. Jian, K. Gao, H. Yu, Effects of different pyrolysis temperatures on the preparation and characteristics of bio-char from rice straw, Acta Sci. Circum. 36 (2016) 1757-1765.
6. M.J. Antal, M. Grønli, The art, science, and technology of charcoal production, Ind. Eng. Chem. Res. 42 (2013) 1619-1640.
7. T.J. Bandosz, K. Block, Effect of pyrolysis temperature and time on catalyticperformance of sewage sludge/industrial sludge-based composite adsorbents, Appl. Catal. B-Environ. 67 (2006) 77-85.
8. A.L. Ahmad, M.M. Loh, J.A. Aziz, Preparation and characterization of activated carbon from oil palm wood and its evaluation on methylene blue adsorption, Dyes. Pigments. 75 (2007) 263-272.
9. I. Ghouma, M. Jeguirim, S. Dorge, L. Limousy, C.M. Ghimbeu, A. Ouederni, Activated carbon prepared by physical activation of olive stones for the removal of NO2, at ambient temperature, CR. Chim. 18 (2015) 63-74.
10. J.M.V. Nabais, P. Nunes, P.J.M. Carrott, M.M.L.R. Carrott, A.M. Garcia, M.A. Diaz-Diez, Production of activated carbons from coffee endocarp by CO2 and steam activation, Fuel Process. Technol. 89 (2008) 262-268.
11. R. Rostamian, M. Heidarpour, S.F. Mousavi, M. Afyuni, Characterization and sodium sorption capacity of biochar and activated carbon prepared from rice husk, J. Agr. Sci. Tech-Iran. 17 (2015) 1057-1069.
12. M.J. Zhao, Synthesis and adsorption properties of modified biochars for Pb2+ removal, Hebei Normal University, 2017.
13. Y. Sun, H. Li, G. Li, B. Gao, Q. Yue, X. Li, Characterization and ciprofloxacin adsorption properties of activated carbons prepared from biomass wastes by H3PO4 activation, Bioresource Technol. 217 (2016) 239-244.

14. S.Y. Oh, D.I. Yoo, Y. Shin, G. Seo, FTIR analysis of cellulose treated with Sodium hydroxide and carbon dioxide, Carbohyd. Res. 340 (2005) 417-428.
15. S. Shin, J. Jang, S.H. Yoon, I. Mochida, A study on the effect of heat treatment on functional groups of pitch based activated carbon fiber using FTIR, Carbon. 35 (1997) 1739-1743.
16. S. Kang, B.S. Xing, Adsorption of dicarboxylic acids by clay minerals as examined by in situ ATR-FTIR and ex situ DRIFT, Langmuir. 23 (2007) 7024-7031.
17. S. Ravulapalli, R. Kunta, Defluoridation studies using active carbon derived from the barks of Ficus racemosa plant, J. Fluorine Chem. 193 (2017) 58-66.

Research progress of water-based release agents

Bo Liang [1,2], Ye Yang [3] and Junping Li [4,*]

[1] School of Aviation and Transportation, Jiangsu College of Engineering and Technology, Nantong 226000, China
[2] Shandong Meichen Industrial Group Co., LTD., Zhucheng 262200, China
[3] Downhole Testing Company of Jianghan Petroleum Engineering Co., Ltd., SINOPEC, Wuhan 430000, China
[4] Changzhou Vocational Institute of Mechatronic Technology, Changzhou 213164, China

Abstract. A mould release agent is an additive that makes the moulded product easy to detach from the mould, it forms a protective film on the product to ensure the surface quality of the mould. Mold release agents are divided into water-based release agents, oil-based release agents and powder-based release agents. Water-based release agents are widely used because they have the advantages of low price, good release effect, green environment protection, smooth casting surface, good air tightness and good automaticity operation. According to the effective release material in the mold, release agent can be divided into silicone oil type, wax and fatty acid type water-based mold release agent. In this paper, the composition, properties and application characteristics of various water-based release agents were analyzed, and the future development direction of water-based release agents was forecasted.

Keyword: Release agent; water - based release agent; silicone oil release agent.

1. Introduction

A release agent is an additive that allows the moulded product to be easily separated from the mould, forming a protective barrier film on the product and ensuring the surface quality of the mould. The principle is that the release agent forms a special insulating film which reduces the adsorption of the product to the mould. This insulating film is only micron thick and highly heat resistant and is a chemically connected film, the main function of which is to reduce the peeling force so that the moulded part can be easily separated intact within a certain temperature range[1-6]. China is a large producer and consumer of plastics and rubber industry, since 2010, the annual production of plastics and rubber has exceeded 60 million tons, the annual production and consumption capacity are ranked first in the world, with the rise of the rubber and plastic industry, the amount of mold release agent is also increasing year by year, so the market size is huge. Mould release agents are widely used in moulded products of elastomers such as rubber and plastics. The specific performance requirements of the release agent are as follows: good release, good continuity of release and the formation of a homogeneous film even for complex shaped moulded objects. Good formability and superior secondary processability: when the release agent is transferred to the formed object, it has no adverse effect on the processed object such as plating, hot press moulding, printing, finishing and bonding. Easy coating, heat resistance and good pollution resistance. Good stability, good physical and chemical stability when used with the compound. Low flammability, low odor, low toxicity and high production efficiency[7-11]。

Water-based release agents are widely used due to their excellent price, good release effect, green environment protection, smooth casting surface, good air tightness and good automaticity of operation[12-16]. After spraying, it is able to produce a thin layer of only micron thickness on the surface of the mould, which not only reduces the impact of the metal liquid on the cavity during pressing and forming, but also reduces the wear of the casting and the cavity, ultimately protecting the cavity and acting as a lubricant. At the same time, it can also adjust the temperature of each part of the mould, so that the internal temperature of the mould remains balanced, thus improving the service life of the mould, therefore, water-based mold release agent instead of oil-based mold release agent has become an inevitable trend. At the same time, it is able to adjust the temperature of each part of the mould, so that the internal temperature of the mould remains balanced, thus increasing the service life of the mould. Therefore, it has become an inevitable trend for water-based release agents to replace oil-based release agents[17]. Depending on the nature of the release agent, mold release agents can be divided into water-based release agents, oil-based release agents and powder-based release agents. Compared to oil-based release agents, water-based release agents are safe, clean, non-polluting, do not damage the surface of the casting, are easy to cool and produce fewer internal pores, and are increasingly used. Water-based release agents have the advantage of

* Corresponding author: Lijunping606@163.com

being less expensive and less costly than powder release agents[18]. Water-based release agents can be classified as silicone, wax or fatty acid based on the effective release substance in the release agent.

2. Silicone oil-based release agents

The term silicone oil is usually used to refer to dimethicone oil [19]. Silicone oils have a lower surface tension than typical organic oils and solvents, are oxidatively stable, chemically inactive and have no physiological activity。Therefore, silicone oil mold release agent has the characteristics of easy to release, does not pollute and corrode the mold and product, use low concentration, can be used in the coating many times, non-toxic, and therefore is widely used in rubber, packaging, plastic synthesis, glass manufacturing, printing production, precision casting and other industries, is the perfect choice of mold release agent[20]. Silicone oils is made by hydrolysis of dimethyldichlorosilane with water to produce a primary polycondensation ring. The cyclosomes are cracked and rectorized to get low cyclosomes, and then the cyclosomes, head capping agent and catalyst are put together to get a mixture of different polymerization degrees, and the low boiling matter can be removed by vacuum distillation to get silicone oil. Silicone oils and their derivatives are suitable materials for polyurethane release agents, which possess excellent release properties, such as aluminium alloy mould release agents consisting of silicone oil, overpressure lubricant, emulsifier and water release agents. Silicone oils and their derivatives have been proven to have excellent release effects for many years. However, if the product is reprocessed, the silicone oil-based release agent will have the disadvantages of less frequent release, easy separation of air bubbles on the product surface and cumbersome post-treatment[6]. Dimethylsilicone oil is a transparent liquid, non-corrosive and physiologically inert. It also has excellent resistance to freezing, heat and oxidation, and can be used for a long time in the temperature range of -50℃ to 200℃. In addition to the properties of methyl silicone oil, methyl hydrogen-containing silicone oil can also be cross-linked at low temperatures under the action of metal salt catalysts and form waterproof films on the surface of various substrates. As the main ingredient of water repellent, anti-adhesive and anti-corrosion agent, and widely used in the surface treatment of textile, leather, paper, metal, glass, wood and other materials. Hydroxyl silicone fluids are methyl silicone fluids capped with hydroxyl silicone fluids. It therefore has special properties due to the active hydroxyl group. It is a processing aid for silicone rubber and can improve the processing properties of silicone rubber. Sewing thread silicone oil is a compound of large molecular weight polysiloxane and liquid paraffin. Therefore, when lubricating sewing threads, it gives them good lubricity, anti-static properties, friction resistance and smoothness. Silicone oil release agents can be divided into solution type silicone oil release agents, oil paste type silicone oil release agents, aerosol type silicone oil release agents and emulsion type silicone oil release agents according to their form. Solution type silicone release agents are formulated by mixing silicone oil and other additives with organic solvents. Depending on the type of silicone oil, its viscosity and the type of additives used, it can be divided into a number of different types. They are mainly used for shell and film casting, die casting and the moulding of rubber and plastic products.

Different kinds of silicone oil are the main component of the solution-type silicone oil release agent, but also can add a variety of wax, clay, MQ-type silicone resin, talc and other ingredients to improve the performance of the release agent, such as dimethyl silicone oil and MQ-type silicone resin mixed with organic solutions to prepare the solution release agent, to improve the mold release durability has very good results. Silicone oil is the main component of the solution-type silicone release agent, but can also add a variety of waxes, clay, MQ-type silicone resin, talc and other ingredients to improve the performance of the release agent. For example, dimethyl silicone oil and MQ type silicone resin mixed with organic solution to prepare the solution release agent, to improve the mold release durability has a very good effect. Paste-type silicone release agents are pastes made by mixing silicone oil with precipitated silica or fumed silica. Compared to oil and emulsion silicone release agents, oil paste release agents are widely used in high temperature and high pressure moulding because of their higher heat resistance and long-lasting release properties. At the same time, the release agent is also suitable for the vulcanisation of bent rubber hoses due to its good lubricating properties. Aerosol silicone release agents consist of methyl phenyl silicone, dimethyl silicone or long chain alkyl silicone as the active ingredient and are sealed in an aerosol can with Freon-11, Freon-22 or Freon replacement. Pure silicone oil release agents are not easy to use, and the release effect is not satisfactory. The solution type release agent also has the danger of fire and poisoning, so the development of emulsion type release agent, and has become the mainstream of the current release agent products. Emulsion release agent can be prepared by mechanical emulsification, that is, silicone oil and water dispersed under the action of emulsifier. It can also be prepared by emulsion polymerization, that is, polysiloxane in the emulsifying agent, catalyst, emulsion polymerization in water system. The emulsifiers used in the latter can be non-ionic, anionic and cationic, but in mechanical emulsification methods, mainly non-ionic emulsifiers are used. The emulsification devices are mainly homogenisers, colloid mills or high-speed stirrers. The silicone oil content of the emulsion products produced is generally 30% to 60% (mass fraction). Further dilute with water to an emulsion with a silicone oil concentration of 0.2% to 1.0% (mass fraction) before use. The diluted emulsion becomes less stable and should be used up as soon as possible. The emulsion release agent has advantages in terms of cost, fire resistance and hygiene, which makes it highly competitive in the market. However, it also has three disadvantages, namely the volatility of water is too poor to be used as a low-temperature mould release agent; poor wettability of the substrate, especially for polymer materials; and the use of emulsifiers reduces the precision of the product. If ordinary hard water is used constantly to dilute the emulsion, scale will form on the surface of the mould, which will further affect the accuracy of the product if it is not cleaned regularly, so it is better to use soft water to dilute the emulsion. In emulsion type products, some are made by adding organic solvents to reduce the amount of emulsifier used to achieve faster water evaporation and reduce the deposition of solids on the mould surface. Wen Jianqiang[21] used methyl silicone oil and methyl phenyl silicone oil as the main raw materials, and stearamide and polyvinyl alcohol were added. After mixing, heating and stirring, emulsification and other steps, developed a kind of silicone oil demoulding agent with excellent demoulding effect, safe and environmental protection, and can extend the service life of moulds, which was suitable for the demoulding of rubber hoses and printing production[22].

3. Wax release agents

A variety of compounds such as oils, waxes, soaps and macromolecular compounds can be used as active ingredients in release agents. There are two main forms, one of which is to make an emulsion of molten paraffin wax in the presence of an emulsifier and then apply it to the formwork for use. The other way is to dissolve the paraffin wax in solvent oil to make a paraffin oil release agent. This release agent has an ideal release effect, but it is not easy to dilute and apply evenly when the temperature is too low; it is flammable, volatile and unsafe to use, and cannot be stored in the open air. Paraffin wax release agents with high wax content often leave residues of paraffin wax on the concrete surface, which affects the further decoration of the mould surface. The mechanism of action of such release agents is isolation and lubrication.

Due to the adhesive nature of the isocyanate in the main polyurethane material, a suitable wax release agent is used to prevent it from bonding to the mould during the moulding process. Wax emulsions can be well used in environmentally friendly waterborne polyurethane release agents, which can reduce the surface tension of the coating and increase the film-forming properties of the release agent on the mould surface, improving the bonding and finishing properties of polyurethane products. This type of release agent, has excellent release performance. Thin film thickness (0.1~μm), high dimensional accuracy of products, dry coating, less adhesion to products, not easy to adhere to moulded products, can be diluted with water for use, environmentally friendly, no VOC emissions, can be applied to more difficult to release epoxy resin products. Waxes have a high melting point, good viscosity and excellent film-forming properties, and contain polar groups such as hydroxyl and carboxyl groups that are easily emulsified, making them a very important class of active ingredients for release agents. The film formation process of emulsions is divided into three processes: filling, fusion and diffusion. The low melting point of paraffin waxes gives them good solubility in polymers and improves their fluidity, helping to complete the three stages of filling, fusion and diffusion and to form continuous and homogeneous films, making waxes a suitable material for film-forming materials[23]. Polyethylene emulsion is a non-ionic AC polymer combination of polyethylene copolymers fused into an emulsion by high pressure. It is a non-toxic, odourless and non-corrosive non-ionic emulsion, easily soluble in water, with good resistance to acidity, alkalinity, electrolysis, bathing with resin, not easy to stratify and coalesce. The emulsion is constant in composition, has good chemical stability and does not delaminate for long term storage. It is widely used in the production of water-based inks, liquid shoe polish, metal release agents and other industries. It is an excellent release agent for a wide range of applications in the polyurethane industry and other special industries.

Zeng Xiangyun [24] et al. developed a polyurethane wax release agent by using release wax, paraffin wax and microcrystalline wax as the main raw materials and adding foam stabilizers and petroleum-based solvents. It has the advantages of good release effect, no pollution, no corrosiveness to the mould surface and can be used repeatedly, etc. It is suitable for the release of polyurethane high resilience foam. Gao Huijiang and Xu Yingying[25] et al. took beeswax and rice bran wax as raw materials, added vegetable oil, silicone oil and talcum powder, etc. After hydrolysis and emulsification, they successfully developed a wax release agent with low cost, reasonable components, and safety and environmental protection, with excellent lubricity and good demoulding properties. The method is reasonable, easy to implement, low cost and suitable for large-scale application in the industry.

4. Fatty acid based release agents

Fatty acids and their soaps are very versatile lubricants and can also be used as release agents. They are available from a wide range of sources, are inexpensive, have good compatibility with many plastics, are heat stabilised and are generally pre-mixed with resin prior to processing and are easy to use. Stearic acid is still one of the most important lubricants because of its low price, comprehensive performance and ease of processing into various metal soaps such as zinc stearate, dull, lead, magnesium, barium, cadmium, aluminium, sodium and lithium are all commonly used lubricants. Fatty acid release agents are mainly composed of high-grade fatty acid esters, fatty acid salts, other additives and water and are widely used in the processing of foam moulded products. The fatty acid derivatives are compatible with the rubber compound and their use is not detrimental to the physical properties of the compound.

Therefore, this type of release agent can promote the mobility of the rubber in the mould cavity during the vulcanisation of SBR and EPDM moulded products, and can also improve the sliding and lubricating properties of rubber seals and gaskets, and can improve the self-lubrication of reciprocating seals. Thus, the release and processing properties of thermoplastic elastomers such as SBR, EPDM, butadiene rubber, acrylate rubber, epichlorohydrin and thermoplastic elastomers can be effectively improved in order to obtain optimum economic efficiency. Used in many other elastomers, it has also been found to improve the flow of the rubber in the mould cavity, reduce operating temperatures and shorten the production cycle time of the product. Jin Minglin and Xu Shi [26] developed a wet molding release agent for magnetic materials by using advanced fatty acids as the main raw material and paraffin as the medium, after heating, stirring and emulsification. The release agent can be evenly distributed on the surface of the mould, which greatly improved the release of the magnetic material blanks and made the surface of the products highly accurate and smooth, and it was a green and environmentally friendly release agent as it is pollution-free and harmless to the health of the operators when used.

5. Conclusion

Mold release agents are divided into water-based release agents, oil-based release agents and powder-based release agents. Water-based release agents are widely used because they have the advantages of low price, good release effect, green environment protection, smooth casting surface, good air tightness and good automaticity operation. According to the effective release material in the mold, release agent can be divided into silicone oil type, wax and fatty acid type water-based mold release agent. The water-based release agent has excellent performance, green environmental protection, good plasticity, suitable price, has a broad market prospect and is widely used. The development and research of water-based release agent have been carried out for many years in foreign countries. Quite a number of products have been produced and the industrial chain is becoming mature. Due to the late start of China's research on water-based release agents, there is no systematic research on water-based release agents in China, and there is still a large gap with foreign countries. The high-end water-based release agent is basically monopolized by foreign countries, with the continuous development of mold release technology, the domestic water-based release agent in some degree has been able to compete with foreign, I believe that in the future, China's mold release agent technology will continue to break through, and then break the monopoly of foreign situation.

Acknowledgments

Thanks to the following funds for supporting this project. Postdoctoral Innovation Project of Shandong Province (202103051), Heilongjiang Provincial Natural Science Foundation of China (LH2020E089), China Postdoctoral Science Foundation (2021M691970), University Nursing Program for Young Scholars with Creative Talents in Heilongjiang Province (UNPYSCT-2018205), Fundamental Research Fundtion for Universities of Heilongjiang Province (2018-KYYWF-1598).

References

1. Amra, Kesrin, Momin, et al. Formulation evaluation of ketoconazole microemulsion-loaded hydrogel with nigella oil as a penetration enhancer[J]. Journal of Cosmetic Dermatology, 2019, 18(6): 1742-1750.
2. Benito J M, Cambiella A, Lobo A, et al. Formulation, characterization and treatment of metalworking oil-in-water emulsions[J]. Clean Technologies and Environmental Policy, 2010, 12(1): 31-41.
3. Beshkar F, Khojasteh H, Salavati-Niasari M. Recyclable magnetic superhydrophobic straw soot sponge for highly efficient oil/water separation[J]. Journal of Colloid and Interface Science, 2017, 497(56): 57-65.
4. Cambiella A, Benito J A, Pazos C, et al. Interfacial properties of oil-in-water emulsions designed to be used as metalworking fluids[J]. Colloids and Surfaces a-Physicochemical and Engineering Aspects, 2007, 305(1-3): 112-119.
5. Cambiella A, Benito J M, Pazos C, et al. The effect of emulsifier concentration on the lubricating properties of oil-in-water emulsions[J]. Tribology Letters, 2006, 22(1): 53-65.
6. Cappelletti G, Ardizzone S, Meroni D, et al. Wettability of bare and fluorinated silanes: A combined approach based on surface free energy evaluations and dipole moment calculations[J]. Journal of Colloid and Interface Science, 2013, 389(557): 284-291.
7. Chaydarreh K C, Shalbafan A, Welling J. Effect of ingredient ratios of rigid polyurethane foam on foam core panels properties[J]. Journal of Applied Polymer Science, 2017, 134(17): 224-675.
8. Cerpnjak K, Zvonar A, Vrecer F, et al. Development of a solid self-microemulsifying drug delivery system (SMEDDS) for solubility enhancement of naproxen[J]. Drug Development and Industrial Pharmacy, 2015, 41(9): 1548-1557.
9. deDamborenea J, Bastidas J M, Vazquez A J. Adsorption and inhibitive properties of four primary aliphatic amines on mild steel in 2M hydrochloric acid[J]. Electrochimica Acta, 1997, 42(3): 455-459.
10. Figueiredo L, Bandeira P, Mendes A, et al. Use of fluoropolymer permanent release coatings for molded polyurethane foam production[J]. Journal of Coatings Technology and Research, 2012, 9(6): 757-764.
11. Sirtl E. Über die reduktion von chlorsilanen mit wasserstoff[J]. Reuschel, 2004, 332(3-4): 113-123.
12. Abou-Taleb N H, El-Sherbiny D T, El-Enany N M, et al. Multiobjective optimization of microemulsion-thin layer chromatography with image processing as analytical platform for determination of drugs in plasma using desirability functions[J]. Journal of Chromatography A, 2020, 1619(33): 4455-5781.
13. Abu-Azzam O, Nasr M. In vitro anti-inflammatory potential of phloretin microemulsion as a new formulation for prospective treatment of vaginitis[J].

Pharmaceutical Development and Technology, 2020, 2(11): 344-556.

14. Albrecht W N, Bryant. Polymer-fume fever associated with smoking and use of a mold-release spray containing polytetrafluoroethylene[J]. Journal of Occupational Medicine Official Publication of the Industrial Medical Associa, 1987, 29(10): 817-819.
15. Bao X J, Lin M, Koh H Z, et al. Preparation of iron oxide and iron oxide/silicon oxide nanoparticles via water-in-oil microemulsion. in Ceramic Nanomaterials and Nanotechnology Ii, DeGuire M. R.,Hu M. Z.,Gogotsi Y.,Lu S. W., Editors. 2004. p. 121-128.
16. Berger-Neto A, Jaccoud-Filho D d S, Wutzki C R, et al. Effect of spray droplet size, spray volume and fungicide on the control of white mold in soybeans[J]. Crop Protection, 1987, 92(Complete): 190-197.
17. Zhu Hui, Li Yuancai, Zhao Wei, et al. Development of aqueous releasing agent for aluminum alloy die casting[J]. Special Casting & Nonferrous Alloys, 2010, 30(04): 342-344+289.
18. Jiaxin W, Kejian L, Qingshan R. Preparation and property study of the aqueous waste engine oil base releasing agent[J]. Fine and Specialty Chemicals, 2014, 22(08): 39-42.
19. Iam-khong P, Muchan P, Sriprang N, et al. Innovation of green chemistry for synthesizing nanoporous and microporous aluminosilicate via microemulsion technique[J]. Chiang Mai Journal of Science, 2020, 47(3): 473-483.
20. Yuguang W, Xiaoying W, Ying L. Progress on silicone oil mold-release agent by silicone high-boiling compound[J]. Silicone Material, 2013, 27(01): 57-59.
21. Jianqiang W. A mold release agent containing methyl silicone oil and its preparation method[P], CN106426677A, 2017.
22. Grande F, Ragno G, Muzzalupo R, et al. Gel formulation of nabumetone and a newly synthesized analog: microemulsion as a photoprotective topical delivery system[J]. Pharmaceutics, 2020, 12(5): 446-664.
23. guojiang W, Guizhi C, Min J. Preparation of wax-emulsion releasing agent for aluminium alloy die casting[J]. Contemporary Chemical Industry, 2014, 43(09): 1695-1697.
24. Yunxiang Z, Jianguo L. Polyurethane wax release agents and their uses[P], CN103419307A, 2013.
25. Gao Huijiang, Yingying X. A safe and environmentally friendly new wax release agent and its preparation method[P], CN105643847A, 2016.
26. Jin MInglin. A wet molding release agent for magnetic materials and its preparation method[P], CN1843655, 2006.

Nonlinear error control of five-axis machining singular region

Liangji Chen [1,2], Zhi Liu [1,*], Zhiguang Li [1] and Zisen Wei [1]

[1] School of Mechanical Engineering, TianGong University, Tianjin, China
[2] College of Mechanical and Control Engineering, GuiLin University of Technology, Guilin, China

Abstract. Aiming at the singular problem of large nonlinear error caused by the extreme change of rotation in five axis NC machining of complex surface. An optimization algorithm of tool path in singular region is proposed. Taking a-c double turntable five axis linkage NC machine tool as an example, firstly, the mathematical models of tool local milling ability, machining bandwidth and tool axis inclination are established. Based on the two constraint models, the mechanism of singularity and the size of singular region are analyzed, and it is proposed to adjust the side angle with the tool contact as the rotation center to avoid the singular region. The modified tool axis vector that does not meet the accuracy is interpolated recursively according to the interpolation principle. This method solves the problem of large nonlinear error caused by the extreme rotation of the rotating axis in the singular region on the basis of meeting the requirement of no curvature interference. At the same time, the processing bandwidth before and after modification is guaranteed to remain unchanged. Simulation results show that the algorithm can effectively detect the singular region and improve the machining quality in the singular region.

1. Introduction

The five-axis machine tool increases two degrees of freedom in the three-axis machine tool,which makes the tool can free cut the workpiece with no attitude, greatly increasing the surface removal rate of the workpiece material and improving the processing quality of the workpiece.However,due to the introduction of rotation axis,it brings special processing problems.Singularity (singular region) is one of the most important problems.In the singular region, the rotation axis will produce discontinuous speed rotation,will produce large nonlinear errors, and even interference,which will cause great damage to the workpiece itself and the machine tool components.Therefore,it is very important to detect the singularity and process the tool position data for improving the machining accuracy of workpiece and protecting the machine tool.

Affouard used polynomial interpolation to modify the tool path in the singular region,but the interpolation algorithm was complex and required a lot of calculation [1].The method of Munlin optimized the rotation axis in the singular region,but failed to fundamentally solve the problem [2]. Wang Feng combined the Jacobian matrix with the variation of each axis to identify the singular region by conditional number,but it is difficult to determine the conditional number accurately in a quantitative way [3-5].Yu Xiaosui proposed to solve the singular problem by adding one more rotating axis,but the machine tool mechanism was complicated and its use was very limited [6].

Based on the above issues,in local can establish cutting tools and milling and processing bandwidth and knife shaft Angle of singular area on the basis of mathematical model of cutter axis vector is optimized,and then five axis machine tool of the modified files for precision detection,did not meet the requirements of precision cutter location data for recursive interpolation,until meet the accuracy requirement.

2. Local millability conditions for five-axis machining

In tool path planning of 5-axis machining,global interference,local interference and tool bottom interference must be taken into account.For local interference,it is necessary to establish the local coordinate system at the tool contact point,and establish the geometric information function relation between the tool and the workpiece in the local coordinate system,and judge whether the local interference occurs by the difference of the two functions.The specific local coordinate system establishment method is as follows:Take the cutting contact as the coordinate origin,the cutting direction F as X-axis,the normal vector of the curved surface cutting contact as z-axis,and use the right hand rule to determine the Y-axis direction.In this coordinate system,the cutter axis vector is represented by the heel deflection Angle,namely $T(\lambda,\omega)$.

Take ring cutter as the example in Fig 1.1 below,establish tool surface function $z=f(x,y)$ and workpiece surface function $z=h(x,y)$,then the difference function is equal to

* Corresponding author: liuzhi2595@163.com

z= f(x,y)-z= h(x,y),when the dupont indication line graph of the difference function at the cutter contact is elliptical,no local interference occurs,and the cutter meets the local millability at the cutting contact.Kmax and Kmin represent the maximum and minimum principal directions of the cutter surface at the cutting contact,Lmax and Lmin represent the maximum and minimum principal directions of the workpiece surface at the cutting contact,are the included Angle of the maximum principal directions of the two surfaces,and h is the residual height.Then, the Dupont indicator line of the difference function is.

$$\begin{aligned}&(L\max\cos^2\theta + k\min\sin^2\theta - L\max)x^2\\&+2(k\max - k\min)\sin\theta\cos\theta xy +\\&(k\max\sin^2\theta + k\min\cos^2\theta - L\min)y^2 = 2h\end{aligned} \tag{1}$$

When

$$\begin{aligned}&k\max + k\min - L\max - L\min > 0\\&-(k\max - L\max)(k\min - L\min)\\&+\sin^2\theta(k\max - k\min)(L\max - L\min) > 0\end{aligned} \tag{2}$$

There is no local interference in the tool contact.Based on the knowledge of differential geometry,the maximum principal curvature of the cutter surface at the cutting contact is:

$$k_{\max} = \frac{1}{a} \tag{3}$$

The minimum principal curvature is:

$$\lambda = \arcsin\frac{\beta}{1-\alpha L_{\max}} \tag{4}$$

When the minimum curvature of the tool surface is equal to the maximum principal curvature of the workpiece surface,the material removal rate is the highest and the machining efficiency is the best.Which kmin = Lmax, then

$$\frac{\sin\lambda}{\beta + \alpha\sin\lambda} = L_{\max} \tag{5}$$

According to the above non-interference conditions,local interference will not occur regardless of the sideslip Angle.In order to facilitate the calculation of subsequent processing bandwidth,let

$$\begin{aligned}a &= L_{\max}\cos^2\theta + k_{\min}\sin^2\theta - L_{\max}\\b &= k_{\max} - k_{\min})\sin\theta\cos\theta\\c &= k_{\max}\sin^2\theta + k_{\min}\cos^2\theta\end{aligned} \tag{6}$$

If the included Angle between the cutting direction and the cartesian coordinate system axis of the tangent plane is, then the processing bandwidth is:

$$W = 2\sqrt{-\frac{2h(a\cos^2\phi + 2b\sin\phi\cos\phi + c\sin^2\phi}{-b^2+ac}} \tag{7}$$

On the basis of the above calculation, the relation between machining bandwidth W and cutter inclination Angle and heel can be obtained.

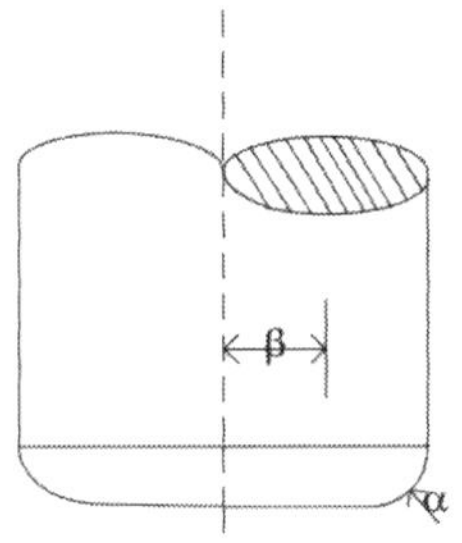

Fig.1 Schematic diagram of ring knife

3. Singularity mechanism analysis of five-axis machining

3.1 Singularity problems in five-axis machining

When i and j in the tool axis vector are both 0,that is, the tool axis vector is (0 0 1),after the post-processing of the machine tool,the degree of freedom of the C-axis is lost,and the tool point at this time is defined as singular point.The corresponding cutter axis is singular axis.In fact,when the included Angle between the tool vector and the singular axis is less than a specific value,the C-axis Angle between the two adjacent tool points may change dramatically,resulting in extreme nonlinear errors and resulting in singular problems.The region swept by the tool axis vector that is less than the Angle is called singular region,as shown in Fig 2 below.

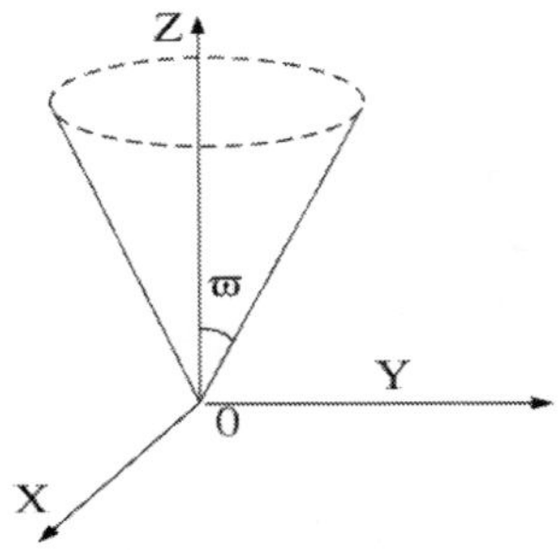

Fig.2 Singular area

3.2 The method of calculating the size of singular region

When the cutter axis vector is (0,0,1), it is the singular point. When the included Angle between the cutter axis vector and the singular axis is very small, the swept conical region is the singular region, as shown in the figure above. Assuming that the cutting depth of the tool is H and the radius of the bottom circle of the vertebral body is R, the bottom circle of the singular cone can be calculated as follows:

$$\begin{cases} \|T\| = i^2 + j^2 + k^2 = 1 \\ \tan\lambda = \sqrt{\dfrac{i^2 + j^2}{k^2}} \\ k = \sqrt{\dfrac{1}{\left(1+\tan^2\lambda\right)}} \\ R = H\tan\lambda \end{cases} \quad (8)$$

K value can be traversed through the calendar, so that the corresponding knife point can be judged. The calculated R value represents the size of the singular circle projected by the singular cone.

3.3 Adjusting strategy of cutter axis vector avoidance in singular region

Based on the above constraints to avoid interference and the relationship between machining bandwidth and tool axis inclination Angle, a strategy was proposed to avoid the singular region by adjusting the sideslip Angle with the tool contact as the rotation center.The specific scheme is as follows:The tool axis vector at each tool contact point on the tool path is projected onto the oxy horizontal plane,and the space vector is planized.For any tool axis vector P(px, py, pz),multiply each coordinate component of the tool axis vector P by 1/ pz to obtain the processed P '(px/pz, py/pz, 1). The Z coordinate is constant, so the z coordinate can be ignored,so the processed projection tool axis vector is P' (px/pz, py/pz).

According to the range of singularity circle projected by singular cone,the cutter position data was taken to judge whether the cutter axis vector entered the singular circle on the bottom surface of singular cone,and the cutter axis vector between the last projection of the cutter axis vector C1 before entering the singular circle and the first one after opening the singular circle C2 was detected.

According to the above constraints of local millability and the function relationship between machining bandwidth and tool axis inclination Angle,when the tool axis vector is adjusted to avoid the singular region,no interference phenomenon must be ensured for each tool contact in the singular region, and the machining bandwidth remains unchanged before and after adjustment.According to the above,when the minimum curvature of the tool surface is equal to the maximum curvature of the workpiece surface at the tool contact point,the meshing effect between the tool and the workpiece is the best,and the machining bandwidth is the maximum.At the same time,no interference occurs when the sideslip Angle is arbitrarily adjusted under the tool attitude.The machining bandwidth and sideslip Angle are symmetric according to the function of machining bandwidth and tool axis inclination Angle.According to the transformation relation of adjacent cutter axis vectors in space, the cutter axis vectors detected in the singular circle can be adjusted to the singular region,so that the adjusted cutter axis vector and the cutter axis vector before adjustment are in a straight line at the center of the singular circle,so that the singular region can be avoided by adjusting the sideshow Angle.At the same time,the adjusted distance G keeps the processing bandwidth unchanged before and after adjustment according to the relationship between the cutter axis vector and the cutter axis pose Angle,as shown in the figure below P1 to P6 are the tool points in the detected singular region.

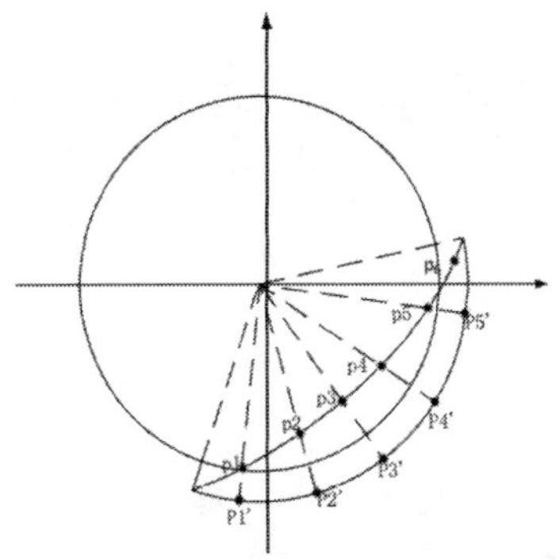

Fig.3 Schematic diagram of tool axis vector avoidance

According to the insertion point,judge whether the adjusted cutter axis vector in the singular region meets the error requirements, and perform recursive interpolation for those that do not meet the error requirements.Insert the processing point according to the Angle of the tool axis vector to determine. The basis for inserting the processing point is:

$$\left|\frac{f(s)-f(m)}{f(s)}\right| \leq \varepsilon_f \tag{9}$$

f(s) is the current tool axis vector, f(m) is the adjacent tool axis vector,εf is the given Angle error.

4. Simulation verification and experimental data analysis

By MATLAB tool path error analysis can be properly through the singular region, the sharp increase of the error is far beyond the limit of the error value. The maximum error value is as high as 0.745mm, as shown in Fig 4 below. Meanwhile, post-processing of the compiled tool position data shows that from tool point 1 to tool point 6,although the variation range of tool axis vector is small, the corresponding change Angle of AXIS A and C is large, as shown in Fig 6.The change of Angle C is 66.7°, and that of axis A is 55.7°.This change will cause the machine tool to produce setbacks, damage the machine tool, for the workpiece processing, it is easy to occur curvature interference, and occur too large nonlinear errors, causing singular problems. Therefore, the tool axis vector should be adjusted in time in tool path planning to avoid the occurrence of strange phenomena.

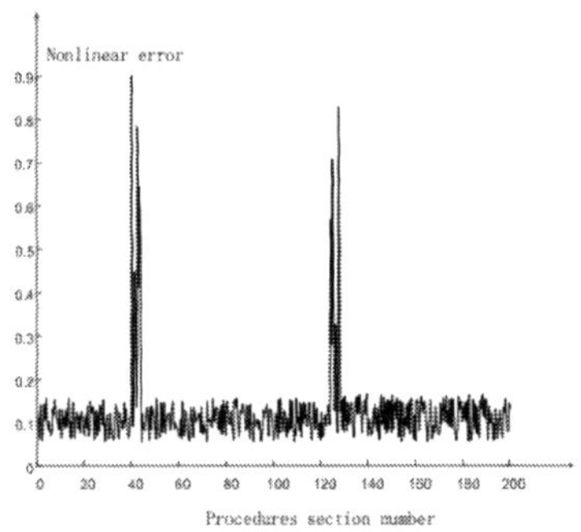

Fig.4 Before optimization

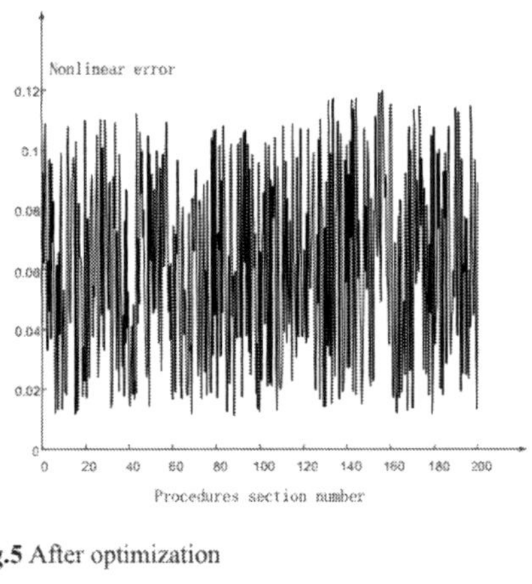

Fig.5 After optimization

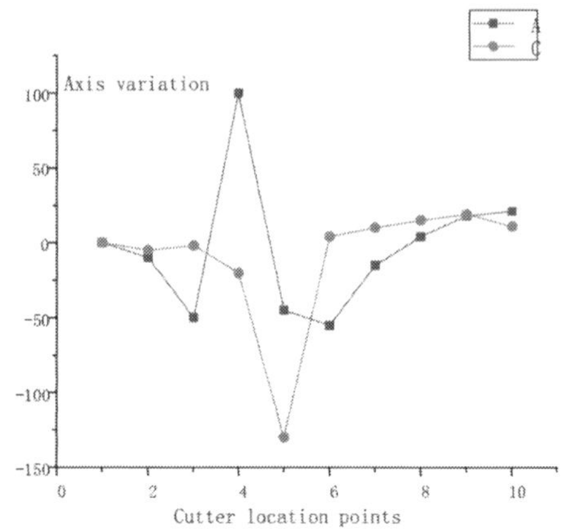

Fig.6 Schematic diagram of change of rotation axis

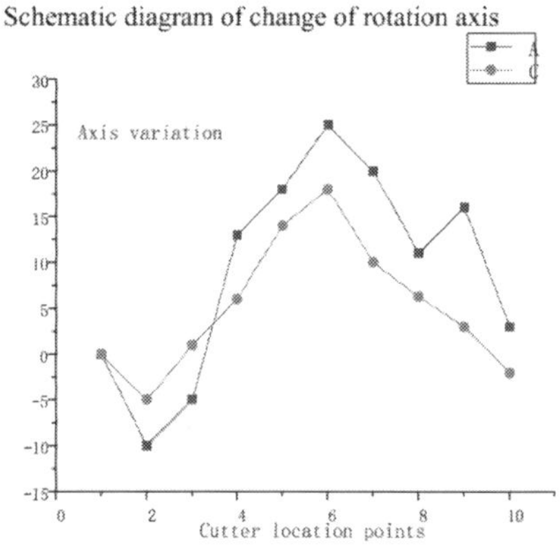

Fig.7 Schematic diagram of rotation axis change after adjustment

Based on the establishment of local milling condition of tool axis vector and the relationship between machining bandwidth and tool axis inclination Angle, the tool axis vector in the singular region is avoided. Compared with the traditional method, the efficiency of machining with the maximum machining bandwidth is greatly improved for the overall tool path. At the same time, the avoidance strategy of singular region has a great improvement over the calculation of simple dense interpolation method.

In terms of machining accuracy, the optimization method proposed in the singular region can control the nonlinear error within a reasonable range,as shown in Fig 5 below. After post-processing, it is verified that the rotation axis of the machine tool changes gently.Meet the requirements of workpiece processing quality.See Fig 7.

5. Conclusion

(1) In this paper,based on the establishment of the function relation between the local milling ability and the machining bandwidth and the tool axis inclination Angle in five-axis machining,a tool axis vector avoidance method is proposed for the problems in the singular region.After modification, error calculation and post detection are carried out.

 https://doi.org/10.1051/matecconf/202235801034

(2) In this way, the processing bandwidth remains unchanged before and after modification, and the processing efficiency and accuracy have been improved.After the post-processing verification of machine tool each axis of the processing stability has a considerable degree of improvement.

Acknowledgments

This work was financially supported by Guangxi Natural Science Foundation under Grant (No. 2021GXNSFAA220019).

References

1. Affouard A, Duc E, Lartigue C, et al.Avoiding 5-axis singularities using tool path deformation[J].International Journal of Machine Tools and Manufacture, 2004, 44(4): 415-425.
2. Munlin M, Makhanov S S,Bohez E L J. Optimization of rotations of a five-axis milling machine near stationary points[J]. Computer-Aided Design, 2004, 36(12): 1117-1128.
3. Sørby K. Inverse kinematics of five-axis machines near singular configurations. International Journal of Machine Tools and Manufacture, 2007, 47(2): 299-30.
4. Eneration for 5-axis machine tools with singularity avoidance[J]. CIRP Annals,2020, 69(1):349-352.
5. Shimojima K, Nakandakari D.Development of Five-axis Machining Center Geometric Deviation Estimation Method using Least Square Method[J]. Journal of the Japan Society for Precision Engineering, 2012, 78(10):869-874.

Optimization method for complex surface partitioning and cutting parameters in 5-axis machining

Liangji Chen [1, 2], Zhiguang Li [1, *], Zhi Liu [1] and Zisen Wei [1]

[1] School of Mechanical Engineering, TianGong University, Tianjin, China
[2] College of Mechanical and Control Engineering, GuiLin University of Technology, Guilin, China

Abstract. For complex surfaces with variable geometric features and high processing requirements, the current stage still uses more conservative machining cutting parameters to generate five-axis machining tool positions, which inevitably leads to some parts not to obtain the best cutting state, and the generated machining tool positions are poorly adapted, so it will produce a great processing deformation error seriously affects the surface quality. In addition, conservative cutting parameters applied to different geometric features of the surface area is very likely to induce dramatic changes in the machining process cutting forces and machine motion parameters, resulting in uneven distribution of machining errors and local super poor, seriously affecting the quality of machining. In five-axis machining, due to the introduction of two rotary axes, the tool vector can be rotated arbitrarily in the spatial domain. When machining complex surfaces, the distance between the tool center coordinates and the oscillation center or rotation center of the five-axis machine will change with the curvature of the surface. surface quality. Therefore, the optimization of tool cutting parameters in 5-axis simultaneous machining is a technical bottleneck that needs to be solved urgently.

1. Introduction

In recent years, many domestic and foreign scholars have made some researches on the optimization of cutting parameters and other issues around complex surfaces with complex morphology and variable curvature. The curvature of complex surfaces, normal vector and other geometric features can reflect all the information of the surface, and the geometric features of complex surfaces themselves have regional differences, so it is easy to carry out regional segmentation by starting from geometric features [1], Optimization have some reference significance.

As the main parameter for evaluating the geometric features of surfaces, curvature has been used by some scholars as the primary judgment criterion for region partitioning of complex surfaces. zheng [2] et al. used boundary projection lines to partition the original processing region to generate a static processing region without interference and constructed a new region boundary based on the reachability analysis method. shen [3] et al. investigated the inverse evaluation of parametric region partitioning of surfaces based on rotational axis motion mechanism, which assigns rotational motion characteristics to each mesh within the parametric region and partitions the region accordingly. All of the above lack partitioning guidelines oriented to the CNC machining process, which leads to the independence of region partitioning and tool trajectory planning and makes it difficult to ensure that all regions of the part meet the accuracy requirements. In view of the advantages of constant cutting forces, many scholars have studied cutting parameter optimization methods oriented to constant cutting force control.In 2015, Moufki [4] et al. analyzed the bevel cutting model using the analytical method, which was applied to the prediction of milling forces for side milling machining. In the same year, Fu [5] et al. considered radial runout for the prediction of milling forces in nickel-aluminum bronze material. However, the above also only considered the geometric information of the tool and the workpiece, the physical information of the cutting state (e.g., cutting force) while not involved, and also lacked the influence of these constraints on cutting stability and surface integrity on the optimization of cutting parameters.

However, most of the studies mentioned above only consider the surface partitioning method and the parametric optimization method based on cutting force modeling, but most of them do not combine them effectively, and it is not comprehensive to consider only one aspect as the optimization objective. In this paper, we propose a five-axis parameter optimization method based on surface partitioning for complex surface machining, based on the prediction model of cutting force, considering the improvement of machining efficiency and surface quality, with the objectives of reducing machining deformation error and cutting force fluctuation, and establishing a multi-objective feed optimization model. The Newton-Raphson algorithm is used to optimize the machining parameters.

* Corresponding author: lizhiguang1001@163.com

2. Partitioning method based on curvature features of complex surfaces

2.1 NURBS surface partitioning guidelines

In this paper, NURBS surfaces are used to describe complex surfaces, and firstly, the surfaces are classified according to the curvature properties of NURBS surfaces. The mean curvature is the convexity of the surface, and the Gaussian curvature is the curvature of the surface. k_H is the mean curvature of the surface, and k_G is the Gaussian curvature of the surface, calculated as follows:

$$k_H = \frac{EN - 2FM + GL}{2\left(EG - F^2\right)} \tag{1}$$

$$k_G = \frac{LN - M^2}{EG - F^2} \tag{2}$$

Where E, F, G are the basic quantities of the first category of NURBS surfaces; L, M, N are the basic quantities of the second category of NURBS surfaces.

2.2 Discretization of complex surfaces and zoning

In order to obtain the point cloud information of the surface, the surface is first discretized. The basic process of surface discretization is to select one isoparametric line in the u-direction and one in the v-direction of the surface, and to discretize these two isoparametric lines according to the size of the surface, and to record the coordinate values of these grid points to form the discretization points of the surface parameter domain. On this basis, the preliminary partitioning of the complex surface is carried out. The NURBS complex surface is firstly created, and the surface is discretized into a 40×40 grid model. On the basis of the discretized grid model, the Gaussian curvature and the mean curvature of each discrete point on the grid are obtained by traversal.

To facilitate subsequent surface partitioning, the clustering centers and boundary vertices of each surface region are numbered, and then each boundary point of the surface region divided on the 2D map is projected onto the surface by coordinate mapping according to its location distribution on the grid for surface partitioning, resulting in coordinate mapping and surface partitioning. After surface mapping, neighboring surface slices of the same type can be combined to reduce the number of partitioning and facilitate subsequent surface partitioning processing.

3. Modeling and calculation of cutting forces

With the analytical method, it is possible to obtain the cutting force mechanically and to obtain the magnitude of the external load in the analytical modeling of the residual stress, and the bevel cutting geometry relationship is shown in the figure below. Figure (1) is the calculation model of milling force for annular tool machining: O_T-$X_TY_TZ_T$ is the tool coordinate system, P is the point on the jth cutting edge, P-$x_0y_0z_0$ is the micro-element coordinate system, and the micro-element coordinate system is continuously transformed with the tool turning angle. Figure (2) shows the bevel cutting model, point P is a point on the cutting edge, and the cutting speed direction corresponding to x_0 at this point corresponds to the direction of the tangential force dF_t.

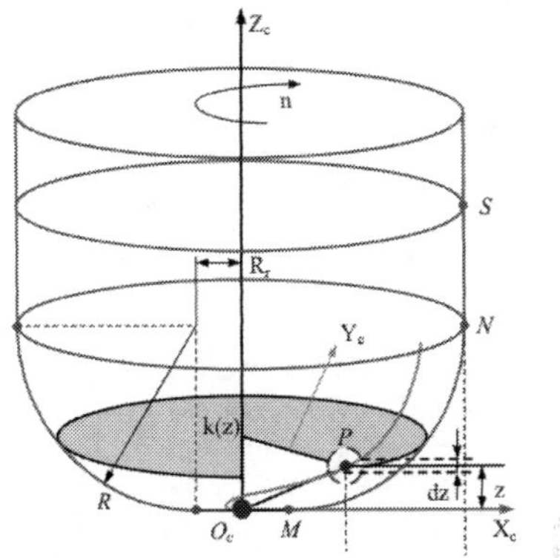

Figure 1. Ring knife geometry model.

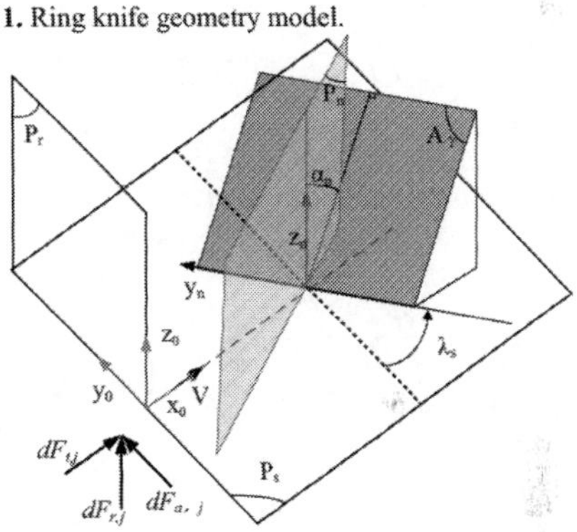

Figure 2. Angled cutting model.

For the micro-element of cutting edge j, the three micro-element cutting force components acting at the point P are:

$$\begin{Bmatrix} dF_{t,j} \\ dF_{r,j} \\ dF_{a,j} \end{Bmatrix} = \begin{bmatrix} \cos\lambda_s \cos\phi_n \cos\eta_{sh} + \sin\lambda_s \sin\eta_s & \cos\lambda_s \sin\phi_n \\ -\sin\phi_n \cos\eta_s & \cos\phi_n \\ \sin\lambda_s \cos\phi_n \cos\eta_{sh} - \cos\lambda_s \sin\eta_s & \sin\lambda_s \sin\phi_n \end{bmatrix} \begin{Bmatrix} dF_s \\ dN_s \end{Bmatrix} \tag{3}$$

After the above equation finds the micro-element force, according to the transformation relationship between the tool coordinate system and the micro-element coordinate system, find the cutting force components dF_x, dF_y, dF_z in the tool coordinate system for the micro element on the jth cutting edge:

$$\begin{bmatrix} dF_{xj}\left[\phi_j(z)\right] \\ dF_{yj}\left[\phi_j(z)\right] \\ dF_{zj}\left[\phi_j(z)\right] \end{bmatrix} = R_2 \begin{bmatrix} dF_{tj}\left[\phi_j(z)\right] \\ dF_{aj}\left[\phi_j(z)\right] \\ dF_{rj}\left[\phi_j(z)\right] \end{bmatrix} \tag{4}$$

where R_2 is the transformation matrix of the micro-element coordinate system P-$x_0y_0z_0$ to the tool coordinate system:

$$R_2 = \begin{bmatrix} -\cos\phi_j & -\sin\phi_j\cos\kappa & -\sin\phi_j\sin\kappa \\ \sin\phi_j & -\cos\phi_j\cos\kappa & -\cos\phi_j\sin\kappa \\ 0 & -\sin\kappa & \cos\kappa \end{bmatrix} \quad (5)$$

Integrate the cutting force component of the micro element in the direction of the tool axis and the number of teeth of the tool axial depth of cut a_p to get the total cutting force magnitude:

$$F_q(\phi) = \sum_{j=1}^{N} \int_0^{a_p} dF_{qj} dz \quad (6)$$

Eight sets of milling tests with different cutting parameters were designed to collect real-time cutting force values and calculate the mean value of cutting force to represent the cutting force of the cutting process. At the same time, different machining parameters were substituted into the cutting force model to calculate the model prediction theoretical force value, and the two were compared to calculate the error. The analysis of the validation results shows that the experimental cutting force values and the predicted force values of the cutting force model basically match, and the overall average error rate is only 4.69%, which verifies that the cutting force prediction model established under different cutting parameters is consistent with the actual machining, and ensures that the predicted force of the cutting force model matches the actual machining cutting force magnitude, which lays the foundation for the subsequent cutting parameter optimization work.

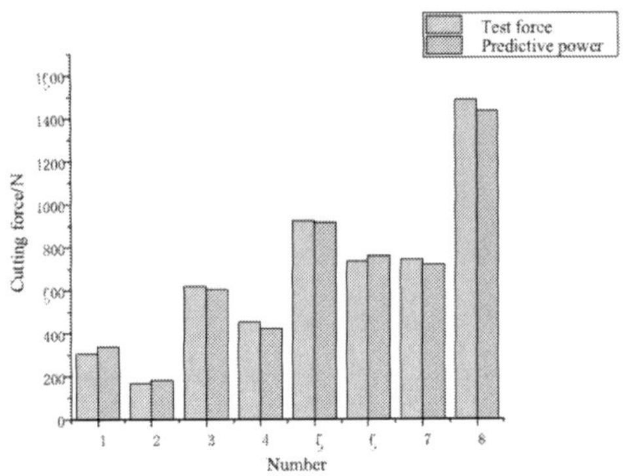

Figure 3. Cutting force model validation diagram.

4. Cutting parameters optimization

Since the constant cutting force ensures the smooth operation of machine and tool and improves the surface quality, the method takes the absolute value of the difference between the given reference cutting force and the maximum cutting force at each tool point as the optimized objective function, and uses the cutting geometry deformation error as the main constraint for complex surface machining. The Newton-Raphson iterative algorithm is used to adjust the cutting parameters in the CAM toolpath file so that the objective function can be minimized. Combined with the prediction model of cutting force to calculate the magnitude of cutting force, so as to establish the mapping relationship between cutting force F and cutting parameter F(ft).

$$F = F(ft), \quad ft = W(n, v_f, a_p, a_e) \quad (7)$$

where n is the rotational speed, v_f is the feed rate, a_p is the depth of cut, and a_e is the depth of cut. For the cutting parameter optimization method oriented to constant cutting force control, the objective function is

$$\min\left|F_{ref} - F_{imax,\ j}\right| \quad (8)$$

Where, F_{ref} is the given reference cutting force; $F_{imaxn,j}$ is the maximum cutting force calculated after the jth iteration of the cutting parameters at the ith tool position point.The process implemented using the Newton-Raphson algorithm is as follows.

Read the initial cutting parameter f_{init} from the toolpath file, set the reference cutting force F_{ref} by the constraints of the machine tool system; read a row of tool points for cutting force calculation, and get the maximum cutting force F_{imax}; solve the new cutting parameter $f_{i,1}$ according to the following formula, and calculate the maximum cutting force $F_{imaxn,j}$ for this tool point under this cutting parameter.

$$f_{i,1} = f_{init} \frac{F_{ref}}{F_{imax}} \quad (9)$$

Solve the cutting parameters according to Eq. and record the optimized cutting parameters for this tool point when the optimization objective of the objective function of Eq. is satisfied.

$$f_{i,2} = f_{i,1} + \frac{(F_{ref} - F_{imax\,n,j})(f_{i,1} - f_{init})}{F_{imax\,n,1} - F_{in}} \quad (10)$$

Select the next tool point and repeat the steps until the calculation of all tool points is finished; the optimized cutting parameters are used to modify the values in the CAM toolpath file to obtain the new toolpath file.

5. Comparison and analysis of results before and after optimization

After optimization, the overall cutting force in the whole machining stage is kept in the area around the given cutting force value of 1000N. After optimization, the real-time cutting force is kept within the area of [-9.57%, 9.78%] of the given cutting force value of Fobj, and the fluctuation interval of cutting force is obviously less than ±10%, which achieves the constant cutting force constraint, and the fluctuation is smaller, and the variance is 10.56, which indicates that the machining process is smoother and ensures the surface quality. The machining surface quality requirement is ensured.

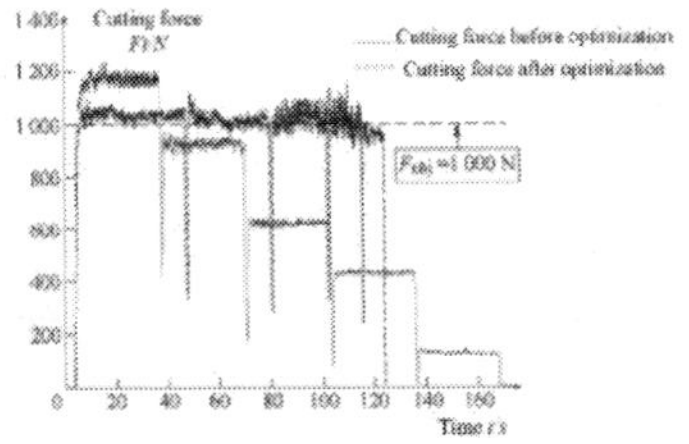

Figure 4. Comparison of cutting force curve before and after optimization.

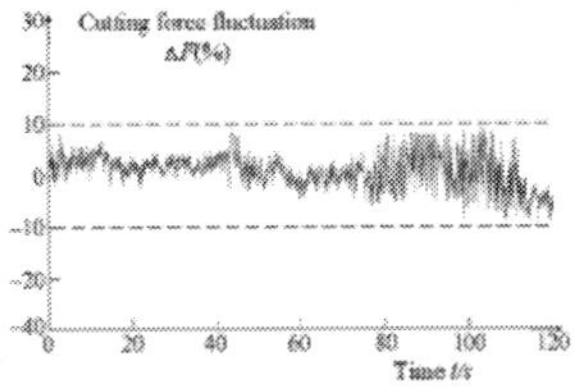

Figure 5. Cutting force fluctuation graph.

Comparing the average profile height before and after optimization, it can be found that the overall surface profile height after optimization is lower than before optimization, and comparing the calculated surface roughness, it can be seen that the average surface roughness Ra before optimization is 4.22μm, and Ra after optimization is 3.53μm, and the surface roughness of machining by optimizing the feed amount is improved by 16.35%, and the overall average surface roughness after optimization is in the range of 2.92~4.98μm, which ensures the surface quality requirement of semi-finishing. The overall average surface roughness after optimization is within the range of 2.92~4.98μm, which ensures the semi-finishing surface quality of the part.

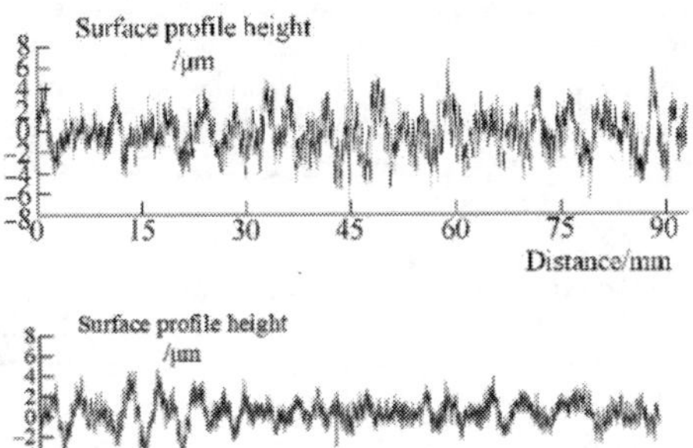

Figure 6. Average surface profile height distribution before and after optimization.

6. Conclusion

This paper addresses the problem that operators often rely on their own experience to select conservative process parameters in five-axis CNC machining of complex surfaces. Effectively avoid excessive cutting force and machining chatter phenomenon, so that the geometric accuracy and machining quality of complex surfaces can be guaranteed, taking into account the cutting stability and part performance constraints to effectively optimize the process parameters, breaking the bottleneck of complex surface machining quality is difficult to ensure. In view of the problem that the overall machining method with uniform process parameters is prone to local machining errors and severe cutting vibrations, the basic theory of area partitioning for CNC machining process is studied, so that the parameters can be adjusted optimally in time to respond to the change of cutting state. The parameters of cutting speed, feed rate, depth of cut, and width of cut are optimally adjusted in time to respond to the change of actual cutting state of the 5-axis machining tool cutting system by combining the control of cutting force and cutting deformation at different machining stages or parts of the workpiece, thus improving the surface quality of the parts.

Acknowledgments

This work was financially supported by Guangxi Natural Science Foundation under Grant (No. 2021GXNSFAA220019).

References

1. Tuong N V, Pokorny P. A practical approach for partitioning free-form surfaces [J]. International Journal of Computer Integrated Manufacturing, 2018, 23(11): 992-1001.
2. Zheng Z J, Zheng G L, Zhou M, et al. An algorithm for precise machining area computation in rough machining of complex 3D pocket [J]. International Journal of Advanced Manufacturing Technology, 2017, 92(1-4): 47-56.
3. Shen H Y, Sun Y F, Zhang L C, et al. A parameter zone subdivision method for rotary axes motion optimization in five-axis toolpath generation using inverse evaluation mechanism[J]. International Journal of Advanced Manufacturing Technology, 2018, 98(9-12):3115-3131.
4. A. Moufki, D.Dudzinski, G.Le Coz.Prediction of cutting forces from an analytical model of oblique cutting, application to peripheral milling of Ti-6Al-4V alloy[J].Int J Adv Manuf Technol, 2015, 81:615–626.
5. Zhongtao Fu, Wenyu Yang, Xuelin Wang, Jürgen Leopold. An analytical force model for ball-end milling using a predictive machining theory considering cutter runout[J]. The International Journal of Advanced Manufacturing Technology, 2016(84):2449–2460.

Effect of surface insulation on thermal stress of concrete

Jianda Xin [1, 2, 3, *], Yi Liu [1, 2, 3], Juan Wang [1, 2, 3], Zhenhong Wang [1, 2, 3], Xiaoming Jiang [4], Dingquan He [4], Huaiyu Jiang [4], Wenqian Hou [1, 2, 3], Shixi Zhang [1, 2, 3]

[1] State Key Laboratory of Simulation and Regulation of Water Cycle in River Basin,China Institute of Water Resources and Hydropower Research, Beijing, China
[2] Key Laboratory of Construction and Safety of Hydraulic Engineering of Ministry of Water Resources, China Institute of Water Resources and Hydropower Research, Beijing, China
[3] Department of Structure and Materials, China Institute of Water Resources and Hydropower Research, Beijing, China
[4] Huadian Tibet Energy Co., Ltd., Lhasa, China

Abstract. For concrete, the appearance of temperature load due to the hydration heat of cementitious materials is inevitable and important to the structural safety of dam. This paper investigates the effect of different surface insulation measures on the thermal stress of dam concrete by means of finite element simulation calculation. The calculation results show that surface insulation is important for lowering the surface thermal stress of concrete. When the equivalent surface coefficient of heat transfer was less than 3kJ/(m2·h·°C), the surface tensile stress can be decreased by 80%~90% for concrete undergoing diurnal temperature variation and cold wave.

Keywords: Mass concrete; Cracking control; Temperature variation; Surface insulation.

1. Introduction

Different from the thin-walled structures/components (beams, columns and slabs) in the industrial and civil construction industry, the hydraulic concrete structures with cross sections larger than 1.0m are highly difficult to construct due to the large amount of cementitious materials and the vulnerability to external environment such as solar radiation and temperature, as well as the complex construction technology of high arch dams. Its construction measures are closely related to its own material characteristics and meteorological conditions in the dam site area [1-3]. Internal temperature can arise after concrete mixing caused by the hydration of cementitious materials such as cement and fly ash. Once the deformation is restraint by the foundation and/or adjacent parts, significant tensile stress inside the structure can be generated [4]. Cracking occurs once the tensile stress in the temperature drop stage reaches the strength of concrete. Therefore, how to effectively reduce the tensile stress of concrete is the most concerned problem in the engineering field.

Generally, for the concrete structure with a section size greater than 1.0m, the adverse factors of too thick section combined with the thermal inertia of concrete materials make it difficult for the internal hydration heat to dissipate through natural heat dissipation, so cooling water pipes are needed for water cooling [5-6]. On the other hand, in order to cope with the large temperature difference (temperature gradient) caused by the sudden change of external temperature, thermal insulation measures should be taken on the surface of the dam body. First cooling by the embedded cooling pipes is carried out at the initial stage of casting to reduce the temperature peak, and later cooling is carried out before the arch grouting to make sure the concrete can reach the target grouting temperature [7]. Due to the low early-age strength of concrete (especially concrete mixed with fly ash), unreasonable first cooling may increase the cracking risk of early-age concrete.

Based on a three-dimensional numerical simulation calculation, this paper investigated the effects of surface insulation measures on the generation and development of concrete thermal stress, quantitatively analyzed the influence law of each factor on dam safety, and finally provided a strong theoretical and technical support for the realization of high quality mass concrete construction.

* Corresponding author: xinjd@iwhr.com

2. Temperature field calculation method

The heat conduction equation of concrete is shown in Equation (1) [8,9]

$$\frac{\partial T}{\partial \tau} = a\left(\frac{\partial^2 T}{\partial^2 x} + \frac{\partial^2 T}{\partial^2 y} + \frac{\partial^2 T}{\partial^2 z}\right) + \frac{Q}{c\rho} \quad (1)$$

where, a is the coefficient of temperature conductivity, m^2/h; T is actual temperature,°C; τ is the age, h; ρ is the density, kg/m^3; c is the specific heat, $kJ/(kg°C)$; Q is the hydration heat of cementitious material, kJ/kg. The adiabatic temperature rise of cementitious material is θ, one can obtain[10]

$$\frac{\partial \theta}{\partial \tau} = \frac{Q}{c\rho} \quad (2)$$

Substituting Equation (2) into Equation (1), one can have

$$\frac{\partial T}{\partial \tau} = a\left(\frac{\partial^2 T}{\partial^2 x} + \frac{\partial^2 T}{\partial^2 y} + \frac{\partial^2 T}{\partial^2 y}\right) + \frac{\partial \theta}{\partial \tau} \quad (3)$$

3. Project summary

3.1 Project and meteorological conditions

Two concrete mixtures for a high concrete arch dam construction is selected for simulation calculation. The arch dam normal water level is 2254.00m, the dam elevation is 2259.00m, the lowest elevation of the dam is 2059.00m, and the maximum height of the dam is 200m. The average annual temperature in the area where the dam site is located is 13.9°C with the extreme maximum temperature being 35.9°C and the minimum temperature being -15.9°C.

3.2 Material performance of concrete

Tables 1 and 2 show the thermal parameters and elastic modulus of dam concrete, respectively.

Table 1. Thermal parameters of dam concrete

	coefficient of temperature conductivity (m^2/h)	coefficient of thermal conductivity (kJ/m·h·°C)	specific heat (kJ/kg·°C)	thermal expansion coefficient (10^{-6}/°C)	adiabatic temperature rise (°C)
C_{180} 35	0.0031	6.01	0.78	9.0	28.0
C_{180} 30	0.0031	5.74	0.75	9.0	26.0

Table 2. Elastic modulus of dam concrete

	Grading	Elastic modulus (GPa)			
		7d	28 d	90 d	180 d
$C_{180}35$	4	17.71	23.04	26.68	31.00
$C_{180}30$	4	15.24	20.43	24.21	28.00

4. Research and analysis of surface insulation measures

4.1 Computational models and boundary conditions

In order to analyze the effect of cold wave and daily temperature difference on the concrete cracking risk of concrete under different surface insulation measures, a refinement model (40 m (length)×20 m (width)×10 m (height)) was prepared. Two different concrete temperature drops (15°C and 20°C), two diurnal temperature drops (6°C and 15°C) and five concrete age (3, 7, 14, 28 and 90d) were adopted. The model has 56320 units and 63068 nodes, as shown in Figure 1.

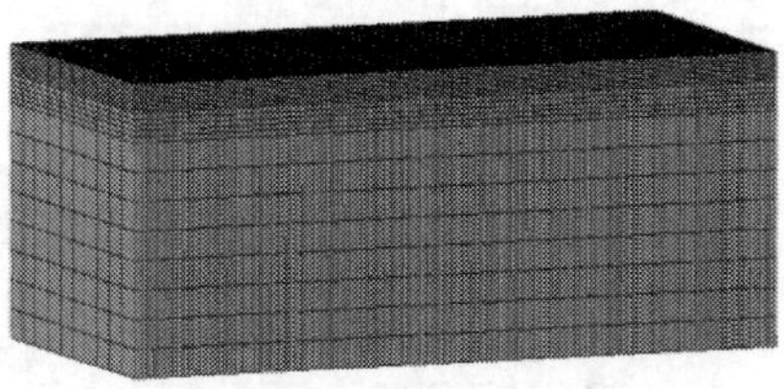

Figure 1. Refined model for surface protection analysis

4.2 Results and discussions

Table 3 and Table 4 show the maximum tensile stress of surface concrete at different ages under the influence of rapid temperature change process based on diurnal temperature difference, respectively. As can be seen from the tables, in the absence of surface protection measures, the maximum surface stress of C35 is 1.09MPa and 1.45MPa when the diurnal temperature drop is 15°C and 20°Crespectively. The maximum stress of concrete decreases to 0.11~0.14MPa when the surface coefficient of heat transfer of insulation material $\beta \leq 3$kJ/(m^2·h·°C). In the absence of surface protection measures, the maximum surface stress of C30 is 1.03MPa and 1.37MPa when the diurnal temperature drop is 15°C and 20°C, respectively. The maximum stress of concrete decreases to 0.10~0.14MPa when $\beta \leq 3$kJ/(m^2·h·°C). To sum up, the maximum reduction magnitude of concrete thermal stress is about 90% when strict surface insulation measures are taken.

Table 5 and Table 6 show the maximum tensile stress of surface concrete under different rapid temperature change processes (6°C and 15°C temperature drop in 2 days). As can be seen from the tables, in the absence of surface protection measures, the maximum surface stress of C35 is 1.30MPa and 3.24MPa when the temperature drop is 6°C and 15°C, respectively. After taking different surface insulation measures, the maximum surface stress level decreases dramatically. When β≤3kJ/(m2·h·°C), the maximum surface stress of concrete can be reduced to 0.26MPa and 0.66MPa when the temperature drop is 6°C and 15°C, respectively. In the absence of surface protection measures, the maximum surface stress of C30 is 1.23MPa and 3.07MPa when the temperature drop is 6°C and 15°C, respectively. When β ≤3kJ/(m2·h·°C), the maximum surface stress of concrete can be reduced to 0.25MPa and 0.63MPa when the temperature drop is 6°C and 15°C, respectively. To sum up, the maximum reduction magnitude of concrete thermal stress is about 80% with strict surface insulation measures.

Table 3. Thermal stress of C35 under diurnal temperature variation (Unit: MPa)

Case	Surface insulation condition (β)	Age (day)				
		3	7	14	28	90
Diurnal temperature variation:15°C	/	0.5	0.64	0.77	0.9	1.09
	10kJ/(m²·h·°C)	0.16	0.2	0.24	0.28	0.34
	5kJ/(m²·h·°C)	0.08	0.1	0.12	0.14	0.17
	3kJ/(m²·h·°C)	0.05	0.06	0.08	0.09	0.11
Diurnal temperature variation:20°C	/	0.66	0.85	1.03	1.2	1.45
	10kJ/(m²·h·°C)	0.21	0.27	0.32	0.38	0.46
	5kJ/(m²·h·°C)	0.1	0.13	0.16	0.19	0.23
	3kJ/(m²·h·°C)	0.07	0.08	0.1	0.12	0.14

Table 4. Thermal stress of C30 under diurnal temperature variation (Unit: MPa)

Case	Surface insulation condition (β)	Age (day)				
		3	7	14	28	90
Diurnal temperature variation:15°C	/	0.47	0.60	0.72	0.85	1.03
	10kJ/(m²·h·°C)	0.15	0.19	0.23	0.27	0.32
	5kJ/(m²·h·°C)	0.07	0.10	0.11	0.13	0.16
	3kJ/(m²·h·°C)	0.05	0.06	0.07	0.08	0.10
Diurnal temperature variation:20°C	/	0.62	0.80	0.97	1.13	1.37
	10kJ/(m²·h·°C)	0.20	0.25	0.30	0.36	0.43
	5kJ/(m²·h·°C)	0.10	0.13	0.15	0.18	0.22
	3kJ/(m²·h·°C)	0.06	0.08	0.10	0.11	0.14

Table 5. Thermal stress of C35 under rapid temperature drop (Unit: MPa)

Case	Surface insulation condition (β)	Age (day)				
		3	7	14	28	90
Temperature drop in 2 days:6°C	/	0.64	0.79	0.93	1.08	1.30
	10kJ/(m²·h·°C)	0.34	0.42	0.49	0.57	0.68
	5kJ/(m²·h·°C)	0.20	0.24	0.29	0.33	0.40
	3kJ/(m²·h·°C)	0.13	0.16	0.19	0.22	0.26
Temperature drop in 2 days:15°C	/	1.59	1.97	2.33	2.69	3.24
	10kJ/(m²·h·°C)	0.86	1.05	1.23	1.42	1.71
	5kJ/(m²·h·°C)	0.50	0.61	0.71	0.83	0.99
	3kJ/(m²·h·°C)	0.33	0.41	0.48	0.55	0.66

Table 6. Thermal stress of C30 under rapid temperature drop (Unit: MPa)

Case	Surface insulation condition (β)	Age (day)				
		3	7	14	28	90
Temperature drop in 2 days:6°C	/	0.60	0.74	0.88	1.02	1.23
	10kJ/(m²·h·°C)	0.32	0.39	0.46	0.54	0.65
	5kJ/(m²·h·°C)	0.19	0.23	0.27	0.31	0.38
	3kJ/(m²·h·°C)	0.13	0.15	0.18	0.21	0.25
Temperature drop in 2 days:15°C	/	1.50	1.85	2.19	2.54	3.07
	10kJ/(m²·h·°C)	0.81	0.99	1.16	1.34	1.62
	5kJ/(m²·h·°C)	0.47	0.57	0.67	0.78	0.94
	3kJ/(m²·h·°C)	0.32	0.38	0.45	0.52	0.63

5. Conclusion

Surface insulation measures can effectively improve the cracking risk of surface concrete under the condition of rapid temperature change. When the equivalent surface coefficient of heat transfer is less than or equal to 3kJ/(m²·h·°C), the maximum thermal stress of concrete decreases by 80%~90% under the condition of rapid temperature variations.

Acknowledgements

This research was supported by the National Natural Science Foundation of China (Grant No. 51779277) and the National Key R&D Program of China (Grant No. 2018YFC0406703). This research was also sponsored by the Special Scientific Fund sponsored by IWHR for Department of Structures and Materials (Grant No. SS0145B612017), and the Fund of State Key Laboratory of Simulation and Regulation of Water Cycle in River Basin of IWHR (Grant No. SKL2020ZY10). Support provided by Huadian Tibet Energy Co., Ltd. (Grant No. 12IJD202000070) is also gratefully acknowledged.

References

1. CHENG NG. Development, prospect and challenge of dam construction in China. Dam and safety, 2001(2):20-22.
2. ZHOU HM. Thermal stress analysis of 300 m high arch dam. China Science and Technology Information, 2010(06): 45-47+63.
3. ZHANG R. Thermal stress simulation analysis and temperature control measures for crack prevention studies on high arch dam. Master dissertation. Wuhan: Wuhan University, 2005.
4. WANG SH, XU P, ZHU BF. Study on thermal stresses of large arch dam during construction. Water Resources and Hydropower Engineering, 2000(7):9-12.
5. ZHU YM, HE JR, XIAO ZF, et al. Calculation and experimental comparison of temperature field of embedded cooling water pipes and their application. Journal of Hohai University (Natural Sciences), 2003(06):626-630.

6. LIU Y, ZHANG GX, YIN HA, et al. Influence of cooling height and water temperature on thermal stress of arch dam. Proceeding of research and progress of concrete materials and temperature control for hydraulic dam,2009:368-373.
7. ZHANG YX, HUANG DH, SONG YP, et al. Analysis of the second stage artificial cooling way of Xiluodu arch dam with numerical simulation method. Journal of Dalian University of Technology, 2003(3):367-371.
8. ZHU BF.Equivalent equation of heat conduction in mass concrete considering the effect of pipe cooling. Journal of Hydraulic Engineering,1991(3):28-34.
9. ZHU BF. Thermal Stresses and Temperature Control of Mass Concrete, Butterworth-Heinemann, Oxford, 2013.
10. ZHU BF. A method for computing the adiabatic temperature rise of concrete considering the effect of the temperature of concrete. Journal of Hydroelectric engineering, 2003(2): 69-73.

Preparation and Thermal Conductivity of FG/PP Composites

Haijun Zhou [1], Xiaolei Zhang [2,*], Mengmeng Zhou [1], Zhengzheng Wang [1], Xiaoqi Chen [1], Fen Zhang [1], Jijun Xiao [2] and Yantao Li [1]

[1] Institute of Energy Source, Hebei Academy of Sciences, Shijiazhuang, 050081, China
[2] Hebei University of Science and Technology, Shijiazhuang, 050018, China

Abstract: The surface modified flake graphite (FG) was filled into polypropylene (PP) and mixed by a double-roll open mill. FG/PP composites with high thermal conductivity were prepared by compression molding. Studies had shown that the modification of FG by coupling agent was beneficial to the dispersion in FG/PP composites. The effect was better when the amount of NDZ-201 was 0.7wt%. With the increase of FG particle size, the thermal conductivity of FG/PP composites increased obviously. Compared with 17μm FG, the average, vertical and parallel thermal conductivity of FG/PP composites prepared with 148μm FG increased by 65.9%, 87.7% and 47.0% respectively. The thermal conductivity of FG/PP composites were improved with the increase of FG content. When FG content was 70wt%, the thermal conductivity increased significantly, the average, vertical and parallel thermal conductivity reached 4.16 W·m-1·K-1, 0.938 W·m-1·K-1 and 18.4 W·m-1·K-1.

Key word: Flake graphite, Polypropylene, Composites, Thermal conductivity

1. Introduction

Polypropylene (PP) is a general plastic with regular structure and high crystallinity. Its working temperature is between -10°C~125 °C, with good chemical stability and excellent mechanical properties. Since there are no free electrons in PP, only atoms, groups and chains can vibrate at room temperature, so the thermal diffusion ability of PP is poor. Its thermal conductivity is only about 0.2W·m^{-1}·K^{-1}, and its thermal conductivity is far lower than that of metal, so its application is limited in heat exchanger, radiator, condensate cooler and other heat transfer equipment [1-5]. Relatively speaking, graphite has good thermal conductivity, high heat resistance and low price, so it has been widely applied in thermal conductive composites. Moreover, China is rich in graphite resources, and it is 90 % of world production. There are many kinds of graphite, which provides us with unique conditions for the study of FG/PP composites [6-8].

The thermal conductivity of Graphite Filled PP composites is attributed to the mechanism of thermal conduction path. Graphite contacts each other to form a three-dimensional thermal conduction path, and then forms a bi-continuous composites with PP. Therefore, graphite is the key factor in the construction of heat conduction path. When the filling amount is low, the composites are coated by polypropylene to form a "sea-island" structure, and the thermal conductivity of the composite has little change. With the increase of the filling amount, the probability of graphite contact in the composite increases. When the filling amount increases to a certain critical value, the heat conduction path is gradually improved, and a three-dimensional heat conduction path of the interpenetrating "sea-sea" structure is formed. Thus, the filling form and amount of graphite become the key to the research of thermal conductive composites [9-11].

In this study, the coupling agent was used to modify the surface of FG to improve its dispersion in PP and further increase the filling amount in FG/PP composites. By designing the molding structure and adjusting the orientation degree of graphite, more complete three-dimensional network heat conduction paths were constructed, so as to meet the use requirements of heat transfer equipment, reduce the amount of metal, control the production cost, save energy, and expand its application in chemical industry, metallurgy, new energy and seawater desalination [12-17].

2. Experimental methods

2.1 Materials

Flake graphite (FG, 99%) with the particle size of 17μm, 27μm, 37μm, 74μm and 148μm were purchased from Hebei Haoteng Trading Co., Ltd; Polypropylene (PP, 230#) was obtained from Sinopec Maoming Petrochemical Company; Silane coupling agents (KH-560, KH-570) and Titanate coupling agent (NDZ-101, NDZ-201) bought from Dinghai Plastic Chemical Co.,

* Corresponding author: xiaoji@hebust.edu.cn

 https://doi.org/10.1051/mateconf/202235801037

Ltd; Absolute ethanol (analytical pure) was purchased from Tianjin Yongda Chemical Reagent Co., Ltd; Antioxidant 1010 and Zinc stearate were obtained from Shanghai Aladdin Biochemical Technology Co., Ltd.

2.2 Equipments

Double-roll open mill (ZG-180) was produced by Dongguan Zhenggong Electromechanical Equipment Technology Co., Ltd; Electric drying oven (101-3AB) was produced by Tianjin Taisite Instrument Co., Ltd; Flat vulcanizer (ZG-200T) was produced by Dongguan Zhenggong Electromechanical Equipment Technology Co., Ltd; Universal prototype (wzy-240) was produced by Chengde Hengtong Test Instrument Co., Ltd; Thermal constant analysing (TPS-2500S) was produced by Sweden Hot disk Co., Ltd.

2.3 Preparation of FG/PP composites

The coupling agent was dissolved in anhydrous ethanol, and then mixed well with FG. The mixture was dried in an oven at 140°C. A certain amount of PP, zinc stearate, antioxidant and treated FG were pre-mixed in a high-speed mill, then mixed in a double-roll open mill, and finally molded by the flat vulcanizer to prepare FG/PP composites.

2.4 Testing methods

The thermal conductivity was tested according to GB / T 32064-2015. The radius of probe was 3.189mm, the test temperature was 25°C, and the sample size was 40×40×3mm。

3. Results and discussion

3.1 Types of coupling agents

The inorganic filler treated by coupling agent has better compatibility with polymer, which can effectively improve the properties of composites. FG was treated with two silane coupling agents and two titanate coupling agents. When FG content of 17μm was 50wt% and the coupling agent dosage (measured by FG content) was 1wt%, the FG/PP composites were prepared and their thermal conductivities were tested (Fig.1). It can be seen from Fig. 1 that the thermal conductivities of FG / PP composites prepared by FG treated with four different types of coupling agents had no obvious change. The coupling agent basically had no effect on the orientation of FG, so the thermal conductivity changed little.

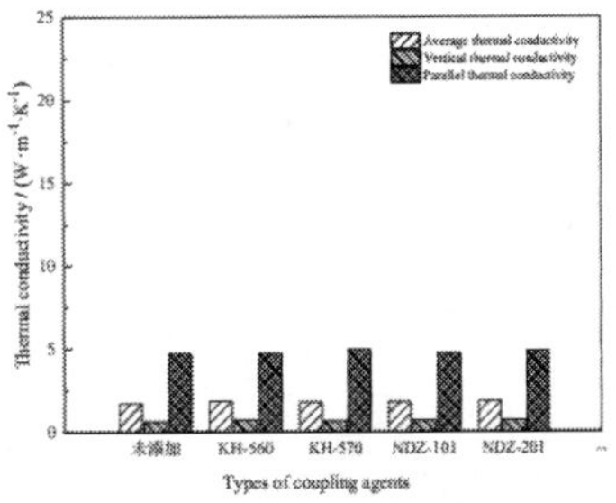

Fig.1 Effect of coupling agent on thermal conductivities of FG/PP composites

3.2 Content of coupling agents

When the FG content of 37μm was 70wt%, the FG/PP composites were prepared by changing the amount of NDZ-201 and their thermal conductivities were tested (Fig.2).

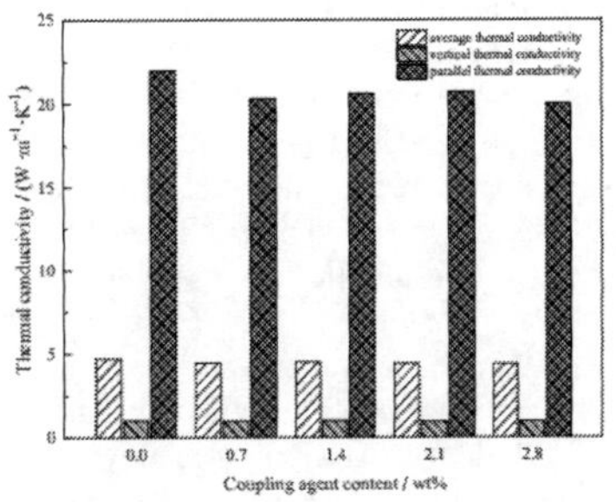

Fig.2 Effect of coupling agent dosage on thermal conductivities of FG/PP composites

As shown in Fig.2, the thermal conductivities of FG/PP composites prepared by FG treated with NDZ-201 were slightly lower than that of composites without coupling agent. The average, vertical and parallel thermal conductivities of the composites without NDZ-201 were $4.73W\cdot m^{-1}\cdot K^{-1}$, $1.01W\cdot m\text{-}1\cdot K^{-1}$ and $22.0W\cdot m^{-1}\cdot K^{-1}$, respectively. When the addition of NDZ-201 was 2.8wt%, the average, vertical and parallel thermal conductivities of the composites were $4.37\ W\cdot m^{-1}\cdot K^{-1}$, $0.951\ W\cdot m^{-1}\cdot K^{-1}$ and $20.0\ W\cdot m^{-1}\cdot K^{-1}$. Thermal conductivities of composites decreased slightly with the increase of coupling agent content. The possible reason is that the thermal conductivity of the coupling agent itself is very low. When the coupling agent is coated on the surface of FG, the internal resistance of the composite increases, which hinders the heat transfer in the composite and reduces the heat conduction flux of the composite.

3.3 Content of FG

In order to investigate the effect of FG content on the properties of the composites, FG/PP composites with different FG content were prepared by 17μm FG and 1wt% NDZ-201 coupling agent. Fig. 3 shows the results.

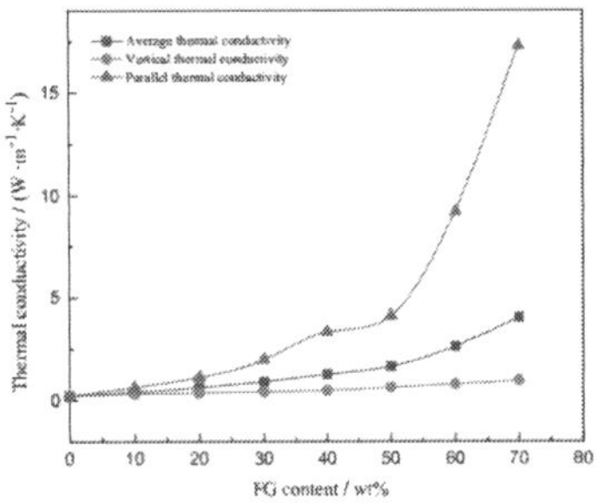

Fig.3 Effect of FG content on thermal conductivities of FG/PP composites

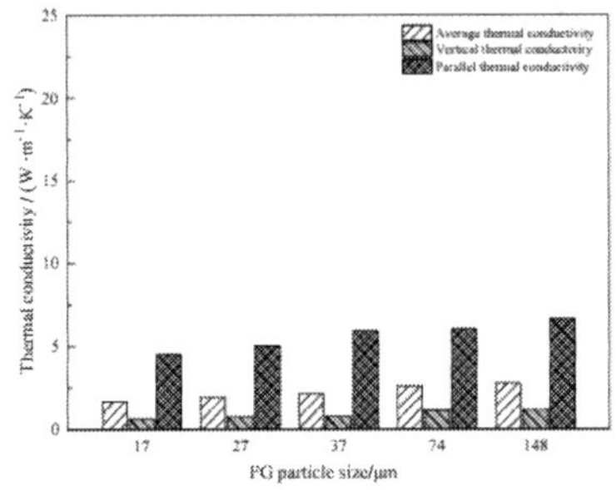

Fig.4 Effect of FG particle size on thermal conductivities of FG/PP composites

The thermal conductivities of the composites increased with the increase of FG content. When the FG content is 10wt%, the average, vertical and parallel thermal conductivity of the FG/PP composites were 0.475 W·m^{-1}·K^{-1}, 0.318 W·m^{-1}·K^{-1} and 0.710 W·m^{-1}·K^{-1}, respectively. When the FG content increased to 70wt%, the average, vertical and parallel thermal conductivity were 4.16W·m^{-1}·K^{-1}, 0.938 W·m^{-1}·K^{-1} and 18.4W·m^{-1}·K^{-1}, which improved by 7.76, 1.95 and 24.9 times, respectively. By contrast, it was found that the parallel thermal conductivities were improved most obviously, which was closely related to the molding method. Before pressing, the molten materials in the mold cavity are unevenly distributed. When pressure is applied in the vertical direction, the graphite sheets in the vertical state are easy to tilt and fall down. At the same time, the molten materials flow to the parts with less materials and lower pressure, which promote the orientation of FG. Therefore, FG is difficult to orientate in the vertical direction and cannot form a good heat conduction path. On the other hand, FG has a high degree of orientation parallel to the surface of the material, and the layers come into contact with each other to form more complete stacking structures, which provides more heat conduction paths for the transfer of heat flux, so the improvement of parallel thermal conductivity of the composites is much greater than that of vertical thermal conductivity. In addition, When FG content was less than 50wt%, the continuous phase in the composite was PP, the thermal resistance was very large, and the thermal conductivity increased slowly. When FG content exceeded 50wt%, FG transformed into continuous phase, and the chance of overlapping between FG was greatly increased, which made it easy to construct more complete thermal conduction paths, thus sharply increasing thermal conductivity.

3.4 Particle size of FG

To explore the effects of FG with different particle sizes on the thermal conductivity of composites, FG with particle sizes of 17, 27, 37, 74 or 148μm was selected to prepare FG/PP composites under the condition of FG content of 50wt%, NDZ-201 dosage of 1wt% and the thermal conductivities were tested (Fig.4).

It can be observed from Fig.4, the thermal conductivities of the FG/PP composites increased with the increase of FG particle size. The average, vertical and parallel thermal conductivities of the composites with FG particle size of 17μm were 1.67 W·m^{-1}·K^{-1}, 0.618 W·m^{-1}·K^{-1} and 4.51 W·m^{-1}·K^{-1}, respectively. When FG particle size was 148μm, the average, vertical and parallel thermal conductivities of the composites were 2.77 W·m^{-1}·K^{-1}, 1.16 W·m^{-1}·K^{-1} and 6.63 W·m^{-1}·K^{-1}, increasing by 65.9%, 87.7% and 47.0%, respectively. The reason is that when the FG particle sizes are small, the particles are coated by PP melt film. During molding, they are not easy to be flattened and crushed; Meanwhile, because the FG particles are small, the specific surface area is large, and there are many PP diaphragms between FG particles, which hinder the formation of heat conduction path, the thermal conductivities are low. On the contrary, when the FG particles are large, the PP melt film between the particles is thick and uneven. FG particles are easy to be separated and contact with each other under compressive shear stress, which provides the possibility to form more complete thermal conductivity network. At the same time, the number of gaps between FG particles is less, which is more conducive to the overlapping contact of uncoated FG particles to form the complete heat conduction path, so as to improve the thermal conductivities of FG/PP composites.

4. Conclusion

In this paper, PP was filled with FG modified by coupling agent, and FG/PP composites with high thermal conductivity were prepared by molding method. The effects of the type and amount of coupling agent, FG content and particle size on the thermal conductivity of the composites were investigated. The results were as follows:

(1). The use of coupling agent NDZ-201 to modify FG was beneficial to the dispersion in the materials, but the thermal conductivities of FG/PP composites decreased slightly.

(2). With the increase of FG content, the thermal conductivities of FG/PP composites increased. When the FG content was more than 50wt%, the thermal

https://doi.org/10.1051/matecconf/202235801037

conductivities of the composites had an obvious critical value, in which the parallel thermal conductivities improved most significantly.

(3). The thermal conductivities of FG / PP composites increased dramatically with the increment of FG particle sizes. Compared with 17μm FG, the average, vertical and horizontal thermal conductivity of the FG/PP composites prepared by148μm FG increased by 65.9%, 87.7% and 47.0%, respectively.

References

1. Li Y, Li J L, Zhu Q S, et al. Research progress in graphene based thermal conductivity materials[J]. Journal of Materials Engineering, 2021,49(11):1-13.
2. Hu M, Yu D, Wei J. Thermal conductivity determination of small polymer samples by differential scanning calorimetry[J]. Polymer Testing, 2007,26(3):333-337.
3. Han Z D, Fina A. Thermal conductivity of carbon nanotubes and their polymer nanocomposites: A review[J]. Progress in Polymer Science, 2011,36(7): 914-944.
4. Chen H, Ginzburg V V, Yang J, et al. Thermal conductivity of polymer-based composites: Fundamentals and applications[J]. Progress in Polymer Science, 2016,59: 41-85.
5. Dialameh L, Yaghoubi M, Abouali O. Nature convection from an array of horizontal rectangular thick fins with short length[J]. Applied Thermal Engineering, 2008,28(17): 2371-2379.
6. Du S Y, Du Y X, Fu Y Z, et al. Preparation and properties of PP/Graphite/AL2O3 thermal conductive composites[J]. Engineering Plastics Application, 2014,42(11):17-22.
7. Xu H G, Peng X T, Zhou J J, et al. Research of thermally conductive composites based on flake graphite filled in polyamide 6/polypropylene[J]. Engineering Plastics Application, 2015,43(9):1-6.
8. Du S Y, Du Y X, Xu B P, et al. Properties of thermally conductive polymers reinforced with graphite mixed in a co-rotating non-symmetric twin-screw[J]. China Plastics, 2015,29(7):86-91.
9. Zhang Z Y, Bai S J, Zang X X, et al. Research progress in factors affecting thermal conductivity of polymeric materials[J]. China Plastics, 2021, 35(9):156-165.
10. Lee G W, Min P, Kim J, et al. Enhanced thermal conductivity of polymer composites filled with hybrid filler[J]. Composites: Part A, 2006,37:727-734.
11. Hu Y, Li R H. Study on Polypropylene composites with high coefficient of thermal conductivity and its preparation[J]. Plastics Science and Technology, 2012, (12):59-63.
12. Zhang L W. Manufacturing, property and application of thermal conductive polymer composites[D]. East China University of Science and Technology, 2010.
13. King J A, Tambling T M, Morrison F A, et al. Effects of carbon fillers on the thermal conductivity of highly filled liquid-crystal polymer based resins[J]. Journal of Applied Polymer Science, 2008, 109(4):2145-2155.
14. Panwar V, Park J O, Park S H, et al. Electrical, dielectric, and electromagnetic shielding properties of polypropylene-graphite composites[J]. Journal of Applied Polymer Science, 2010,115(3):1306-1314.
15. Wang Z Z, Liu D J, Zhou H J, et al. Preparation and properties of polypropylene/flake graphite thermal conductive composites[J]. Engineering Plastics Application,2020,48(1):30-34+45.
16. Wu J W, Ji X Y, Yang J D, et al. Preparation and properties of polypropylene / graphite composites with high thermal conductivity[J]. Engineering Plastics Application,2016,44(08):36-40.
17. Xu H G, Peng X T, Zhou J J, et al. Research of thermally conductive composites based on flake graphite filled in polyamide 6 / polypropylene[J]. Engineering Plastics Application,2015,43(09):1-6.

Statics and Modal Analysis of Main Frame of Composite Board Detaching Sleeve Equipment

Wenchang Zhu, Guangzhong Hu *, Qiuyue Bin, Junjie Liao and Haoming Xu

School of Mechanical Engineering, Sichuan University of Science and Engineer, Sichuan, China

Abstract. A new type of special machine tool for stripper sleeve was developed according to the stripper sleeve process of composite board. In order to ensure the reliability of the mainframe structure and the performance of the linear guide mechanism, the static structural analysis and modal analysis of the mainframe are carried out. The results show that the maximum stress of the supporting beam under the main frame is about 48.18 MPa, the maximum strain is 0.00023, and the maximum stress of the four-legged column is about 5.35 MPa, which is full of strength and stiffness design requirements. The end-cutting mechanism is installed on the beam through the installation plate. The deformation of the beam has a great influence on the operation accuracy of the linear guide mechanism and the cutting accuracy of the end-cutting mechanism. The maximum displacement in the X and Y directions of the installation plate is 0.014 mm, 0.030 mm and 0.197 mm, respectively, which meets the normal use requirements and cutting accuracy requirements of the linear guide mechanism. The first six natural frequencies of the mainframe are 2.963 – 47.137 Hz. In order to make the mainframe have good dynamic characteristics, avoid resonance and meet the end-cutting requirements of the composite board, the motor speed should be controlled between 300 – 493 r / min.

Keywords. Mainframe of stripper equipment; Linear guide mechanism; Strength and rigidity; Dynamic characteristics

1. Introduction

A kind of core board material is closely combined with the metal frame and two layers of steel sheet to form a metal composite board [1]. Before use, it is necessary to remove the outer steel sleeve composed of the steel sheet and the metal frame. The composite board stripper equipment is a new type of equipment to remove the outer steel sleeve of the metal composite board. The designed mainframe, as the support structure of the stripper equipment, has a large size, total length of 3.57 m, width of 2.5 m, and height of 2.4 m. It is mainly used to withstand the gravity load generated by the end-cutting mechanism, the reaction force generated by the leveling mechanism, and the influence of motor excitation. Under the influence of these complex loads, the mainframe will deform, resulting in changes in the cutting accuracy of the end-cutting mechanism and the motion accuracy of the linear guide mechanism, affecting the performance of the entire mechanical equipment. Therefore, the mainframe should not only have sufficient strength and stiffness, but also have good dynamic characteristics. Many domestic and foreign scholars have done a lot of research on the static and dynamic performance of mechanical structure. P. Thejasree [2] carried out the finite element analysis on the of crankshaft under alternating load, and studied the influence of crankshaft stress on the reliability and life of engine, which provided theoretical support for the lightweight design of crankshaft . Hou Yangguo [3] proposed a conforming band method based on spline FSM, which is suitable for the static analysis of thin-walled structures with transverse stiffeners. The finite element method was used to carry out the static analysis and buckling analysis of thin-walled structures with transverse stiffeners, and the accuracy of the composite band method was verified. Khondaker Sakil Ahmed [4] used the finite element analysis method to study and analyze the strength, stiffness and fatigue life of a drilling active lifting structure in practical application, and obtained the influence of environmental factors on its structural performance, which provided guidance for the design of offshore jack-up drilling platform in the future. Leonardo B. Saint Martin [5] used the operational modal analysis technology to extract the modal parameters of a rotating machinery structure. Compared with the experimental modal analysis technology, the accuracy of the operational modal analysis technology was verified, which provided a theoretical basis for the prediction and analysis of structural modal parameters by the operational modal analysis. Han Fei [6] proposed a modal analysis method of double-beam system based on dynamic stiffness, and used a method based on dynamic stiffness

* Corresponding author: 568092170@qq.com

to study the influence of elastic support on the dynamic characteristics of mechanical system and the change law of dynamic stiffness in frequency domain, which can quickly obtain the information of frequency and modal shape of the system.

Most of the above studies used the finite element analysis method to carry out the finite element analysis of the mechanical structure, and explored whether the static and dynamic performance of the frame structure met the requirements, so as to verify the rationality of the structural design. However, the relevant standards of non-standard machine tools were not studied. In this paper, the finite element analysis of the mainframe of the composite board stripper sleeve equipment is carried out to explore the strength and stiffness characteristics of the overall structure of the main frame under static load and whether the displacement of the guide slider mechanism caused by the deformation of the mainframe meets the relevant standards of the non-standard machine tool, and whether the deformation of the beam leads to the cutting accuracy of the equipment can meet the requirements. In addition to meeting the static strength requirements of the mainframe structure, the dynamic characteristics of the structure should be further studied to avoid the resonance of the mainframe by avoiding the natural frequency of the mainframe, which provides an important theoretical basis for the reliability of the overall structural performance of the stripper equipment.

2. Working Principle and Structure Design of Mainframe

During the working process, the composite board is sent to the workbench, the positioning mechanism makes the composite board move to the middle of the workbench, the leveling mechanism starts, and the bending composite board is flattened. The positioning mechanism and the leveling mechanism are installed on the lower support beam of the mainframe, and the end-cutting mechanism of the beams on both sides of the mainframe is installed to complete the cutting work at both ends of the composite board.

The mainframe of the composite board stripper equipment can effectively absorb the static load generated by the leveling device when it works, and also has good dynamic characteristics under the normal working state of the end-cutting mechanism. Therefore, it is determined that the mainframe of the composite stripper equipment is mainly composed of leg columns, lower beams, beams, lower support beams and upper bending beams. The end-cutting mechanism is installed on the crossbeams on both sides of the mainframe through the linear guide and the mounting plate. The leveling mechanism is mainly composed of multiple cylinders and installed on the crossbeams under the mainframe to bear the reaction force generated in the flattening process of the composite plate. The mainframe structure design is shown in Fig. 1.

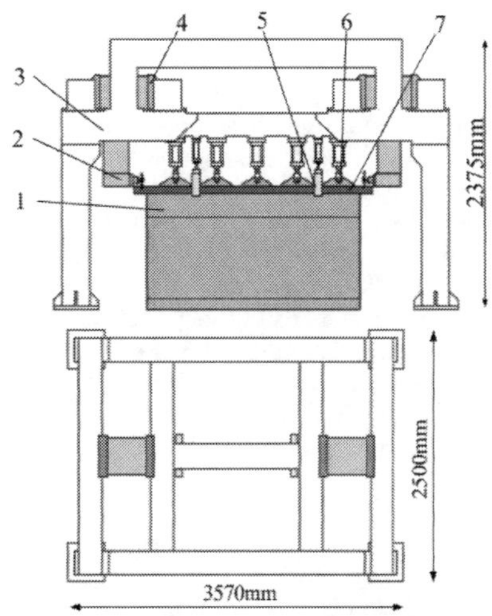

1.Workbrnch 2. Cutting mechanism 3.Mainframe 4. Installation plate of end-cutting mechanism 5. Positioning mechanism 6. Levelling mechanism 7. Composite board

Figure 1. Mainframe structure design diagram

3. Static Analysis of Mainframe

3.1 Mainframe finite element model establishment

According to the actual size of the composite board stripper equipment, SolidWorks software was used to establish the overall three-dimensional model, and then the mainframe part was extracted. The chamfer, hole forming, thread hole and bolt structure included in the mainframe model were simplified, which was convenient for subsequent grid division and analysis and calculation. The simplified model is shown in Fig. 2. The mainframe material is Q235 and its material properties [7] are shown in Table 1.

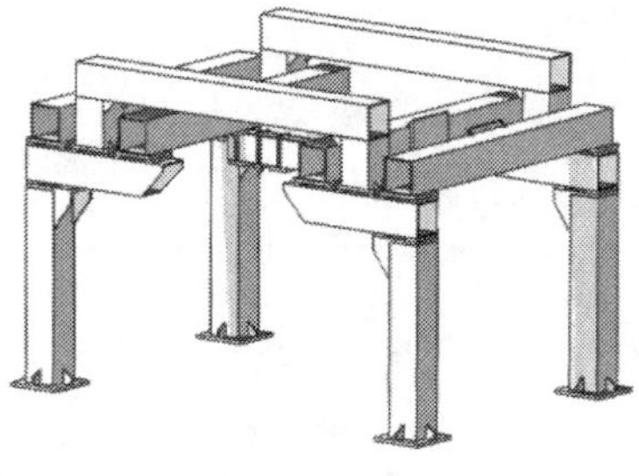

Figure 2. Mainframe analysis model

Table 1. Material Properties of Mainframe.

Material	Density/ $kg{\cdot}m^{-3}$	Young's modulus/MPa	Poisson's ratio
Q235	7860	$2.1{\times}10^5$	0.3

 https://doi.org/10.1051/matecconf/202235801038

The mainframe is a welded part. In order to reduce the error of calculation results, the welded joints of the mainframe structure should be connected by solder joints. The local amplification diagram of the solder joint connection is shown in Fig. 3. The contact surface type of the welded joint is set as friction contact, and the friction coefficient is 0.2. The mainframe is mainly divided by hexahedral mesh, and a small number of irregular or complex contact parts are segmented to divide tetrahedral mesh, which is conducive to reducing calculation time and improving calculation accuracy [8]. The mesh size of the finite element model of the mainframe is controlled to 40 mm. The total number of mesh divided is 85068, and the number of nodes is 27454. The meshing effect is shown in Fig. 4.

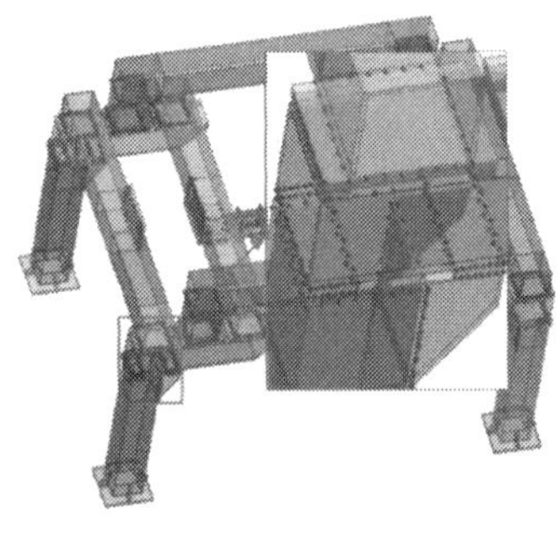

Figure 3. Local magnification of welding spot

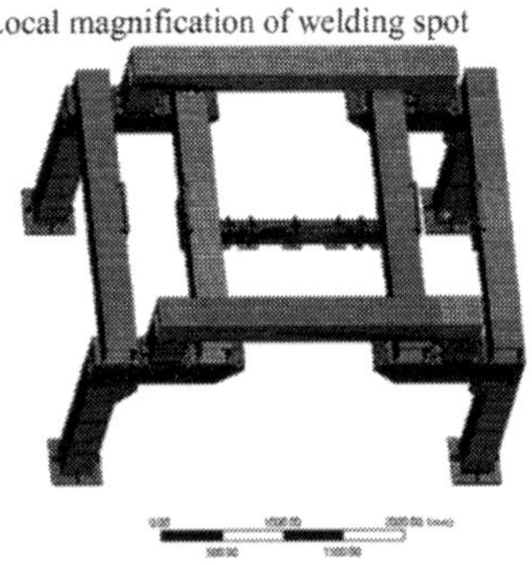

Figure 4. Grid division of mainframe

The bottom foot of one column of the mainframe is fixed on the ground, and the remaining three feet can produce translational displacement in the horizontal direction ; due to the large quality of the mainframe itself, the influence of its own gravity factor should be considered. End-cutting mechanisms with mass of about 704.3 kg were installed on both sides, and leveling and positioning devices with mass of about 26.9 kg, 14.7 kg, 21.6 kg, 21.6 kg, 21.6 kg, 14.7 kg and 21.6 kg were installed on the middle beam; the pressure is 0.7MPa and the cylinder diameter is 100mm. After calculation, the cylinder generates about 0.23 MPa reverse pressure on the lower support beam when it works. According to the above constraint and load distribution law, the corresponding boundary conditions are added to the mainframe, as shown in Fig. 5.

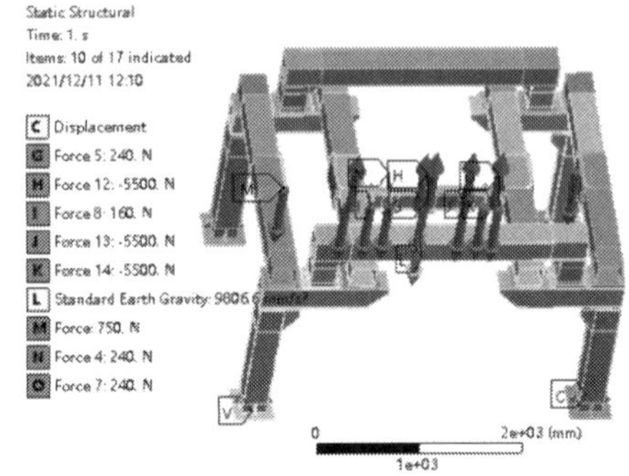

Figure 5. Load and constraint of mainframe

3.2 Analysis of Result

After the solution, the equivalent stress, strain and deformation distribution cloud diagrams of the overall structure of the mainframe are extracted, as shown in Fig. 6, Fig. 7 and Fig. 8.

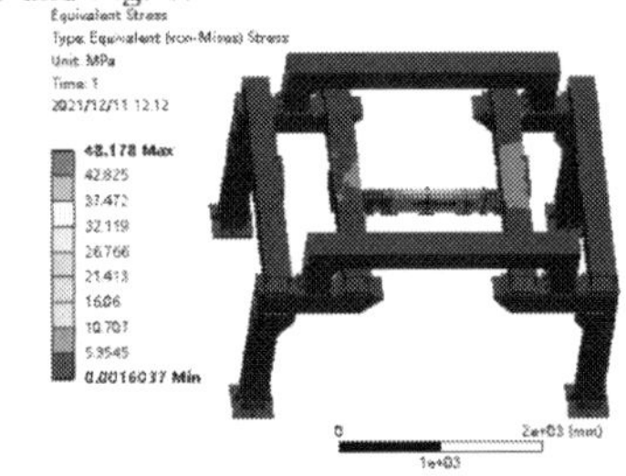

Figure 6. Stress cloud diagram of mainframe

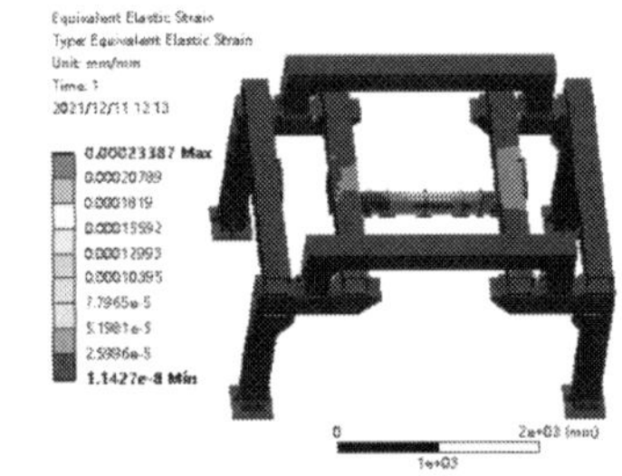

Figure 7. Strain cloud diagram of mainframe

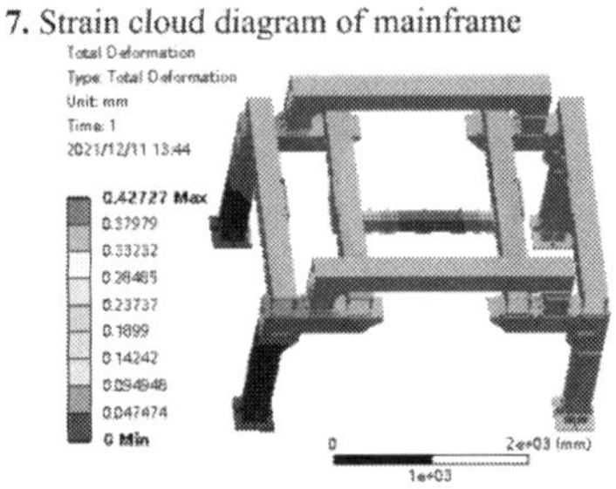

Figure 8. Deformation cloud diagram of mainframe

From the above results, it can be seen that the stress of the mainframe is mainly concentrated in the middle part of the lower support beam. The maximum stress is about 48.18 MPa, the maximum deformation is about 0.43 mm, the maximum strain is about 0.00023, and the maximum stress of the column is about 5.35 MPa. Referring to *Modern Mechanical Design Manual* [9], the strength and stiffness of the main frame are checked, which meet the design requirements. The specific results are shown in Table 2.

Table 2. Strength and Stiffness Checking Results of Mainframe

Check the project	Result	Requirements	Whether it meets
Maximum stress of support beam /MPa	48.18	≤50	Yes
Maximum stress of four-legged column /MPa	5.35	≤30	Yes
Maximum strain of mainframe	0.00023	0.002~0.0004	Yes

The cutting mechanism is installed on the installation plate of the beams at both ends of the mainframe, and the installation schematic diagram is shown in Fig. 9.

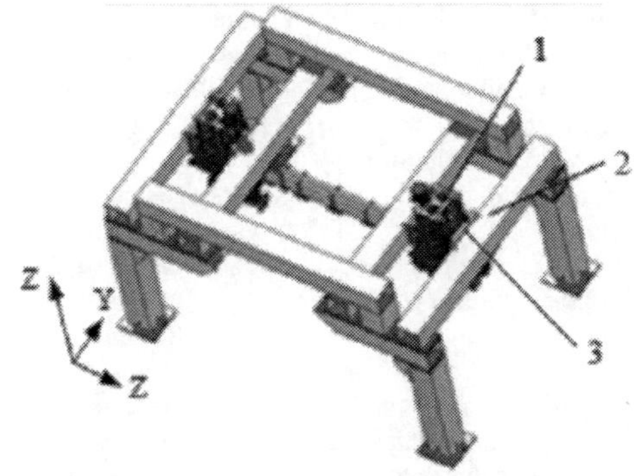

1. End-cutting mechanism 2. beam 3. Installation plate

Figure 9. Installation diagram of end-cutting mechanism

When the mainframe is loaded, the deformation of the beams at both ends will lead to the displacement of the installation plate. Due to the linear guide mechanism is installed in the end-cutting mechanism, without considering the error between the linear guide and the end-cutting mechanism itself, the displacement of the installation plate directly affects the service life of the linear guide and the cutting accuracy of the whole equipment. The X and Y direction displacement cloud images of the installation plate are extracted as shown in Figures 10 and 11.

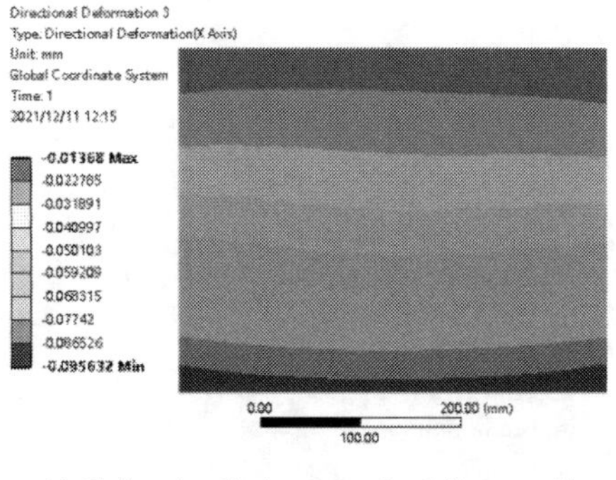

Figure 10. X direction displacement cloud diagram of installed plate

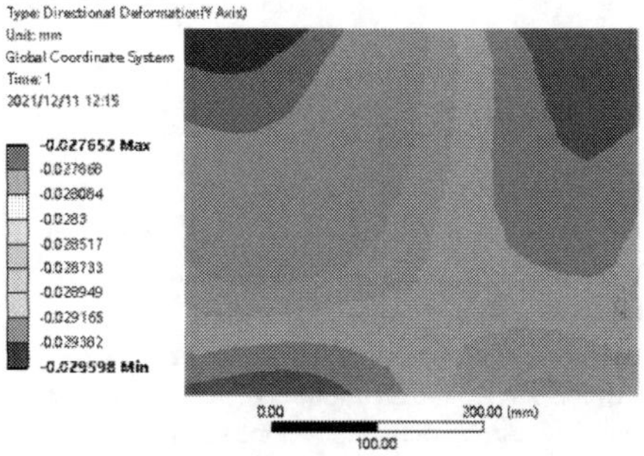

Figure 11. Y direction displacement cloud diagram of installed plate

The analysis shows that the displacement of the installation plate in the X direction is about 0.014 mm, and the displacement in the Y direction is about 0.030 mm. According to *Modern Mechanical Design Manual* [9], the displacement of the bottom surface of the linear guide can not exceed 0.15 mm, and the displacement of the side of the guide can not exceed 0.10 mm, which meets the motion requirements of the linear guide. At the same time, the installation plate displacement also affects the cutting accuracy of the equipment. According to the milling accuracy range of the milling machine, it is 0.54 mm, which meets the cutting accuracy requirements.

4. Modal Analysis of Mainframe

Modal analysis is mainly used to study the dynamic characteristics of the structure. Modal is the inherent vibration characteristics of the mechanical structure, and each mode has a specific natural frequency and modal shape [10, 11]. In order to avoid the resonance of the mainframe under the excitation of the motor and affect the reliability of the structure of the stripper equipment, it is necessary to carry out the modal analysis of the mainframe, study the dynamic characteristics of the mainframe, and understand the change law of the natural frequency and vibration mode of each order of the mainframe.

The excitation source of the mainframe is mainly generated by the motor of the end-cutting mechanism. When the motor works steadily, it can be regarded as the rotor moving in a uniform circular velocity. The relationship [12] between the generated excitation frequency $f(Hz)$ and its own rotational speed $n(r / \min)$ is :

$$f = \frac{1}{60} n \tag{1}$$

The speed range of the motor is 0 ~ 2000r / min, and the excitation frequency range is 0 ~ 33.3Hz calculated by (1) formula. According to *Mechanical Vibration and Noise* [13], the natural frequency of the mainframe has the following relationship with the excitation frequency :

$$\bar{\omega} = \frac{\omega}{\omega_n} \tag{2}$$

Formula : $\bar{\omega}$ frequency ratio ; ω_n the natural frequency of the frame ; ω for excitation frequency.

When $\omega \to \omega_n$, that is, $\bar{\omega} \to 1$, the mainframe resonates. In engineering application, the interval of $0.75 < \bar{\omega} < 1.25$ is taken as the resonance interval [14]. In order to ensure that some natural frequencies and vibration modes are not omitted in the study of the mainframe mode, the extraction of the modal frequency interval of the mainframe should be appropriately broadened to 0 – 50 Hz. The results of the first six modal natural frequencies and the corresponding modal characteristics are shown in Table 3.

Table 3. Natural Frequency Result Table of Mainframe

Number	Natural frequency /Hz	Modal characteristics
1	2.963	Left and right deformation of mainframe around fixed foot column occurs
2	10.955	Mainframe deformed diagonally along fixed foot column
3	17.653	Torsion of mainframe along fixed foot column
4	31.139	Mainframe twists up and down along fixed foot and deforms left and right
5	40.167	Inward deformation of left two columns of mainframe
6	47.137	Outward deformation of two columns in front of mainframe

It can be seen from Table 4 that the first six natural frequencies of the mainframe are in the range of 2.963 – 47.137, which is the same as the range of external excitation frequency, indicating that the mainframe will resonate under the external excitation frequency, and the low-order frequency has a great influence on the dynamic performance of the mainframe structure. In order to avoid the resonance phenomenon of the main frame, the excitation frequency should be controlled to avoid the natural frequency of the main frame, and the speed of the motor should be controlled between 222 and 493 r / min to avoid the natural frequency of the main frame. According to the experimental test of the cutting process of the composite board, when the motor speed reaches 300 r / min, it can meet the cutting requirements of the composite board. In order to avoid the natural frequency of the mainframe structure, avoid the resonance phenomenon of the mainframe and meet the cutting quality of the composite board, the motor speed range should be selected between 300 and 493 r / min.

5. Conclusion

In this paper, the mechanical device of composite stripper sleeve is developed. The finite element method is used to analyze the structural performance of the mainframe. The mainframe meets the requirements of static structural design. In order to avoid the resonance of the mainframe, the motor speed range of the end-cutting mechanism is determined to be 300 ~ 493 r / min, which provides a theoretical basis for the reliability of the overall structural performance of the composite board stripper sleeve equipment. The specific conclusions are as follows :

(1) Through the static analysis of the mainframe, it can be seen that there is stress concentration in the middle part of the lower support beam, the maximum stress value is about 48.18 MPa, the maximum deformation is about 0.43 mm, the maximum strain is about 0.00023, and the maximum stress on the four-legged column is about 5.35 MPa, which meets the strength and stiffness requirements of the mainframe. The maximum displacement of the mounting plate in X and Y directions is about 0.014 mm, and 0.030 mm, which meets the requirements of linear guide and equipment cutting accuracy.

(2) The excitation frequency range of the end-cutting mechanism installed on the mainframe is about 5 – 33.3 Hz. In order to study the dynamic performance of the mainframe structure, the modal analysis is carried out. The first six natural frequencies of the mainframe are extracted to be 2.963 – 47.137 Hz. In order to avoid the natural frequency of the mainframe structure, prevent the over-resonance of the mainframe, and meet the cutting requirements of the composite board, the motor speed should be controlled between 300 – 493 r / min.

Acknowledgments

This work was financially supported by Anhui Province Key Laboratory of Intelligent Manufacturing of Construction Machinery Opening Project (Project number: IMCM2021KF03) fund.

References

1. L. Q. Wang, R. R. Liu, Y. Chen, et al. China Patent CN104409121A. (2015).
2. P. Thejasree, G. Dileep Kumar, S. Leela Prasanna Lakshmi, Modelling and Analysis of Crankshaft for

passenger car using ANSYS, Materials today: proceedings. 10 (2017) 11292-11299.

3. G. Y. Hou, Z. J. Li, J. H. Gong, A compound strip method for static and buckling analyses of thin-walled members with transverse stiffeners, Thin-Walled Structures. 165 (2021) 107829.
4. K. S. Ahmed, A. K. Keng, Stress and stiffness analysis of a 7-teeth pinion/rack jacking system of an Offshore jack-up rig, Engineering Failure Analysis. 115 (2020) 104623.
5. Leonardo B. Saint Martin, Leonardo L. Gusmão, Tiago H. Machado, et al, Operational modal analysis application to support structure identification under rotating machinery unbalance, Engineering Structures. 249 (2021) 113344.
6. F. Han, D. H. Dan, Z. C. Deng, A dynamic stiffness-based modal analysis method for a double-beam system with elastic supports, Mechanical Systems and Signal Processing. 146 (2021) 106978.
7. F. F. Gu, Z. S. Zhu, X. Z. Yan, Design and Strength check of Large Blow Molding Machine Rack, Physics Procedia. 22 (2011) 252-257.
8. G. J. Kim, H. G. Kwak, C. K. Jin, Three-dimensional equivalent static analysis for design of submerged floating tunnel, Marine Structures. 80 (2021) 103080.
9. D. T. Qin, L. Y. Xie, Modern Mechanical Design Manual, first ed., Chemical Industry Press, Beijing, 2011.
10. J. Zhou, J. Y. Su, Examples of Finite Element Analysis : Dynamis, first ed., Posts and Telecom Press, Beijing, 2019.
11. T. Ding, S. C. Yu, Frame construction mode numerical simulation analysis based on ANSYS, Journal of Liaoning Technical University (Natural Science) , 29 (2010) 43-44.
12. J. J. Liu, L. Ma, W. Xiang, et al, Static, Modal Analysis and Structural Optimization of 9ZM-5.0 Feeding-ramie Shredder Frame, Plant Fiber Science in China, 42 (2020) 8.
13. M. Zhao, Mechanical Vibration and Noise, first ed., Science Press, Beijing, 2004.
14. S. Y. Lou, Vibration Behavior Analysis and Test Verification Based on 3WQF80-10 Engine Driving Unmanned Helicopter Frame, Agricultural Equipment and Technology, 2018.

Design and optimization of elevator car bottom structure

Naiqiu Huang, Mengmeng Song *, Shungen Xiao, Zhipeng Zheng

School of information, Mechanical and Electrical Engineering, Ningde Normal University, Ningde, China

Abstract Taking the elevator car bottom structure as the research object, the car bottom sheet metal process structure model was established by the three-dimensional design software SolidWorks. The part connection and support were considered in the simulation, and the mesh division finite element analysis was carried out to study static stress characteristics of the car bottom structure under load. The car bottom structure is optimized by introducing crisscross weldments according to the characteristics. The simulation results show that all the performances are improved, so as to illustrate the effectiveness of structural optimization, which provides theoretical basis for car bottom structure design.

Key words: Car bottom; design; simulation; optimization.

1. Introduction

The car is an important part that passengers directly contact when riding the elevator. The bottom structure of the car carries the load in the car and the weight of the car. The design of the car bottom structure affects the safety performance, carrying capacity and riding experience. At present, there are only a few researches on the optimization of the elevator car bottom, mainly including the frame structure, material processing and force analysis of the elevator car carried out by major elevator manufacturers [1]. Li Yingjie studied th, e mechanical law of the car bottom platform under various working conditions by means of numerical analysis, and proposed that the quality of the car bottom can be reduced under the premise of satisfying safety and reliability [2].Liu Dazhu et al. used finite element and modal analysis methods to analyze key components such as the car, and proposed that the traditional car structure design could be optimized and improved and its rationality was verified [3]. Cai Jian et al. applied the research methods of structural statics and dynamics and proposed to reduce the material cost of the car frame to meet the overall safety under extreme working conditions [4].It can be seen from the above that the current industry research on the optimization of the car structure has achieved great results, but it mainly focuses on the analysis of the overall structure of the car and the optimization of the parameters of the mathematical model of the car bottom. There are few studies on the simulation optimization of the elevator car bottom. Therefore, in this paper, the model of the car bottom is established by modeling software, and the finite element static analysis method is used for analysis. According to the simulation analysis results of the car bottom structure, an optimization plan is proposed.

2. Car bottom design

In order to facilitate quantitative analysis, combined with the actual. In this paper, the research is carried out from the perspective of the actual production of elevators by manufacturing enterprises, taking the car bottom of an elevator with a main specification parameter of 1640 mm×1160 mm×3660 mm and a rated bearing capacity of 630 kg as an example. Referring to the conventional car bottom design scheme of the current manufacturing enterprises, refer to the various manufacturing parameters [5]. Based on the C-shaped steel shown in Figure 1, the car bottom design is carried out.

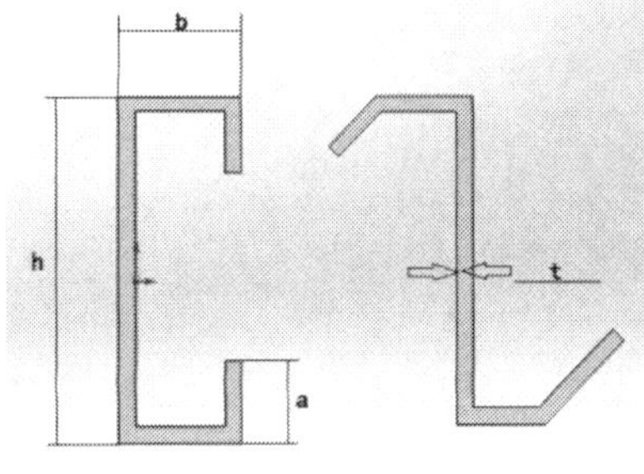

Figure 1. Outline drawing of C-shaped steel.

The car bottom is a rectangular structure, and the car support bracket and the car support C-shaped steel are used as the main load-bearing frame. Select the "Ω" type car bottom reinforcing rib with wall thickness t=2 mm[6], so that the bearing surface of the car bottom is as large as possible, and the two sides are fixed on the C-shaped groove of the bracket. The car trailer bracket adopts an

* Corresponding author: 544824964@qq.com

asymmetric C-shaped structure, one end is matched with the C-shaped groove, and the other end is matched with the reinforcing rib bottom to achieve the effect of supporting the bottom. The reinforcing rib bottom cooperates with the C-shaped steel of the car support to increase the bearing capacity of the car bottom surface. The third-layer steel structure is C-shaped steel for the car support, which increases the rigidity of the connection between the car support bracket and the bottom of the reinforcing rib. To sum up, in order to eliminate interference, all parts except the car bottom structure are compressed, and the intuitive car bottom structure as depicted in Figure 2 is obtained. In order to fit the frame, the dimensions of the parts of the car bottom and the C-shaped steel are defined in the same way: "h" is the height of the contour, "bL" and "bS" are the widths of both sides, "a" is the height of the folded edge of the parallel contour, "t" is the wall thickness of the part, and "l" is the part Length, specific dimensions are as follows:
C-shaped steel: h=110 mm, bL=67 mm, bS=47 mm, a=0 mm, t=3 mm, l=1636 mm; Car bottom reinforcement: h=85 mm, bL=73 mm, bS=73 mm, a=28 mm, t=2 mm, l=1050 mm; Car support bracket: h=100 mm, bL=105 mm, bS=30 mm, a=28 mm, t=5 mm, l=1630 mm; Rib bottom: h=108 mm, bL=37 mm, bS=37 mm, a=17 mm, t=3 mm, l=1640 mm; C-shaped steel for car support: h=120 mm, bL=50 mm, bS=50 mm, a=20 mm, t=3 mm, l=1130 mm.

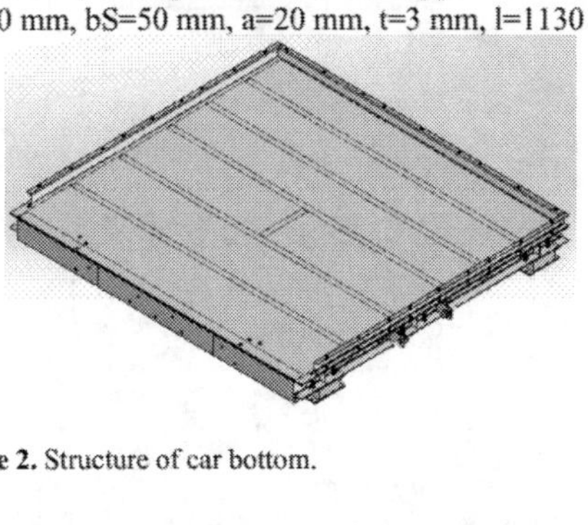

Figure 2. Structure of car bottom.

3. Simulation analysis of car bottom

3.1 Car bottom simulation

Simulation, as a simulation analysis plug-in in SolidWorks software, can perform static stress analysis on various components in mechanical design. Parts and assemblies to be analyzed can be selected locally or globally before performing a static stress analysis. To define component contact, contact sets must be defined to form connections before analysis. Fixtures describe how the model is supported, including fixed geometry, fixed hinges, elastic supports, and advanced fixtures. Generally a static stress analysis consists of at least one fixture and one external load. For the forces acting on the part, including various vector loads such as force, torque, pressure, etc. External loads should be added in the analysis process, the action analysis of no-load is invalid. The entire part must also be meshed before the static stress analysis, dividing the part into thousands of simple mesh elements. When performing control variable analysis, pay attention to the consistency of grid density, otherwise the analysis results will be affected. After running the static stress analysis, results options such as strain force, displacement, and strain appear to complete the finite element analysis of the part assembly [7-9].

3.2 Car bottom analysis

In the designed car bottom model, the structural model is simplified, and the selected Q235A material is applied to all. After the bottom plate is hidden, the rest of the parts are regarded as steel beams, and all the beam node groups are set, and the number of nodes is calculated to be 53. Use a global contact connection, use a fixed geometry fixture to make the joints hinged, and set all 53 joints to be hinged. The bearing capacity is 630 kg, that is, a force of 6174 N is applied to the bottom of the car as presented in Figure 3.Meshing using the default mesh density resulted in a total of 9105 meshes as shown in Figure 4.

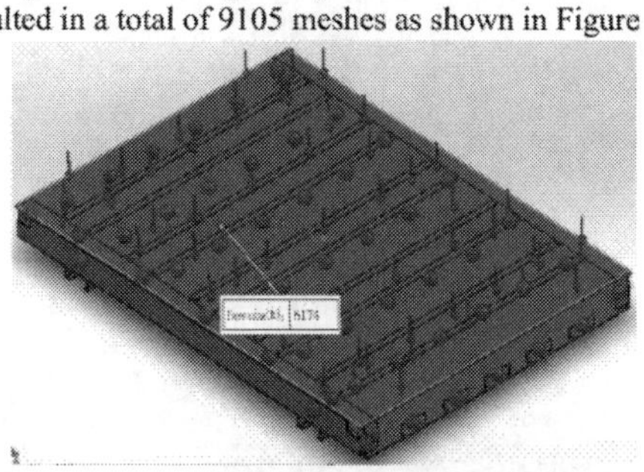

Figure 3. Load distribution diagram.

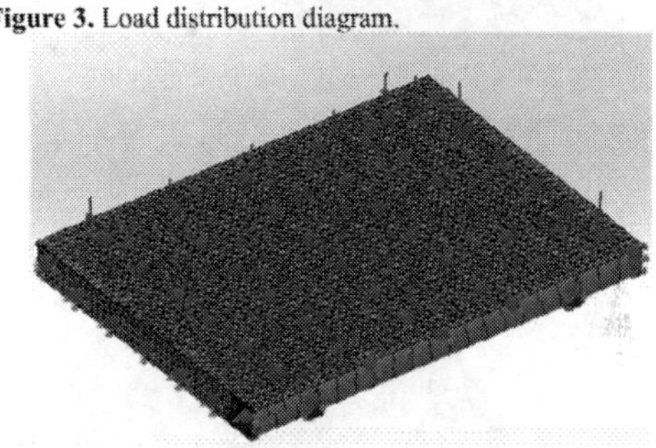

Figure 4. Mesh division diagram.

The running result is shown in Figure 5, and It is known that: The minimum displacement is 1.00e-30 mm, and the maximum displacement is 4.34e-02 mm; The minimum strain is 0.00e+00 mm/mm, and the maximum strain is 4.20e-05 mm/mm; Yield force is 2.35e+08 N/m2, minimum stress is 4.31e+01 N/m2, maximum stress is 1.08e+07 N/m2. The minimum safety factor (FOS) is 2.18e+01, and the maximum safety factor is 5.45e+06.

https://doi.org/10.1051/matecconf/202235801039

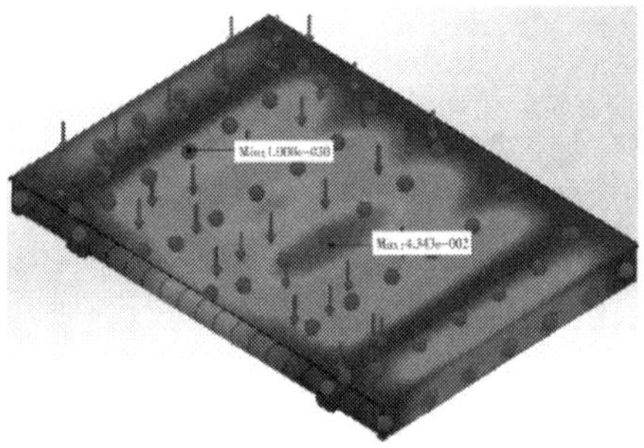

(a) Displacement

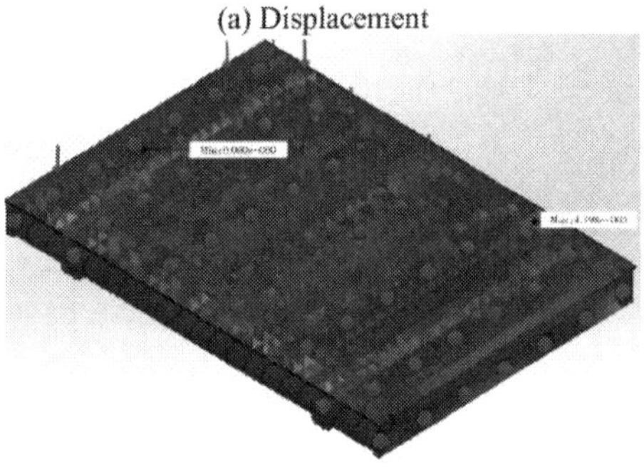

(b) Strain

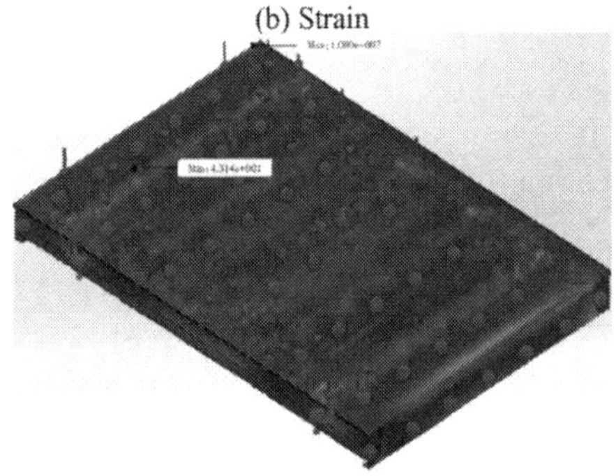

(c) Stress

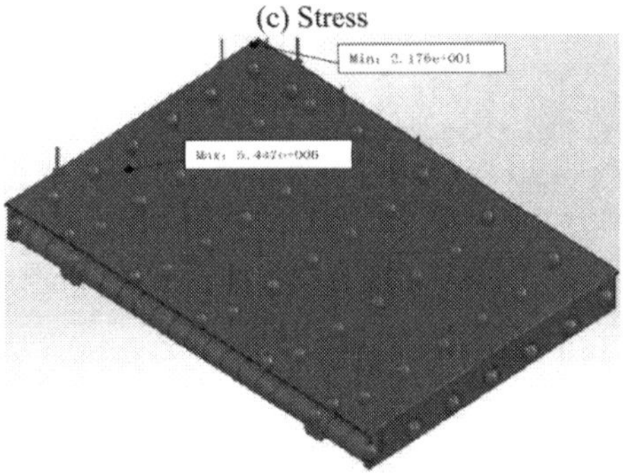

(d) Safety factor

Figure 5. Simulation and analysis results of the designed car bottom.

4. Car bottom optimization

4.1 Car Bottom Modeling Optimization

According to the above-mentioned static stress analysis of the designed car bottom, from the position where the maximum stress, strain and displacement occur, it is known that the optimized car bottom structure can be optimized from the structure and position of the car bottom reinforcing rib contacted by the car floor. In the case of ensuring that the weight of the car bottom structure does not increase, the welding structure is used to replace the car bottom reinforcing ribs to design a more stable load-bearing structure. According to the position distance of the original C-shaped steel, set the intersecting weldment at the diagonal of the two brackets. The midpoints of the two intersecting lines are connected to form a rectangle, and the components are square tubes with a size of 40 mm × 40 mm × 4 mm, which are matched with the end face of the C-shaped steel to generate a cross-shaped cross structure as shown in Figure 6. In the designed car bottom model, delete the car bottom reinforcing ribs, compress the shock-absorbing pad, insert the cross-shaped cross structure, and assemble as shown in Figure 7.

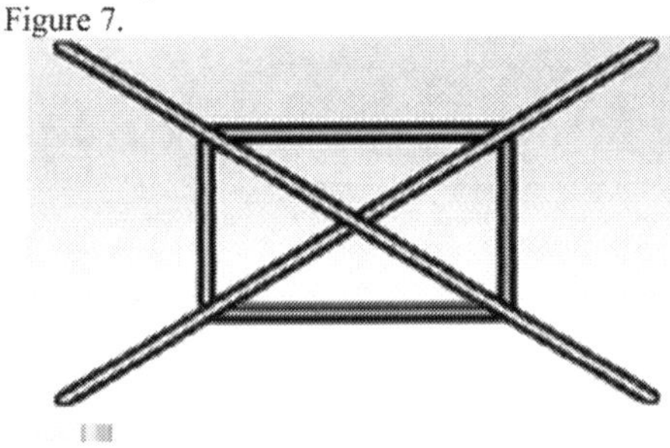

Figure 6. Cross-shaped structure.

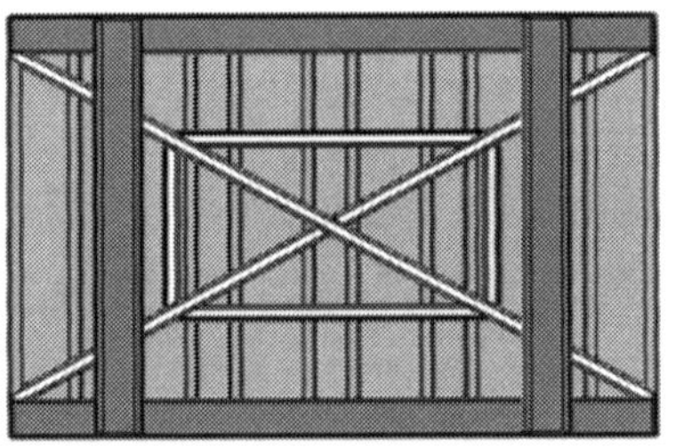

Figure 7. Optimized assembly drawing.

4.2 Optimization analysis of car bottom

In the optimized car bottom assembly model, the parts are regarded as beams, Q235A material is selected, and fixtures and loads are added to the calculation nodes. Mesh and run the analysis.

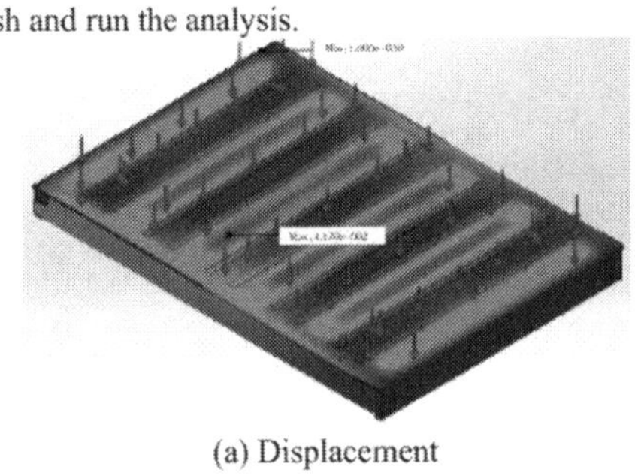

(a) Displacement

https://doi.org/10.1051/matecconf/202235801039

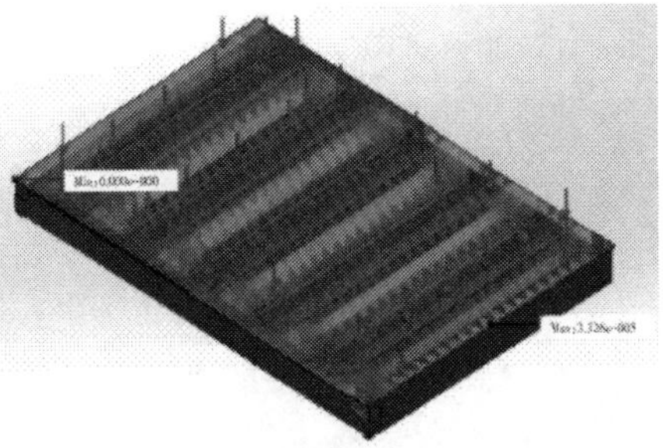

(b) Strain

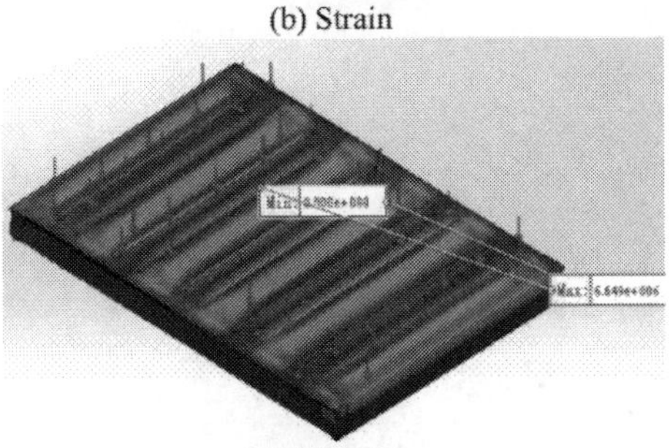

(c) Stress

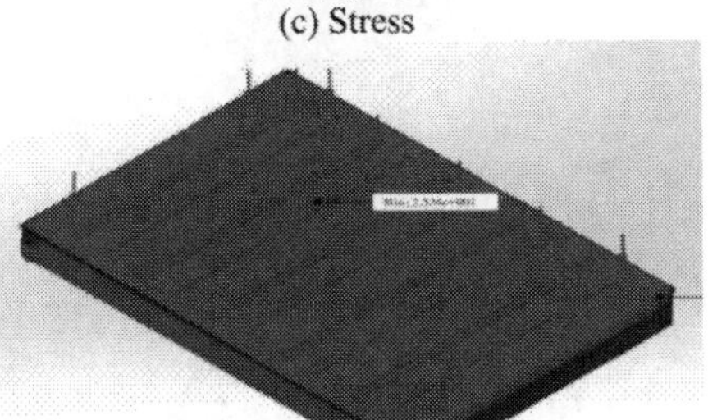

(d) Safety factor

Figure 8. Optimal car bottom simulation analysis result diagram.

The running result is shown in Figure 8, and It is known that: The minimum displacement is 1.00e-30 mm, and the maximum displacement is 4.17e-02 mm; The minimum strain is 0.00e+00 mm/mm, and the maximum strain is 3.33e-05 mm/mm; Yield force is 2.35e+08 N/m2, minimum stress is 0.00e+00 N/m2, maximum stress is 6.65e+06 N/m2. The minimum safety factor (FOS) is 3.53e+01, and the maximum safety factor is 1.00e+16.

4.3 Car bottom optimization comparison

Through the systematic simulation of the car bottom, the static stress analysis of the optimized car bottom structure is carried out, and the results of stress, strain, displacement, and safety factor are obtained. Table 1 shows the analysis results compared with the designed car bottom analysis results.

Table 1. Comparison of the analysis results between the optimized car bottom and the designed car bottom.

Project	Stress / (N/ m^2)		Strain / (mm/mm)		Displacement /mm		FOS	
	Min	Max	Min	Max	Min	Max	Min	Max
Design	4.31 e+01	1.08 e+07	0.00 e+00	4.20 e-05	1.00 e-30	4.34 e-02	2.18 e+01	5.45 e+06
Optimi zation	0.00 e+00	6.65 e+06	0.00 e+00	3.33 e-05	1.00 e-30	4.17 e-02	3.53 e+01	+∞
Differe nce	- 4.31 e+01	- 4.15 e+06	0.00 e+00	- 8.70 e-6	0.00 e+00	- 1.70 e-03	1.35 e+01	+∞
Result	excel lent	excel lent	—	exce llent	—	exce llent	excel lent	excel lent

Notes: 1. Stress (N/m^2) is the internal force per unit area. The greater the stress at that point, the more dangerous. 2. Strain (mm/mm) is the amount of displacement per unit length. The greater the strain of a point along a certain direction, the greater the deformation of the point along that direction.3. Displacement (mm) is the distance that a point within a member moves in a certain direction. 4. FOS is the ratio of the yield strength of the material to the actual stress.

It can be seen from the comparison that the isosceles trapezoid structure with diagonal cross and midpoint reinforcement can enhance the stability of the car bottom and reduce the extrusion of the car bottom plate caused by the car load. The performance of the optimized car bottom in terms of stress, strain, displacement, FOS, etc. is obviously improved compared with the designed car bottom. Under the same conditions, the maximum stress is reduced by 38.4%, the maximum displacement is reduced by 3.9%, and the minimum safety factor is increased by 61.9%. At the same time, the base plate structural support adopts the national standard square tube to facilitate the production and assembly of manufacturing enterprises.

5. Conclusions

Build a model and simulate the structural design around the conventional car bottom of the manufacturing enterprise, and analyze the static state, then find out the characteristics of the car bottom structure under load. According to the characteristic research, it is shown that there is a structural change plan that increases the minimum safety factor of the car bottom by more than 50% without changing the factors that affect the simulation results of the car bottom such as quality, material, connection, etc.

Acknowledgments

This paper was supported by the following research projects: by the Natural Science Foundation of Fujian Province (Grant No. 2020J01432), Research and Innovation Team of of Ningde Normal University (Grant No. 2020T02), Ningde Normal University Serving Local Special Project(Grant No.2018ZX401).These supports are gratefully acknowledged.

References

1. Q.S.WANG, H.Y.ZHAO, Structural analysis of the elevator car frame and platform, Mechanical Research and Application. 1 (2015)133-134.
2. Y.J.LI, Strength analysis and optimization design of elevator platform based on ANSYS Workbench, Mechanical and Electrical Engineering Technology. 7(2021)227-230 .
3. D.Z.LIU, Research on structure design and analysis of a new type of household elevator.Suzhou:Suzhou university,2015.
4. J.CAI, Research and analysis of elevator structural mechanics base on ANSYS Workbench.Suzhou:Suzhou university,2016.
5. X.K.JIAN, L.WANG,Y.N.ZHAO, M.G.LI and H.C.LI, Preliminary study of lift car parametric design, one family redevelopment based on VB.NET Solidworks, China Elevator. 15(2020) 26-31.
6. Y.GAO, D.Q.JING, Y.F.KOU, Static analysis and fatigue damage research of elevator car reinforcement bar, China Elevator.1(2020):6-9.
7. Y.D.CHEN, Z.Q.BAO, H.J.REN and Y.X.WANG, Finite element analysis for product design based on SolidWorks Simulation, Computer Technology And Development.9(2012):177-180.
8. Q.S.LI, L.J.GUO,W.B.WANG, J.Z.FAN and R.R.SU. Fatigue analysis of head board of die castingmachine based on Solidworks Simulation, Journal of Mechanical & Electrical Engineering. 10(2010): 47-50.
9. H.B.JIN, X.JI Xin, Based on SolidWorks simulation finite element analysis of piston and rod of hydraulic cylinder, Journal of Machine Design.37(2020):176-179.

Stress analysis of pipelines with small misalignment under internal pressure using fast modelling system ABAP

Sunting Yan [1, 2, *], Zhongqiang Liu [1, 2], Jiansheng Hu [1, 2], Yonggui Chen [1, 2], Xiaoli Shen [3], Guoyang Teng [1, 2]

[1] Zhejiang Academy of special equipment science China
[2] Key Laboratory of Special Equipment Safety Testing Technology of Zhejiang Province China
[3] Zhejiang institute of mechanical& electrical engineering co. LTD., China

Abstract. This paper presents a numerical finite element analysis of pipelines with small misalignment under internal pressure using a developed fast pipeline modelling system ABAP using Python development environment of ABAQUS. The effect of small misalignment is shown to affect at most 10% of stress levels. Extensive parametric analyses have been conducted on the pipe length, pipe thickness. Pipe length has trival effect on non-uniform stress distribution when length/diameter ratio is greater than 1. For thinner pipe, misalignment amplifies the circumferential non-uniformity more significantly than thicker pipe. This paper serves to enhance the understanding of pipes with small installation misalignment defects under pure internal pressure.

1. Introduction

Pipelines under internal pressure may experience burst failure and lead to economic loss and threats to safety of humans. The assessment of integrity of pipelines with various kinds of defects is pivotal to maintain safe operation.

There are various researches on these aspects. For example, Chen et al.[1] presented an analytical method to calculate the burst pressure of long corroded pipelines under internal pressure. The existence of corrosion flaws reduces the burst pressure. Furthermore Jin et al.[2] conducted integrity assessment of the pipelines containing an isolated corrosion pit by the similar analytical method. Ye et al.[3] innovatively presented an analytical large deformation collapse model of pipelines under excessive external pressure by including the effect of plastic hardening of materials. Similarly Yan et al.[4] studied the collapse failure and equal potential energy load of shallow cylindrical shells. A beam-model was developed by Yan et al.[5] to investigate the corrosion-induced thickness loss on the critical collapse pressure through the tangent modulus method to account for plasticity effect. Netto et al.[6] also studied the burst pressure of pipelines when initial thickness loss was considered. Finally Ye et al.[7] and Yan et al.[8] presented analytical investigations of corroded pipes and shallow cylindrical shells. The thickness non-uniformity was shown to affect significantly the collapse pressure.

From above, the thickness-loss defect's effect on critical pressure has been well understood. But in some occasional case, especially during installation, pipes may be welded imprecisely and it leads to some misalignment defect as illustrated in Fig.1. The assessment of the severity of the misalignment defect is critical for safety maintenance of the entire piping system. Thus this paper presents some finite element analyses of pipelines with misalignment defect under internal pressure based on our previously developed fast piping modelling system ABAP. Extensive parametric analyses(based on linear perturbation analysis in ABAQUS) are conducted to illustrate the effect of thickness, misalignment and pipe lengths on the stress distribution.

Figure 1. A realistic example of pipes with small misalignment

* Corresponding author: yansunting@zju.edu.cn

https://doi.org/10.1051/matecconf/202235801040

2. Finite element model description

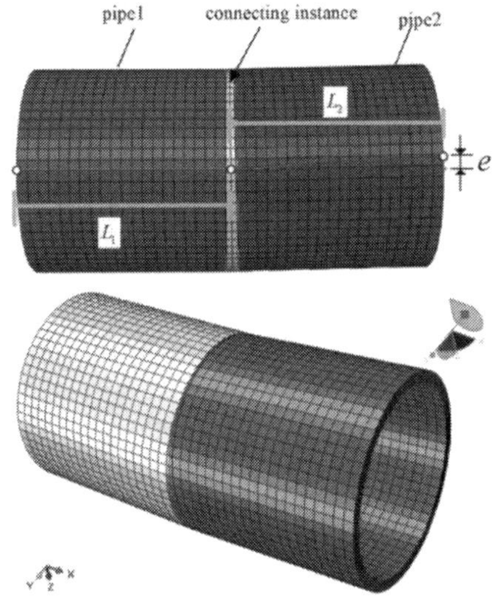

Figure 2. Model parameters of two pipes with misalignment

Fig.2 shows the finite element models and parameters. The model consists of two pipes indexed as pipe1 and pipe2 with lengths L_1, L_2, outer diameter $D_1, D_2 = 1\text{m}$ and thicknesses t_1, t_2 respectively. The misalignment is modelled by the parameter e defined as the deflection distance between center point of right end cross-section of pipe2 and the expected center point if pipe1 and pipe2 are in straightline. Thus e / L_2 determines the misalignment angle. To connect pipes with misalignment, ABAP system(see section 4) automatically detects the misalignment defect and generats a connecting instance based on orphan mesh method in ABAQUS. The Dofs are fully coupled between connecting instance and pipes. Internal pressure is imposed on the internal surfaces of this assembly with end cap effect considered. The end cap effect is modelled by binding the nodes of two end cross-sections to reference points at centers through BEAM type constraints in ABAQUS and imposing forces normal to the bound cross-sections at ends of magnitudes equal to $p_i A_i$ where p_i is internal pressure magnitude and A_i is the internal area defined as $A_i = \pi / 4 \cdot (D_1 - 2t_1)^2 = \pi / 4 \cdot (D_2 - 2t_2)^2$. Note that in this paper $D_1 = D_2, t_1 = t_2$. By linear perturbation analysis, without loss of generality, internal pressure is fixed as 1MPa.

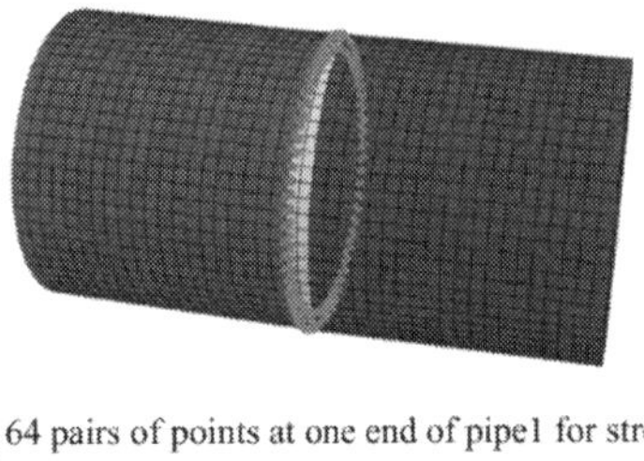

Figure 3. 64 pairs of points at one end of pipe1 for stress linearization

After calculation, we use linear stress linearization functionality in ABAQUS to use 64(64 elements are uniformly assigned in circumferential direction) pairs of points circumferentially to extract stress information along each line linked by each pair(see Fig.3 and Fig.5 for circumferential indices). Fig.4 shows the Mises contour when $L_1 = L_2 = 1\text{m}, t_1 = t_2 = 0.05\text{m}, e / L_2 = 0.07$ under 1MPa internal pressure. The mises stress level is higher at index 32 and lower at index 0. Fig. 5 shows the Mises stress contour when pipe is thicker with $t_1 = t_2 = 0.1\text{m}$ and similarly the convex part has lower stress level.

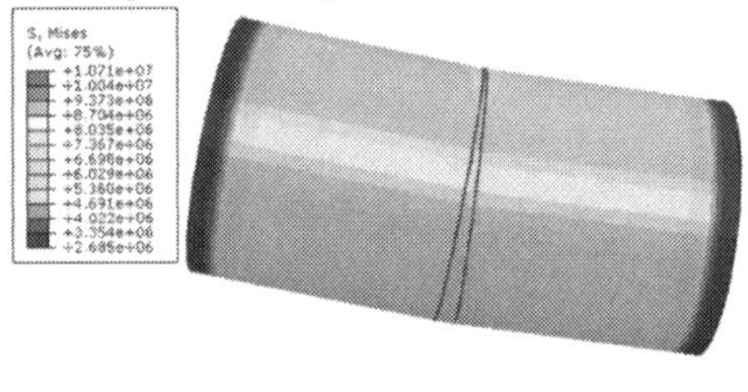

Figure 4. Mises stress contour when
$L_1 = L_2 = 1\text{m}, D_1 = D_2 = 1\text{m}, t_1 = t_2 = 0.05\text{m}, e / L_2 = 0.07$

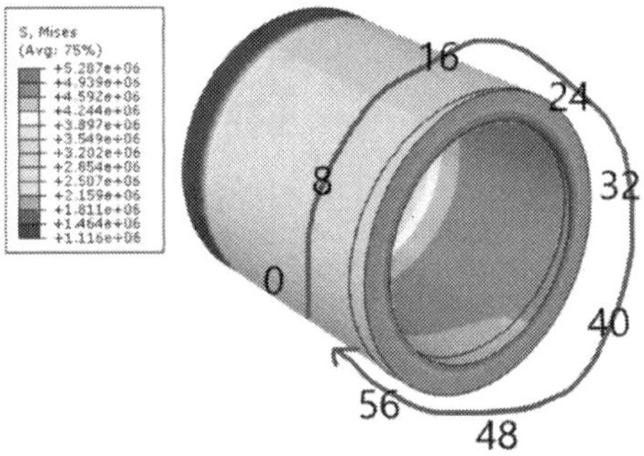

Figure 5. Mises stress contour when
$L_1 = L_2 = 1\text{m}, D_1 = D_2 = 1\text{m}, t_1 = t_2 = 0.1\text{m}, e / L_2 = 0.07$

3. Parametric analysis on circumferential stress distribution

Fig. 6 shows the Tresca stress and Mises stress of linearized stress along each pair of points. By stress linearization the averaged stress along thickness is called membrane stress and the linearized stress at one end of thickness is called membrane+bending1(the other is called membrane+bending2 and in this analysis membrane+bending1 stress is always greater than membrane+bending2). Tresca stress of membrane+bending1 is largest and Tresca stress of membrane is lower. By definition Tresca stress is always larger than Mises stress. The abscissa from 0 to 63 means circumferential indices as in Fig.5. As index increases from 0 to 32, all stresses increase and by symmetry as index increases from 32 to 63, all stress decreases.

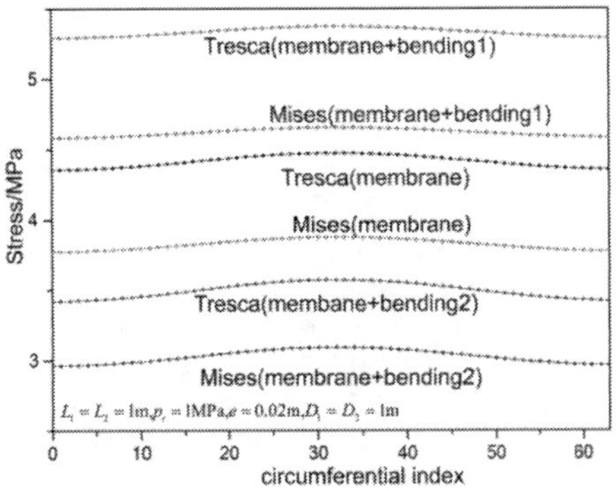

Figure 6. Tresca stress and Mises stress along circumferential direction

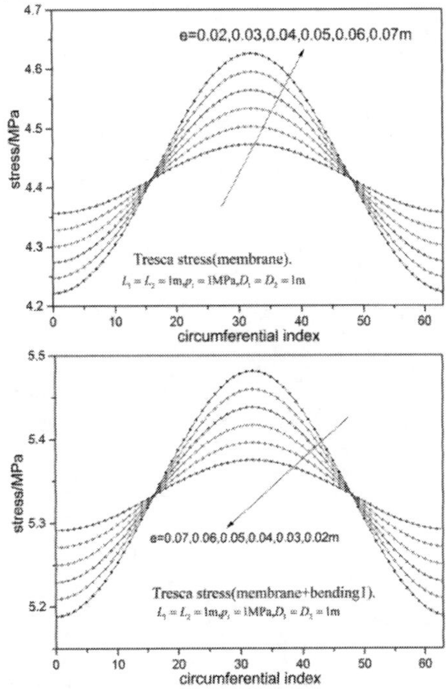

Figure 7. Effect of misalignment parameter e on the Tresca stress distribution

From Fig.7, the Tresca stress(membrane) and Tresca stress(membrane+bending1) are presented for various misalignment parameter e. As e increases, the effect of circumferential index is increasing. For example, at index 0, for e=0.02, the Tresca(membrane) stress is about 4.35MPa and at index 32, the corresponding stress is about 4.47MPa. However for e=0.07, the stress increases from about 4.22MPa to about 4.63MPa. The trend is similar for membrane+bending1 case. This analysis shows the significant effect of misalignment parameter e.

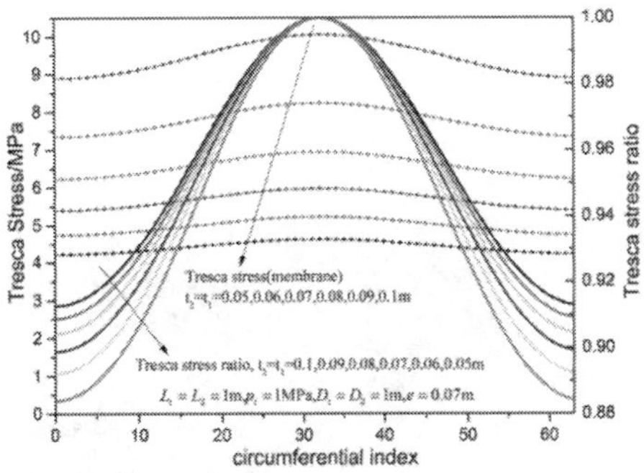

Figure 8. Tresca stress for various thickness levels

Fig. 8 shows the Tresca stress(membrane) for various thicknesses. The Tresca stress ratio is defined as the Tresca stress(membrane) divided by maximum Tresca stress(membrane) for all circumferential indices. As thickness decreases the non-uniformity of stress distribution increases significantly. For example when $t_1 = t_2 = 0.05\text{m}$, ratio has minimumvalue about 0.88 and when $t_1 = t_2 = 0.1\text{m}$ the ratio has minimum value about 0.91. This shows that thinner pipe is more significantly affected by misalignment defect. Fig. 9 presents the effect of pipe lengths on Tresca stress distribution by studying circumferential distribution of Tresca stress(membrane) for $L_1 = L_2 = 1,2,3,4,5,6\text{m}$ and fixed $e / L_2 = 0.07$, $t_1 = t_2 = 0.05, 0.1\text{m}$. All lines coincide indicating that pipe length has trivial effect on stress distribution.

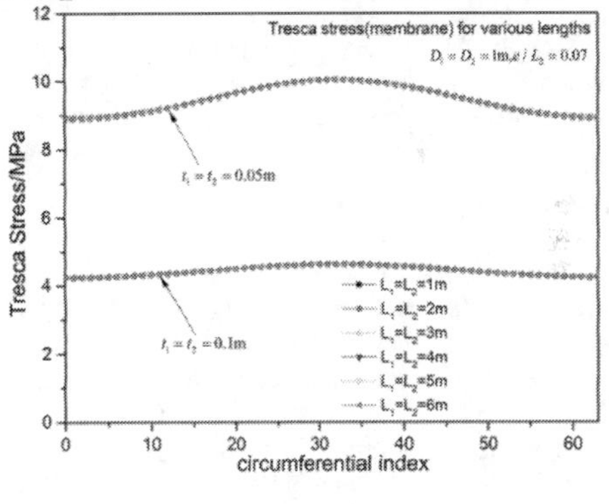

Figure 9. Tresca stress distribution for various lengths

4. Brief introduction of ABAP system

ABAP(namely ABAqus & Python) system was developed by Python by the first author and the kernel is based on secondary python development environment of ABAQUS software. The developed GUI interface is able to input the node locations, pipe connectivity information and geometric parameters. The kernel is responsible to accept the GUI-generated command text, interpret the text, build geometric pipe model, mesh the pipes by structural grids, assembly the pipes and impose the load to submit the job. All parametric analyses in this paper are based on this system quite conveniently without the tedious process of repeatedly building finite element models manually.

Fig.10 is the GUI of ABAP system supporting preview of the model.

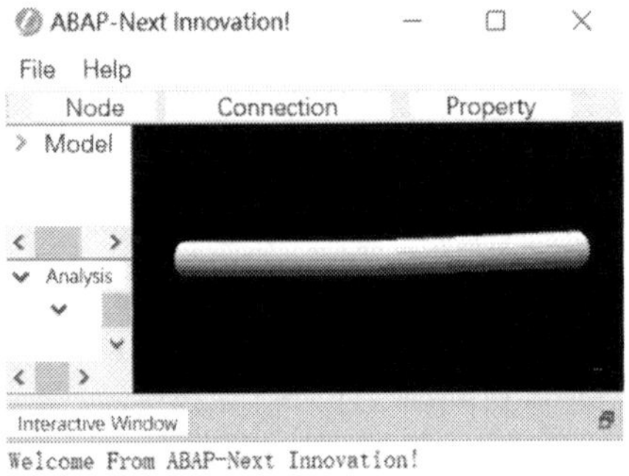

Figure 10. Interface of ABAP system and preview of pipes with small misalignment

5. Conclusions

(1) Under internal pressure the pipe misalignment defect slightly affects the circumferential stress distribution. For example, a 4-degree misalignment(corresponding to e/L2=0.07) leads to about 10% variation of stress circumferentially.
(2) The misalignment's effect is more significant when pipe is thinner.
(3) The pipe length has trivial effect on stress distribution circunferentially.
(4) All parametric analyses can be conveniently conducted by ABAP system without manually repeatedly building FEA models. This shows the effectiveness of ABAP system in enhancing the analysis efficiency.

Acknowledgments

This work was financially supported by natural science fundation of Zhejiang Province LQ21E050001, Science and technology plan project of Zhejiang Bureau of Quality and Technical Supervision (NO.20200307), Eyas Program Incubation Project of Zhejiang Provincial Administration for Market Regulation(No. CY2022221) and Public Project of Zhejiang Province LGG19E060004.

References

1. Chen ZF, Wang W, Yang H, Yan ST, Jin ZJ. On the effect of long corrosion defect and axial tension on the burst pressure of subsea pipelines. Applied Ocean Research. 2021 Jun 1; 111:102637.
2. Jin, Z.J., Qiu, C., Chen, Z., Yan, S.T. and Shen, X.L., 2020. Integrity assessment of the pipelines containing an isolated corrosion pit. Engineering Failure Analysis,113, p.104539.
3. Ye, H., Yan, S.T., Shen, X.L., Chen, Z.F. and Jin, Z.J., 2019. On the response of postbuckling of rings with strain hardening under external pressure. International Journal of Solids and Structures, 167, pp.127-141.
4. Yan, S.T., Shen, X., Chen, Z. and Jin, Z., 2018. Symmetric snap-through and equal potential energy load of non-uniform shallow arch under a concentrated load considering imperfection effect. International Journal of Mechanical Sciences, 146, pp.152-179.
5. Yan, S.T., Shen, X.L., Jin, Z.J. and Ye, H., 2016. On elastic-plastic collapse of subsea pipelines under external hydrostatic pressure and denting force. Applied Ocean Research,58, pp.305-321.
6. Netto, T.A., Ferraz, U.S. and Estefen, S.F., 2005. The effect of corrosion defects on the burst pressure of pipelines. Journal of constructional steel research,61(8): 1185-1204.
7. Ye, H., Qian, J., Yan, S., Jiang, C., Wei, L. and Jin, Z.J., 2016. Limit bending moment for pipes with two circumferential flaws under combined internal pressure and bending. International Journal of Mechanical Sciences, 106, pp.319-330.
8. Yan, S.T., Shen, X. and Jin, Z., 2019. Static and dynamic symmetric snap-through of non-uniform shallow arch under a pair of end moments considering critical slowing-down effect. Proceedings of the Institution of Mechanical Engineers, Part C: Journal of Mechanical Engineering Science, 233(16), pp.5735-5762.

Study on Serviceability of Transition Section Between Road and Tunnel

Chonghui Zhong [1], Bingwei Cheng [1,2], Linhao Wang [1,2], Shijie Yu [1,2]

[1] China Railway No.3 Bureau Group Investment Co., Ltd. Taiyuan, Shanxi,030006, China
[2] Tianshui Sanyangchuan Tunnel Traffic Engineering Co., Ltd. Tianshui, Gansu,741030, China

Abstract: The determination of allowable differential settlement in bridge transition is a key problem to prevent vehicle jump at bridge head, but there are few theoretical research achievements in this aspect atroduction home and abroad. In this paper, four different structures of the road surface of The Sanyangchuan tunnel and the lead project are studied. The allowable differential settlement of asphalt pavement is calculated by asphalt pavement-layer system, and the allowable differential settlement is calculated by Ladan Lasse transform.

Keywords: ITD; Trip; Deflection value; Post-construction settlement.

1. Introduction

Highways play a very important role in the development of China's transportation industry, and in recent years, bridge head jumping has become a prominent problem in the construction of highways. Bridge head jumping has a very obvious impact on the safety, stability and comfort of driving, which will cause damage to the car as well as the safety of driving if it is serious. Therefore, it is significant for highway construction to analyze and discuss the problem of bridge head jumping and propose targeted prevention and control measures.

In this paper, we mainly study the jumping problem of pavement transition section of Sanyangchuan Tunnel and lead project in Tianshui City. The asphalt pavement is subjected to the calculation of permanent deformation and bending and settlement values; the small bridge deck is subjected to the calculation of post-work settlement. The theory of asphalt pavement design in China is the theory of multi-layer elastic laminate system, which adopts the principle of bicircular vertical homogeneous load action[1], and at the same time, in order to make the actual road use condition and calculation reasonable, the contact conditions between the layers are set as a completely continuous system when calculating the road surface bending and settlement and layer bottom tensile stress[2-4].

Many domestic and foreign information to the approach road embankment of the allowable post-work settlement as a bridge head jump settlement control index. The Japanese Road Association[5] recommends that the allowable residual settlement of bridge approach roads should not exceed 10-30 cm within 3 years, and it is pointed out in the U.S. highway study "Soft Foundation Treatment of Highway Embankments"[6] that: during the economic life of a highway (10-15 years), the allowable post-work settlement of general embankments is 30-60 cm (1-2 feet), but the allowable post-work settlement of bridge approaches is 15-30 cm (0.5-1 ft).

2. Background

This project is located in Maiji District of Tianshui City, which can connect the main urban area of Tianshui City and Sanyangchuan New District, and also as a part of Luojiagou - Sanyangchuan Tunnel - Magan Highway Passage, which can connect Lianhuo Expressway, Xihuang Avenue of Tianshui main urban area, Chengji Avenue and Sanyangchuan New District in the long term.Technical indicators are executed according to the Technical Standards for Highway Engineering[7] (JTG B01-2014) issued by the Ministry of Transportation.

2.1 Meteorological conditions

Tianshui City is located in the northern temperate zone, a semi-arid continental climate. The natural area is divided into the arid zone in the Yang Mountain in the north, the river valley in the middle and the humid zone in the shallow mountains in the south. The general climatic characteristics of Tianshui are: rapid warming in spring; no scorching heat in summer; continuous rain in autumn; and no severe cold in winter. The climate is mild, with four distinct seasons, sufficient sunshine and moderate precipitation. The average multi-year temperature in Tianshui is 10.9°C, the extreme high temperature is 38.3°C, the extreme minimum temperature is -19.2°C, and the average temperature of the coldest month is about -

2°C; the average multi-year precipitation is 580.0mm.The annual temperature distribution in Tianshui is shown in Figure 1.

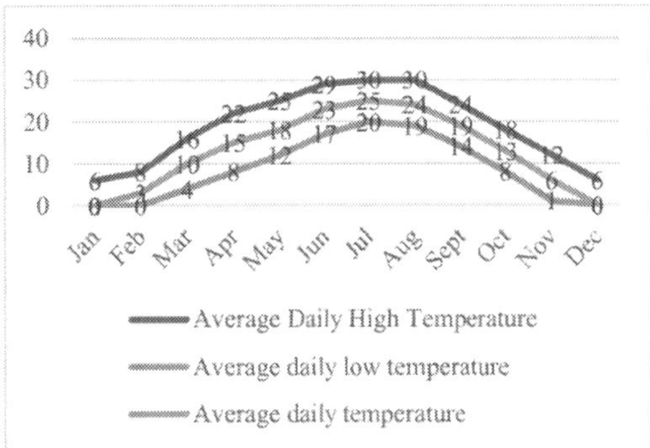

Figure 1. The annual temperature distribution in Tianshui

2.2 Geomorphology,Hydrogeology

The project traverses through the valley area of the river through the river, the north mountain of the river in the loess mount area to the south bank of the Wei River, the geomorphology is erosion accumulation valley landform and denudation structure hilly landform, the topography is undulating, the relative height difference of 200 ~ 400 m. The type of groundwater along the line can be divided into two types of pore water and bedrock fracture water according to their conditions and aquifer nature.

3. Simulation

3.1 Mainline and interchange ramp pavement structure

(1) Traffic volume calculation

Class I highway; target reliability index of 1.28; annual average daily traffic volume of 1500 vehicles/day in both directions for large passenger cars and trucks in the initial year; design service life of pavement of 15 years; 15 years from opening to the first time for rutting repair[8-12].

Traffic volume annual average growth rate of 6.5%, direction factor 0.55, lane factor 0.5, integral truck proportion 25%, semi-trailer truck proportion 35%, the initial year design lane large passenger cars and trucks annual average daily traffic volume of 412/day, the design life of the design lane cumulative large passenger car and truck traffic volume 3636514, so the pavement traffic load level for the light traffic load Grade.

Axle load calculation:

When the fatigue cracking of asphalt mixture layer and inorganic bonding material stabilization layer is tested, the cumulative number of equivalent design axle loads on the design lane during the design life is 5383818 times and 2.184496E+08 times, respectively.

When calculating the permanent deformation of the asphalt mixture layer: the cumulative number of equivalent design axle loads on the design lane during the period from opening to the first rutting repair is 5383818 times.

When calculating the vertical compressive strain on the top surface of the roadbed: the cumulative number of equivalent design axle loads acting on the design lane during the design life is 8130594 times[13-14].

(2) Pavement structure design and calculation

The number of layers of pavement structure is 6, the design axle load is 100kN standard axle load, the design layer of pavement is 4, and the starting thickness of design layer is 320mm. the parameters of each structural layer are shown in Table 1. below.

Table 1. Structural layer parameters

Layer	Name of structural layer material	Thickness(mm)	Modulus(MPa)	Poisson's ratio	Inorganic bonding material stabilization class material bending and tensile strength(MPa)	Asphalt mixture rutting permanent deformation amount(mm)
1	Fine grain type AC-13C	40	11000	0.25	-	1.5
2	Medium Grain Type AC-20C	60	10000	0.25	-	2.5
3	Coarse Grain Type AC-25	80	10000	0.25	-	2.5
4	Cement stabilized gravel (5% cement admixture)	180	7500	0.25	1.4	-
5	Cement stabilized gravel (3.5% cement admixture)	180	7500	0.25	1.4	-
6	Cement-lime gravel soil	180	3000	0.4	0.8	-

The fatigue cracking test for each inorganic bonding material stabilized layer is shown in Table 2.

Table 2. Fatigue cracking test for each inorganic bonding material stabilized layer

	Level 4	Level 5	Level 6
Design layer thicknessH(4)(mm)	320	320	320
Seasonal permafrost area adjustment factorKA	0.8	1.8	0.8
Temperature adjustment factorKT2	1.06	1.06	1.06
Field integrated correction factorKC	-1.157	-1.39	-0.83
Layer bottom tensile stress σ (MPa)	0.029	0.065	0.041
Fatigue cracking life of inorganic bonding material stabilization layerNF2(Axis)	9.37924E+10	2.610822E+10	6.964421E+09
The cumulative number of equivalent design axle load actions on the design lane during the design service lifeNZB2(Axis)	2.184496E+08	2.184496E+08	2.184496E+08

(3) asphalt mixture layer permanent deformation calculation

The asphalt mixture layer permanent deformation equivalent temperature TPEF = 19.6 °C, the opening to

the first time for rutting repair period on the design lane equivalent design axle load cumulative number of times NZB3 = 5383818 axle times, the asphalt mixture layer permanent deformation test number of layers N = 6, 1 to 6 layers RAI were 0.34mm, 0.61mm, 1mm, 0.72mm, 0.44mm, 0.55mm. RA = 3.66mm, RAR = 15mm, so the asphalt mixture layer permanent deformation to meet the specification requirements.
(4) Calculate the accepted bending and settlement values of the designed pavement structure
Dry and wet cycles or freeze-thaw cycles under the conditions of soil modulus reduction factor KAT = 0.8, the top surface of the road acceptance bending value LG = 59.8 (0.01mm), the road surface acceptance bending value LA = 8.2 (0.01mm).
The first layer of asphalt mixture rutting test dynamic stability technical requirements for 5139 times/mm; the second layer of asphalt mixture rutting test dynamic stability technical requirements for 2412 times/mm; the third layer of asphalt mixture rutting test dynamic stability technical requirements for 2412 times/mm.
The minimum frost protection thickness of the pavement structure is 500 mm, and the calculation results show that the total thickness of the pavement structure meets the frost protection requirements.

3.2 Tunnel Pavement Structure

(1) Traffic volume calculation
The data is the same as the traffic volume calculation data of the pavement structure of the main line and interchange ramps.
(2) Pavement structure design and calculation
The number of layers of pavement structure is 6, the design axle load is 100kN standard axle load, the design layer of pavement is 4, and the starting thickness of design layer is 320mm. the parameters of each structural layer are shown in Table 3 below.

Table 3. Structural layer parameters

La yer	Name	Modulus (mm)	Modulus (MPa)	Poiss on's ratio	tensile strengt(MPa)	deform ation amount (mm)
1	AC-13C	40	11000	0.25	-	1.5
2	AC-20C	60	10000	0.25	-	2.5
3	Cement stabilized steel slag	220	10000	0.25	1.4	-
4	Cement stabilized gravel	-	7500	0.25	1.4	-

The fatigue cracking test for each inorganic bonding material stabilized layer is shown in Table 4:

Table 4. Fatigue cracking test for each inorganic bonding material stabilized layer

	Level 3	Level 4
Design layer thicknessH(4)(mm)	320	320
Seasonal permafrost area adjustment factorKA	0.8	1.8
Temperature adjustment factorKT2	0.888	0.888
Field integrated correction factorKC	-0.24	-1.362
Layer bottom tensile stressσ(MPa)	0.041	0.118
NF2(Axis)	7.217371E+11	1.116348E+10
NZB2(Axis)	2.184496E+08	2.184496E+08

(3) asphalt mixture layer permanent deformation calculation
TPEF = 19.6 °C, NZB3 = 5383818 axle times, the asphalt mixture layer permanent deformation test number of layers N = 5, 1 to 5 layers of asphalt mixture layer permanent deformation are 0.18mm, 0.49mm, 0.89mm, 0.66mm, 0.41mm. RA=2.63mm, RAR=15mm, so the permanent deformation of the asphalt mixture layer meets the specification requirements.
(4) Calculate the acceptance bending and settlement value of the designed pavement structure
KAT = 0.8, LG = 59.8 (0.01mm), LA = 9.3 (0.01mm).

3.3 Bridge deck pavement structure

Set up the lap board of the road and bridge transition section, the lap board will rotate around the abutment, although the settlement of the abutment is smaller than the settlement of the approach road embankment, but can not be ignored. In the figure, x is the length of the hitching plate; S_r^b and S_r^x are the post-work settlement of the foundation of the bridge deck and the hitching plate away from the soil under the end of the platform (referred to as the post-work settlement of the hitching plate); i indicates the longitudinal slope of the bridge deck, hitching plate and approach road when completed, and the longitudinal slope of the approach road when the soil settlement becomes stable;i_b'andi$'$are the longitudinal slope of the bridge deck and hitching plate when the soil settlement of the approach road becomes stable[15-16], respectively. As shown in Figure 2.

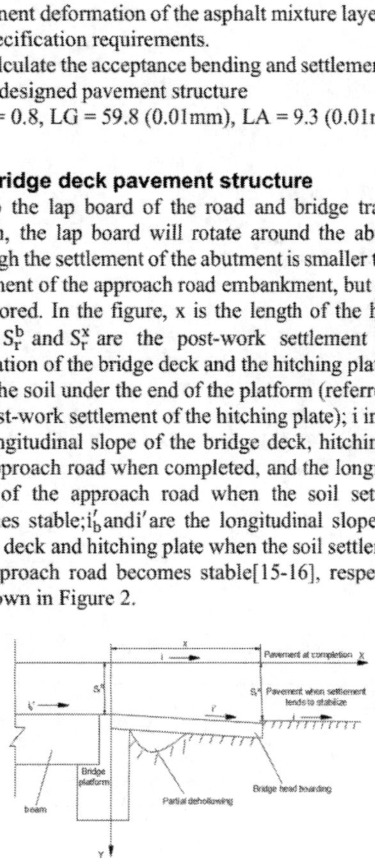

Figure 2. Analysis of differential settlement at the bridge head with a lap plate

https://doi.org/10.1051/matecconf/202235801041

From the geometric relationship shown in Figure 1.1, the longitudinal slope change of the bridge deck, i.e. the differential settlement between the abutment and the pier on the other side of the side span and the ratio between the span diameter of the side span and the slope difference between the deck and the lap slab are calculated as shown in Equation 1 and 2, respectively:

$$\Delta i_1 = i_b' - i \quad (1)$$

$$\Delta i_2 = i' - i_b' \quad (2)$$

Then the longitudinal slope change value before and after the settlement of the lap slab due to post-work settlement is calculated as shown in Equation 3:

$$\Delta i = \Delta i_1 + \Delta i_2 \quad (3)$$

Then the post-work settlement of the soil under the lap slab away from the end of the platform is calculated as shown in Equation 4:

$$S_r^x = S_r^b + x \cdot \Delta i \quad (4)$$

And S_r^b calculated as shown in Equation 5, for low, medium and high compressibility saturated soils α_1, the foundation soil type coefficients are 0.4. 0.7 and 0.85, respectively; $\alpha_2$is for bridge foundation type coefficients, for pile foundations $\alpha_2 = 1/500$; for expanded foundations $\alpha_2 = 1/300$; L is the bridge side span span diameter, still calculated as 25m when the span diameter is less than 25m[17].

$$S_r^b = \alpha_1 \alpha_2 L \quad (5)$$

Therefore, the after-work settlement of the bridge deck of this small bridge is calculated as shown in Equation 6.

$$S_r^x = 0.7 \cdot \frac{1}{500} \cdot 25 + 15 \cdot 0.2\% = 0.065cm \quad (6)$$

4. Summary

4.1 Data Comparison

Based on the previous parameter calculations for the four different pavement structures, the permanent deformation and bending values for the mainline and interchange ramp pavement structures and tunnel pavement structures are shown in Table 5.A comparison of the permanent deformation and bending values of the mainline and interchange ramp pavement structure and the tunnel pavement structure is shown in Figure 5.

Table 5. Mainline pavement and tunnel pavement permanent deformation and bending and settlement values

	Mainline	Tunnel Pavement	Difference
RAI(1)	0.34mm	0.18mm	0.16mm
RAI(2)	0.61mm	0.49mm	0.12mm
RAI(3)	1mm	0.89mm	0.11mm
RAI(4)	0.72mm	0.66mm	0.06mm
RAI(5)	0.44mm	0.41mm	0.03mm
RAI(6)	0.55mm	-	-
RA	3.66mm	2.63mm	1.03mm
LA	8.2(0.01mm)	9.3(0.01mm)	1.1(0.01mm)
Total bending value	68.0(0.01mm)	69.1(0.01mm)	-

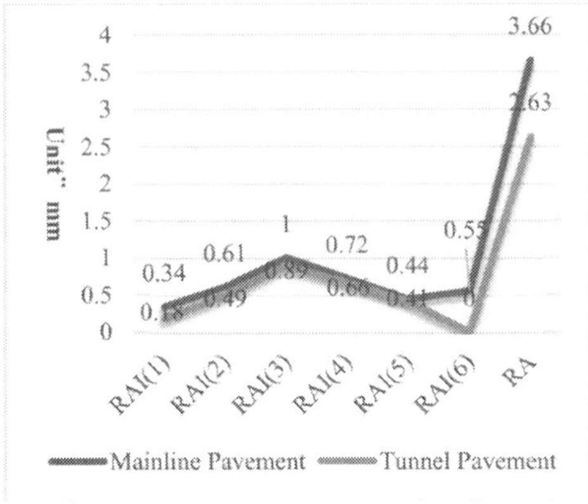

Figure 3. A comparison of the permanent deformation and bending values of the mainline and interchange ramp pavement structure and the tunnel pavement structure

The data in Table 1 shows that the difference in permanent deformation between the main line and the tunnel is small, so the transition section of the two areas will not have the problem of jumping; Chongqing Highway Science Institute of the Ministry of Communications believes that for the advanced pavement pavement within 20 years, the allowable post-work settlement of the section adjacent to the artificial structure is 10-20cm more appropriate[18]. And the road of the small bridge deck settlement after work for 0.65mm, the main line and tunnel pavement of the total bending value of 68.0mm and 69.1mm, and the small bridge deck settlement difference of 67.35mm and 68.45mm, respectively. are less than 10cm, so to meet the requirements of smooth vehicle driving; and the toll station square pavement structure for the concrete structure, its settlement is generally negligible, and The main line bending value is 68.0mm, and the settlement difference with the toll station square pavement is about 6.8cm, which satisfies the requirement of smooth driving of vehicles.

5. Conclusion

(1) The difference in permanent deformation between the main line and the tunnel is 1.03mm, and the difference in bending and settlement values is 1.1mm, both of which are small, so there will be no jumping problem in the transition section area between the two;

(2) the road of the small bridge deck after work settlement of 0.65mm, the main line and tunnel road surface of the total bending value of 68.0mm and 69.1mm, and the small bridge deck settlement difference of 67.35mm and 68.45mm respectively. are less than 10cm, so meet the requirements of smooth vehicle driving requirements;.

(3) And the toll station square pavement structure is concrete structure, its settlement is generally negligible, and the main line bending settlement value is 68.0mm, and the settlement difference of the toll station square pavement is about 6.8cm, which meets the requirements of smooth vehicle driving.

References

1. Liu Xiao, Li Yantao, Hu Peng. Problems and suggestions in asphalt pavement design standards in China [J]. Jinan:Shandong Institute of Transportation, 2017:55-56.
2. Huang F. Research on the theory and method of semi-flexible composite pavement structure design [D]. Chongqing:Chongqing Jiaotong University.2008:57-59.
3. Miao Dongsheng. Research on the design of special road surface of Wuyi Road BRT project in Changzhou [D]. Nanjing:Nanjing University of Technology. 2008:2-5.
4. Liu, Edward. Empirical design system and mechanics[J]. Urban Construction Theory Research: Electronic Edition.2011:96-97.
5. Japan Road Association, translated by Cai Enjie. Technical guide to the treatment of soft ground for road geotechnical. Beijing:People's Traffic Press.1984:7-8.
6. Ye Mianshu. Study on the prevention and control measures of bridge head jumping, Annex I - Study on the post-work settlement criteria of the roadbed after the bridge platform. Institute of Transportation Engineering, Southeast University.
7. Technical Standards for Highway Engineering (JTG B01-2014). Beijing: People's Traffic Publishing House.
8. Industry Standard of the People's Republic of China (Liu Boying), JTG D50-2017. Design Specification for Asphalt Pavement of Highways [S]. Beijing: General Office of the Ministry of Transport. 2017.
9. Teng XQ. Road surface engineering [M]. Chengdu: Southwest Jiaotong University Press. 2013:156-198.
10. Liu Liping. New roadbed pavement engineering [M]. Shanghai:Tongji University Press.2019.
11. Sha Qinglin. High-grade highway semi-rigid base asphalt pavement [M]. Beijing: People's Traffic Publishing House. 1998:45-85.
12. Industry standard of the People's Republic of China (Liu Boying), JTG D50-2017. design specification for highway asphalt pavement [S]. Beijing:General Office of the Ministry of Transport.2017.
13. Cao Q. Structural mechanical analysis of flexible base asphalt pavement under heavy load [D]. Hefei:Anhui University of Construction,2017:39-57.
14. Gao Yaliang. Research on structural design of asphalt pavement under heavy traffic[D]. Qinhuangdao: Yanshan University.2016:15-45.
15. Specification for the design of highway bridges and culverts (CJTJ024-85). Beijing: People's Traffic Publishing House.
16. Lai Guolin. Study report of bridge head jumping prevention and control measures of Annex II a proposal on the allowable value of settlement slope difference of bridge head lap plate []. Institute of Transportation Engineering, Southeast University.
17. Ye Mianshu. Study on post-work settlement control standard for bridge head approach. Journal of Southeast University. 1997:12-17.
18. Chongqing Highway Research Institute, Ministry of Communications. Experimental study on the construction of highway embankment on soft foundation in Xingang. In: Dredging Highway Engineering Technology (33-4-1), 1985.

Analysis of Mechanical Properties of Loess with Cement Incorporation Ratio

Baogui Rong [1, 2], Qingbo Liu [1], Gang Mei [1, 2], Bingwei Cheng [1, 2]

[1] China Railway No.3 Bureau Group Investment Co., Ltd. Taiyuan, Shanxi, 030006 China
[2] Tianshui Sanyangchuan Tunnel Traffic Engineering Co., Ltd. Tianshui, Gansu,741030 China

Abstract: Taking loess as the research object, the influence of cement content on mechanical properties of loess was studied. Firstly, the basic physical parameters of loess were obtained through laboratory physical index test, and improved by using Portland cement. Secondly, under different incorporation ratios (3%, 5%, 7%, 9%), heavy compaction test and liquid-plastic limit test were carried out respectively to obtain mechanical parameters of cement-improved loess. Finally, the shear strength parameters and compressive strength of the improved soil samples are obtained by unconsolidation undrained triaxial shear test and unconfined compressive strength test. The results show that the liquid limit, plastic limit and plastic index of loess improved by adding cement increase compared with the original soil sample, indicating that the binding force between soil particles is improved and the plasticity is enhanced. The variation range of dry density and water content has no obvious relationship with the content of cement. At the same time, the compressibility of soil is reduced and the collapsibility of soil is eliminated. With the increase of age, the strength of cement-modified soil also increases. Through unconsolidated and undrained triaxial shear tests and unconfined compressive strength tests, the strength of loess improved significantly after adding cement, and the strength of soil-cement increased with the increase of cement content, and water stability was obviously improved.

Keywords: Loess; Cement incorporation ratio; Mechanical properties.

1. Introduction

Loess refers to the yellow silt deposits transported by wind during the Quaternary Geological Age. There is a certain regularity in geographical distribution, mainly distributed in Shaanxi, Henan, Gansu and other regions. In soil classification, loess is often classified as "special soil", which is mainly attributed to two aspects. On the one hand, loess not only has strong hydrophilicity, adheres to the particle skeleton, but also has swelling and shrinking properties. Due to water swelling, the soil structure is softened, and the bonding capacity between particles is weakened, which improves the compressibility and strength of the soil. reduce. On the other hand, loess in its natural state has the characteristics of large void ratio and low density. When it encounters water, it will collapse rapidly, drop in strength, and increase in deformation[1].

Loess is widely distributed in China, especially in collapsible areas. Once it meets rain weather, a large amount of water penetrates into the soil, making the pore water pressure in the soil rise continuously, while the strength of the soil decreases continuously, which is easy to induce landslides, road subgrade collapse, uneven settlement of the foundation and other engineering accidents. Therefore, a large number of scholars have carried out research on the mechanical properties of loess. Du Jingfang[2] studied the variation rules of compression characteristics, shear strength and permeability of loess under dry-wet cycle through direct shear test. The experimental results show that the shear strength and its parameters decrease with the increase of the number of drying and wetting cycles and moisture content, and increase with the increase of moisture content. Wei Yao et al.[3] analyzed the influence of freezing temperature on mechanical properties of freezing-thawing loess by testing its unconfined compressive strength, shear strength and compressibility under different freezing temperatures. The test results show that the unconfined compressive strength of loess with different moisture content decreases after freezing-thawing at different freezing temperatures, and the cohesion of soil with different moisture content decreases with the decrease of freezing temperature. Fei Xiaoou et al.[4] conducted experimental studies on mechanical properties such as pore pressure dissipation and different water content in order to study the failure mechanism of loess foundation in the dump of open-pit coal mine under large area loading. The results show that the shear strength of loess decreases greatly with the increase of water content, and the change of water content and cohesion is log-linear. The dissipation degree of pore water pressure decreases with the increase of confining pressure, and the dissipation coefficient increases with the increase of water content, and presents the embrittlement failure phenomenon at low confining pressure. Shoubang Sun et al.[5] compared the vertical and horizontal shear strength differences of intact Q_4 loess samples through triaxial and direct shear tests. The results show that the shear strength of surrounding rock of Q_4 loess has obvious anisotropy. The internal friction angles of loess are basically the same in different directions. The vertical strength is generally greater than the horizontal strength. Leng Yanqiu et al.[6] studied the variation rules of structural strength, deformation modulus and strength parameters of intact loess under different moisture content through measuring instrument test and triaxial shear test. It is found that the yield stress, initial deformation modulus and cohesion of structure are sensitive to the water content. With the increase of initial water content, the yield stress, initial deformation modulus and cohesion of structure decrease by power function or exponential function. Kui Liu et al.[7] studied the effects of dry density and water content on mechanical properties of loess through shear strength and compressibility tests, and the results showed that: the lower the water content, the greater the impact of dry density change on cohesion. With the same dry density, the internal friction Angle of compacted loess decreases with the increase of water content. Xiangtian Xu et al.[8] studied the influence of water content on strength, stiffness and damage characteristics of frozen loess through triaxial compression and cyclic loading and unloading tests. The test results show that moisture content of frozen soil is an important factor affecting damage parameters during deformation. The effects of confining pressure on mechanical and damage properties of frozen loess are sensitive, indicating that the effects of water content and confining pressure are coupled. Taking loess as the research object, Shi Chang 'an[9] studied the influence of freezing-thawing cycle on mechanical properties of loess through triaxial test. The results show that the strength of soil decreases with the increase of the number of freezing-thawing cycles, and the influence on the strength decreases with the increase of the number of freezing-thawing cycles. When the water content is constant, the cohesion of loess decreases and the internal friction Angle decreases with the increase of freezing-thawing cycles.

In recent years, loess is the foundation of the industrial development in northwest China, as well as the source and basic conditions of building materials for highway transportation, water conservancy facilities and civil and industrial construction. In order to improve the strength of loess and meet the characteristics of compressibility, permeability and durability, the improved method is often used to meet the needs of practical engineering[10]. Therefore, a large number of scholars have carried out research on soil improvement. Wang Yimin et al.[11] analyzed the physical and mechanical properties of cement-improved loess based on the strict requirements for filling of high-speed railway roadbed and laboratory tests, and studied the influence of different cement content and curing age on strength growth. The results show that the maximum dry density and compression coefficient of Q_3 cement-improved loess decrease with the increase of incorporation ratio, and the unconfined compressive strength increases with the increase of incorporation ratio. In dry-wet cycle, 5% cement-improved loess has the best stability. Ma Xuening et al.[12] studied the compaction characteristics, compression characteristics, strength characteristics and influencing factors of cement-improved loess through a large number of geotechnical tests and theoretical analysis. The results show that the shear strength and unconfined compressive strength of cement-improved loess are affected by density, mass ratio and age, and the strength value increases with the increase of mass fraction of cement content. Zhang Yabin et al.[13] studied the compaction characteristics, strength characteristics and dry-wet cycling effects of cement-improved loess through a large number of laboratory tests. The results show that with the increase of cement content, the optimal water content of cement-improved loess increases and the maximum dry density decreases. When the cement content is low, the strength of cement-modified loess increases obviously. The strength growth rate of cement-improved loess slows down when the cement content is higher. With the increase of the number of dry and wet cycles, the strength of cement-improved loess will slightly decrease or slightly increase, but the range of change is not drastic, and its strength tends to be stable. Jia Ding Wang et al.[14] studied the influence of cement mixing ratio on the dynamic characteristics of cement-improved loess through dynamic triaxial test. The results show that with the increase of cement content, the dynamic characteristics of modified loess become better, and there is no optimal cement content. Taking loess as the research object, He Xiaoliang et al.[15] used GDS dynamic triaxial apparatus to conduct progressive triaxial test on loess after cement improvement, and studied the variation law of cement improvement loess deformation with cement ratio under locomotive vibration load. The results show that with the increase of cement proportion, the vibration subsidence of loess decreases obviously. The vibration subsidence of cement-improved loess increases gradually with the increase of vibration times. When the vibration times exceeds 3000 times, the vibration subsidence basically becomes stable. Guo Xing et al.[16] took loess as the research object and carried out cement soil improvement test on it. The test results show that the maximum dry density, optimal water content and compression coefficient of the improved soil are affected by the cement content. With the increase of cement content, the maximum dry density and compression coefficient decrease while the optimal water content increases. Lang Weijun et al.[17] took highway subgrade in Loess region of Qinghai as the research object and analyzed water content, compactness and settlement observation of cement improved subgrade through tests. The results show that 6% cement-soil ratio meets the construction control requirements, indicating that the construction of cement improved roadbed in loess area has good applicability and reliability. Jiang Yingjun et al.[18] studied the influence of cement dosage, compaction coefficient and health preservation age on the stability of cement improved loess subgrade in order to reveal the factors influencing the stability of cement

 https://doi.org/10.1051/matecconf/202235801042

improved loess subgrade. The results show that the compressive strength of the improved loess increases linearly with the increase of cement dosage or compaction coefficient under dry-wet and freeze-thaw conditions, respectively. When the number of dry-wet cycles is more than 15 times, the number of freeze-thaw cycles is more than 12 times, and the cement dose is more than 3%, the compressive strength of the improved loess tends to be stable. Necmi Yarbasi et al. [19] respectively conducted freezing-thawing cycle tests on improved soils mixed with lime, cement and fly ash, and from the test results and data, they believed that the frost resistance and durability of the three improved soils were effectively improved.

By summarizing the research results of the above scholars, it is not difficult to find that the improvement of loess with cement can improve its water stability, improve its strength, reduce its permeability, and avoid the occurrence of engineering accidents. At the same time, it has important theoretical and practical significance in engineering. In this paper, the influence of cement content on mechanical properties of loess was studied by compaction test, liquid-plastic limit test, consolidation test, collapsibility test, triaxial test and unconfined compressive strength test.

2. Physical Property

Loess is non-saline soil, and its physical parameters are shown in Table 1 through indoor parameter test. Grade 32.5 ordinary Portland cement is used for cement, and the physical and mechanical indexes are shown in Table 2.

Table 1 Physical parameters of loess

Liquid limit/%	Plastic limit/%	Plasticity index	Maximum dry density/(g/cm^3)	Optimum water content/%
29.2	19.8	9.4	1.880	14.3

Table 2 Physical and mechanical indexes of ordinary Portland cement

Fineness	Ignition loss	Stability	Setting time		Compressive strength/MPa		Flexural strength/MPa	
			Initial set	Final set	3d	28d	3d	28d
≤ 10%	≤ 5.0%	Normal	≥ 45min	≤ 10h	11.0	32.5	2.5	5.5

3. Mechanical Analysis

The methods of loess improvement mainly focus on physical improvement and chemical improvement. Physical improvement is to join the coarse particle in loess and the use of dynamic compaction, the compaction pile method such as physical soil compaction method to change the internal grain size distribution, particle arrangement and contact, increase the contact area between particles and reduce the pore volume, make the internal structure of soil tend to be more compact, thus achieve permeability, such as improving strength improvement purpose. Chemical modification is to improve the stability and strength of loess by adding a certain modifier to the loess, using the effect of modifier and water and the chemical reaction with soluble salt in the loess, to produce cementing material to fill the pores, change the soil particle composition and spatial structure, so as to improve the stability and strength of loess engineering properties. The commonly used improvers are mainly cement, lime, fly ash and other materials [20-23] to reduce the permeability of loess and improve its strength.

Cement as a powdery hydraulic inorganic cementitious material, mixed with water into slurry, hardening in air or hardening in water, and the sand, stone and other materials firmly cemented together. Therefore, cement as a modifier, not only low cost, economic and environmental protection, and significant improvement effect, is widely used in construction engineering, water conservancy engineering and traffic engineering and other special construction in the engineering field. Since this paper studies the influence of cement content on mechanical properties of loess, it will be analyzed through the following tests [24] :

(1) Heavy compaction test and liquid plastic limit test were carried out by adding cement into loess at the rate of 3%, 5%, 7% and 9% respectively.

(2) Improved soil specimens with corresponding density were prepared by pressing 95% compacting method and kept in a constant temperature and humidity curing box (chamber) with a temperature of 20±2°C and humidity >90%. After keeping in good health, their compression coefficient, compression modulus and collapsibility coefficient were measured.

(3) Modified soil specimens with the corresponding density were prepared by pressing 95% compacty by hydrostatic method. The specimens were kept in a constant temperature and humidity curing box (chamber) with a temperature of 20±2°C and humidity >90%, and their unconsolidated and undrained triaxial shear strength was measured after 28 days of curing.

(4) 39.1×80 mm specimens were prepared by pressing 95% compactness method, kept in a constant temperature and humidity curing box (chamber) with a temperature of 20±2°C and humidity >90%, and their water-soaked and unconfined compressive strength without water-soaked for 7 days were measured (water-soaked means that the specimens were taken out one day before they reached age. The specimen was immersed in water at a temperature of 20 ±2°C for 24h).

3.1 Beating Experiment

Beating experiment is a basic method to study the compactability of fine grained soil in laboratory. The dry density and moisture content of the soil samples were measured by heavy-duty compactor. By changing the moisture content and repeating the test, the compaction curve is obtained and the maximum dry density and optimal water content of soil samples are obtained, among which the maximum dry density and optimal water content are the necessary conditions for controlling the quality of subgrade filling. The results of compaction test are summarized in Table 3.

Table 3 Results of compaction test and liquid plastic limit test

Soil sample	Liquid limit/ %	Plastic limit/ %	Plasticity index	Maximum dry density/ (g/cm^3)	Optimum water content/ %	Dry density /(g/cm^3)	Dry unit weight / (kN/m^3)
3% cement soil	35.7	22.7	13.0	1.900	14.3	1.805	17.689
5% cement soil	36.5	22.8	13.7	1.902	14.3	1.807	17.708
7% cement soil	37.1	23.6	13.5	1.880	14.4	1.786	17.503
9% cement soil	39.5	25.8	13.7	1.900	14.3	1.805	17.689

3.2 Liquid Limit-Plastic Limit Joint Test

Liquid limit is the limit water content of soil sample when it transitions from plastic state to flowing state. Plastic limit is the limit water content of soil sample from plastic state to semi-solid state. Liquid and plastic limits are important physical characteristics of clay and reflect the degree of interaction between soil particles and water. In engineering practice, the exact value of liquid-plastic limit index directly affects the determination of soil name and bearing capacity of corresponding soil foundation, and indirectly reflects the engineering properties of soil samples. The liquid limit, plastic limit and plastic index of the modified filler were obtained through the liquid plastic limit test. The results of liquid plastic limit test are summarized in Table 3, in which the dry density is 95% of the maximum dry density.

It can be seen from Table 3 that the change of maximum dry density and optimal water content is not obvious after the loess is mixed with different cement content. The liquid limit and plastic limit of the improved soil increased with the increase of cement content, and both increased compared with the original soil sample, indicating that the adhesion between soil particles was improved and the plasticity was enhanced after the addition of cement. The plasticity of the improved loess with 9% cement content was the strongest.

3.3 Consolidation Test, Loess Collapsibility Test

Consolidation test refers to the compression test to determine the saturated clay sample under the condition of confining pressure. Compressibility of soil is mainly caused by the reduction of pore volume. Therefore, the relationship between pressure and sample deformation or pore ratio is measured through consolidation test, so as to calculate the compression coefficient and compression modulus of soil sample. The test results are shown in Table 4.

Loess collapsibility is the process of loess compression and collapsibility deformation under certain pressure and water immersion. The collapsibility coefficient of loess can be obtained through loess collapsibility test, and the results are shown in Table 4.

Table 4 Results of consolidation test and loess collapsibility test

Soil sample	Degree of compaction /%	Curing age: 1d			Curing age: 7d		
		Compression coefficient /MPa-1	Compression modulus /MPa	Collapsible coefficient	Compression coefficient /MPa^{-1}	Compression modulus /MPa	Collapsible coefficient
Original soil sample		0.18	10.6	0.026			
Compaction of soil samples	95	0.10	15.0	0.000	0.09	16.2	0.000
3% cement soil	95	0.06	24.9	0.000	0.05	30.2	0.000
5% cement soil	95	0.06	24.9	0.000	0.05	30.1	0.000
7% cement soil	95	0.06	25.1	0.000	0.05	30.9	0.000
9% cement soil	95	0.07	21.3	0.000	0.05	30.2	0.000

It can be seen from Table 4 that, compared with the original soil sample, the compression coefficient of the improved soil sample decreases and the compression modulus increases after adding a certain proportion of cement, indicating that the compressibility of the improved soil sample decreases. Meanwhile, the collapsibility of the improved soil sample is eliminated, and the relationship between the change range and the dosage is not obvious. With the increase of health age, the compression coefficient decreases and the compression modulus increases with the same cement content, indicating that the strength of cement improved soil increases slightly.

3.4 Unconsolidated and undrained triaxial shear test

The unconsolidated and undrained shear test, also known as the fast shear test, does not allow the sample to drain during the process of applying confining pressure and increasing axial pressure until the specimen is broken. The shear strength parameters, namely c and φ, can be obtained through the test. The index parameters of soil shear strength are an important index to determine the bearing capacity of soil. Therefore, it is of great significance to accurately measure the index parameters of soil shear strength for engineering design and construction. The results of unconsolidated and undrained triaxial shear test are summarized in Table 5.

It can be seen from Table 5 that the modified loess specimen prepared by pressing 95% compaction was kept in a curing box (chamber) with a temperature of 20±2°C and humidity >90% for 28 days, and the internal friction Angle and cohesion increased with the increase of cement content. Among them, the shear strength parameters of the modified loess with 9% cement content were the maximum.

Table 5 Results of unconsolidated and undrained triaxial shear test

Soil sample	Degree of compaction/ %	Curing age/d	Internal friction angle/ (°)	Cohesion/ (kPa)
Original soil sample			31.2	21
Compaction of soil samples	95	28	35.6	34
3% cement soil	95	28	34.3	214
5% cement soil	95	28	30.5	437
7% cement soil	95	28	42.4	442
9% cement soil	95	28	44.7	553

3.5 Unconfined Compressive Strenght Test

The unconfined compressive strength test is a special case of the triaxial test, that is, the soil sample is subjected to the pressure test under the condition of no lateral limitation, so as to obtain the ultimate strength of resisting axial pressure, that is, the unconfined compressive strength. The unconfined compressive strength test results are summarized in Table 6.

Table 6 Unconfined compressive strength test results

Soil sample	Degree of compaction/%	Curing age/ d	Unconfined compressive strength during immersion/kPa	Unconfined compressive strength without flooding /kPa	Strength degradation /%
Original soil sample			0	100	100
Compaction of soil samples	95	7	0	387	100
3% cement soil	95	7	816	1360	40
5% cement soil	95	7	999	1461	32
7% cement soil	95	7	1341	1651	19
9% cement soil	95	7	1559	1919	19

As can be seen from table 6, under the condition of 95% compaction degree, the unconfined compressive strength of cement-improved soil under standard curing, dry-wet cycle and saturation conditions for 7 days increases significantly with the increase of cement content, and the attenuation rate of strength decreases gradually, that is, the water stability gradually increases. After soaking for 24h, the unconfined compressive strength attenuates by 19% ~ 40%, except that the strength attenuating rate of the modified soil mixed with 3% cement is slightly higher, the other attenuation rates are relatively small. Among them, the improved soil sample with 9% cement content has the maximum unconfined compressive strength and the minimum strength attenuation, both after soaking and without soaking, indicating the strongest water stability.

4. Conclusion

In this paper, loess is taken as the research object. Firstly, the basic physical parameters of loess are obtained through laboratory physical parameter test. Secondly, the loess was modified by adding different doses of cement (3%, 5%, 7%, 9%). The mechanical parameters of the loess were measured by compaction test, liquid-plastic limit test, consolidation test, collapsibility test, triaxial test and unconfined compressive strength test. Finally, the influence of cement content on mechanical properties of loess is studied through data analysis. The results show that:

(1) The liquid limit, plastic limit and plastic index of the loess improved by adding cement increased compared with the original soil sample, indicating that the binding force between soil particles was improved and the plasticity was enhanced. The variation range of dry density and water content had no obvious relationship with the content of cement.

(2) After adding cement to loess, the compressibility of soil is reduced and the collapsibility is eliminated; With the increase of age, the strength of cement-modified soil also increases.

(3) The unconsolidated and undrained triaxial shear test shows that the strength of loess improved significantly after adding cement, and the strength of soil-cement increases with the increase of adding cement.

(4) The unconfined compressive strength test shows that the water stability of loess improved by adding cement is obviously improved.

In conclusion, after adding a certain proportion of ordinary Portland cement into the loess, the properties of the soil are greatly improved, which can improve the strength and water stability of the soil, reduce the compressibility of the soil, and eliminate the soil collapsibility.

References

1. Zhang Xuejiao. Experimental study on permeability and mechanical properties of lime improved loess [D]. Guangdong University of Technology,2020.
2. Du Jingfang. Study on mechanical properties and slope stability of remolded loess under dry-wet cycle [D]. Taiyuan University of Technology,2020.
3. Wei Yao, Yang Genshe, Ye Wanjun. Journal of Yangtze River scientific research institute, 2018, 35(08).

4. Fei Xiaoou, Miao Haibin, Wang Jianguo. Experimental study on mechanical properties of loess under large area loading [J]. Coal mine safety, 2016, 47(12).
5. Shoubang Sun,Xiaoping Cao. Study on Physical and Mechanical Properties of Q4 Loess in Lanzhou Rail Transit Area[J]. Recent Patents on Engineering, 2021,15(4).
6. Leng Yanqiu,Peng Jianbing,Wang Shuo,Lu Feng. Development of water sensitivity index of loess from its mechanical properties[J]. Engineering Geology, 2021,280.
7. Kui Liu,Liu Kui,Yuan Kang ze,Wu Kai. Effect of Water Content and Dry Density on Mechanical Properties of Compaction Loess[J]. IOP Conference Series: Earth and Environmental Science, 2020, 598(1).
8. Xiangtian Xu,Qionglin Li,Ying Lai,Wentai Pang,Ruiping Zhang. Effect of moisture content on mechanical and damage behavior of frozen loess under triaxial condition along with different confining pressures[J]. Cold Regions Science and Technology, 2018.
9. Shi Chang'an. Experimental study on the effect of freezing-thawing cycle on mechanical properties of loess [D]. Xijing University,2019.
10. Li Lei. Experimental study on mechanical properties and structure of loess modified by silica powder [D]. Xi 'an University of Technology,2019.
11. Wang Yimin, Liang Bo, Ma Xuening, Huang Zhijun. Experimental study on Cement Improved Loess in high-speed Railway Subgrade [J]. Journal of Lanzhou Jiaotong University,2005(04).
12. Ma Xuening, LIANG Bo. Geotechnical Engineering Technology,2005(05).
13. Zhang Yabin, Yang Youhai. Shanxi architecture, 2013, 39(25).
14. Jia Ding Wang,Shu Jun Peng,Wan Li Xie. Study on the Cement-Improved Loess under the Vibratory Load by Dynamic Tests[J]. Advanced Materials Research, 2014, 2837.
15. He Xiaoliang, Liu Xiaomin, Wang Zhishuo. Groundwater, 2014, 36(02).
16. Guo Xing, Zhang Qiqi, Liu Borong, Zeng Youjiang. Groundwater, 2015, 37(03).
17. Lang Weijun, Dong Cuo, Zhu Chunming. Research on construction technology of cement improved roadbed in Loess area of Qinghai Province [J]. Science and Technology Innovation and Application, 2017(09): 251.
18. Jiang Yingjun, Wang Hanyue, Qiao Huaiyu, Yue Weimin, Dong Xin. Science technology and engineering, 2020, 20(35).
19. Necmi Yarbaşı,Ekrem Kalkan,Suat Akbulut. Modification of the geotechnical properties, as influenced by freeze–thaw, of granular soils with waste additives[J]. Cold Regions Science and Technology, 2006, 48(1).
20. Xue Huike. Experimental analysis on the effect of fly ash, lime and cement to improve loess filling [J]. Science and Technology Vision,2016(14).
21. Alireza Tabarsa, Nima Latifi,Christopher L. Meehan, Kalehiwot Nega Manahiloh. Laboratory investigation and field evaluation of loess improvement using nanoclay – A sustainable material for construction[J]. Construction and Building Materials, 2018, 158.
22. Brian G Palmer,Tuncer B Edil,Craig H Benson. Liners for waste containment constructed with class F and C fly ashes[J]. Journal of Hazardous Materials, 2000, 76(2).
23. Phongthorn Julphunthong. Chemical Stabilization of Loess in Northeast Thailand Using the Mixture of Calcined Marble Dust Waste and Sugarcane Bagasse Ash Waste[J]. Geotechnical Engineering Journal of the SEAGS & AGSSEA,2015,46(1).
24. Ji lei. Performance test of loess improvement soil [J]. Railway survey, 2009,35(05).

Analysis of Mechanical Properties of Loess with Lime Incorporation Ratio

Tiehai Wu [1], Zhaofeng Jia [1], Baogui Rong [1, 2], Laishuan Zhao [1, 2]

[1] China Railway No.3 Bureau Group Investment Co., Ltd. Taiyuan, Shanxi,030006, China
[2] Tianshui Sanyangchuan Tunnel Traffic Engineering Co., Ltd. Tianshui, Gansu,741030, China

Abstract: Taking loess as the research object, the effect of lime content on mechanical properties of loess was studied. Firstly, the basic physical parameters of loess were obtained by laboratory physical index test and improved by lime. Secondly, the mechanical parameters of lime-modified loess were obtained by heavy compaction test and liquid-plastic limit test under different incorporation ratios (3%, 5%, 7%, 9%). Finally, the shear strength parameters and compressive strength of improved soil samples were obtained by unconsolidated and undrained triaxial shear test and unconfined compressive strength test respectively. The results showed that the liquid limit, plastic limit and plastic index of the loess improved by adding lime increased compared with the original soil sample, indicating that the binding force between soil particles was improved and the plasticity was enhanced. The variation range of dry density and water content had no obvious relationship with the dosage of lime. At the same time, the compressibility of soil is reduced and the collapsibility of soil is eliminated. With the increase of age, the strength of lime-modified soil also increases. Through unconsolidated and undrained triaxial shear tests and unconfined compressive strength tests, the strength of the loess improved by adding lime is significantly improved, and the strength of soil-cement increases with the increase of the dosage, and the water stability is obviously improved.

Keywords: Loess; Lime incorporation ratio; Mechanical properties.

1. Introduction

Loess refers to the yellow silt deposits transported by wind during the Quaternary Geological Age. There is a certain regularity in geographical distribution, mainly distributed in Shaanxi, Henan, Gansu and other regions. In soil classification, loess is often classified as "special soil", which is mainly attributed to two aspects. On the one hand, loess not only has strong hydrophilicity, adheres to the particle skeleton, but also has swelling and shrinking properties. Due to water swelling, the soil structure is softened, and the bonding capacity between particles is weakened, which improves the compressibility and strength of the soil. reduce. On the other hand, loess in its natural state has the characteristics of large void ratio and low density. When it encounters water, it will collapse rapidly, drop in strength, and increase in deformation[1].

Loess is widely distributed in China, especially in collapsible areas. Once it meets rain weather, a large amount of water penetrates into the soil, making the pore water pressure in the soil rise continuously, while the strength of the soil decreases continuously, which is easy to induce landslides, road subgrade collapse, uneven settlement of the foundation and other engineering accidents. Therefore, a large number of scholars have carried out research on the mechanical properties of loess. Du Jingfang[2] studied the variation rules of compression characteristics, shear strength and permeability of loess under dry-wet cycle through direct shear test. The experimental results show that the shear strength and its parameters decrease with the increase of the number of drying and wetting cycles and moisture content, and increase with the increase of moisture content. Wei Yao et al.[3] analyzed the influence of freezing temperature on mechanical properties of freezing-thawing loess by testing its unconfined compressive strength, shear strength and compressibility under different freezing temperatures. The test results show that the unconfined compressive strength of loess with different moisture content decreases after freezing-thawing at different freezing temperatures, and the cohesion of soil with different moisture content decreases with the decrease of freezing temperature. Fei Xiaoou et al.[4] conducted experimental studies on mechanical properties such as pore pressure dissipation and different water content in order to study the failure mechanism of loess foundation in the dump of open-pit coal mine under large area loading. The results show that the shear strength of loess decreases greatly with the increase of water content, and the change of water content and cohesion is log-linear. The dissipation degree of pore water pressure decreases with the increase of confining pressure, and the dissipation coefficient increases with the increase of water content, and presents the embrittlement failure phenomenon at low confining pressure. Shoubang Sun et al.[5] compared the vertical and horizontal shear strength differences of intact Q_4 loess samples through triaxial and direct shear tests. The results show that the shear strength of surrounding rock of Q_4 loess has obvious anisotropy. The internal friction angles of loess are basically the same in different directions. The vertical strength is generally greater than the horizontal strength. Leng Yanqiu et al.[6] studied the variation rules of structural strength, deformation modulus and strength parameters of intact loess under different moisture content through measuring instrument test and triaxial shear test. It is found that the yield stress, initial deformation modulus and cohesion of structure are sensitive to the water content. With the increase of initial water content, the yield stress, initial deformation modulus and cohesion of structure decrease by power function or exponential function. Kui Liu et al.[7] studied the effects of dry density and water content on mechanical properties of loess through shear strength and compressibility tests, and the results showed that: the lower the water content, the greater the impact of dry density change on cohesion. With the same dry density, the internal friction Angle of compacted loess decreases with the increase of water content. Xiangtian Xu et al.[8] studied the influence of water content on strength, stiffness and damage characteristics of frozen loess through triaxial compression and cyclic loading and unloading tests. The test results show that moisture content of frozen soil is an important factor affecting damage parameters during deformation. The effects of confining pressure on mechanical and damage properties of frozen loess are sensitive, indicating that the effects of water content and confining pressure are coupled. Taking loess as the research object, Shi Chang 'an[9] studied the influence of freezing-thawing cycle on mechanical properties of loess through triaxial test. The results show that the strength of soil decreases with the increase of the number of freezing-thawing cycles, and the influence on the strength decreases with the increase of the number of freezing-thawing cycles. When the water content is constant, the cohesion of loess decreases and the internal friction Angle decreases with the increase of freezing-thawing cycles.

In recent years, loess is the foundation of the industrial development in northwest China, as well as the source and basic conditions of building materials for highway transportation, water conservancy facilities and civil and industrial construction. In order to improve the strength of loess and meet the characteristics of compressibility, permeability and durability, the improved method is often used to meet the needs of practical engineering[10]. Therefore, a large number of scholars have carried out research on soil improvement. Yang Mei et al.[11], through the performance test of improving loess with lime, showed that adding lime into loess can effectively improve the properties and structure of loess, reduce the permeability and compressibility of loess, control the occurrence of its collapsibility, partially or completely eliminate the collapsibility, and thus improve the strength of loess. Tao Fujin[12] studied the physical properties, strength characteristics, water stability and strength variation rule of lime-improved loess under dry-wet cycles through laboratory tests and field tests. The results show that the control range of water content of lime-improved loess becomes wider with the increase of ash content. The shear strength of lime-modified soil increases with the increase of compaction coefficient. The compressibility decreases with the increase of compaction coefficient, mixing ratio and age. The cohesion of lime-improved soil increases with the increase of wetting and drying cycles, and its strength also increases significantly. Zhou Jianji et al.[13] took loess as the research object and analyzed the influence of different incorporation ratios of lime on improved loess performance through consolidation test. The experimental results show that with the increase of lime incorporation ratio, the compression coefficient of improved loess decreases first and then increases, and the permeability coefficient decreases first and then increases. The addition of lime can effectively improve the properties of loess, and the maximum improvement can be achieved when the lime incorporation ratio is 7%, so the best lime incorporation ratio is 7%. Wang Bo[14] studied the physical properties, stress-strain relationship and compression characteristics of lime-improved loess through indoor geotechnical tests and theoretical analysis. The results show that the maximum dry density of lime-improved soil decreases and the optimal water content increases with the increase of lime mixing ratio. The stress-strain curve of lime-modified loess has obvious stress peak point, and the lime-modified loess shows obvious brittle failure. Mr.li[15] through the red bed mudstone and lime improved soil compaction test, then consider the optimal moisture content and saturation moisture content without confining pressure, CBR and direct shear mechanics test, the results show that with the increase of the ratio of lime, improvement of red mudstone layer of optimum moisture content increases gradually, while the maximum dry density decreases. Wang Lixia et al.[16] explored the optimal incorporation ratio of lime-improved silty sand in the seasonally frozen zone under the condition of high temperature curing, based on the conventional triaxial unconsolidated and undrained (UU) shear test of lime-improved silty sand in the seasonally frozen zone with different freeze-thaw cycles. The results show that compared with plain soil, the cohesion, internal friction Angle and shear strength of lime-improved silty sand are significantly increased, and the increase is the largest when the lime incorporation ratio is 6%. Compared with lime improved silt, the cohesion, internal friction Angle and shear strength of fly ash/lime improved silt are improved to a certain extent, and the improvement is most obvious when the ratio of fly ash/lime improved silt is 4：1. Wang Lifeng et al.[17] discussed the yield characteristics of lime-improved soil through triaxial test, and established the elastic-plastic constitutive model of lime-improved soil. The results show that the improved lime soil with 10% cement ratio has higher shear strength. Liang Zhichao et al.[18] used unsaturated consolidation apparatus to conduct compression tests on lime improved loess, and studied soil-water characteristics, compression characteristics, humidification deformation and collapsibility deformation characteristics of lime improved loess with different lime content and dry density. The results show that the lime content has little effect on the soil-water characteristic curve of unsaturated loess, but the dry density has a great effect on the soil-water characteristic curve of lime-modified loess. Both the compression index and yield stress increase with the increase of suction. Under the same vertical pressure, the greater the initial suction, the greater the coefficient of collapsibility. Xu Zhijian[19] studied the strength and physical-mechanical variation of soft soil with different lime incorporation ratio, compaction degree and curing age. The results show that the improved soft soil with lime incorporation ratio of 6% has good road performance and can meet the requirements of design and economic benefit. Wei Jie[20] studied the settlement characteristics of lime-improved loess as subgrade material through tests on parameters such as unconfined compressive strength, springback modulus and compression consolidation. The results show that when the lime incorporation ratio is 6%, the workability and economy of lime soil are the best, and the compaction degree, bending modulus and rebound modulus of the roadbed meet the design requirements, which can effectively reduce the roadbed settlement.

By summarizing the research results of the above scholars, it is not difficult to find that the use of lime to improve the loess water stability, improve its strength, reduce its permeability, avoid the occurrence of engineering accidents, at the same time, has important theoretical and practical significance in engineering. In this paper, the influence of lime incorporation ratio on mechanical properties of loess was studied by compaction test, liquid-plastic limit test, consolidation test, collapsibility test, triaxial test and unconfined compressive strength test.

2. Physical Property

Loess is non-saline soil, and its physical parameters are shown in Table 1 through indoor parameter test. Grade ii calcareous quicklime is used for lime, and quicklime is used for self-digestion of hydrated lime. The physical and mechanical indexes are shown in Table 2.

Table 1 Physical parameters of loess

Liquid limit/%	Plastic limit/%	Plasticity index	Maximum dry density/(g/cm^3)	Optimum water content/%
29.2	19.8	9.4	1.880	14.3

Table 2 Physical and mechanical indexes of lime

Available calcium and magnesium oxide content	Water content	0.17mm Sieve allowance of square hole sieve	0.125mm Sieve allowance of square hole sieve
$\geq 60\%$	$\leq 4\%$	$\leq 1\%$	$\leq 20\%$

3. Mechanical Analysis

The methods of loess improvement mainly focus on physical improvement and chemical improvement. Physical improvement is to join the coarse particle in loess and the use of dynamic compaction, the compaction pile method such as physical soil compaction method to change the internal grain size distribution, particle arrangement and contact, increase the contact area between particles and reduce the pore volume, make the internal structure of soil tend to be more compact, thus achieve permeability, such as improving strength improvement purpose. Chemical modification is to improve the stability and strength of loess by adding a certain modifier to the loess, using the effect of modifier and water and the chemical reaction with soluble salt in the loess, to produce cementing material to fill the pores, change the soil particle composition and spatial structure, so as to improve the stability and strength of loess engineering properties. The commonly used improvers are mainly cement, lime, fly ash and other materials[21-24] to reduce the permeability of loess and improve the strength of loess.

Lime is a kind of gaseous inorganic cementitious material with calcium oxide as the main component. It is made of limestone, dolomite, chalk, shell and other products with high calcium carbonate content, calcined at 900~1100°C. Therefore, lime as a modifier, not only low cost, economic and environmental protection, and significant improvement effect, is widely used in construction engineering, water conservancy engineering and traffic engineering and other special construction engineering field. Since this paper studies the influence of lime incorporation ratio on mechanical properties of loess, the following tests will be carried out to analyze it[25] :

(1) Heavy compaction test and liquid plastic limit test were carried out by adding lime to the loess at the rate of 3%, 5%, 7% and 9% respectively.

(2) Improved soil specimens with corresponding density were prepared by pressing 95% compacting method and kept in a constant temperature and humidity curing box (chamber) with a temperature of 20±2°C and humidity >90%. After keeping in good health, their compression coefficient, compression modulus and collapsibility coefficient were measured.

(3) Modified soil specimens with the corresponding density were prepared by pressing 95% compacty by hydrostatic method. The specimens were kept in a constant temperature and humidity curing box (chamber) with a temperature of 20±2°C and humidity >90%, and their unconsolidated and undrained triaxial shear strength was measured after 28 days of curing.

(4) 39.1×80 mm specimens were prepared by pressing 95% compactness method, kept in a constant temperature and humidity curing box (chamber) with a temperature of 20 ±2°C and humidity >90%, and their water-soaked and unconfined compressive strength without water-soaked for 7 days were measured (water-soaked means that the specimens were taken out one day before they reached age. The specimen was immersed in water at a temperature of 20 ±2°C for 24h).

3.1 Beating Experiment

Beating experiment is a basic method to study the compactability of fine grained soil in laboratory. The dry density and moisture content of the soil samples were measured by heavy-duty compactor. By changing the moisture content and repeating the test, the compaction curve is obtained and the maximum dry density and optimal water content of soil samples are obtained, among which the maximum dry density and optimal water content are the necessary conditions for controlling the quality of subgrade filling. The results of compaction test are summarized in Table 3.

Table 3 Results of compaction test and liquid plastic limit test

Soil sample	Liquid limit/ %	Plastic limit/ %	Plasticity index	Maximum dry density/ (g/cm^3)	Optimum water content/ %	Dry density /(g/cm^3)	Dry unit weight / (kN/m^3)
3% lime soil	36.6	22.9	13.7	1.760	15.6	1.672	16.386
5% lime soil	36.1	22.5	13.6	1.772	15.9	1.683	16.497
7% lime soil	36.0	23.5	12.5	1.740	15.8	1.653	16.199
9% lime soil	36.1	23.8	12.3	1.753	15.8	1.665	16.320

3.2 Liquid Limit-Plastic Limit Joint Test

Liquid limit is the limit water content of soil sample when it transitions from plastic state to flowing state. Plastic limit is the limit water content of soil sample from plastic state to semi-solid state. Liquid and plastic limits are important physical characteristics of clay and reflect the degree of interaction between soil particles and water. In engineering practice, the exact value of liquid-plastic limit index directly affects the determination of soil name and bearing capacity of corresponding soil foundation, and indirectly reflects the engineering properties of soil samples. The liquid limit, plastic limit and plastic index of the modified filler were obtained through the liquid plastic limit test. The results of liquid plastic limit test are summarized in Table 3, in which the dry density is 95% of the maximum dry density.

It can be seen from Table 3 that the maximum dry density and optimal water content do not change significantly after the addition of lime with different incorporation ratios in loess, and the maximum dry density and water content are found at 5% lime incorporation ratio. The liquid limit and plastic limit of the improved soil decreased with the increase of lime incorporation ratio, and both increased compared with the original soil sample, indicating that the adhesion between soil particles was improved and the plasticity was enhanced after the addition of lime. The plasticity of the improved loess with 3% lime incorporation ratio was the strongest.

3.3 Consolidation Test, Loess Collapsibility Test

Consolidation test refers to the compression test to determine the saturated clay sample under the condition of confining pressure. Compressibility of soil is mainly caused by the reduction of pore volume. Therefore, the relationship between pressure and sample deformation or pore ratio is measured through consolidation test, so as to calculate the compression coefficient and compression modulus of soil sample. The test results are shown in Table 4.

Loess collapsibility is the process of loess compression and collapsibility deformation under certain pressure and water immersion. The collapsibility coefficient of loess can be obtained through loess collapsibility test, and the results are shown in Table 4.

Table 4 Results of consolidation test and loess collapsibility test

Soil sample	Degree of compaction /%	Curing age: 1d			Curing age: 7d		
		Compression coefficient /MPa-1	Compression modulus /MPa	Collapsible coefficient	Compression coefficient /MPa^{-1}	Compression modulus /MPa	Collapsible coefficient
Original soil sample		0.18	10.6	0.026			
Compaction of soil samples	95	0.10	15.0	0.000	0.09	16.2	0.000
3% lime soil	95	0.08	20.3	0.000	0.08	20.1	0.000
5% lime soil	95	0.08	20.0	0.000	0.08	20.3	0.000
7% lime soil	95	0.08	20.6	0.000	0.06	27.0	0.000
9% lime soil	95	0.08	23.1	0.000	0.07	22.9	0.000

As can be seen from Table 4, compared with the original soil sample, the compression coefficient of the improved soil sample decreases and the compression modulus increases, indicating that its compressibility decreases. Meanwhile, the collapsibility of the improved soil sample is eliminated, and the relationship between the change range and the incorporation ratio is not obvious. With the increase of health age, the compression coefficient decreases and the compression modulus increases with the same incorporation ratio of lime-improved soil, indicating that the strength of lime-improved soil increases slightly.

3.4 Unconsolidated and undrained triaxial shear test

The unconsolidated and undrained shear test, also known as the fast shear test, does not allow the sample to drain during the process of applying confining pressure and increasing axial pressure until the specimen is broken. The shear strength parameters, namely c and φ, can be obtained through the test. The index parameters of soil shear strength are an important index to determine the bearing capacity of soil. Therefore, it is of great significance to accurately measure the index parameters of soil shear strength for engineering design and construction. The results of unconsolidated and undrained triaxial shear test are summarized in Table 5.

Table 5 Results of unconsolidated and undrained triaxial shear test

Soil sample	Degree of compaction/%	Curing age/d	Internal friction angle/(°)	Cohesion/(kPa)
Original soil sample			31.2	21
Compaction of soil samples	95	28	35.6	34
3% lime soil	95	28	36.5	84
5% lime soil	95	28	36.9	222
7% lime soil	95	28	39.7	125
9% lime soil	95	28	45.0	121

As can be seen from Table 5, the improved soil specimen prepared by pressing 95% compaction was kept in a curing box (chamber) with a temperature of 20±2°C and humidity >90% for 28 days. The internal friction Angle increased with the increase of cement incorporation ratio, and the cohesion showed a trend of first increasing and then decreasing. Among them, the cohesion was the largest at 5% lime incorporation ratio.

3.5 Unconfined Compressive Strenght Test

Table 6 Unconfined compressive strength test results

Soil sample	Degree of compaction/ %	Curing age/d	Unconfined compressive strength during immersion/kPa	Unconfined compressive strength without flooding/kPa	Strength degradation /%
Original soil sample			0	100	100
Compaction of soil samples	95	7	0	387	100
3% lime soil	95	7	337	605	44
5% lime soil	95	7	426	782	46
7% lime soil	95	7	416	843	51
9% lime soil	95	7	378	855	56

The unconfined compressive strength test is a special case of the triaxial test, that is, the soil sample is subjected to the pressure test under the condition of no lateral limitation, so as to obtain the ultimate strength of resisting axial pressure, that is, the unconfined compressive strength. The unconfined compressive strength test results are summarized in Table 6.

From table 6 unconfined compressive strength test results can be seen that the degree of compaction (95%) cases, lime improved soil under the condition of standard curing and dry-wet circulation and saturated 7 d of unconfined compressive strength increase significantly with the increase of the cement mortar, the water stability are improved obviously, flooding after 24 h the unconfined compressive strength attenuation 44% ~ 56%, Except that the strength attenuation rate of modified soil mixed with 9% lime is slightly higher, the other attenuation rates are relatively small. Among them, the improved soil sample with 3% cement incorporation ratio has the lowest strength attenuation, indicating that water stability is the strongest.

4. Conclusion

In this paper, loess is taken as the research object. Firstly, the basic physical parameters of loess are obtained through laboratory physical parameter test. Secondly, the loess was modified by adding different doses of lime (3%, 5%, 7%, 9%). The mechanical parameters of the loess were measured by compaction test, liquid-plastic limit test, consolidation test, collapsibility test, triaxial test and unconfined compressive strength test. Finally, the influence of lime incorporation ratio on mechanical properties of loess is studied through data analysis. The results show that:

(1) Compared with the original soil sample, the liquid limit, plastic limit and plastic index of the loess improved by adding lime increased, indicating that the binding force between soil particles was improved and the plasticity was enhanced. The variation range of dry density and water content had no obvious relationship with the incorporation ratio.

(2) After adding lime to the loess, the compressibility of the soil is reduced and the collapsibility of soil is eliminated; With the increase of age, the strength of lime-modified soil also increases.

(3) Unconsolidated and undrained triaxial shear tests show that the strength of loess improved by adding lime;

(4) The unconfined compressive strength test shows that the water stability of loess improved by adding lime is obviously improved.

In conclusion, after adding a certain proportion of lime into the loess, the properties of the soil are greatly improved, which can improve the strength and water stability of the soil, reduce the compressibility of the soil and eliminate the soil collapsibility.

Reference

1. Zhang Xuejiao. Experimental study on permeability and mechanical properties of lime improved loess [D]. Guangdong University of Technology,2020.
2. Du Jingfang. Study on mechanical properties and slope stability of remolded loess under dry-wet cycle [D]. Taiyuan University of Technology,2020.
3. Wei Yao, Yang Genshe, Ye Wanjun. Journal of Yangtze river scientific research institute, 2018,35(08):61-66.
4. Fei Xiaoou, Miao Haibin, Wang Jianguo. Experimental study on mechanical properties of loess under large area loading [J]. Coal mine safety, 2016,47(12).
5. Shoubang Sun,Xiaoping Cao. Study on Physical and Mechanical Properties of Q4 Loess in Lanzhou Rail Transit Area [J]. Recent Patents on Engineering, 2021,15(4).
6. Leng Yanqiu,Peng Jianbing,Wang Shuo,Lu Feng. Development of water sensitivity index of loess from its mechanical properties[J]. Engineering Geology, 2021,280.
7. Kui Liu,Liu Kui,Yuan Kang ze,Wu Kai. Effect of Water Content and Dry Density on Mechanical

Properties of Compaction Loess[J]. IOP Conference Series: Earth and Environmental Science, 2020, 598(1).

8. Xiangtian Xu, Qionglin Li,Ying Lai,Wentai Pang,Ruiping Zhang. Effect of moisture content on mechanical and damage behavior of frozen loess under triaxial condition along with different confining pressures[J]. Cold Regions Science and Technology,2018.
9. Shi Chang'an. Experimental study on the effect of freezing-thawing cycle on mechanical properties of loess [D]. Xijing University,2019.
10. Li Lei. Experimental study on mechanical properties and structure of loess modified by silica powder [D]. Xi 'an University of Technology, 2019.
11. Yang Mei, Liu Yonghong. Experimental study on the performance of loess modified by adding lime [J]. Subgrade Engineering,2007(01).
12. Tao Fujin. Experimental study on lime modified loess filler for passenger dedicated line [J]. Journal of lanzhou jiaotong university,2011,30(06).
13. Zhou Jianji, Liang Jiyun, Zhang Fangyu, Wu Wanjiong, Yan Xude. Experimental Study on Engineering Characteristics of Lime Modified Loess [J]. Railway Construction, 2014(09).
14. WANG Bo. Experimental Study on Lime Improved Soil of Highway Subgrade [J]. West Transportation Science and Technology, 2016(03).
15. Li Junfeng. Experimental study on mechanical properties of red mudstone soil modified by lime under different water content [J]. Urban Roads, Bridges and Flood Control, 2021(04).
16. Wang Lixia, Han Bing, Zhang Yunlong, Wang Jing, Sun Shuang, Zhang Xinyue. Journal of jilin jianzhu university, 2020, 37(01).
17. Wang Lifeng, Xie Xirong, Wang Tian. Mechanical properties and constitutive model of lime-improved soil [J]. Chinese science and technology bulletin, 2019, 35(11).
18. Liang Zhichao, Hu Zaiqiang, Guo Jing, Wang Kai, Feng Zhe, Zheng Haicheng. Journal of hydroelectric engineering, 2020,39(03).
19. Xu Zhijian. Experimental study on the Performance of lime improved silty soft soil Road [J]. Traffic Construction and Management,2019(03).
20. Wei Jie. Experimental study and engineering application of lime improved loess performance [J]. Heilongjiang Transportation Science and Technology, 201, 44(02).
21. Xue Huike. Experimental analysis on the effect of fly ash, lime and cement to improve loess filling [J]. Science and Technology Vision,2016(14).
22. Alireza Tabarsa,Nima Latifi,Christopher L. Meehan, Kalehiwot Nega Manahiloh. Laboratory investigation and field evaluation of loess improvement using nanoclay – A sustainable material for construction[J]. Construction and Building Materials, 2018.
23. Brian G Palmer,Tuncer B Edil,Craig H Benson. Liners for waste containment constructed with class F and C fly ashes[J]. Journal of Hazardous Materials,2000,76(2).
24. Phongthorn Julphunthong. Chemical Stabilization of Loess in Northeast Thailand Using the Mixture of Calcined Marble Dust Waste and Sugarcane Bagasse Ash Waste[J]. Geotechnical Engineering Journal of the SEAGS & AGSSEA,2015,46(1).
25. Ji lei. Performance test of loess improvement soil [J]. Railway survey, 2009, 35(05).

Effect of different replacement depth on slope stability of roadbed

Zhaofeng Jia [1], Yu Liu [1], Baogui Rong [1, 2], Pingyu Ling [1, 2]

[1] China Railway No.3 Bureau Group Investment Co., Ltd. Taiyuan, Shanxi,030006 China
[2] Tianshui Sanyangchuan Tunnel Traffic Engineering Co., Ltd. Tianshui, Gansu,741030 China

Abstract: Combined with the engineering example of Sanyangchuan tunnel and lead project in Tianshui City, firstly, the cement improved soil and lime improved soil with better mechanical properties were selected by experiment. Secondly, based on Geostudio software, a numerical calculation model is established to analyze the influence of different replacement materials and different replacement depths on the stability of the roadbed slope for different replacement ranges.

Keywords: Roadbed slope; Improved soil; Replacement filling method.

1. Introduction

The stability of the slope of the roadbed is the key to the quality of the whole road construction, and whether its structure and quality are stable and qualified plays a decisive role in the whole project. There is a soft soil geographical construction foundation, low bearing capacity, not easy to construct, and often occur uneven settlement phenomenon, to a certain extent to reduce the length of the road, the construction of the road in the soft soil foundation, easy to trigger a variety of road hazards. [1] The standard definition of soft ground in the industry is a weak soil layer with low strength and relatively high compression, small permeability, and the presence of certain organic materials in these soils. In highway design and construction, the formation of soft ground is mainly clay and powder soil with more fine particles, which will lead to settlement if the water table rises or the stability of the building is insufficient. [2-3] The characteristics of soft soil roadbed are divided into the following points: First, the soil is loose, and the clay and powder soil contains more fine particles. Secondly, it has large voids, high water content and high compressibility. Due to the above two characteristics, resulting in the construction of road and bridge soft ground strength and stiffness are too low and prone to uneven settlement phenomenon, has become one of the difficulties in the construction of buildings. If under construction or after the completion of the geological influence, easy to trigger the foundation subsidence and collapse of the problem, more serious will also be a threat to the personal safety of construction personnel.

The replacement method is widely used in highway projects and has many advantages, the most important being that it can strengthen the roadbed and improve the construction quality of the whole project. The replacement method is also called the soil replacement method. The so-called replacement method refers to the reinforcement of the soft foundation of the highway. The specific process of the replacement method is to use a combination of manual and mechanical equipment to clean up the weak soil layer in the roadbed, and then use solid materials for filling and compaction, usually using hard soil and stones for filling in layers, thus enhancing the bearing capacity of the roadbed. Most of the time, natural gravels and flakes are used to fill the roadbeds because they are easy to find and relatively cost effective. [4-8] However, in determining the depth of the replacement fill, due to many influencing factors, it is often based on empirical method and lacks a quantitative calculation basis. According to the relevant literature, the quantitative calculation methods include the roadbed load working area method, the foundation bearing capacity method, etc. [9]

In this paper, we firstly compare the sexual mechanics experiments of cement and lime with different dosing amounts to select the suitable dosing materials, and then conduct numerical simulation based on Geostudio software to simulate and calculate the influence of different replacement materials and different replacement depths on the stability of roadbed slope.

2. Project Background

2.1 Geographical location and traffic conditions

Tianshui Sanyangchuan tunnel and lead project is located in Tianshui City, Maiji District, the tunnel inlet is located in Dengjiazhuang, before connecting to the northern end of the Tianxiu Bridge ramp, and through the ramp to

Chengji Avenue; tunnel exit is located in Qiaohe New Village, through the interchange and Magan Highway connection, the tunnel import and export end are close to the existing road, convenient traffic.

2.2 Meteorology

Tianshui City is located in the northern temperate zone and has a semi-arid continental climate. The natural area is divided into the arid zone in the Yang Mountain in the north, the river valley in the middle and the humid zone in the shallow mountains in the south. The general climatic characteristics of Tianshui are: rapid warming in spring; no scorching heat in summer; continuous rain in autumn; and no severe cold in winter. The climate is mild, with four distinct seasons, sufficient sunshine and moderate precipitation. According to the weather observation data of Tianshui weather station, the multi-year average temperature of Tianshui is 10.9 ℃, the extreme maximum temperature is 38.3℃, the extreme minimum temperature is -19.2℃, and the coldest monthly average temperature is -2 ℃; the multi-year average precipitation is 580.0mm, and the highest record since the historical data was recorded in 2003 was 810mm. The distribution of precipitation is uneven, mainly concentrated in July-September, accounting for about 65% of the annual precipitation. According to the meteorological data of Qinzhou District, the maximum precipitation in the district is 286.6mm in a row, 113mm in a day and 57.3mm in an hour.

July-September, which corresponds to the period of concentrated precipitation, is also the period of frequent occurrence of landslide, landslide and debris flow, accounting for 70%, 80% and 90% of the total frequency respectively. The average annual precipitation day is 99 days.

The average maximum evaporation is 190.6mm (August); the average minimum is 34.0mm (January); the average annual evaporation is 1314.9mm. The maximum annual snow thickness is 15cm, the maximum permafrost depth is 61cm, and the prevailing wind direction is east.

2.3 Geomorphology

The project traverses the valley area of the river through the river, the north mountain of the river through the loess mount area to the south bank of the Wei River, the geomorphology is erosion accumulation valley landform and denudation structure hilly landform, the topography is undulating, the relative height difference is 200 ~ 400m.

(1) Erosion and accumulation valley

It is distributed at both ends of the project area, and the starting point of the line is located in the area of the valley step of the river by the river, and the end of the line is located in the area of the valley step of the river by the Wei River. Due to the erosion of water and artificial transformation, the form of the order surface has been unclear, and it is actually close to the foot of the loess hills. The Weihe River valley is a long east-west spreading valley with large width changes, and the Weihe River valley at the end of the project area is open, with a width of 1 km. The river floodplain and terrace landforms in the valley are typical and open.

(2) Stripping structure hilly landform

The hills are distributed in the middle of the project area and are loess hills, with the beams extending from east to west, elevation 1112~1534m, relative height difference 422m, slope gradient 10~35°. The base of the mountain is Neoproterozoic mudstone and Precambrian granite gneiss, overlain by the Fourth Series loess. The main characteristics are that the top of the mountain is beam-shaped, the top is narrower, the slope of the beam is slope gully development, cut deeper, the valley is mostly "V" valley, and the local gully slope is upright. The area is a landslide and debris flow development area, and geological hazards are very developed.

3. Mechanical properties of the modified soil

The original soil of the roadbed slope is loess, which is a non-saline soil, and its physical properties are shown in Table 1.

Table 1 Physical properties of loess

Liquid limit/ %	Plastic limit/ %	Plasticity Index	Maximum dry density/ (g/cm^3)	Optimum moisture content/ %
29.2	19.8	9.4	1.880	14.3

3.1 Cement improved soil

The cement is 32.5 grade ordinary silicate cement, and the physical and mechanical indexes are shown in Table 2.

Table 2 Physical and mechanical indexes of ordinary silicate cement

Fineness	Burning loss	Adequacy	Coagulation time		Compressive strength/MPa		Flexural strength/MPa	
			First condensation	Final condensation	3d	28d	3d	28d
≤ 10%	≤ 5.0%	Qualified	≥ 45min	≤ 10h	11.0	32.5	2.5	5.5

The cement was mixed into the loess at 3%, 5%, 7% and 9%, respectively, and heavy compaction tests and liquid-plastic limit tests were conducted. The mechanical properties of the cement-improved soil selected by comparing the experimental results with the 9% admixture are as follows.

Table 3 Compaction test and liquid-plastic limit test results

Soil sample	Liquid limit /%	Plastic limit /%	Plasticity Index	Maximum dry density /(g/cm^3)	Optimum moisture content/%	Dry Density/(g/cm^3)	Dry heaviness /(kN/m^3)
9% hydro mulch	39.5	25.8	13.7	1.900	14.3	1.805	17.689

 https://doi.org/10.1051/matecconf/202235801044

Table 4 Unconsolidated undrained triaxial shear test results

Soil sample	Compaction/%	Age of maintenance/d	φ/(°)	c /(kPa)
9% hydromulch	95	28	44.7	553

3.2 Lime amended soil

The lime was made of Class II calcareous quicklime, and the quicklime was made of quicklime with its own dissipation, and the physical and mechanical indexes are shown in Table 5.

Table 5 Lime physical and mechanical indexes

Effective calcium plus magnesium oxide content	Water content	Sieve allowance of 0.17mm square hole sieve	Sieve allowance of 0.125mm square hole sieve
≥ 60%	≤ 4%	≤ 1%	≤ 20%

Lime was mixed into the loess at 3%, 5%, 7% and 9%, respectively, and heavy compaction tests and liquid-plastic limit tests were conducted. The mechanical properties of the cement-amended soils selected by comparing the experimental results with the 5% admixture are as follows.

Table 6 Compaction test and boundary moisture content test results

Soil sample	Liquid limit /%	Plastic limit /%	Plasticity Index	Maximum dry density /(g/ cm^3)	Optimum moisture content t/%	Dry Density /(g/ cm^3)	Dry heaviness/ (kN/m^3)
5% lime soil	36.1	22.5	13.6	1.772	15.9	1.683	16.497

Table 7 Unconsolidated undrained triaxial shear test results

Soil sample	Compaction/%	Age of maintenance/d	φ/(°)	c /(kPa)
5% lime soil	95	28	36.9	222

4. Numerical simulation model

The numerical simulation model based on the roadbed slope of the project is established as follows.

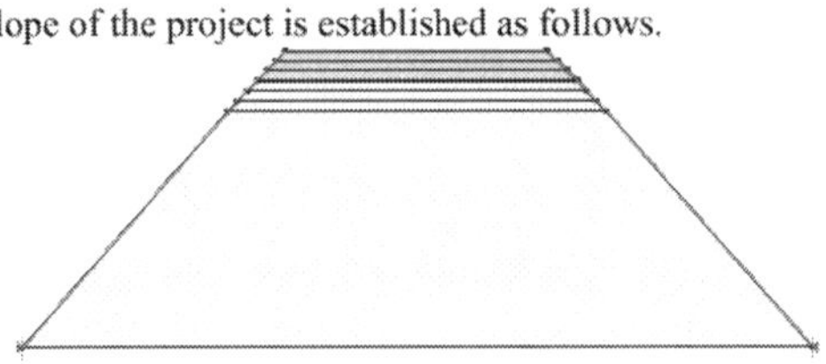

Fig.1 Numerical simulation model

After inputting the soil properties of the roadbed slope, six areas were set up at 0.5m intervals in the range of 0.5-3m to analyze the changes of the stability of the roadbed slope when using 9% dosed cement improved soil and 5% dosed lime improved soil with the replacement range of 0.5m, 1m, 1.5m, 2m, 2.5m and 3m respectively.

The variation curve of the safety factor with the range of the replacement fill when using 9% blended cement improved soil is obtained as follows.

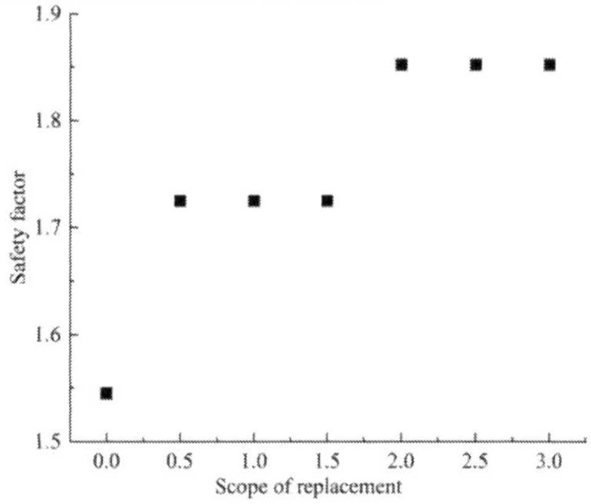

Fig. 2 Variation curve of safety factor (9% blended cement improved soil)

Analyzing the above figure, it can be seen that 9% cementitious improved soil can effectively improve the safety coefficient of the roadbed slope, and the safety coefficient of the roadbed slope is 1.725 when the replacement range is 0.5-1.5m, and the safety coefficient of the roadbed slope is 1.852 when the replacement range is 2-3m. Considering the economic benefits and road safety, when using 9% cementitious improved soil to replace the roadbed, the replacement range of 0.5m is A better quality choice.

The variation curve of the safety factor with the range of replacement fill when using 5% lime amended soil is obtained as follows.

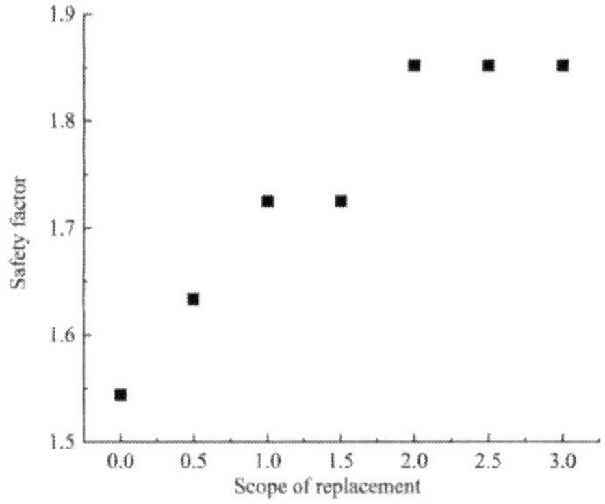

Fig.3 Variation curve of safety factor (5% lime amended soil)

Analyzing the above figure, it can be seen that 5% lime amended soil can effectively improve the safety coefficient of roadbed slope, and the safety coefficient of roadbed slope is 1.633 when the replacement range is 0.5m, 1.725 when the replacement range is 1-1.5m, and 1.852 when the replacement range is 2-3m. Considering the economic benefits and road safety, it is better to use 5% lime amended soil when the replacement range is 1m. In consideration of economic benefits and road safety, when using 5% lime improved soil for roadbed replacement, the replacement range of 1m is a better choice.

5. Conclusions

Through the combination of indoor experiments, the cement improved soil and lime improved soil with better mechanical properties were selected, and the numerical calculation model was established based on Geostudio software to analyze the influence of different replacement materials and different replacement depths on the stability of the slope of the roadbed, and the following conclusions were drawn.

(1) In the comparison of the performance of improved soil with different doping amounts, 9% cement improved soil and 5% lime improved soil have better mechanical properties than other improved soils, and are more suitable as improved soil for roadbed slope replacement.

(2) Both 9% cement-improved soil and 5% lime improved soil can effectively improve the stability of the slope of the roadbed, and the improvement performance of both of them is similar in the numerical simulation results.

(3) The safety coefficient of the slope of the roadbed changes when the 9% cement-improved soil is replaced by 0.5m and 2m, and considering the economic and safety factors, the replacement range of 0.5m is a more economic and safe choice.

(4) When 5% lime improved soil is used to fill the slope of the roadbed, the safety coefficient of the slope of the roadbed changes when the range of replacement is 0.5m, 1m, 2m, and the range of replacement is 1m is a more economical and safe choice.

References

1. Wan Tigui, Wan Qilong. Practice of roadbed replacement technology in highway engineering[J]. Traffic World, 2018(36):72-73.
2. Wu Shuai. Application of replacement fill construction process in the treatment of soft foundation of highway [J]. Jiangxi building materials, 2016(18):166+171.
3. Luo Xun. Experimental discussion on the impact of urban road material selection on the environment and improvement [J]. Urban geography,2015(06):188.
4. Su Jianguo. A brief discussion on the application of roadbed replacement technology in highway construction [J]. Henan building materials, 2016(03): 124.
5. Ren Kaixuan. Discussion on the application of roadbed replacement technology in highway construction [J]. Information technology construction, 2016(02):367+369.
6. Gou Yajun. Analysis of the application of roadbed replacement technology in highway construction [J]. Construction knowledge, 2016(04).
7. Yu Panpan. Analysis of the application of roadbed replacement technology in highway construction [J]. Jiangxi building materials, 2015(23):163-165.
8. Pei Gang. Discussion on the application of roadbed replacement technology in highway construction [J]. Information technology construction,2016(4):155.
9. Xiao Huade. Application of HPDS in the calculation of roadbed replacement depth [J]. Transportation standardization, 2011(18):97-99.

Research Progress in Characterization Methods of Anti-corrosion and Wear-resistant Polyurethane Coatings

Xinghui Jia [1, 2], Xiwen Song [1], Jinjuan Fan [2, *], Wei Sun [2], Changkui Liu [2]

[1] School of Materials and Metallurgy, Inner Mongolia University of Science and Technology, Baotou 014000, China.
[2] AECC Beijing Institute of Aeronautical Materials, Beijing 100095, China

Abstract: Polyurethane is a coating material widely used in the field of anti-corrosion and wear resistance. At present, a lot of research work has been carried out in the research and development of polyurethane materials and process improvement. However, there are few researches on the characterization methods of polyurethane coatings, and suitable characterization methods are selected. Conducive to the improvement and application of materials. This article summarizes the characterization methods of polyurethane coatings from the aspects of morphology, structure, mechanical properties, electrochemical properties and thermodynamic properties. It is found that there is no standardized and unified evaluation standard for the electrochemical characterization methods of polyurethane; polyurethane as a high molecular polymer has a certain sensitivity to the service environment. Humidity and temperature have a great influence on its physical and chemical properties. Conventional characterization methods cannot perform microstructure characterization under the stress state of the material and the service environment. The establishment of standardized and unified electrochemical evaluation standards for polyurethane; the ability to characterize polyurethane under service conditions is an important research direction for improving the characterization method.

Key words: Polyurethane; Coating materials; Characterization methods.

1. Introduction

Polyurethane [1,2] is made of polyisocyanate and polyether polyol or polyester polyol or small molecular polyol, polyamine or water and other chain extenders or crosslinking agents. The produced polymer has a unique soft and hard segment structure and a microphase separation structure. It is a polymer material with excellent performance. Compared with traditional metal and ceramic-based anti-wear composite materials, polyurethane coatings have better processing properties, wear resistance, erosion resistance [3 , 4], low temperature flexibility [5], and lower manufacturing costs. It is used in protection fields such as rain [6, 7], sand control [8], UV protection [9], and has broad development prospects. At present, domestic scientific researchers and related units have carried out research and improvement on the production and synthesis process of polyurethane, and have achieved obvious development: Zhang Weigang [10] prepared graphene-modified PU/Al composite coating by scratch coating method, which significantly improved the coating. Stability to salt water corrosion; Qiao Longyun [11] used halloysite nanotubes (HNTs) to encapsulate the inorganic corrosion inhibitor Ce3+, and prepared a controlled release nanocontainer, which was doped into a polyurethane base material to enhance the durability of the coating.

Although domestic scientific researchers have done a lot of research and development on the corrosion and wear resistance of polyurethane in recent years, my country's polyurethane industry started relatively late, and some of the core technologies in the polyurethane field are still in the hands of a few large companies[12]. Skilled application of various characterization methods to analyze anti-corrosion and wear-resistant polyurethane is an important means to improve the international competitiveness of my country's polyurethane products. Only by fully understanding the properties of materials and establishing the relationship between structure, composition and performance can we better analyze the raw materials and product properties of polyurethane, control the polyurethane production process, explore the failure mechanism of coatings, and complete my country The transformation from a big country of polyurethane to a strong country of polyurethane, therefore, this article will briefly summarize the analysis and characterization methods of anti-corrosion and wear-resistant polyurethane coatings.

2. Morphology analysis method

In the process of in-depth research on the corrosion and wear resistance of polyurethane, scanning electron microscope (SEM) is the most widely used morphology observation tool. The analysis principle of SEM is mainly to use electron beams to scan the surface of the sample to be tested. The difference in morphology stimulates different morphological information, and the information is studied to obtain the microstructure morphology of the sample [13,14]. Analyzing the morphology of polyurethane before and after erosion by SEM and comparing the erosion resistance of polyurethane coatings with different compositions can help us to screen coating materials with better erosion resistance. Zhong Ping [15] used SEM to compare Observation of different samples before and after the erosion found that the polyurethane/polyurea hybrid system coating itself has a large number of bubbles, resulting in weak erosion resistance, while the pure polyurea coating itself has fewer bubble defects, erosion resistance and anti-erosion performance. The micro-cutting performance is better than that of polyurethane/polyurea hybrid system coatings. Using SEM to observe the morphology before and after the erosion to study the erosion behavior is helpful to explore the erosion mechanism of erosion particles. Gao Jin [16] studied the erosion behavior of sand particles with different sizes through the morphological changes of polyurethane coatings. :Under the same flow velocity conditions, most sand particles with larger diameters will directly impact, and most sand particles with smaller diameters will obliquely rub the coating. Using SEM to observe the corrosion degree of the polyurethane aging process can study the failure mechanism of the coating in the corresponding environment, and then predict the service life of the coating. Liu Jinhe[17] simulated the desert atmospheric environment on the acrylic polyurethane coating The layer is aged, and the microscopic morphology before and after the aging is observed by SEM. Before aging, the coating has no obvious defects under low magnification, and there are microscopic defects such as pores under high magnification. The anti-wind sand erosion performance of the coating is reduced, which weakens the coating of the pigment, causing a large number of white pigment particles to precipitate on the surface of the polyurethane and more erosion pits; Hao Yunhong [18] used SEM to observe the microscopic damage morphology of polyurethane coatings with different aging times, as shown in Figure 1. After multiple cycles of high and low temperature cyclic aging, cracks, peeling, and a large number of polyurethane coatings appeared. The temperature aging seriously reduces the smoothness of the polyurethane coating surface, which reduces the erosion resistance of the coating and shortens the service life of the coating.

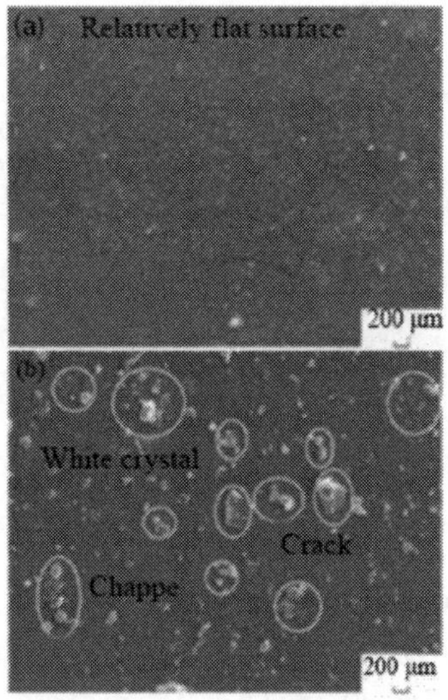

Figure 1 Microscopic damage morphology of polyurethane coating (a not aged, b aged 50 times)

SEM has high magnification and spatial resolution, and the imaging is full of stereoscopic effect, and can be used in conjunction with X-ray energy spectrometer devices [19]. However, SEM also has many problems when processing polyurethane samples. The electron beam of SEM will damage the polyurethane samples. Polyurethane is a non-conductive material. In order to make the polyurethane surface emit secondary electrons, it is necessary to spray the surface of the polyurethane scanning electron microscope sample. A layer of precious metal particles, which will increase the difficulty of the operation process. If you do not need to study the failure mechanism of the coating, only observe the flatness of the coating surface and judge whether the coating fails, you can choose to use an Optical Microscope, Li Xu [20] use OM to observe different PTMGs PUA coating prepared with a molar ratio of N-330, it is found that polyurethane prepolymer with high N-330 content has a high degree of branching, which will lead to poor compatibility between the prepolymer and acrylic resin, which is not conducive to the formation of a good surface condition. coating. Cui Xiaofei [21] used OM to observe the aging morphology of the coating as shown in Figure 2. As the aging cycle increases, the smoothness of the coating surface decreases severely. During the 8-9 cycle of aging, the bubbles are serious and corrosives are precipitated, which can be preliminarily judged. Coating failure.

 https://doi.org/10.1051/matecconf/202235801045

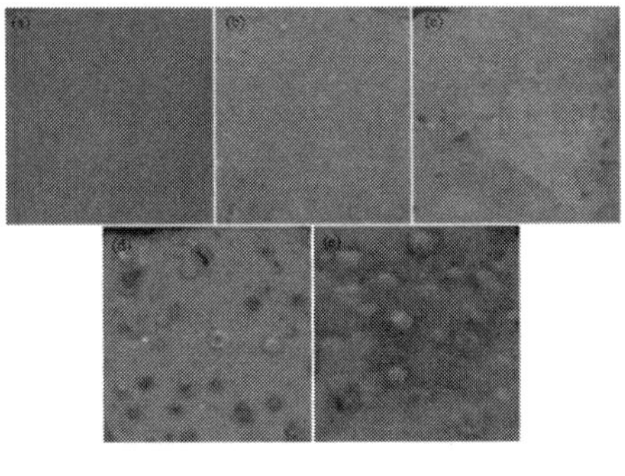

Figure 2 Surface aging morphology of coatings in different aging cycles

3. Structural analysis method

1) Fourier infrared absorption spectrum

Fourier Transform Infrared Spectroscopy can qualitatively identify and quantitatively analyze the molecular structure according to the position and intensity of the infrared absorption peak [22,23, 24, 25]. Fourier infrared spectroscopy can be used to analyze the changes of polyurethane molecular structure and functional groups, which is an effective method for the modification of polyurethane coatings. The mechanism of the polyurethane modification process is more complicated. According to the corresponding relationship between the absorption spectrum and the substances in each stage of the reaction, the group changes in the substance can be inferred. According to the position and relative intensity of the infrared characteristic absorption peak, each group in the sample can be analyzed. The content of points. Liu Gang [26] In the study of porous nano SiO2/graphene oxide modified polyurethane coating materials, the analysis of graphene oxide modified polyurethane by Fourier infrared spectroscopy showed that: phthalic anhydride polyester polyol and polyphenyl polyisocyanate The high-viscosity polyurethane prepared as a raw material was successfully synthesized and there are obvious characteristic peaks in it, making the product molecular structure mostly benzene rings to form a higher-strength modified polyurethane; compared to the simple graphene oxide infrared spectrum curve, graphene oxide The hydroxyl group, carboxyl group and epoxy group are obviously present, which can effectively complete the modification of polyphenyl polyurethane by graphene oxide. In determining the degree of crosslinking and curing of polyurethane, infrared spectroscopy is also an important analysis tool [27]. -NCO group has a sharp absorption peak at 2300cm-1, and the absorption peak of -NCO group will inevitably decrease with the curing process of polyurethane. In the case of a known control sample, infrared spectroscopy can be used to make a preliminary judgment of the curing speed.

Fourier infrared absorption spectroscopy is often used to analyze the reasons for the failure of polyurethane coatings. The construction of polyurethane on the contaminated substrate is an important reason for the failure of coatings. In general, as long as we use microscopy to observe the surface of the coating delamination Contaminants can be found, but some contaminants, such as lubricating oil, small molecule curing agents, etc., are difficult to observe with the naked eye and need to be analyzed by infrared spectroscopy. In addition, infrared spectroscopy can also be used to determine the mixing ratio of polyurethane materials. Polyurethane is a two-component coating material. Only by strictly controlling the mixing ratio can the polyurethane coating material with the best performance be obtained. The variety of cyanate resins and the multiplicity of -NCO group reactions greatly increase the difficulty of infrared spectroscopy to determine the mixing ratio.

2) Raman spectroscopy

The theoretical basis for the qualitative analysis of the molecular structure of Raman spectra is that the Raman shift depends on the change of the molecular vibrational energy level. Different chemical bonds or functional groups have different vibration modes, which determine the energy change between their energy levels [28,29,30]. Fourier infrared absorption spectroscopy has the advantages of high light flux, low noise, fast measurement speed, high resolution, high wavenumber accuracy, and wide spectral range. It mainly reflects the functional groups of molecules. Raman spectroscopy mainly reflects the skeleton of molecules. Complement the results. Like infrared absorption spectroscopy, Raman spectroscopy can also be used in the modification of polyurethane. Hao Weihong [31] used Raman spectroscopy to study the resistance of modified graphene oxide/polyurethane coatings to wind and sand erosion. The ID/IG of GO was about 1.02, while the ID/IG of FGO was about 1.18. Functional groups will be grafted on the GO surface, which will increase GO defects and disorder, and the ratio of ID/IG will increase, indicating that the modification will graft functional groups on the surface of graphene and increase defects, and GO modification will be successful.

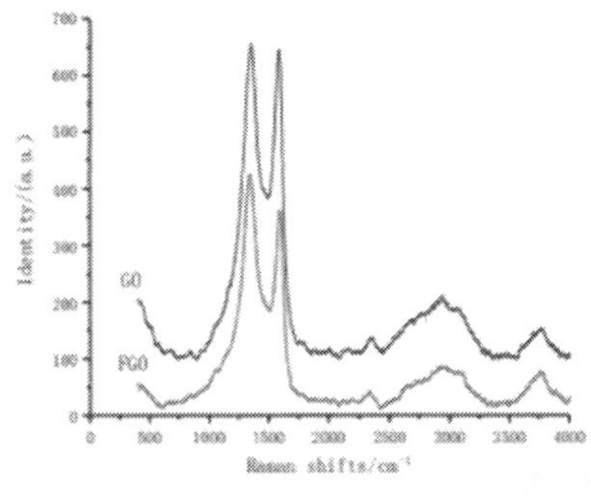

Figure 3 Raman spectra of GO and FGO

Raman spectroscopy can also detect the progress of the polyurethane production reaction. In the process of industrial production of polyurethane prepolymers, the offline detection of the content of -NCO to judge the reaction process has huge safety hazards, and the specific situation cannot be grasped in time. Lu Jinxing [32] Using

Raman spectroscopy and multivariate analysis methods based on angle measurement, a quantitative analysis model of -NCO groups in the synthesis process of polyurethane prepolymers was established, and a method for online monitoring of -NCO content was provided.

3) XRD phase analysis

XRD (X-ray diffraction) phase analysis is a method for analyzing and determining the existence of each component based on the diffraction effect of polycrystalline samples on X-rays. Raman spectroscopy and Fourier infrared spectroscopy are important research methods to study polyurethane materials from a molecular perspective, but the factors that determine the properties of polyurethane materials are not only the chemical composition of the molecules, but also the way related atoms combine into molecules in space. In the research of anti-corrosion and wear-resistant polyurethane, XRD is mainly used to determine the existence of modified substances in polyurethane. Hao Weihong. [31] used XRD to analyze the crystal structure of FGO before and after modification and found that the order of the GO lattice structure was destroyed after modification, and the Si-O-Si bond grafted on the GO surface and increased The functional group increases the interlayer spacing, which is beneficial to improve the dispersibility of FGO. Ye Yinping [33] used XRD diffractometer to characterize nano-Ni powders when preparing Ni/polyurethane nanocomposite coatings. From the diffraction peaks, it can be determined that the Ni nanocrystals have a face-centered cubic structure. Wide and 2θ angles can calculate that the initial average grain size of Ni nanoparticles is 10nm, and the final Ni nanoparticles prepared are uniformly distributed in the polyurethane, and the particle size is 50nm, indicating that the magnetic and magnetic properties of the Ni nanoparticles are Surface activity produces agglomeration.

4. Test method of mechanical properties

1) Macroscopic mechanics

When testing the macro-mechanical properties of polyurethane, the universal testing machine is an indispensable testing tool. The universal testing machine is a macro-mechanics testing instrument with high precision and stable performance that integrates stretching and shearing functions. It is mainly used to test the tensile strength and tear strength of polyurethane. Tensile strength is an important parameter affecting the wear resistance of polyurethane coatings [34]. Polyurethane is a typical soft-hard segment block polymer. Capacitiveness is closely related. Li Shuai [35] studied the relationship between the content of soft and hard segments and the tensile strength and found that: when the content of hard segments is low, as the content of hard segments increases, the physical cross-linking points inside the material gradually increase, and orderly hydrogen bonding With the increase of the degree, the tensile strength and elongation at break show an increasing trend; when the content of the hard segment is higher, the rigidity of the molecular chain becomes larger and the overall hydrogen bonding degree decreases significantly, which makes the orientation of the molecular chain difficult during the stretching process. Decrease in strength and elongation at break; Yang Yilin [36] and others used the tensile strength to characterize the self-healing ability of polyurethane. The crosslink density increased with the increase of the amount of HDI trimer, which increased the hard segment content of the prepared polyurethane material and improved the tensile strength performance. Wang Xinggang [37] conducted a tensile experiment on butyl polyurethane, and the results showed that polyisobutylene is a highly saturated molecular chain structure and cannot form intermolecular forces such as hydrogen bonds, which leads to an increase in the content of soft segments, which makes butyl polyurethane The tensile strength decreases.

At present, there are many testing methods for the tear strength of polyurethane, but there are few reports on the tear strength of polyurethane alone. Tear strength and tensile strength are commonly used to characterize the mechanical properties of polyurethane together, and then characterize the ratio of soft and hard segments, crystalline properties and properties of polyurethane. compatibility. In the process of tearing experiment, the selected test standard is different, the shape of the sample and the tensile speed will also be different. The shape of the sample has a particularly significant effect on the tear strength. For a right-angled shape, the tear strength with cuts is lower than without cuts. The stretching speed also has a certain influence on the tearing behavior. It shows brittle failure at a higher stretching speed, and shows elastic failure at a lower stretching speed. It is generally concluded that the tearing strength will vary with The stretching speed increases and decreases.

2) Micromechanics

Nanoindentation technology makes the load change continuously through the control of the program, and measures the depth of the indentation at the same time. It has the characteristics of simple operation, low applied load and high displacement resolution of the monitoring sensor. Therefore, it is in the measurement of the mechanics of polyurethane film materials. It has unique advantages in performance and can test the mechanical properties of polyurethane on the nanometer scale. Carroll [38] et al. used the nanoindentation method to show that reducing the indentation storage modulus of the polyurethane coating is beneficial to increase the shock wave dissipation. The literature on using nanoindentation technology to characterize polyurethane coatings is relatively rare at present, but this technology can be used as a reference for the research of polyurethane in the testing of other materials. Yan Xiaoxin [39] used the nanoindentation test technology to study and showed that with the increase of CB content, the elastic modulus and hardness of CB/PP composites increase. Among them, the elastic modulus increases significantly, and the hardness increase has obvious three-stage characteristics.

5. Electrochemical analysis method

1) Electrochemical impedance spectroscopy
Electrochemical Impedance Spectroscopy (EIS) is an electrochemical testing technique with a wide frequency range and little disturbance to the system. The structure and performance of the sample are analyzed by measuring the impedance shape of the sample under different frequency sinusoidal AC voltages. It is one of the important technical means to evaluate whether the polyurethane coating fails [40]. It is believed that 10Ω·cm2 is the critical value of whether the organic coating has protective performance. Cao Chunan and Zhang Jianqing [41, 42] established 6 different equivalent circuit models to analyze the coating system, and realized EIS data analysis by nonlinear least squares fitting. Liu Jinhe [17] and others used acrylic polyurethane coating to simulate the accelerated test of the desert atmosphere and used EIS technology to find that after the 8-cycle accelerated test, the low-frequency impedance modulus is close to the critical value, and the coating basically loses its protective performance. ,As shown in Figure 4. Song et al. [43] found that multi-walled carbon nanotubes (MWCNTs) can enhance the adhesion of acrylic polyurethane coatings on the surface of carbon steel through EIS technology and delay the peeling rate of the coatings. Zhu et al. [44] used EIS to find that during the aging process of acrylic polyurethane coatings in 3.5% NaCl solution and ultraviolet radiation, the rate of decrease in coating resistance and the rate of increase in coating capacitance and porosity were lower than that of aliphatic polyurethane coatings.

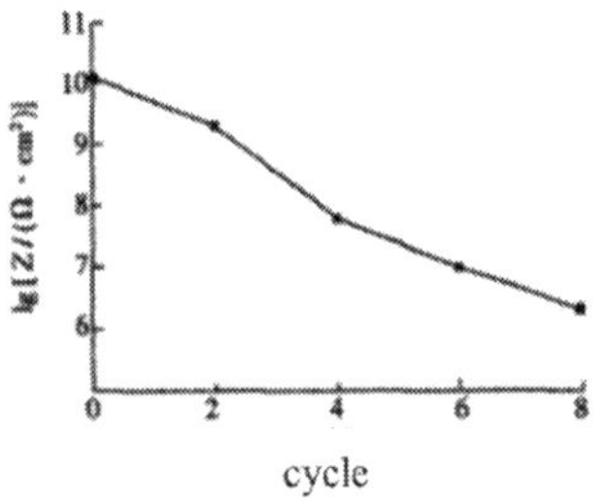

Figure 4 Variation of coating impedance with test cycle

2) Electrochemical noise
Electrochemical Noise Method (ENM) is a technology that uses the phenomenon of random and spontaneous fluctuations of potential or current on the electrode surface to reflect the information of the system from quantitative change to qualitative change. Iverson [45, 46] discovered electrochemical noise for the first time, and it has made great progress in coating protection in recent years. Coating microcracks, coating blistering and changes in ambient temperature are the main factors that cause electrochemical noise. Jamali et al. [47] used the same working electrode to evaluate the anti-corrosion performance of different organic coatings by electrochemical noise and electrochemical impedance spectroscopy, and the results of the two test methods were consistent. Although electrochemical noise technology is relatively rare in the research reports of corrosion-resistant polyurethane coatings, ENM, as an in-situ, non-destructive testing technology, has huge application potential in the research of polyurethane coatings.

6. Thermodynamic analysis method

Thermodynamic analysis is an analytical method for testing the relationship between the properties of sample materials and temperature under program-controlled temperature. In recent years, thermodynamic analysis technology and its development have greatly promoted the in-depth study of polymer coating materials. As the temperature changes, the structure and properties of polyurethane will also change. Therefore, the use of thermal analysis technology to study the evolution of the physical and chemical properties of polyurethane with temperature changes has unique technical advantages. The thermodynamic analysis methods commonly used in the research of polyurethane mainly include differential scanning calorimetry (DSC) and thermogravimetric analysis (TG).
1) Differential scanning calorimetry
Differential scanning calorimetry is a technology that measures the heat flow of the test volume relative to the reference material with temperature under the control of the program. DSC is an important tool for studying the polymerization, curing, cross-linking and decomposition reactions of anti-corrosion and wear-resistant polyurethane coatings. It is mainly used to analyze the change of the glass transition temperature Tg of polyurethane, When Dong [48] studied CB/TPU nano-synthetic coatings, they used DSC observation and found that the glass transition temperature of the coating increased with the increase of carbon black content. Li Tingting [49] found through DSC that the increase of the hard segment content in the polycarbonate-type high-hardness polyurethane elastomer increases the degree of cross-linking of the polyurethane, increases the rigidity of the molecular backbone and affects the molecular motion of the soft polyurethane segment, which significantly weakens the soft and hard segments. The force between the two degrees increases the energy required for the soft segment to reach the activated state, which in turn increases the glass transition temperature of the polyurethane.
2) Thermogravimetric analysis
Thermogravimetric analysis can measure the relationship between the quality of polyurethane and temperature under program control temperature. This method is highly quantitative and can accurately measure the change and rate of polyurethane quality, and understand the thermal stability of polyurethane by observing the weight loss of polyurethane.
When Dong [48] studied CB/TPU nano-synthetic coatings, compared with pure TPU using TGA analysis, when the mass loss of TPU-12CB during thermal decomposition is 5% and 10%, the corresponding temperature increases respectively. Observing the DTG curve at 16.65°C and 13.56°C, it is concluded that the maximum thermal decomposition temperature of TPU-

12CB is 36.32°C higher than that of pure TPU, and the thermal stability of the coating is improved.

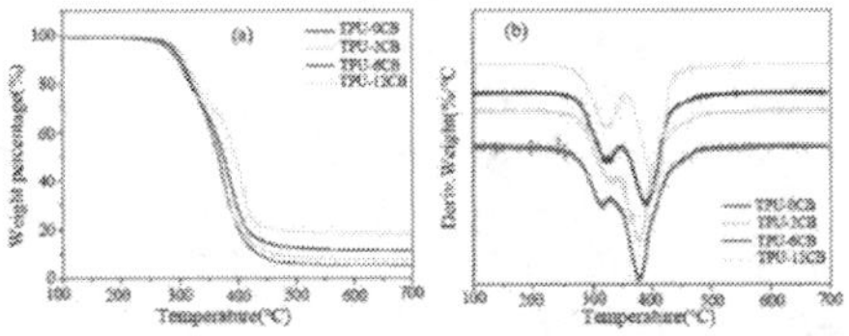

Figure 5 (a) TGA and (b) DTG curve of the tested sample

Guo Mingjie [50] and others have shown that when the content of MWCNTs is about 20wt%, the temperature at which the weight loss of mMMCs/WPU composite reaches 10% of its final value increases by 20°C, and the weight loss reaches half of its final value. The temperature at that time increased by 29°C. After being modified by mMMCs nanocomposite, the thermal stability of WPU has been significantly improved.

7. Outlook

Polyurethane as an excellent anti-corrosion and wear-resistant polymer coating material has important research significance in various industrial fields. This article reviews some of the commonly used characterization methods in the research of anti-corrosion and wear-resistant polyurethane coatings, such as the use of SEM to observe the micro-morphology of polyurethane, analysis of erosion and aging mechanisms; use of Fourier infrared absorption spectroscopy, Raman spectroscopy, and XRD Phase analysis is used to study the reaction process and material modification of polyurethane; the universal testing machine is used to measure the tensile strength and tear strength of the polyurethane material, and then the content of the soft and hard segments of the polyurethane is characterized, etc., the conventional polyurethane characterization method has been developed relatively mature, Its future research work should focus on the following two aspects:

(1) Standardization of electrochemical evaluation standards. As an efficient characterization method for analyzing the structure and performance of polyurethane, electrochemical characterization technology has not yet established a standardized and unified evaluation standard, which is not conducive to the exchange and comparison of scientific research and production. Therefore, the establishment of standardization and uniformity for different purposes of polyurethane The standard of electrochemical evaluation is of great significance.

(2) Characterization of polyurethane in service state. Polyurethane as a high molecular polymer has a certain sensitivity to the service environment. Humidity and temperature have a great influence on its physical and chemical properties. Conventional characterization methods cannot perform microstructure characterization under the stress state of the material and the service environment. Therefore, it can Characterization in the service state of polyurethane is an important research direction to improve the characterization method.

References

1. RAJPUT S D, HUNDIWALE D G, MAHULIKAR P P, et al. Fatty acids based transparent polyurethane films and coatings[J]. Progress in Organic Coatings,2014,77(9):1360-1368.
2. SCRINZI E, ROSSI S, DEFLORIAN F, et al. Evaluation of aesthetic durability of waterborne polyurethane coatings applied on wood for interior applications[J]. Progress in Organic Coatings, 2011,72(1/2):81-87.
3. ZHANG J, TU W, DAI Z. Synthesis and characterization of transparent and high impact resistance polyurethane coatings based on polyester polyols and isocyanate trimers[J]. Progress in Organic Coatings, 2012,75(4):579-583.
4. SUN WEI, XING CHAO, TANG XIAOBO, et al. Comparative study on the degradation of a zin-rich epoxy primer/acrylic polyurethane coating in different simulated atmospheric solutions. J.Coat.Technol.Res.,2021,18(2):397-413.
5. ESLAMI R, BAGHERI R, HASHEMZADEH Y, et al. Optical and mechanical properties of transparent acrylic based polyurethane nano Silica composite coatings[J]. Progress in Organic Coatings, 2014, 77(7):1184-1190.
6. JOKT-KORPELA F, KARVINEN J, PAIVANR ANTA B, et al. Hydrophobic and oleophobic anti-reflective polyacrylate coatings[J]. Microelectronic Engineering, 2014,114:38-46.
7. XU KAIXIANG, YANG JIANJUN. The research situation and the progress of functional waterborne polyurethane building coatings [J]. Polyurethane Industry, 2016, 31(2):1-4.
8. JIN ZHENG, QINGHUA ZHANG, ZHENHUAN LUO, et al. Progress on the protective coating layer materials of wind turbine blades[J]. Polymer Materials Science and Engineering, 2012(11):182-186.
9. WANG SHULUN, ZHAI WEN, TANG ZHIYONG, et al. Preparation and properties of the new ageing resistance polyurethane elastomer[J]. Polyurethane Industry, 2008, 23(4):24-27.
10. ZHANG WEIGANG, LU MIN. Effect of graphene on salt water resistance of polyurethane/al composite coating[J]. Surface Technology, 2019, 48(03): 163-167.
11. QIAO LONGYUN. Research on enhancing corrosion resistance of polyurethane coating by nano-container packaging corrosion inhibitor[D]. Huazhong University of Science and Technology, 2017.
12. LU GANG. The present situation and development trend of polyurethane coating[J]. Chemical Industry, 2013,31(01):23-26.
13. LING YAN, ZHONG JIAOLI, TANG XIAOSHAN, et al. The principle and application of scanning

electron microscope[J]. Shandong Chemical Industry, 2018, 47(09):78-79+83.

14. ZHOU GUANJUN. Discussion on detection techniques for seal failure of airplane fuel tanks[J]. Failure Analysis and Prevention,2020,15(01):62-68.
15. ZHONG PIN, PENG ENGAO, LI JIAN, et al. Study of erosion behavior of polyurethane-urea[J]. Tribology, 2007(05):447-450.
16. GAO JIN, ZHANG YONG, WANG JIAN, et al. Study on coatings erosion of wind turbine blades under sand-carrying wind[J]. Acta Energeia Solaris Sinica,2020,41(07):367-371.
17. LIU JINHE, ZHANG YAQIN, LIU MUYI, et al. Study on accelerated test of organic coating in desert atmospheric environment[J]. Surface Technology, 2014,43(04):64-67+96.
18. HAO YUNHONG, LI JIE, TIAN XULE, et al. Effect of temperature aging on wind erosion resistance of steel structure coatings[J]. Journal of Building Materials, 2020,23(02):407-413.
19. YU LIFANG, YANG ZHIJUN, ZHOU YONGZHANG, et al. Overview of scanning electron microscope and environmental scanning electron microscope in the field of geosciences[J]. Postgraduate Journal of Sun Yat-sen University (Natural Science, Medical Edition), 2008(01):54-61.
20. LI XU, ZHANG YUNSHENG, ZUO YAN, et al. Wind-blown sand resistant polyurethane acrylate coatings for aerospace transparency[J]. Polyurethane Industry,2016,31(06):40-43.
21. CUI XIAOFEI, TAN XIAOMING, WANG DE, et al. Assessment of aging performance of polyurethane coating for 7B04 al-alloy with an accelerated testing spectrum[J]. Journal of Chinese Society for Corrosion and Protection,2018,38(01):74-80.
22. HAYES, P.A., VAHUR, S., LEITO, I. ATR-FTIR spectroscopy and quantitative multivariate analysis of paints and coating materials[J]. Spectrochim Acta A Mol Biomol Spectrosc,2014,133: 207-213.
23. POPA, A.M., ANGELONI, S., BÜRGI, T., et al., Dynamic perspective on the function of thermoresponsive nanopores from in situ AFM and ATR-IR investigations[J]. Langmuir, 2010, 26(19):15356-15365.
24. ZHANG, W.R., ZHU, T.T., SMITH, R., et al., An investigation on the melamine self-condensation in polyester/melamine organic coating[J]. Progress in Organic Coatings, 2010, 69(4):376-383.
25. REN HAIPING, WANG JIANYU, LI JIAGE, et al. A novel method for the wavelength measurement of infrared light source[J]. China Pharmaceutical Affairs,2010,24(08):807-808+811.
26. LIU GANG, OU BAOLI, ZHAO XINXIN, et al. Preparation of porous nano sio2/graphene oxide modified polyurethane coating materials[J]. Journal of Ceramics,2020,41(04):531-536.
27. YANG QI. The application of infrared spectrum analysis in coating failure mechanism [J]. Aviation Maintenance & Engineering,2015(12):84-87.
28. ZHANG, W.R., ZHU, T.T., SMITH, R., et al. A non-destructive study on the degradation of polymer coating II: Modelling of degradation depth profiles[J]. Polymer Testing,2012, 31 (8):1100-1104.
29. ZHANG, W.R., ZHU, T.T., SMITH, R., et al. A non-destructive study on the degradation of polymer coating I: Step-scan photoacoustic FTIR and confocal Raman microscopy depth profiling[J]. Polymer Testing,2012,31(7):855-863.
30. YAMANAKA, T., NAKAGAWA, H., TSUBOUCHI, S., et al. In situ Raman spectroscopic studies on concentration change of electrolyte salt in a lithium ion model battery with closely faced graphite composite and LiCoO2 composite electrodes by using an ultrafine microprobe[J]. Electrochimica Acta, 2017, 234: 93-98.
31. HAO YUNHONG, MA SIHAN, LI JIE, et al. Study on sand erosion resistance of graphene oxide/polyurethane coatings[J/OL]. Journal of Building Materials:1-14[2021-01-06].
32. http://kns.cnki.net/kcms/detail/31.1764.TU.20200413.1957.006.html.
33. LV JINXING, SU HUI1, YAO ZHIXIANG. et al. On - line monitoring of —NCO percentage in polyurethane prepolymer during polymerization via raman spectroscopy [J]. Paint & Coatings Industry,2015,45(08):57-60+66.
34. YE YINPING, ZHANG YONGSHENG, SUN XIAOJUN, et al. Preparation and tribological behavior of Ni / polyurethane nanocomposite coatings [J]. Tribology,2003(02):104-107.
35. SUN WEI, FAN JINJUAN. Research progress on the solid particle erosion of polyurethane coatings[J]. Equipment Environmental Engineering,2020,17(12):67-75.
36. SHUAI LI, JUN ZHANG, JIANJUN CHEN. et al. Synthesis and characterization of polycarbonate-derived shape memory polyurethane[J]. Polymer Materials Science and Engineering,2019,35(01):1-6.
37. YANG YILIN, LU XUN, WANG WEIWEI, et al. Preparation and characterization of thermally reversible self-healing polyurethane elastomer[J]. Journal of Materials Engineering,2017,45(08):1-8.
38. WANG XINGGANG, HE XIAORONG. Study on synthesis and mechanical properties of new type butyl polyurethane elastomer[J]. Science and Technology Innovation,2020,33:29-30.
39. A.O'CARROLL, M.HARDIMAN, E.F.TOBIN, T.M.YOUNG. Correlation of the rain erosion performance of polymers to mechanical and surface properties measured using nanoindentation[J]. Wear,2018,8(412-413):38-48.
40. YAN XIAOXIN, CAI HONGLEI, ZHANG ANJIANG, et al. Study on the mechanical properties of carbon black/polypropylene composites based on

nanoindentation technology[J]. Polymer Bulletin,2020(06):35-42.

41. ZHOU LIXIN, CHENG JIANG, YANG ZHUORU, et al. Methods for study and evaluation of anticorrosion performance of organic coatings[J]. Corrosion Science and Protection Technology, 2004(06):375-377+380.
42. ZHANG JIANQIANG, CAO CHUNAM. Electrochemical impedance spectroscopy data analysis by using a mathematical model for impedance of irreversible electrode reaction[J]. Journal of Chinese Society of Corrosion and Protection,1993(01):27-34.
43. ZHANG JIANQIANG, SUN GUOQING, CAO CHUNAM. EIS data analysis for evaluation of protective property of organic coatings[J]. Corrosion Science and Protection Technique,1994(04):318-325.
44. SONG,D.,YIN,Z.,LIU,F., et al. Effect of carbon nanotubes on the corrosion resistance of water-borne acrylic coatings[J]. Progress in Organic Coatings, 2017, 110: 182-186.
45. ZHU, Y., XIONG, J., TANG, Y., et al., EIS study on failure process of two polyurethane composite coatings[J]. Progress in Organic Coatings, 2010, 69 (1): 7-11.
46. ZHANG JIANQING, ZHANG ZHAO, WANG JIANMING, et al. Analysis and application of electrochemical noise I theory of electrochemical noise analysis[J]. Journal of Chinese Society of Corrosion and Protection,2001(05):55-65.
47. ZHANG JIANQING, ZHANG ZHAO, WANG JIANMING, et al. Analysis and application of electrochemical noise technology II . application of electrochemical noise[J]. Journal of Chinese Society of Corrosion and Protection,2002(04):50-57.
48. JAMALI, S.S., MILLS, D.J., SYKES, J.M., Measuring electrochemical noise of a single working electrode for assessing corrosion resistance of polymer coated metals[J]. Progress in Organic Coatings, 2014, 77 (3): 733-741.
49. MENGYAO DONG, QIANG LI, HU LIU, et al. Thermoplastic polyurethane-carbon black nanocomposite coating: Fabrication and solid particle erosion resistance [J]. 2018,
50. LI TINGTING, LV XIAOJIAN, SHEN ZHAOYU, et al. Synthesis and properties of polycarbonate-type polyurethane elastomer with high hardness[J]. Thermosetting Resin,2021,36(01):14-18.
51. GUO MINGJIE. Study on Preparation and Performance of Inorganic Nanomaterials Modified Polyurethane [D]. South China University of Technology,2019.

https://doi.org/10.1051/matecconf/202235801046

Preparation and Characterization of High Temperature Resistant Polyimide Films

Zhirui Li, Yong Hou, Pengchang Ma and Wenhua Zhao*

Zhongshan Polytechnic, Zhongshan, China

Abstract. To prepare high temperature resistant polyimide thin films, we have discussed the synthesis of high-purity diamine precursors in this study. We have prepared polyamide acid solutions first via two-step solution-based polymerization of heat-resisting diamine and dianhydride, and then obtain polyimides after dehydration during a high temperature curing process. The thermal, mechanical and hygroscopic properties of four polyimides thin films were tested and compared.

1. Introduction

Polyimides (PI) have been widely used. For example, they can be used as the flexible substrate of thin-film solar cells in the form of thin films [1-3]. They can also be used in the form of high temperature protection coatings for aerospace engineering [4, 5]. It has been the focus of researchers to synthesize polyimides with excellent comprehensive properties. Via molecule structural design, researchers discovered that the introduction of certain functional groups, such as imidazole ring, oxazole ring, pyridine, etc., can enhance the thermal endurance of polyimides [6].
To prepare high temperature resistant polyimide thin films, we have discussed the synthesis of high-purity diamine precursors in this study. We have prepared polyamide acid solutions first via two-step solution-based polymerization of heat-resisting diamine and dianhydride, and then obtain polyimides after dehydration during a high temperature curing process. The thermal, mechanical and hygroscopic properties of four polyimides thin films were tested and compared.

2. Experimental

As shown in Figure 1, the common method to produce polyimide thin films in the lab is a two-step procedure: first, under the protection of inert gases such as N2, diamine and dianhydride undergo low temperature polymerization to polyamide acid (PAA) in polar aprotic solvents (DMAc, DMF, etc.) under even stirring; then PAA is applied onto a substrate via a coater, forming films of 200~500 um in thickness; the coated substrate is transferred to a hot plate and allowed to slowly pre-dry at 30~60℃ in order to ensure the smoothness of the film; finally, it is transferred to a vacuum oven and heated in multiple stages under the protection of inert gases such as N2, in which hot pressing amination and cyclodehydration take place (the amination temperatures of different thin films vary between 300 and 400℃ to ensure complete dehydration and optimal film properties).

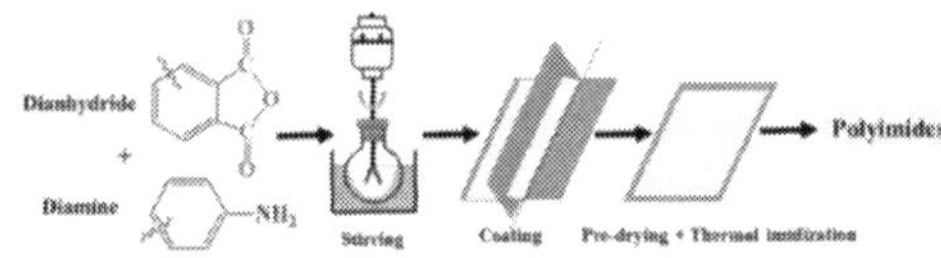

Figure 1. Schematics of the classic two-step synthesis of polyimide thin films.

3. Results and discussions

According to previous reports, the characteristic peaks of the vibration absorption bands of aromatic imides are around 1780 cm-1, 1720 cm-1, 1380 cm-1 and 725 cm-1 [82], for the four polyimides The infrared transmission experiment of the film is carried out, and the infrared spectrum obtained is shown in Figure 2. The four films are all at 1780 cm-1 (respectively at 1778 cm-1, 1176 cm-1, 1176 cm-1, and 1777 cm-1.), 1720 cm-1 (respectively at 1731 cm-1, 1728 cm-1, 1737 cm-1, 1730 cm-1) and 725 cm-1 (respectively at 735 cm-1, 742 cm-1, 727 cm -1, 733 cm-1) near the first strong peak of the carbonyl C=O in the imide ring, which is caused by the large dipole moment of the carbonyl group, the stretching and bending vibration of the carbonyl group; Near 1380 cm-1 (respectively at 1380 cm-1, 1377 cm-1, 1382 cm-1, 1374 cm-1), there are more CN stretching vibration peaks in the imide ring, all at 3480 cm-1 (The stretching vibration absorption peaks of NH appear near 3380 cm-1, 3483 cm-1, 3495 cm-1, 3482 cm-1, respectively. The existence of these characteristic absorption peaks proves the aromaticity in the prepared polyimide film. The successful synthesis of imide groups, namely the dehydration and ring closure of polyamic acid. The infrared spectrum, on the other hand, can also reflect the

* Corresponding author: 63614802@qq.com

effect of the curing temperature we set on the degree of thermal imidization.

In the infrared spectra of the above four films, we did not find the characteristic peak of the intermediate polyamic acid solution. Generally, carbonyl-C=O (carboxy-COOH) has a characteristic absorption peak at 1720~1706 cm-1, and CO (The carboxyl-COOH) has a characteristic stretch peak at 1320~1210 cm-1. From this perspective, it can be seen that the intermediate PAA solution has a high degree of imidization. This result also verifies the rationality of the laboratory setting curing temperature (BPDA-PPD, PMDA-ODA, 350 ℃ ; BAPBBI-BPDA, BAPBBOA-BPDA, 370 ℃).

Compared with the infrared spectrum of the BPDA-PPD film, we can see that the PMDA-ODA film exhibits unique strong peaks at 1166 cm-1 and 1120 cm-1. This is because the molecular chain skeleton structure of the PMDA-ODA film Containing COC bond, that is, the peak generated by the deformation and stretching vibration of the -O- bond, is not in the infrared spectrum of the BPDA-PPD film; compared to the polyimide film containing bibenzoxazole groups, it can be seen The infrared spectrum of the polyimide film containing bibenzimidazole group has a relatively strong peak at 3610 cm-1. This peak is due to the hydrogen bond formed between the NH bond molecules in the diamine structure. And the resulting peaks correspond to the infrared spectra of the two diamine monomers we are analyzing.

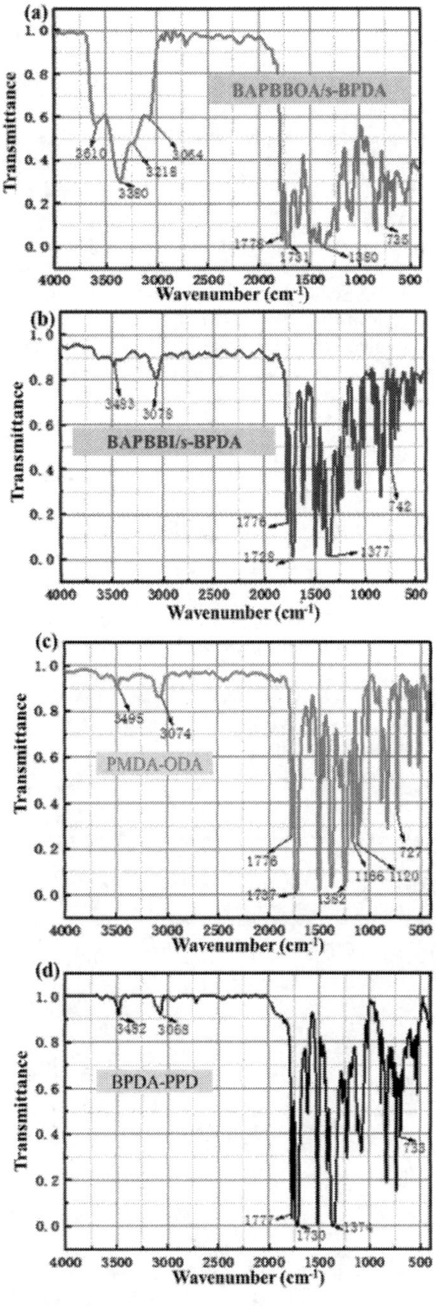

Figure 2. (a)- (d) FTIR spectrums of BAPBBI / s-BPDA, BAPBBOA / s-BPDA FTIR spectrum of PMDA / ODA, and BPDA / PPD thin films.

 https://doi.org/10.1051/matecconf/202235801046

As mentioned above, when used as the flexible substrate of CIGS thin-film solar cells, polyimide thin films need to meet high standards in thermal stability and resistance. The thermal stability of polymers is usually represented by Td in chemistry, and obtained via analyzing thermogravimetric curves experimentally. Due to the great challenge in finding the precise initial temperature drop point, Td5% is often used. Tg is often used to represent the thermal resistance in physics, which is generally obtained via analyzing the peak point of DMA or TMA curves (glass transition temperature). In this part, the thermogravimetric curves are collected with TGA (). The details are as follows: during the TGA tests, the thin film samples are cut to strips (5-10 mg) and placed in crucibles; set the measurement parameters, with temperature at 25-800 °C, ramp rate at 10 °C / min under N2 protection. The detailed result analysis is as follows:
From the TGA curves of four types of polyimide thin films, it can be clearly seen that film containing dibenzimidazole groups has Tg5% of 581 ℃, film containing dibenzoxazole groups of 585 ℃. The synthesized DuPont and Ube films in the lab have lower thermal stability, with Tg5% of 505 and 560 ℃, respectively. It is apparent that polyimides with dibenzimidazole and dibenzoxazole groups, which have rigid molecular chain skeleton and strong intermolecular forces, have relatively higher thermal stability and can basically meet the requirement for temperature in the production of thin-film solar cells.

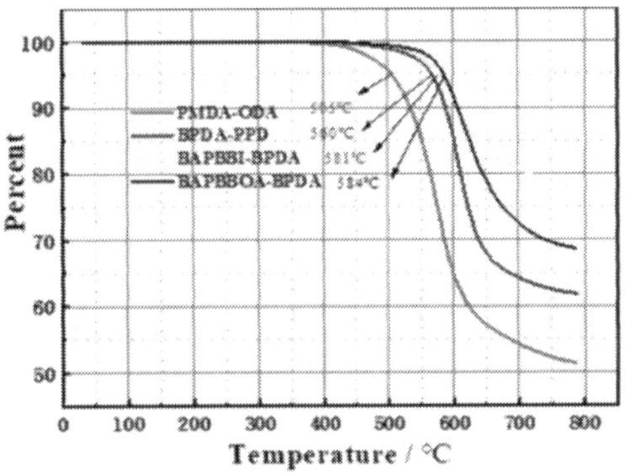

Figure 3. TGA curves of four types of polyimide thin films.

In the process of preparing thin-film solar cells, the polyimide thin-film substrate also needs to meet the physical heat resistance, which can be a measure of the time duration of the heating temperature of the thin-film material in the process of preparing the solar cell substrate. The higher the heat resistance of the polyimide film, the more it can withstand the long-term high-temperature environment preparation process, which plays an important role in the production and preparation of solar cells. In this part, the Q800 of American TA Company is used for dynamic mechanical analysis. The specific experimental steps are as follows: In the dynamic mechanical analysis (DMA) of two polyimide films containing bibenzimidazole and oxazole groups, first cut Take a 6.35 mm×6.35 mm (length×width) polyimide film with a thickness of about 30 um; then set the measurement parameters, select the stretching mode to measure, the temperature range is 25-500 ℃, the heating rate is 5 ℃ / min, nitrogen atmosphere protection. The specific experimental results are analyzed as follows.
It can be seen from Figure 4 that the glass transitions of bibenzoxazole and imidazole polyimide films are around 339 ℃ and 421 ℃ respectively. Both polyimide films have higher energy storage energy. At the glass transition point, the change in energy storage energy is caused by the rapid movement of the molecular chain. The existence of the biaromatic ring makes the molecular chains of the bibenzimidazole group and the bibenzoxazole group have higher rigidity. At the same time, the N atom in the bibenzimidazole diamine and the bibenzoxazole diamine The O atoms can form hydrogen bonds between molecular chains and within molecular chains to further enhance the intermolecular force, which explains the higher heat resistance of polyimide films containing these two functional groups.

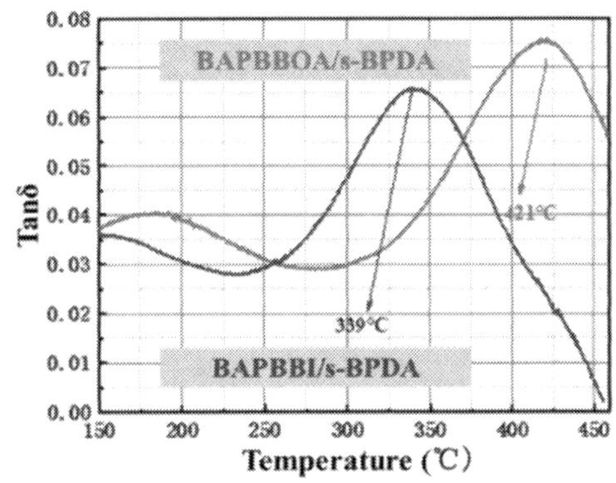

Figure 4. TMA curves of dibenzimidazole and dibenzoxazole thin films.

4. Conclusion

In this study, we synthesize four polyimide thin films, two of which are commercial ingredients. The other two high temperature resistant polyimide films are prepared in our lab with high temperature resistant diamine precursors and diphenyl tetracarboxylic anhydride via a two-step solution-based method, that is, polycondensation followed by cyclodehydration. The FTIR spectra, mechanical, thermal and water absorption properties of four films are characterized and the results show: Polyimide thin films made in the lab via reaction with diamine precursors (BAPBBI, BAPBBOA) and diphenyl tetracarboxylic anhydride possess excellent thermal properties, with Td5% of 581 and 584 °C and Tg of 421 and 339 °C for BAPBBI-BPDA and BAPBBOA-BPDA, respectively, outperforming commercial thin films made in the lab.

Acknowledgments

This work was financially supported by the Science and Technology Major Project of Guangdong Provincial Education Department No. 2019GZDXM009, the Science and Technology Major Project of Guangdong Provincial Education Department No. 2020ZDZX2081， Provincial special funds for science and technology (Major Project) of Guangdong Province No. 2020sdr007 and the Science and Technology Major Project of Zhongshan Polytechnic No. KYA2002.

References

1. P. Ma, W. Li, C. Yi, C. Dai, H. Luo, and C. Yang, "Investigation of Mo films deposited on high temperature polyimide substrate by magnetron sputtering for flexible CIGS thin film solar cells application," AIP Advances, 9, (2019) 045024,.
2. S. Shi, L. Yao, P. Ma, Y. Jiao, X. Zheng, D. Ning, M. Chen, F. Sui, H. Liu, and C. Yang, "Recent progress in high temperature resistance PI substrate with low CTE for CIGS thin film solar cells," Materials Today Energy, (2021) 100640.
3. C. Yi, W. Li, S. Shi, K. He, P. Ma, M. Chen, and C. Yang, "High-temperature-resistant and colorless polyimide: Preparations, properties, and applications," Solar Energy, 195, (2020) 340-354.
4. P. Ma, C. Dai, H. Wang, Z. Li, H. Liu, W. Li, and C. Yang, "A review on high temperature resistant polyimide films: Heterocyclic structures and nanocomposites," Composites Communications, 16, (2019) 84-93.
5. A. Périchaud, R. Iskakov, A. Kurbatov, T. Akhmetov, O. Prokohdko, I. V. Razumovskaya, S. L. Bazhenov, P. Apel, V. Y. Voytekunas, and M. Abadie, "Auto-reparation of polyimide film coatings for aerospace applications challenges & perspectives," in High Performance Polymers–Polyimides Based–From Chemistry to Applications, (2012) 215-244.
6. T. Okada, R. Ishige, and S. Ando, "Effects of chain packing and structural isomerism on the anisotropic linear and volumetric thermal expansion behaviors of polyimide films," Polymer, 146, (2018) 386-395.

Research on Photo Catalytic Performance of TiO_2 Film Doped with Li^+

Jing Zhang [1,*], Zhengxing Huang [1], Liming Zheng[1], Xiaobo Tian [1], Wengang Yin [1] and Shihua Ding [2]

[1] Armed Police College of CAPF, Chengdu, China
[2] School of Materials Science and Engineering, Xihua University, Chengdu, China

Abstract. Based on sol-gel method and spin coating process, TiO_2 film doped with Li^+ is prepared by utilizing analytical pure butyl titanate, lithium nitrate, absolute ethanol, glacial acetic acid, acetylacetone and deionized water as raw material. The Li^+ ion which has equal radius as Ti^{4+} ion is chosen in experiments for doping and modification. Experimental results show certain increase of catalytic efficiency in TiO_2 film doped with Li^+. However, the increase level of catalytic efficiency gradually shows some decrease when the dopant amount changes from 0.025mol to 0.1mol. Catalytic efficiency is raised to 25.7% when the dopant amount is 0.025mol.

1. Introduction

In 1970s, A.Fujishima etc. found that continuous redox reaction could occur on the surface of irradiated TiO_2[1], since then, researches on photo catalytic materials represented by TiO_2 are widely launched. Titanium dioxide is extensively used in areas of environment protection, sensing, solar battery, etc. since its excellent photo catalytic and photoelectric characteristics [2~5]. Titanium dioxide film produced by different process or method has diversified performances in aspects of structure, component, width, orientation, etc[6~7]. Sol-gel method [8] needs simple equipment and low reaction temperature. No vacuum condition is required. Moreover, the obtained film has advantages such as uniform composition, high purity and low difficulty of doping. Therefore, sol-gel is a usual preparation method of TiO_2 film. $Ti(OR)_4$ is used as the main raw material, which is then changed into sol by hydrolysis and polycondensation reaction in the organic medium. The above-mentioned sol is then coated on basal body by spin coating or pulling method. The leftover organic material and water are eliminated by dry roasting. At last, TiO_2 film is obtained. Currently, low efficiency of photo catalytic is the main obstacle which blocks nano-scale TiO_2 film spread and application. By producing film after ion doping in nano-scale TiO_2 film, efficiency of photo catalytic and visible light activity are effectively increased[7,9].

2. Experimental sample preparation

Sol-gel technique is adopted in our experiments. Analytical pure butyl titanate, lithium nitrate, absolute ethanol, glacial acetic acid, acetylacetone and deionized water are prepared as raw material. Homogeneous canary yellow sol which is transparent is produced. The PH value of prepared sol ranges from 4.8 to 5.2. In the system of $(C_4H_9O)_4Ti\text{-}C_2H_5OH\text{-}H_2O$, alkoxide of titanium has hydrolysis reaction with water and $Ti(OH)_x(OR)_y$ monomer is then formed. Equation 1 shows related reaction.

$$Ti(OR)_4 + XH_2O \rightarrow Ti(OH)_x(OR)_{4-x} + XROH(x<4) \quad (1)$$

The generated $Ti(OH)_4$ coalescences and forms amorphous structure. Poly condensation reaction is then occurred in acid solution.

$$\text{-Ti-OH} + \text{OH-Ti-} \rightarrow \text{-Ti-O-Ti-} + H_2O \quad (2)$$

or

$$\text{-Ti-OH} + \text{RO-Ti-} \rightarrow \text{-Ti-O-Ti-} + ROH \quad (3)$$

Titanium dioxide is obtained after organic material and water molecule are removed in high-temperature process. Theoretically, the mole stoichiometric ratio of butyl titanate and water is $n[Ti(OR)_4/H_2O]=1:4$. In our experiments, water is overdosed compared with butyl titanate, so as to promote full hydrolysis of butyl titanate. The preparation technique and process of doped ionic sol is shown in Figure 1. According related operation steps, doped ionic sol will be produced. A solution： 1/3 absolute anhydrous ethanol is mixed with butyl titanate and then stirred strongly for 10 min. Acetyl acetone and glacial acetic acid are gradually dripped in at the same time. The mixture is stirred for 10 min. Certain additive is mixed together and stir is going on. B solution： 2/3 absolute anhydrous ethanol is mixed with deionized water and ions for doping, the mixture is stirred strongly for 30 min. Thereafter, B solution is slowly added into A solution and stirred strongly for another 2h. Doped ion sol

* Corresponding author: 13408519980@163.com

is then obtained and placed in dry clean sealed container. Ageing happens in room temperature of 25°C.
Single crystal polishing silicon wafer is adopted in our experiments as base material. Gel film is then produced by spin coating method is applied after the sol which has 48h ageing. After heat treatment, the solvent in the gel was evaporated, and polymerization between colloidal particles occurs. Finally, the required TiO_2 film sample is produced.

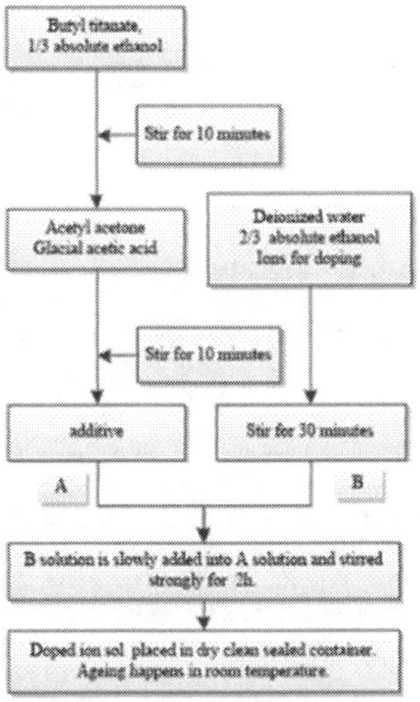

Figure 1. Process of Sol preparation with doping

3. Test and analysis

3.1 Photo catalysis performance of pure TiO_2 film

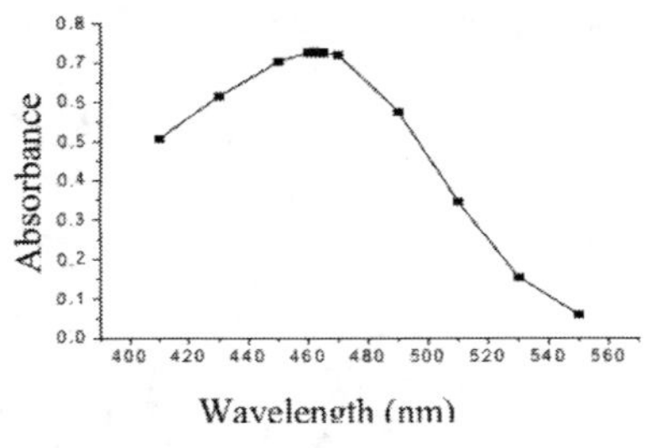

Figure 2. Absorbance of methyl orange

Photo catalysis activity of TiO_2 film is expressed by ultraviolet visible spectrophotometry in experiments. Unico WFJ7200 spectrophotometer is used. First, Absorbance curve of methyl orange with concentration of 10mg/L under different visible wavelength ranges from 400nm to 550nm is tested by utilizing ultraviolet visible spectrophotometry. The absorption wavelength in the range of visible light of adopted sample methyl orange solution is determined in photo catalytic activity test. Figure 2 shows related results. In the range of 400nm to 550nm, absorbance of methyl orange increases first and then decreases. At the 462nm point, maximum absorbance appears. Therefore, in experiment for testing degradation rate of photo catalytic degradation of methyl orange, wavelength of visible spectrophotometer is fixed to 462nm. In our experiments, prepared methyl orange solution is placed in darkroom. Photo catalytic test is applied after 30 min.
Photo catalytic degradation of methyl orange is applied in obtained pure TiO_2 film, the concentration of methyl orange solution is 10mg/L. 20W ultraviolet lamp which is the only light source in degradation experimental environment is utilized as ultraviolet light source. The 20mm×20mm film sample is then placed into 10ml methyl orange solution. The film faces ultraviolet lamp and methyl orange solution is stirred continually.
Photo catalytic performance test is applied to pure TiO_2 film which is heat-treated in different temperature from 500°C to 800°C in experiments. Experimental results are given in Figure 3. Maximum value of degradation rate is 15.9% and appears at 600°C. Value of degradation rate at 500°C, 700°C and 800°C are 14.1%, 12.6% and 12.2%, respectively. Heat treatment with 600°C are used in photo catalytic experiment.

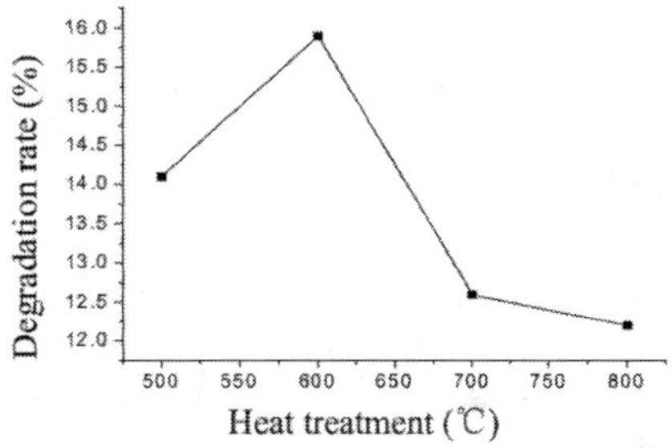

Figure 3. Photocatalytic performance of film with different heat treatment temperatures

3.2 Photo catalytic performance of TiO_2 film doped with Li^+ ion

Modification study on TiO_2 film is implemented. Effective metal ion doping should require two conditions[10]: First, dopant could capture electrons and holes at the same time for the effective separation between the two. Second, the captured electrons and holes can be released and drifted to reaction interface. The Li^+ ion which has similar radius as Ti^{4+} ion is chosen in experiments for doping.

Table 1. Parameters of doping ions

Doping ions	Coordination number	Ionic radius(Å)
Li^+	6	0.76

The influence of different dopant amount on TiO_2 photo catalytic performance is studied. Parameters of doped ion are given in Tab.I. Ti^{4+} ion is 6-coordinate with ion radius of 0.605Å. The selected Li^+ ion has a 6-coordinate radius of 0.76Å.

Photo catalytic experiment is applied on Li^+ doped TiO_2 film. The relation between UV irradiation time and degradation rate is shown in Figure 4. Li^+ dopes 0.025mol with UV irradiation 30 min and degradation rate is 11.8%. Degradation rate changes to 19.8%, 22.7% and 25.7% when irradiation time changes to 60 min, 90 min and 120 min, respectively. It is thus clear that as long as illumination time extends, catalytic efficiency increases evidently. Under comprehensive consideration of experiment and time, illumination time of photo catalytic experiment is fixed to 120 min.

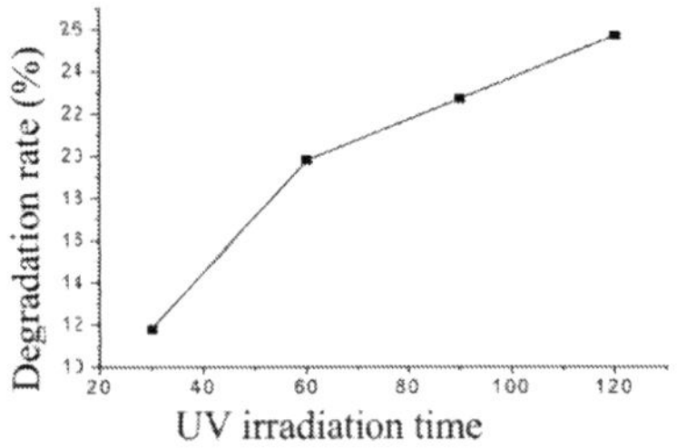

Figure 4. Photocatalytic performance of Li+ doped films at different times

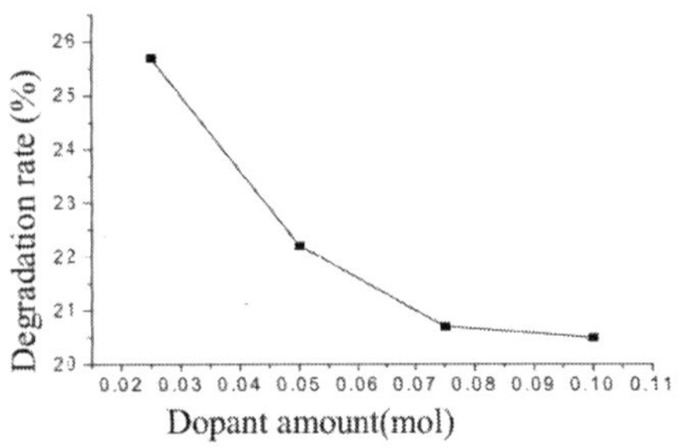

Figure 5. Photocatalytic performance of Li+ doped films at different qulity

Photo catalytic experiment is applied on Li^+ doped TiO_2 film. 20W ultraviolet irradiation is applied for 120 min. As shown in Figure 5, compared with film which has no dopant, the catalytic rate raises from 15.9% to a larger value. However, as dopant amount rises from 0.025mol to 0.1mol, the increase level of catalytic efficiency gradually shows some decrease. Catalytic efficiency is 25.7%, 22.2%, 20.7% and 20.5% when dopant amount is 0.025mol, 0.05mol, 0.075mol and 0.1mol, respectively.

By utilizing bede 01 X-ray diffraction(XRD) made in UK, XRD test and analysis are applied on the film. Related parameters are set as follows: λ=1.5406Å, scanning speed is 18°/min, step is 0.06°. XRD images of TiO_2 film doped with Li^+ are given in Figure 6. As dopant amount changes from 0.025mol to 0.1mol, second phase does not appear. After Li^+ ions are mixed with TiO_2, two kinds of status usually exist: Li^+ ions enter into crystal lattice of TiO_2 and interact with TiO_2 crystal lattice. The interaction results in micro local defects in TiO_2 crystal lattice and lattice field distortion occurs near the defects.

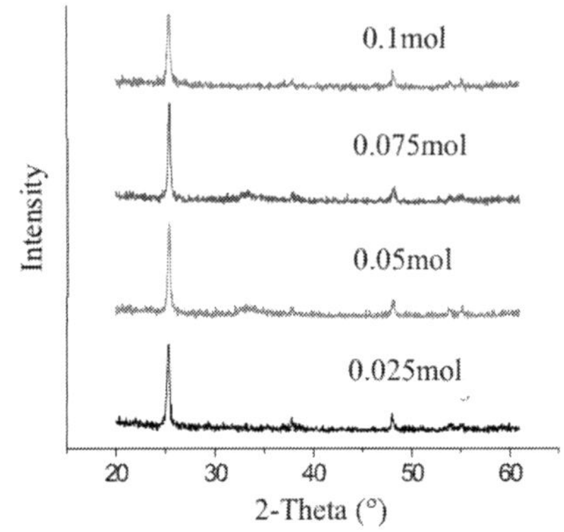

Figure 6. XRD of Li+ doped films at 600°C heat treatment

Under light conditions, more electrons are inspired to conduction band, number of electron-hole pairs increases, so photo catalytic activity of TiO_2 increases. In the other case, Li^+ ions exist between gaps and become traps for capturing electrons or holes. The recombination of electrons and holes is restrained so as to enhance photo catalytic activity of TiO_2.

The 6-coordinate radius of Li^+ and Ti^{4+} ions is 0.76Å and 0.605Å, respectively. Since Li^+ ions have larger radius, they are more likely to exist in the form of interstitial ions. However, electron hole concentration of valence band increases as Li^+ ions are excessive, meanwhile, surface potential barrier is reduced, transition layer is thickened, the ability for space charge layer to separate electron-hole decreases, distance between trap bits for trapping carrier becomes smaller, the doped ions evolve into reunite center of electrons and holes, which increases carrier recombination rate and decreases photo catalytic degradation efficiency. Thus catalytic efficiency moves down when amount of Li^+ ions rises up, as shown in Figure 5.

4. Conclusions

Photo catalysis activity of TiO_2 film is expressed by ultraviolet visible spectrophotometry. The wavelength is fixed to 462nm. Configured methyl orange solution with concentration of 10mg/L is placed in darkroom for 30min. After that, photo catalytic experiment is implemented. TiO_2 film with dopant Li^+ ions is then under heat-treatment of 600°C. Compared with TiO_2 film with no dopant, TiO_2 film with dopant Li^+ ions increases catalytic rate of methyl orange solution. However, the increase level of catalytic efficiency gradually shows some decrease when the dopant amount changes from 0.025mol to 0.1mol. The reason is excessive Li^+ ions enhance carrier recombination rate, which decreases photo catalytic degradation efficiency.

Acknowledgments

This work was financially supported by the research of "Real-time Digital Camp Security Monitoring System Based on Internet of Things" No. 2017FZ0011.
This work was supported by the research of Automatic target training and assessment system of explosion-proof projectile throwing based on UWB technology No. 2021KJ01.

References

1. Fujishima A, Honda K. Electrochemical photolysis of water at a semiconductor electrode [J]. Nature,1972,238(5358):37-38.
2. Anpo M, Takeuchi M. The design and development of highly reac-tive titanium oxide photocatalysts operating under visible light irra-diation[J]. Journal of Catalysis, 2003, 216(1): 505-516.
3. Fujishima A, Rao T N, Tryk D A. Titanium dioxide photocataly-sis [J]. Journal of Photochemistry & Photobiology C: Photochemistry Reviews, 2000, 1(1): 1-21.
4. Ma Ke, Zhu Yan, et a1. The preparation technique of nanometer TiO2 film and the application in environment protection. Applied Chemical Industry. 2012,(08):1426.
5. O'Regan B, Grätzel M.A low-cost, high-efficiency solar cell based on dye-sensitized colloidal TiO2 films[J]. Nature, 1991, 353(6346): 737-740.
6. Xiao Huaqing, Wang Kunping, Qiu Junjie, et a1. The Preparation and Characterization of TiO2 Film Prepared by a Sol-gel Method [J]. Journal of Chongqing University of Technology (Natural Science), 2018(11):93—99.
7. Zhang Jun, Dong Jie, et al. Preparation of Fe3+-doped TiO2 thin film on glass and its photocatalytic degradation performance for methyl orange. Electropating & Finishing, 2017,(22) :1234.
8. Shangguan W, Yoshida A, CHen M. Physicoch emical properties and photocatalytic hydrogen evolution of TiO2 films prepared by sol gel processes Solar Energy Materials &Solar Cells, 2003, (80): 433-441.
9. Zhang Yong-lin, Zhao Chen, et al. Preparation and Properties of Zn+ Doped TiO2 Composite Thin Films. Journal of synthetic crystals. 2014(01):0019.
10. Yaqi Hou, Daming Zhuang, et al. Factors affecting photocatalytic performance of TiO2 film and ways to improve it. [J]. Chemical industry and environmental protection, 2004, 24(3):180—184.

Influence of Surface Crack Propagation on Ring-Shell Structure

Haichao Lin

School of Mechanical Engineering, Jiangsu University of Science and Technology, Zhenjiang 212003, Jiangsu, China

Abstract. In this paper, the surface cracks of deep-sea ring-shell structures are studied. In this paper, the geometric and physical parameters of 1000m underwater titanium alloy ring shell structure were analyzed. Firstly, the mathematical model of its relevant contour curve was obtained by combining with the parameter equation. Based on the mathematical model, the finite element model was established and the force analysis under static equilibrium condition was carried out. Secondly, the stress maximum point was solved by finite element analysis, and the initial crack was added at the stress maximum point. The stress intensity factor method was used to analyze the stress intensity factor and crack propagation. Finally, according to the physical parameter load, stress ratio, morphology coefficient and different Angle analysis, the influence factors that interfere with crack growth and the dynamic trend of crack morphology were evaluated. Aiming at the study of dynamic performance of cracked ring-shell structure, the crack characteristic curves under different complex working conditions were analyzed to deepen the research in related fields and provide some scientific reference for the study of related ring-shell structure.

Keywords: Circular ring shell; surface crack; stress intensity factor; initial crack; crack growth rate.

1. Introduction

Human never cease to explore the ocean and move towards the vast ocean. Throughout history, the development of marine resources should not only solve the problem of how to mine, but also solve the problem of research and design of the development of survey equipment. At present, the development of submersibles is constantly meeting the development needs of the times, and the development of multi-functional, safe and efficient deep-sea submersibles is a trend in the field of deep-sea research. The studies on the traditional spherical, cylindrical and conical pressure-resistant structures are relatively complete, while the circular ring pressure-resistant shell has a good hydrodynamic shape, which can overcome the shortcomings of the traditional cylindrical shell, at preset, it is an indispensable pressure-resistant shell structure of deep-sea submersibles. As far as the main structure of the submersible, the circular ring shell structure not only provides enough buoyancy, its safety is the basis for the normal operation of the submersibles. Therefore, the systematic research on the structural safety of the circular ring shell is an important guarantee of the stable and reliable operation of the underwater submersible.

The shell body of circular ring shell structure is welded by two stamping end shells along the equator. Due to stress concentration, small cracks often occur around the welding toe formed in the process of manufacturing and processing, with the increase of the service times of the structure, the cracks will continue to expand and threaten the safety of the circular ring shell structure. The surface elliptical crack is the most common defect form in the circular ring shell structure, and the crack is a typical three-dimensional crack. It is a common method to solve the stress intensity factor (SIF) and simplify the analysis of the fatigue crack problem via numerical analysis. Wang Qingfeng [1] et al. took the rectangular plate with the single crack on the surface as the research object, and studied the influence of interference crack on the stress intensity factor of the crack on the surface of foundation. Xie Yuzhi, Ai Shumin [2], et al. used Franc3D to simulate the surface crack propagation problem at the welding joints on the plate. Stein and Mcelman [3] studied the buckling performance of circular ring shell structures without initial defects. D. Redekop and Y. M. Zhang [4, 5] et al. studied the linear elastic strength and stability of circular ring shell structure. Flügge and Sobel [6, 7] carried out buckling and stability analysis of circular ring shell for the buckling problem of circular ring shell structures. At present, the studies on cracks are mostly aimed at various types of plates, pressure vessels, wing or bracket aerospace structures, conical or spherical shell bodies, tube joints and other structures, the studies on circular ring shells mainly concentrate on the buckling characteristics, and there are few studies on the influence of crack propagation.

In this paper, the geometric model and finite element model of the circular ring shell structure were built, the circular ring shell structure was subjected to uniform

* Corresponding author: 18361651387@163.com

https://doi.org/10.1051/matecconf/202235801048

external pressure and residual stress caused by welding, which caused instability and uncertainty of the cracking position, therefore, the studies on the fatigue crack propagation law near the equator of the circular ring shell structure had a certain reference basis. Therefore, the surface semi-elliptical crack was inserted at the welding of circular ring shell structure. Theoretical and numerical studies on the stress intensity factor of surface cracks were carried out. The stress intensity factor method was used to solve the stress intensity factor of surface cracks under the combined action of seawater external pressure and welding residual stress. Furthermore, the influence factor of crack growth rate was studied, which provided references for the engineering application of the pressure shell of the submersible circular ring shell structure.

2. Building of the Finite Element Model of the Circular Ring Shell

2.1 Structural parameters

At present, in many current submersibles, their main structure usually adopts the circular ring shell structure. The underwater space station with the circular ring shell structure was designed in accordance with the conceptual model of the underwater space station proposed by Ross, the geometric parameters of the structural model are shown in Table.1, the circular ring shell structure model built by 3D UG software is shown in Fig.1.

Table.1 Geometric parameters of the circular ring shell structure model

name	parameter
radius of gyration R (mm)	7100
circle radius of section a (mm)	2700
shell thickness t (mm)	34

Fig.1 Circular ring shell structure

The material used in the deep-sea circular ring shell structure is high-strength titanium alloy Ti-6Al-4V, and the material performance parameters are shown in Table.2.

Table.2 Material performance parameters of titanium alloy Ti-6Al-4V

elastic modulus E/GPa	Poisson's ratio μ	yield strength σ_s /MPa
110	0.3	830

2.2 Finite element model

Theoretically, the circular ring shell structure is not constrained under external pressure, in order to eliminate the rigid displacement of the model without hindering the relative deformation, according to the studies on the constraints of the spherical shell and the theoretical constraints of the deep-sea circular ring shell structure by the China Classification Society, the symmetric boundary condition or the anti-symmetric boundary condition were applied to the analysis of the symmetric structure, and the four-point constraint mode with 90° symmetry was set up, and the symmetric nodes corresponded to the x-y and x-z displacement constraints, as shown in Fig.2.

The built 3D model was imported into the finite element software, and the force analysis of the circular ring shell structure model under static balance condition was carried out, the total number of divided units is 8104, the number of element nodes was 11662, and the element type was tetrahedral mesh. The tetrahedron was mainly located in the overall structure, mainly because the tetrahedral mesh generation algorithm was stronger than the hexahedral mesh generation algorithm in automation, efficiency, reliability and geometric versatility.

 https://doi.org/10.1051/matecconf/202235801048

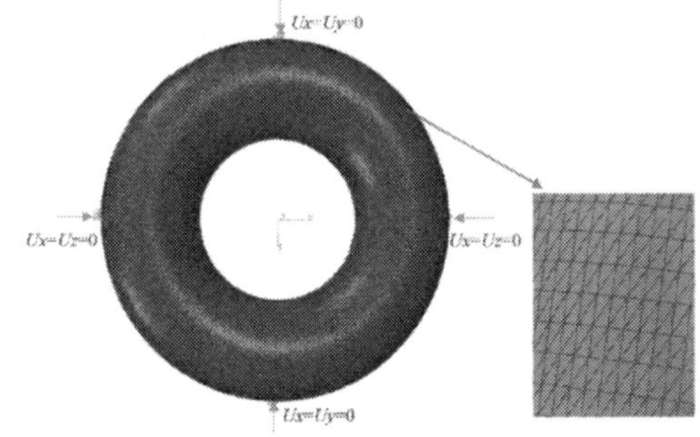

Fig.2 Meshing and boundary constraint patterns of the numerical model

3. Stress Analysis

The overall stamping forming process of the circular ring shell is to connect the two hemispheres to form the whole circular ring shell through equator welding [8]. The circular ring shell obtained by this process has less welding seam, and its uniformity and consistency are relatively better. However, the welding residual tensile-compressive stress inevitably exists in the welding structure, and the tensile strength of the structure is affected by the welding residual tensile stress. Therefore, it is very important to study the influence of residual stress of equator welding seam on the whole circular ring shell. Due to the high cost of the overall test and research of the circular ring shell, and the welding residual stress of the internal shell is generally difficult to test, and because the overall equatorial welding seam of the circular ring shell also belongs to butt welding, therefore, it is very valuable to carry out the numerical simulation of residual stress of titanium alloy butt welding thick plate.

3.1 Analysis of welding residual stress

The general force loading method, the displacement loading method and the predefined temperature field loading method are the three main methods to realize residual stress loading. Compared with the other two methods, the predefined temperature field loading method is used for simulation analysis, the analysis results are the most reasonable, therefore, this paper adopts the predefined temperature field loading method to load residual stress. The overall process is as follows: first, the model was built in accordance with the basic geometric dimensions, and the area materials in the model were defined as the isotropic linear elastic materials, the expansion coefficient in the thickness direction of the welding area was defined, the residual stress in Fig.3 was loaded into the model to obtain the stress distribution along the thickness of the pipe wall, set the load step to be 1000 steps and the sub-step to be step 5, setting a reasonable step size can not only ensure the accuracy of the stress solution, but also can avoid the problem of non-convergence caused by the excessive gradient of the temperature field due to the excessively large load step. (Namely the last step) the calculated welding residual deformation, it means that the residual deformation after welding can be simulated via the welding process of finite element simulation, according to Fig.3, the residual deformation after welding includes shrinkage deformation and angular deformation. Since the welding seam is sealed welding, the welding order is different, if first welding, then shrinks first, if welding late, then the shrinks late, it will inevitably cause a certain bending deformation, and this is consistent with the welding residual deformation after the simulation.

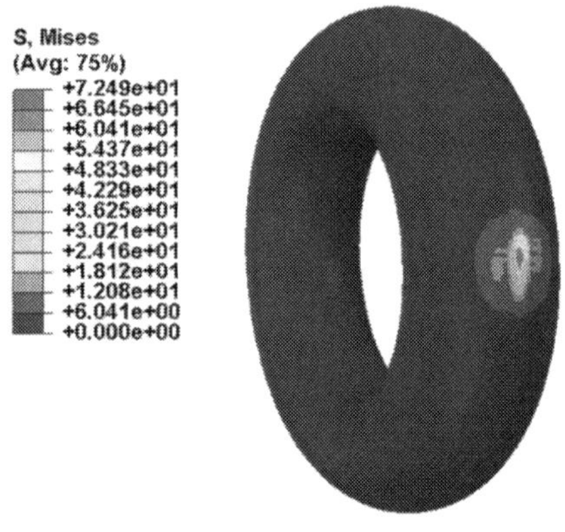

(a) equivalent stress diagram of welding area at the initial position

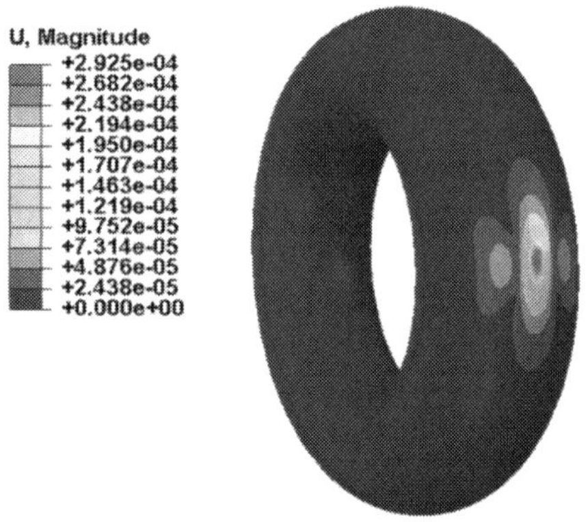

(b) equivalent deformation diagram of welding area at the initial position

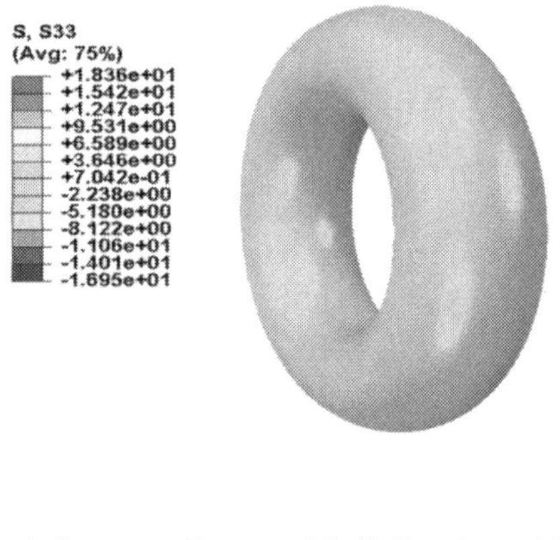

(c) equivalent stress diagram of the Z-direction welding area at the initial position

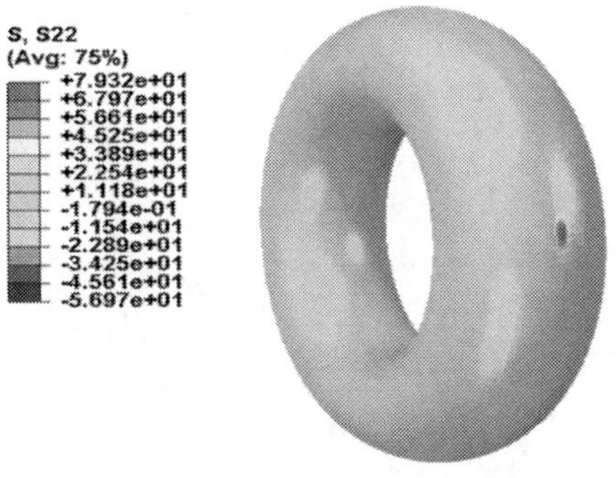

(d) equivalent stress diagram of the Y-direction welding area at the initial position

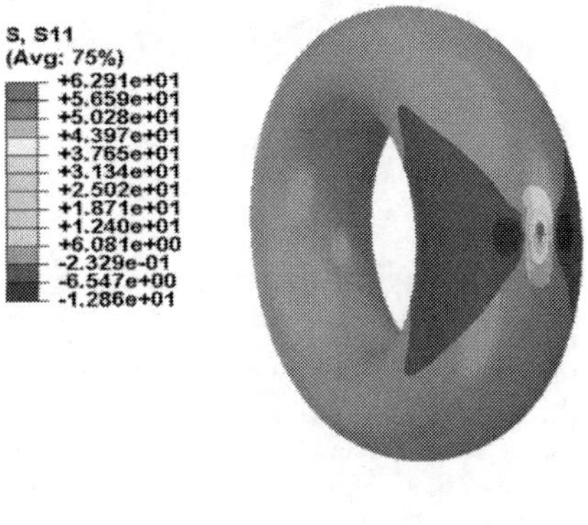

(e) equivalent stress diagram of the X-direction welding area at the initial position

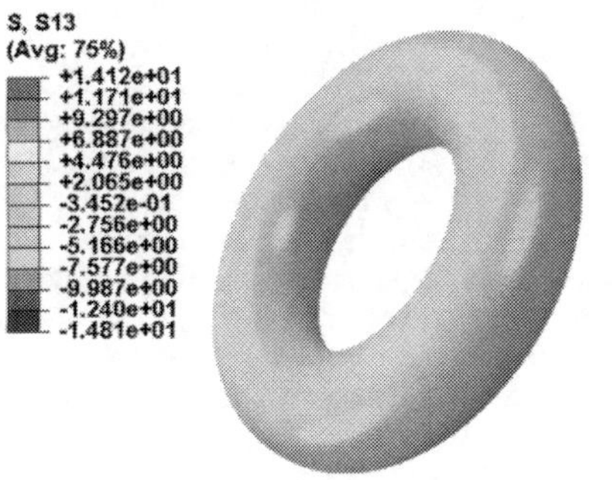

(f) equivalent stress diagram of the X-Z direction welding area at the initial position

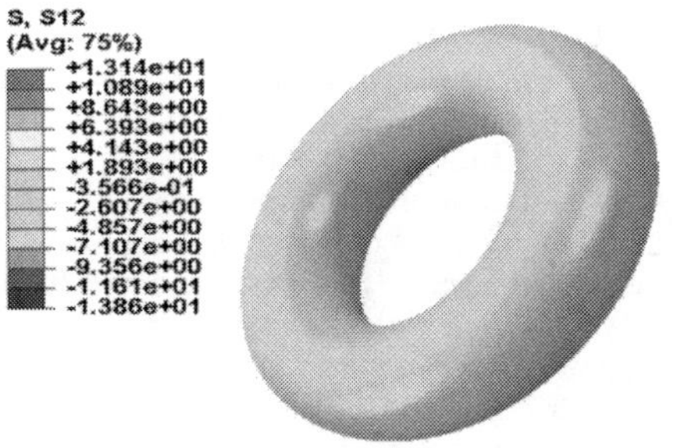

(g) equivalent stress diagram of the X-Y direction welding area at the initial position

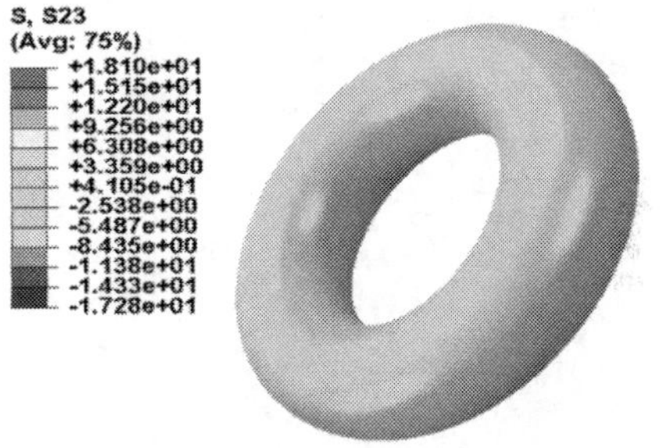

(h) equivalent stress diagram of the Y-Z direction welding area at the initial position

Fig.3 Welding force analysis diagram of ring shell

According to the above figure, we can know the longitudinal residual stress along the welding seam on the upper surface. The maximum equivalent stress is 72MPa, mainly concentrated in the Y direction, and the maximum equivalent strain is 0.0029mm. X, Y., Z are all mainly tensile stress, and the maximum value is about 62.9MPa, 79.3MPa, 18.4MPa near the center of the welding seam. There are compressive stress and tensile stress in three directions, the compressive stress appears at the first welding end of the welding seam, and quickly change over to the tensile stress, present the increasing trend, we can also know that the longitudinal residual stress at the welding seam on the upper surface is significantly greater than transverse residual stress. In the heat-affected area parallel to the welding direction, the longitudinal residual stress Y is greater than the X and transverse residual stress. Y presents the decreasing trend in stress in comparison with the welding seam, the corresponding longitudinal residual stress corresponds to the decreasing magnitude of X, the magnitude of the Y direction is the largest, and it has changed over to the compressive stress. At the end of the welding plate parallel to the direction of the welding seam, the longitudinal residual stress X and the transverse residual stress Y are smaller than those at the welding

 https://doi.org/10.1051/matecconf/202235801048

seam, while the longitudinal residual stress has mostly become compressive stress.

The maximum stress on the surface of the welding site is 72MPa, and the residual stress reaches the maximum tensile stress on the external surface of the crack, and begins to present state of compressive stress when it decreases to the quarter of the thickness along the thickness direction, and increases along the thickness direction. Due to the rigorous design of the pressure-resistant shell body, when the crack expands to one-tenth of the thickness, it is considered that the service life of the submersible pressure-resistant spherical shell has been reached, and further calculation is no longer performed, at this time, the state of residual stress has not changed, the distribution law of residual stress can be easily obtained by referring to the distribution law of the above residual stress, the residual stress is added to the crack surface through FRANC3D software, and its stress model is shown in Fig.4.

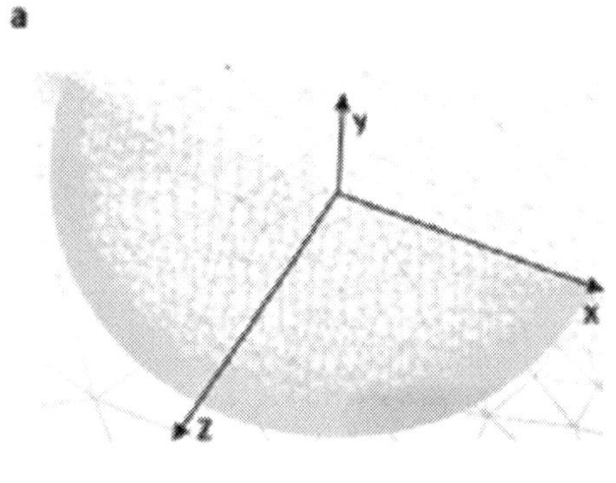

Fig.4 Residual stress model on crack surface

3.2 Analysis of compressive stress

For the diving process of any pressure-resistant shell [9], the calculated pressure PS obtained is shown in formula (1).

$$P_S = \rho g H \quad (1)$$

In the formula: ρ——the density of seawater; g——gravity acceleration; H——calculated depth.

Their values are $\rho = 1.025\text{g/cm}^3$, $g = 9.8\text{m/s}^2$, $H = 1000\text{m}$.

The deep-sea circular shell structure is subjected to the uniform external load of seawater when it operates 1000m underwater, theoretically, the model is constrained by no limit, so the free degree is not limited, and the external pressure is directly added. According to formula (1), the pressure of the circular ring shell structure is 10MPa when it operates 1000m underwater, so the uniformly distributed pressure of 10MPa is applied to the external surface of the shell body in the finite element model.

The pressure-resistant shell of the circular shell structure is subjected to the uniform external load of seawater when it operates 1000m underwater, and the force under the deep-water pressure is shown in Fig.5. The static strength meets the requirements, and the stress is mainly concentrated on the inside of the circular ting shell structure. The yield strength is 830MPa, the maximum equivalent stress is 509MPa, the static strength meets the design requirements, and the strain is mainly concentrated on the circular surface.

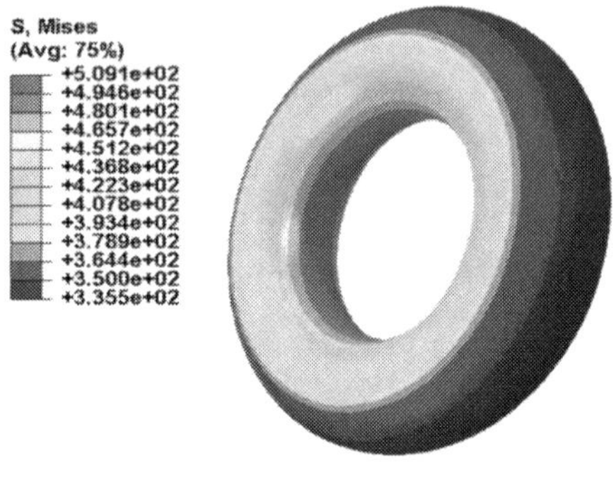

Fig.5 Stress analysis diagram of ring shell structure

4. Influencing Factors of Crack Propagation Characteristics

4.1 Influence of load on crack propagation characteristics

The surface cracks generally adopt semi-elliptical cracks in engineering practice. In this paper, the elliptical crack with the depth to length ratio of 1/4 was adopted, and the semi-elliptical surface crack with the crack depth of 4 mm was taken. The sub-model structure was set up, the single crack was set at the welding seam position of the circular ring[10-12], the initial crack was introduced into the model, and the built crack node data used the joint analysis method in FRANC3D, and the calculation module of the software ABAQUS was used, and various nodes of the crack front were selected to build the point set, the sub-models were meshed again, and the crack mesh was introduced. In order to ensure the calculation accuracy, the mesh at the crack tip will be automatically encrypted. It is convenient to divide the singular meshes around the crack tip below. The numerical model of circular ring shell structure which introduces surface crack is shown in Fig.6.

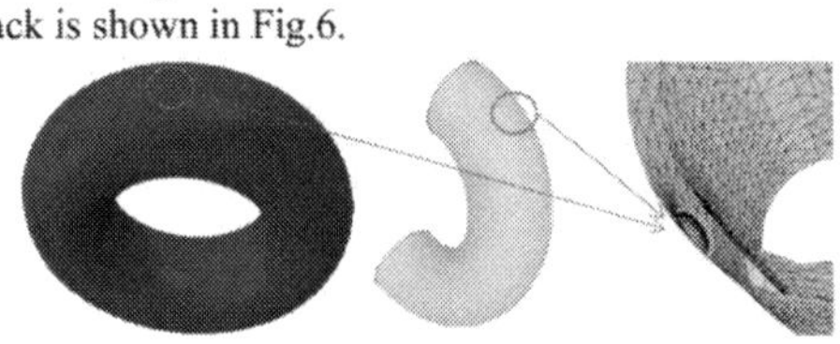

Fig.6 Finite element model with surface crack

In order to study the influence of load on crack propagation, loads of 36KN, 27KN, and 18KN were applied on the circular ring shell structure, respectively, the stress ratio was 0.1, and the sample thickness of 34mm remained unchanged, through control variable method, the influence of load on crack propagation was studied only by changing the load, the stress intensity factor is an important factor to measure the fracture of the material, in order to make the simulation more accurate, six winding integrals were set along the thickness direction, the integral errors of the first two windings is relatively large and were excluded, and the average value of the last four sets of data was taken as the calculated stress intensity factor, the relationship among the crack propagation length, the number of cycles and the stress intensity factor is shown in Fig.7, and the analysis data is shown in Table.3.

Table 3 Analysis parameters

crack propagation (mm)	stress intensity factor ($Mpa\sqrt{mm}$)			cycle times		
	18 (KN)	27(KN)	36(KN)	18(KN)	27(KN)	36(KN)
13	67	100	134	5698	3018	1846
16	67	113	151	5698	16973	9999
19	67	131	173	5698	29758	17937
21	97	144	197	73310	36135	23068
24	111	167	224	87670	43434	26827
27	140	202	269	101210	49541	30472

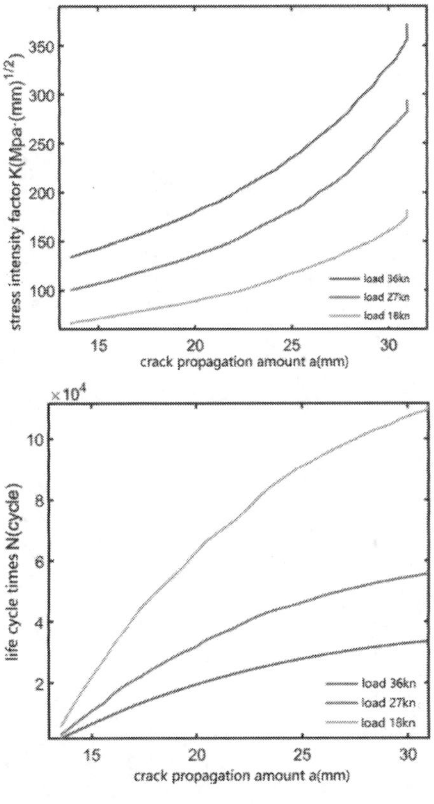

Fig.7 Geometrical parameters under load

According to Fig.7, we can know that the load is proportional to the propagation rate and the stress intensity factor, as the load increases, the crack propagation rate increases, and the growth rate of the stress intensity factor also increases, on the contrary, the load is inversely proportional to the cycle times, the larger the load, the smaller the cycle times, according to the calculated stress intensity factor, the result of K_I was found to be the largest. The stress intensity factor determines the crack type, so the cracks are mainly I-type cracks, when the cracks start to expand, the relatively serious opening displacement occurs, but there is also a little sliding displacement. According to the stress intensity factor, it has been judged that the main mode of crack is mainly type I, type II as supplement.

According to the relevant characteristics of the materials, it can be seen from the analysis diagram that the maximum stress intensity factor KI of the titanium alloy Ti-6Al-4V in the FRANC3D analysis result meets the requirements and conforms to the fracture criterion. The stress intensity factor in the middle of the crack tip is larger than that at the edge, what happens is mainly because the tensile stress in the middle of the crack is relatively large.

As the load continues to increase, the propagation rate, stress intensity factor, and cycle times are nonlinear change, but the crack length approaches a fixed value, and the change is small.

4.2 Influence of stress ratio on crack propagation characteristics

In order to study the influence of stress ratio on crack growth, the stress ratios of 0.1, 0.2 and 0.5 were used for the samples, respectively, these three stress ratios could well show the influence of stress ratio on crack propagation, by controlling variables, the thickness of the sample was 34mm, the load size was 36KN, and the stress ratios were 0.1, 0.2, 0.5, respectively, the relationship between the crack propagation length and the cycle times, and the relationship between the crack length and the stress intensity factor were obtained, as shown in Fig.8, and the analysis data are shown in Table 4.

It can be seen from the curve diagram that under the same △K, the change of the stress ratio has little influence on the fatigue crack propagation rate of the paris area, as long as the △K is the same, the value of the crack propagation length will be very close, but the cycle times decrease gradually with the increase of the stress ratio. Under the same crack propagation length, the cycle times also decreases gradually with the increase of the stress ratio. Under the same cycle times, the crack propagation length and stress intensity factor also increase gradually with the increase of the stress ratio. The larger the stress ratio, the less the cycle times when reaching the same fatigue crack length, namely, the greater the fatigue crack propagation rate. The increase of stress ratio will lead to the reduction of crack tip strain and strain rate, thus increasing the hydrogen and hydrogen embrittlement hydrogen embrittlement susceptibility into the low alloy steel, so in the range of medium △K value, the crack propagation rate will increase with the increase of stress ratio.

Under the same stress ratio, the stress intensity factor is proportional to the crack length and the cycle times, and the crack length is proportional to times.

Table.4 Analysis parameters

crack propagation (mm)	stress intensity factor (MPa$\sqrt{mm}$)			cycle times		
	0.1	0.2	0.5	0.1	0.2	0.5
13	134	134	139	1846	2124	5063
16	151	151	152	9999	12665	29027
19	173	174	174	17937	21977	49395
21	190	189	191	21789	22347	60620
24	233	226	226	26872	32912	73760
27	269	264	270	31165	29914	84990

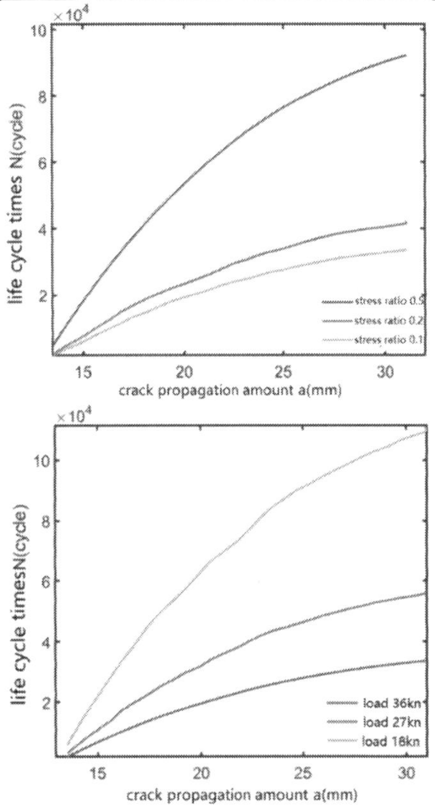

Fig.8 Geometrical parameters under stress ratio

4.3 Influence of shape factor on crack propagation characteristics

According to the provisions of GB/T 6398-2017 [9], the stress intensity factor of the standard sample is calculated by the following formula:

$$K_1 = \frac{F}{BW^{1/2}} g\left(\frac{a}{w}\right) \quad (2)$$

In the formula, $g\left(\frac{a}{w}\right)$ represents the shape factor of the standard sample, it is calculated in accordance with the formula

$$g\left(\frac{a}{w}\right) = \frac{(2+\alpha)(0.886+4.64\alpha-13.32\alpha^2+14.72\alpha^3-5.6\alpha^4)}{(1-\alpha)^{3/2}} \quad (3)$$

In the formula, when $0.2 \leq a/w \leq 1.0$, equation is effective.

The influence of the t and p stress intensity factors had been studied above, what follows will further study the influence of the shape factor on the crack propagation, Zhang Xing gave the dimensionless calculation results of rectangular compact tensile specimens containing notched cracks when shape factors are different under the isotropic and orthotropic conditions, the influence of different thicknesses and loads on the stress intensity factor with the change of the shape factor has not been deeply studied, the results are shown in the figure below, the stress ratio changes, and the crack propagation is not affected when shape factor is the same, the maximum stress intensity factor at the crack tip is selected as the fracture stress intensity factor under this shape factor. The relationship between the crack propagation length and the cycle times, and the relationship between the crack length and the stress intensity factor were obtained, as shown in Fig.9, and the analysis data are shown in Table.5.

As can be seen from Fig. 9, under the same load, the larger the shape factor, the larger the stress intensity factor, under the same shape factor, the larger the load, the larger the stress intensity factor. Under the same thickness, the larger the shape factor, the larger the stress intensity factor, under the same shape factor, the larger the load, the larger the stress intensity factor, under the same shape factor, the larger the thickness, the smaller the stress intensity factor.

Table.5 Analysis parameters

crack propagation (mm)	36KN		
	6.1(mm)	10(mm)	12.2(mm)
0.26	83	52	43
0.31	91	62	50
0.36	109	70	52
0.41	132	76	59
0.46	146	84	70
0.51	168	104	81

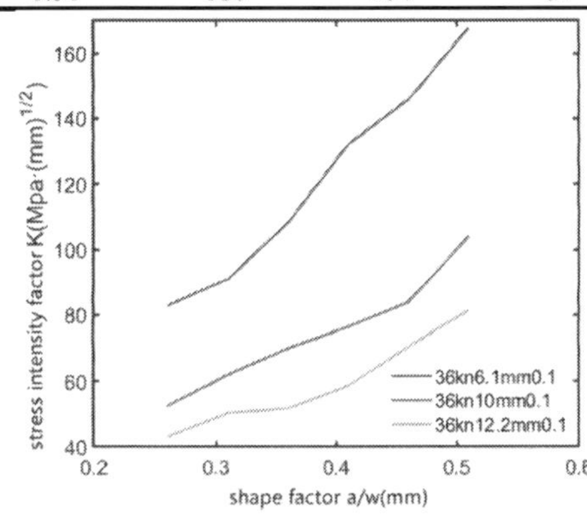

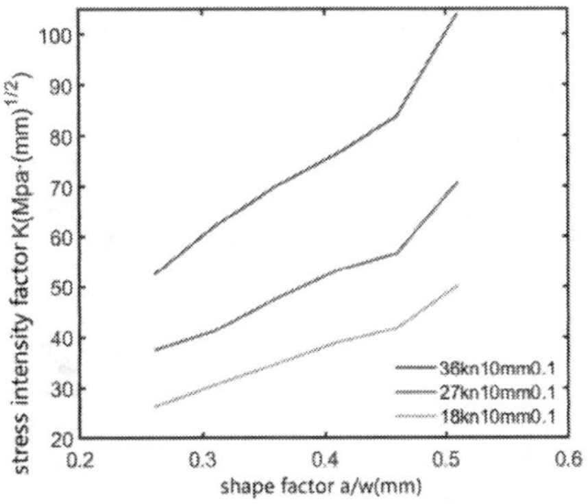

Fig.9 Geometrical parameters under shape factor

4.4 Influence of initial cracks at different angles on crack propagation characteristics

In this paper, three representative initial cracks of 0°, 30°, and 60° were prefabricated, all the initial crack length was 2mm, the load was 36KN, the thickness was 10mm, and the stress ratio was 0.1, the change of the cracking stress intensity factor of crack tip with crack cycle times was studied when the cracks were at different initial angles, as shown in Fig.10, the analysis data are shown in Table 5.

Under the same angle, the crack propagation rate is faster than the increase rate of the stress intensity factor, the crack propagation length is also larger than the growth rate of cycle number, and the crack propagation rate is smaller than the growth rate of the stress intensity factor, under different angles and the same propagation length, the cycles times increase with the increase of the angle, the values are similar at 0° and 30°, and there is a significant change at 60°. The larger the angle, the less obvious the trend of crack propagation length, under different angles and the same cycle times, the stress intensity factor decreases with the increase of the angle, the value is similar at 0° and 30°, similarly, there is also a significant change at 60°. Under different angles and the same crack length, the angle has little influence on the stress intensity factor, and the change is not large, under different angles, the values are approximate. When the crack propagation length is the same, the influence of the angle on the stress intensity factor is relatively small, and the influence of the angle on the crack propagation rate is relatively large.

Judging from the longitudinal direction, the angle continues to increase, the values are similar at 0° and 30°, and the influence is not significant, and there is a significant change at 60°. From the analysis of force angle, under the same load, 0° and 30°, the stress intensity factor of the crack tip at the stress concentration position, combined with the calculation formula of the crack stress intensity factor, the influence of crack length and stress intensity factor at the 0° and 30° crack propagation area is relatively small, the influence is great at 60°, it has a great influence on the crack propagation rate at 60°.

Table.6 Analysis parameters

crack propagation (mm)	stress intensity factor (Mpa$\sqrt{mm}$)			cycle times		
	0°	30°	60°	0°	30°	60°
13	142	138	139	3175	3394	10451
16	162	161	160	11472	12223	18281
19	185	186	186	17506	18842	24899
21	209	205	211	21862	22347	29240
24	243	247	247	25969	26614	33221
27	292	294	290	28922	29914	36031

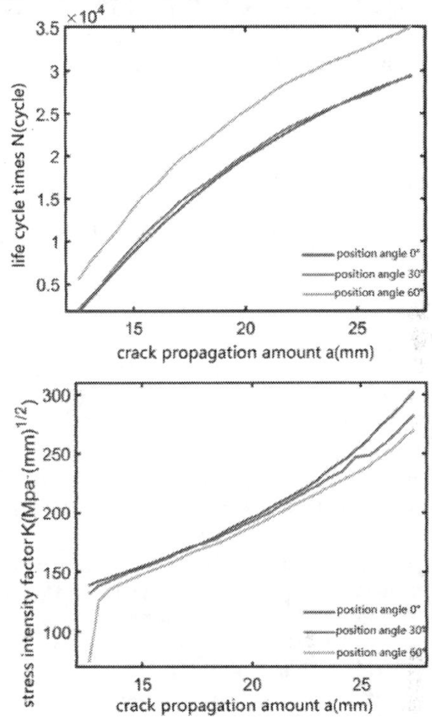

Fig.10 Geometric parameters at different angles

5. Conclusion

In this paper, the research status of the circular ring shell structure was mainly analyzed, the research method for crack propagation was puts forward based on this, through the numerical simulation analysis of the generated crack shape, the titanium alloy Ti-6Al-4V as the main material, through the joint analysis of ABAQUS and FRANC3D software, the simulation analysis of 17mm thickness and 34mm thickness under different load conditions was carried out, and the change of the crack in the thickness direction was obtained, and the following laws were obtained through the analysis:

1 It was found that the load was proportional to the propagation rate and the stress intensity factor, as the load increased, the crack propagation rate increased, the growth rate of the stress intensity factor also increased, on the contrary, the load was inversely proportional to the cycles times.

2 Under the same stress ratio, the stress intensity factor was proportional to the crack length and the cycle times, and the crack length was proportional to the cycle times. As the thickness increased, the static strength of the material specimen also increased, the corresponding stress also increased. The stress intensity factor and the cycle times are also increasing.

3. Under the same load, the larger the shape factor, the larger the stress intensity factor, under the same shape factor, the larger the thickness, the smaller the stress intensity factor. Under the same angle, the crack propagation rate was faster than the increase rate of the stress intensity factor, the crack propagation length was also larger than the growth rate of cycle times, and the crack propagation rate was smaller than the growth rate of the stress intensity factor.

References

1. Wang Qingfeng. Study on Residual Strength of New Type Composite Pipe Under Crack Damage[J]. Journal of Jiangsu University of Science and Technology (, 2018, 40(05):623-624.
2. Ai Shumin, Yu Ming, Cheng Xiaoming, et al. Analysis and Application of Three-Dimensional Crack Growth Based on Franc3D[J]. Journal of Mechanical Strength, 2018, 40(01):251-254.
3. D. Redekop, B. Xu, Y. M. Zhang. Stability of a toroidal fluid-containing shell[J]. International Journal of Pressure Vessels and Piping, 1999, 76(9): 575-581
4. YM Zhang, P. Mirfakhraei, B. Xu, et al. A computer program for the elastostatics of a toroidal shell using the differential quadrature method[J]. International Journal of Pressure Vessels and Piping, 1998, 75(13): 919-929.
5. W. Flügge, L. H. Sobel. Stability of shells of revolution: General theory and application to the torus[J]. Lockheed Missiles and Space Company Report, 1965, 6(12):65-75.
6. W. Flügge, L. H. Sobel, Stability of toroidal shells under uniform external pressure[J]. American Institute of Aeronautics and Astronautics, 1967, (5): 425-431.
7. Jie Zhiyu, Li Yadong, Wei Xing, et al. Numerical Study of Mixed-mode Stress Intensity Factor of Surface Crack of Welded Joint[J]. Journal of the China Railway Society, 2017, 39(02):127-133.
8. Tang Fei. Buckling Behavior of Composite Egg-shaped Shell under Hydrostatic Pressure [D]. Dalian University of Technology, 2019.
9. Zhang Sujie. Analysis on the Effects of Material and Structural Defects on the Property of Deep-sea Manned Spherical Hull[D]. Jiangsu University of Science and Technology, 2018.
10. Tang Fei. Buckling Behavior of Composite Egg-shaped Shell under Hydrostatic Pressure[D]. Dalian University of Technology, 2019.
11. Tu Sihao. Application of XFEM in Fatigue Crack Growth Analysis of Pressure Vessel[D]. Zhejiang University of Technology, 2017.
12. Priscilla L, Chin. Stress Analysis, Crack Propagation and Stress Intensity Factor Computation of a Ti-6Al-4V Aerospace Bracket using ANSYS and FRANC3D Rensselaer Polytechnic Institute, 2011:11.

https://doi.org/10.1051/matecconf/202235801049

Kinematic Simulation of Automatic Conveying and Stamping Machine based on UGNX

Wenyue Liang and Xianghui Zhan*

Zhuhai College of Science and Technology, Zhuhai, China

Abstract: This paper designed a kind of automatic conveying and stamping machine, which was mainly used for drawing thin sheet metal. UGNX software was used to design the motion law of the mechanism and the simulation mechanism. The time displacement diagram obtained by theoretical calculation and simulation was compared to verify the correctness of theoretical analysis and the rationality of the design. The comparison shows that the device meets the requirements of theoretical calculation. The device has the characteristics of simple structure, high efficiency and easy operation, which provides a reference for the subsequent realization of conveying and stamping devices with different requirements.

1. Introduction

Automatic conveying and stamping machine, is a kind of automatic material conveying equipment, compared with the traditional conveyer operation, it is more simple, higher degree of automation, widely used in light, heavy industry parts stamping, is an important device. The type of automatic conveyer on the current market is relatively single, the market is still not perfect.

This paper designs a machine, which make guide rod institution with swing rod institution together to be as the stamping institution, crank slider institution as the conveying institution. It has the characteristics of easy design, simple structure and good conformity with the actual production requirements. It provides the reference for different production requirements.

2. Automatic conveying and stamping machine overall design scheme

2.1 The design requirements

In this paper, the automatic conveying and stamping machine for metal sheet billet from the side to stay processing position. Upper die with larger speed nearly blank first and then down at a constant speed, forming work, continue downward billet products will launch of upper die cavity after, finally back quickly, to complete a working cycle.Its requirements are:

(1) Stamping execution component have the movement characteristics as following： close to the workpiece with a fast speed, uniform down the stretch and fast return the beginning position.

(2) Fine forming products productivity about 30 pieces per minute.

(3) On mobile total distance of 280 mm, the drawing schedule in the central part of total travel, about 100 mm.

(4) Travel speed ratio coefficient is 1.2 or higher.

(5) The transporting material maximum distance of 200 mm.

2.2 Design flow chart

The design of the automatic conveying and stamping machine flow chart is shown in Fig. 1[1].

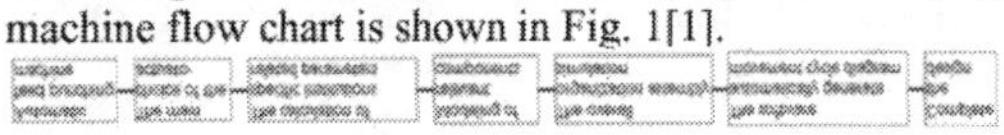

Figure 1. Project flow chart

3. Main components of the design and its modeling

3.1 Conveying institution design and modeling

Conveying institution include: rack, crank, connecting rod, guide rod, nut, driven gear, etc. Conveying institution are mainly consisted of crank connecting rod institution and guide bar institution[2].The institution has the character of reciprocating motion, can realize the request of conveying.According to the change of the crank and connecting rod length, can realize different conveying requirements.

Conveying institution diagram as shown in Fig. 2.

Using UGNX to conveying institution are modeled as shown in Fig. 3.

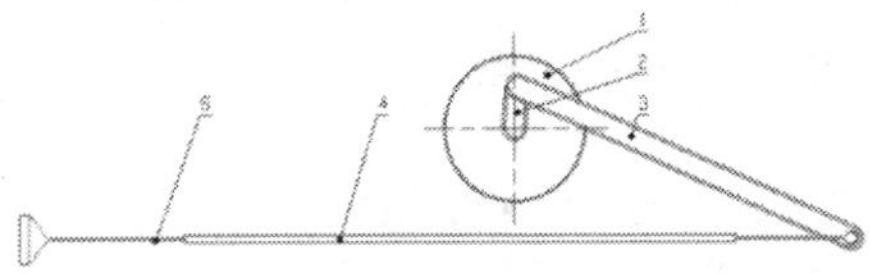

1 - gear; 2 - crank; 3 - connecting rod; 4 - rack; 5 - guide bar

Figure 2. The sketch of conveying institution

* Corresponding author: zhanxianghui@zcst.edu.cn

Figure 3. Modeling for conveying institution

3.2 Stamping design and modeling of the institution

Stamping institution mainly includes: rack, gear, shaft, rocker, guide rod, punch, slider, driven gear, etc.Stamping institution is mainly composed of slider-crank mechanism of guide bar mechanism.The institution is quickly approaching, uniform motion, fast return movement characteristics, can well realize the request of stamping[3]. Stamping institution diagram as shown in Fig. 4.
Using UGNX to stamping institution are modeled as shown in Fig. 5.

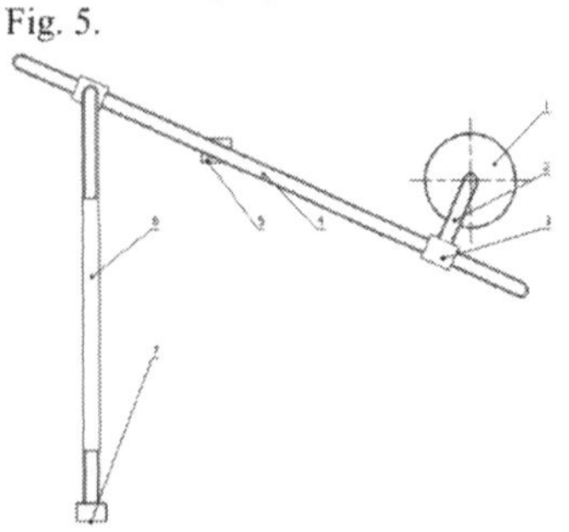

1 - gear; 2 - crank; 3 - slide block; 4 - rocker; 5and6 - rack; 7 - the punch

Figure 4. The sketch of stamping institution

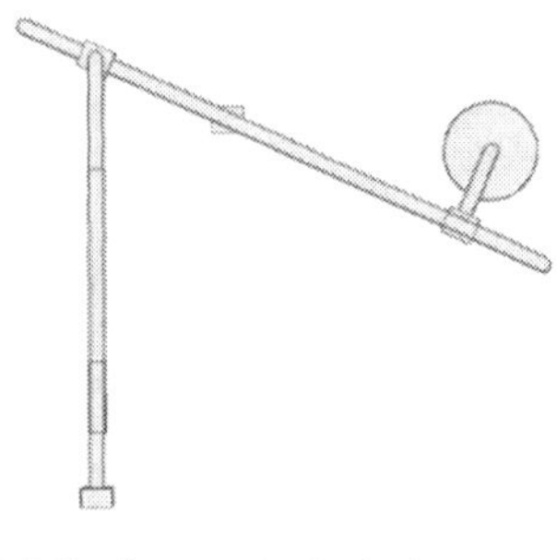

Figure 5. Modeling for stamping institution

4. The model establishing of Automatic conveying and stamping machine

4.1 The overall structure

The parts assembled of automatic conveying and stamping machine model as shown in Fig. 6.
The basic composition of institution is as follows: including rack, gear, shaft, the slider, rocker, guide rod, connecting rod, etc.In the initial drop-down menu choice movement simulation, new simulation in sports navigator, into the movement simulation interface[4].

As shown in Fig. 6 is the automatic conveying and stamping machine,which has entered the simulation interface.Its conveying institution is a offset slider-crank mechanism, which has simple structure, convenient operation, can be a very good implementation movement back and forth[5].Stamping institution for rocker guide bar mechanism, this mechanism can achieve good fast approaching, drawing at a constant speed, fast return work requirements[6].
Transmission part as input by the gear wheel 3, gear 2 and gear 5 as a driven wheel, respectively independent input to convey institution and stamping institution, to achieve the overall transmission. The number of teeth for the gear 3 is 240; The number of teeth for the gear 2 and gear 5 is 120, so the total transmission ratio of 1:1.

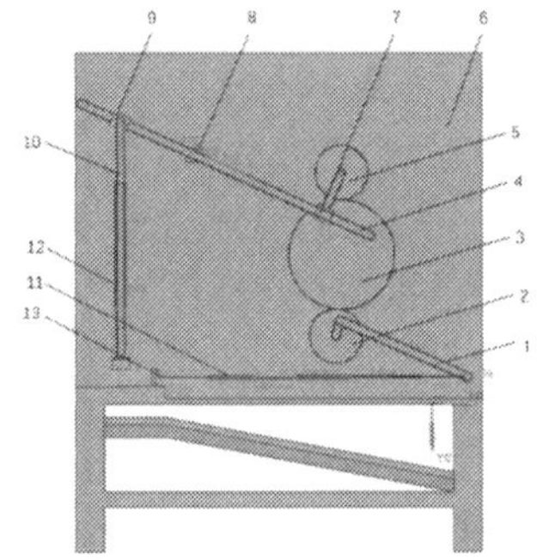

1 - connecting rod; 2 - driven wheel 1; 3 - input wheel; 4 - rocker; 5 - driven wheel 2; 6 - rack; 7 - slide block; 8 - rack; 9 - slide block; 10 - guide rod; 11 - rack; 12 - rack; 13 - guide rod

Figure 6. The overall structure of the automatic conveying and stamping machine Modeling

4.2 Define the connecting rod

Definition of connecting rod general method are following: defines the same motion components for the same connecting rod;Static components are defined as fixed connecting rod;Of motion simulation is independent components can be omitted.
The rack 6、8、11、12, respectively defined as fixed connecting rod L001, L002, L003, L004;The input gear 3 defined as connecting rod L006;Driven gear 2 and the crank have the same motion, therefore, defined as connecting rod L005;On the driven gear 5 and its crank due to have the same motion, therefore, defined as connecting rod L007;Connecting rod 1 defined as L008;Guide bar 13 is defined as the connecting rod L009;Rocker defined as connecting rod L010;Sliding block 7 and 9 respectively defined as connecting rod L011 and L012; Guide bar 10 is defined as the connecting rod L013 guide rod.

4.3 Create motion pair

Create a general method of kinematic pair: typically create a kinematic pair between two connecting rods; To collide in the movement of connecting rod, should create as a 3 dimensional contact vice, can appear otherwise simulation animation connecting rod phenomenon of

mutual penetration, is inconsistent with the actual.The fixed connecting rod kinematic pair, can be without selecting the second connecting rod, UGNX is fixed it by default.Sports navigator of the connecting rod corresponding to the green words, kinematic pair corresponding to the orange icon on the right.
As shown in Fig. 6, L001-L004 respective set to the fixed vice J001-J004; Give L005 relativing to the ground a rotation vice, called rotation vice J005; Giving L006 relativing to the ground a rotation vice, called rotation vice J006, adding a 180 degree constant speed drive on it, make its can be rotated in the cycle for a week. Give L007 relativing to the ground a rotation vice, called rotation vice J007;L005 relative L008 set as rotation vice J008;L009 relativing L008 set as rotation vice J009;L009 relativing L004 set as sliding vice J009;L007 relativing L012 set as rotation vice J010; L010 relativing L0011 set as sliding vice J014;L010 relative L002 set as rotation vice J011;L010 relativing L012 set as sliding vice J015;L013 relativing L012 set as rotation vice J012;L013 relativing L003 set as sliding vice J016;Gear 5 relativing gear 3 set as 3 dimensional tangent vice G001;Gear 2 relativing to gear 3 set as 3 dimensional tangent vice G002;Guide bar 13 relativing rack 12 set as 3 dimensional line contact vice G003;Sliding block 7 relativing rocker 4 set as 3 dimensional line contact vice G004;Sliding block 9 relativing rack 11 set as 3 dimensional line contact vice G005;Guide bar 10 relativing rack 12 set as 3 dimensional line contact vice G006;Between gear 2 and gear 3 adding a gear vice J017;Between gear 5 and 3 adding a gear vice J018.

4.4 Automatic conveying and stamping machine with the holistic movement

As the rotation of the gear 3 will drive the rotation of gear 2, which drives the swinging rod 3 to move and swing, thereby making guide bar 13 to sliding,which can realize the function of conveying. After rotating a week later, the gear 3 will return to the starting position, to complete a cycle.
As the rotation of the gear 3 will drive the rotation of gear 5, which drives the swinging rod 4 to move and swing, thereby making guide bar 10 to sliding, which can realize the function of stamping. After rotating a week later, the gear 3 will return to the starting position, to complete a cycle.

5. Automatic conveying and stamping machine of the kinematics simulation

Motion analysis is carried out in accordance with the input the time and the number of dynamic simulation analysis. Click the movement simulation animation icon, select a cycle as the research, the time set to 2 second, count to 500 steps, ADAMS solver after calculating, can directly see the simulation process of automatic conveying and stamping machine[7]. Because of the connecting rod can't for generating charts, selection of connecting rod L009 in conveying institution as a research object, aautomatically generated in the mapping function of time - - displacement speed, time - acceleration diagram, as shown in Fig. 7 is the convey movement of the institution loop diagram;
Selection of connecting rod L013 in stamping institution as a research object, automatically generated in the mapping mechanism of time - - speed, time - - displacement speed, and time- - acceleration diagram, as shown in Fig 8 is the stamping institution motion cycle chart.(the red line as the time of displacement curve and the blue line speed of time displacement curve and the green line acceleration of the displacement curve of time.)
Choose connecting rod L009 and L013, automatic generation of its time - - displacement diagram, as shown in Fig. 9. From Fig. 9 you can see the whole machine from 0 second began to move.0 second moment, conveying institution, located at 107.5 mm on workpiece, and at the same time stamping institution at the workpiece distance 280 mm, 0 to 0.5 second, conveying institution to push billet to processing position, stamping institution is quickly approaching billet.When 0.5 second moment, complete convey and return immediately. At the same time, stamping institution encounter blank, start to drawing the blank. 0 to 0.8 second, keeps uniform drawing stamping mechanism, conveying institution continue to stay away from the workpiece.
0.8 second moment, stamping institution to complete the drawing 0.8 to 1.2 second, stamping institution to continue downward movement, after drawing blank will launch processing position.The conveying institution continue to stay away from the workpiece 1.2 second moment, stamping institution will completely blank off a processing location, then immediately return. 1.2 to 2 second,stamping institution from processing position, and return to furthest distance artifacts in the 2 second moment distance is 280 mm.While conveying mechanism in 1.6 second moment reaches position for 200 mm, then immediately return back to the starting location, prepare for the next conveying .

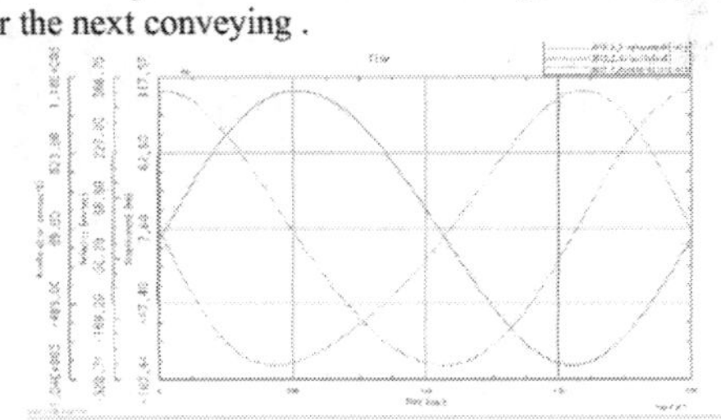

Figure 7. Conveying institution loop cycle chart

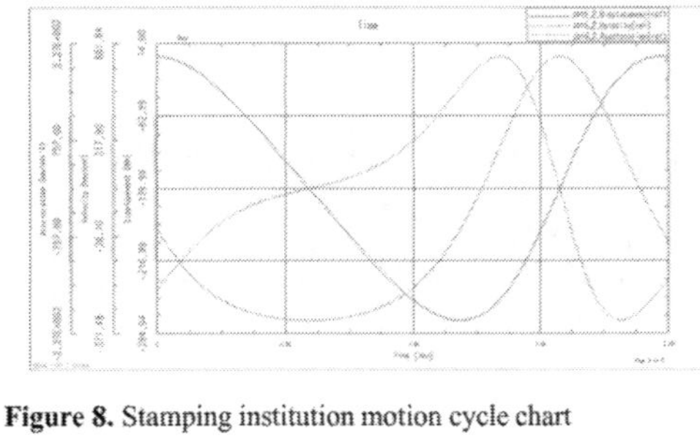

Figure 8. Stamping institution motion cycle chart

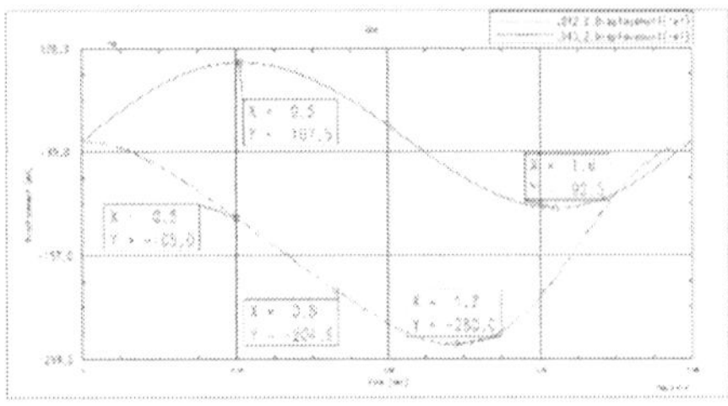

Figure 9. The simulation time displacement cycle chart

6. Conclusion

In this paper, the accuracy of theoretical analysis is verified by modeling and simulation. The design of automatic conveying and stamping machine has the advantages of simple and compact structure and high matching precision. The motion simulation function of UGNX software, can dynamics and kinematics analysis to carry out on the body.
The results show that this method can easily find out the cause of the failure in the operate, or sensitive factors, so as to take effective measures to ensure the normal operation of the automatic conveying and stamping machine.

References

1. Jiacheng Chen,Miaoxing Li,Haifeng Zhao,Motion Simulation of Energy Storage Mechanism Based on UG NX4.0[J], China Academic Journal Electronic Publish House. 1007-3175(2012)01-0060-03:60-61.
2. Wang Hua,Jianfeng Zhang, Design and Motion analysis of a New Type of Jujube Pitting Machine[J], China Academic Journal Electronic Publish House. 1005-1295.2020.05.011:50-54.
3. LU Yuhao, Shen Debo, Li Jie Zhu Junile, Li Gang, Lu Xinjian, Simulation Analysis and Scale Optimization of a New Type of Multi-Link Stamping Mechanism [J], MECHANICAL&ELECTRICAL ENGINEERING TECHNOLOGY. 201811276022Z: 80-84.
4. Information on https://v.youku.com/v_show/id_XNDU5MDE4NjU2OA==.html
5. Zhang Feng, Zhang Li, Offset slider-crank mechanism motion simulation analysis [J], The China science and technology information. 2017.01.026:70-71.
6. Chenchen Wang, Based on the UG NX8.0 crank rocker mechanism motion simulation and analysis [J], China Academic Journal Electronic Publish House. 2018.19.004:5-6.
7. Xiehua Zeng, Li Shan, Dongming Sun, The dynamic simulation analysis of mechanical system based on UG[J], China Academic Journal Electronic Publish House.2005, 32:61-63.

Effect of ultrasonic wave on coating and corrosion of ZK60 magnesium alloy

Nana Zhang, Zhibo Tong*, Hongyan Li, Shichang Liu, Qing Li, Qiuyan Xie, Jianbo Ding

Yangtze Normal University, the faculty of materials science and engineering, Chongqing, 408100 China

Abstract. In this paper, hydroxyapatite coating was successfully deposited on ZK60 magnesium alloy surface by ultrasound-assisted electrodeposition method, and the macroscopic structure, chemical composition and corrosion properties of ZK60 magnesium alloy sample and HA coating were analyzed. The results show that the quality of ZK60 magnesium alloy surface coating is improved by ultrasonic power, and the HA coating is uniform, smooth and compact. XRD and FTIR analysis show that the main composition of the coating obtained by ultrasonic assisted electrodeposition is hydroxyapatite. The results show that ultrasonic treatment is beneficial to the corrosion resistance of SBF solution.

Keywords: Ultrasonic; ZK60 magnesium alloy; Coating; corrosive.

1. Introduction

It has been reported [1] that more than 2 million artificial joint replacement operations are performed every year, and the number continues to increase rapidly as time goes by, which makes the demand for metal materials to replace human bones growing. Among them, traditional orthopedic implant materials that appeared earlier, such as 316L stainless steel, cobalt-chromium alloy, Ti-6Al-4V, etc., will generate products that have toxic effects on cells in the human body, even cause immune rejection and inhibit the growth of osteoblasts [2]. In recent years, some of the rise of new biodegradable materials is more suitable for materials for medical applications, including ZK60 magnesium alloy as a kind of Mg -Zn- Zr alloy, with high strength, good biocompatibility and widely research [3], the in vitro degradation of performance and biological response experiments also show that ZK60 magnesium alloy can be used as a biological material [4,5]. However, despite the ZK60 magnesium alloy as human bone material has many advantages, however, the degradation of magnesium in the body too fast speed, resulting in graft site, alkaline and hydrogen release speed too fast, the formation of airbags, affect tissue healing, therefore, how to slow down and control the degradation of magnesium alloys become the key to its clinical application [5].

At present, the preparation of HA coating on magnesium alloy surface by scholars are widely applied to improve the corrosion resistance of magnesium alloy, its main methods include electrochemical plating, conversion coating, anodizing and deposition process and so on, these methods can effectively balance on the mechanical properties of the magnesium alloy and biological properties of HA, make its application in bone implant material [6]. In addition, the study of ultrasound-assisted phosphate [7] on nickel base [8] copper base [9] shows that at low temperature, ultrasound-assisted reaction can improve ion diffusion, current efficiency and reduce cathode polarization due to cavitation, thus speeding up the reaction speed. Moreover, ultrasonic also has a positive effect on the second phase doped metal coating [10]. In addition, a large number of studies [11,12] show that, the addition of ultrasonic treatment to the electrodeposition process helps to eliminate bubbles and produce a smooth coating. However, there are few reports on the use of ultrasonic assisted electrodeposition technology to obtain uniform HA coating and improve the corrosion resistance of ZK60 magnesium alloy.

Therefore, ZK60 magnesium alloy in Mg-Zn-Zr alloy was selected in this study to explore the corrosive effect of ultrasonic assisted electrodeposition on HA coating on its surface. The HA coating with different ultrasonic power and its changing behavior in bionic anti-corrosion research were studied, and its influence on biological mineralization ability was further studied in human body simulation fluid (SBF).

2. Experiment

2.1 Material

The reagents used in the experiment were pure chemical reagents for Chinese medicine analysis, and were used directly without special treatment. The base material is ZK60 magnesium alloy, after cutting it into samples with a size of 5cm×5cm×1cm, Grind and polish them step by step with sand paper of size 600, 800, 1000, 1200, 1500 and 2000 respectively. A piece of 0.5 mol/L citric acid

* Corresponding author: 404545023@qq.com

corroded for 2 min was used as the cathode, and then a piece of uncorroded sample was selected. Ultrasonic cleaning with ethanol, acetone and deionized water for 10 min, and air drying as anode. The electrolyte used for electrodeposition is composed of $CaCl_2$, KH_2PO_4 analytical pure reagent and ultra-pure water. After the configuration, the concentration of Ca^{2+} and PO_4^- is 0.04mol/L and 0.024mol/L, making the Ca/P ratio 1.67. Then add 0.5mL/L hydrogen peroxide prepared in 2000mL volumetric bottle, constant volume, 1500mL was taken as the solution required for an ultrasonic electrodeposition [12].

After electrodeposition of HA coating on ZK60 magnesium alloy, the bionic corrosion resistance of ZK60 magnesium alloy was tested by foaming in human bionic solution SBF solution. SBF solution is supersaturated for apatite, therefore, improper preparation method may lead to hydroxyapatite precipitation in the solution [12]. In this paper, according to the given quality in Table 1, reagents were added successively from left to right to prepare SBF solution. The configuration process was completed on the magnetic stirrer, and the next reagent must be added after the reagent was completely dissolved each time. Meanwhile, the internal temperature of the solution should be kept at 37°C during the dissolution process. After the solution was dissolved, the pH of the solution was adjusted to 7.4±0.1 by Tris and HCl buffer solution.

Table 1 Chemical composition of SBF solution /g

NaCl	$NaHCO_3$	KCl	$K_2HPO_4\cdot 3H_2O$	$MgCl_2\cdot 6H_2O$	$CaCl_2$	Na_2SO_4	Tris	HCl
8.036	0.352	0.225	0.230	0.311	0.293	0.072	6.063	5~6 mL

2.2 Experimental Process

The prepared ZK60 magnesium alloy was cleaned and sealed with adhesive tape to make the area of the exposed part of the reaction 1cm*1cm. The area of the exposed part of the reaction was put into 1500 ml electrolyte with electrodes 20 mm apart. Then, the constant piezoelectric deposition experiment was carried out under the constant current and constant voltage power supply (Shenzhen Solid Measurement Electronic Technology Co., LTD., NPS605W), and the experimental voltage was 1.5V. The schematic diagram of the experimental device is shown in Fig.1 During the reaction process, the electrolyte was placed in a low temperature constant temperature tank, the reaction temperature was kept at 60°C, and the electrochemical deposition time was 1500 s. In order to explore the influence of ultrasonic wave on electrodeposition, 2s was treated by ultrasonic wave and 2s was closed (Wuxi Woxin Instrument Manufacturing Co., LTD., V0Shin-501D), and the influence rules of different powers (0%, 15%, 25%, 40%, 80%) on coating and corrosion were studied.

In the bionic anti-corrosion performance experiment of SBF solution, half of the electrodeposited ZK60 magnesium alloy sample was sealed with tape, and then the sample was put into SBF solution (40 mL) for reaction. Comparison has been made. During the reaction, the container was sealed to avoid pollution or reaction with air. Then the whole product was placed in a water bath (MPLR-702) to maintain 37±1°C, and a new solution was changed every day during the reaction process to ensure the availability of fresh reaction ions. After seven days, the sample was removed from the solution, washed with ultrapure water, and dried in a drying oven at 80°C for 24 h, and then analyzed by XRD and FTIR.

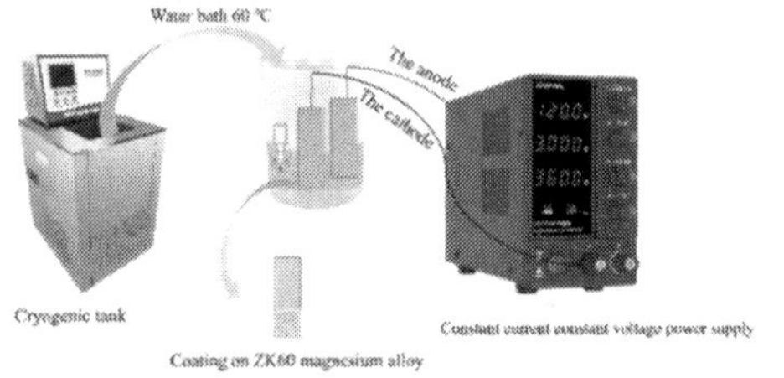

Fig.1. Schematic diagram of ultrasonic assisted electrodeposition process

2.3 Characterization

Shimazu 6100Lab X-ray diffractometer was used to detect the sample phase. The tube voltage was 40kV, the tube current was 100mA, the continuous scanning range was 10-90 °, the step size was 0.02°, and the scanning speed was 10°/min. Infrared spectral analysis (FTIR) was performed by Thermo Fisher is5, with an infrared spectral scanning range of 400-2000 cm^{-1} and spectral resolution of 4cm^{-1}.

3. Results and Discussion

Fig.2(a) shows the macroscopic morphology of HA coating electrodeposited on ZK60 magnesium alloy at different ultrasonic powers, visible, in the absence of ultrasound (that is, the ultrasonic power was 0%), only a small amount of ZK60 magnesium alloy surface HA coating on the majority of coating fall off in the electrolyte, it shows that the adhesion strength of the electrodeposition coating on ZK60 magnesium alloy is not very good, And with the increase of ultrasonic power, the quality of coating on the surface of the ZK60 increases, visible ultrasound is helpful to improve the surface of hydroxyapatite in ZK60 deposits, in addition, the ZK60 surface coatings are also more level off, confirmed that it has been reported in the HA material surface smooth, easy to induce cell adhesion, and convex concave surface, cell adhesion is poor [13], therefore, Ultrasonic wave is beneficial to the deposition of HA coating and the quality of biomimetic material.

The sample of electrodeposited ZK60 magnesium alloy soaked in SBF solution for seven days is shown in Fig.2(b). The upper part of the red line is divided into the reaction area, and the lower part of the red line is divided into the tape sealing part. From the perspective of the reaction region, there is only a very small amount of crystal on ZK60 magnesium alloy without ultrasonic electrodeposition, while with the increase of ultrasonic power, the coating surface thickness becomes larger and larger. From 5.92 mg/cm^2 without ultrasonic wave to 23.32 mg/cm2, 27.52 mg/cm^2, 28.42 mg/cm^2,

30.52mg/cm^2 (corresponding to 15%, 25%, 40%, 80% of ultrasonic power respectively), the surface morphology is more and more dense, visible, Ultrasonic treatment is beneficial to the biocompatibility of ZK60 magnesium alloy coating. The sealed part of the tape (the part below the red line) is still reacting where the tape is not firmly sealed due to the HA coating which has been electrodeposited and the strong agitation of ultrasonic waves. From corrosion, no ultrasonic electrodeposition ZK60 magnesium alloy surface is uneven, severe corrosion and corrosion were not only at the bottom of the gap, and the central part is corroded, and with the increase of ultrasonic power, hydroxyapatite coating is obvious thickening, also some improved at the bottom of the corrosion rate, and a central part in addition to ultrasonic power 40%, also to have certain to improve, This could have been caused by accidentally knocking off the coating on the central part during the experiment.

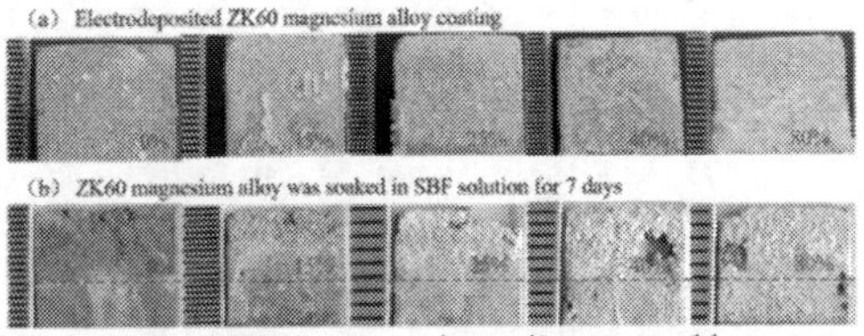

Fig.2. ZK60 magnesium alloy coated by electrodeposition and soaked in SBF solution for 7 days

The coating was scraped off the ZK60 magnesium alloy and spread on the surface of silicon powder for X-ray diffraction analysis, and the potassium bromide tablet for Fourier transform infrared spectroscopy analysis, the results are shown in Fig.3. Fig.3(a) shows the XRD patterns of electrodeposited coating on ZK60 magnesium alloy. The diffraction peak 2θ = 25.9° and 32.9° are HA characteristics [6]. It can be seen that without ultrasonic treatment, the HA peak of coating is not obvious, and the crystallinity of hydroxyapatite after ultrasonic treatment is significantly enhanced. Fig.4 is a Fourier infrared image. The wavelengths of 474, 564, 602, 972, 1031, 1086 and 1095 cm^{-1} are characteristic peaks of PO_4^{3-} [12,14], while 634cm^{-1} is characteristic peak of hydroxyl (OH) [15]. Its peak is generally thought to be produced by the presence of PO_4^{3-} and OH^- in hydroxyapatite $Ca_{10}(PO_4)_6(OH)_2$. As can be seen from Fig.3(b), the introduction of ultrasonic wave significantly enhanced the infrared characteristic vibration peak of hydroxyapatite, which was consistent with the XRD examination results, and there was weak hydrogen phosphate (HPO_4^{2-}) groups at absorption peaks 525, 1075 and 1110cm^{-1} [16]. This is generally believed to be caused by the formation of the calcium-phosphate phase (OCP) [17], due to the decomposition of hydroxyl radicals in water during ultrasonic treatment, which in turn promotes the formation of the HA phase rather than other, less stable OCP phases. In addition, bands related to adsorbed water were also found at map 1625 and 1672cm^{-1} [18]. The infrared spectra of 870, 1391 and 1400 to 1500cm^{-1} show CO_3^{2-} [19], which is because the electrolyte is exposed to CO2 in the air during the deposition process, and the dissolved CO_3^{2-} reacts with Ca^{2+} in the electrolyte to generate a small amount of $CaCO_3$. $CaCO_3$ does not affect the binding of hydroxyapatite coating to cells [20].

In the process of electrodeposition without ultrasonic treatment, equation (1)~(4) will occur in water during electrolysis, and a large amount of hydrogen will be precipitated at the cathode [20], so that the bonding force of the formed hydroxyapatite coating is not strong, and it is easy to fall off into the solution. In the case of electrical deposition at the same voltage, ultrasonic wave has a strong stirring effect, which makes hydrogen discharge faster and reduces the blocking effect of bubbles, thus making HA coating uniform and smooth with dense surface, which is similar to the dense needle-like structure formed in literature [21]. At the same time, the hydroxyl radicals on ZK60 magnesium alloy forming hydroxyapatite as shown in formula (5), when the consumption of alcohol is more near cathode ion to supplement, can affect the crystallization of hydroxyapatite, and the role of water in ultrasonic processing will produce hydroxyl, able to replenish the hydroxyl free radicals[18], effectively promote the electrodeposition, Improve surface roughness [22].

$$2H_2O+2e^- \rightarrow H_2\uparrow+2OH^{-1} \quad (1)$$

$$H_2O_2+2e^- \rightarrow 2OH^{-1} \quad (2)$$

$$H_2PO_4^-+e^- \rightarrow HPO_4^{2-}+1/2H_2\uparrow \quad (3)$$

$$HPO_4^-+e^- \rightarrow PO_4^{3-}+1/2H_2\uparrow \quad (4)$$

$$10Ca^{2+}+6PO_4^{3-}+2OH^{-1} \rightarrow Ca_{10}(PO_4)_6(OH)_2 \quad (5)$$

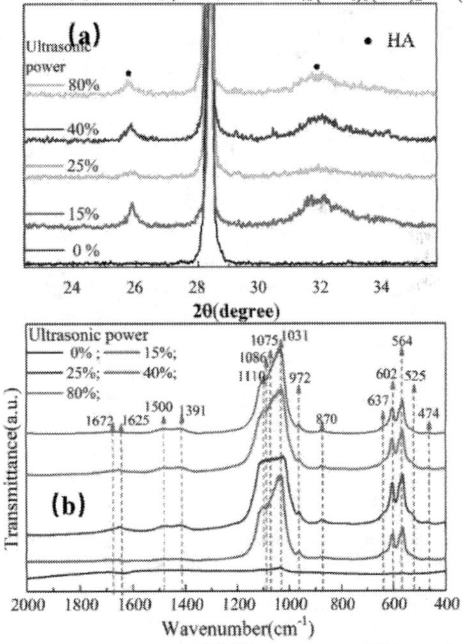

Fig.3. XRD patterns and FTIR spectrum of HA coating prepared at different ultrasonic powers

4. Conclusion

(1) Ultrasonic power is beneficial to increase the surface coating quality of ZK60 magnesium alloy, making the HA coating uniform, smooth and compact.

(2) XRD and FTIR analysis show that the coating obtained by ultrasonic-assisted electrodeposition is mainly composed of hydroxyapatite.
(3) The results of SBF solution blistering showed that ultrasonic treatment was beneficial to the corrosion resistance of samples.

References

1. Zeng, R., et al., Progress and Challenge for Magnesium Alloys as Biomaterials. Advanced Engineering Materials, 2008. 10(8): p. B3-B14.
2. Puleo, D.A. and W.W. Huh, Acute toxicity of metal ions in cultures of osteogenic cells derived from bone marrow stromal cells. Journal of Applied Biomaterials, 1995. 6(2): p. 109-116.
3. Witte, F., et al., Degradable biomaterials based on magnesium corrosion. Current Opinion in Solid State and Materials Science, 2008. 12(5-6): p. 63-72.
4. X.N. Gu, N. Li, Y.F. Zheng, et al. In vitro degradation performance and biological response of a Mg–Zn–Zr alloy[J]. Materials Science and Engineering: B, 2011, 176(20):1778-1784.
5. HORNBERGER H, VIRTANEN S, BOCCACCINI A R. Biomedical coatings on magnesium alloys—A review[J]. Acta Biomaterialia, 2012, 8(7): 2442−2455.
6. Hua Shuai et al., Preparation and corrosion resistance of hydroxyapatite coatings on magnesium alloys by electrochemical deposition. Functional Materials, 2017. 48(11): 11216-11220.
7. Sheng, M., et al., Ultrasonic irradiation and its application for improving the corrosion resistance of phosphate coatings on aluminum alloys. Ultrasonics Sonochemistry, 2010. 17(1): p. 21-25.
8. I. Tudela, Y. Zhang, M. Pal, I. Kerr, T.J. Mason, A.J. Cobley, Ultrasound-assisted electrodeposition of nickel: effect of ultrasonic power on the characteristics of thin coatings, Surf. Coat. Technol. 264 (2015) 49–59.
9. Y.L. Yang, Y.D. Wang, Y. Ren, C.S. He, J.N. Deng, J. Nan, J.G. Chen, L. Zuo, Singlewalled carbon nanotube-reinforced copper composite coatings prepared by electrodeposition under ultrasonic field, 62, 2008, pp. 47–50.
10. I. Tudela, Y. Zhang, M. Pal, I. Kerr, A.J. Cobley, Ultrasound-assisted electrodeposition of composite coatings with particles, Surf. Coat. Technol. 259 (2014) 363–373.
11. J. Liu, L. Yang, Z. Song, C. Xu, Microstructures and capacitance performance of MnO2 films fabricated by ultrasonic-assisted electrodeposition,Appl.Surf.Sci.478(2019)94–102.
12. Li, T., et al., Effects of ultrasonic treatment and current density on the properties of hydroxyapatite coating via electrodeposition and its in vitro biomineralization behavior. Materials Science and Engineering: C, 2019. 105: p. 110062.
13. T. Iwamoto, Y. Hieda, Y. Kogai, Effect of hydroxyapatite surface morphology on cell adhesion, Mater. Sci. Eng. C Mater. Biol. Appl. 69 (2016) 1263–1267.
14. Fathyunes, L. and J. Khalil-Allafi, Characterization and corrosion behavior of graphene oxide-hydroxyapatite composite coating applied by ultrasound-assisted pulse electrodeposition. Ceramics International, 2017. 43(16): p. 13885-13894.
15. M. Manoj, D. Mangalaraj, N. Ponpandian, C. Viswanathan, Core–shell hydroxyapatite/Mg nanostructures: surfactant free facile synthesis, characterization and their in vitro cell viability studies against leukaemia cancer cells (K562), RSC Adv. 5(2015) 48705–48711.
16. M.R. Etminanfar, J. Khalil-Allafi, On the electrodeposition of Ca-P coatings onnitinol alloy: a comparison between different surface modification methods, J.Mater. Eng. Perform. 25 (2016) 466–473. (20)
17. Lin, D. and X. Wang, Electrodeposition of hydroxyapatite coating on Co Ni Cr Mo substrate in dilute solution. Surface and Coatings Technology, 2010. 204(20): p. 3205-3213.
18. D. Gopi, E. Shinyjoy, A. Karthika, S. Nithiya, L. Kavitha, D. Rajeswari, T. Tang,Single walled carbon nanotubes reinforced mineralized hydroxyapatite composite coatings on titanium for improved biocompatible implant applications, RSC Adv. 5(2015) 36766–36778.
19. F. Marashi-Najafi, J. Khalil-Allafi, M.R. Etminanfar, Biocompatibility of hydroxyapatite coatings deposited by pulse electrodeposition technique on the Nitinol superelastic alloy, Mater. Sci. Eng. C 76 (2017) 278–286.
20. Yi Yuejun. Study on preparation of Hydroxyapatite coating on Medical magnesium alloy [D]. Central South University,2012.
21. L. Yan, Y. Xiang, J. Yu, Y. Wang, W. Cui, Fabrication of antibacterial and antiwear hydroxyapatite coatings via in situ chitosan-mediated pulse electrochemical deposition, ACS Appl. Mater. Interfaces 9 (5) (2017) 5023–5030.
22. I. Tudela, Y. Zhang, M. Pal, I. Kerr, A.J. Cobley, Ultrasound-assisted electrodeposition of composite coatings with particles, Surface & Coatings Technology 259 (2014) 363–373.

Modulating anionic activities in layered Li-rich cathode materials with inverse spinel $MnFe_2O_4$ coating

Liying Bao [1,2], Xinyu Zhu [1,2], Ning Li [1,2,*], Yongjian Li [1,2], Lifeng Xu [1,2], Lai Chen [1,2], Duanyun Cao [1,2], Yun Lu [1,2], Yuefeng Su [1,2,*], Youyou Fang [1,2], Chenxing Yang [1,2], Ran An [1,2]

[1] Beijing Key Laboratory of Environmental Science and Engineering, School of Material Science and Engineering, Beijing Institute of Technology, Beijing 100081, China
[2] Beijing Institute of Technology Chongqing Innovation Center, Chongqing, 401120, China

Abstract: Layered Li-rich cathode oxides can provide high specific capacity due to oxygen anion redox involving in charge compensation process during cycling, but there is a severe structural transition from layered to spinel accompanying with irreversible oxygen loss during cycling, which lead to electrochemical degradation. The current researches show that the irreversible oxygen evolution reaction of layered lithium-rich materials in the first cycle mainly comes from the surface lattice oxygen, so the surface modification by the materials with more stable structure is one of the effective ways to improve the electrochemical performance of layered lithium-rich materials. In this paper, we report a modified layered lithium-rich cathode material by surface coating of inverse spinel $MnFe_2O_4$. The inverse spinel has strong polarization effect on anion migration due to its different atoms occupying octahedral sites from layered structure and it can also modulate the Fermi level and stretching the O-O bond, thereby increasing the energy barrier for surface oxygen oxidization. Furthermore, the three-dimensional connected tunnel structure of the inverse spinel also makes the surface layer of the material have a faster lithium ion transferring rate, and a large number of lithium storable vacancies inside of it improved the Li+ intercalation efficiency, initial coulombic efficiency and rate performance. Compared with the unmodified sample, the modified material coated with 2% $MnFe_2O_4$ has superior cycle stability and outstanding rate performance. It is hoped our work would provide the knowledge for the future development of high-performance cathode materials for Li-ion batteries.

1. Introduction

Layered Li-rich oxides have become one of the most promising cathode candidates for next-generation high-energy-density Li-ion batteries due to their high specific discharge capacity (250 mAh/g) and low cost as well as environmental benignity. Different from the traditional layered cathode materials, such as $LiCoO_2$ and NCM cathode materials, where the capacities are mainly provided by transition metal redox for charge compensation, the additional high specific capacity of layered lithium-rich cathode materials mainly comes from the anionic oxygen redox ($O^{2-/n-}$)[1-3]. However, a major bottleneck in the practical application of layered lithium-rich cathode materials is the severe capacity and voltage fade, caused by the irreversible migration of transition metal ions leading to phase transform from layered to the spinel structure during cycling[4]. In addition, the side reaction occurs in the electrode and the electrolyte, and the oxygen evolution at surface caused by the deep oxidation of anionic oxygen, are the two other important reasons for the failure of the layered lithium-rich cathode material. At present, researchers have developed various surface protective layers to reduce the side reactions occurs in the electrode and the electrolyte and slow down the erosion of the electrode material by the electrolyte, including oxides (Al_2O_3[5], TiO_2[6], ZrO_2[7]), phosphates ($AlPO_4$[8], Li_3PO_4[9], $LiFePO_4$[10]) and fluorides (AlF_3[11], LiF[12], MgF_2[13]). However, the oxygen evolution caused by the deep oxidation of anionic oxygen still remains, which can not only bring about the safety problem of the battery, but also reduce the coulombic efficiency in the first cycle. More importantly, it can also deteriorate surface mass and charge transfer during the subsequent electrochemical reaction, thus, it is crucially needed to regulate the oxygen activities for the layered Li-rich materials.

The oxygen evolution reaction during initial charging process of the layered lithium-rich cathode material mainly includes the following processes: First, the undercoordinated Li-O bonds and the TM-O bonds near surface are broken, and then a large number of oxygen anions migrate to the surface and lose electrons to form new O-O bonds and release oxygen[14-15]. Therefore, to inhibit oxygen evolution of layered lithium-rich cathode materials, we can increase the binding energy of TM-O through surface doping with high-electronegative transition metal elements, thus making oxygen evolution thermodynamically difficult. At the same time, preventing the migration of oxygen ions to the surface is the second

* Corresponding author: ningli@bit.edu.cn

effective way to alleviate the oxygen evolution of layered lithium-rich cathode, and the migration process of anions is related to the effect of the external electric field[16, 17] and the formation of oxygen vacancies on the surface during synthesis and cycling[18]. Therefore, the driving force for the migration of oxygen anions can be weakened by coating the dielectric polarized materials on the surface of layered lithium-rich cathode material; and reducing the oxygen vacancies on the surface of the material by optimizing synthesis process. Since both layered lithium-rich oxides and spinel oxides belongs to cubic oxygen close-packing structure[19-21] and Mn-based spinel oxides are also widely used to catalyze the oxygen reduction reaction, so spinel oxides can be used as the coating agents to improve the cycle stability, rate capability, and most importantly to reduce the irreversible oxygen release of layered lithium-rich cathode materials. In general, spinel can be written as $\{A\}[B_2]X_4$ which is a cubic close-packed structure consisting of X^{2-} ions. A atom occupies tetrahedral positions and B atom occupy octahedral positions. If the A atoms exchange positions with half of the B atoms, then $\{A\}[B_2]X_4$ would transform from an ortho-spinel structure to an inverse-spinel structure—$\{B\}[AB]X_4$[22, 23]. In the Mn-Fe-based spinel, Mn atoms and half of Fe atoms co-occupy the octahedral sites, and the other half of Fe atoms occupy tetrahedral sites. Here, we report a co-precipitation method to build a $MnFe_2O_4$ coating layer on the surface of Li-rich layered oxide after annealing at 550 °C. Through systematically electrochemical and structural characterization, it is confirmed that $MnFe_2O_4$ has been successfully coated on surface of layered Li-rich cathode $Li_{1.2}Mn_{0.6}Ni_{0.2}O_2$ (labeled as LMO), and this coating modification can greatly enhance the initial coulombic efficiency, reducing the irreversible oxygen loss, and improving the cycle stability and rate performance. It is hoped this work should shed the light on the future cathode materials modification for advanced Li-ion batteries.

2. Experiment

Synthesis of $Li_{1.2}Mn_{0.6}Ni_{0.2}O_2$: According to the chemical formula, the stoichiometric metal salts $LiC_2H_3O_2.2H_2O$, $MnC_4H_6O_4.4H_2O$, $NiC_4H_6O_4.4H_2O$ were weighed (5% excess of lithium salt), then deionized water was added and the mixture was stirred to form a clear solution. In addition, the citric acid was dissolved into deionized water and added in the mixture, and the ammonia was added to adjust the pH to 7. Finally, the mixture solution was heated at 80 °C until to form a green gel. The obtained gel was scraped and dried in an oven, and then calcined in a muffle furnace at 450 °C for 5 hours and at 900 °C for 12 hours with a heating rate of 5 °C/minute.

Spinel $MnFe_2O_4$ coating modification: 1.0 gram of the pristine LMO material was dispersed in deionized water, which is marked as solution A. Then stoichiometric anhydrous $FeCl_3$ and $MnSO_4.H_2O$ were dissolved in deionized water with stirring, which is marked as solution B. The solution B was dropwise added into the solution A and stirred uniformly. Then NaOH was added to -adjust pH value to ~ 11, and after reacting for 2 hours, the mixed solution was then centrifuged and dried in an oven. Finally, the dry powder was calcinated in a muffle furnace at 550 °C for 5h with heating rate of 5 °C/min.

Structure characterization and electrochemical tests: The crystal structure of the sample was characterized by X-ray diffraction (XRD), 2θ = 10–90 °, 10 °/min, step size of 0.02 °. The morphology and elemental composition of the samples was characterized by scanning electron microscopy (SEM-EDS). All samples were assembled into a coin cell with using lithium metal as the anode, where the electrode was the uniform mass mixture for the ratio of 8(active material):1(acetylene black):1(PVDF). Here, the average mass loading of the active cathode material on carbon-coated aluminum foil is 2.0~3.0 mg/cm^{-2}. The cathodes and anodes are separated by two two Celgard 2500films, and 1 mol $LiPF_6$ was dissolved in EC/DEC (1:1 v/v) as the electrolyte. The cells were cycled by a LAND device at 30 °C between 2–4.8 V (nominal specific capacity of 250 mAh/g).

3. Results and discussion

In order to characterize the structure of layered lithium-rich cathode material before and after coating, the XRD patterns of all the samples were collected. Figure 1 shows the X-ray diffraction patterns of the as-synthesized pristine material LMO and the modified materials with coating ratios (molar ratio) of 1.5%, 2%, and 4% (labeled as LMO1.5, LMO2, and LMO4). It can be seen that all the patterns of the pristine layered lithium-rich material can be well indexed with layered structure with R3-m symmetry, except the superlattice peak at 20-25° which is related to the $LiMn_6$ or $LiMn_5Ni$ ordering in the TM slabs [24, 25]. The diffraction peaks of the three modified samples are similar to the pristine material LMO, indicating that the crystal structures of the modified samples have not changed significantly after surface coating. However, the LMO1.5 sample has an obvious spinel structure impurity phase, and the super-lattice peaks of the three modified samples near 20-25 ° gradually weaken with the increasing of the coating amount. The decreasing of intensity ratio I003/I104 (LMO-1.31, LMO1.5-1.33, LMO2-1.14, LMO4-1.21) of the modified samples also indicates the increasing in the degree of Li/Ni mixing after surface coating, while the appropriate Li/Ni mixing may stabilize the layered structure and the improve the cycle stability.

Figure 2 shows the scanning electron microscope (SEM) images of the pristine LMO and the sample with a coating amount of 2%. The morphologies of all samples present as 100-300 nm primary particles with different degree of agglkomeration. It can be seen from Fig.2a that as the coating amount increases, the surface of primary particles gradually becomes rough, which indicates that the surface of the material is successfully covered by coating agent. In order to investigate the uniformity of coating, we performed EDS mapping tests on the pristine material LMO(Fig.2b) and the modified material LMO2(Fig.2c). The EDS mapping results show that in the pristine material LMO, Mn, Ni and O are uniformly dispersed. As for the modified sample LMO2, the Fe, Mn and Ni are uniform dispersed as well. It can be concluded that the coating layer has been successfully coated on the surface of the LMO sample.

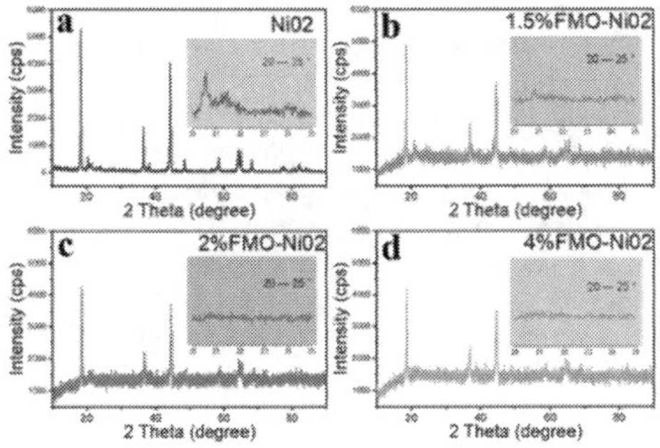

Figure 1. XRD pattern of the four samples; (a)—LMO pristine; (b)—LMO1.5; (c)—LMO2; (d)—LMO4.

In order to investigate the effect of MFO coating on the electrochemical performance of layered Li-rich material, Figure 3 shows the first cycle curve and the corresponding dQ/dV curve of the pristine LMO and the modified samples at a current density of 0.1 C, and Table 1 shows the first charge-discharge electrochemical data. It can be seen that the LMO, LMO1.5, LMO2 and LMO4 samples can yield initial charge capacities of 326.7 mAh/g, 240.6 mAh/g, 298.4 mAh/g, and 273.7 mAh/g, and the gradual decrease of charge capaicty with the increase of MFO coating, indicates that the initial oxygen release was alleviated. In addition, with the increase of the coating layer, the irreversible capacity in the first cycle can also be effectively reduced and the initial coulombic efficiency is increased from 81.5% of the pristine LMO to 96.4% of LMO4. The initial charge-discharge curves of all samples are shown in Fig. 3a. The charge curves are composed of the slope region and a plateau at around 4.5 V, corresponding to the oxidation peaks of Ni and O respectively as shown in the dQ/dV plots in Fig. 3b. In Figure 3b, there is also an obvious sharp reduction peak around 2.8V, corresponding to the Mn^{3+}/Mn^{4+} redox couple of the inverse spinel $MnFe_2O_4$ structure. Combined with the EDS spectrum, it further confirms that the inverse spinel $MnFe_2O_4$ is successfully coated on the surface of the material. More importantly, it can be seen that with increasing of coating amount, the potential of the plateau increases, corresponding to the right shift of O oxidization peak in the dQ/dV plots (Fig. 3b). This phenomenon infers that the oxidation of oxygen becomes difficult, that is, the irreversible release of oxygen is alleviated.[22, 23]It is believed that the inverse spinel $MnFe_2O_4$ coating layer promotes the charge polarization effect and prevent the migration of oxygen anion to surface. Meanwhile, the inverse spinel has stronger binding energy to oxygen anion to prevent the formation of the O-O dimer at the surface. Therefore, the coating of inverse spinel $MnFe_2O_4$ alleviate the oxygen evolution reaction both kinetically and thermodynamically. In addition, lithium-deficient spinel $MnFe_2O_4$, can provide lithium ion vacancies to accommodate many extra lithium ions, and thus improve the initial coulombic efficiency. The discharge curve of the sample is a gradually decreasing slope. It can be seen an obvious plateau appears below the discharge 3.0 V among the LMO2 and LMO4 samples.

Figure 2. (a)SEM patterns of the modified samples; (b)SEM patterns and corresponding EDS spectroscopy of LMO; (c)EDS spectroscopy of LMO2.

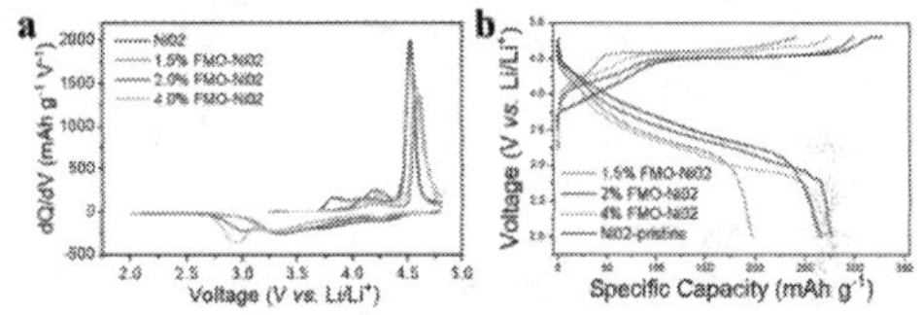

Figure 3. The first charge-discharge curves of the four samples (b) and corresponding dQ/dV curves (a).

Table 1. The electrochemical data of first cycle for the four samples.

	Charge capacity mAh/g	Discharge capacity mAh/g	ICE	Irreversible capacity mAh/g
LMO1.5	240.6	197.4	82.06%	43.2
LMO2	298.4	276.9	92.82%	21.5
LMO4	273.7	263.8	96.37%	9.9
LMO-Ni02	326.7	266.4	81.52%	60.3

Figure 4 shows the charge-discharge curves of the pristine LMO and the modified samples under 0.1 C current density for 50 cycles. With the increase of cycling number, all samples show capacity decay. The discharge capacity of the pristine material LMO decreased from 266.4 mAh/g to 194.6 mAh/g after 50 cycles, and the capacity retention is only 73.5%. Among the modified samples, the LMO1.5 sample did not form a good coating structure and showed rapid capacity decay and poor capacity retention while the LMO2 and LMO4 samples shows outstanding cycle performance. The LMO2 sample shows the best capacity retention, from the initial capacity of 276.9 mAh/g to the 236.9 mAh/g after 50 cycles with capacity retention of 85.6%. In addition, a small plateau around 2.8 V remains on the discharge curves of LMO2 and LMO4 samples, indicating the formation of a stable inverse spinel structure coating layer[26, 27].

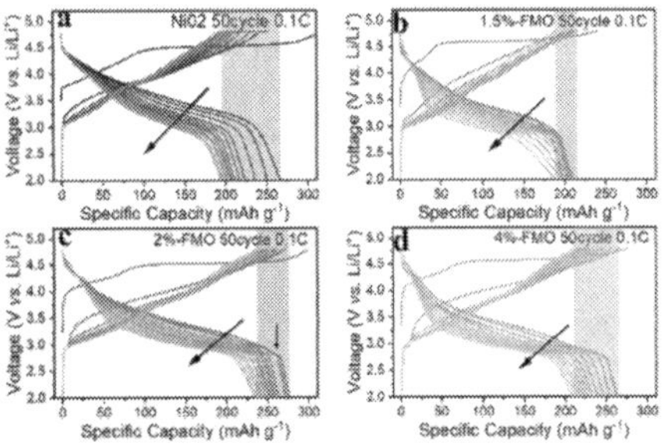

Figure 4. The charge-discharge curves of 50 cycles at a current density of 0.1 C (a)LMO、(b)LMO1.5、(c)LMO2、(d)LMO4.

In order to study the effect of the inverse spinel coating on the rate performance of the material, Figure 6 shows the rate performance test of all the samples at different current densities. Under 0.3 C, 0.5 C, 1 C, 3 C, 5 C and 10 C rates, the pristine LMO shows discharge capacities of 218 mAh/g, 206 mAh/g, 190 mAh/g, 149 mAh/g, 110 mAh/g, 38 mAh/g, respectively. After surface coating with the Li^+ conductive inverse spinel, the modified samples display superior rate performance The LMO2 sample shows the best rate performance, and exhibits discharge specific capacities of 257 mAh/g, 252 mAh/g, 243 mAh/g, 218 mAh/g, 189 mAh/g, and 113 mAh/g at the current density of 0.3 C, 0.5 C, 1 C, 3 C, 5 C, 10 C, respectively. It is indicated that the inverse spinel structure can provide a three-dimensional lithium ion transport channel, which accelerates the interfacial lithium ions mass transfer from the electrode to the electrolyte.

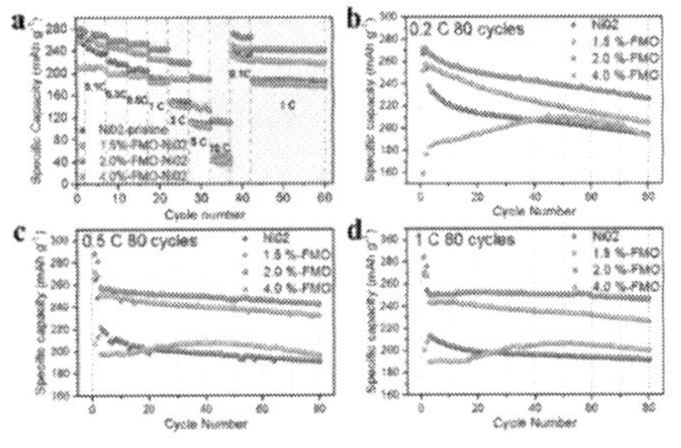

Figure 5. (a)The rate performance of the four samples; (b-d) The discharge capacity curves of the four samples at the current density of 0.2 C, 0.5 C and 1 C for 80 cycles.

Figures 5b-d shows the rate cycling performance of all the samples at 0.2 C, 0.5 C and 1 C. At different rate, except for that of LMO1.5, the capacities of all other samples gradually decrease with the cycling number. The modified samples LMO2 and LMO4 show much higher capacity and capacity retention than the pristine materials at all different rates. As shown in Fig. 6b–d, the pristine LMO have initial capacities of 238 mAh/g, 221 mAh/g and 213 mAh/g at 0.2 C, 0.5 C, and 1 C, respectively. After 80 cycles, it shows the discharge capacity of 194 mAh/g, 191 mAh/g and 192 mAh/g with the capacity retention are 81.5%, 86.4%, and 90.1%, respectively. After modification, the initial capacities of the sample LMO2 greatly increase to 267 mAh/g, 257 mAh/g, 251 mAh/g at 0.2 C, 0.5 C, and 1 C, respectively. And after 80 cycles it shows the capacities of 226 mAh/g, 242 mAh/g and 246 mAh/g with the capacity retention are 84.6%, 94.2%, and 98.0%. The significant enhancement in the discharge specific capacity and capacity retention of the modified materials is mainly due to that the stable inverse spinel coating structure on the surface enables the substantial improvement in the lithium ion conductivity of the material. In addition, it can be seen that the excess inverse spinel coating would not further increase the rate performance, which can be attributed to that the obtained incompletely crystallized inverse spinel coating would increase the surface charge transfer resistance by low-temperature processing. Therefore, only a suitable coating amount can optimize the rate performance of lithium-rich manganese-based cathode material.

4. Conclusion

In summary, we have successfully prepared $MnFe_2O_4$ inverse spinel-coated layered Li-rich cathode materials, and electric field effect of the inverse spinel surface coating layer can suppress the outward migration of O anions, and its stronger binding energy to oxygen anion can also slow down the anion migration and oxygen gas release. In addition, the oxygen oxidation reaction in the first cycle become thermodynamically difficult and its kinetic has been enhanced after coating. With these merits of inverse spinel surface coating, the layered Li-rich cathodes show greatly reduced the irreversible initial capacity and improved initial coulombic efficiency (LMO2-92.82%, LMO-81.52%), and the LMO2 sample shows best capacity retention of 85.6% capacity retention after 50 cycles at 0.1C and outstanding rate capability of 267 mAh/g, 257 mAh/g, 251 mAh/g at 0.2 C, 0.5 C, and 1 C with high capacity retention). Our results suggest surface coating with inverse spinel enable to tune the oxygen activities in Li-rich layered oxides, and such finding should provide valuable knowledge for development of the high-performance cathode materials for the next-generation Li-ion batteries.

Acknowledgements

This work was supported by National Key R&D Program of China 2021YFC2902905, Chongqing Talents Plan for Young Talents CQYC202005032, National Natural Science Foundation of China (22109010, 21875022, 22179008), N. Li, L. Chen and D. Cao acknowledge the support from Beijing Institute of Technology Research Fund Program for Young Scholars.

References

1. Assat, G.; Tarascon, J.-M. Nature Energy 2018, 3, (5), 373-386.
2. Assat, G.; Foix, D.; Delacourt, C.; Iadecola, A.; Dedryvere, R.; Tarascon, J. M. Nat Commun 2017, 8, (1), 2219.

3. Rozier, P.; Tarascon, J. M. Journal of The Electrochemical Society 2015, 162, (14), A2490-A2499.
4. Ates, M. N.; Jia, Q.; Shah, A.; Busnaina, A.; Mukerjee, S.; Abraham, K. M. Journal of The Electrochemical Society 2013, 161, (3), A290-A301.
5. Kobayashi, G.; Irii, Y.; Matsumoto, F.; Ito, A.; Ohsawa, Y.; Yamamoto, S.; Cui, Y.; Son, J.-Y.; Sato, Y. Journal of Power Sources 2016, 303, 250-256.
6. Kim, S.-J.; Kim, M.-C.; Kwak, D.-H.; Kim, D.-M.; Lee, G.-H.; Choe, H.-S.; Park, K.-W. Journal of Power Sources 2016, 304, 119-127.
7. Wang, Z.; Liu, E.; Guo, L.; Shi, C.; He, C.; Li, J.; Zhao, N. Surface and Coatings Technology 2013, 235, 570-576.
8. Wu, F.; Zhang, X.; Zhao, T.; Li, L.; Xie, M.; Chen, R. ACS Appl Mater Interfaces 2015, 7, (6), 3773-81.
9. Chen, D.; Zheng, F.; Li, L.; Chen, M.; Zhong, X.; Li, W.; Lu, L. Journal of Power Sources 2017, 341, 147-155.
10. He, L.; Xu, J. M.; Wang, Y. J.; Zhang, C. J. ACTA PHYSICO-CHIMICA SINICA 2017, 33, (8), 1605-1613.
11. Li, G. R.; Feng, X.; Ding, Y.; Ye, S. H.; Gao, X. P. Electrochimica Acta 2012, 78, 308-315.
12. Fu, C.; Wang, J.; Wang, J.; Meng, L.; Zhang, W.; Li, X.; Li, L. Journal of Materials Chemistry A 2019, 7, (40), 23149-23161.
13. Sun, S.; Wan, N.; Wu, Q.; Zhang, X.; Pan, D.; Bai, Y.; Lu, X. Solid State Ionics 2015, 278, 85-90.
14. Lin, F.; Markus, I. M.; Nordlund, D.; Weng, T.-C.; Asta, M. D.; Xin, H. L.; Doeff, M. M. Nature Communications 2014, 5, (1).
15. Strehle, B.; Kleiner, K.; Jung, R.; Chesneau, F.; Mendez, M.; Gasteiger, H. A.; Piana, M. Journal of The Electrochemical Society 2017, 164, (2), A400-A406.
16. Hu, E.; Yu, X.; Lin, R.; Bi, X.; Lu, J.; Bak, S.; Nam, K.-W.; Xin, H. L.; Jaye, C.; Fischer, D. A.; Amine, K.; Yang, X.-Q. Nature Energy 2018, 3, (8), 690-698.
17. Li, X.; Qiao, Y.; Guo, S.; Xu, Z.; Zhu, H.; Zhang, X.; Yuan, Y.; He, P.; Ishida, M.; Zhou, H. Adv Mater 2018, 30, (14), e1705197.
18. Li, Q.; Ning, D.; Zhou, D.; An, K.; Wong, D.; Zhang, L.; Chen, Z.; Schuck, G.; Schulz, C.; Xu, Z.; Schumacher, G.; Liu, X. Journal of Materials Chemistry A 2020, 8, (16), 7733-7745.
19. Peng, J.; Li, Y.; Chen, Z.; Liang, G.; Hu, S.; Zhou, T.; Zheng, F.; Pan, Q.; Wang, H.; Li, Q.; Liu, J.; Guo, Z. ACS Nano 2021.
20. Zhang, W.; Sun, Y.; Deng, H.; Ma, J.; Zeng, Y.; Zhu, Z.; Lv, Z.; Xia, H.; Ge, X.; Cao, S.; Xiao, Y.; Xi, S.; Du, Y.; Cao, A.; Chen, X. Adv Mater 2020, 32, (19), e2000496.
21. Luo, D.; Ding, X.; Fan, J.; Zhang, Z.; Liu, P.; Yang, X.; Guo, J.; Sun, S.; Lin, Z. Angew Chem Int Ed Engl 2020, 59, (51), 23061-23066.
22. Béjar, J.; Álvarez-Contreras, L.; Ledesma-García, J.; Arjona, N.; Arriaga, L. G. Journal of Electroanalytical Chemistry 2019, 847.
23. Li, C.; Han, X.; Cheng, F.; Hu, Y.; Chen, C.; Chen, J. Nat Commun 2015, 6, 7345.
24. Wang, G.; Yi, L.; Yu, R.; Wang, X.; Wang, Y.; Liu, Z.; Wu, B.; Liu, M.; Zhang, X.; Yang, X.; Xiong, X.; Liu, M. ACS Appl Mater Interfaces 2017, 9, (30), 25358-25368.
25. Hy, S.; Felix, F.; Rick, J.; Su, W. N.; Hwang, B. J. J Am Chem Soc 2014, 136, (3), 999-1007.
26. Eum, D.; Kim, B.; Song, J. H.; Park, H.; Jang, H. Y.; Kim, S. J.; Cho, S. P.; Lee, M. H.; Heo, J. H.; Park, J.; Ko, Y.; Park, S. K.; Kim, J.; Oh, K.; Kim, D. H.; Kang, S. J.; Kang, K. Nat Mater 2022.
27. Pei, Y.; Xu, C. Y.; Xiao, Y. C.; Chen, Q.; Huang, B.; Li, B.; Li, S.; Zhen, L.; Cao, G. Advanced Functional Materials 2017, 27, (7).

Research status and prospect of dissimilar light alloy welding technology

Xiaohui He [1, a], Bin Liu [2, b], Xiong Zhou [2, c], Cheng Li [1, d], Xuhui Zhang [3, e], Junjie Chang [3, f]

[1] Chongqing Vocational Institute of Engineering, Chongqing 402260, China
[2] Chongqing Xinxiya Aluminum Industry Group, Chongqing 402284, China
[3] Chongqing Longyu Precision Copper Pipe Co., Ltd, Chongqing 402284, China

Abstract: Magnesium alloys, aluminum alloys and titanium alloys are currently widely used light alloy materials with good prospects. The welding of Mg/Al/Ti dissimilar light alloys has a wide range of application requirements in aerospace, automobile manufacturing, electronic information and other fields, and is a hot topic of current research. This article summarizes the welding research results of Mg/Al/Ti dissimilar light alloys in recent years, outlines the current research status of welding technology of Mg/Al/Ti dissimilar light alloys, analyzes the common problems and solutions in welding, and proposes further Carry out the research of Mg/Al/Ti welding technology and basic theory, look forward to the effective ways to improve the welding performance of Mg/Al/Ti dissimilar light alloys and future research directions, and promote the promotion of Mg/Al/Ti dissimilar light alloy joints The practicality of it is of great significance.

Keywords: Magnesium alloy; aluminum alloy; titanium alloy; welding technology.

1. Introduction

With the rapid development of science and technology, higher requirements are put forward for the lightweight of metal structural materials. While meeting the design cost requirements, the performance advantages of the materials must be fully reflected. The applications of the three light alloys of aluminum, magnesium, and titanium are extensive and cross-cutting, and meet the current requirements for lightweight. Therefore, it is necessary to weld the three to form composite structural parts with complementary advantages. At present, research in this area has received widespread attention. For example, magnesium-aluminum composite structural parts have the characteristics of low density and high specific strength. Under the condition that the strength is ensured, the weight of the structural parts can be effectively reduced and the goal of light weighting can be achieved. Titanium-aluminum composite structural parts can not only take advantage of the high strength of titanium alloy, but also show the light weight characteristics of aluminum alloy. The sandwich structure of the American NASA YF-12 fighter wing is a composite structure. It has two parts, one is the skin, mainly made of titanium alloy, which is used to enhance the strength of the aircraft structure and ensure the limit of vibration resistance, and the other is the sandwich, which is made of aluminum. Alloys are mainly used to reduce the weight of aircraft; the new composite material automobile exhaust system developed by the German Titan company has a weight reduction of 40% compared to the steel exhaust system, and can be used in service at temperatures exceeding 800°C [1].

With the development of many industrial branches in the direction of lightweight, the effective connection of magnesium/aluminum/titanium dissimilar light alloys by welding has broad application prospects. Dissimilar light alloy welded joints with strong interface bonding can not only give full play to the respective performance advantages of dissimilar light alloys, but also reduce structural weight, save materials, and balance material performance and economic effects.

2. Weldability analysis of Mg/Al/Ti

In the welding of dissimilar light alloy materials, melting point, atomic radius, crystal structure, thermal conductivity and linear expansion coefficient are the main influencing factors. According to the physical properties of the three light alloys of magnesium/aluminum/titanium, first, magnesium alloys and aluminum alloys have lower melting points, while titanium alloys have a higher melting point of about 1000°C. Magnesium alloys and aluminum alloys with low melting points, When the temperature exceeds 660°C, it is in a molten state, and the high melting point titanium alloy is still mainly solid, which makes the magnesium alloy and aluminum alloy with low melting point easy to evaporate and lose the alloy elements during welding, which makes it impossible to weld the joint. Second, the thermal conductivity and linear expansion coefficients of the three light alloys are greatly different, which makes the deformation capacity of the light alloys in the heating and cooling processes different. When the molten pool is crystallized, large residual stresses will be generated in the joints after welding. One stress cannot be eliminated by heat treatment. For the metals on both sides of the weld, the stress state is different, which easily leads to cracks in the weld, and even the phenomenon of separation between the weld and the base metal. Third, the three light alloys have huge differences in lattice constants, lattice structures, atomic radii, etc., resulting in metallurgical incompatibility between light alloys, low mutual solubility, and low mutual solubility, which makes it very difficult to form a fusion zone. Fourth, the three light alloys are all typical active metals, which easily react with oxygen to produce oxides such as MgO, Al2O3, and TiO2. In particular, Al2O3 has a very compact structure and a very high melting point, reaching 2050°C. Removal is very difficult. During the welding process, the oxides produced will seriously affect the bonding between the magnesium/aluminum/titanium dissimilar light alloys, and cause defects such as inclusions and cracks in the joint area, resulting in poor joint bonding performance and difficulty in welding. In addition, combined with the Mg-Al binary phase diagram (Figure 1), we can conclude that magnesium and aluminum can easily form a large amount of Mg-Al brittle intermetallic compounds such as Mg2Al3 and Mg17A112 during welding. Under different high temperature conditions, titanium and aluminum react, producing many Ti-Al brittle intermetallic compounds, such as TiAl. The intermetallic compound lowers the bonding strength of the welded joint, causing the joint to be prone to brittle fracture. Therefore, the above analysis shows that the weldability of the magnesium/aluminum/titanium dissimilar light alloy is poor and the welding is very difficult [2].

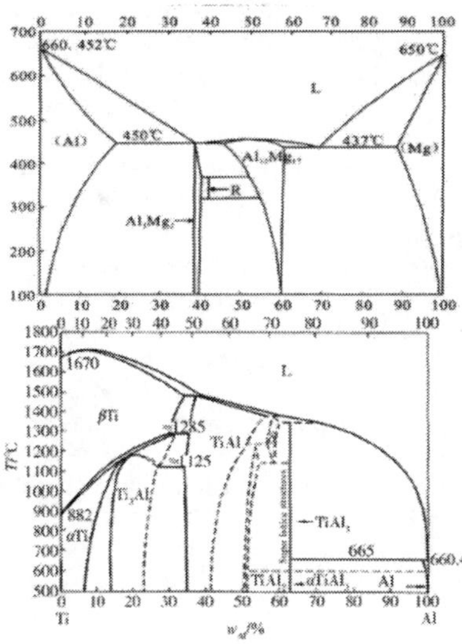

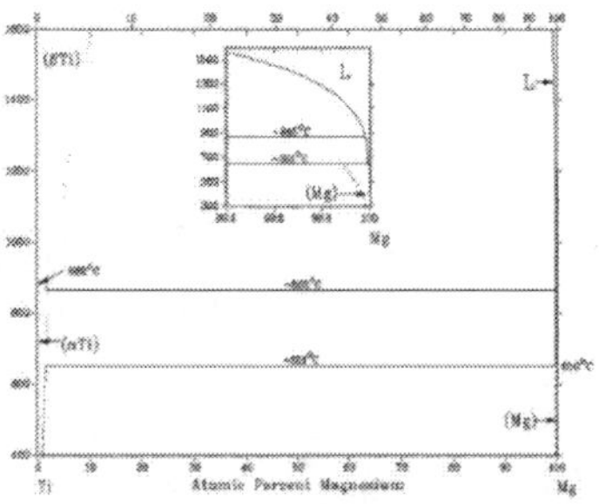

Fig. 1 Binary phase diagram between Mg、Al and Ti

3. Dissimilar light alloy welding technology of Mg and Al

At this stage, for magnesium alloys, the welding methods commonly used for magnesium alloys and aluminum alloys are fusion welding and solid welding. Among them, the most effective welding method is fusion welding, such as tungsten argon arc welding (TIG), electron beam welding (EBW).) And laser welding (LAW), etc. Ben-Artzy and Munitz et al. [3] investigated the weldability of 6063 aluminum alloy with AM50 and AZ31 through TIG and EBW, and came to the following conclusions, the application of TIG will produce a heat-affected zone around the magnesium alloy, which is around AZ31 magnesium alloy. The width of the heat-affected zone is several millimeters, and the width of the heat-affected zone around AM50 magnesium alloy is hundreds of microns, and there is a eutectic structure at the joint fusion position. The core position of the fusion zone is γ-$Mg_{17}(Al,Zn)_{12}$ The phase is dominant, and a continuous brittle phase is formed near the grain boundary. The application of EBW will not produce a heat-affected zone around the magnesium alloy, but there will be β-$Mg_{17}(Al,Zn)_{12}$ phases in the joint fusion zone. Peng Liu et al. [4] used TIG to weld magnesium (230) and aluminum (1060). They found that there are brittle intermetallic compounds of $Mg_{17}Al_{12}$ and Mg_2Al_3 in the fusion zone around the Mg welded joint. For the fusion zone, it has two Part, one is the melting crystallization zone, and the other is the semi-melting crystallization zone. For the melting zone, it has two kinds of structure, one is columnar crystal, and the other is columnar dendritic structure..Rattana Borrisutthekuld et al. [5] used LAW to weld magnesium and aluminum alloys, because many brittle intermetallic compounds, such as $Mg_{17}Al_{12}$, were formed during the welding process. The thickness of the inter-compound layer is reduced to enable them to be effectively overlapped. Takehiko et al.[6] compared and analyzed the tensile properties of magnesium alloy and aluminum alloy spot welded joints, and found that for resistance spot welded joints, its strength is only 1/4 to 1/5 of the strength of the same magnesium alloy. The thickness of the intermetallic compound layer between Mg and Ag decreases with the increase of the thickness of the added Ag metal foil, and the thickness of the eutectic reaction layer between Ag and Al increases with the increase of the thickness of the Ag metal foil, thereby improving the joint strength . Liu Xujing et al. [7] used laser-TIG composite heat source to weld different types of magnesium alloys and aluminum alloys. The study found that laser-TIG composite heat source welding is fast and can effectively stir the molten pool, thereby improving the welding of two light alloys. The microstructure shows that the intermetallic compound of the welded joint is transformed from uninterrupted layered to diffuse.

Solid phase welding uses physical effects such as friction, pressure or thermal diffusion to overcome the roughness of the two connecting surfaces, and remove oxide films or other impurities, so that the atoms on the two connecting surfaces can approach each other to the lattice distance, and then Achieve solid-state connections. At this stage, there are many solid-phase welding methods, such as diffusion welding, friction stir welding (FSW) and explosive welding. Solid phase welding is used to weld Mg/Al dissimilar metals, which overcomes defects such as cracks, deformation, and pores caused by fusion welding, and can obtain joints with excellent performance. Li Yajiang's research group [8-10] deeply analyzed the influence of vacuum diffusion welding process on the microstructure and mechanical properties of magnesium/aluminum light alloy welded joints. Through research, it is concluded that under the best process parameters such as heating temperature and holding time, a dense Mg/Al dissimilar metal diffusion welding joint can be obtained. The interface transition zone formed by the diffusion of Mg/Al atoms is mainly composed of three parts. One is the Mg_2Al_3 phase layer, the second is the Mg-Al phase layer, and the third is the Mg_3Al_2 phase layer. These new features have an important impact on the performance of the joint. Liu Peng et al. deeply explored the vacuum diffusion welding method of Mg/Al dissimilar light alloy. The study found that through effective control of vacuum diffusion welding process parameters, such as holding time, heating temperature, and welding pressure, a welded joint with good performance can be obtained. The maximum shear strength can reach 18.94MPa. For Mg/Al diffusion welded joint area , It has three phase layers, one is Mg_2Al_3 phase, which is the transition zone on the aluminum side, the other is Mg-Al phase, which is the intermediate diffusion zone, and the third is Mg_3Al_2 phase, which is the transition zone on the magnesium side. The hardness of the different phase layers is very large. The difference seriously affects the performance of the joint. Zhao Limin et al. used zinc interlayer to connect Mg/Al by diffusion welding at 365°C. Through research, it is concluded that there are mainly three layers in the interface area of magnesium-aluminum welded joints, one is the aluminum-zinc reaction layer, the other is the insufficient diffusion layer, and the third is the zinc-magnesium reaction diffusion layer. The addition of zinc interlayer effectively prevents the diffusion between magnesium and aluminum. The shear strength of zinc interlayer magnesium-aluminum diffusion welded joints far exceeds the shear strength of magnesium-aluminum direct vacuum diffusion welded joints. Fracture analysis shows that the fracture occurs in the zinc-magnesium reactive diffusion layer.

In addition, in recent years, many new welding methods have been produced, such as electromagnetic pulse welding, high-energy pulse laser welding, multi-beam laser welding, etc., which have effectively avoided various defects caused by welding of magnesium and aluminum dissimilar light alloys. Good seam mechanical properties. However, no matter what kind of welding method, it will be restricted under the established welding situation. How to effectively play the role of different welding methods is one of the topics that scientific researchers must pay attention to in the future.

4. Dissimilar light alloy welding technology of Mg and Ti

The connection of magnesium alloy and titanium alloy is restricted by the huge difference in properties between the two, and it is difficult to effectively connect the Mg/Ti alloy through the previous fusion welding process. At this stage, most of the welding research related to Mg/Ti alloys focuses on the following three aspects. First, special welding methods are used, such as transient liquid phase diffusion welding (TLP), laser welding, explosive welding, ultrasonic spot welding (USW), etc. Even for arc welding, optimization methods such as cold metal transition technology (CMT) or ultrasonic assistance are used; second, other alloy elements are added to effectively adjust the metallurgical reaction during the welding process; third, special welding methods are combined with other elements The organic connection can effectively solve the welding problem of Mg/Ti alloy. Xiong Jiangtao et al. used aluminum foil as an intermediate layer to carry out transient liquid phase diffusion welding of Mg/Ti alloy, and thoroughly explored its connection principle. GAO et al. used laser welding technology to realize Mg/Ti alloy butt welding and revealed its connection principle. Relevant personnel from Kumamoto University in Japan and Komazawa University in Japan used magnesium alloy sheet as the intermediate layer, and used explosive welding technology to connect TP270 titanium alloy plate and AZ31 magnesium alloy plate underwater. When the explosive device tilt angle α is 30° When the distance D between the explosive and the sample is between 5-35mm, a magnesium-titanium alloy joint with excellent performance can be obtained. Daxin Ren et al. of Dalian University of Technology realized the ultrasonic spot welding (USW) connection of Mg/Ti dissimilar alloys. Within a certain process range, with the increase of ultrasonic energy, the overall strength of the joint showed an upward trend. Lan Shaojuan et al. used heterogeneous ultrasonic welding to conduct a large number of experiments on magnesium and titanium alloys to study the influence of welding parameters on welding results, analyze the formation principle of solder joints, and optimize the ultrasonic welding process for future research on ultrasonic welding technology. Provided for reference. Japanese scholar Watanabe T et al. took magnesium and titanium alloy sheets as examples, carried out friction stir welding on them, and carried out in-depth research on the microstructure of the interface diffusion zone and the mechanical properties of the joints, and realized the comparison between ZK60 magnesium alloy and titanium alloy sheets. Metallurgical connection between. Masayuki Aonuma et al. performed FSW welding on three magnesium alloys with different Al content and Ti, respectively, and analyzed the influence of Al element on the interface structure of welded joints. The results showed that the thickness of the Ti-Al compound layer at the interface varies with the content of the magnesium base material. The amount of A1 increases, and the tensile strength of the joint decreases with the increase of the amount of A1 in the magnesium base metal. R. Cao et al. used CMT technology to realize the welding and brazing connection of titanium and magnesium alloys. Through research, it is concluded that with the cooperation of AZ61 magnesium alloy welding wire, the connection mode between magnesium and titanium is higher, and the strength of the joint is higher. Although the research on the welding technology of Mg/Ti dissimilar light alloys has made some progress, it is still in the initial stage. The influence of welding parameters and material properties on the welding effect, and the formation mechanism of the welding joints need to be further studied. Especially in view of the special property differences between Mg/Ti alloys, a reliable connection of Mg/Ti can be achieved only by applying various special welding methods with other elements to adjust the interface reaction.

5. Dissimilar light alloy welding technology of Al and Ti

At this stage, there are many welding methods for Ti/Al alloys, such as argon tungsten arc welding, laser welding, friction stir welding, diffusion welding, and fusion brazing. Ma Zhipeng et al. used tungsten argon arc welding to connect TC4 titanium alloy and aluminum alloy. Through research, it was found that when the welding current is 110-120A, the tensile strength of the joint reaches the maximum value of 158MPa, and the weld structure is mainly composed of A1-Si eutectic phase and $TiAl_3$ intermetallic compound composition. When the welding current exceeds 140A, the thickness of the intermetallic compound at the joint weld increases to 300μm, resulting in a continuous decrease in the tensile strength of the joint. Foreign scholars Vaidya et al. studied the 4KW split beam YAG laser welding Ti_6Al_4V titanium alloy and 6056 aluminum alloy. The laser beam irradiates the aluminum alloy to melt and wrap it on the titanium alloy to directly form a welded joint. Only $TiAl_3$ is formed at the interface of the welded joint. Phase, the maximum tensile strength can reach 255MPa, and the joints are all broken in the base material. Chen Yuhua et al. used friction stir welding to butt TC1 titanium alloy and LF6 aluminum alloy. Through research, it is concluded that the interface between the titanium alloy base material and the weld nugget is not flat, and there are white granular intermetallic compounds at the boundary position, while the interface between the aluminum alloy base material and the weld nugget is relatively flat. There are two types of Ti in the aluminum base metal in the weld nugget area. -Al intermetallic compound. Yao Wei et al. used diffusion welding to join 5A06/TA2 dissimilar light alloys. Research has shown that when the welding temperature is higher than 525°C, the $Al_{18}Ti_2Mg_3$ type mesophase appears in the joint. The strength of the welded joint is closely related to the new phase area and is related to solid solution metallurgy. The bonding zone is closely connected. With the growth of the new phase zone of the diffusion reaction, the bonding strength continues to decrease, and with the increase of the solid solution metallurgical bonding zone, the bonding strength continues to increase.

The physical and chemical properties of titanium/aluminum dissimilar alloys are quite different, and it is difficult to obtain welded joints with excellent properties. Although the current welding technology can solve welding defects such as missing alloy elements, high temperature oxidation, pores and cracks, it cannot solve the generation of brittle intermetallic compounds. The type, quantity and distribution of intermetallic compounds will have a direct impact on the performance of the joint. Therefore, controlling the production of brittle intermetallic compounds during welding is still an important issue in the welding of titanium/aluminum dissimilar alloys.

6. Conclusion

At present, many scholars have conducted a lot of research on the welding methods of Mg/Al/Ti dissimilar light alloys. Among them, welding technologies such as fusion welding, brazing, diffusion welding, explosive welding, friction stir welding, electromagnetic pulse welding, etc. but there are still certain problems and limitations. Among them, brazing requires a vacuum working environment, which will be restricted by the size and shape of the workpiece. Brazing technology, brazing material design, etc. have not formed a mature theoretical system. The strength of brazing joints still needs to be improved; diffusion welding requires high parts assembly, expensive equipment, and long heat preservation time, which is difficult to meet industrial production requirements. Need; Fusion welding is prone to produce intermetallic chemicals, which can easily deteriorate the performance of the joint; the explosive speed in explosive welding is not stable, and the quality of the obtained weld is not good; friction stir welding has high requirements for the joint form and is only suitable for plate welding; Electromagnetic pulse welding is still in the initial stage of research and development, and the low power of the welding machine limits its application areas.

Although the welding technology of dissimilar light alloys has achieved phased results, the current research results can only reduce the impact of intermetallic compounds on joint performance, and affect the formation of different intermetallic compounds at the welding interface. The mechanism, distribution law and its correlation with joint performance still need to be further studied; on the other hand, there are still many limitations in welding technology that need to resolve. Therefore, the welding technology of dissimilar light alloys will still be the focus of future research.

Acknowledgements

Supported by the Science and Technology Research Program of Chongqing Municipal Education Commission (Grant No.KJQN202103407).

References

1. Lan Tian, Dong Peng, Xiao Rongshi. Experimental analysis of laser deep penetration brazing of aluminum/titanium dissimilar alloys[J]. Transactions of the China Welding Institution, 2010, 31(8): 109~113.
2. Shi H, Chen K, Liang Z Y, Dong F B, Yu T W, Dong X P, Zhang L T. Intermetallic compounds in the banded structure and their effect on mechanical properties of Al/Mg dissimilar friction stir welding joints[J]. Journal of Materials Science & Technology, 2017, 33 (4): 359~366.
3. BEN-ARTZY A, MUNITZ A, KOHN G, et al. Joining of light hybrid constructions made of magnesium and aluminum alloys[A].Magnesium Technology[C]. Seattle, Washington: TMS, 2002.295-302.
4. Liu Peng, Li Yajiang, Geng Haoran et al. Materials Letters[J], 2007, 61(6): 1288.
5. Rattana Borrisutthekul, Yukio Miyashita, Yoshiharu Mutoh. Dissimilar material laser welding between magnesium alloy AZ31B and aluminum alloy A5052-O[J]. Science and Technology of Advanced Materials, 2005, 6: 199~204.
6. Takehiko W, Yoshitaka S. Journal of Japan Institute of Light Metals[J], 2004, 54(7): 293.
7. Liu Xujing, Liu Liming, Wang Heng et al. Welding Journal [J], 2005, 26(8): 31.
8. Liu Peng, Li Yajiang, Wang Juan, et al. Microstructure of the interface area of vacuum diffusion welding of Mg/Al dissimilar materials[J]. Transactions of the China Welding Institution, 2004, 25(5): 5~8.
9. Liu P, Li Y J, Geng H R, et al. A study of phase constitution near the interface of Mg/Al vacuum diffusion bonding[J]. Materials Letters, 2005,59 (16): 2001~2005.
10. Liu P, Li Y J, Geng H R, et al. Investigation of interfacial structure of Mg/Al vacuum diffusion-bonded joint[J]. Vacuum, 2006, 80(5):395~399.

Effect of different active materials on WAAM depth of penetration and weld morphology

Yunfeng Huang [1, 2], Xuezhi Shi [1, 2, *]

[1] School of mechanical engineering, Zhejiang Ocean University, Zhoushan, China
[2] Zhoushan Haida Science and Technology Research Institute Co. Ltd, Zhoushan, China

Abstract: The Gas Metal Arc welding process was used to prepare single-track and single-track multilayer samples. The effects of CaF2 and TiO2 on the penetration width and depth of samples are studied. The results show that the addition of these two active agents will reduce the surface morphology quality of single-track, but it will deepen the penetration depth of single-track. Compared with sample of CaF2, sample of TiO2 can maintain a better single track morphology while increasing the penetration. The ratio of effective melting width to maximum melting width is close to that without adding.

1. Introduction

Wire arc additive manufacturing (WAAM) technology uses electric arc as the manufacturing heat source and the metal welding Wire as the filling metal material. It adopts arc welding equipment and process methods to prepare full weld metal layer by layer to form a part structure. The arc welding equipment and its process method are used to prepare the fully welded metal layer by layer to form the part structure. It has the characteristics of high forming efficiency and low manufacturing cost [1-2].
Since this technology can realize the direct and rapid manufacturing of metal parts, it has gained more and more attention in the field of additive manufacturing. WAAM technology is mainly divided into the following three categories: Plasma arc welding (PAW), Gas tungsten arc welding (GTAW) and Gas metal arc welding (GMAW). GMAW technology is widely used due to its relatively simple equipment and high deposition rate[1-2].
Although GMAW has greater advantages, it generates more heat in the additive manufacturing process. Therefore, in the Marangoni effect, the metal material flows outward from the center of the molten pool during the melting process, which enlarges the welding surface, making the depth of penetration (DOP) smaller, resulting in a wider heat affected zone (HAZ).
Using activated material to increase the DOP is a common practice. However, use of activated material in MIG welding to increase the DOP is less investigated. So, in this work, the effect of different kinds of activated fluxes was studied on DOP during MIG welding. Activated material of CaF_2 and TiO_2 were used during MIG welding to identify their effect on single-track and single-track multilayer samples of mild steel for ships.

2. Experimentai materials and method

Q235B mild steel plates were used as base material and ER50-6 mild steel wire was used as Melting material. The nominal chemical composition of base material and melting material are mentioned in Table 1.
GMAW forming system and activated flux are show in figure 1. GMAW was carried out with and without activated flux.
In order to determine the appropriate input parameters (voltage, feed rate, welding speed and shielding gas flow), a large number of experimental studies have been carried out. Through a large number of experiments, these parameters have been optimized. And the suitable process parameters have been determined. The process parameter values are shown in Table 2.

* Corresponding author: shixuezhi@zjou.edu.cn

Table 1. Chemical composition of wire(wt %).

Alloy	C	Si	Mn	S	P	Fe
ER50-6	≤0.22	0.12-0.30	0.03-0.65	≤0.045	≤0.045	Bal.
Q235B	≤0.15	≤1.15	≤1.85	≤0.035	≤0.025	Bal.

Table 2. Wire arc additive manufacturing process parameters.

Process parameters	Without adding	Adding CaF_2	Adding TiO_2
Current （A）	160	160	160
Voltage (V)	24	24	24
Scan speed (mm/s)	10	10	10
Drying temperature (℃)	0	200	200
Ratio to alcohol	0	1.2：1	1.2：1

3. Experimental results and discussion and analysis

3.1 Result & analysis for single-layer and single-track for with and without activated flux

The influence of different activated fluxes on the surface morphology is shown in Figure 1. It can be seen from Figure 1 that the DOP of TiO_2 activated flux is significantly greater than the DOP of CaF_2 and without activated flux.

According Table 3, we can see the melting width without activated flux are larger than the activated flux of CaF_2 and TiO_2 and the melting depth and depth to width ratio without activated flux are smaller than the activated flux of CaF_2 .

The result shows that the DOP with activated flux is improved compared with welding without activated flux and different activators have different effects on the geometric dimensions of the weld. The chemical composition of the activator plays an important role in changing the geometry of the weld.

This is because the metal flow of the arc welding seam is affected by the Marangoni effect, buoyancy, surface tension and Lorentz force.

In the Marangoni effect, due to the melting of the flux material, the weld metal liquid flows outward from the center of the molten pool to widen the weld surface, resulting in a decrease in DOP. The use of activated flux can reverse the Marangoni effect. In the reverse Marangoni effect, the weld metal flows from the outside to the inside, resulting in a higher DOP[3]. At the same time, the influence of oxides on penetration is greater than that of halides[4].

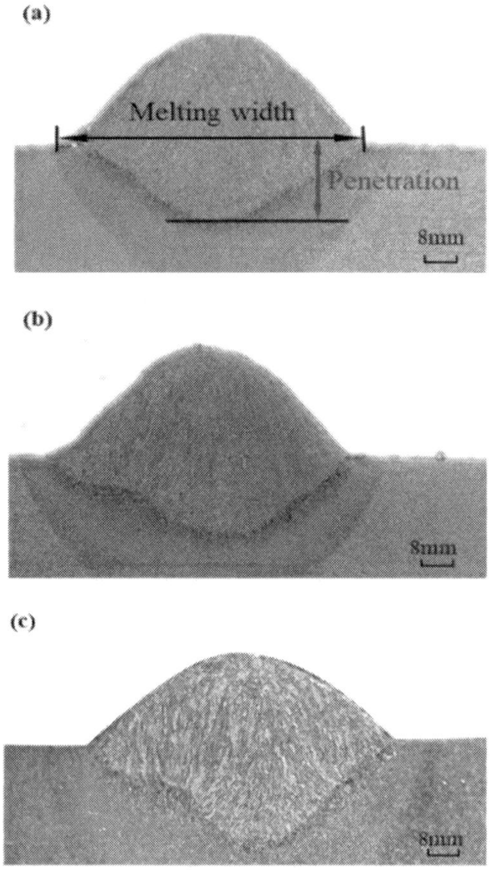

(a) Without Flux (b) Adding CaF_2 (c) Adding TiO_2.

Fig.1 Single-track comparison without adding , adding CaF_2 and adding TiO_2

Table 3. The melting width, melting height, and depth to width ratio of test pieces under different processes.

Flux	Depth (D) (mm)	Width (W) (mm)	Depth to width ratio (D/W)
Without Flux	1.91	6.56	0.29
Adding CaF_2	2.02	6.68	0.30
Adding TiO_2	2.49	6.94	0.36

3.2 Result & analysis for multi-layer and single-track for with and without activated flux

Figure 2 is the section diagram of the specimen formed by multi-layer single-track deposition after edm cutting, inlaying, grinding, polishing and corrosion. Table 3 shows the maximum melting width, effective melting width and effective ratio of the specimens under different processes. It can be seen from the data in the table that the maximum ratio of effective width to maximum width without adding indicates the best molding quality. The activated flux can increase the maximum melting width of samples, especially the activated flux of TiO_2. The ratio of effective width to maximum width is close to that of without adding. The activated flux of CaF_2 has less increase in melting width, the lowest ratio of effective width to maximum width and the worst molding quality.

This is due to the influence of TiO_2 activated flux on the arc shape and arc voltage of tungsten inert gas welding. Therefore, the decrease of the average surface tension and the change of the combined temperature dependence of the surface tension have no significant correlation with the arc performance. The increase of oxygen content in molten pool is the main reason for the change of surface tension gradient and the increase of penetration of molten pool[5].

The main action principle of activated flux of CaF_2 is that Ca ion and F ion are ionized during the molding process. The ionization of Ca ion increases the arc temperature, while the ionization of F ion makes the arc shrink[6]. This is because CaF2 can greatly increase the arc temperature. As the arc temperature increases, the absolute value of average surface tension decreases sharply and the penetration depth increases. Arc heating affects the stability of additive manufacturing, produces a large amount of splash, and affects the quality of parts[7-8].

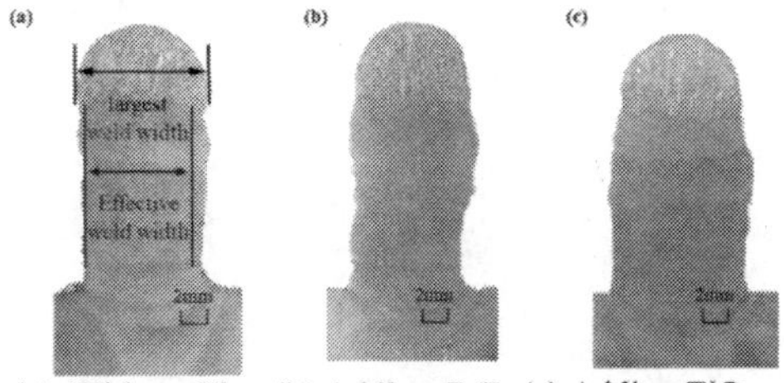

(a) Without Flux (b) Adding CaF_2 (c) Adding TiO_2.

Fig.2 Multilayer comparison without Flux , adding CaF_2 and TiO_2

Table 4. The largest melting width, Effective melting width, and ratio of test pieces under different processes.

Craft	largest weld width (mm)	Effective weld width (mm)	Ratio
Without Flux	9.48	8.68	0.91
Adding CaF_2	9.77	7.71	0.79
Adding TiO_2	10.55	9.32	0.88

4. Conclusion

1. The experimental results show that the penetration depth of the weld bead is greatly increased after the activation material is added, and the morphology quality of the penetration single-crack is reduced.
2. CaF_2 ionizes F ions during the arc additive manufacturing molding process, which will affect the arc morphology and arc temperature, and have a greater impact on the morphology of the sample. Therefore, it should be avoided as an additive during the molding process.
3. TiO_2 activated flux increases the penetration depth by increasing the oxygen content in the molten pool, and it has little effect on the arc morphology and molten pool temperature. Therefore, compared with CaF_2 activated flux, TiO_2 activated flux has better molding quality and is suitable to be used as an additive to expand the penetration depth.

Acknowledgements

The work was financially supported by the Zhoushan Science and Technology Project (2019C21014).

References

1. Willlams S W, Martina F, Addison A C, et al. Wire + arc additive manufacturing [J]. Materials Science and Technology, 2015, 32(7): 641-647.
2. Ding D, Pan Z, Cuiuri D, et al. Wire-feed additive manufacturing of metal components: technologies,

developments and future interests [J]. International Journal of Advanced Manufacturing Technology, 2015, 81(4): 465-481.

3. Baghel A , Sharma C , Rathee S , et al. Influence of activated flux on micro-structural and mechanical properties of AISI 1018 during MIG welding[J]. Materials Today: Proceedings, 2021(1).
4. Zhang Ruihua, Fan Ding, YU Shurong. Development of Active Agent for A-TiG Welding of Low Carbon Steel [J]. Transactions of the China Welding Institution,2003(02):16-18+3..
5. Qing-ming L I, Wang X H, Effect of Activating flux on Arc Shape and Arc Voltage of Tungsten inert gas welding[J]. Journal of The Chinese Society for Nonferrous Metals, 2007.
6. MARYA, EDWARDS, G. R . Chloride contributions in flux-assisted GTA welding of magnesium alloys.2002.
7. Chunkai Li,Yu Shi,YuFen Gu,Ding Fan,Ming Zhu. Effects of different activating fluxes on the surface tension of molten metal in gas tungsten arc welding[J]. Journal of Manufacturing Processes, 2018,32:
8. Leconte,Paillard,Chapelle,Henrion,Saindrenan. Effects of flux containing fluorides on TIG welding process[J]. Science and Technology of Welding and Joining,2007,12(2):

Study on the organization and properties of medium-Mn steel with V

Hongru Yang [1], Xiaoyan Liu [1, 2, *]

[1] School of Materials Science and Engineering, Hebei University of Engineering, Handan, Hebei, 056038, China
[2] Hebei Engineering Research Centre for Rare Earth Permanent Magnetic Materials & Applications, Handan, Hebei, 056038, China

Abstract: V is an important strengthening element in alloy steels, and its addition to the steel significantly improves the overall performance of the material. This paper describes the principle of V for optimizing the properties of medium-Mn steels. The excellent properties of V-containing Mn steels and their applications in practical industrial production are presented.

Key words: Medium-Mn Steel, V, Organization, Properties.

1. Introduction

Medium-Mn steels are mainly characterized by a low carbon medium alloy with a manganese content maintained between 3% ~ 12%[1]. Compared to the previous high-strength steel, the content of alloying elements has been reduced, so the cost has dropped significantly. Both C and Mn are the main strongly austenitic stabilizing elements in steel, and they significantly enlarge the austenite region, lower the A1 and A3 lines, and reduce the austenite transformation temperature, thus acting as grain refinement [2]. Among them, Mn can reduce the critical quenching rate of steel and improve the hardenability of steel[3]. At the same time, it can make the organization of the steel fine and uniform, and avoid the carbide agglomerates in the carburized layer. However, if the Mn content is too high, it will affect the welding performance of the material and thus have an impact on the safety of the steel structure. The composition of low-carbon medium manganese steel is designed so that the steel can be weldable and tough on the basis of maintaining high strength, which is beneficial to the application of the material in actual production.

In recent years, medium manganese steels have become the mainstay of the third generation of automotive steels by modifying them in different degrees and in different ways[4]. Professor Morris[5] in the USA studied that high strength plasticity can be obtained in mild steel by modulating its organization to ultrafine grain ferrite and austenite. In 2007, a researcher added "reverse-transformation austenite" to a steel containing about 6 wt.% Mn to further develop the properties of medium manganese steels. With the further research on the modification of medium manganese steel, its excellent performance is receiving more and more extensive attention. With the rapid development of China's further economic construction, the ship and marine industries are now ushering in the rapid development, and the further development of the marine industry for the requirements of ship and marine engineering steel has also increased. The ocean is a very harsh corrosive environment for various materials, and the sea water itself is a strong electrolyte. The content of atmospheric moisture and sodium chloride in the marine environment is very high, coupled with the influence of sunlight, sea breeze, including environmental temperature and humidity, will form a very harsh corrosive environment[6-11]. Due to the high concentration of sodium chloride and through the collaboration of various factors, a corrosive water film will adhere to the steel surface, and some highly corrosive substances in the environment will also be incorporated into the water film[12,13], further enhancing the corrosiveness and causing further damage to the steel, resulting in a corrosion failure situation. The medium-Mn steel can give full play to its excellent mechanical properties as a steel for marine engineering, while its corrosion resistance needs to be studied.

* Corresponding author: x918y@126.com

Alloying elements can improve the properties of metals to varying degrees. Mn is good for enhancing the strength of the steel, but the large amount of Mn will lead to the formation of a large amount of Mn-rich compounds on the surface of the rust layer of medium manganese steel after corrosion[14], which is not conducive to the corrosion resistance of the steel itself, and the addition of other corrosion-resistant alloying elements will improve this problem. Among the alloying elements, Cr, Ni, Cu, V, etc. can improve the corrosion resistance of the material, and adding such elements to Mn steels will improve the corrosion resistance of Mn steels from the perspective of composition, and to maintain the mechanical properties of the material itself.V can refine the grain and improve the strength, toughness and welding properties. If it is present in the form of carbide, it reduces its hardenability. Tarek[15] developed a medium manganese steel containing Cr-Ni-V-Cu and investigated its corrosion resistance. A manganese steel exhibiting high corrosion resistance and excellent strength and good plastic toughness was obtained by the development of medium manganese steel alloying. Among the alloying elements mentioned above, N plays a role in enhancing pitting resistance in addition to the addition of Cr to obtain the expected high corrosion resistance, and the V alloy can enhance the denseness of the rust layer by refining the grain[16,17]. Grain refinement will contribute to a denser rust layer after corrosion, and a dense rust layer will give better protection to the substrate, thus enhancing the corrosion resistance of Mn steels[18,19].

In this paper, the organization and properties of manganese steels containing V are investigated.

2. The role of element V in medium-Mn steel

Adding V to medium manganese steel will form the intermediate phase VC[4]. When the N content in the steel is low, VC has a high solubility in austenite, usually at 1000 °C. When the austenitic phase ferrite transforms, VC tends to precipitate in the interphase and strengthen the steel at the same time. The precipitated VC phase is distributed linearly at the phase boundaries, and these lines are almost in parallel distribution. In addition to the strengthening effect of precipitation, there is a role in this process to optimize the steel properties by increasing the density of the precipitated phase through grain refinement. Fine-grain strengthening and precipitation strengthening are the main toughening modes of V alloying. Haiping Wu[20] found through their study that the addition of V with a mass fraction of 0.1-0.2% to medium-Mn steel resulted in less wear than high manganese steel at lower impact loads.

The VC precipitated in V-containing manganese steel is not only uniformly distributed in the austenite matrix in a large number of fine particles, but also has the presence of some large particles. VC is an important V alloy additive that can effectively improve the performance of steel. Adding it to medium manganese steel can not only improve the strength and toughness of steel, but also make steel with good weld ability, eliminate inclusions in steel extension and other effects.

3. Effect of V on the properties of medium-Mn steel

V is an important strengthening element in alloy steel[21]. The nature of VC formed by solidification in V-containing Mn steels is very stable, difficult to dissolve, hard and high melting point, uniformly distributed on the matrix tissue, and has a very obvious strengthening effect on Mn steels.Shizhong Wei[22] found that the vanadium to the carbon ratio and absolute content have an important effect on the morphology of vanadium carbide. Xiaoqin Cha[23] investigated the effect of vanadium on the fatigue properties of 38MnS steel using bending fatigue experiments in the hot forged and annealed states. It was shown that the fatigue resistance of 38MnS steel was effectively enhanced due to the addition of V.

VC is a refractory second phase particle, and it is difficult to dissolve completely in the austenite of medium manganese steel during solid solution treatment. The precipitation of vanadium-carbon compounds along the grain boundaries will effectively organize the grain boundary diffusion, thus acting to inhibit the grain boundary movement. This leads to the austenite grain growth received to prevent, will play a fine crystal strengthening effect on the medium manganese steel, improve the strength and toughness of the steel.

VC is extremely hard (2650 HV) and also has a high melting point, which when present in steel as second-phase particles not only increases the overall hardness of the steel, but also significantly improves the wear resistance of the steel[24]. Due to the unique properties of VC, V is considered to be one of the most significant alloying elements affecting the wear resistance of steel.

In the medium-Mn steel matrix, VC is uniformly and diffusely distributed in the tissue, which increases the overall deformation resistance of the tissue of the matrix, thus improving the yield strength of the material. In addition, it also improves the work-hardening ability of the medium manganese steel under stress, which enhances the overall performance of the material and enhances the wear resistance. Among the alloying elements, the effect of V on the yield strength of medium manganese steel is the most significant[25].

4. Conclusion

The presence of many VC particles, which are nano-sized particles and are uniformly distributed throughout the structure of Mn steels, can significantly improve the hardness and fracture toughness of the base alloy. VC is considered to be the alloying element with the greatest influence on the wear resistance and yield strength of steel. In summary, medium manganese steels containing V have better overall performance and can better cope with the actual production requirements.

Acknowledgments

Authors thank the foundation of Science and Technology research in Hebei Universities (ZD2018213), the Science and Technology Research and Development Projects of Handan City (19422111008-20) for the financial support of investigations.

References

1. Lee, Han. Current opinion in medium manganese steel[J]. Materials Science and Technology, 2015, 31(7): 843.
2. D.Z. Zhang. Research on Microstructure-Properties and Marine Corrosion Behavior of 690MPa Grade Medium-Mn Steel for Offshore Platform[D]. Shenyang: Northeastern University, 2016.
3. W.X. Song. Metal Science (Second Edition) [M]. Beijing: Metallurgical Industry Press, 2004.
4. M. Lei. Study of Microstructure and Mechanical Properties of V-alloyed Medium Mn Steel[D]. Beijing: Beijing Jiaotong University, 2019.
5. M. Niikura, J.W. Morris. Thermal processing of ferritic 5Mn steel for toughness at cryogenic temperatures[J]. Metallurgical Transactions A, 1980, 11(9): 1531-1540.
6. Y.T. Li, B.R. Hou. Corrosion Law of Steel in Shallow Sea Petroleum Development Area of Shengli Oilfield[J]. Studia Marina Sinica, 2006(47): 101-108.
7. K.M.A. Hossain, S.M. Easa, M. Lachemi. Evaluation of the effect of marine salts on urban built infrastructure[J]. Building and Environment, 2009, 44(4): 713-722.
8. L. Li, X.L. Xu, F.F. Ai, et al. Effect of Technology and Microstructure on Corrosion Behavior of Bridge Steels in Simulated Ocean Splash Zone Environment[J]. Corrosion & Protection, 2013, 34(06): 503-506.
9. B.R. Hou, D. Xifang, L.C. Shui. Corrosion behavior of steel in the seawater-air-sea transition interface zone[J]. Oceanologia et Limnologia Sinica, 1995(05): 514-519.
10. X.L. Cui, Z.H. Ma, L. Zhang, et al. Analysis of low-alloy steel rust layer in real sea splash zone (abstract)[J]. Corrosion Science and Protection Technology, 1995(03): 253-254.
11. X.R. Zhu. Research progress of rust in marine environment[J]. Total Corrosion Control, 1998(02): 24-27.
12. K.V.S. Ramana, S. Kaliappan, N. Ramanathan, et al. Characterization of rust phases formed on low carbon steel exposed to natural marine environment of Chennai harbour-South India[J]. Materials and Corrosion, 2007, 58(11): 873-880.
13. T. Ishikawa, K. Takeuchi, K. Kandori, et al. Transformation of γ-FeOOH to α-FeOOH in acidic solutions containing metal ions[J]. Colloids & Surfaces A Physicochemical & Engineering Aspects, 2005, 266(3): 155-159.
14. Z. Liu, X. Lian, T. Liu, et al. Effects of rare earth elements on corrosion behaviors of low carbon steels and weathering steels[J]. Materials and Corrosion, 2020, 71(2): 258-266.
15. A. Tarek, F.X. Guo, S. Simon, et al. Development of a Cr-Ni-V-N Medium Manganese Steel with Balanced Mechanical and Corrosion Properties[J]. Metals, 2019, 9(6): 705.
16. L.C. Feng, Y.Q. He, B. Qiao, et al. Corrosion and Protection of Metal and Alloy in Marine Environment[J]. Hot Working Technology, 2013, 42(24): 13-17.
17. W.W. Peng, W.D. Zeng, W.Q. Yan, et al. Effect of tempering process on microstructure and toughness of Aermet100 ultrahigh strength steel[J]. Transactions of Materials and Heat Treatment, 2013, 34(06): 58-61.
18. P. Montoya, I. Díaz, N. Granizo, et al. An study on accelerated corrosion testing of weathering steel[J]. Materials Chemistry and Physics, 2013(142): 220-228.
19. H. Luo, X.G. Li, C.F. Dong. Research on Atmosphere Exposure in Tropical Marine and Accelerated Corrosion Test of 304 Stainless Steel[J]. Journal of Chinese Society for Corrosion and Protection, 2013, 33(03): 193-198.
20. H.P. Wu. Effects of Vanadium and Titanium on the Microstructure and Properites of As-cast Medium Manganese Steel[J]. Foundry Equipment & Technology, 2011(6): 12-13.
21. J.B. Deng. Application and Research Progress of Vanadium in Metal lurgy[J]. Ferro-Alloys, 2011, 42(2): 45-48.
22. S.Z. Wei, L.J. Xu, J.H. Zhu, et al. Effect of Carbon and Vanadium Content on Microstructure and Mechanical Properties of High Vanadium High Speed Steel[J]. Journal of Iron and Steel Research, 2005, 17(5): 6.
23. X.Q. Cha, W.J. Hui, Q.L. Yong, et al. Effect of Vanadium on the Fatigue Properties of Microalloyed Medium-Carbon Steels[J]. Acta Metal lurgica Sinica, 2007, 43(7): 719-723.
24. Y.J. Ma, Y. Wang. Analysis of Wear Resistance of High Carbon and High Vanadium High Speed Steel

and High Chromium Cast Iron[J]. China High-Tech Enterprises, 2007(4): 101.

25. Z.Z. Yuan, S. Chou, Y. Kuang. Countermeasure of Improving High Manganese Steel Casting Metallurgical Quality[J]. Foundry Technology, 2003(05): 380-382.

In Situ Preparation of Thermotropic Liquid Crystalline Polyester and Nylon 10T

Ze Rong [1, a], Shiyu Chen [2, b], Hanwen Xu [2, c], Xiaoping Yang [3, d], Haojing Wang [4, e] and Yuanzhe Li [5, f, *]

[1] Chool of Mathematical and Physical Sciences, University College London, London, WC1E 6BT, United Kingdom
[2] Chool of Polymer Materials and Engineering, Sichuan University, Chengdu, 610065, China
[3] Chool of Chemistry and Chemical Engineering, Xiamen University, Xiamen, 361005, China
[4] Chool of Chemical and Biomedical Engineering, Nanyang Technological University, Singapore, 637459, Singapore
[5] Chool of Materials Science and Engineering, Nanyang Technological University, Singapore, 639798, Singapore

Abstract. Poly-decamethylene terephthalamide (PA10T) is one of high temperature special engineering plastics, which is widely used in aviation automobile, and electronic industries. However, its bad processability, poor crystallinity, and toughness may affect its further applications. In this work, the melting polycondensation synthesis of thermotropic liquid crystalline polyester (TLCP) prepolymer was explored. This synthesized TLCP was then added into PA10T via in-situ compounding, to obtain the poly-decamethylene terephthalamide / thermotropic liquid crystalline polyester composites (TLCP/PA10T). The TLCP was found easy to be oriented, which can reduce the entanglement of PA10T and improve the processing properties of PA10T. Moreover, the anisotropy under tensile conditions after the TLCP orientation can form microfibers and improve the mechanical properties of the composites. The relationship between crystallization properties, tensile strength and TLCP content of TLCP/PA10T composites was also studied. Such TLCP/PA10T composites can serve as a high temperature special engineering plastic in industrial manufacturing.

Keywords: Melt polymerization, Thermotropic liquid crystalline polyester (TLCP), Poly (decamethylene terephthalamide) (PA10T), Crystallization properties, Mechanical properties.

1. Introduction

Poly (decamethylene terephthalamide) (PA10T) is a semi-aromatic nylon made by condensation of decylenediamine and terephthalic acid. The long flexible diamine chain in the molecular chain makes the macromolecule have certain flexibility, so that it has a high crystallization rate and crystallinity, while its low water absorption rate, so that its products have good mechanical properties [1,2]. Compared with conventional aliphatic nylon materials, semi-aromatic nylons have more rigid molecular chains and restricted molecular chain movement which resulting in higher glass transition temperature (T_g) and better dimensional stability, mechanical properties [3], corrosion resistance, and thermal properties. Due to these excellent properties, PA10T is more suitable for certain small electronic components, such as LED reflection bracket and connector than conventional aliphatic nylon. Also, the modified products of PA10T can also be applied to some chemical reagents or heat-resistant environments, such as water treatment, Nano Molding Technology (NMT) and engine parts [4]. However, semi-aromatic nylon has the problems of crystallization, poor toughness and processability due to high melt viscosity [5]. To solve these problems of semi-aromatic nylon, it can be modified to obtain a modified material of semi-aromatic nylon with good processing performance without losing its excellent mechanical properties and other advantages. The modification by adding inorganic fillers such as glass fibers (GF) [6] can enhance the mechanical properties, but at the same time, it will make the material less flowable and thus difficult to process.

Many studies have shown that thermotropic liquid crystal polymers (TLCP) can be used as a processing aid for thermoplastics to significantly reduce the melt viscosity of the blend system. Isayev [7] found that the 1.0 wt. % TLCP could significantly improve the processability of polycarbonate (PC). Chang [8] found that TLCP can effectively improve the processability of polyphenylene ether/polystyrene alloy. When the content of TLCP is more than 5.0 wt. %, compatibilizers must be added to improve the compatibility between TLCP and thermoplastic in order to obtain satisfactory blending. On the other hand, TLCP has good flowability, low melt viscosity, and excellent mechanical properties, and it is also close to the processing temperature of semi-aromatic nylons such as PA10T. Therefore, TLCP has been regarded as one of proper polymers for modification of

* Corresponding author: yuanzhe001@e.ntu.edu.sg

PA10T. In the modification of PA10T by TLCP, TLCP is easily oriented, which can reduce the entanglement of PA10T and thus improve the processing properties of PA10T. On the other hand, TLCP orientation is anisotropic in tensile conditions and forms microfibers, improving the mechanical properties of the composite.

In this work, TLCP prepolymer was successfully synthesized by melt polycondensation, after which, the poly-decamethylene terephthalamide / thermotropic liquid crystalline polyester composite (TLCP/PA10T) was obtained by adding TLCP into PA10T through in situ compounding. The mechanical properties, thermal properties and crystallization behavior of TLCP/PA10T composites with different mass fractions of TLCP have been studied. The reasons for the differences in properties of TLCP/PA10T composites with different mass fractions of TLCP have been discussed.

2. Materials and Method

2.1 Materials and Devices

P-hydroxybenzoic acid (HBA), terephthalic acid (TA) and isophthalic acid (IA) were purchased from Sinopharm Chemical Reagent Co., Ltd. (Chengdu, China), 4,4'-dihydroxybenzene (BP) was purchased from Sigma-Aldrich (Shanghai, China), sulphuric acid was purchased from XiLONG Scientific Co., Ltd. (Chengdu, China), acetic anhydride was purchased from Chron Chemicals Co., Ltd. (Chengdu, China), methylimidazole was purchased from Adamas Reagent, Ltd. (Shanghai, China), pentafluorophenol was purchased from Quanzhou Qianda Technology Co., Ltd. (Chengdu, China), PA10T prepolymer was used as purchased from Sichuan Lang Di New Material Co., Ltd (Deyang, China).

The polymerization reaction device was a self-designed high temperature and high-pressure reactor. The reactor was equipped with a stirring device, which could adjust the stirring speed. The reactor was electrically heated, with a temperature controller to control the real-time temperature of the reaction through a thermometer. The reaction kettle was connected with a vacuum pump and a nitrogen cylinder to take away the by-product acetic acid and provide inert gas protection.

2.2 Preparation of 4- Acetyloxybenzoic Acid (ABA)

All 30 g of p-hydroxybenzoic acid (HBA), 60 g of acetic anhydride and 0.09 g of sulfuric acid were added into a three-neck flask. After vacuumizing, nitrogen was introduced and heated to 80 °C for 2 h. After the reaction, the reaction system was poured into ice water, the system was stirred until all the precipitates were precipitated, the precipitates were filtered, washed with deionized water for 3 to 4 times until the pH value of the system was between 4 - 5 to obtain 4-acetoxybenzoic acid (ABA). The reaction equation was indicated in Figure 1(a).

2.3 Preparation of 4,4'- Diacetoxybiphenylene (BPA)

Typically, the BPA was firstly synthesized by adding 30 g of hydroquinone (BP), 60 g of acetic anhydride and 0.09 g of sulfuric acid into a three-necked flask. After vacuumizing, nitrogen was introduced and heated to 80 °C for 2 h. After the reaction, the reaction system was poured into ice water, the system was stirred until all the precipitates were precipitated, the precipitates were filtered, washed with deionized water for 3 to 4 times until the pH value of the system was neutral to obtain 4,4'-diacetoxybiphenyl (BPA). The reaction equations were indicated in Figure 1(b).

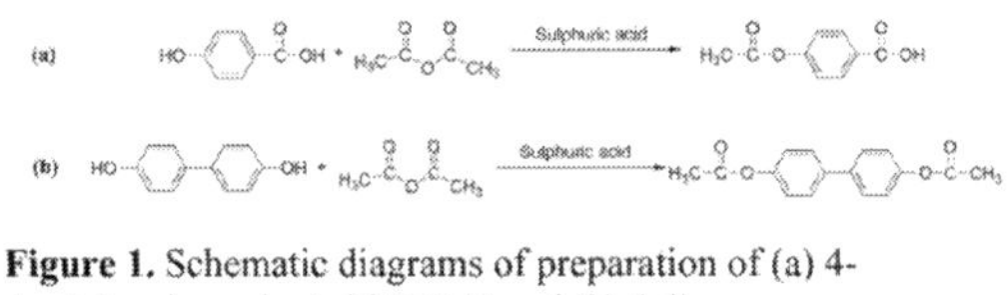

Figure 1. Schematic diagrams of preparation of (a) 4-Acetyloxybenzoic Acid (ABA) and (b) 4,4'-diacetoxybiphenylene (BPA).

2.4 Synthesis of TLCP

ABA and ABP monomer raw materials were weighted and mixed according to a certain proportion. Then the prepared catalyst methylimidazole aqueous solution was added into the reactor as 0.5 wt.%. The vacuum pump was used to remove the air in the reactor, nitrogen was injected for more than 5 times, then the reactor was heated to 170 °C for 1 h at the heating rate of 0.8 °C min-1, until 240 °C. After heating, the exhaust port was opened to remove the small molecule by-product acetic acid, the reaction system was cooled down to 100 °C for 1h. The reactor was opened and added an appropriate amount of catalyst methylimidazole aqueous solution again, the air in the reactor was removed by using the vacuum pump, nitrogen was injected for more than 5 times, then the reactor was raised to 240 °C, the stirring device was turned on at the speed of 300 rpm min^{-1}. The temperature was raised to 300 °C at the rate of 0.6 °C min-1, and finally the temperature was raised to 320 °C at the rate of 0.5 °C min-1, stop the reaction [9].

2.5 Synthesis of TLCP/PA10T Composites

PA10T prepolymer and TLCP were weighed and mixed in different proportions and were put into polymerization reactor. The vacuum pump was used to remove the air in the reactor, nitrogen was injected for more than 5 times, the temperature of the reactor was set to 305 °C, the stirring device was turned on at the speed of 300 rpm min^{-1},the vacuum pump was turned on to vacuum, stay for 1 h, the temperature was raised to 315 °C, stay for 1 h, the temperature was raise to 325 °C, stay for 1 h, raised the temperature to 335 °C, and the reaction was closed after 30 min [10]. The composite splines were prepared according to the requirements and specifications of ASTM D882-09 Standard. There were five groups of PA10T, 2.0 wt.% TLCP/PA10T, 5.0 wt.% TLCP/PA10T, 10.0 wt.% TLCP/PA10T and 20.0 wt.% TLCP/PA10T, with five dumbbell splines (20.0 mm × 2.0 mm × 0.2 mm) in each group.

2.6 Characterizations

The samples were characterized by total reflection infrared spectroscopy using Nicolet iS50 Fourier transform infrared spectroscopy of Thermo Fisher Company. Differential scanning thermal analysis was performed using NETZSCH DSC 214 differential thermal analyzer. Thermogravimetric analysis was performed using NETZSCH TG 209F3 thermogravimetric analyzer. The crystallization of TLCP was analyzed by Ultima IV X-ray diffractometer of RIGAKU company. The texture structure of polymer samples was observed under Leica DM4P polarizing microscope. 5967 universal material testing machines of Instron company was used to test mechanical properties, test rate 5 mm min^{-1}.

3. Results and Discussions

3.1 FTIR Characterizations of TLCP Prepolymer and TLCP/PA10T Composites

The FTIR spectra of TLCP prepolymer is shown in Figure 2(a), the stretching vibration peak of C-H is displayed at 2925 cm^{-1}. C=C benzene ring skeleton vibration peaks are displayed at 1601 cm^{-1}, 1492 cm^{-1} and 1425 cm^{-1} which indicate the presence of benzene ring in the synthesized product. Characteristic peaks of para disubstituted benzene are displayed at 902 cm^{-1}, 880 cm^{-1}, and 690 cm^{-1}. The ester carbonyl stretching vibration peak of phenol is observed at 1737 cm^{-1}. The characteristic absorption peak of acetate of phenol at 1264 cm^{-1} is the absorption peak of biphenyl. FTIR spectra prove that the synthesized TLCP is consistent with the citations [11].

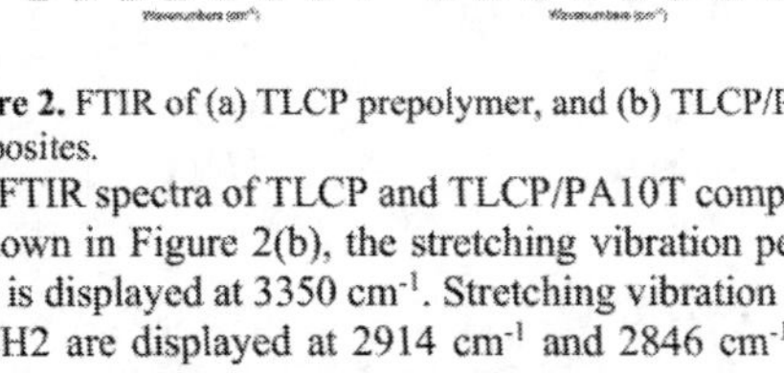

Figure 2. FTIR of (a) TLCP prepolymer, and (b) TLCP/PA10T composites.

The FTIR spectra of TLCP and TLCP/PA10T composites is shown in Figure 2(b), the stretching vibration peak of N-H is displayed at 3350 cm^{-1}. Stretching vibration peaks of CH2 are displayed at 2914 cm^{-1} and 2846 cm^{-1}. The stretching vibration peak of C=O bond is displayed at 1625 cm^{-1}. The peak displayed at 1530 cm^{-1} is the combined peak of N-H bond in-plane bending and C-N bond stretching vibration. The peak displayed at 1297 cm^{-1} is the combined peak of C-H bond in-plane bending and C-N bond stretching vibration. Peaks displayed at 860 cm^{-1} and 728 cm^{-1} are bending vibration absorption peaks in the C-H bond plane of benzene ring. The FTIR spectra indicates that the products are PA10T, which is corresponded with the citations [12].

3.2 Thermal Properties of TLCP Prepolymer and TLCP/PA10T Composites

The synthesized TLCP prepolymer was analyzed and tested by differential thermal analyzer. The sample was measured cyclically at the heating rate of 20 °C min-1 under the protection of nitrogen within 5.0 mg, and the heating range was 25 °C to 330 °C. The sample was thermogravimetric analyzed by thermogravimetric analyzer. The test was carried out under nitrogen protection, and the temperature rise range was 25 - 600 °C. The thermal history of the sample was eliminated before the test, and the DSC curve is shown in Figure 3(a). As the product is TLCP prepolymer, compared with the commercial TLCP products, due to its small molecular weight and poor thermal stability, the melting point of TLCP prepolymer is about 265 °C according to the DSC curve, and the thermal decomposition temperature of TLCP prepolymer is about 500 °C according to the TGA curve. The TGA curve of TLCP prepolymer is shown in Figure 3(b). Figure 3(c) shows the TGA curves of PA10T and TLCP/PA10T composites. The analysis shows that comparing with PA10T, the thermogravimetric temperature of TLCP/PA10T compounded with TLCP decreases slightly with the increase of TLCP content, and the degree of reduction is almost negligible. Therefore, it could be seen that the addition amount of TLCP TLCP/PA10T almost does not affect the thermal stability of the composites. Table 1 indicates the specific data where T_5 represents 5 % of thermal weight loss.

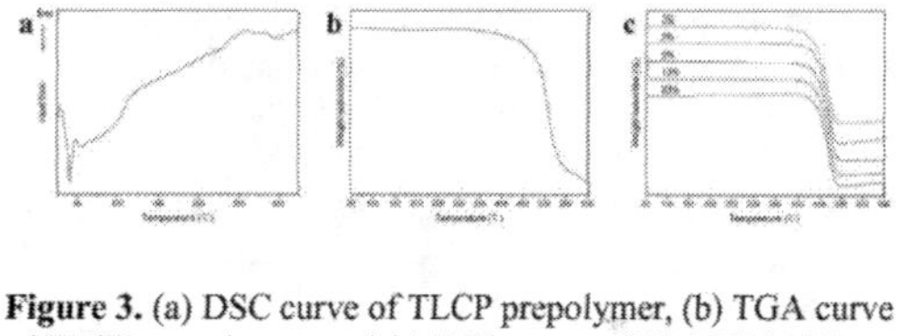

Figure 3. (a) DSC curve of TLCP prepolymer, (b) TGA curve of TLCP prepolymer, and (c) TGA curve of TLCP/PA10T composites.

Table 1. TLCP/PA10T composites T_5.

Mass fraction of TLCP in composite (%)	T_5 (°C)
0	434.79
2	429.64
5	426.43
10	421.70
20	409.66

3.3 Crystallization properties of TLCP Prepolymer and TLCP/PA10T Composites

The synthesized TLCP prepolymer was characterized by Ultima IV X-ray diffractometer of RIGAKU company in Japan. Figure 4(a) is its X-ray diffraction pattern. The X-ray diffraction pattern shows that the TLCP prepolymer is in 2θ. There is an obvious diffraction peak at about 20 °, which is consistent with the citations [12]. Figure 4(c) shows the DSC second temperature rise curve of PA10T, TLCP and TLCP/PA10T composites. According to the analysis of the figure, TLCP does not melt obviously, while TLCP/PA10T composite is similar to PA10T, and there is an obvious crystallization peak near the crystallization temperature of PA10T [13]. According to the analysis of the figure, the melting point of TLCP/PA10T composites decreases slowly with the increase of TLCP content. According to DSC (Figure 4(d)) the crystallization properties of TLCP/PA10T composites increase first and then decrease with the increase of TLCP content in the composites. When the proportion of TLCP is relatively small, on the one hand, TLCP can play the role of nucleation in the crystallization process, as a nucleating agent to crystallize PA10T. Although TLCP destroyed the regularity of PA10T, the relatively rigid molecular chain of TLCP and the role of nucleation limit the movement of PA10T molecular chain segment, TLCP plays a dominant role as a nucleating agent. Therefore, the content of TLCP increases gradually, and the crystallinity of TLCP/PA10T composites is higher. When the mass fraction of TLCP is increased to 5.0 wt. %, the crystallization performance of TLCP reaches the best. When the TLCP is greater than 5.0 wt. %, due to the increase of the proportion of TLCP, more and more TLCP formed microfibril structure. Therefore, the role of nucleating agent is gradually weakened, and the role of hindering the movement of PA10T segment is gradually dominant, showing a trend of decreasing crystallinity. From X-ray diffraction, 5.0 wt.% TLCP/PA10T has the largest crystallinity and the best crystallinity, which confirms the above analysis of crystallization behavior. In addition, it can be seen from Figure 4(b) that there is a diffraction peak with smaller peak intensity at high angle and low angle respectively in TLCP/PA10T compared with pure PA10T, and it can also be seen that the crystal forms of TLCP and PA10T are different.

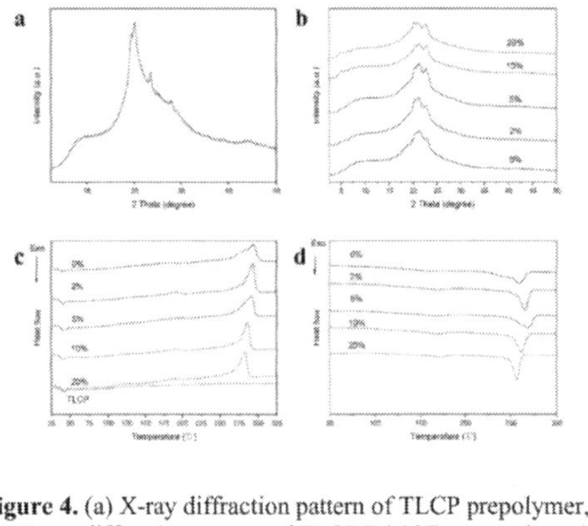

Figure 4. (a) X-ray diffraction pattern of TLCP prepolymer, (b) X-ray diffraction pattern of TLCP/PA10T composite, (c) DSC heating curves of PA10T and TLCP/PA10T composites, and (d) DSC cooling curves of PA10T and TLCP/PA10T composites.

TLCP prepolymer sample was placed on a glass slide, heated, melted, and pressed with another glass slide on the heated state, and observed after cooling. As shown in Figure 5(a), a typical nematic schlieren texture is observed. Polymerized TLCP sample is placed on a glass slide, heated, melted and pressed with another glass slide to an appropriate thickness and slid it in the same direction on the heated state, and then it is placed under a polarizing microscope for observation and photography after cooling. As shown in Figure 5(b) and (c), TLCP forms a typical nematic band texture, showing stripes oriented along the direction of shear stress. It is speculated that this kind of band texture is formed by many microfibers oriented along the direction of shear stress, regularly arranged and periodically bent into sawtooth shape. It can be inferred from the analysis of polarizing microscope photos that TLCP is composed of many ordered microcrystalline regions with random orientation before shear. When the TLCP is subjected to shear, the ordered microcrystalline regions with random orientation are aligned in the stress direction of shear. After the shear action stops, the highly oriented state gradually returns to the ordered microcrystalline region structure with regular.

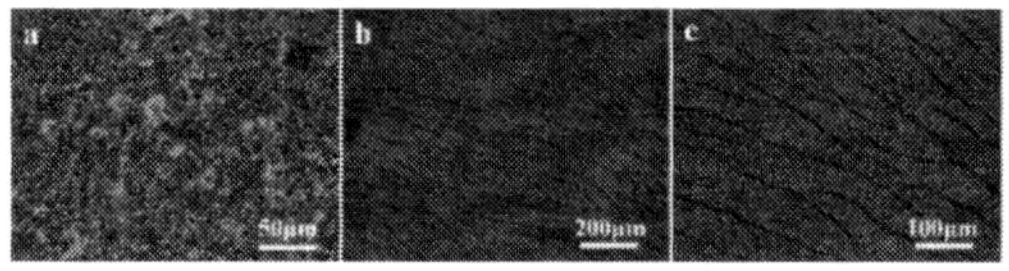

Figure 5. (a) Texture structure of TLCP prepolymer, and (b), (c) TLCP prepolymer texture (shear).

3.4 Mechanical properties of TLCP and TLCP/PA10T

As indicated in Figure 6, the tensile strength of TLCP/PA10T composite spline first increased but then decreased with the increase of TLCP content of the composite, and the tensile strength reached the maximum at 5.0 wt. %, which was 68 MPa. Based on the above analysis, the mechanical properties of the composite can be enhanced since there are few microfibers formed by a small amount of TLCP added into the PA10T. With the gradual increase of TLCP content, the tensile strength of the microfibril gradually increase until the TLCP content exceeds 5.0 wt.%, which is due to the domination of the poor compatibility between PA10T and TLCP and the compatibility became worse. The test results of mechanical properties of composite splines also confirmed the previous analysis.

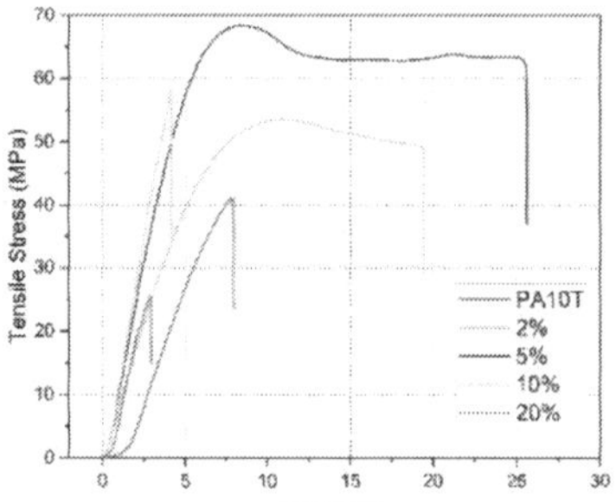

Figure 6. Stress-strain Diagram of PA10T and TLCP/PA10T composites.

4. Conclusion

Thermotropic liquid crystalline polyester (TLCP) prepolymer was demonstrated to be firstly synthesized and PA10T was successfully compounded in situ with TLCP prepolymer to explore the appropriate synthesis process of TLCP/PA10T melt polymerization. The effects of different mass fractions of TLCP on the thermal, crystalline, and mechanical properties of TLCP/PA10T composites were characterized in this work. The mechanical and crystalline properties of PA10T can be adjusted by varying the mass fraction of TLCP content. The TLCP/PA10T showed optimized mechanical and crystalline properties with TLCP content at its 5.0 wt.% and then deteriorated with the decrease of TLCP content. The tensile strength of 5.0 wt.% TLCP/PA10T composites reached 68 MPa which is also the optimal sample in this work. Results showed that TLCP/PA10T composites has better properties compared with conventional PA10T and can be widely used in aerospace, automotive industry, electronic industry and other fields.

Acknowledgments

This work has been supported by Asia Education Institution (SG) since 2020, Project number: AEI210802Z.

References

1. Chang, D. Y., & Chang, F. C. J. J. o. a. p. s. In situ reactive compatibilized noryl/LCP blends. 56(9), 1015-1028.
2. Chen, B.-K., Tsay, S.-Y., & Chen, J.-Y. Synthesis and properties of liquid crystalline polymers with low Tm and broad mesophase temperature ranges. Polymer, 46(20), 8624-8633.
3. Huang, B., & Walsh, J. J. Solid-phase polymerization mechanism of poly(ethylene terephthalate) affected by gas flow velocity and particle size. Polymer, 39(26), 6991-6999.
4. Isayev, A. I., & Modic, M. J. P. C. Self-Reinforced melt processible polymer composites: Extrusion, compression, and injection molding. 8(3), 158-175.
5. Liu, P., Wu, T., Ye, G., & Xu, J. Novel polyarylates containing aryl ether units: synthesis, characterization and properties. Polymer International, 62(5), 751-758.
6. Novitsky, T. F., Lange, C. A., Mathias, L. J., Osborn, S., Ayotte, R., & Manning, S. J. P. Eutectic melting behavior of polyamide 10, T-co-6,T and 12,T-co-6,T copolyterephthalamides. 51(11), 2417-2425.
7. Pan, X., Chi, Z., Cheng, D., Jin, M., & Bu, H. Solid-State Polymerization of a Liquid Crystalline Copolyester Derived from 2,6-Naphthalene Dicarboxylic Acid, Terephthalic Acid, 4-Acetoxybenzoic Acid and Hydroquinone Diacetate. Journal of Macromolecular Science, Part B, 44(2), 249-259.
8. Pisharath, S., & Wong, S. C. Development of the morphology and crystalline state due to hybridization of reinforced toughened nylon containing a liquid-crystalline polymer. Journal of Polymer Science Part B-Polymer Physics, 41(6), 549-559.
9. Ruggiero, V., & Acierno, D. Effects of the addition of small amounts of a thermotropic liquid crystalline polymer on the processing characteristics of polyphenylene oxide-polyamide alloys. Advances in Polymer Technology, 26(2), 109-120.
10. Wang, J., Li, L., He, Y., Song, H., Chen, X., & Guo, J. The effect of thermo-oxidative ageing on crystallization, dynamic and static mechanical properties of long glass fibre-reinforced polyamide 10T composites. 5(6), 172029.
11. Wang, W. Z., Liu, B. Y., Li, R. X., Zhang, Y. H. J. G. C. K. Y. G. P. M. S., & Engineering. Synthesis and physical and mechanical properties of poly(decanediamine p-phenylenedacetic amides). 25(4), 144-146.
12. Ward, Y., & Young, R. J. Deformation studies of thermotropic aromatic copolyesters using NIR Raman spectroscopy. Polymer, 42(18), 7857-7863.
13. Zou, G., Wang, P., Feng, W., Ren, Z., & Ji, J. Poly (decamethylene terephthalamide) copolymerized with long-chain alkyl dodecanedioic acid: Toward bio-based polymer and improved performances. 135(31), 46531.

Preparation and Properties of Sn-TiO_2 nanocomposite coatings

Shengping Zhang

School of Materials Science and Engineering, Jiangsu University of Science and Technology, Zhenjiang, China

Abstract. Sn-TiO_2 nanocomposite coatings were prepared by sol-gel enhanced electrodeposition technique and the effect of c concentration on the performance of the composite coatings was investigated by adding different volumes of TiO_2 sol to the electrolyte. The microscopic morphology and corrosion resistance of the coatings were characterized by XRD tests, SEM tests, hardness tests and a series of electrochemical tests. From the test results, it was found that the coating had the highest hardness value and the best corrosion resistance when the TiO_2 sol was added at 15 mL/L; however, when the TiO_2 sol concentration was too large, the corrosion resistance and hardness value of the Sn-TiO_2 composite coating decreased significantly due to the destruction of the crystal structure and coating structure by the agglomerated TiO_2 nanoparticles in the coating.

Keywords: Nanocomposite; TiO_2 sol; electrodeposition;coating.

1. Introduction

Nowadays, with the development of the world economy, there is an increasing demand for high-performance coating materials[1-3]. How to improve the quality of coatings has become a top priority in the field of coating research. As the surface of iron and steel devices is easy to rust, this problem limits its further application in the industrial field. Generally, new coatings can be prepared on the surface of iron and steel devices by electrodeposition technology to enhance the corrosion resistance of iron and steel devices. The coatings prepared by electrodeposition technology should have good corrosion resistance and mechanical properties, such as Zn[4], Ni[5], Cu[6] and so on. Because of its good corrosion resistance and solderability, tin coating is widely used in various industrial fields as a protective coating for all kinds of metal devices. However, in a relatively harsh environment, the protective effect of a single tin coating on the metal substrate is limited, how to further improve the quality of tin coating has become a hot issue in the field of surface engineering technology.

At present, most of the research is to improve the coating quality of tin coating by doping metal elements and nanoparticles. In recent years, a new TiO_2-sol-enhanced electrodeposition technology has been developed, TiO_2 nanoparticles will be formed in situ in the process of electrodeposition, and this sol-enhanced electrodeposition technology can improve the dispersion of nanoparticles in the coating, thus can better improve the quality of the coating[7]. However, there is no research on Sn-TiO_2 nano-coating at present. In order to prove the potential of Sn-TiO_2 nanocomposite coatings, Sn-TiO_2 nanocomposite coatings were prepared by TiO_2-sol enhanced electrodeposition, and the effects of different concentrations of TiO_2 sol on tin coatings were systematically studied.

2. Experimental

In this paper, Fe sheets were used as the substrate, and Fe substrate was acid-washed using concentrated hydrochloric acid before the preparation of the coating, with the aim of removing the corrosion products surviving on the surface of the iron sheets. In our experiments, we prepared Fe/Sn-TiO_2 composite coatings by DC electrodeposition method, using the methanesulfonic acid system solution as the electrolyte (the main components are MSA, MSASn and gelatin, respectively), and slowly drop a certain amount of TiO_2 sol into the plating solution before plating .The coating was prepared as follows: the tin plate (40 mm × 40 mm) was used as the anode of the electrolytic cell, and the treated Fe sheet was used as the cathode of the electrolytic cell, under 30 mA/cm^2 current density During the electroplating process, the electrolyte was stirred at a constant speed using a magnetic stirrer, the plating temperature was controlled at 30°C, and the plating time was 20 min; the prepared samples were rinsed with anhydrous ethanol and deionized water and then transferred to a drying oven at 100°C to obtain the Fe/Sn-TiO_2 composite coating.

Microhardness tester was used to test the hardness of the sample, and 50g charge for 10s was applied to each indentation on the sample.The crystal structure was identified by X-ray Diffraction(XRD, D2 Phaser, Bruker, Germany) at a step size of 2°.The surface morphology was

https://doi.org/10.1051/matecconf/202235801056

deserved by a scanning electron microscope (SEM, Phenom proX, Netherlands).The electrochemical analyses were tested by the electrochemical workstation(CS2350H, Chenhua Instrument,Shanghai). A three-electrode system was used in the whole electrochemical test, Fe/Sn-TiO_2 (10mm×10mm) as working electrode, platinum plate (20mm×20mm) as counter electrode, saturated calomel electrode (SCE) as reference electrode, and the 3.5%wt NaCl solution used as electrolyte solution. In the dynamic potential scan test, the potential scan range is -0.5V~0.5V, and the scan speed is 0.5mV/s; in the electrochemical impedance test, the frequency range is 0.01~100000Hz.

3. Result and Discussion

Figure 1 shows the crystal structure of the Sn-TiO_2 coating. The positions of all the diffraction peaks in the figure are consistent with the PDF card of the tetragonal Sn (PDF:88-2625). Due to the small amount of TiO_2 sol, there is no diffraction peak of TiO_2 in the XRD pattern. The results show that with the increase of the amount of TiO_2 sol, the intensity of the diffraction peak decreases and the FWHM increases, indicating that the size of the crystal decreases gradually. However, when the addition of TiO_2 sol exceeds that of 15mL/L, the intensity of the diffraction peak increases, indicating that the addition of excessive TiO_2 sol will lead to grain coarsening.

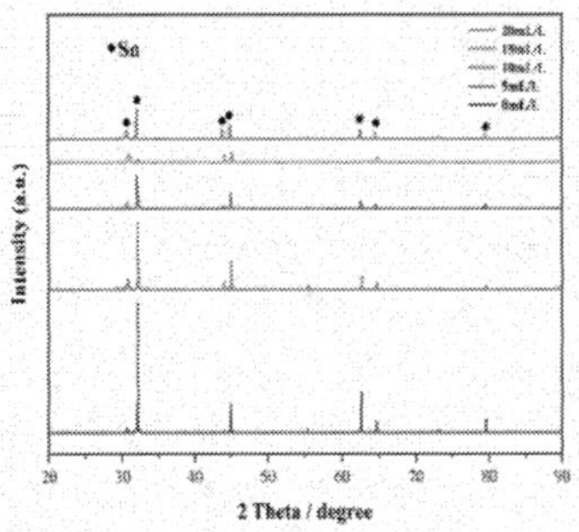

Fig. 1. XRD patterns of Sn–TiO_2 nanocomposite coatings prepared at different TiO_2 sol addition.

Figure 2 depicts the surface morphology of Fe/Sn coating and the coating prepared by adding different volumes of TiO_2 sol. It can be seen from the figure that when the addition of TiO_2 sol is 0Ml/L, the particle size on the surface of Sn coating is larger, at the same time, the surface structure of Sn coating is loose and there are large pores. With the increase of the amount of TiO_2 sol, the number of pores on the coating surface decreased gradually, and the coating surface was obviously refined, this is because the dispersed TiO_2 nanoparticles can make up for the coating defects and improve the coating quality. When the addition of TiO_2 sol is 15mLl/L, the particle size on the surface of the coating is the smallest and the arrangement is the most uniform. It can be seen from figure 2 (e) that when the addition of TiO2 sol is further increased, the particle size on the surface of the coating is different, and the crystal clusters are easy to be found, and the quality of the coating is poor.

Figure 3 shows the 3D surface topography of the coating prepared by adding different volumes of TiO_2 sol. When the addition of TiO_2 sol is 15mL/L, the coating has a smooth and dense surface structure, but when the addition of TiO_2 sol is too large, the surface of the coating becomes rough. This is due to the fact that when the amount of TiO_2 sol is appropriate, the dispersion of TiO_2 nanoparticles is better, and the coating quality is improved; while when the amount of TiO_2 sol is too large, TiO_2 nanoparticles are easy to agglomerate, which reduces the quality of the coating[8].

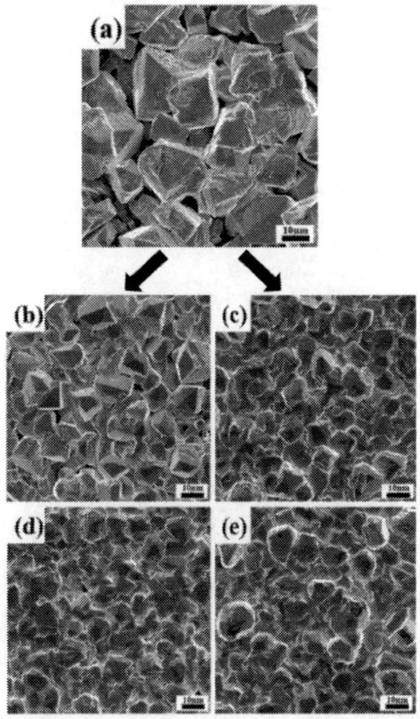

Fig.2. SEM images of Sn-TiO_2 Nano-Coatings prepared by adding different volumes of TiO_2 sol (a) 0mL/L (b) 5mL/L (c) 10mL/L (d) 15mL/L (e) 20mL/L

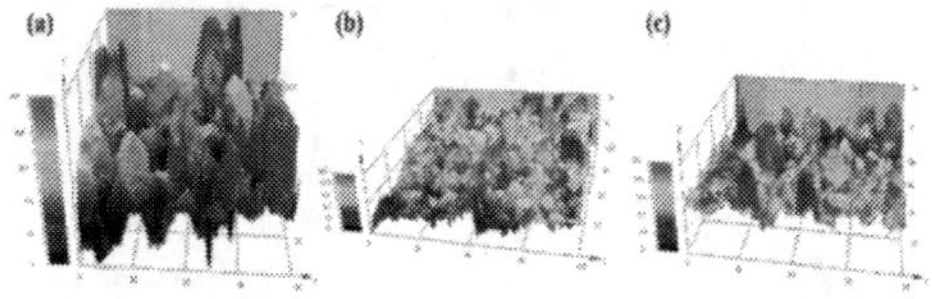

Fig.3. 3D images of Sn-TiO_2 Nano-Coatings prepared by adding different volumes of TiO_2 sol (a) 0mL/L(b) 15mL/L (c) 20mL/L

The hardness of the Sn coating prepared by adding different concentrations of TiO_2 sol is tested, and the results are shown in the figure 4. The results of the bar chart show that with the increase of the amount of TiO_2 sol, the hardness of the coating increases from 67HK (pure tin coating) to 167HK (15mL/L TiO_2 sol), but when the amount of TiO_2 sol continues to increase, the hardness of the coating decreases, indicating that excessive TiO_2 sol will damage the hardness of the coating. The results show that the dispersed TiO_2 nanoparticles can effectively improve the hardness of tin coating.

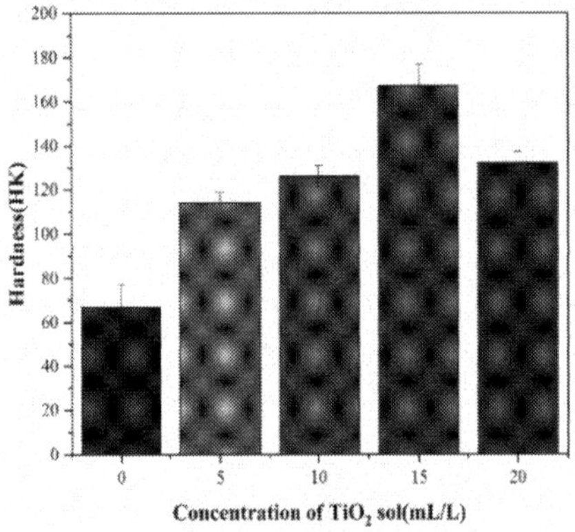

Fig.4. Hardness values of different Sn-TiO_2 Nano-Coatings

Figure 5 shows the Tafel curve of the coating prepared by adding different volumes of TiO_2 sol and iron substrate. Through the fitting calculation of the polarization curve, the corrosion parameters (corrosion potential, corrosion current density and corrosion rate) are shown in Table 1. The corrosion resistance is related to the corrosion potential and corrosion current density. The coating with higher corrosion potential and lower corrosion current density has better corrosion resistance. It can be seen from the table that the corrosion resistance of Fe/Sn is obviously higher than that of Fe matrix, and the corrosion rate is up to 4.9×10^{-2}mm/a. With the increase of the addition of TiO_2 sol, the corrosion rate of the coating decreased gradually, and when the addition of TiO_2 sol was 15mL/L, the corrosion potential of the coating was the highest (- 0.55V), the corrosion potential was the lowest (6.9×10^{-7}A/cm2), and the corrosion rate reached 8.1×10^{-3}mm. It is proved that the corrosion resistance of the coating is the best, and when the addition of TiO_2 sol continues to increase, the corrosion resistance of the coating becomes worse, which is related to a large number of agglomerated nanoparticles[9].

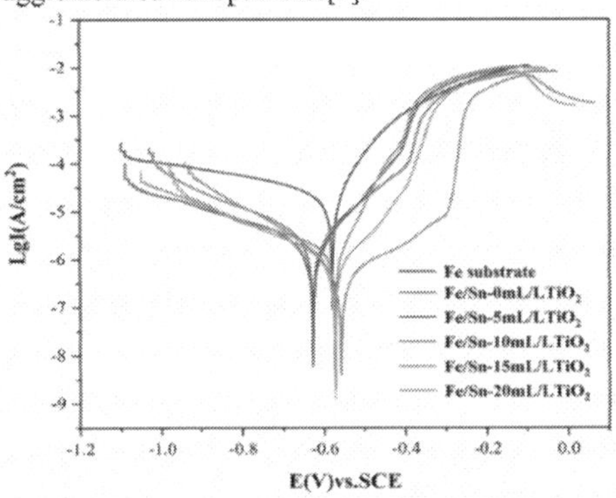

Fig.5. Tafel curve of Sn-TiO_2 Nano-Coating

Table 1 Tafel fitting results of Sn-TiO_2 Nano-Coating

Sample name	E_{corr}(V)	I_{corr}(A/cm^2)	Corrosion rate(mm/a)
Fe substrate	-0.58	3.13×10^{-5}	3.68×10^{-1}
Fe/Sn-0mL/LTiO_2	-0.63	4.17×10^{-6}	4.9×10^{-2}
Fe/Sn-5mL/LTiO_2	-0.63	2.0×10^{-6}	2.4×10^{-2}
Fe/Sn-10mL/LTiO_2	-0.57	1.42×10^{-6}	1.7×10^{-2}
Fe/Sn-15mL/LTiO_2	-0.55	6.9×10^{-7}	8.1×10^{-3}
Fe/Sn-20mL/LTiO_2	-0.57	9.2×10^{-7}	1.1×10^{-2}

Figure 6 shows the Nyquist images of iron substrates and different coatings in 3.5%wt NaCl solution. The capacitive rings in the image show different ring radii, which is related to the change of polarization resistance in the corrosion process. The larger the ring radius is, the higher the charge transfer impedance is and the better the corrosion resistance of the coating is. The equivalent circuit can be fitted according to the Nyquist image, as shown in figure 6. R_s and R_{ct} are solution impedance and charge transfer impedance respectively, and CPE is the original of constant phase angle. The results show that the R_{ct} value increases gradually with the increase of TiO_2 sol addition, and when the TiO_2 sol addition is 15mL/L, the Rct value is the highest (100.3kΩ/cm^2), indicating that it can better resist the corrosion of the external environment.

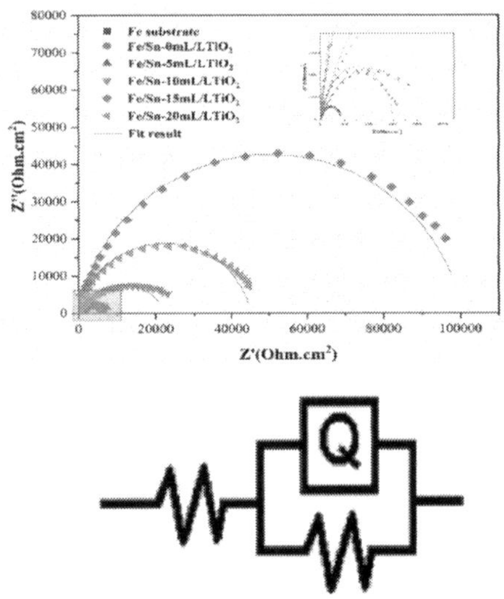

Fig.6. Nuquist image of Sn-TiO_2 Nano-Coating and Equivalent circuit diagram

4. Conclusion

In this study, a method for the preparation of Sn-TiO_2 composite coatings by sol-gel enhanced electrodeposition was pioneered. It is concluded that with the increase of TiO_2 sol addition, the grain refinement occurs and the coating structure is more dense and flat, and when the TiO_2 sol addition is 15 mL/L, the hardness of the coating is the highest and the corrosion resistance is the best; however, when too much TiO_2 sol is added, it leads to more defects in the coating, which is not conducive to the improvement of the mechanical properties and corrosion resistance of the coating.

References

1. Zhang H, Yu M , Liang C , et al. Effect of SiC Deposition Behavior on Microstructure and Mechanical Properties of Cold-Sprayed Al5056/SiC Composite Coatings[J]. Journal of Thermal Spray Technology, 2021(2).
2. Di W, Yao Y, X Shan, et al. ZrO2-doped YTaO4 as Potential CMAS-Resistant Materials for Thermal Barrier Coatings Application[J]. Journal of the American Ceramic Society, 2021.
3. Hui W, Hpa B, Yla B, et al. Effect of Selective Pre-oxidation on the Growth and Corrosion Resistance of Hot-Dip Zinc Coating of Nodular Cast Iron[J]. Corrosion Science, 2021.
4. Ramanauskas R, Quintana P, Maldonado L, et al. Corrosion resistance and microstructure of electrodeposited Zn and Zn alloy coatings[J]. Surface & Coatings Technology, 1997, 92(1-2):16-21.
5. Liu Y, Zhao Q. Study of electroless Ni–Cu–P coatings and their anti-corrosion properties[J]. Applied Surface Science, 2004, 228(1):57-62.
6. [Shan C L, Huang X M, Yu-Cheng W U, et al. Effect of Cu Content in Electroless Ni-Cu-P Coatings on Their Microstructure and Properties[J]. Chinese Journal of Applied Chemistry, 2008.
7. Wang Y, Gao W, He Z, et al. Cu–Sn–Zn nanocomposite coatings prepared by TiO2 sol-enhanced electrodeposition[J]. Journal of Applied Electrochemistry, 2020, 50(14).
8. He Z, Zhang S, Yin L, et al. Cu–TiO2 nanocomposite coatings prepared from sol-enhanced electrodeposition[J]. International Journal of Modern Physics B, 2019:2040038.
9. Zhang W , Di C, Qiao Y, et al. Preparation and properties of duplex Ni-P-TiO2/Ni nanocomposite coatings[J]. International Journal of Modern Physics B, 2018, 33(1n03):1940019.

Research Progress in the Preparation of Phosphor for White Light Emitting Diode

Xueye Li *

Beijing Key Laboratory of Materials Utilization of Nonmetallic Minerals and Solid Wastes, National Laboratory of Mineral Materials, School of Materials Science and Technology, China University of Geosciences, Beijing 100083, China

Abstract. As a solid-state lighting source, white light emitting diode (LED) has become the most promising green lighting energy source in the 21st century with its three advantages of energy-saving, environmental-friendly and green-oriented. The study summarized the research progress in the preparation of phosphor for white light emitting diode, analyzed and compared the effects of different materials and synthesis methods on its luminescence performance. The future development trend was also prospected.

Keywords: White LED; phosphor; rare earth luminescent materials.

1. Research Background

1.1 The development of light sources in lighting

The development of modern human illuminate lamp is marked by the following important periods. In 1882, Thomas Edison from the United States made a carbon fiber incandescent lamp. In 1898, the British chemist Ramsay accidentally invented the neon lamp when checking whether rare gases conduct electricity. In 1906, the high-pressure mercury lamp brought mankind into the era of gas discharge lamps. In 1962, GE's Holonyak and Bevacqua et al. invented red LEDs using gallium arsenide phosphide (GaAsP) material[1]. In the 1970s, gallium phosphide (GaP) green LEDs and silicon carbide (SiC) yellow LEDs emerged to supplement the orange, yellow and green parts of the LED light spectrum; but at the same time, the LED luminous efficiency was pretty low, and it lacks the blue light sources in the RGB trichromes. As a result, it was always difficult to make the breakthrough in terms of white LEDs[2]. In 1993, Shuji Nakamura and others of Nichia Chemical Company (Nichia) first developed gallium nitride (GaN) blue LEDs, and used indium gallium nitride (InGaN) to prepare ultraviolet, blue and green LEDs with ultra-high brightness; and used aluminum gallium indium phosphide (AlGaInP) to prepare ultra-high brightness red and yellow LEDs [3]. In 1996, the first white LED was ultimately made by using gallium nitride (GaN) blue LED + Y3Al5O12：Ce3 +yellow phosphors[4]. Since then, the modern light sources in lighting have gone through three stages: incandescent, neon and gas discharge lamps. And then it entered the fourth stage - the white LED as the leading lighting source.

1.2 The advantages of white LEDs

LED is now the main light source of semiconductor lighting; and compared with other daily lamps such as incandescent and tungsten halogen lamps, white LED lamps have many advantages[5--6].

In terms of energy consumption, white LEDs have the advantages of using low-voltage power supplies, low energy consumption, and high photoelectric conversion efficiency.

In terms of application, white LEDs have high stability, short response time, the ability to emit light in multiple colors, a wide operating voltage range, and flexible materials that can be embedded in any plane.

In terms of environmental protection, the average service life of white LEDs is over 100,000 hours, which is over a hundred times more than that of ordinary light sources; and compared to incandescent lamps, which contain harmful elements such as mercury, white LEDs are made without elements harmful to humans and the environment, and are therefore considered one of the most popular new energy sources.

2. The Implementation Methods and Principles of White LEDs

The main structures of LEDs are a diode composed of a p-type semiconductor and an n-type semiconductor. The carriers of the p-type semiconductor are negatively charged electrons and the carriers of the n-type semiconductor are positively charged holes. When a forward voltage is input, the negatively charged electrons move to the vicinity of the p-n junction and compound with the positively charged holes to produce photons[7].

* Corresponding author: 1003191321@cugb.edu.cn

The bandgap energy E_g of electrons and holes and the wavelength of the radiated light λ satisfy $E_g=hc/\lambda$, which also determines that the light emitted by LEDs is monochromatic and cannot emit a continuous spectrum of white light. Therefore, to obtain white light through the LED, it is necessary to mix the single-color LED light source with other light-emitting materials to form white light. The implementation methods of white LED can be divided into single-chip type and multi-chip type according to the number of chips.

Single-chip white LED refers to the use of an LED chip to generate light to excite other fluorescent materials to emit light so that a variety of color light mix to form white light, the most commonly used are the following three methods [8].

(1) Blue LEDs are coated with yellow phosphors.

(2) Blue LEDs are coated with green and red phosphors.

(3) UV LEDs are coated with RGB tricolor phosphors or a single white phosphor to produce white light.

Multi-chip white LEDs refer to the use of two or more LED chips mixed to form white light, and this method does not require other fluorescent materials.

Table 1 The Implementation Methods and Principles of White LEDs

Implementation method	Number of chips	The way of mixture	The principles of lighting
Single-chip type	1	Blue LEDs+yellow phosphors	Phosphors photoluminescence, mixed with monochromatic light from LED to produce white light
		Blue LED+green&red phosphors	
		UV LEDs+RGB tricolor phosphors/white phosphors	
Multi-chip type	2	Blue LEDs+yellow LEDs	p-n junction electroluminescence, producing complementary color light
	3	RGB tricolor LED combination	

3. The Role of Phosphors and the Selection Principles

Phosphors play a key role in the process of producing white light. It is mainly composed of two parts: the matrix lattice and the activator. The matrix lattice commonly has an aluminate system, a silicate system, a molybdate system, a tungstate system, a sulfide system, a nitride and nitrogen oxide system, etc., whose function is to absorb the excitation energy (LED emitted light). The activator is mainly rare earth-like metal ions. In the process of phosphor preparation, it is often at high temperatures to the matrix lattice doping activator and produce impurity defects and thus defective energy levels. Due to the nature of the 4f shell layer electrons of rare earth-like metal ions, when the activator receives the activation energy transferred from the matrix lattice, the electrons in the 4f shell layer will undergo an energy level jump and be excited by the light mixed with the light emitted from the LED, thus achieving white light.

The selection principles of phosphors for white LEDs are usually divided into two aspects: in terms of preparation, the preparation of phosphor requires a simple process, lower cost, and produces fewer toxic substances; in terms of performance, they usually regard luminous efficiency, CIE color coordinates, color temperature, color rendering index, and other parameters as measurements. In addition, the selection of phosphor also needs to meet the following principles: the excitation spectrum of the phosphor and the LED emission spectrum should match; the effective excitation position of phosphor should be wide; and phosphor quenching temperature should be high and thermal stability should be good[9].

The yellow phosphor with blue LEDs is currently the most widely used and the most advanced technology to achieve the white LED. The traditional yellow phosphor is mostly used yttrium aluminum garnet ($Y_3Al_5O_{12}$) as the matrix lattice. Nichia company initially applied the high-temperature solid-phase synthesis of $Y_3Al_5O_{12}$: Ce^{3+}(YAG:Ce) to achieve the white LED; it is also the mainstream of the current industrial production solution. This method has a simple structure and low cost, but the disadvantages are the high temperature required for synthesis, the uneven particle size of the product, the partial absence of red light in the synthesized phosphor, and the unsatisfactory color rendering, which makes it difficult to meet the needs of low color temperature lighting[10].

4. The Preparation Methods of Phosphors

From the perspective of preparation, the phosphor synthesized by the traditional high-temperature solid-phase method has the disadvantages of large particle size and difficult encapsulation. Therefore, various synthesis methods, such as the co-precipitation method, sol-gel method, hydrothermal method, microwave radiation synthesis method and spray pyrolysis method, have been tried. The use of co-precipitation method can improve the particle size of phosphors, making it easy to encapsulate, and it is convenient to control the process parameters to get different sizes and shapes of phosphors. The disadvantages of the co-precipitation method are a slight decrease in luminescence intensity and agglomeration of phosphors, which can be improved by adding surfactants[11]. The molten salt-assisted co-precipitation method is based on the co-precipitation method and introduces high-temperature molten salts such as fluorine salts or halide salts as solvents and reaction media, which prepares samples with a smaller particle size and a cubic shape and better phosphor color saturation, improving the phenomenon of red light deficiency[12]. The calcination method is an extension of the traditional high-temperature solid-phase method with a lower synthesis temperature and a smaller particle size of the synthesized phosphor, but the luminescence of the phosphor is greatly affected. So the overall effect is not as good as the high-temperature solid-phase method[13]. The precursors prepared using the sol-gel method have the advantages of high purity, small particle size, and low synthesis temperature, but the disadvantage is the high cost of the production process and certain environmental pollution problems[14]. The hydrothermal method is the use of a high-pressure reactor to prepare precursors. The method synthesis temperature is low but phosphor luminescence performance is not as high as the high-temperature solid-phase method. The microwave method is an emerging synthesis method. It uses polar molecules in the microwave effect of dielectric loss, so that the material heats up to complete the reaction, greatly reducing the reaction energy consumption and speeding up the reaction. Therefore, it has some application prospects[15]. The spray pyrolysis method mainly uses the aerosol process of precursors to prepare spherical particles with a uniform particle size distribution and a large specific surface area. The special advantage of the spray pyrolysis method over other synthesis methods is that the size and size morphology of the luminescent material can be controlled. Spray pyrolysis is a simple technology, environmentally friendly production process, and low cost, so it is expected to replace the high-temperature solid-phase method as the mainstream method for phosphor synthesis in the future[16].

5. Blue-light-excited Phosphors

Blue LEDs can excite yellow phosphors or red-and-green mixed phosphors to produce white light. Yellow powder excited by blue LEDs has high luminous efficiency but poor color rendering. Red-and-green mixed phosphors excited by blue LEDs can make up for the lack of red light, but still exist many problems. Therefore, people continue to explore new materials to make white LEDs perform better.

5.1 Yellow phosphors

The most commonly used material for yellow phosphor is $Y_3A_{15}O_{12}$: Ce^{3+}(YAG:Ce). On this basis, co-doping other ions is often the method to improve the luminous performance. By co-doping with rare-earth ions to replace some of the yttrium ions in YAG, different ratios of YAG:Ce^{3+},Ln (Ln=Pr^{3+},Sm^{3+} ,Gd^{3+}, Tb^{3+}) are prepared, which can change the emission wavelength of yellow powder and improve the color rendering performance[17]. It was found that when the doping concentration is small, the CIE coordinates move toward the red region, which improves the color rendering to some extent and can meet the demand for warm illumination. But when the doping concentration is larger, the concentration quenching effect is gradually enhanced, which can cause a sharp decrease in the intensity of the emission peak and affect the luminous efficiency. Zhang J et al. compensated for the missing part of red light by doping Cr^{3+} into $Y_3A_{15}O_{12}$: Ce^{3+} and using red light excited by Cr^{3+}[18]. Yu found that the luminescence intensity was increased by 185% when Li^+ was introduced with a doping amount of 0.06; and by 170% when Zn^{2+} was introduced with a doping amount of 0.1. The emission spectrum peak was red-shifted by 35 nm and the color rendering index was 81.7, which was increased by about 17.7%[19].

There are also many other substrates of yellow powders that have exhibited good performances during the study of new materials. For example, $Ca_3Tb_7BO_4$ (SiO_4) $_5$O in silicate system with apatite[20] as the matrix doped with Tb particles; $LiBa_2Ga$ (P_2O_7) $_2$: $Dy^{3+}$and Sr_8MgAl (PO_4) $_7$: xEu^{2+} in phosphate matrix lattice, the color rendering index of the latter has reached 90.1[21-22]. Among the nitride based phosphors, Ca-α-SiAlON:Eu^{2+} emits orange-yellow light at 590 nm and β-SiAlON: Eu^{2+} emits yellow-green light at about 550 nm, and the modified Li-α-SiAlON: Eu^{2+} has better luminous efficiency and can meet the illumination needs of warm white light. The advantage of nitride phosphors is that they have good thermal and chemical stability, but the disadvantage is that the preparation conditions are harsher and are subject to subsequent development[23].

5.2 Red phosphors

To obtain white LEDs with low color temperature and high color rendering index, the method of coating blue LEDs with red and green phosphors is gradually coming into the limelight. The existing sulfide system of red phosphor luminous efficiency is low with poor stability, so the development of new red phosphors is of great significance.

At present, red phosphors are broadly divided into

sulfide systems, nitride systems and silicate systems.

Among the sulfide systems, SrS:Eu^{2+} and CaS:Eu^{2+} are more mature phosphors with emission wavelengths of 620 nm and 650 nm, respectively, and have high luminous efficiency and can meet the demand for low color temperature lighting. On this basis, the compounding of the two into $(Ca_{1-x}Sr_x)S$:Eu^{2+} further expands the excitation bandwidth, and by changing Sr/Ca, the wavelength of the emitted red light can also be regulated. Han X B et al. improved the emission intensity of phosphors by replacing Ca^{2+} with Mg^{2+}[24]. However, phosphors of sulfide systems are still more difficult to solve the problem of light decay, so sulfide phosphors will be gradually eliminated during subsequent studies[25]. In the nitride system, $M_2Si_5N_8$: Eu^{2+}

(M=Ca,Sr,Ba) series phosphors can emit red light with a maximum wavelength of 680 nm[26]. They have good temperature quenching properties and low degree of thermal attenuation. Furthermore, they can also be used to prepare long afterglow materials by co-doping with Nd^{3+}, Sm^{3+}, Dy^{3+} ions. $Sr_xCa_{1-x}AlSiN_3$:Eu^{2+} phosphors, developed by Watanabe H, have more excellent temperature quenching properties compared to $M_2Si_5N_8$: Eu^{2+} [27]. The development of nitride phosphors is still limited by the harsh production conditions and high production costs. The phosphors of silicate systems are gradually gaining attention due to the cheap and easy availability of raw materials and a simple synthesis process. Those with Eu^{3+} as an activator include Li_2ZnSiO_4: Eu^{3+}, $Sr_3B_2SiO_8$: Eu^{3+}, etc. Those with Eu^{2+} as activator[28] include $(Ba_{1.2}Ca_{0.74-x})_2SiO_4$: xEu^{2+}.

5.3 Green phosphors

Green phosphors mainly use Eu^{2+} as the activator, and the green phosphors studied earlier were mainly sulfide systems, including Ga_2S_3 : Eu^{2+}、MGa_2S_4 : Eu^{2+}(M= Ca、Sr、Ba)[29], etc., but because of the poor stability and toxicity of sulfides, their promotion and application were limited. Subsequently,$(M_{1-2x/v}Yb_x)_{m/v}Si_{12-m-n}Al_{m+n}O_nN_{16-n}$(M=Ca、Li、Mg) phosphors with Yb as activator, $Ca_3Cs_2Si_3O_{12}$: Ce^{3+} and M_2SiO_4:Eu(M=Ba、Sr、Ca) with Ce^{3+} as the activator were gradually developed[30、31]. All these have achieved good results and can be applied in the production of daily white LEDs. The green phosphors of nitride systems such as Ca-α-Sialon：Yb and $Sr_3Si_{13}Al_3O_2N_{21}$：Eu developed in recent years have higher quantum efficiency and excitation intensity and have certain prospects for application.

6. UV-excited phosphors

The color of white LED made by UV-excited phosphor is only determined by the phosphor itself, which has overcome the problems of poor color rendering index and that blue LED scheme color temperature is high. Initially, people used UV LED coated RGB tricolor phosphor to produce white light, but a variety of phosphors mixed ratio not only increases the difficulty of packaging but also faces the problem of color re-absorption, which largely affects the luminous efficiency and color reproduction. So the development of a single substrate of white phosphor is of great importance.
Single matrix white phosphor, according to the type of activation ion doping, can be divided into two kinds: the first is the white phosphors with only one kind of activation ion-doped. Its light-emitting principle is the use of light-emitting matrix lattice and activation ions mixed to produce white light or regulate the location of activation ions in the lattice, to make it light at different peak positions. The second is a variety of activation ion co-doped white phosphor. Its light-emitting principle is to choose the right number of ions ratio so that a variety of color mixing to achieve white light.

6.1 Single-ion-doped white phosphors

Light-emitting matrix lattice, commonly including vanadate (VO_4^{3-}), molybdate (MoO_4^{2-}), tungstate (WO_4^{2-}), etc., usually emits blue-green light with the emission of red light activation ion Eu^{3+} to obtain white light, such as the vanadate single phosphor $Ca_2NaMg_2V_3O_{12}$:Eu^{3+} developed by Setlur et al. of GE, $ZnWO_4$:Eu^{3+} developed by Chen et al., and $Li_3Ba_2Y_3(WO_4)_8$: Tb^{3+}, Eu^{3+}developed by Tian et al.
The white light emission principle of single-ion-doped white light phosphors is that when the activating ion is in different lattice points of the matrix lattice, it produces broadband emission with the peak position in different regions. And by changing the composition of the matrix, the position of the emission peak is adjusted to obtain white light[32]. The absorption and emission bands of Eu^{2+} are relatively wide, and there exists a single 5d1 electron of Eu^{2+} which is susceptible to external influence, so Eu^{2+} can form multiple luminescent centers in the lattice of different compounds and is commonly used as an activator for single-ion-doped white light phosphors. There are mainly silicate systems, phosphate systems and borate systems. Silicate system phosphors have the advantages of high stability, UV shock resistance, and high photoconversion efficiency[33].
Kim et al. in 2004 reported a silicate system phosphor $M_3MgSi_2O_8$:Eu^{2+},Mn^{2+} (M=Ba, Sr, Ca) with excellent color temperature of 3600 K and a color rendering index of 95 by adjusting the ion ratios. The color coordinates of (0.33,0.34) are very close to natural white light[34]. Yang et al. developed Sr_2MgSiO_5:Eu^{2+}, Mn^{2+} on this basis with better excitation in the red light part[35]. Phosphate system phosphors have good physical and chemical stability and high conversion efficiency. Luan et al. synthesized $SrMg_2(PO_4)_2$: Eu^{2+}, Tb^{3+} phosphor with color coordinates of (0.33,0.34) and color temperature up to 5510 K[36]. The advantages of borate system phosphors are low synthesis temperature and easy synthesis of high brightness luminophores. The main results are $Ba_2Ca(B_3O_6)_2$:Eu^{2+}, Mn^{2+} synthesized by Xiao et al. and $Sr_3Y_2(BO_3)_4$:Dy^{3+}, $LiBaBO_3$:Ce^{3+}, Eu^{2+}, etc. synthesized by Li et al.

6.2 Co-doped systems of white light phosphors

The advantage of white light phosphors achieved by co-doping two or more activator ions is that the doping concentration of activator ions can be regulated, thus changing the intensity ratios of different color emission peaks, which can meet the needs of different applications. The common co-doping systems are Eu^{2+}/Ce^{3+}, Eu^{2+}/Mn^{2+}, Ce^{3+}/Tb^{3+}, $Bi^{3+}/Eu^{3+}/Tb^{3+}$. The main problem among them is still the general lack of red-light part, so that the color rendering index is not high. While Eu^{2+}/Mn^{2+} has better luminescence performance compared to other co-doped systems due to the energy transfer efficiency between activators, thus making the Eu^{2+}/Mn^{2+} co-activated phosphors have obvious emission in the red region. At this stage, Eu^{2+}/Mn^{2+} co-doped phosphors are mainly based on the aluminate system and silicate system. The phosphors of aluminate system have the advantages of high quantum conversion efficiency, good color rendering and a wide excitation range[37]. The main results include $La_{0.827}Al_{11.9}O_{19.09}$: Eu^{2+},Mn^{2+} developed by Won, $Ba_2MgSi_{2-x}Al_xO_7$:0.1Eu^{2+},0.1Mn^{2+} developed by Shen, $CaAl_2O_4$: Eu^{2+},Mn^{2+}developed by Xu, etc. The phosphors of silicate system are mainly alkaline earth halosilicates which have the advantages of chemical stability, high luminescence and low synthesis temperature. The main results include the $Ca_2SiO_3Cl_2$: Eu^{2+}, Mn^{2+} series of chlorosilicate phosphors synthesized by Yang et al, $Ba_2SiO_3Cl_2$: Eu^{2+}, Mn^{2+}synthesized by Shen et al, and the recent $Ca_4Si_2O_7F_2$: Eu^{2+}, Mn^{2+} synthesized by Huang et al.

7. Conclusion and Prospects

After decades of development of rare-earth fluorescent materials, the technologies have become sophisticated and various new fluorescent materials and preparation methods are emerging. But considering various factors such as economic efficiency, synthesis technology and practical performance, it is difficult to find a new solution to completely replace the traditional method of coating

YAG: Ce^{3+} yellow phosphor for blue LEDs in a short time.

China's rare earth resources is accounted for 41.36%. Therefore, China is a veritable power of rare earth resources, and is also the production and consumption of rare earth light-emitting materials of the first country. But China's overall research level of rare earth light-emitting materials is still in a backward state; most of the technology patents and core intellectual property rights are monopolized by Japan, the United States and other places units. Our researchers should develop innovative and independent capabilities and find a research path suitable for themselves. To fill the gap in the field of fluorescent materials in China, we should start from the following aspects.

(1) To conduct in-depth study of the basic theory, such as rare earth examples of 4f electronic layer of the law of motion, sensitization process, energy transfer process, luminescence quenching process, etc.

(2) To look for new ways to prepare rare earth luminescent materials, such as nitride phosphors with excellent performance (but due to the harsh synthesis conditions, it has not been put into widespread application).

(3) To increase the research in thermal stability as well as conversion efficiency, as our country tends to pay more attention to the color coordinates, color temperature, and color rendering index of rare earth fluorescent materials.

(4) To develop new applications of rare earth fluorescent materials, such as light conversion materials, long afterglow materials, nano-emitting materials, etc.

References

1. HOLONYAK N J, BEVACQUA S F, Coherent (visible) light emission from Ga(As1-xPx) junctions [J]. Appl. Phys. Lett. 1962, 1: 82-83.
2. Hu, X.D. LED to Illuminate the World [J], College Physics, 2015, 34: 19-24.
3. [3] Li, L.L. Synthesis and Luminescence Properties of Rare-earth Doped Molybdate and Tungstate Phosphors [D]. Jilin University, 2015.
4. SHIMIZU Y, SAKANO K, NOGUCHI Y, et al. Light emitting device having a nitride compound semiconductor and a phosphor containing a garnet fluorescent material:US, 5998925 [P].1997-7-29.
5. JIA Y, LÜ W, GUO N, et al. Utilizing Tb3+ as an energy transfer bridge to connect Eu2+-Sm3+ luminescent centers: Realization of efficient Sm3+ red emission under near-UV excitation[J]. Chem. Commun. 2013, 49: 2664-2666.
6. Wu, D.N. Synthesis and photoluminescence properties of SrMoO4 ： Eu3+/Dy3+ single-host phosphors[J]. Journal of Optoelectronics·Laser, 2020(2): 161-167.
7. Liu J, Li W W, Sun K. Progress in Research on White Light Emitting Diode and Its Phosphors[J]. Materials Reviews, 2007(08):116-120.
8. Yang, L.L. Synthesis and properties of new LED fluorescent materials[D]. Shanghai Institute of Technology, 2021.
9. Liu F, Pan Z W, Zhang J H, Wang X J. White LED-based Solid State Lighting Using Phosphors as Spectral Converters[J]. Journal of the Chinese Society of Rare Earths, 2017, 35(02):171-182.
10. Hong, G.Y., Research Progress of Rare Earth Luminescent Materials[J]. Journal of Synthetic Crystals, 2015, 44(10):2641-2651.
11. Ruan, Z. Research of the YAG Phosphors for White LEDs Preparation Process[D]. University of Electronic Science and Technology of China, 2015.
12. Zhu, J.M. Research of Synthesis and Properties of Rare-earth Co-doped YAG:Ce3+ Phosphors for High-Color Rendering White LEDs[D]. Guilin University Of Electronic Technology, 2015.
13. Zhang H, Su C, Han H, et al., Synthesis of neodymium-doped yttrium aluminum garnet (Nd:YAG) nano-sized powders by low temperature combustion[J]. Journal of Rare Earths, 2005, 23(3):304-308
14. Zhang Z T, Zhang J Y. Inorganic Luminescent Materials and the Applications [M]. Beijing: Chemical Industry Press, 2005, 186-192
15. Zhang X F, Tu H M. Microwave Synthesis of Luminescent Materials[J]. Journal of Hebei Normal University, 2002(06):609-613.
16. Zhou Y H, Lin J, Yu M, Zhang H J. Progress in Luminescent Materials Prepared by Spray Pyrolysis Process[J]. Chinese Journal of Luminescence, 2002(05):503-508.
17. Li, J.Y. Rare Earth Luminescent Materials and the Applications[M]. Beijing: Chemical Industry Press, 2003.
18. Ma R, Ma C, Zhang J, et al. Energy Transfer Properties and Enhanced Color Rendering Index of Chromaticity Tunable Green-Yellow-Red-Emitting Y3Al5O12:Ce3+, Cr3+ Phosphors for White Light-Emitting Diodes[J], 2017, (2): 454-467.
19. Yu, Z.T. Preparation of Non-luminous Ion Doped YAG:Ce Phosphors and Application in Warm White LEDs[D]. Harbin Institute of Technology, 2018.
20. Liu W, Tian T, Huang LW, Xu JY, Zhang Y. Synthesis and Luminescent Properties of Ca3Tb7BO4(SiO4)5O Yellow Fluorescent Material [C]//Selected Papers from the China Rare Earth Society 2020 Annual Conference and Jiangxi (Ganzhou) Rare Earth Resources Green Development and Efficient Utilization Conference
21. Zhao D, Liu W, Deng L, Li Y N. Synthesis and Luminescent Properties of LiBa_2Ga(P_2O_7)_2:Dy~(3+) Yellow Fluorescent Material [J]. Journal of Liaocheng University (Social Science Edition), 2022, 35(01):78-83
22. Zhou J C, Chen M T, Lai Y Q, Shi C, Wu D W, Wu Q S. Synthesis and Luminescence Properties of Broadband Emitting Yellow Phosphor Sr8MgAl

(PO4)7: xEu2+ for White Light-Emitting Diode[J]. Chinese Journal of Inorganic Chemistry, 2021,37(07):1237-1244.

23. Xie, W.J. Research on Silicon-Based Oxynitride Phosphors for Phosphor-Converted White LED[D]. University of Science and Technology of China, 2010.
24. Han X B, Feng Y Z, Peng G H, et al. Synthesis of high brightness red long persistent phosphor (Sr1-xCax) S: Eu, Dy by Sol-Gel thermolysis method[J]. Advanced Materials Research, 2013.652/653/654:581-584.
25. Tang Y, Fu R L, Cao B, Zhang P F, Yang F. Luminescent Properties of Blue Light Activated Red Phosphors for White LED[J]. Journal of Nanjing University of Aeronautics & Astronautics, 2016, 48(01):58-66. DOI:10.16356/j.1005-2615.2016.01.010.
26. Li Y Q, van Steen J E J, van Krevel J W H, et al. Luminescence properties of red-emitting M2Si5N8: Eu2+ (M=Ca, Sr, Ba) LED conversion phosphors[J]. Journal of Alloys and Compounds, 2006, 417(1/2):273-279.
27. Watanabe H, Kijima N. Crystal structure and luminescence properties of SrxCa1-xAlSiN3: Eu2+ mixed nitride phosphors[J]. Journal of Alloys and Compounds, 2009, 475(1/2):434-439.
28. Choi N S, Park K W, Park B W, et al. Eu2+-Mn2+ energy transfer in white light emitting T-phase (Ba,Ca)2SiO4: xEu2+,Mn2+ phosphor[J]. Luminescence, 2010, 130(4):560-566.
29. Xu J, Zhang X M, Zhang J H, et al. Synthesis and Luminescence Properties of MGa2S4：Eu2+(M=Ca、Sr、Ba) and EuGa2S4 Phosphors [J]. Acta Scientiarum Naturalium Universitatis Sunyatseni, 2004, 3: 49-51.
30. SSHIMOMURA Y, KURUSHIMA T, SHIGEIWA M, et al. Redshift of green photoluminescence of Ca3Sc2Si3O12: Ce3+ phosphor by charge compensatory additives [J]. Journal of the Electrochemical Society, 2008, 155 (2) :J45-J49.
31. XIE R J, LI Y Q, HIROSAKI N, et al. Nitride phosphors and solid-state lighting [M]. USA: CRC Press, 2011.
32. LIU J, JIANG M, MEI Y M, WU Z C, KUANG S P. Single-substrate white light phosphors for white light-emitting diodes[J]. Progress in Chemistry, 2013, 25(12):2068-2079.
33. Kim J S, Joe P E, Choi J C, et al. Warm-white-light emitting diode utilizing a single-phase full-collor Ba3MgSi2O8: Eu2+, Mn2+ phosphor [J]. Apply Phys Lett, 2004, 84(15):2931.
34. Sun X Y, Zhang J H, Zhang X, et al. A Single White Phosphor Suitable for Near Ultraviolet Excitation Applied to New Generation White LED Lighting[J]. Chinese Journal of Luminescence, 2005, 26(3): 404.
35. Yang Z P, Liu Y F, Xiong Z J, et al. LUMINESCENT PROPERTIES OF THE SINGLE WHITE EMITTING PHOSPHOR Sr2MgSiO5:(Eu2+, Mn2+) [J]. Journal of the Chinese Ceramic Society, 2006, 34(10);1195.
36. Luan L, Guo C F, Ding X. A novel full-color emission W-LED phosphor based on SrMg_2(PO_4)_2 [J]. Journal of Functional Materials and Devices, 2010, 16(1): 52.
37. ZHANG X M, ZHU Y H, WANG H B, JIN Y J. Progress in Research on Aluminate Red Phosphors Materials[J]. Materials Reports, 2008, 22(S1):270-273.

Theory study of H_2O adsorbed TiO_2 (110) surface: structural and electronic properties

Siying Zhong [1, a, *], Shaoyi Wu [1, b], Xiaohong Chen [2, c] and Kailai Xu [3, d]

[1] Department of Applied Physics, School of Physics, University of Electronic Science and Technology of China, Chengdu 610054, People's Republic of China
[2] School of Sciences and Research Center for Advanced Computation, Xihua University, Chengdu 610039, People's Republic of China
[3] Key Laboratory of Green Chemistry & Technology of Ministry of Education, Sichuan University, Chengdu 610064, People's Republic of China

Abstract. There are three main crystal forms of TiO_2 in nature: rutile, anatase and brookite. In this paper, the GGA-PBE method of density functional theory was used to study the H_2O molecules adsorbed (110) surface of these three kinds of titanium dioxide. H_2O molecules tend to polymerize into dimmers due to the increase of H-O bond length on the surface of rutile than anatase. The lowest adsorption energies of H_2O molecules adsorbed on these three structures are −5.479, −0.085 and 4.278 eV, respectively. H_2O molecule is most likely to be adsorbed on the (110) surface of anatase. As for H_2O molecules adsorbed in the (110) surface of these structures, rutile exhibits the smallest changes of both host TiO_2 and H_2O molecules.

Keywords: TiO_2; DFT; Density of states; Adsorption energy.

1. Introduction

Water is the lubricant of life and a necessary condition for the existence and development of human life on earth. For water molecules adsorbed on the material surface, the physical and chemical properties of the material and water molecules would change somewhat, which may be useful for material modification in materials science. It also helps certain oxides become catalysts for certain processes (especially photoelectric transformation) [1]. Since Fujishima and Honda [2] discovered experimentally in 1972 that titanium dioxide electrodes catalyze the dissociation of water and produce hydrogen under ultraviolet light irradiation, many physics and chemists have devoted themselves to understanding the mechanism of this novel catalytic process [3-10]. Vittadini et al. [11] calculated the adsorption structure and adsorption energy of water on the anatase (101) and (001) surfaces using the DFT method. Schaub et al. [12] studied the water dissociated and adsorbed at the oxygen defect sites on (110) surface of rutile with combination of STM experiments and DFT calculations. By using the CASTEP method, Lindan et al. [13-14] revealed that water on (110) surface of rutile exists not only at the surface of the adsorption but also with the non-dissociation adsorption method, and the potential energy changes of the adsorption process were also calculated and analyzed.

In order to clarify the above problems, in this work, the plane wave pseudopotential density functional method (CASTEP) is adopted for the studies of adsorption of water molecules on the (110) surface of three forms of TiO_2, which would likely enrich the current highlight of the theoretical researches on TiO_2 clusters with adsorbed water molecules.

2. Computational methods

Calculations are based on the spin-polarized DFT using generalized gradient approximation (GGA) for exchange-correlation potential prescribed by Perdew, Burke and Ernzerh of (PBE)[15], implemented in CASTEP package. The Grimme's DFT-D [16] correction is introduced for the accurate treatment of van der Waals dispersion. A Fermi smearing of 0.005 Ha and a global orbital cutoff of 10 eV are employed. For Brillouin zone (BZ) sampling, a $3 \times 3 \times 3$ k - point mesh is applied, since convergence with respect to the number of k - points is vital to generate reliable results. Convergences in energy, force and displacement are set as 2×10^{-5} eV/atom, 0.05 eV/Å and 0.002 Å, respectively.

3. Results and discussions

3.1 Structural optimization of H_2O molecule

Water is an inorganic substance and its constituent elements are hydrogen and oxygen. The hydrogen bonds of water molecules have the same number, so the interaction between hydrogen bonds could form multiple hydrogen bonds in three-dimensional space. The

* Corresponding author: zhongsiying68@163.com

https://doi.org/10.1051/matecconf/202235801058

abnormal interaction between water molecules makes water have some special physical properties. A dynamic hydrogen bond network could be formed between water molecules. The existence of liquid water molecules is mainly in the form of molecules $(H_2O)_n$. It can be seen from Figure 1 that there are 4 water molecules connected near each water molecule.

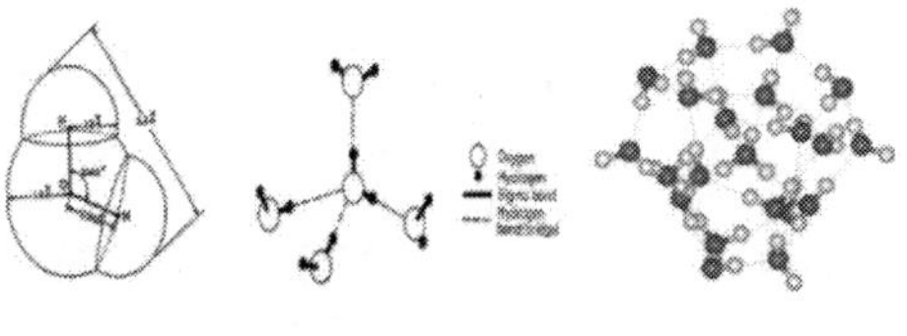

Figure 1: The structure of H2O molecule

The structural data of H_2O molecule are calculated and showed in Table 1, and the density of states for H_2O molecule is also displayed in Fig.2.

Table 1: Original structure data of H_2O molecule

	H_2O
d_{H1-O1}(Å)	0.975
d_{H2-O1}(Å)	0.975
d_{H1-H2}(Å)	1.540
Θ_{H-O-H}(°)	104.086
Enthalpy(eV)	-468.592
Frequency(cm^{-1})	5009.959
Energy(eV)	-467.899

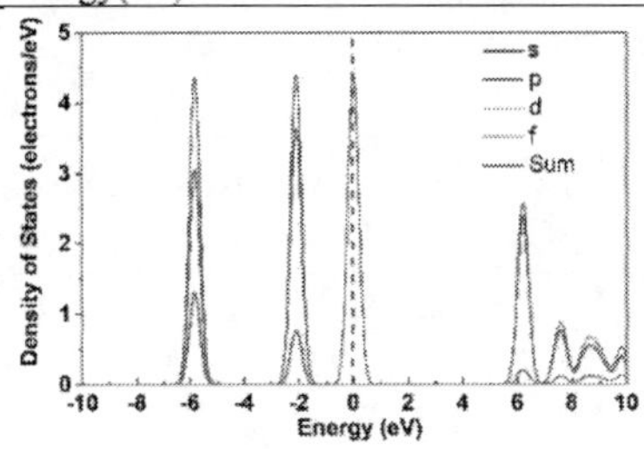

Figure 2 That a peak locates at around −2 eV

It can be seen from Figure 2 that a peak locates at around −2 eV, indicating the corresponding electronic state arising from the oxygen 2s and 2p orbitals. There is no charge state in the position from -17 to -7 eV, corresponding to the forbidden energy region. In the middle of -7 to 1 eV, there are three obvious peaks, accounting for three allowed energy regions. Moreover, the electronic states in this energy range originate mainly from oxygen 2p orbital. It is worth mentioning that the electronic states near 0 eV come from only the oxygen 2p orbital. However, there is no electronic state distribution in the range of -5 to -3 eV and around -1.5 eV, indicating two forbidden regions. In general, there are 7 peaks in the water molecule, associated with 7 allowed regions. In addition to a more obvious dense distribution of charge states around -18 eV, other electronic states locate mainly in the ranges of -7 to 1 eV and 5 to 11 eV.

3.2 Structural optimization of TiO_2

There are three crystal structures of TiO_2 in nature: rutile, anatase and brookite. Consisting of interconnected TiO_2 octahedra, rutile and anatase exhibit higher catalytic activity, especially the latter. But there is also a difference between the two structures in the way of the octahedra connecting with each other. The three crystal structures are shown in Figure 3.

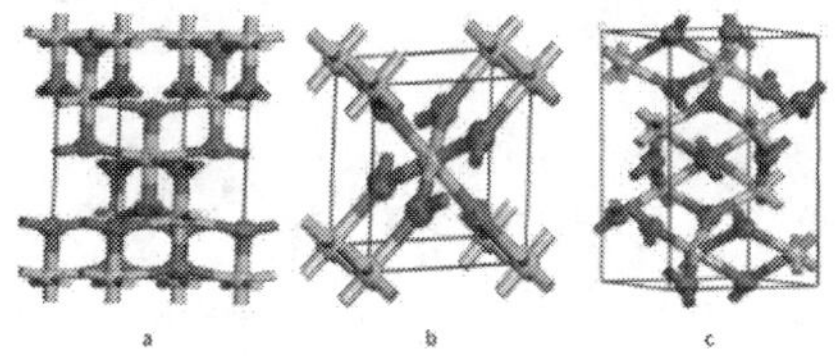

Figure 3: The structure of TiO_2. (a) anatase, (b) rutile, (c) brookite

There are two ways to connect octahedral, i.e., coedge and common vertex, as shown in Figure 4.

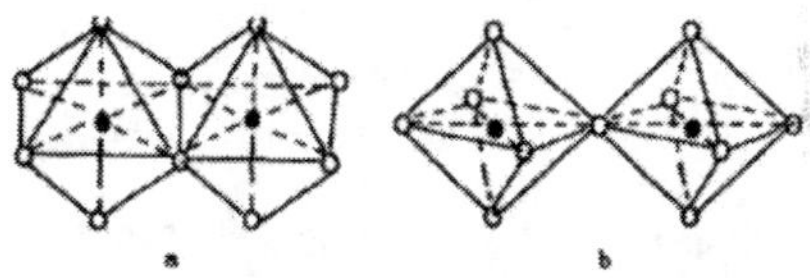

Figure 4: The connection method of TiO2. (a) coedge, (b) common vertex

The Ti-O-Ti angle, distances between Ti-O atoms, enthalpy and frequency of these three structures are shown in Table 2.

Table 2: The three crystal structures of TiO_2

	TiO_2			
	θ_{Ti-O} (°)	d_{Ti-O}(Å)	Enthalpy(eV)	frequency(cm^{-1})
rutule	90	2.046	-4961.862	519.849
anatase	77.588	2.004	-4962.006	563.967
brookite	85.325	2.208	-1984.806	867.207

In the three structures, the angle between the Ti-O-Ti link of the rutile structure is 90°, larger than those for both anatase and brookite. The lengt of the Ti-O bond in rutile is longer than that of anatase, which may indicate that the structure can be more compact around the Ti^{4+} in the former. As for the enthalpy and frequency, anatase and rutile are similarly small, whereas brookite has much higher enthalpy and frequency.

3.3 H_2O molecule adsorbed on TiO_2(110)

There are many possibility positions of the adsorption of H_2O molecules on TiO_2(110). Three possible positions (above, in and below) are calculated for H_2O molecules adsorbed on (110) surface of TiO_2, as shown in Figure 5.

 https://doi.org/10.1051/matecconf/202235801058

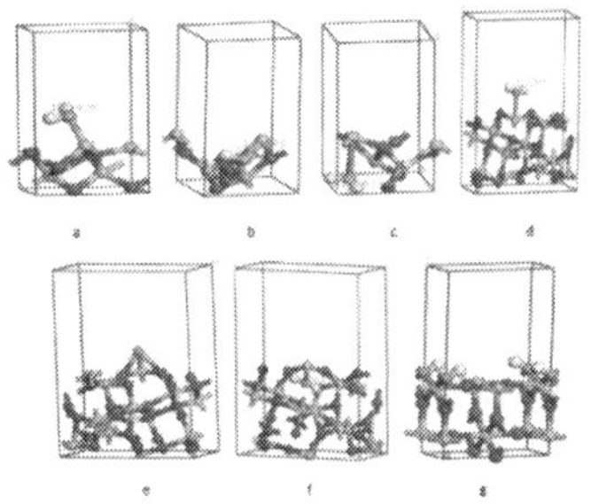

Figure 5: The structure of H2O molecule adsorbed on (a-c) anatase, (d-f) brookite and (g) rutile TiO2(110) surface.

Using density functional theory (DFT) plane wave pseudopotential method, it could be clearly understanding the various changes in the structure of TiO_2 with the adsorbed H_2O molecules. The useful data, i.e., hydrogen-oxygen bond length, hydrogen-hydrogen bond length, Ti-O-Ti angle, distance between Ti-O atoms, enthalpy, and adsorption energy for the three kinds of TiO_2 structures are displayed in Table 3, 4 and 5, respectively. H_2O adsorbed on (110) surface of anatase with three possible positions, and all the H-O bonds of water molecules are elongated. The H-H bond increases significantly only when H_2O is adsorbed within the crystal. Comparing the adsorption energy, it can be seen that while H_2O adsorbed above the surface and in the titanium dioxide crystal are negative. While the adsorption energy of H_2O adsorbed below the surface of titanium dioxide is positive, indicating that H_2O is least likely to be adsorbed below the (110) surface of TiO_2.

When H_2O was adsorbed on the structure of brookite, the change of the H_2O molecule adsorbed below the surface of TiO_2 is not obvious, but the changes of the H_2O molecule adsorbed above the (110) surface of titanium dioxide are compared. Comparing the changes of titanium dioxide, it was found that the situation is just the opposite. The adsorption of H_2O molecules below TiO_2(110) makes the structure of titanium dioxide atoms change greatly. It can be seen that the adsorption energy adsorbed in the titanium dioxide crystals is negative, indicating that this adsorption position is favorable, while the adsorption energy of H_2O molecule below TiO_2(110) is positive, means the H_2O molecule is least likely to be adsorbed with this situation.

In the rutile structure, the adsorption of H_2O molecule does not significantly change the structure of titanium dioxide, but regardless of the angle or the distance between titanium and oxygen atoms are reduced, which means the titanium dioxide atoms are closer together while H_2O molecule titanium adsorbed on it. Similarly, the changes of the H_2O molecule themselves is not obvious. It is worth to mention that the structure of H_2O molecule is basically not distorted while it adsorbed above TiO_2(110) surface. The adsorption energy of H_2O molecule adsorbed above the (110) surface of rutile is 4.278 eV, which shows that this kind of adsorption is impossible.

Table 3: H_2O molecule adsorbed on (110) surface of anatase.

	anatase TiO_2(110)						
	d_{H1-O1} (Å)	d_{H1-O1} (Å)	d_{H1-H2} (Å)	θ_{Ti-O} (°)	d_{Ti-O} (Å)	enthalpy (eV)	E_{ads} (eV)
initial				77.588	2.004	−4962.006	
relaxation				125.057	1.990	−7438.898	
above	1.002	0.981	1.619	115.932	2.174	−9511.773	−0.935
in	2.448	0.978	2.497	108.671	1.797	−9512.663	−5.497
below	1.014	0.977	1.618	52.865	3.456	−9511.195	8.417

Table 4: H_2O molecule adsorbed on (110) surface of brookite.

	brookite TiO_2(110)						
	d_{H1-O1} (Å)	d_{H1-O1} (Å)	d_{H1-H2} (Å)	θ_{Ti-O} (°)	d_{Ti-O} (Å)	enthalpy (eV)	E_{ads} (eV)
initial				76.058	2.230	−1984.006	
relaxation				122.190	1.890	−2976.733	
in	2.449	0.979	2.448	76.765	2.360	−3023.647	9.811
below	0.978	0.979	0.157	66.749	3.552	−3023.605	−0.085

Table 5: H_2O molecule adsorbed on (110) surface of rutile.

	rutile TiO_2(110)						
	d_{H1-O1} (Å)	d_{H1-O1} (Å)	d_{H1-H2} (Å)	θ_{Ti-O} (°)	d_{Ti-O} (Å)	enthalpy (eV)	E_{ads} (eV)
initial				90.000	1.956	−4961.802	
relaxation				79.913	1.978	−9483.536	
above	0.978	0.979	1.612	68.727	1.682	−9953.024	4.278

4. Conclusions

Structural and electronic properties as well as the activity of H_2O molecule adsorbed on (110) surfaces of three forms of TiO_2 were calculated by DFT method of GGA-PBE. Based on the analysis, the following conclusions could be drawn:

(1) For the three forms of titanium dioxide, anatase may exhibit the best photocatalytic activity. Both rutile and anatase are tetragonal crystal systems, with the local rhombic symmetry for the host Ti^{4+} sites, while brookite belongs to orthorhombic structure.

(2) As TiO_2 bulks are cleaved from (110) surface, the symmetries are reduced to C222 (D2-6), P21 (C2-2) and PMM2 (C2V-1), respectively, for anatase,

brookite and rutile. At the same time, this operation also makes the arrangements of Ti and O atoms more compact, and the bonds between these two kinds of atoms are also strengthened.

(3) For H_2O molecule adsorbed on anatase structure, the possible adsorption position is the center of the system. Under the influence of external factors, the H_2O molecule adsorbed above the (110) surface prefers to decomposing. In the brookite structure, H_2O molecule tends to be adsorbed below the surface, and H_2O molecule is also easiest to be decomposed.

References

1. Lars Erik Walle, Davide Ragazzon, Anne Borg, et al, Photoemission studies of water dissociation on rutile TiO2(110): Aspects on experimental procedures and the influence of steps Lars, Applied Surface Science, 303 (2014) 245-249.
2. Fujishima A, Honda K. Electorchemical photolysis of water at a semiconductor electrode, Nature, 238 (1972) 37-38.
3. Tan S, Feng H, Ji Y, et al. Observation of Photocatalytic Dissociation of Water on Terminal Ti Sites of TiO2(110) Surface. J. Am. Chem. Soc, 134 (2012) 9978-9985.
4. Cheng J, Sulpizi M, VandeVondele J, et al. Hole Localization and Thermochemistry of Oxidative Dehydrogenation of A queous Rutile TiO2(110). ChemCatChem, 4 (2012) 636-640.
5. Cheng J, Vande Vondele J, Sprik M, et al. Identifying Trapped Electronic Holes at the Aqueous TiO2 Interface. J. Phys. Chem. C, 118 (2014) 5437-5444.
6. Nakamura R, Nakato Y. Primary Intermediates of Oxygen Photoevolution Reaction on TiO2 (Rutile) Particles, Revealed by in Situ FTIR Absorption and Photoluminescence Measurements. J. Am. Chem. Soc, 126 (2004) 1290-1298.
7. Imanishi A, Okamura T, Ohashi N, et al. Mechanism of Water Photooxidation Reaction at Atomically Flat TiO2(Rutile) (110) and (100) Surfaces: Dependence on Solution Ph. J. Am. Chem. Soc, 129 (2007) 11569-11578.
8. Waegele M M, Chen X, Herlihy D, et al. How Surface Potential Determines the Kinetics of the First Hole Transfer of Photocatalytic Water Oxidation. J. Am. Chem. Soc, 136 (2014)10632-10639.
9. Cheng J, Liu X, Kattirtzi J A, VandeVondele J, et al. Aligning Electronic and Protonic Energy Levels of Proton-Coupled Electron Transfer in Water Oxidation on Aqueous TiO2. Angew. Chem. Int. Ed, 53 (2014)12046-12050.
10. Whitesides GM, Crabtree GW. Don't Forget Long-Term Fundamental Research in Energy. Science, 315 (2007)796-798.
11. Vittadini A, Selloni F, et al. Structure and Energetics of Water Adsorbed at TiO2 Anatase （101） and （001） Surfaces. Phys. Rev. Lett, 81(1998) 954-956.
12. Schaub R, Thostrup N, et al. Oxygen Vacancies as Active Sites for Water Dissociation on Rutile TiO2(110). Phys. Rev. Lett, 87 (2001) 266104-266106.
13. Philip JD Lindan, Harrison NM. Mixed Dissociative and Molecular Adsorption of Water on the Rutile (110) Surface. Phys. Rev. Lett, 80 (1998) 762-765.
14. Philip JD Lindan, Changjun Zhang. Exothermic water dissociation on the rutile TiO2 （110） surface. Phys. Rev. Lett, 72 (2005) 0754391-0754397.
15. Perdew, J. P.; Burke, K.; Ernzerhof, M. Generalized Gradient Approximation Made Simple. PHY S I CAL REV I EW LETTERS, 77 (1996) 3865-3868.
16. GRIMME, S., Semiempirical Gga-Type Density Functional Constructed with a Long-Range Dispersion Correction. J. Comput. Chem, 27 (2006)1787-1799.

Research on influencing factors of fatigue performance of cruciform welded joints on offshore platforms

Da Li [1], Yunlong Su [1,*], Xueping Bai [1], Chentao Peng [2], Lianyong Xu [2]

[1] CNOOC Research Institute, Beijing, China
[2] School of Materials Science and Engineering, Tianjin University, Beijing, China

Abstract. In this study, the fatigue life of load-carrying cruciform welded joints under various plate thicknesses and degrees of penetration was performed. The fatigue assessment results showed a significant "thickness effect." The fatigue limit of the 25 mm-thick sample was 13.4% lower than that of the 16 mm-thick fatigue sample. Essentially, the fatigue life of the weld toe does not change with the degree of penetration. Plate thickness is the key factor affecting the fatigue life. The thickness effect is determined by both the stress gradient along the plate thickness and the degree of surface stress concentration. The correction result based on the effective notch stress approach is more accurate than the thickness correction formula in the ABS standard, which underestimates the thickness effect of load-carrying cruciform welded joints and may lead to danger. The penetration ratio a/t affects the fatigue failure position of load-carrying cruciform welded joints. The fatigue design of load-carrying cruciform welded joints can be performed by appropriately increasing the degree of penetration while achieving the damage at the weld toe.

Keywords: Cruciform welded joints, fatigue strength, thickness effect, degree of penetration.

1. Introduction

Structural fatigue is one of the most important concerns in marine ships, and the fatigue life of key nodes in hulls and offshore platforms has always been crucial. Welding is one of the common connection forms in offshore steel structures [1]. As the welding parts are often used at key nodes with geometric discontinuities, their fatigue performance has great engineering significance for the overall fatigue resistance design of steel structures.

As a common structure in ships and offshore operations, fatigue failure in cruciform welded joints usually starts from the weld roots or toes. The crack propagates through the weld or base metal under cyclic loading, resulting in fatigue failure at the joints. Depending on the applied forces, cruciform welded joints can be divided into load-carrying and non-load-carrying cruciform welded joints [2].

Based on the structural stress approach, K Kinoshita et al. [3] studied the influence of the length of the welding leg and gap depth, i.e., the penetration length, on the fatigue performance of non-load-carrying cruciform welded joints. Ngoula et al. [4] corrected the crack propagation behavior of cruciform welded joints by considering the effects of residual stress and weld toe geometry. Liao et al. [5] used three-dimensional numerical simulation to analyze the effect of low temperature on the crack propagation life of non-load-carrying cruciform welded joints. Andud [6] studied the effects of high-frequency mechanical shock treatment on the fatigue life of cruciform welded joints. There are still uncertainties regarding the factors affecting the fatigue strength; therefore, the study of the factors affecting the fatigue performance is crucial in structural design.

In this study, we tested the fatigue performance of load-carrying cruciform welded joints under different plate thicknesses and degrees of penetration, and plotted the S-N curve. The influence of plate thickness and degree of penetration on fatigue performance was analyzed using the values of fatigue limit. Combined with finite element analysis, the influence pattern of plate thickness and degree of penetration on the fatigue strength of load-carrying cruciform welded joints has been discussed in depth.in their paper.

2. Experimental design

All manuscripts must be in English, also the table and figure texts, otherwise we cannot publish your paper. Please keep a second copy of your manuscript in your office.

When receiving the paper, we assume that the corresponding authors grant us the copyright to use the paper for the book or journal in question. Should authors use tables or figures from other Publications, they must ask the corresponding publishers to grant them the right to publish this material in their paper.

* Corresponding author: suyl11@cnooc.com.cn

https://doi.org/10.1051/matecconf/202235801059

2.1 Material selection

EH36 common steel was selected as the material for offshore platforms. Its material properties and chemical composition are shown in Tables 1 and 2.

Table 1. Chemical composition of EH36(wt.%).

Material	C	Si	Mn	P	S	Cu	Cr	Ni
EH36	0.16	0.28	1.33	0.021	0.011	0.25	0.15	0.29

Table 2. Mechanical properties of EH36.

Material	Yield strength (MPa)	Tensile Strength (MPa)	Elongation (%)	Elastic Modulus (MPa)
EH36	347	580	28.5%	2.06×10^3

2.2 Sample preparation

According to the welding manual on common offshore platforms, the welding process uses flux-cored arc welding (FCAW) as the base, and submerged-arc welding (SAW) as the filling and cover. The design of the sample is shown in Figure 1. The thickness of the main plate in the two samples was 16 mm and 25 mm. The joint was a cruciform welding joint processed through wire cutting.

Table 3. Welding processing conditions

Welds	Method	Current (A)	Voltage (V)	Gas	Weld speed (mm/min)
Bottom	GMAW	127-155	18-22	CO_2	170-280
Filled	SAW	400-650	26-33	/	400-600
Cover	SAW	400-650	26-33	/	400-600

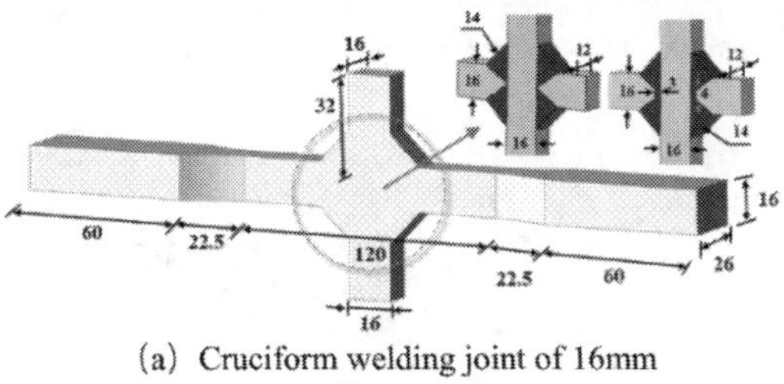

(a) Cruciform welding joint of 16mm

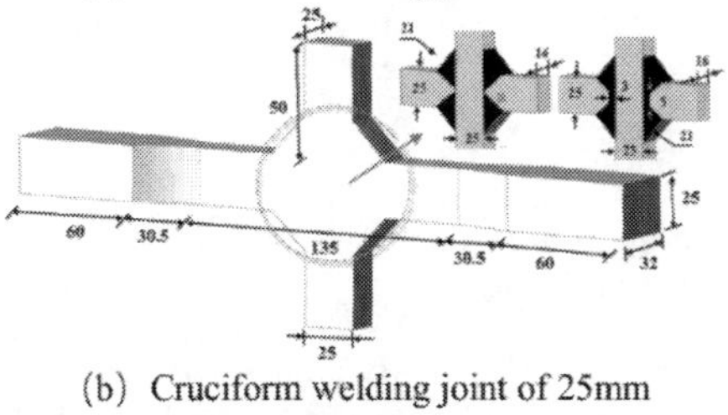

(b) Cruciform welding joint of 25mm

Figure 1. Fatigue specimen configurations for cruciform welding joint with different thickness

2.3 Test equipment and procedures

Referring to GB/T4337-2015 [7], SDS high-frequency fatigue testing machine was used as the test equipment with a stress ratio R of 0.1. A sine wave was applied through constant amplitude loading with a test frequency of 120 Hz. The frequency of the testing machine decreases when the sample cracks. The test ends automatically when the frequency is lower than 10 Hz, and the crack length is approximately half of the thickness of the main plate. The uniaxial tensile fatigue performance test was performed on four types of samples using the group method. Each group of samples was subjected to fatigue tests at 5 to 6 stress levels.

3. Results and Discussion

3.1 Fatigue macroscopic quality

The cruciform welded joints used in the fatigue tests were polished, and the weld appearance quality was inspected using a macroscopic metallographic approach. The metallographic test is a direct and objective method to assess the quality of weld forming [8], which can be used to observe invisible welding defects. The results of the metallographic test are shown in Figure 2. It can be seen that the welding seam is well formed, and there are no welding defects, such as pores, slag inclusions, and incomplete penetration. We measured the unpenetrated depth and gap size of the partially penetrated welded samples. It was observed that the average unpenetrated length of the 16 mm thick partially melted sample was 4.47 mm with an average gap width of 0.5 mm; and the average unpenetrated length of the 25 mm thick partially melted sample was 4.73 mm with an average gap width of 0.51 mm.

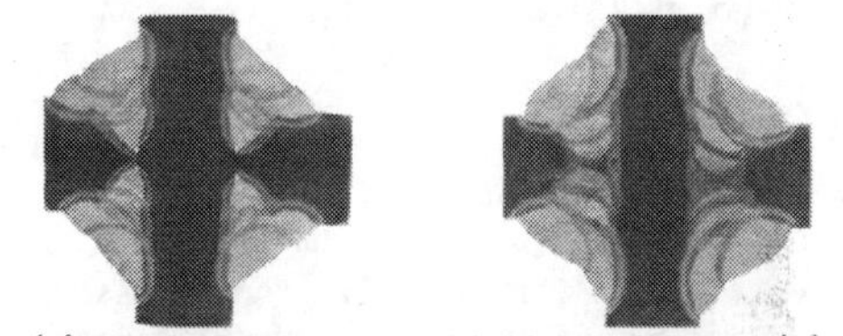

(a) Partial penetration welded joint of 16mm (b) Full penetration welded joint of 16mm

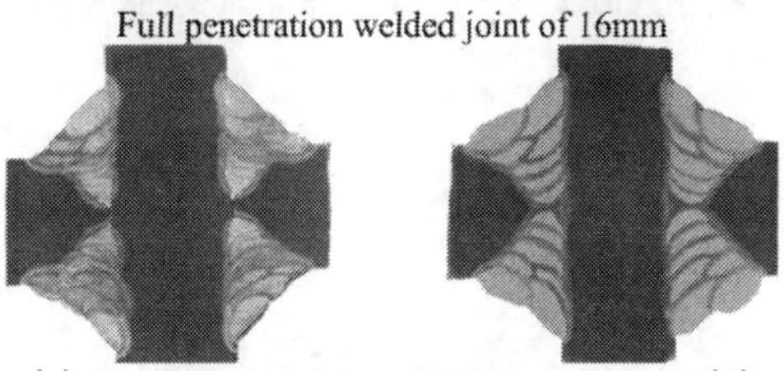

(c) Partial penetration welded joint of 25mm (d) Full penetration welded joint of 25mm

Figure 2. Macroscopic quality inspection results of fatigue specimens.

3.2 Fatigue test results

The failure modes of the load-carrying cruciform welded joints are usually weld root failure and weld toe failure [9]. The experimental results showed that for all the tested samples, the cracks originated at weld toes and propagated along the thickness direction perpendicular to the main plate until the final failure. The fatigue failure mode is shown in Figure 3.

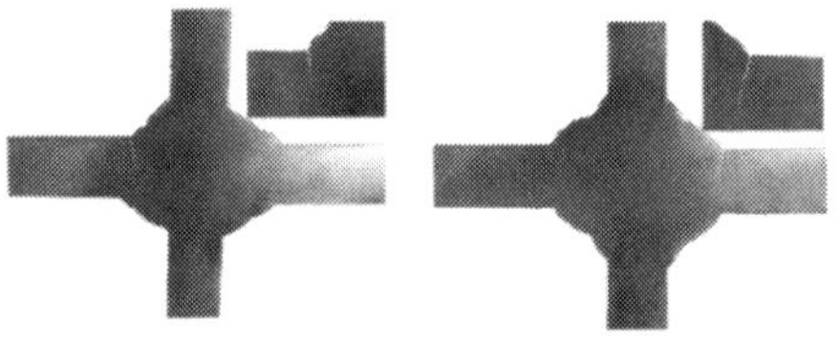

(a) Partial penetration welded joint of 16mm (b) Full penetration welded joint of 16mm

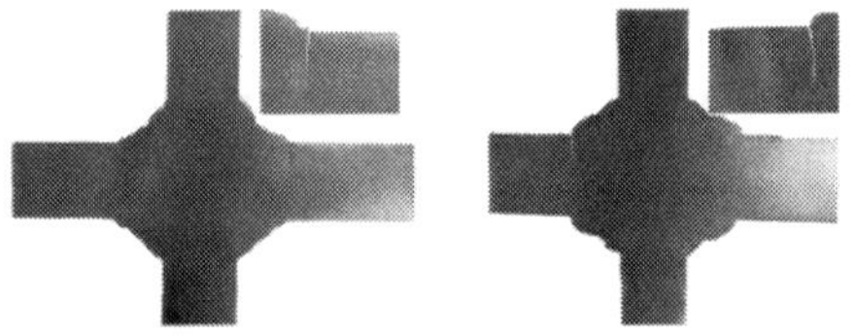

(c) Partial penetration welded joint of 25mm (d) Full penetration welded joint of 25mm

Figure 3. Failure modes of fatigue sample.

According to the results of the fatigue test and nominal stress method in IIW [10] (International Institute of Welding), the test data were processed, as in Eq. (1), to obtain the corresponding median S-N curve. The median S-N curve was determined using the method of least squares, where the slope of the regression was set to 3.0. The fatigue strengths corresponding to the fatigue life of two million cycles were compared.

$$\log\Delta\sigma=-1/m\log N+1/m\log C \quad (1)$$

where $\Delta\sigma$ is the stress range, N is the cycle numbers, and m and C are parameters.

The median S-N curves under the four conditions are shown in Figure 4. Each point in the graph represents the average value of fatigue life at the corresponding stress level. All, but one, data points are well distributed on both sides of the fitting curve or coincide with the curve, which proves the accuracy of the test data.

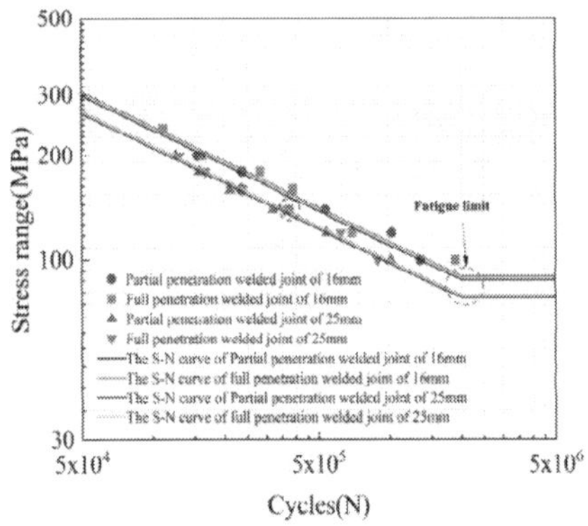

Figure 4. Median S-N curves of different plate thicknesses and penetration degrees.

The fatigue performance of the 16 mm thick cruciform welded joints was significantly better than that of the 25 mm thick cruciform welded joints. The fatigue limits of the partial penetration cruciform welded joints were 86.1 MPa and 77.7 MPa at thicknesses of 16 mm and 25 mm, respectively. The fatigue limits of the full penetration cruciform welded joints at thicknesses of 16 mm and 25 mm were 89.8 MPa and 79.2 MPa; the difference between the fatigue limits of the two was 10.8 % and 13.4 %, respectively. The results showed that plate thickness is a key factor affecting the life as the increase in the plate thickness reduces the fatigue performance of welded joints.

The degree of penetration determines the extent of fusion of the attachment plate and the main plate. Part of the penetration has an unfused area at the junction of the two plates, which causes geometric discontinuity at the weld root and leads to stress concentration. In the load-carrying cruciform welded joint, the weld root and weld toe are the two key weak parts susceptible to fatigue. The degree of stress concentration determines the occurrence of fatigue failure [11], and the strength of stress concentration on the two parts determines the location of fatigue failure.

In our tests, the fatigue failure positions were all located at the weld toes. It can be seen from Figure 4 that the fatigue failure life at the weld toe is essentially the same between the partial penetration and full penetration samples, and their fatigue curves mostly overlap each other. The fatigue limits of the 16 mm thick partial penetration and full penetration cruciform welded joints were 86.1 MPa and 89.8 MPa, respectively. The fatigue limits of the 25 mm thick partial penetration and full penetration cruciform welded joints were 77.7 MPa and 79.2 MPa, respectively. The difference in the fatigue limits of the two was 3.1 % and 1.9 %, respectively.

Table 3. Fatigue limit values of welded joints under different conditions

Penetration	Thickness (mm)	Fatigue limit (MPa)
Partial	16	86.1
Full	16	89.8
Partial	25	77.7
Full	25	79.2

3.3 Influence of plate thickness and degree of penetration

To analyze the influence of plate thickness and degree of penetration on fatigue life further, ABAQUS software was used to model four types of welded joints based on their actual sizes. Owing to the symmetry of the structure, a 1/4 model was established. The stress at the tip of the weld toe had singularity and local micro-constraint effects. A virtual radius of 1 mm was constructed at the weld toe using the effective notch stress method. The effective notch stress method was used as suggested by IIW [10]. It is based on the studies of multiple researchers [12,13], and is published as the guideline. The notch was idealized through the concept of mean stress [14], and the notch stress calculation coefficient was obtained through the local maximum principal stress at the radius based on the simulated reference radius.

Wang Q [15] studied the grid sensitivity of the notch stress method. Notch stress varied with the element sizes when the number of elements was no more than 20, however, it remained almost unchanged when the number exceeded 20. We used a grid size of 0.1 mm for the fatigue weak spot of welding toes and weld roots, and performed the finite element analysis based on the linear elasticity theory. The model and grid refinement are shown in Figure 5. The grid refinement at the weld toe was consistent between the partial penetration and full penetration welded joints; therefore, the grid refinement at the partial penetration weld toe is not shown.

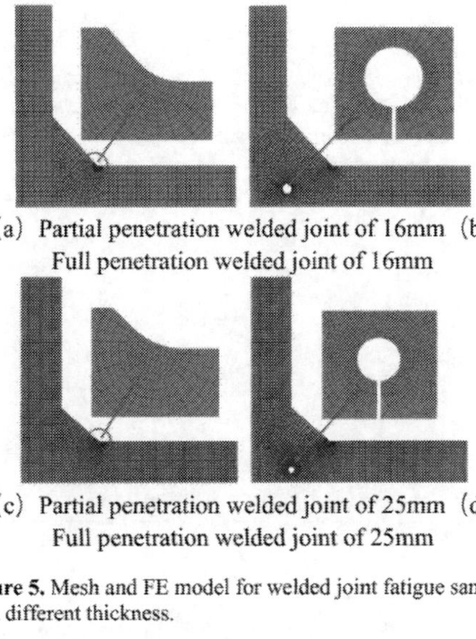

(a) Partial penetration welded joint of 16mm (b) Full penetration welded joint of 16mm

(c) Partial penetration welded joint of 25mm (d) Full penetration welded joint of 25mm

Figure 5. Mesh and FE model for welded joint fatigue samples with different thickness.

The maximum principal stress at the notch was extracted as the notch stress concentration factor, and the results are shown in Table 4.

It can be seen from the results that the increase in the plate thickness leads to a significant increase in the notch stress concentration factor at the weld toe, and the notch stress concentration factor at the weld toe of the partially penetrated welded joint is larger than that of the weld root. Changes in the degree of penetration had no distinct effects on the stress concentration factor at the weld toe, which was consistent with the test results.

Table 4. Results of notch stress concentration coefficient.

Penetration	Thickness (mm)	Notch stress concentration factor	
		Weld toe	Weld root
Partial	16	2.58	2.11
Full	16	2.64	\
Partial	25	2.76	2.24
Full	25	2.81	\

3.3.1 Influence of plate thickness.

All manuscripts must be in English, also the table and figure texts, otherwise we cannot publish your paper. Please keep a second copy of your manuscript in your office. Please keep a second copy of your manuscript in your office. When receiving the paper, we assume that the corresponding authors grant us the copyright to use the paper for the book or journal in question. Should authors use tables or figures from other Publications, they must ask the corresponding publishers to grant them the right to publish this material in their paper.

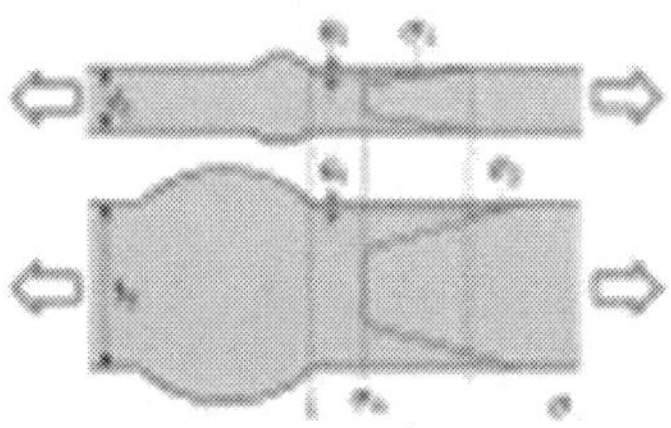

Figure 6. Schematic diagram of thickness effect

ABS standard introduced a thickness correction formula based on statistical results, which corrects the welded structure with a thickness greater than 22 mm in the form of a power function. For untreated welded joints, the correction factor is 0.25.

$$k=\left(\frac{T}{22}\right)^{0.25} \tag{2}$$

Where k is the thickness factor for fatigue strength and T means the real thickness of welded joint.

In the effective notch stress method, owing to the existence of the virtual notch, the notch stress concentration factor is the average stress from the surface to the depth of 1 mm from the sharp notch of the weld toe. The notch stress concentration factor at the weld toe radius includes the change in the degree of stress concentration and stress gradient at the weld toe caused by the increase in plate thickness. The results obtained using the effective notch stress method already cover the thickness effect.

The fatigue test data were processed using the thickness correction formula and the effective notch stress method, as shown in Figures 7 and 8. The figures show three S-N curves under the survival rates of 2.3 %, 50 %, and 97.7 % to reflect the dispersion degree of the corrected results. The corresponding dispersion bandwidth was calculated.

https://doi.org/10.1051/matecconf/202235801059

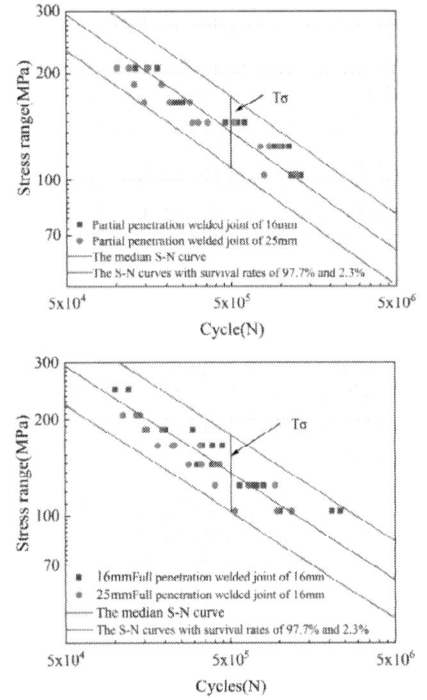

(a) Correction results of partial penetration with different thicknesses (b) Correction results of full penetration with different thicknesses

Figure 7. Correction result of thickness correction formula

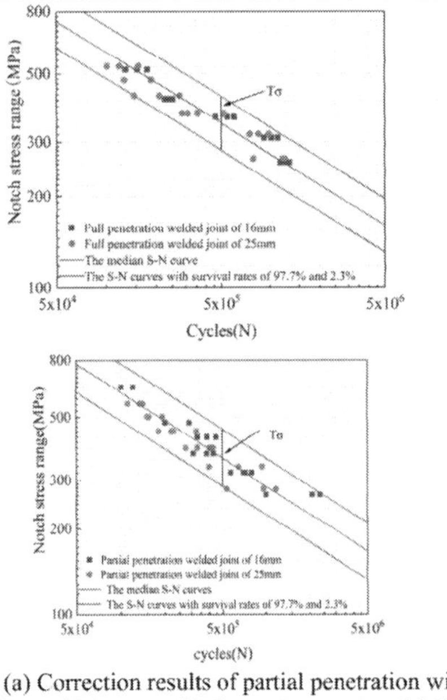

(a) Correction results of partial penetration with different thicknesses (b) Correction results of full penetration with different thicknesses

Figure 8. Correction results of effective Notch Stress Method.

Based on the thickness correction formula, the calculated value under the thickness of 25 mm was 1.032. However, as shown in Figure 2, the fatigue life correction results of welded joints are not ideal. The distribution of data points under different thicknesses is relatively discrete. Under high stress, the distribution of data points under different thicknesses is relatively concentrated, whereas under low stress they are relatively discrete. However, they still show a distinct thickness effect. The calculated dispersion bandwidth was $T\sigma_1 = 1:1.59$ and $T\sigma_2 = 1:1.73$.

Compared with the thickness correction factor, the correction result of the effective notch stress method is more accurate. The distribution of data points under different thicknesses is well patterned. The difference between the test data is small, which effectively aggregates the fatigue life under the two plate thicknesses onto the same S-N curve, as shown in Figure 3. The calculated dispersion bandwidths were $T\sigma_3 = 1:1.44$ and $T\sigma_4 = 1:1.56$.

The load-carrying cruciform welded joint was enlarged in equal proportion to obtain the fatigue strength reduction curve under different thicknesses. The correction results were compared, as shown in Figure 9.

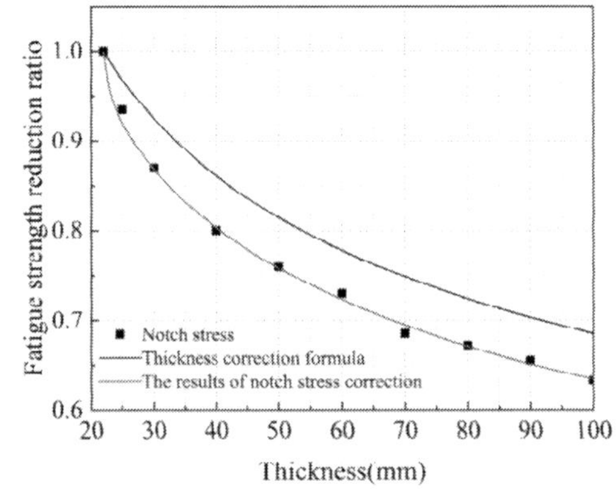

Figure 9. Fatigue strength reduction curve under different thickness.

It was found that the thickness effect of the welded joints enlarged in equal scale was distinct [18], whereas the thickness effect of the welded joints with non-proportional enlargement was ambiguous. In the out-of-plane gusset structure commonly used in the ocean [19], when the thickness of the main plate changes, the thickness correction coefficient index is in the range 0.1–0.14. Toshiaki [20] found that HFMI could effectively improve the fatigue performance of thick plates. The thickness effect is a function of the plate thickness in addition to being the result of the combined effect of various factors, such as the welding structure, joint form, and weld size. It is more accurate to use "size effect" [21] instead of thickness effect.

The thickness effect is determined by the stress gradient along the plate thickness and the degree of surface stress concentration. Compared with non-proportional magnification, proportional magnification leads to an increase in the size of the attached plate and the weld. The surface stress concentration at the weld toe increases, which leads to greater stress gradient along the thickness of the main plate. In the high stress area, the average stress of the proportionally scaled welded joint is greater than that of the non-proportionally scaled welded joint, resulting in significant thickness effect. When the degree of stress concentration at the weld toe in the welded structure is not sensitive to the increase in the plate thickness, even when the plate thickness changes, the stress gradient generated along the thickness direction and the change in the average stress are relatively small.

The notch stress method has a larger correction on the fatigue strength reduction ratio of the load-carrying cruciform welded joints. The calculation result of the thickness correction formula proposed in the ABS standard is underestimated, and there is a certain risk in the fatigue design. Owing to its own characteristics, the notch stress concentration factor includes the influence of welding structure, joint form, and weld geometry, so that the fatigue life can be more accurately corrected when the geometry of the welded structure changes.

3.3.2 Influence of the degree of penetration.

The penetration ratio a/t is defined as the ratio of penetration length 'a' to plate thickness 't.' The relationship between the stress concentration factor of the notch at the root and toe of the load-carrying cruciform welded joint and penetration ratio a/t is plotted, as shown in Figure 10.

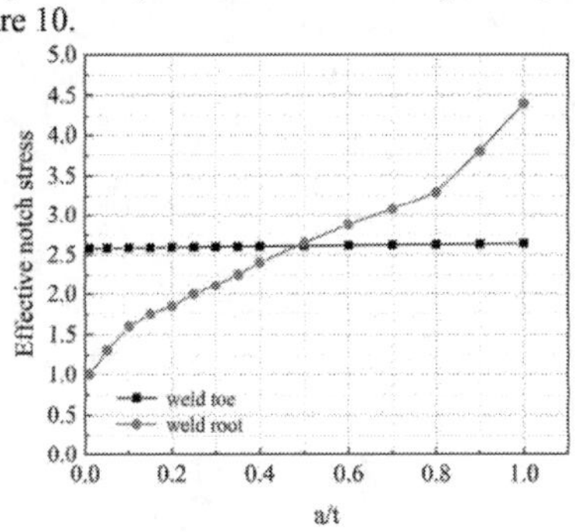

Figure 10. Variation of calculation coefficient of notch stress at weld toe and weld root with a/t.

With an increase in a/t, the notch stress concentration factor at the weld root increases rapidly. The notch stress concentration factor at the weld root is more sensitive to the change in a/t, which increases from 1 to approximately 4.4. The notch stress concentration factor at the weld toe essentially does not change with a/t. The change in the penetration ratio a/t has no distinct effect on the fatigue life at the weld toe of the load-carrying cruciform welded joint.

When $a/t < 0.48$, the stress concentration factor of the notch at the weld toe is considerably higher than that at the weld root. The high stress concentration at the weld toe results in fatigue failure of the weld toe. When $a/t = 0.48$, the notch stress concentration factors at the weld toe and the weld root are equal, and the notch stress concentration factor curves of the two intersect at one point. When $a/t > 0.48$, the stress concentration at the weld root is relatively higher, and the weld root becomes the fatigue failure point. When a/t=1, which is a zero-penetration fillet weld, the notch stress concentration factor at the weld root is considerably higher than that at the weld toe. The effective notch stress method can clearly distinguish weld root and weld toe failure. The results show that the degree of penetration can be appropriately increased while ensuring the fatigue life [22], and the amount of welding consumables can be saved.

4. Conclusion

The following conclusions can be drawn through the fatigue tests of the commonly used load-carrying cruciform welded joints in the ocean under different plate thicknesses and degrees of penetration, as well as the finite element analysis of the influence of the plate thickness and degree of penetration on the fatigue life:

(1) The load-carrying cruciform welded joint exhibited a distinct thickness effect, and the fatigue limit decreased by 13.4%. The fatigue performance of welded joints reduced significantly with the increase in the plate thickness. The degree of penetration had no distinct effect on the fatigue life of the toe of the load-carrying cruciform welded joint. The fatigue limits of the partial penetration and full penetration welded joints differed by only 3%.

(2) The thickness effect is determined by the stress gradient along the plate thickness direction and the degree of surface stress concentration. The thickness correction formula in the ABS standard is not suitable for load-carrying cruciform welded joints. The thickness effect is a function of the plate thickness in addition to being the result of the joint action of various factors, such as welding structure, joint form, and weld size. The effective notch stress method can make an accurate correction for the thickness effect.

(3) The notch stress concentration factor at the weld root is sensitive to changes in the penetration ratio a/t, and the notch stress concentration factor at the weld toe is not affected by the penetration ratio. At $a/t < 0.48$, fatigue failure occurred at the weld toe; and at $a/t \geq 0.48$, the weld root became the most fatigue-sensitive point. While achieving the damage at the weld toe, we can appropriately increase the degree of penetration to save the amount of welding wire.

Acknowledgments

This work was financially supported by Lingshui Semi-submersible Production Platform Research Project (LSZX-2020-HN-01-04).

References

1. Shin W S, Chang K H, Muzaffer S. Fatigue analysis of cruciform welded joint with weld penetration defects, J. Engineering Failure Analysis, 2021, 120:105111.
2. Chang K H, Lee C H, Jang G C, et al. OS4-7-4 Weld geometry effect on the fatigue strength of non-load-

carrying fillet welded cruciform joints, J. The Japan Society of Mechanical Engineers, 2017.

3. Kinoshita K, Arakawa S. Influence of incomplete penetration size on fatigue crack initiation point of load-carrying cruciform welded joints, J. Kou kouzou rombunshuu, 2011, 16:65-70.
4. Ngoula D T, Beier H T, Vormwald M. Fatigue crack growth in cruciform welded joints: Influence of residual stresses and of the weld toe geometry, J. International Journal of Fatigue, 2016, 101(PT.2):253-262.
5. Liao Xiaowei, Wang Yuanqing, et al. Fatigue performance of cross-shaped non-force-transmitting fillet weld joints in low temperature environment, J. Journal of Zhejiang University: Engineering Science, 2020, 54(10):9.
6. Andud D, Saidin S, Manurung Y. Fatigue Life Behaviour of Transverse Fillet Weld and Transverse Fillet on Weld of the HSLA S460G2+M Followed by HFMI/PIT, J. Applied Mechanics and Materials, 2020, 899:126-134.
7. General Administration of Quality Supervision, Inspection and Quarantine of the People's Republic of China, China National Standardization Administration. GB/T 24176—2009 Metal Materials Fatigue Test Data Statistical Scheme and Analysis Method, S. Beijing: China Standard Press, 2016.
8. Peng Y, Dai Z, Chen J, et al. Fatigue behaviour of load-carrying fillet-welded cruciform joints of austenitic stainless steel, J. Journal of Constructional Steel Research, 2021, 184(8):106798.
9. Zhao H, Li X, Seng T L. Fracture analysis of load-carrying cruciform welded joint with a surface crack at weld toe, J. Engineering Fracture Mechanics, 2020, 241.
10. Hob Bac Her A F. The new IIW recommendations for fatigue assessment of welded joints and components-A comprehensive code recently updated, J. International Journal of Fatigue, 2009, 31(1):50-58.4+161-162.
11. Zhang M, Gou G, Hang Z, et al. Effect of stress concentration on the fatigue strength of A7N01S-T5 welded joints, J. International Journal of Modern Physics B, 2017, 31(16-19):1744047.
12. Ayg M. Fatigue evaluation of welded details-using the finite element method, J. Chalmers University of Technology, 2013.
13. Xz A, Sx B, Yz A, et al. Fatigue Reliability Analysis of Metro Bogie Frame Based on Effective Notch Stress Method. 2021.
14. H. Neuber. Über die berücksichtigung der spannungskonzentration bei festigkeitsberechnungen, Konstruktion, 20 (7) (1968), pp. 245-251.
15. Wang Q, Ji B, Fu Z, et al. Evaluation of crack propagation and fatigue strength of rib-to-deck welds based on effective notch stress method, J. Construction and Building Materials, 2019, 201(MAR.20):51-61.
16. Boni L, Lanciotti A, Polese C. "Size effect" in the fatigue behavior of Friction Stir Welded plates, J. International Journal of Fatigue, 2015, 80(nov.):238-245.
17. Yagi J, Tomita Y, Machida S, et al. Thickness effect criterion for fatigue strength evaluation of welded steel structures, J. Journal of the Japan Society of Naval Architects & Ocean Engineers, 2009, 115(1):58-65.
18. Shiratsuchi T, Osawa N. Investigation of thickness and bead profile effects on fatigue strength of welded joints based on relative stress gradient, J. International Journal of Fatigue, 2020, 134:105520.
19. Tetsuo, Okada, Masashi, et al. An analytical and experimental study on the thickness effect of fatigue strength in large-scale-welded models, J. Welding in the World Journal of the International Institute of Welding Journal of the International Institute of Welding, 2014.
20. Toshiaki I, Toshio N, Yoshihisa T, et al. Thickness Effect on Fatigue Strength of Welded Joint Improved by HFMI, J. Quarterly Journal of the Japan Welding Society, 2016, 34(4):249-259.
21. Lotsberg I. Assessment of the Size Effect in Fatigue Analysis of Butt Welds and Cruciform Joints. American Society of Mechanical Engineers, 2014.
22. Rohani H, Dabiri M, Ahola A, et al. Re-evaluation of weld root fatigue strength for load-carrying fillet welded joints using the notch stress concept, J. International Journal of Fatigue, 2020, 144(6).

Dielectric Properties of Polypropylene by Deashing Method for DC Polymer Film Capacitors

Zhaoliang Xing [1], Chong Zhang [1], Xiying Dai [1], Fei Li [1], Yu Xiao [1] and Jian Li [2,*]

[1] State Key laboratory of Advanced Power Transmission Technology, Global Energy Interconnection Research Institute Co., Ltd., Beijing 102209, China
[2] College of Chemical Engineering and Material Science, Tianjin University of Science and Technology, Tianjin 300457, China

Abstract. In this paper, a novel deashing method was proposed to prepare polypropylene (PP) materials with different ash contents (80 ~ 560 ppm). Effects of the ash on dielectric and energy storage characteristics of PP in polymer film capacitors were studied. The experimental results reveal that a low content of ash will help to improve the dielectric properties. Compared to the sample with 560 ppm of ash content, the PP film with 60 ppm ash possesses a lower DC conductivity (1.85×10^{-14} S/m). After the purification, the decline of the ash content in PP prevents the local electric field distortion and enhances the breakdown strength, which is beneficial for the improvement of the discharge energy density. Meanwhile, the reduction in the number of carriers ionized by the ash weakens the transport behavior and reduces the conductivity. It is concluded that the PP with high purity shows a great potential in polymer film capacitors.

1. Introduction

Polymer film capacitors play an important role in modern electronics and electrical systems, such as power capacitors in high voltage direct current (HVDC) projects. With the penetration of the high-level and large-capacity HVDC project, the requirements for capacitors with high reliability and small volume have increased [1-3].The excellent dielectric performance is the prerequisite for the capacitor to operate stably, including a high breakdown strength and low dielectric loss [4]. And the outstanding energy storage capability is beneficial to the trend of small-size and light-weight for capacitors. Theoretically, PP films as the dielectric insulation of the capacitor possess satisfactory electrical performances due to molecular chains without any polar groups [5-7]. However, PP materials contain various proportions of the ash (residual catalysts and additives) introduced by manufacturers during the production, which significantly affect the dielectric performance and the operational life time of films.

PP materials are polymerized from propylene using the Ziegler-Natta catalyst, which mainly comprises of a major catalyst (compounds of $TiCl_4$and $MgCl_2$), an activated catalyst (such as triethylaluminum) and an electron donor withSi. After the polymerization, the catalyst will remain in the PP matrix as the solid.The ash content is used to describe the residual catalyst, which is defined as the mass ratio of residual inorganic oxides after the calcination to the original material[8-9].The ash would introduce lots of polar functional groups, such as carbonyl (-C=O), hydroxyl (-OH), vinyl and conjugated double bond, which were partly responsible for the reduction ofthe capacitor lifetime [10]. It is found in [4] that the dielectric loss was impurity-controlled and any potential increase was mainly due to the effect of residual catalysts. Based on the functional density theory, the research results indicate that the ash wouldlead to the electron density interference between polymer backbones andcharge carriers, resulting in a significant increase of conductivity and obvious decline of breakdown strength [11-12].

Up to now, most of researchesaboutthe effects of theash on dielectric properties have been carried out using the simulation method, and there lacks the verification by the experimental tests. In particular, few studies were done to comprehensively optimize the dielectric and energy storage properties of PP filmsfor capacitors [13-16]. This paper makes an effort to build the relationship between the ash and dielectricas well as energy storage performance, which helps to further understand the role of the ash in PP matrix and reveal the insulation failure mechanism.PP materials were synthesized using a Ziegler-Natta catalyst in the laboratory.An effectivedeashing method was proposed toprepare PP with different ash contents(80 ~ 560 ppm) and a calcination device was used to detect the ash content. The crystallization characteristics was investigated by differential scanning calorimetry (DSC). The electrical properties tests include the breakdown strength and DC conductivity. The deashingmethod showsa great potential in the production of insulation materialswith high purity.

* Corresponding author: lijian@tust.edu.cn

2. Experimental

2.1 Materials Preparation of sMAO

PP materials with different ash contents were obtained by washing the powder with the suitable solvent. PP powder was synthesized by anionic polymerization using a catalyst in the laboratory. The Ziegler-Natta catalyst mainly comprises of a Ti and Mg compound, an organo-aluminium compound and an external electron donor with Si. Ethanol and N-hexane were chosen as solvents. Polar ethanol is compatible with polar ash in the PP matrix and can effectively remove impurities. Nonpolar N-hexane can swell and loosen molecular chains of PP, which helps solvents to enter the interior and achieve better deashing effect.

Table 1. Washing process parameters.

Parameters / Samples	Ethanol (mL)	N-hexane (mL)	Washing frequency
PP5	-	-	-
PP4	200	0	1
PP3	0	200	1
PP2	100	100	1
PP1	100	100	3

The deashing (washing) device was designed and shown in Figure1a, mainly including a 3-necked round flask, a magnetic stirrer and a thermostat. The washing process is as follows:

(a) The atmosphere without moisture and oxygen was created by vacuuming and filling nitrogen in the 3-necked round flask, which can effectively avoid the chemical reaction and oxidation of impurities.

(b) About 40g of PP powder and 200 mL of the solvent were added into the flask, both of them were fully mixed through magnetic stirring for 2h at 80 °C.

(c) The mixture after stirring was filtered, centrifuged and dried until the PP powder was clean and pure. And the PP powder after the deashing treatment was obtained.

PP with different ash contents were produced by controlling washing process parameters, the details are shown in Table 1.The washing frequency is number of times of repeating the washing process of (b). According to the different process parameters, the obtained PP powder are respectively named byPP5,PP4, PP3, PP2 and PP1.PP5 is the original material after polymerization and has not been washed.

The calcination method was used to verify the washing effect and measure the ash content in PP, the device is shown in Figure 1b. The calcination process includes:

(a)40g of PP powder was putted in a quartz crucible. PP powder gradually melted, degraded and carbonized under the constant heating of an electric stove. The organic components were removed and inorganic components as residues remained in oxidation state (such as Al_2O_3, TiO_2, SiO_2 and MgO).

(b) After the calcination for 1 ~ 2 hour, the quality of the residue was measured by the electronic balance (with the accuracy of ~0.0001g).

(c) Repeatthe process of (a) and (b) until the difference of the last two testing is less than 0.0002g. The ash content is the original PP. The ash content of PP1- PP5 are shown in Table 2.

In this section, five kinds of PP with different ash contents (80 ~ 560 ppm) are obtained through the washing process.

Table 2 Test samples and the ash content.

Samples	PP1	PP2	PP3	PP4	PP5
Ash content (ppm)	80	120	280	400	560

A certain amount of PP powder was melted at 190 °C for 5 min, hot pressed at 24 MPa for 10 minutes, and then cooled to 120 °C with the rate of 15 °C/min. The films with the thickness of 25 ~ 30 μm were prepared.

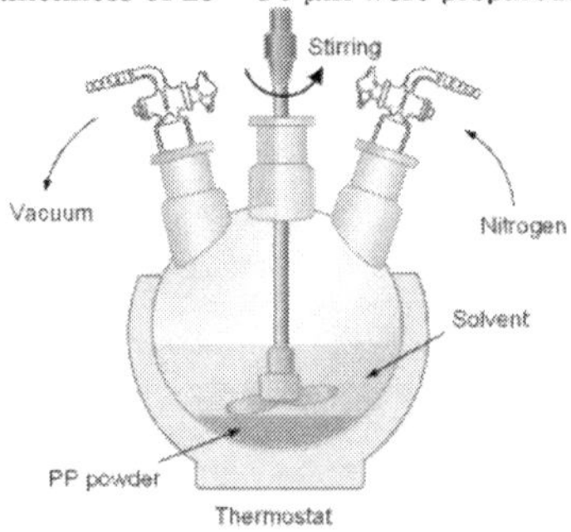

(a)

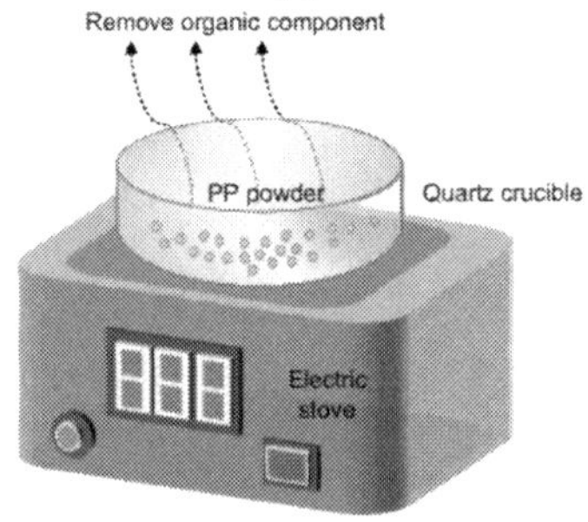

(b)

Figure 1. Thedevices of (a) washing and(b) the ash content measurement for PP powder.

2.2 Characterization

Thermograms were obtained using a PerkinElmer Diamond differential scanning calorimeter (DSC) system. PE productions were heated from 25 to 250 °C at a heating rate of 10 °C /min under constant purging of N_2 at 20 mL/min.

Samples were placed at 60 °C for 1h to remove moisture before the test. In order to investigate the effects of the ash on electric properties, the breakdown strength and DC conductivity were tested. The breakdown strength was tested using the ball-plate electrode in oil.

3. Results and discussion

PP is a typical semi-crystalline polymer, which is composed of crystalline and amorphous regions. PP molecules fold, pack and arrange around the nucleation center to form the high-density crystalline region [4]. DSC was used to study the crystallization characteristic, samples were put in the liquid nitrogen and the testing temperature increased from 20 to 250 °C with the rate of 10 °C/min.The results are shown in Figure 3. The melting curve indicates the melting temperature of samples are between 165.3 to 167.6 °C, the relative error is within 1.43 %, which means the ash has almost no influence on the melting temperature.

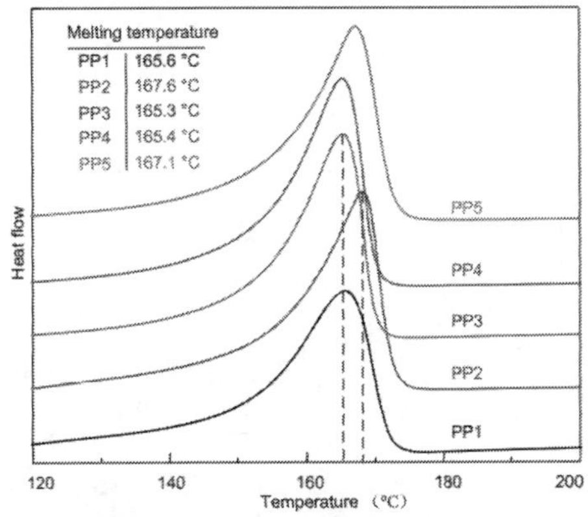

Figure 2. The melting curve of DSC test.

The crystallinity (the volume ratio of crystalline regions) can be defined as:

$$X_m = \frac{\Delta H_m}{\Delta H_m^0} \quad (1)$$

X_m represents the crystallinity, ΔH_m is the absorbed heat (melting enthalpy) by the sample during the heating process, and ΔH_m^0 is melting enthalpy when the material fully crystallizes.

Based on the melting curve in Figure 3 and Equation (1), the crystallinity iscalculated and shown in Figure 4. Traditionally, the researchers have agreed with the conclusion that when the content of additivesin PP is high, which will hinder the crystallization of PP and cause the decrease of the crystallinity [14-15]. But in the research, an interesting phenomenon is found. As shown in Figure 4, the crystallinity increases sharply with the ash content increases from 80 to 400 ppm. And it reaches a steady state from 400 to 560 ppm. In other words, the paper gets an opposite conclusion contrary to conventional wisdom. The phenomenon can be explained by the nucleating agent. The common nucleating agents include organic and inorganic materials, such as aromatic compounds and rare earth compounds [16]. The ash in PP may play a role of the nucleating agent. During the crystallization process, part of the ash increases the number of nucleation centers, leading to an increase of the crystallinity. But the other part of the ash is separated from the crystalline region, preventing the growth of the crystal, which resulting in the decrease of the crystallinity [4]. When the ash content is low, the ash mainly promotes the crystallization. And the effect is opposite at higher content.

Therefore, the paper believes that the low content of the ash can promote the crystallinity of PP.

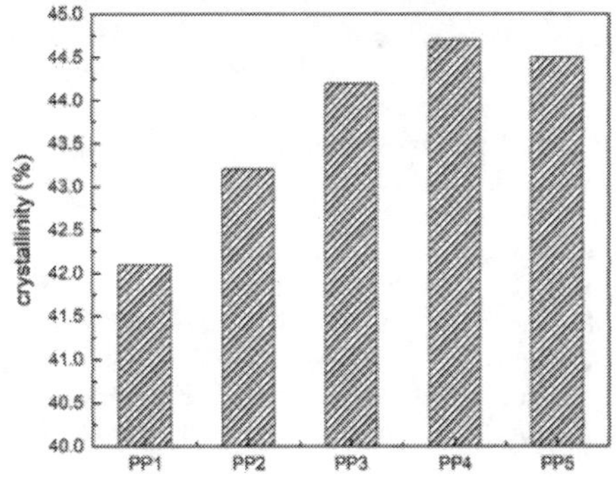

Figure 3. The crystallinity of test samples.

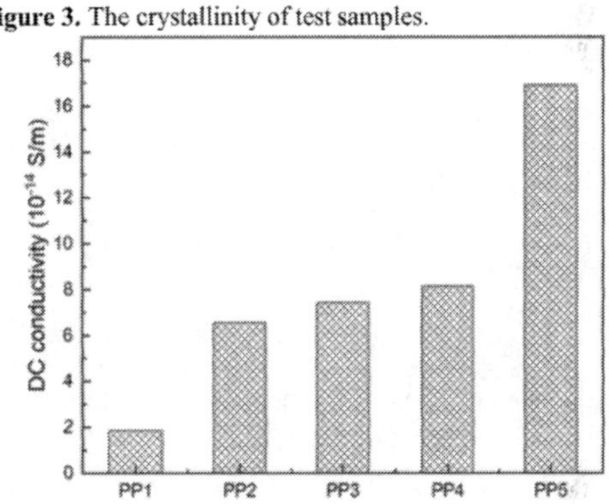

Figure 4. DC conductivity of test samples.

DC conductivityat 25°Cis described in Figure 5.It can be seen that the conductivity increases as the increase of the ash content.PP1 has the lowest conductivity (1.85×10^{-14} S/m) among all the samples. The conductivity of PP2, PP3 and PP4 is similar and the range varies from 6.56 ~ 8.14×10^{-14}S/m, which increases rapidly to 1.69×10^{-13}S/m for PP5 (almost 16 times higher than that of PP1).The growth amplitude can reach an order of magnitude.

For insulation materials with thewide band gap (about 7eV for PP), the conductivity mechanisms at room temperature mainly arethe electronic and ionic conduction [23]. The carriers generated by impurities and injected from electrodes can jump over a barrier from one trap site to another, forming a steady conduction current under the electrical field. It is found that thecrystallinityis improved by the ash,the high crystallinityis beneficial to decrease the conductivity. But more ashis separated from thecrystalline regions for samples with high ash content. These impurities (such $TiCl_4$ and $MgCl_2$) at amorphous regions generate carriers by ionization, contributing to the ionic conduction and enhancingthe conductivity.

Therefore, the crystalline and amorphous regions together affect the conductivity, the highercrystallinity does not mean the lower conductivity, but also depends on the ashat the amorphous region.

4. Conclusion

In the study, PP materials with different ash contents(80 ~ 560 ppm) were prepared by the deashing method. The dielectric and energy storage characteristics of PP films in capacitors were studied, respectively. The high content of impurities in PP materials does not cause the decrease of the crystallinity. During the crystallization process, part of the ash plays a role of the nucleation center, and enhances the crystallinity. And the other part exists in amorphous regions, which has a great influence on the dielectric property of films. As the ash contentin creases, the DC conductivity of films show adeterioration trend. The development of materials with high purity is an ideal way to ensure the long-term reliable and stable operation of capacitors and can realize the miniaturization of capacitors.

Acknowledgments

The authors thank the financial support from the Project of Global Energy Interconnection Research Institute (5500-201958321A-0-0-00)

References

1. C. Lu et al, "Online Degradation of Biaxial-Orientated Polypropylene Film from HVDC Filter Capacitors," IEEE Trans. Dielectr. Electr. Insul., vol. 26, no. 1, pp. 26-23, Feb. 2019.
2. M. Ritamäkiet al, "Large-area Approach to Evaluate DC Electro-thermal Ageing Behavior of BOPP Thin Films for Capacitor Insulation Systems," IEEE Trans. Dielectr. Electr. Insul., vol. 24, no. 2, pp. 826-836, Apr. 2017.
3. Y. Zhang et al, "EnhancedElectric polarization and breakdown strength in the all-organic sandwich-structuredpoly(vinylidene fluoride)-based dielectric film for high energy density capacitor," APL Mater. vol. 5, no. 7, pp. 1-6, Jul. 2017.
4. M. Ritamäkiet al, "Performance metrics for a modern BOPP capacitor film," IEEE Trans. Dielectr. Electr. Insul., vol. 26, no. 4, pp. 1229-1237, Aug. 2019.
5. C .Chenet al, "Designing of Ferroelectric/Linear Dielectric Bilayer Films: AnEffective Way to Improve the Energy Storage Performances ofPolymer-Based Capacitors,"J Phys Chem C, vol. 124, no. 11, pp. 5920-5927, Mar. 2019.
6. Y. K. Zhuet al, "High Energy Density Polymer Dielectrics Interlayered by Assembled Boron Nitride Nanosheets," Adv.Energy mater., vol. 9, no. 40, pp. 1-10, Oct. 2019.
7. H. M. Umranet al, "Ageing: Causes and Effects on the Reliability of Polypropylene Film Used for HVDC Capacitor," IEEE Access, vol. 8, pp. 40413-40430, 2020.
8. Alain. standaertet al, "Process for the production of propylene polymers having low ash content," US 2009/0312507A1, Dec. 17, 2009.
9. Alain. standaertet al, "Process for the production of a propylene polymer having a broad molecular weight distribution and a low ash content," US 2011/0034649A1, Feb. 10, 2011.
10. H. Nguyen et al, "Structural and Electronic Properties ofDefect-Free and Defect-Containing Polypropylene: A ComputationalStudy by van der Waals Density-Functional Method," J. Phys. Stat. Solidi(b), vol. 255, no. 3, pp. 1-7, 2017.
11. M. E. Stournaraet al, "A first principles investigation of isotactic polypropylene," J. Mater. Sci., vol. 45, no. 2, pp. 443–447, Oct. 2010.
12. A. Huzayyin, et al, "Density functional analysis of chemical impurities in dielectric polyethylene," IEEE Trans. Dielectr. Electr. Insul., vol. 17, no. 3, pp. 926-930, Jun. 2010.
13. F. H. Zhenget al, "Space Charge Characterization in Biaxially OrientedPolypropylene Films,"IEEE Trans. Dielectr. Electr. Insul., vol. 23, no. 5, pp. 3102-3107, Oct. 2016.
14. A. Tariq et al, "Study of thermal, morphological, barrier and viscoelastic properties of PP grafted with maleic anhydride (PP-g-MAH) and PET blends," J.Polym. Res., vol. 27, no. 10, pp. 1-10, Sep. 2020.
15. M.Streibl et al, "The Influence of Nanoparticles and their Functionalizationon the Dielectric Properties of Biaxially OrientedPolypropylene for Power Capacitors,"IEEE Trans. Dielectr. Electr. Insul., vol. 27, no. 2, pp. 468-475, Apr. 2020.
16. W. P. Li et al, "High-energy-density dielectric films based on polyvinylidene fluoride and aromatic polythiourea for capacitors," J. Mater. Chem. A, vol. 2, no. 38, pp. 15803-15807, Jul. 2014.

https://doi.org/10.1051/matecconf/202235801061

Experimental and Simulation Research of Belt Extrusion Process Based on Simplified Chip Model

Qi Liu *, Yuwei Wang, Na Zhao and Wenfang Zhu

Northwest Institute of Mechanical and Electrical Engineering, Xianyang, China

Abstract. The projectile extrusion process is an extremely important part of the initial trajectory of the artillery launch. The differentiation of the extrusion process brings about the stability problem of the extrusion process, which directly affects the stability and safety of the artillery, ballistic and projectile launching process. sex. Since the simplified chip intrusion resistance model does not pay attention to the projectile movement process and the formation process of the elastic belt groove, and only studies the change law of the intrusion resistance, the model is relatively simple and the parameters are easy to determine. In this paper, a quasi-static test device for the extrusion process of the elastic belt is established. The model is calibrated with the simplified chip resistance model with the intrusion process. Finally, the model is used to calculate the resistance along the process of a large-caliber howitzer projectile extrusion process, and it is compared with the embedded resistance test results of a large-caliber howitzer., the results show that in the whole process of extrusion, the relative error between the calculated value of the total extrusion resistance and the experimental value is basically below 10%, and the maximum relative error is 10.98%, which verifies the accuracy of the model.

key word: belt, extrusion, simplified chip, extrusion resistance.

1. Introduction

The squeeze-in period is defined as the time from the start of projectile movement until the tape is fully embedded in the rifling [1]. The squeeze-in process is an extremely important part of the launch trajectory of the artillery. In the process of artillery firing, there are a large number of problems related to the extrusion process, such as projectile initial velocity error, projectile jamming reliability and artillery barrel life. With the development of high-performance launch and new principle technology, the impact of the projectile extrusion process on the entire launch performance has become more and more prominent [2]. For example, due to the increasingly complex firing conditions and requirements of high-performance artillery, the differences in the projectile extrusion process due to different loading parameters will eventually significantly affect the muzzle velocity stability and even cause abnormal chamber pressure; There is a dynamic overlap between the projectile jamming and the extrusion process in new launching principles and technologies such as technology, which makes the extrusion process more complicated. Because the time and stroke of the extrusion process are very short, it is difficult to measure by experiments. Therefore, the classical internal ballistics usually adopts the assumption that the projectile starts to move when the extrusion pressure is reached, and ignores the extrusion process of the elastic belt [3]. However, the research on the extrusion process of the ammunition belt is very important. It is to define the wear relationship between the ammunition belt and the inner bore and the rifling, to check the matching degree of the ammunition belt and the sloped bore, to explore the mechanism of the ballistic peak in the artillery and related to the life of the artillery. The preliminary work on the problems such as the clipping of the bullet belt is indispensable, and it also has an important impact on the shooting density [4]. The importance and complexity of the extrusion process make it a hot and difficult point in the research of launch technology. Many scholars are devoted to the model research of the extrusion process of the elastic belt, such as the Johnson-Cook constitutive model [5]. With the help of three-dimensional simulation analysis software such as ABAQUS studies the large deformation of the elastic band, the plastic deformation flow process of the elastic band material, the shear failure process, and the grooving process [6], and then studies the projectile motion law, extrusion pressure, extrusion resistance change law. However, so far, there is no complete theory about the squeeze-in period in the ballistic literature at home and abroad, and scholars have not reached a consensus on the physical meaning of the squeeze-in pressure value, let alone a satisfactory squeeze-in resistance calculation. Methods [7-9]. With the help of three-dimensional

* Corresponding author: shamozhiying-67@163.com

https://doi.org/10.1051/matecconf/202235801061

simulation software to calculate the intrusion resistance, the modeling process is complicated, the number of parameters is large and difficult to accurately determine, and the accuracy of the calculation results is difficult to grasp. In view of the above problems, this paper establishes a quasi-static test device for the extrusion process of the elastic belt. Data, the model calibration was carried out on the simplified chip resistance model of the belt extrusion process. The calibration model was used to calculate the resistance along the process of a large-caliber howitzer projectile extrusion process, and compared with the embedded resistance test results of a large-caliber howitzer. By comparison, the results show that in the whole process of extrusion, the relative error between the calculated value of the total extrusion resistance and the experimental value is basically below 10%, and the maximum relative error is 10.98%, which verifies the accuracy of the model.

2. Experimental research on the quasi-static extrusion process of the belt

2.1 Introduction of Test Bench

2.1.1 Tested Product

The tested items are elastic belts of different structure sizes and materials. The design drawings of standard elastic belts (1#, material brass H96, first top surface width 4mm, chip groove width 2.5mm, belt forward inclination angle 20°) and The physical map is shown in Figure 1, and the material properties of different material elastic belts are shown in Table 1.

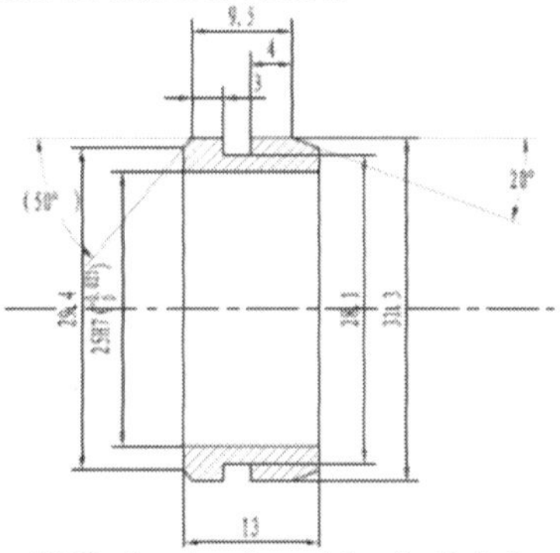

(1) Design drawing of the elastic belt

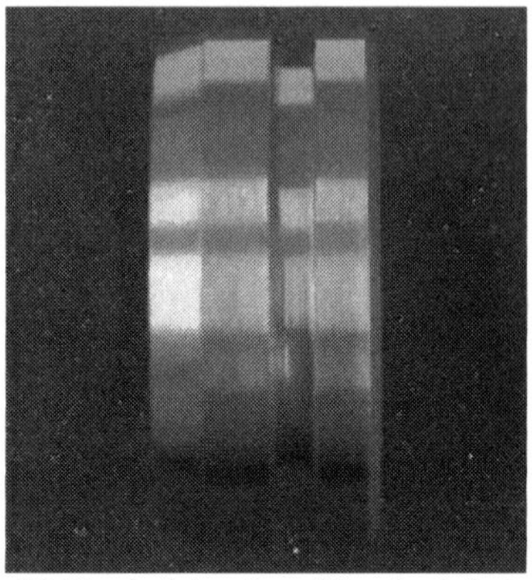

(2) Physical drawing of the elastic belt

Figure 1 Standard belt diagram

Table 1 Material properties of different elastic bands

material	material properties				
	density $\rho(kg/m^3)$	elastic modulus $E(GPa)$	poisson's ratio μ	strength limit $[\sigma](MPa)$	strength criterion
Brass H96	8900	115	0.34	320	Tensile strength
Duralumin 2A-12	2770	69	0.33	502	Tensile strength
Nylon 1010	1150	8.3	0.28	60	Tensile strength

2.1.2 Participating Products

The test items of this test bench are mainly simulated projectiles, truncated barrels and supports. Among them, the simulated projectile consists of a warhead, a tail and a simulated belt, as shown in Figure 2. When assembling, first put the elastic belt on the end of the warhead from the rear end, and then fasten the tail and the warhead through threads, and at the same time perform the axial positioning of the elastic belt, and then snap the simulated projectile into the rifling barrel as a whole. Since the warhead and the tail are separated and connected by threads, the elastic belts of different materials and different structural parameters can be replaced, and the influence of different material elastic belts and different elastic belt structure sizes on the intrusion resistance can be studied.

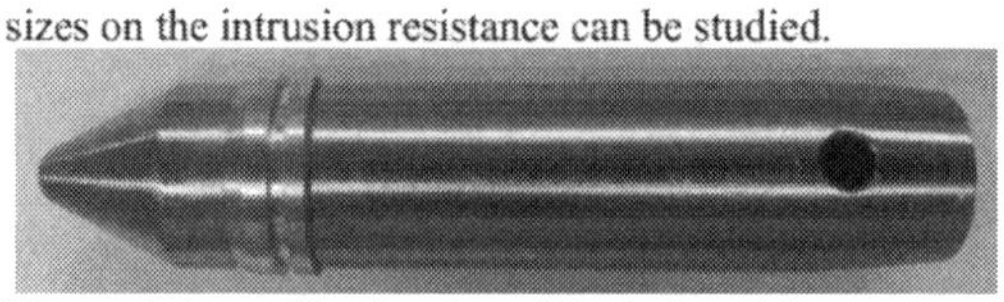

Figure 2 Assembly drawing of the belt and the simulated projectile

The truncated barrel is designed according to the structure and size of a small diameter barrel, as shown in Figure 3. The truncated barrel includes a complete sloped bore portion, a rifling start portion and a part of a full rifling portion. When assembling, the simulated projectile is snapped into the truncated barrel as a whole (apply a preload of about 200N). During the squeeze-in test, the design length of the tail should ensure that the notching of the elastic belt is completed and that the elastic belt part of the simulated projectile can completely pass through the truncated barrel. The support body is made of gun steel. Its function is to provide support for the shortening of the barrel during the extrusion process of the elastic belt, to ensure that the extrusion direction is along the axial direction of the barrel, and to provide a channel for simulating the movement of the projectile and play a soft recovery role. The whole test The physical drawing of the device is shown in Figure 4.

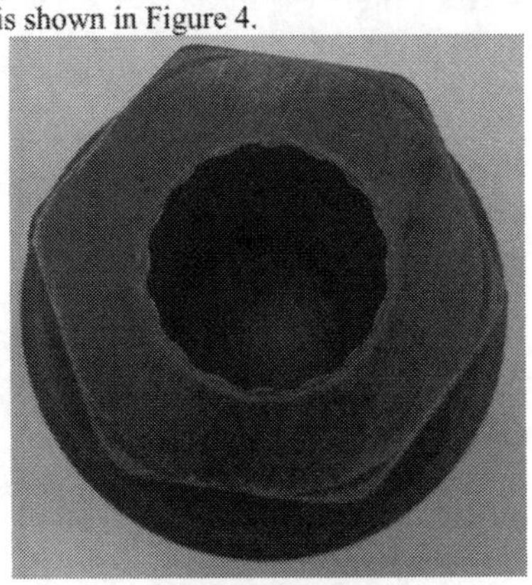

Fig.3 The actual picture of the truncated body tube

Fig.4 The actual picture of the assembly of the extrusion test device

2.1.3 Power Loading Mechanism

The loading mechanism of the intrusion force is a computer-controlled universal testing machine, and its main parameters are: model: 105D; maximum test force: 100KN; displacement resolution: 0.025μm; test force resolution: 1/500000FS; sampling frequency 30HZ.

2.1.4 Test Plan

The experimental protocol is shown in Table 2.

Table 2 Test Embodiments

Test item	Crowding condition				Test quantity		Test content
	material	top width (mm)	Chip flute width (mm)	Front inclination angle (°)	Single	accumulative	
band material Comparison	H96 Brass	4	2.5	20	3	3	F-L、F-t、L-t、Weighing、Ordinary Photography (Video)
	duralumin 2A-12	4	2.5	20	3	6	
	Nylon1010	4	2.5	20	3	9	
first top surface wide contrast	H96 Brass	6	2.5	20	3	12	
chip flute width degree comparison	H96 Brass	4	3.5	20	3	15	
tape forward angle contrast	H96 Brass	4	2.5	30	3	18	

2.2 Test Results

2.2.1 Comparison of Extrusion Resistance Curves of Different Belt Materials

Figure 5 is a comparison of the typical extrusion resistance curves of the three elastic belt materials. It can be seen from Figure 5 that the extrusion resistance curve of the standard brass H96 elastic belt is obviously double-peaked, and the resistance peak value when passing through the second top surface Higher than the peak value of the first top surface; the extrusion resistance curve of the duralumin 2A-12 elastic belt is single-peaked, and the resistance curve has an inflection point when passing through the first top surface; the extrusion resistance curve of the nylon Nylon1010 elastic belt is double-peaked. And the first peak is higher than the second peak. It can be seen from the three extrusion resistance curves that the peak extrusion resistance is duralumin 2A-12>brass H96>nylon 1010.

https://doi.org/10.1051/matecconf/202235801061

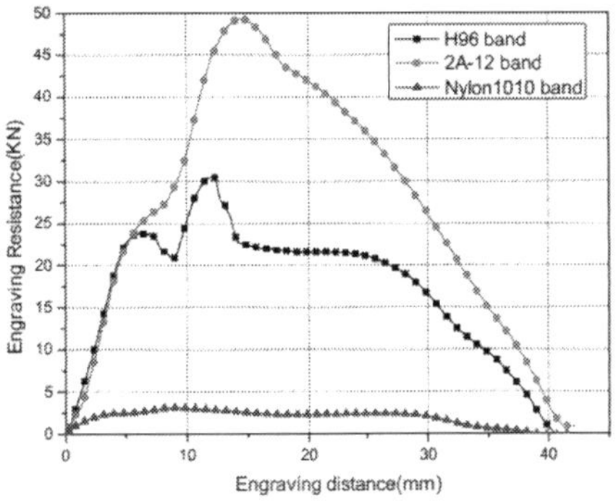

Figure 5 Comparison of the intrusion resistance of belts of three materials curves of differe nt first top surface wide belts

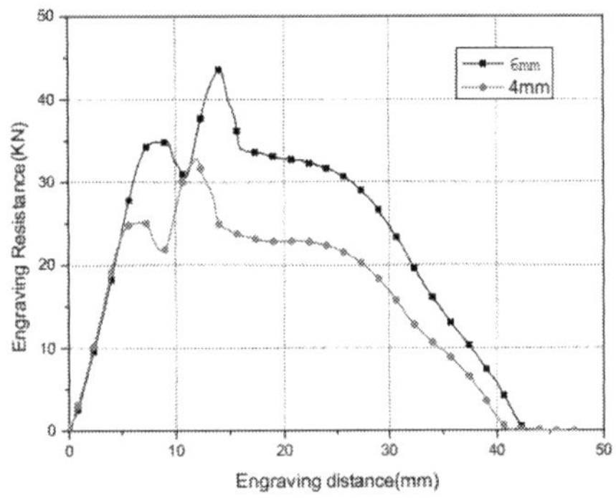

Figure 6 Comparison of the intrusion resistance

2.2.2 Comparison of Extrusion Resistance Curves of Different Belt Structures

Figure 6 is a comparison of the resistance curves of the elastic belts with different first top surface widths; Figure 7 is a comparison of the resistance curves of the elastic belts with different chip groove widths; Comparison of resistance curves.

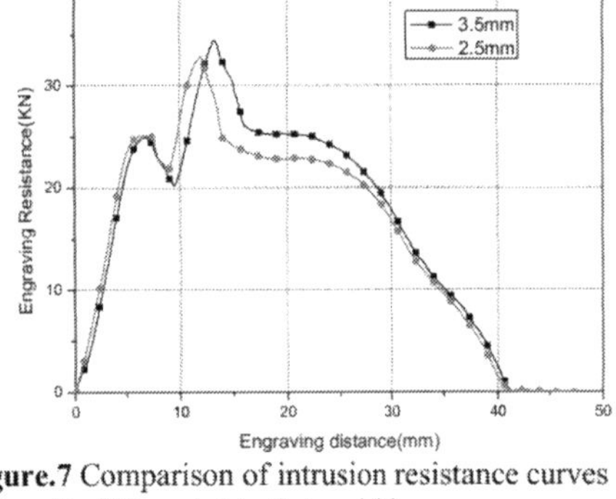

Figure.7 Comparison of intrusion resistance curves of belts with different chip flute widths

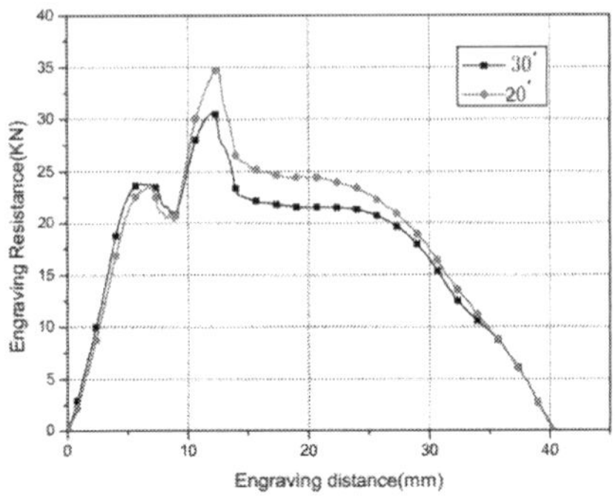

Figure.8 Comparison of intrusion resistance curves of belts with different forward inclination angles

It can be seen from Figure 6 that the peak value of the extrusion curve of the first top surface width of 6mm lags behind that of the 4mm elastic belt, which is in line with the structural characteristics of the elastic belt, and the extrusion resistance of the elastic belt with a first top surface width of 6mm The peak value is higher, which is mainly because the width of the first top surface of the elastic band increases, the width of the material chip becomes larger, and the contact area between the elastic band and the rifling increases, so the two peaks of the extrusion resistance curve increase; from Figure 7, it can be seen that It can be seen that there is little difference in the peak resistance of the elastic band with different chip flute widths. The difference is that the second peak resistance of the 3.5mm chip flute width lags behind that of 2.5mm, which is also caused by the different structure of the elastic band; from the figure 8 It can be seen that the peaks of the two resistance curves of different belt forward inclination angles are quite different. The peak value of the resistance curve of the elastic belt with a forward inclination angle of 20° is about 5KN higher than that of the elastic belt with a 30° angle. This is mainly due to the The width of the elastic belt with a forward inclination angle of 20° is larger, and the contact area with the barrel is larger, so the peak resistance value is also larger. It can be seen from Fig. 6-Fig. 8 that when the structural parameters of the elastic band are different, the intrusion resistance curves show an obvious double-peak shape.

3. Research on simplified chip extrusion resistance model

According to the structural parameters of a small-caliber barrel, projectile, and elastic belt, as well as the material parameters of brass H96 (see Table 3-Table 4), the contact area between the barrel and the elastic belt in different extrusion processes can be obtained.

Table 3 Strength coefficient values of brass H96

ε_0	0	0.05	0.1	0.15	0.2	0.3	0.4	0.5	0.6
$\sigma_\tau^* / (kgf / mm^2)$	8.5	16	21	25	27	30	32	33.5	34.7
$\overline{n}$	14.7	5.93	3.43	1.36	0.89	0.47	0.28	0.18	0.12

Table 4 Material properties of brass H96

material	material properties				
	density $\rho(kg/m^3)$	elastic modulus $E(GPa)$	poisson's ratio μ	strength limit $[\sigma](MPa)$	strength criterion
Brass H96	8900	115	0.34	320	tensile strength

According to the mathematical model of the contact stress, the contact stress acting on the contact area can be obtained. Based on the quasi-static intrusion test data of the elastic belt obtained by the test, the prediction model of the intrusion resistance in the simplified chip process is obtained by multi-parameter fitting:

$$F_{BP} = \mu \cos\alpha_k (1+\frac{tg\alpha_k}{\mu})\frac{\overline{m}}{\gamma}\left\{\left(1+\gamma\frac{\sigma_s}{\overline{m}}\right)\left[1-\frac{tg\alpha_k}{h}(x-H)\right]^n-1\right\}S_k \tag{1}$$

Among them, F_{BP} is the sum of the embedded resistance. Assuming that the contact stress on the contact area between the elastic band and the barrel is uniform at the same extrusion stage, the formula (1) can be further simplified as:

$$F_i = \mu_i \cos\alpha_{k,i}(1+\frac{tg\alpha_{k,i}}{\mu_i})\sigma_{cp,i} \cdot S_i \tag{2}$$

When the insertion section is composed of a smooth groove, a rifling slope and a cylindrical section (of the rifling), the insertion resistance is the sum of three components: the first slope resistance F_1, the second slope resistance F_2 and the cylindrical section resistance F_3.

4. Research on predicition squeeze-in resistance of a large-caliber howitzer

In order to verify the accuracy of the established simplified chip extrusion resistance model, a large-caliber howitzer was taken as the research object, and the extrusion resistance of the howitzer projectile extrusion process was calculated. The contact area and contact resistance results calculated according to the structural parameters of the howitzer barrel and the projectile are compared with the test results as shown in Table 5.

Table 5 Comparison of calculation results and test results of contact area and contact resistance of a certain caliber howitzer

Feature section	x / mm	S_k / mm^2			F_{BP} / kgf		
		Test value	Calculated value	Relative error (%)	Test value	Calculated value	Relative error (%)
0	0	0	0	0	0	0	0
0.5H1	3.79	1484	1419	-4.38	5968	6179	3.54
H1	7.58	2968	2720	-8.356	11380	12630	10.98
0.5Hnp	9.50	3720	3635	-2.285	14453	15172	4.97
0.5(H1 + Hnp)	13.3	5193	5530	6.49	20515	21966	7.07
Hnp	19.0	7522	7077	-5.916	29597	31457	6.284
Hnp+0.5 H1	22.8	7612	7453	-2.089	29669	30572	3.0436
H1+ Hnp	26.6	7718	7949	2.99	30306	30596	0.5969
H1+0.5 (H2+ Hnp)	37.6	8090	8107	0.21	30441	30645	0.6701
H1+ H2	48.6	8463	8615	1.80	30837	31765	3.0094
H2+0.5 (H1+ Hnp)	54.3	8616	8860	2.83	31530	32215	2.1725
H2+ Hnp	60.0	8711	9110	4.58	32155	32320	0.5131
H2+ Hnp+0.5 H1	63.8	8739	9260	5.96	32523	32920	1.2207
H1+ H2+ Hnp	67.6	8741	9409	7.64	32847	33980	3.4493

It can be seen from the comparison between the calculated value and the test value in Table 5 that the absolute value of the relative error of the total contact area does not exceed 10%, and the absolute value of the maximum relative error is 8.356%; the relative error of the intrusion resistance along the way is basically below 10% , the maximum relative error is 10.98%, which proves that the established intrusion model to simplify the chipping process is accurate.

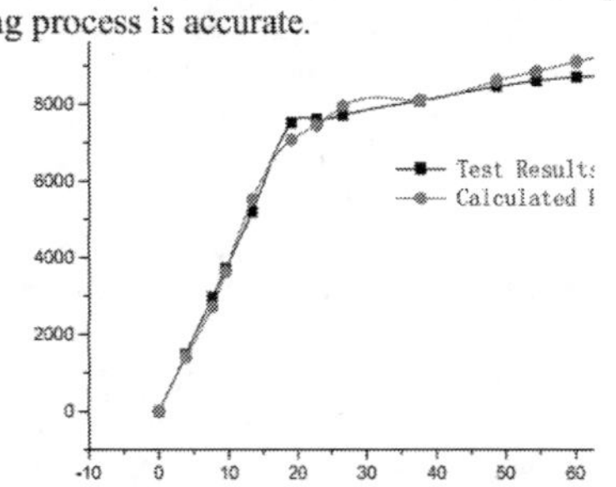

(a) The comparison between the test value and the calculated value of the contact area

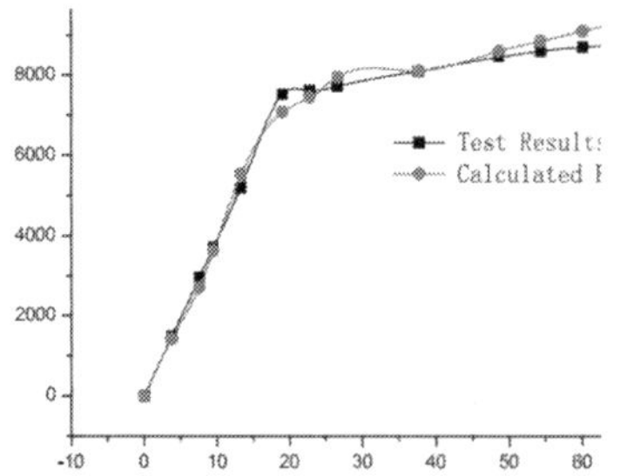

(b) The comparison between the test value and the calculated value of the intrusion resistance

Figure 9. Comparison of calculation results and test results of a large-caliber howitzer during the extrusion process

5. Conclusion

In this paper, a quasi-static test device for the extrusion process of the elastic belt is established. The simplified chip resistance model of the belt extrusion process is calibrated. The recalibrated model is used to calculate the resistance along the path of a large-caliber howitzer projectile extrusion process, and the calculation results are compared with the test results of the embedded resistance. draw the following conclusions:

(1) The extrusion resistance curves of different materials are quite different. The extrusion resistance curve of the standard brass H96 elastic belt is obviously double-peaked, and the peak resistance value when passing through the second top surface is higher than the peak value of the first top surface; The extrusion resistance curve of the duralumin 2A-12 elastic belt has a single peak shape. When passing through the first top surface, the resistance curve has an inflection point; the extrusion resistance curve of the nylon Nylon1010 elastic belt has a double peak shape, and the first peak is higher than the second peak. peak.

(2) The shape of the elastic belts with different structural parameters is not very different after extrusion, the shape of the elastic belt grooves is relatively regular, and the extrusion resistance curves show an obvious double-peak shape, and the extrusion resistance curve of the elastic belt with different structural parameters peak The difference is mainly caused by the structure size.

(3) Under the same working conditions, the extrusion resistance curve is consistent during the extrusion process of the elastic belt, and the peak consistency of the extrusion resistance is better, but the extrusion resistance curves under different working conditions (such as different structural parameters) are quite different. .

(4) Using the simplified chip extrusion resistance model established in this paper, in the whole extrusion process, the relative error between the calculated value of the total extrusion resistance and the test value is basically below 10%, and the maximum relative error is 10.98%. the accuracy of the model.

References

1. Johnson GR, Cook WH. Fracture characteristics of three metals subjected to various strains, strain rates, temperatures and pressures. Engineering Fracture Mechanics, 1985, 21 (1): 31-48.
2. Li, H., et al., Ductile fracture: Experiments and computations. International Journal of Plasticity, 2011. 27(2): p. 147-180.
3. Xu S, Chen X, Feng X. Some type of artillery shells squeeze into the numerical simulation and analysis [J]. Sichuan Ordnance Engineering, 2011.32. No.8:14-15.
4. E.B. Culbanov, Yang Jingrong, Zhou Yanhuang. Ballistics and squeeze-in pressure calculation during squeeze-in period [M]. Beijing: National Defense Industry Press, 1997.
5. Nahshon, K., Z. Xue, A modified Gurson model and its application to punch-out experiments. Engineering Fracture Mechanics, 2009. 76(8): p. 997-1009.
6. Xue, Z., et al., Calibration procedures for a computational model of ductile fracture. Engineering Fracture Mechanics, 2010. 77(3): p. 492-509.
7. Wang P, Yang G, Ge J et al. Based on the Johnson-Cook constitutive model of the elastic belt extrusion process numerical simulation [J]. Journal of Ballistics, 2015.27. No.2:55-56.
8. Sun Q, Yang G, Wang Peng et al. A large-caliber howitzer cannonball jamming process numerical simulation research [J]. Journal of Ordnance Engineering, 2015.36. No.2:206-207.
9. Nahshon, K., J.W. Hutchinson, Modification of the Gurson Model for shear failure. European Journal of Mechanics - A/Solids, 2008. 27(1): p. 1-17.

Experimental and Simulation Research on the Quasi-dynamic Extrusion Process of the Belt

Qi Liu *, Yuwei Wang, Wenfang Zh and Na Zhao

Northwest Institute of Mechanical and Electrical Engineering, Xianyang, China

Abstract. The extrusion process of the elastic belt is a process in which the elastic belt and the inner bore of the barrel undergo complex contact and collision, and the rifling gradually penetrates the elastic belt, so that the material of the elastic belt continues to fail and destroy. It has high instantaneous, strong impact, high temperature, high-speed change There is a big gap between the existing theory and the actual squeeze-in process, and it is urgent to carry out research work on this problem. In this paper, a quasi-dynamic test device for the extrusion process of the elastic belt is established. On this basis, the extrusion history of a small-caliber projectile under different belt materials, different extrusion speeds, different extrusion force loading rates and different chamber pressure loading rates is established. The experimental comparison research was carried out. Based on the experimental data, the model calibration was carried out for the simplified chip resistance model of the belt extrusion process, and the calibration model was used to test the extrusion resistance of a newly designed large-caliber and howitzer belt extrusion process. The prediction research is carried out, which provides a reference for the optimal design of internal ballistics and charges, projectiles and barrels, and the coupling design of projectiles, guns and charges in a system.

key word: Belt, extrusion, quasi-dynamic, experiment

1. Introduction

When the artillery automatic weapon is fired, the projectile moves along the barrel under the action of high temperature and high pressure gunpowder and gas, and interacts with the barrel. The extrusion process of the elastic belt has an important impact on the internal ballistic cycle of the barrel weapon. In the classical internal ballistics, friction work and projectile rotational energy are often considered as secondary work and are included in the internal ballistic energy balance equation, and the extrusion process is ignored. When the pressure reaches the starting pressure, the projectile starts to move, which is quite different from the actual situation [1].

The extrusion process of the elastic belt is actually a process of contact and collision between the elastic belt and the inner bore of the barrel, the rifling gradually penetrates the elastic belt, and the material of the elastic belt continues to fail and destroy. The laws of advance resistance and elastic belt deformation are very complex [2-4]. Due to many difficulties, there is still a big gap between the existing research theories and the actual extrusion process, and it is urgent to carry out research work on the extrusion process.

The importance and complexity of the extrusion process make it a hot and difficult point in the research of launch technology. Many scholars are devoted to the model research of the extrusion process of the elastic belt, such as the Johnson-Cook constitutive model [5]. With the help of three-dimensional simulation analysis software such as ABAQUS studies the large deformation of the elastic band, the plastic deformation flow process of the elastic band material, the shear failure process, and the groove formation process [6], and then study the projectile motion law, extrusion pressure, extrusion resistance change law. However, so far, there is no complete theory about the squeeze-in period in the ballistic literature at home and abroad, and scholars have not reached a consensus on the physical meaning of the squeeze-in pressure value, let alone a satisfactory squeeze-in resistance calculation. Methods [7-9]. With the help of 3D simulation software to calculate the intrusion resistance, the modeling process is complex, the number of parameters is large and difficult to determine, and the accuracy of the calculation results is difficult to grasp. In view of the above problems, this paper establishes a quasi-dynamic test device for the extrusion process of the elastic belt. Based on the experimental data, the simplified chip resistance model of the belt extrusion process was model calibrated, and the calibrated model was applied to the extrusion process of a newly designed large-caliber plus howitzer belt. The intrusion resistance is predicted and researched, which provides a reference for the optimal design of internal ballistics and charges, projectiles and

* Corresponding author: shamozhiying-67@163.com

barrels, and the coupling design of projectiles, guns and charges in a systematic way.

2. Experimental research on the quasi-dynamic extrusion process of the belt

2.1 Introduction of Test Bench

2.1.1 Tested Product

The tested items are elastic belts of different structure sizes and materials. The design drawings of standard elastic belts (1#, material brass H96, first top surface width 4mm, chip groove width 2.5mm, belt forward inclination angle 20°) and The physical map is shown in Figure 1, and the material properties of different material elastic belts are shown in Table 1.

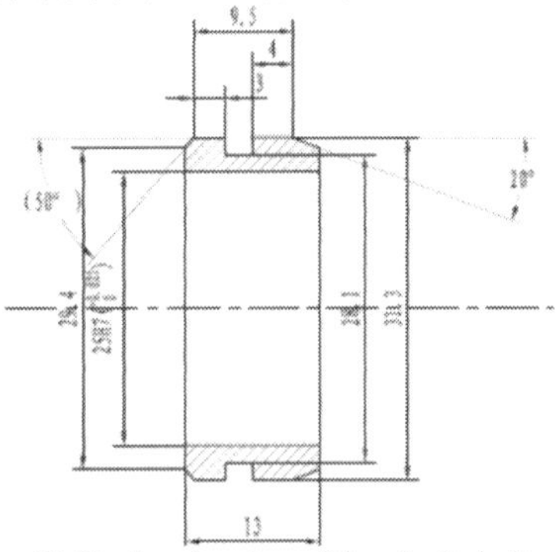

(1) Design drawing of the elastic belt

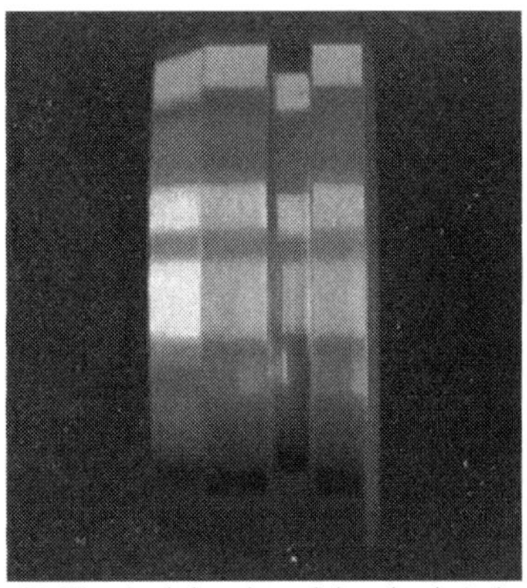

(2) Physical drawing of the elastic belt

Figure 1 Standard belt diagram

Table 1 Material properties of different elastic bands

material	material properties				
	density $\rho(kg/m^3)$	elastic modulus $E(GPa)$	poisson's ratio μ	strength limit $[\sigma](MPa)$	strength criterion
Brass H96	8900	115	0.34	320	Tensile strength
Duralumin 2A-12	2770	69	0.33	502	Tensile strength
Nylon 1010	1150	8.3	0.28	60	Tensile strength

2.1.2 Participating Products

The test items of this test bench are mainly simulated projectiles, truncated barrels and supports. Among them, the simulated projectile consists of a warhead, a tail and a simulated belt, as shown in Figure 2. When assembling, first put the elastic belt on the end of the warhead from the rear end, and then fasten the tail and the warhead through threads, and at the same time perform the axial positioning of the elastic belt, and then snap the simulated projectile into the rifling barrel as a whole. Since the warhead and the tail are separated and connected by threads, the elastic belts of different materials and different structural parameters can be replaced, and the influence of different material elastic belts and different elastic belt structure sizes on the intrusion resistance can be studied.

Figure 2 Assembly drawing of the belt and the simulated projectile

The truncated barrel is designed according to the structure and size of a small diameter barrel, as shown in Figure 3. The truncated barrel includes a complete sloped bore portion, a rifling start portion and a part of a full rifling portion. When assembling, the simulated projectile is snapped into the truncated barrel as a whole (apply a preload of about 200N). During the squeeze-in test, the design length of the tail should ensure that the notching of the elastic belt is completed and that the elastic belt part of the simulated projectile can completely pass through the truncated barrel. The support body is made of gun steel. Its function is to provide support for the shortening of the barrel during the extrusion process of the elastic belt, to ensure that the extrusion direction is along the axial direction of the barrel, and to provide a channel for simulating the movement of the projectile and play a soft recovery role. The whole test The physical drawing of the device is shown in Figure 4.

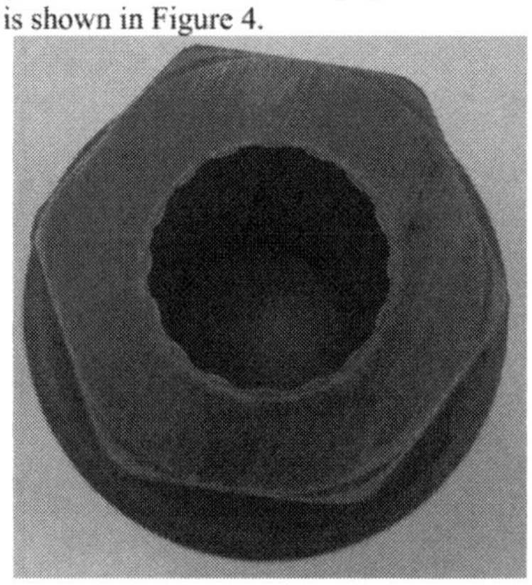

Figure. 3 The actual picture of the truncated body tube

Figure. 4 The actual picture of the assembly of the extrusion test device

2.1.3 Power Loading Mechanism

The loading mechanism of the intrusion force is a computer-controlled universal testing machine, and its main parameters are: model: 105D; maximum test force: 100KN; displacement resolution: 0.025μm; test force resolution: 1/500000FS; sampling frequency 30HZ.

2.1.4 Test Plan

The experimental protocol is shown in Table 2.

Table 2 Test Embodiments

test item	squeeze working condition				test quantity		test content
	band material	extrusion force loading rate (KN/s)	chamber pressure loading rate (MPa/s)	extrusion speed (mm/s)	single	accumulative	
material comparison	H96 brass	—	—	—	3	3	F-L、F-t、L-t、weighing、ordinary photography (video)
	duralumin 2A-12	—	—	—	3	6	
	nylon1010	—	—	—	3	9	
squeeze speed contrast	—	2	—	—	3	12	
	—	4	—	—	3	15	
	—	6	—	—	3	18	
	—	8	—	—	3	21	
intrusion force loading rate comparison	—	—	0.5	—	3	24	
	—	—	1	—	3	27	
	—	—	1.5	—	3	30	
	—	—	2	—	3	33	
	—	—	2.5	—	3	36	
chamber pressure loading rate comparison	—	—	—	20	2	38	
	—	—	—	30	2	40	
	—	—	—	40	2	42	

2.2 Analysis of Test Results

2.2.1 Comparison of the Extrusion Process of Different Belt Materials

Figure 5 shows the comparison of elastic belts of different materials before and after extrusion. It can be seen from Figure 5 that the elastic belt of nylon Nylon1010 only undergoes elastic deformation of the material before and after extrusion, and no chips are formed; while the duralumin 2A-12 material The hardness is relatively high. Although the chip deformation occurs, the chip material of the elastic belt is not adhered to the elastic belt and the simulated projectile. Soft, copper scraps accumulate in the chip slot after the extrusion is completed, and the copper hanging on the rifling is more serious. When the projectile passes through the fully-rifled barrel, there are obvious traces of copper hanging on the outer wall of the projectile circumference. This phenomenon can also be explained. , using the brass H96 material belt, when designing the charge, the copper remover should be used as the charge accessory.

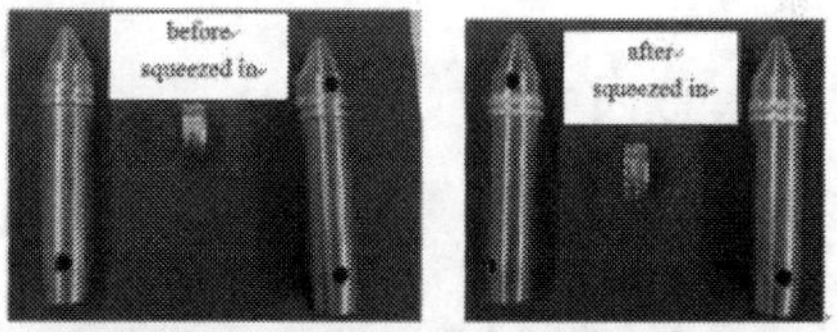

(1) Comparison before and after the brass H96 material belt is squeezed in

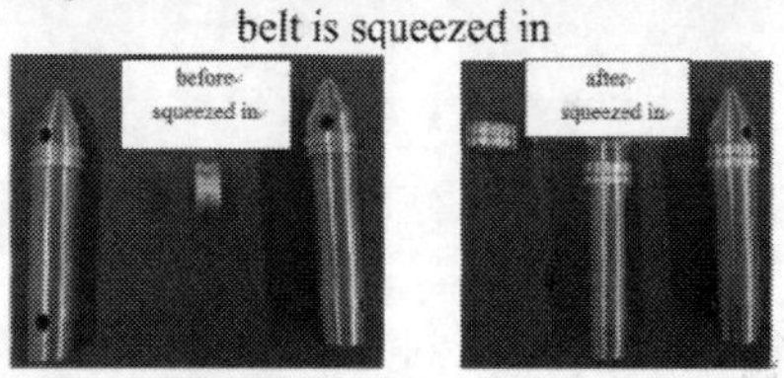

(2) Comparison of before and after extrusion of duralumin 2A-12 material belt

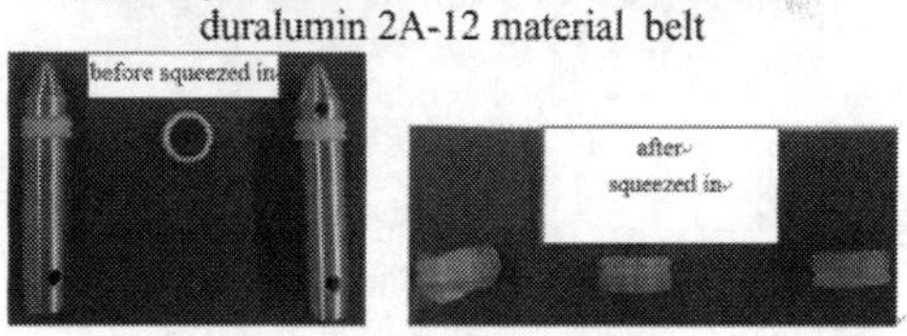

(3) Comparison of nylon Nylon1010 material belt before and after extrusion

Figure 5 Comparison of elastic belts of different materials before and after extrusion

Figure 6 is a comparison of the typical extrusion resistance curves of the three elastic belt materials. It can be seen from Figure 6 that the extrusion resistance curve of the standard brass H96 elastic belt is obviously bimodal, and the resistance peak value when passing through the second top surface Higher than the peak value of the first top surface; the extrusion resistance curve of the duralumin 2A-12 elastic belt is single-peaked, and the resistance curve has an inflection point when passing through the first top surface; the extrusion resistance curve of the nylon Nylon1010 elastic belt is double-peaked. And the first peak is higher than the second peak. It can be seen from the three extrusion resistance curves that the peak extrusion resistance is duralumin 2A-12>brass H96>nylon Nylon1010.

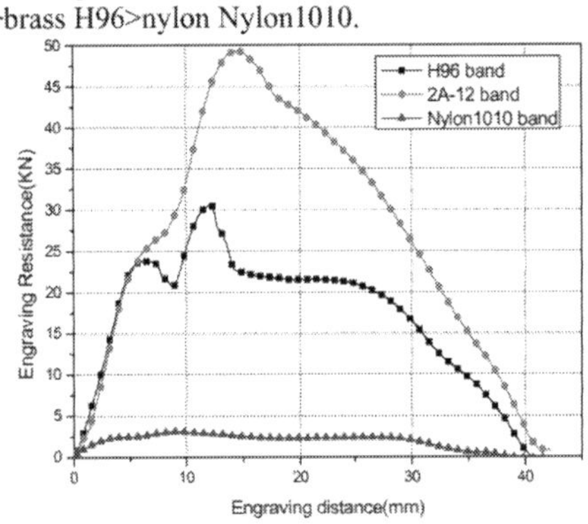

Figure 6 Comparison of intrusion resistance curves at of belts of three materials

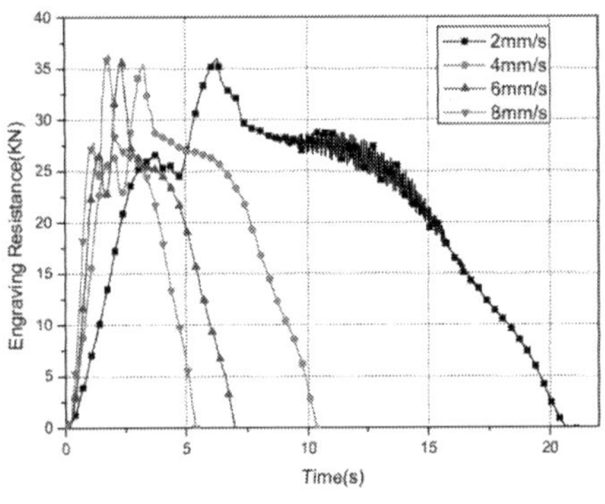

Figure 7 Comparison of intrusion resistance different extrusion speeds

2.2.2 Comparison of Extrusion Resistance Curves at Different Extrusion Speeds

Set the extrusion speed to 2mm/s, 4mm/s, 6mm/s, and 8mm/s respectively. Figure 7 shows the comparison of the extrusion resistance curves at different extrusion speeds. It can be seen from Figure 7 that at different extrusion speeds The shape of the extrusion resistance curve is similar, and the peak value of the resistance curve is almost the same, and the difference is only in the completion time of extrusion.

2.2.3 Comparison of Intrusion Resistance Curves at Different Intrusion Force Loading Rates

Figure 8 shows the comparison of the intrusion resistance curves when the loading pressure at the bottom of the projectile increases according to the law of 500N/s-2500N/s. It can be seen from Figure 8 that with the increase of the intrusion dynamic loading rate, the peak value of the intrusion resistance curve There is a slight downward trend, and with the decrease of the extrusion dynamic loading rate, the extrusion process of the elastic belt becomes more and more unstable, and the extrusion resistance curve shows obvious oscillations.

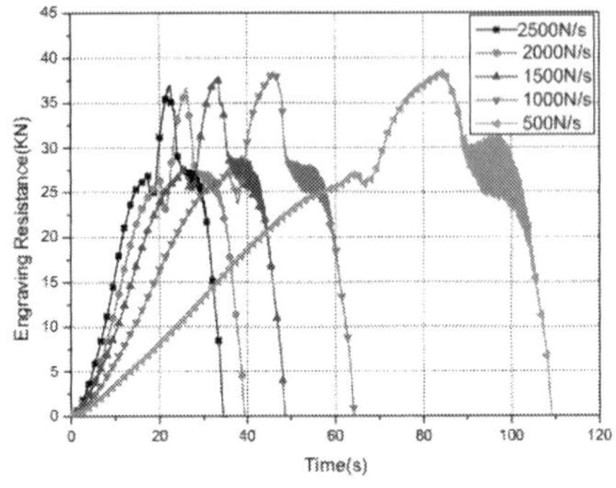

Figure.8 Comparison of intrusion resistance curves under different cavity pressure loading rates

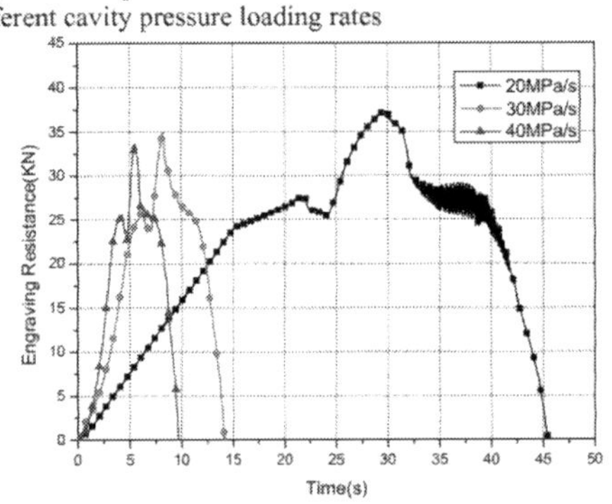

Figure.9 Comparison of intrusion resistance curves under different intrusion dynamic loading rates

2.2.4 Comparison of Intrusion Resistance Curves at Different Chamber Pressure Loading Rates

Figure 9 shows the comparison of the intrusion resistance curves when the chamber pressure loading rate increases according to the law of 20MPa/s-40MPa/s respectively. The peak value of the extrusion resistance curve decreased from 37.5KN at 40MPa/s to 28KN at 20MPa/s, a decrease of about 25.3%, and with the decrease of the chamber pressure loading rate, the extrusion process of the elastic belt became more and more unstable, and the extrusion process became more and more unstable. The resistance curve shows obvious oscillation at 20MPa/s, which is consistent with the changing trend under different extrusion pressure loading rates.

3. Research on simplified chip extrusion resistance model

3.1 Mathematical Model of Contact Stress

Two-dimensional deformation is described by the plastic theory equations:

$$\frac{\partial\sigma_x}{\partial x}+\frac{\partial\tau_{xy}}{\partial y}=0$$
$$\frac{\partial\sigma_y}{\partial y}+\frac{\partial\tau_{xy}}{\partial x}=0$$
$$(\sigma_x-\sigma_y)^2+4\tau_{xy}^2=\frac{4}{3}\sigma_\Phi^2 \quad (1)$$
$$u_x=\lambda_1(\sigma_x-\sigma_{cp})$$
$$u_y=\lambda_1(\sigma_y-\sigma_{cp})$$
$$\dot{\gamma}_{xy}=2\lambda_1\tau_{xy}$$

By solving these equations, the contact stress calculation model is obtained:

$$\sigma_k=\frac{\bar{m}}{\gamma}\left[\left(1+\gamma\frac{\sigma_{k,0}}{\bar{m}}\right)\left(1-\frac{tg\alpha_k}{h_0}x\right)^{\gamma}-1\right] \quad (2)$$

The contact stress is described as follows for different body sections:

(1) First slope:

$$\sigma_k=\frac{\bar{m}}{\gamma_1}\left[\left(1+\gamma_1\frac{\sigma_{k,0}}{\bar{m}}\right)\left(1-\frac{tg\alpha_{k,1}}{h_0}\cdot x\right)^{\gamma_1}-1\right] \quad (3)$$

(2) Second slope :

$$\sigma_k=\frac{\bar{m}}{\gamma_2}\left\{\left(1+\gamma_2\frac{\sigma_{k,1}}{\bar{m}}\right)\left[1-\frac{tg\alpha_{k,2}}{h_1}(x-H_1)\right]^{\gamma_2}-1\right\} \quad (4)$$

(3) After entering the cylindrical part of the barrel (IV area), the contact stress is constant, on the negative line, and on the positive line:

$$\sigma_{k,1}^1=\sigma_{k,1}-\bar{m}\cdot\alpha_{k,1} \quad (5)$$

$$\sigma_{k,2}^1=\sigma_{k,2}-\bar{m}\cdot\alpha_{k,2} \quad (6)$$

According to the structural parameters of a small-caliber artillery barrel, projectile, and belt, as well as the material parameters of brass H96 (Table 3-Table 4), the contact area S_i between the barrel and the belt can be obtained. According to the contact area between the elastic belt and the inner bore of the barrel and the contact stress acting on the area, the following contact resistance model is obtained:

$$F_{BP}=\mu\cos\alpha_k(1+\frac{tg\alpha_k}{\mu})\int_0^{S_k}\sigma_k\cdot dS_k \quad (7)$$

Among them, F_{BP} is the sum of the embedded resistance. Assuming that the contact stress on the contact area between the elastic band and the barrel is uniform at the same slope, (10) can be further simplified as:

$$F_i=\mu_i\cos\alpha_{k,i}(1+\frac{tg\alpha_{k,i}}{\mu_i})\sigma_{cp,i}\cdot S_i \quad (8)$$

The total embedded resistance is determined by the sum of the stroke resistance of each segment: $F_{BP}=\sum F_i$. When the insertion section is composed of a smooth groove, a rifling slope and a cylindrical section (of the rifling), the insertion resistance is the sum of three components: the first slope resistance F_1, the second slope resistance F_2 and the cylindrical section resistance F_3. According to the quasi-static intrusion test data of the elastic belt in the previous section, the prediction model of the intrusion resistance in the simplified chip process is obtained through multivariate regression of parameters. The strength parameters of brass H96 are shown in Table 3-4:

$$F_{BP}=\mu\cos\alpha_k(1+\frac{tg\alpha_k}{\mu})\frac{\bar{m}}{\gamma}\left\{\left(1+\gamma\frac{\sigma_k}{\bar{m}}\right)\left[1-\frac{tg\alpha_k}{h}(x-H)\right]^{\gamma_2}-1\right\}S_k \quad (9)$$

Table 3 Strength coefficient values of brass H96

ε_0	.0	0.05	0.1	0.15	0.2	0.3	0.4	0.5	0.6
$\sigma_s^*/(kgf/mm^2)$	8.5	16	21	25	27	30	32	33.5	34.7
$\bar{n}$	14.7	5.93	3.43	1.36	0.89	0.47	0.28	0.18	0.12

Table 4 Material properties of brass H96

material	material properties				
	density $\rho(kg/m^3)$	elastic modulus $E(GPa)$	poisson's ratio μ	strength limit $[\sigma](MPa)$	strength criterion
Brass H96	8900	115	0.34	320	tensile strength

4. Research on predicition of squeeze-in resistance of a large-caliber howitzer

On the basis of simplifying the calibration of the intrusion model in the chip process, this model is used to study the intrusion resistance of a newly designed large-caliber and howitzer projectile in the intrusion process. The structural parameters and material properties used in the calculation are shown in Table 5. , the calculated contact area changes with the extrusion process are shown in Table 6, and the extrusion resistance and the extrusion component force with the extrusion process are shown in Table 7.

Table 5 Structural parameters and material properties of a large-caliber howitzer

name	symbol	unit	value
belt outer diameter	d_0	mm	163.8
yin wire diameter	d_H	mm	159.54
projectile stripping belt outer diameter	d_k	mm	153.8
band width	Hnp	mm	36.458
sunline tooth width	a	mm	3.81
yin line tooth width	B	mm	6.332
number of rifling	n_H	1	48
first slope angle	$\alpha_{k,1}$	rad	0.0495
second slope angle	$\alpha_{k,2}$	rad	0.018
the first slope bore considers the tangential stress action coefficient	ξ_1	1	1.0
the second slope bore considers the tangential stress action coefficient	ξ_2	1	1.0
plastic coefficient of elastic band material	$2K$	kgf/mm^2	22
material strength factor	$\overline{n}$	1	4.5
the friction coefficient of the first groove	μ_1	1	0.093
the friction coefficient of the second groove	μ_2	1	0.2
gunpowder gas pressure acting on the belt	P_n	MPa	0
first slope height	H1	mm	11.85
second slope height	H2	mm	89.15
rifle height	t	mm	2.3

Table 6 Variation of the contact area with the insertion stroke during the extrusion process of a large-caliber plus howitzer projectile

feature section	x/mm	S_k/mm	S_1		S_2		S_3	
			mm^2	%	mm^2	%	mm^2	%
0	0	0	0	0	0	0	0	0
0.5H1	5.925	1485	1485	100	0	0	0	0
H1	11.85	4591	4591	100	0	0	0	0
0.5Hnp	18.229	1257	1257	100	0	0	0	0
0.5(H1 + Hnp)	24.154	3296	3296	100	0	0	0	0
Hnp	36.458	2513	2513	100	0	0	0	0
Hnp+0.5 H1	42.383	5980	5603	93.7	185.5	3.1	191	3.19
H1+ Hnp	48.308	6722	0	0	593	8.82	6129	91.2
H1+0.5 (H2+ Hnp)	74.654	7970	0	0	1237	15.5	6733	84.5
H1+ H2	101	8218	0	0	1237	15.1	6981	84.9
H2+0.5 (H1+ Hnp)	113.304	8094	0	0	1237	15.3	6857	84.7
H2+ Hnp+0.5 H1	131.533	8321	0	0	927.6	11.1	7393	88.8
H1+ H2+ Hnp	137.458	8446	0	0	0	0	8446	100
> H1+ H2+ Hnp	138	8446	0	0	0	0	8446	100

Table 7 Variation of total intrusion resistance and resistance component of a certain caliber plus howitzer with embedded stroke

x/mm	F_{ZP}/kgf	F_1		F_2		F_3	
		kgf	%	kgf	%	kgf	%
0	0	0	0	0	0	0	0
5.925	2138	2138	100	0	0	0	0
11.85	2269	2269	100	0	0	0	0
18.229	4276	4276	100	0	0	0	0
24.154	4682	4682	100	0	0	0	0
36.458	5027	5027	100	0	0	0	0
42.383	7629	4798	62.9	577	7.56	2254	29.54
48.308	8071	0	0	3844	47.63	4227	52.37
74.654	8475	0	0	3844	45.36	4631	54.64
101	8574	0	0	3844	44.83	4730	55.17
113.304	8578	0	0	3844	44.81	4733	55.18
131.533	8646	0	0	2883	33.34	5763	66.66
137.458	8668	0	0	0	0	8668	100
138	8668	0	0	0	0	8668	100

The calculation results of Table 6 and Table 7 are shown in Figure 10. It can be seen that the contact area continues to increase with the extrusion process. At the beginning of the extrusion process, the contact area increases rapidly, and then grows slowly until the end of the extrusion process. , the contact area almost no longer increases, indicating that the extrusion process is completed, and the elastic belts have been embedded in the rifling of the cylindrical part of the barrel; and the extrusion resistance also shows a similar change rule to the contact area, that is, in the early stage of extrusion, the total contact resistance is rapid. Growth, followed by slow growth, to the basic level of the late extrusion curve, indicating that the extrusion resistance is no longer increasing, and the entire belt has entered the cylindrical rifling area. It can also be seen from Figure 10 that the extrusion force F1 in the first groove area is established first, and it is also the main part of the total extrusion resistance in the early stage of extrusion, but when the elastic belt extrusion distance exceeds about 37mm (this distance It is about the sum of the height of the first groove and the width of the elastic belt), and it quickly decays to 0, indicating that the elastic belt has completely passed the first groove area; at this time, the second groove is squeezed into the resistance component F2 and the cylindrical rifling area is squeezed The advancing resistance component force F3 has been established, and the extrusion component force of the second slope bore area only exists in the range of about 20mm to 75mm of the extrusion process, that is, only at this stage, the second slope bore and the bullet have contact stress . The contact stress of the rifling in the cylindrical part of the barrel is established when the extrusion process reaches about 20mm, and it increases continuously with the extrusion process, indicating that the contact area between the elastic belt and the rifling continues to increase. When the extrusion process is completed, the extrusion process is completed. The resistance component F3 basically no longer increases. From the extrusion resistance curve of the entire projectile extrusion process, the bottom pressure converted by the maximum extrusion resistance is 4.81MPa, which is much

https://doi.org/10.1051/matecconf/202235801062

smaller than the starting pressure of 20MPa defined by the classical internal ballistics.

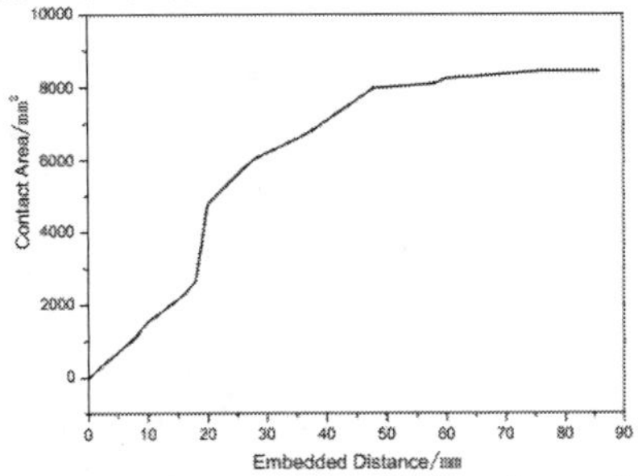

(a) Curve of contact area with insertion stroke

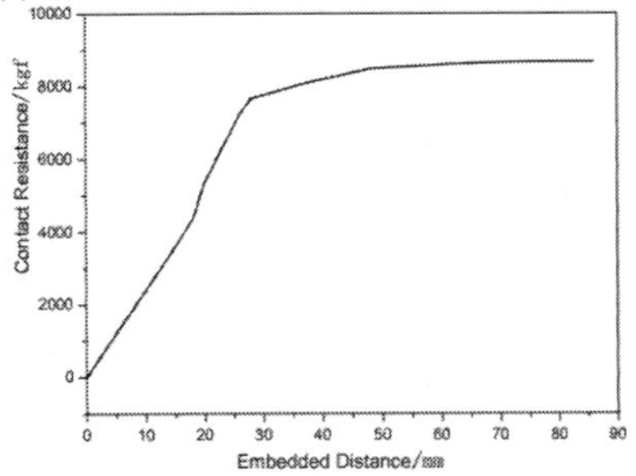

(b) Curve of insertion resistancewith insertion stroke

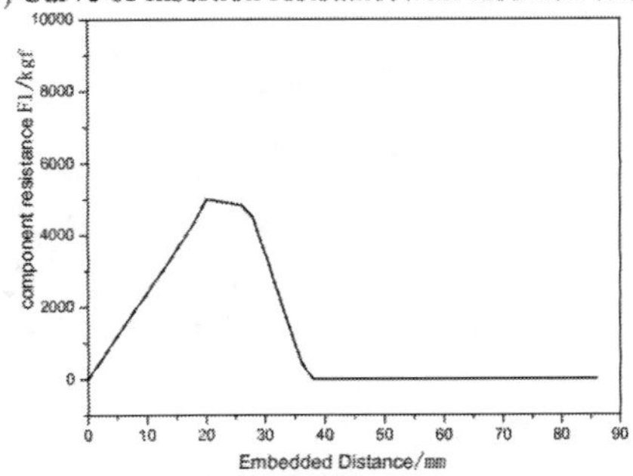

(c) Curve of the resistance force component F1 with respect to the insertion stroke

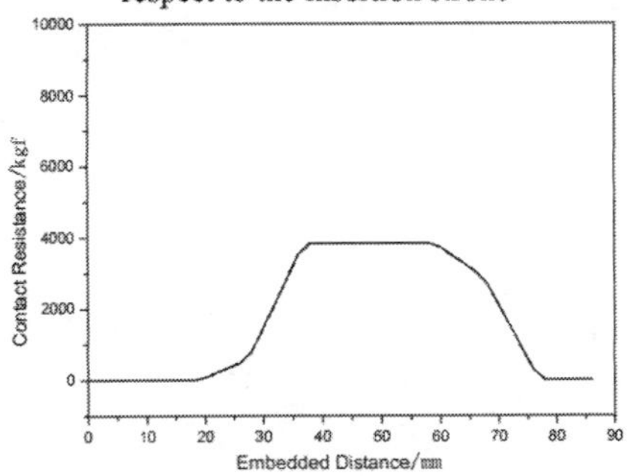

(d) Curve of the resistance componentF2 with the insertion stroke

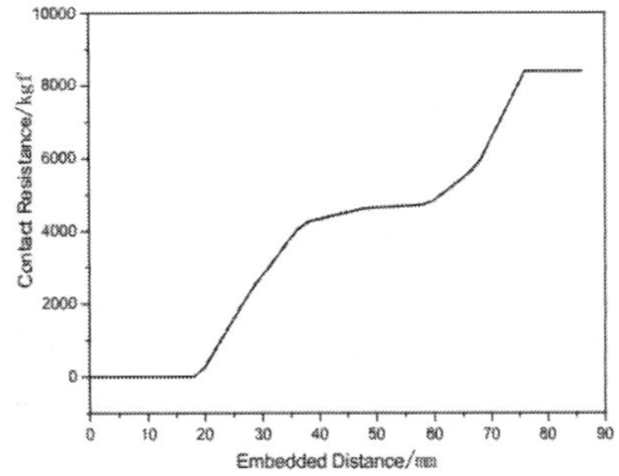

(e) Curve of the resistance force F3 with the insertion stroke

Figure 10 Contact area and squeezing resistance change with insertion stroke

5. Conclusion

In this paper, a quasi-dynamic test device for the extrusion process of the elastic belt is established. On this basis, the extrusion history of a small-caliber projectile under different belt materials, different extrusion speeds, different extrusion force loading rates and different chamber pressure loading rates is established. The experimental comparison research was carried out. Based on the experimental data, the model calibration was carried out on the simplified chip resistance model of the belt intrusion process by using the multivariate parameter regression method. The process intrusion resistance curve was predicted and studied, and the following conclusions were obtained:

(1) The material dynamic behavior of elastic belts of different materials is quite different during the extrusion process. The nylon Nylon1010 elastic belt only undergoes plastic deformation, while the duralumin 2A-12 elastic belt produces brittle chips, and the brass H96 elastic belt produces plasticity at the same time. Deformation, brittle chips and viscoplastic deformation, therefore, the phenomenon of copper hanging on the rifling is more serious.

(2) The shape of the intrusion resistance curve is similar at different intrusion speeds, and the peak value of the resistance curve is almost the same, the difference is only in the completion time of the intrusion; with the increase of the intrusion dynamic loading rate, the peak value of the intrusion resistance curve is slightly different. With the decrease of the loading rate of the extrusion force, the extrusion process of the elastic belt becomes more and more unstable, and the extrusion resistance curve shows obvious oscillation; with the increase of the loading rate of the chamber pressure, the peak value of the extrusion resistance curve The peak value of the extrusion resistance curve decreased from 37.5KN at 40MPa/s to 28KN at 20MPa/s, a decrease of about 25.3%, and with the decrease of the chamber pressure loading rate, the extrusion process of the elastic belt became more and more unstable. , the extrusion resistance curve shows obvious oscillation at 20MPa/s.

(3) Under the same working conditions, the extrusion resistance curve is consistent during the extrusion process

of the elastic belt, and the peak consistency of extrusion resistance is good, but under different working conditions (when the structural parameters and extrusion force loading methods are different) The resistance curves are quite different.
(4) From the extrusion resistance curve of the entire projectile extrusion process, the bottom pressure converted by the maximum extrusion resistance is 4.81MPa, which is much smaller than the starting pressure defined by the classical inner ballistics 20MPa.

References

1. WAND Peng, YANU Guolai, GE Jianli, et al. Numerical simulation of rotating band engraving process based on Johnson-Cook constitutive model[J]. Journal of Ballistics, 2015,27(2):55-61. (in Chinese)
2. MA Mingdi, CUI Wanshan, ZENG Zhiyin, et al. Engraving process analysis of projectiles based on coupling of FEM and SPH[J]. Journal of vibration and shock, 2015,346(6):146-150. (in Chinese)
3. MA Mingdi, CUI Wanshan, ZENG Zhiyin, et al. Dynamic Response Analysis of Chamber Throat in the Engraving Process of Heavy Caliber Gun Projectile, Journal of North University of China(Natural Science Edition) , 2014, 35(3):263-269. (in Chinese)
4. WANG Liang-kuan, Liu Yi, Zhou Yu, Liu Tie-jun, Diwu Qiang-Qiang, Numerical Simulation of the Engraving Process of Pure Iron Rotating Band [J].Journal of Gun Launch & Control, 2018,39(152), 04 52-55+61. (in Chinese)
5. E.B. Culbanov, Yang Jingrong, Zhou Yanhuang. Ballistics and squeeze-in pressure calculation during squeeze-in period [M]. Beijing: National Defense Industry Press, 1997.
6. ZHUANG Zhuo, Nonlinear Finite Elements for Continua and Structures[M]. Tsinghua University Press, 2002. (in Chinese)
7. G.R. Liu, M.B. Liu, Smoothed Particle Hydrodynamics: A Meshfree Particle Method, World Scientific Publishing, 2003
8. Luca Massidda, Yacine Kadi, SPH simulation of liquid metal target dynamics, Nuclear Engineering and Design 240 (2010) 940–946
9. G. R. Johnson and W. H. Cook, A constitutive model and data for metals subjected to large strains, high strain rates and high temperatures. Proc. 7th Int. Symp. on Ballistics, pp. 541-547. The Hague, The Netherlands (April 1983).

Study on Recrystallization Process of Poly(DL-lactide)/Poly(ethylene glycol) Diblock Copolymer

Xiaoqi Chen[1], Lu Bai[2], Chengyu Li[2], Jia Tian[2], Haijun Zhou[1], Mengmeng Zhou[1], Jijun Xiao[2]* and Yantao Li[1]

[1] Institute of Energy Source, Hebei Academy of Sciences, Shijiazhuang, China
[2] School of Materials Science and Engineering, Hebei University of Science and Technology; Hebei Engineering Laboratory of Aviation Lightweight Composite Materials and Processing Technology; Hebei Key Laboratory of Material Near-Net Forming Technology, Hebei Key Laboratory of Flexible Functional Materials, Shijiazhuang, 050018, China

Abstract: Poly (DL-lactide)/Poly (ethylene glycol) diblock copolymer is successfully synthesized with mPEG and DL-LA as main raw materials, and the residual monomers of the polymer were removed by recrystallization. FT-IR is used to characterize the structure of the polymer. Gas chromatography is used to determine the residual monomers and residual solvent of the polymer. The solubility of polymer in different solvents was determined, and the effect of different solvent systems on the removal of residual monomers in copolymer were discussed. The results showed that the Poly(DL-lactide)/Poly(ethylene glycol) diblock copolymer was recrystallized with acetone/ isopropyl ether solvent system has the best ability to remove residual monomers, and the residual monomers of the obtained polymer are less than 0.01 wt%, the total residual solvent is only 0.24 wt%, and the productivity is as high as 95.5 wt%.

key word: Methoxypolyethylene glycols, DL-lactide, residual monomer, recrystallization

1. Introduction

Polyethylene glycol (PEG) is a widely used polyether polymer compound with excellent biocompatibility, blood compatibility, good hydrophilicity and softness. The introduction of PEG into the molecular structure of PLA can improve the disadvantages of PLA, such as poor hydrophilicity and compliance, and difficult control of degradation cycle, which is one of the main research directions at present. However, at present, Poly(DL-lactide)/Poly(ethylene glycol) diblock copolymerization is mainly obtained by solution polymerization or direct melt polymerization, and there may be unreacted monomers such as lactide, lactic acid, PEG, poly(lactic acid) initiated by trace water in the system, oxidation substances and other impurities in the polymer, which affect its use effect, so it is necessary to recrystallize the polymer to remove impurities and improve the purity of the sample.[1-6]

In this paper, polylactic acid - polyethylene glycol block copolymer was prepared by direct melt polymerization with polyethylene glycol monoformaldehyde and racemic lactide as the main raw materials. The solubility of block copolymer and racemic lactide in various solvents was studied, and the effect of solvent system on the residual monomer and yield was discussed. Because recrystallization can introduce organic solvents into the polymer, resulting in the polymer has a certain toxicity, so the solvent residue in the re-crystallized polymer is monitored.

2. Experimental

2.1 Materials

Racemic lactide (DL-LA) purity >99.5%, provided by jinan daigang biotechnology limited public Company. Poly(ethylene glycol) methyl ether (mPEG, Mn=2000), ethyl acetate, toluene, dichloromethane, acetone, absolute ethanol, diethyl ether, n-hexane, isopropyl ether, stannous octanoate ($SnOct_2$), etc. are all analytically pure (AR), which are provided by Sinopharm Chemical Reagent Co., Ltd. All raw materials must be dried before used.

* Corresponding author: xiaojj@hebust.edu.cn

 https://doi.org/10.1051/matecconf/202235801063

2.2 Instrument

Vacuum drying oven, Shanghai Senxin Laboratory Instrument; Fourier transform infrared spectrometer, model FT-IR 100, Perkin Elmer Company; Gas chromatograph, model 6890A, Agilent Technologies Inc.

2.3 Preparation of PLA-mPEG diblock copolymer

Adding DL-LA, mPEG and $SnOct_2$ into an ampoule bottle with a rotor according to a certain proportion, sealed it under a vacuum environment, then reacted at 145 °C for 24 h and cooled to obtain crude PLA-mPEG diblock copolymer. The polymerization equation was shown in Figure 1.

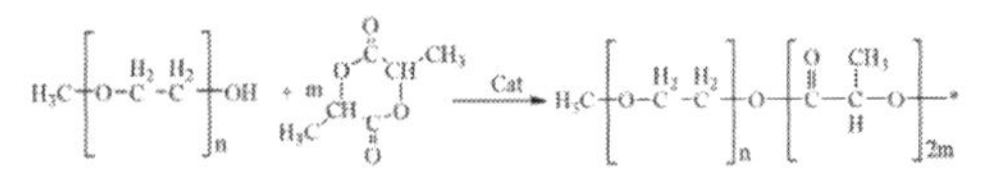

Figure 1. Polymerization equation

2.4 Recrystallization of PLA-mPEG diblock copolymer

A certain amount of crude PLA-mPEG diblock copolymer is added into the prefabricated organic solvent with the ratio of 1g/5 ml. After it is completely dissolved, a poor solvent is slowly added into the PLA-mPEG diblock copolymer solution, making the solution become turbid gradually, which the total calculated amount is 5-10 times the solution volume, and the PLA-mPEG diblock copolymer is slowly precipitated. Then, it is precipitated, filtered and dried at 45°C in a vacuum environment to obtain the high-purity PLA-mPEG diblock copolymer.

2.5 Performance characterization

Fourier transform infrared spectroscopy (FT-IR): The molecular structure of polymer was determined by FT-IR. Residual lactide monomers: Measured by gas chromatograph, determined according to General Rule 0521 of Chinese Pharmacopoeia (2020 edition), lactide content was calculated by peak area according to external standard method. Solvent residue: According to the second method of Chinese Pharmacopoeia General Rule 0861 (2020 edition), the solvent content was calculated by peak area according to the external standard method.

3. Result and discussion

3.1 Molecular structure characterization

The molecular structure of the prepared PLA-mPEG diblock copolymer was characterized by FT-IR, and the results were shown in Figure 2.

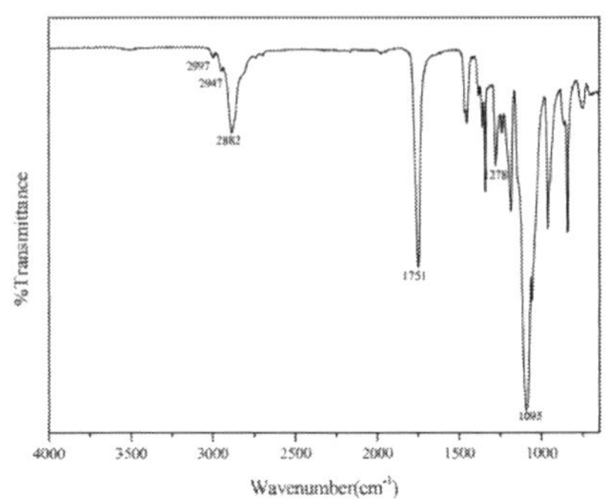

Figure 2. FI-IR of PLA-mPEG Diblock Copolymer

As shown in Figure 2, the characteristic peak of stretching vibration at 2997 cm^{-1} is $-CH_3$, the characteristic peak of stretching vibration at 2947 cm^{-1} is -CH and the characteristic absorption peak at 2882 cm^{-1} is $-CH_2$. At 1095 cm^{-1} and 1278 cm^{-1}, there are asymmetric stretching vibration absorption peaks and symmetric stretching vibration absorption peaks of C-O-C respectively. At 1751 cm^{-1}, the telescopic vibration absorption peak of C=O appears. No vibration absorption peak of the ring skeleton is found at 934 cm^{-1}, which indicates that there is no ring structure. Since the residual monomers are small and difficult to monitor by infrared spectroscopy, further determination of the residual monomers by gas chromatography is required.

3.2 Solubility of PLA-MPEG diblock copolymer

In this paper, the solubility of polymer and racemic lactide in different solvents is compared, as shown in Table 1.

Table 1. The solubility of polymer and racemic lactide in different solvents

Number	Solvent type	PELA/(g/ml)	DL-LA/(g/ml)
1	dichloromethane	0.5	0.1
2	acetone	0.2	0.2
3	toluene	0.07	0.02
4	absolute ethanol	0.01	0.02
5	ethyl acetate	0.08	0.07
6	diethyl ether	<0.01	<0.01
7	isopropyl ether	<0.01	<0.01
8	n-hexane	<0.01	<0.01

As shown in Table 1, PLA-mPEG diblock copolymer and DL-LA have high solubility in methylene chloride and acetone, among which acetone has the strongest ability to dissolve dl-lactide. The solubility of PLA-mPEG diblock copolymer is not too high, so it is suitable for being a good solvent for recrystallization. The solubility of diethyl ether, isopropyl ether and n-hexane on PLA-mPEG diblock copolymer and DL-LA is not high, so it is suitable for recrystallization of poor solvents. Combined with the toxicity of various solvents and the degree of difficulty in the removal of polymer and other factors, choose dichloromethane and acetone the most recrystallized good solvent, choose n-hexane, diethyl ether and isopropyl ether as the poor solvent recrystallization, the next recrystallization test, select the best solution system.

3.3 Recrystallization of PLA-MPEG diblock copolymer

The content of residual monomers in PLA-MPEG diblock copolymer obtained by direct melt polymerization were as high as 1.89 wt%. Recrystallization was used to remove the residual monomers in the copolymer. Firstly, the polymer was completely dissolved in a good solvent, and then poor solvent was added to slowly precipitation to obtain a high-purity PLA-mPEG diblock copolymer. The effects of several different solvent systems on the removal of unreacted monomers from copolymers were compared, as shown in Table 2.

Table 2. Comparison of properties of PLA-mPEG block copolymer after recrystallization

Good solvent	Poor solvent	Appearance	LA residues /(wt%)	Total residual solvent /(wt%)	Productivity /(wt%)
dichloromethane	n-hexane	dark brown	0.55	0.57	90.0
dichloromethane	diethyl ether	brown	0.36	0.40	81.1
dichloromethane	isopropyl ether	buff	0.03	0.22	70.0
acetone	n-hexane	brown	0.55	0.56	88.9
acetone	diethyl ether	light brown	0.34	0.32	82.3
acetone	isopropyl ether	buff	<0.01	0.24	95.5

It can be seen from Table 2 that the residual monomers of polymer can be well removed by recrystallization. Among them, poly (DL-Lactide)/poly (ethylene glycol) diblock copolymer obtained by recrystallization in acetone/isopropyl ether solvent system has the highest purity, and lactide residue and total solvent residue are lowest, and the residual monomers are less than 0.01 wt%, which almost completely removed, and the yield is the highest, up to 95.5wt%; Methylene chloride/isopropyl ether has the second highest ability to remove residual monomers, but the lowest yield rate is only 70%. It can also be seen from Table 2 that whether the good solvent is of dichloromethane or acetone, the poor solvents use isopropane ether, both the residual monomers in PLA-mPEG diblock copolymer can be well removed. Considering the appearance and cost of the product, the best recrystallization system is acetone/isopropyl ether solvent system.

4. Conclusion

In this paper, PLA-mPEG diblock copolymer was successfully synthesized and residual monomers were removed by recrystallization. Acetone/isopropyl ether solvent system and dichloromethane/isopropyl ether solvent system had the best effect on the removal of residual monomers from PLA-mPEG diblock copolymer, and the total amount of residual solvent was the least, the content of residual monomers less than 0.01 wt% and 0.03 wt%, and the total residual solvent content was 0.24 wt% and 0.22 wt%. However, the productivity of PLA-mPEG diblock copolymer obtained by recrystallization of acetone/isopropyl ether solvent system was as high as 95.5 wt%, while the productivity of PLA-mPEG diblock copolymer obtained by recrystallization of methylene chloride/isopropyl ether solvent system was only 70 wt%.

References

1. Li L, Cao Z Q, Bao R Y, et al. Poly(l-lactic acid)-polyethylene glycol-poly(l-lactic acid) triblock copolymer: a novel macromolecular plasticizer to enhance the crystallization of poly(l-lactic acid)[J]. European Polymer Journal, 2017, 97: 272-281.
2. Zhou D, Shao J, Sun J, et al. Effect of the different architectures and molecular weights on stereocomplex in enantiomeric polylactides-b-MPEG block copolymers[J]. Polymer, 2017, 123: 49-54.
3. Tsuji H. Poly (lactic acid) stereocomplexes: A decade of progress[J]. Advanced drug delivery reviews, 2016, 107: 97-135.
4. Sun P, Xiao Y, Di Q, et al. Transferrin receptor-targeted PEG-PLA polymeric micelles for chemotherapy against glioblastoma multiforme[J]. International Journal of Nanomedicine, 2020, 15: 6673.
5. Han L, Yu C, Zhou J, et al. Enantiomeric blends of high-molecular-weight poly (lactic acid)/poly (ethylene glycol) triblock copolymers: Enhanced

stereocomplexation and thermomechanical properties[J]. Polymer. 2016, 103: 376-386.

6. Liu Y, Shao J, Sun J, et al. Improved mechanical and thermal properties of PLLA by solvent blending with PDLA-b-PEG-b-PDLA[J]. Polymer degradation and stability, 2014, 101: 10-17.

https://doi.org/10.1051/matecconf/202235801064

Synthesis of a copolymerization and the anti-high temperature property in oil well cement slurry fluid

Wenbin Wang [1,2,*], Xiangchen Wan [1,2], Kai Tang [1,2]

[1] Drilling and Production Engineering Technology Research Institute of CNPC Chuanqing Drilling Engineering Co., Ltd, 710018, Xi'an, China
[2] National Engineering Laboratory for Exploration and Development of Low Permeability Oil and Gas Fields, 70018, Xi'an, China

Abstracts. Four monomers including AMPS, acrylic acid (AA), N,N-dimethylacrylamide (DMAA), and sodium allylsulfonate (AS) were used to synthesize quaternary fluid loss reducer (FRW) through copolymerization. The optimal reaction conditions were optimized by orthogonal experiment method, and the water loss reduction performance of FRW was evaluated. The results show that the optimal synthesis conditions for FRW are that the molar ratio of AMPS:AA:DMAA:AS is 4.2:1.8:3.5:0.5, the amount of initiator added is 0.3% of the total monomer mass, the monomer mass percentage is 40%, The reaction temperature was 50°C, and the pH was 7. The performance evaluation results show that FRW has good fluid loss control ability. Under the condition of adding 2%, the fluid loss is 141mL when the temperature is 220°C, which has no negative impact on the comprehensive performance of cement slurry and has strong practicability.

Key words: Fluid loss reducer; copolymer; orthogonal experiment.

1. Introduction

The main function of the oil well cement slurry fluid loss additive is to prevent or reduce the rate of the mixing water in the cement slurry from filtering to the formation[1,2]. With the gradual depletion of shallow oil and gas resources, the exploration and development targets have shifted to deeper oil and gas reservoirs[3-5]. The cementing process will not only face the problem of high temperature and high pressure, but also may encounter high salinity[6,7]. Stratum of formation water. This poses a new challenge to the performance of oil well fluid loss additives. At present, the fluid loss agent used in my country is mainly a copolymer of 2-acrylamido-2-methylpropanesulfonic acid (AMPS) and acrylamide (AM) as the main monomers [8,9,10]. This kind of fluid loss agent has low temperature resistance and salt resistance, and is easy to decompose at high temperature, causing excessive retardation of the slurry, and also reducing its ability to control fluid loss[11,12,13]. In order to solve the above problems, based on the summary of previous research results, four monomers including AMPS, acrylic acid (AA), N,N-dimethylacrylamide (DMAA) and sodium allyl sulfonate (AS) were selected. A quaternary fluid loss agent (FRW) was synthesized through copolymerization, and its structure and performance were tested.

2. Materials and methods

2.1 Materials

AMPS, AA, DMAA, AS, sodium bisulfite, ammonium persulfate, and sodium hydroxide are all analytically pure; sodium chloride is industrial grade; dispersant (SXY) and defoamer (BP-1A) are purchased from Chengdu Chuanfeng Chemical Engineering Co., Ltd.; The retarder (BXR-300L) was developed by Dagang Oilfield; Quartz sand was purchased from Henan Huarong Silica Fume Material Co., Ltd.; Jiajiang G-grade cement was purchased from Sichuan Jiajiang Weiju Feicheng Cement Co., Ltd..

2.2 Synthesis method of fluid loss agent

According to a certain molar ratio, weigh 4 kinds of monomers, AMPS, AA, DMAA, AS, etc., and dissolve them in a certain amount of distilled water, then transfer them into a 4-neck flask, and adjust the pH value with 30% sodium hydroxide solution by mass fraction. Blow in nitrogen and heat to the reaction temperature, then add initiator (ammonium persulfate: sodium bisulfite=1:1), keep the temperature constant, stir for a certain period of time, and then obtain the quaternary copolymerization fluid loss agent (FRW).
FRW performance test

* Corresponding author: zhangjian_gcy@cnpc.com.cn

 https://doi.org/10.1051/matecconf/202235801064

The performance evaluation of fluid loss agent is carried out in accordance with my country's oil and gas industry standard SY/T5504.2-2005 oil well cement admixture evaluation method. The cement slurry formula is: Jiajiang Grade G cement + 35% quartz sand (adding at an experimental temperature greater than or equal to 110°C) + 5% microsilica + 0.75% dispersant (SXY) + x% FRW + x% retarder (BXR-300L) + x% NaCl tap water. The x% in the formula represents the mass percentage of the additive in the cement, and the slurry is formulated according to the water-solid ratio of 0.44.

3. Results and discussion

3.1 FRW synthesis experiment

3.1.1 Effect of monomer ratio on fluid loss reduction performance

FRW performance is mainly determined by the number and ratio of functional groups, so the monomer ratio is one of the key factors that determine FRW performance. According to the literature investigation and preliminary investigation, the following synthesis conditions are used: the initiator is 0.4% of the total monomer mass, the monomer mass fraction is 40%, the experimental temperature is 50°C, the experimental pH value is 7.0, and the reaction time For 4h, 5 groups of different monomer ratio experiments were designed (Table 1). Then test its water loss performance at a FRW addition of 2% and an experimental temperature of 150°C.

Table 1 Performance table of FRW samples at 220°C under different monomer ratios

Numbering	Monomer molar ratio				Filter loss /mL
	AMPS	AA	DMAA	AS	
1	5	1	3.5	0.5	112
2	4.5	1	4	0.5	96
3	4	1.5	4	0.5	73
4	4	2	3.5	0.5	50
5	4.2	1.8	3.5	0.5	52

It can be seen from Table 1 that the FRW samples with the proportions numbered 4 and 5 have low fluid loss and close to each other after adding cement slurry. Although the fluid loss of No. 4 is less than that of No. 5, it is considered that if the content of carboxyl groups in FRW is higher, it will cross-link with Ca^{2+} and Al^{3+} produced by cement hydration, resulting in increased slurry consistency and thickening. The "wrapped heart" phenomenon will also appear in the experiment [14]. Considering the fluid loss of cement slurry and the properties of the slurry, the optimal monomer ratio relationship is determined to be AMPS: AA: DMAA: AS=4.2: 1.8: 3.5: 0.5.

3.2 Optimization of synthesis conditions

On the basis of determining the best monomer unchanged, a set of orthogonal experiments was designed, taking the fluid loss of the cement slurry as the evaluation standard, and considering its influence on the strength and performance of the cement stone, and optimizing the synthesis conditions. The factors of the orthogonal experiment are shown in Table 2. The 24h strength test condition is 21MPa×150°C in water bath.

Table 2 Orthogonal experiment factors and level table

temperature /°C	Initiator mass fraction	pH value	Monomer mass fraction	Nitrogen gas time /min	Reaction time /h	Filter loss /mL	Cement stone strength /MPa
30	0.2%	5	20%	10	2	65.4	18.4
40	0.3%	6	25%	20	3	52.3	21.2
50	0.4%	7	30%	30	4	49.6	21.5
60	0.5%	8	40%	40	5	51.2	18.9
70	0.6%	9	45%	50	6	74.1	18.6

Through the analysis of the orthogonal experiment results, the optimal synthesis conditions were determined as follows: the initiator is 0.3% of the total monomer mass, the monomer mass percentage is 40%, the reaction temperature is 50°C, the pH value is 7, and the nitrogen flow time is 30min, the reaction time is 5h.

3.3 FRW fluid loss reduction performance evaluation

3.3.1 High temperature resistance

One of the most important performance indicators of FRW is high temperature resistance. Therefore, the above formula will be used to test the high temperature resistance of FRW. The addition amount of FRW is 1.0% and 2.0%, and the addition amount of retarder is 2.0%. The experimental results are shown in Figure 1.

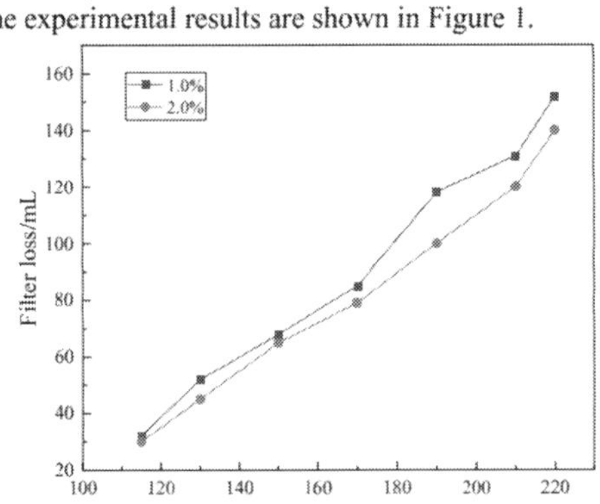

Figure 1 The relationship between the fluid loss of cement slurry and temperature

It can be seen from Figure 1 that although the fluid loss of cement slurry gradually increases as the temperature increases, the fluid loss is 152 mL when the temperature is 220°C. As the amount of FRW increases, the fluid loss of cement slurry decreases. For example, when the temperature is 220°C and the FRW increases from 1.0% to 2.0%, the fluid loss decreases by 8%.

3.3.2 Salt resistance

Stratums with high salinity often greatly reduce the effectiveness of fluid loss additives. In order to test the salt resistance of FRW, cement slurries were prepared with different concentrations of NaCl solution to investigate the salt resistance of FRW. The addition amount of FRW is 2%, the addition amount of retarder is 2%, and the curing condition of cement stone is 150°C×21MPa×48h water bath curing. The experimental results are shown in Table 4.

Table 3 Table of the influence of different salt concentrations on the performance of cement slurry

NaCl concentration	Filter loss /mL	Water separation /mL	Cement stone strength /MPa
10%	40.1	1.1	31.2
15%	42.1	1.1	32.1
20%	43.7	1.3	29.6
25%	43.9	1.4	28.3
30%	45.2	1.8	26.9
35%	45.9	2.5	25.2

It can be seen from Table 3 that with the increase of the salt content in the cement slurry, the fluid loss and water separation of the cement slurry increase slightly, but the increase is small, indicating that FRW has good salt resistance. The morphology of C-S-H gel changes, and the strength of cement stone gradually decreases. But in the case of saturated brine, the strength still reaches 20.5MPa, which is very different from the general low strength of saturated brine cement slurry.

3.3.3 The influence of FRW on cement slurry engineering performance

The fluid loss agent not only achieves the function of reducing fluid loss, but also meets other performance requirements of the cement slurry, such as rheology and mechanical properties. The effect of FRW on the engineering performance of cement slurry is studied here. The curing condition of cement stone is 150°C×21MPa×48h. The results are shown in Table 4.

Table 4 The influence table of FRW on cement slurry engineering performance

FRW increase	Liquidity /cm	Water separation /mL	Filter loss /mL	Fluidity index	Thickening factor /Pa·s^n	Thickening time /min	Cement stone strength /MPa
0	21.2	3	942	0.83	0.38	229	28.3
1%	20.3	0.8	110	0.81	0.34	235	27.6
2%	19.8	0	42	0.82	0.36	232	26.9

It can be seen from Table 4 that as the amount of FRW increases, the fluidity index (n) of the slurry decreases slightly, and the consistency coefficient (K) increases slightly, but overall the value of n is greater than 0.80 and the value of K is less than 0.40Pa•s, indicating that the cement slurry has excellent rheological properties, which is very beneficial for improving the displacement efficiency of the eccentric annulus. The water separation rate of the cement slurry is 0, and the filtration loss is less than 50mL, indicating that the slurry stability is good. The thickening time of cement slurry and the strength of cement stone did not change much with the increase of FRW, indicating that FRW would not decompose at high temperature.

4. Conclusions

In the experiment, a quaternary high-temperature and salt-resistant FRW sample was synthesized by the solution copolymerization method, and the best synthesis conditions were optimized: the molar ratio of AMPS:AA:DMAA:AS was 4.2:1.8:3.5:0.5, and the initiator was added as a monomer. 0.3% of the total mass, 40% of the monomer mass, reaction temperature of 50°C, and pH value of 7. At the same time, FRW has good fluid loss control ability. Under the condition of 2% addition, the fluid loss is 141mL when the temperature is 220 °C, and it has no negative impact on the comprehensive performance of the cement slurry and is practical.

Acknowledgments

The work was supported financially by Shaanxi Provincial Key Research and Development Program (2019GY-136), Key Scientific Research Program of Shaanxi Provincial Department of Education (21JY035) and Youth Innovation Team of Shaanxi University. And we thank the work of Modern Analysis and Testing Center of Xi`an Shiyou University.

References

1. Siming Yan. Synthesis and mechanism study of temperature-resistant fluid loss reducer for oil well

cement[J]. Advances in Cement Research, 2017, 29(5): 183-193.

2. Vassilios C. Kelessidis, Christina Tsamantaki, Athanasios Michalakis, George E. Christidis, Pagona Makri, Kassiani Papanicolaou, Antonios Foscolos. Greek lignites as additives for controlling filtration properties of water–bentonite suspensions at high temperatures[J]. Fuel, 2007, 86(7): 1112-1121.
3. Knapen E and Van Gemert D. Cement hydration and microstructure formation in the presence of water-soluble polymers[J]. Cement and Concrete Research, 2009, 39(1): 6-13.
4. Lu Hongsheng. Yuan Xiap, Huang Zhiyu. Synthesis and performance evaluation of quaternary polymerization fluid loss additive AMPF[J]. Fine Chemical Industry. 2013, 30(11)1256-1250.
5. Li Gao, Fu Qiang, Yu Hanghang, et al. Synthesis of quaternary copolymerization high temperature and saltresistantoil well ement fuid loss aditive[J]. Natural Gas Industry, 2018, 38(12): 96-10.
6. Xiao Q, Rao P, Xiao W, Liu X and Zhang W. Preparation of a novel forpolymer as fluid loss additive for high temperature oil well cementing[J]. Russian Journal of Applied Chemistry, 2014, 87(9): 1377–1381.
7. Xu J, Hu R, Teng C, Chang L and Zhou X, The well cement technology in ultra-deep horizontal well Yuanba-103H[J]. Journal of Y angtze University (Natural Science Edition), 2013, 10(8): 64–67.
8. Song W, Wang Q, Hou Y, et al. AMPS/DMAM/FA - A Graft Copolymer Humic Acid Fluid Loss Reducer for Well Cementing[J]. Drilling Fluid and Completion Fluid. 2017, 34(1): 101-105
9. Liu Jingli, Hao Hujun, Li Xumei, et al. Study on graft modified PVA-A cement slurry fluid loss reducer[J]. Drilling Fluid & completion Fluid, 2014, 31(5): 78-80.
10. Yu Yongjin, Liu Shuoqiong, Liu Liwen, et al. Preparation and evaluation of high temperature cement slurry loss reduction additive DRF-120L[J]. Oil Drilling & Production Technology, 2011, 33(3): 24-27.
11. Xia X, Feng Y, Guo J, et al. Zwitterionic copolymer for controlling fluit loss in oilwell cementing: preparation, characterization, and working mechanism[J]. Polymer Engineering and Science, 2017, 57(1): 78-88 .
12. Liu Fei,Liu Xuepeng.Xia Chengy u,et al. Synthesis and performance evaluation of high temperature resistant cement slurry fluid loss additive SCF-1[J]. Chemistry of Petroleum and N atural Gas, 2016, 45(4): 64-67
13. Lu Jiahan, Guo Juan, Liang Haixiang. Development and application of a crosslinking agent for cementing polymer fluid loss reducer [J]. Henan Chemical Industry, 2021, 38(8): 18-21

 https://doi.org/10.1051/matecconf/202235801065

Finite Element Simulation of Graphene Thermal Conductivity Framework

Jun Yang
College of Chemistry and Chemical Engineering, China University of Petroleum, Qingdao, China

Abstract. With the rapid development of electronic devices towards miniaturization and high integration, the problem of overheating and heat dissipation forces people to find thermal conductive materials with good comprehensive properties to meet the needs of development. Inspired by the three-dimensional boron nitride network, this paper creatively proposed the construction of a composite thermal conductive material with 3D graphene network structure and PDMS as filling matrix, and carried out finite element analysis and simulation with COMSOL. The results show that the thermal conductivity of the composite thermal conductive material with graphene 3D network structure is significantly better than that of the single thermal conductive material without graphene and the thermal conductive material with random distribution of graphene, and most of the heat flow forms a connected thermal conductive path along the heat transfer direction in the graphene of 3D network composite. It can be seen that the prepared composites with 3D thermal conduction network structure have excellent thermal conductivity, and will have broad market prospects in the fields of heat dissipation materials, computers, smart phones, laser weapons and so on.

1. Introduction

With the rapid development of technology and the continuous expansion of industrial demand, as well as the development of electronic devices towards miniaturization and high integration, the problem of overheating and heat dissipation has become a huge obstacle to the development of electronic devices[1]. At present, there are many kinds of high thermal conductivity materials studied, and they can be roughly divided into several categories. One is metal based thermal conductivity materials. Metal is one of the widely used heat conducting materials at present. However, due to its poor corrosion resistance and electrical insulation, it can not meet the heat dissipation requirements of electrical insulation and chemical corrosion. The second type is ceramic thermal conductive materials. Ceramics are inorganic materials with the advantages of high hardness, high temperature resistance and high pressure resistance among common engineering materials. Among them, ceramics such as aluminum nitride and silicon nitride have excellent thermal conductivity. However, their large-scale application is seriously limited due to their high processing cost and poor impact resistance [2].
Whether the thermal conductivity of thermal conductive materials is excellent is very important in industries with high requirements for heat dissipation performance. At present, there is an urgent need to research and develop new heat conducting materials with good comprehensive properties to meet the needs of the development of industry, science and technology and national defense, so people turn their attention to polymer materials. Polymer materials have excellent plasticity, electrical insulation, corrosion resistance and easy processing. They are widely used in many fields. However, most polymer materials have low thermal conductivity, which greatly limits their large-scale application, so polymer based thermal conductive materials came into being[3][4][5]. That is, some high thermal conductivity materials are compounded in the polymer matrix to improve its comprehensive properties, so as to promote the development of thermal conductivity materials in the industrial field.
Hexagonal boron nitride has the advantages of high thermal conductivity, good insulation performance, good chemical stability and industrial production. It is considered to be one of the most ideal thermal conductive fillers. However, hexagonal boron nitride polymer thermal conductive composites have not been applied yet. The two key factors causing this problem are the poor dispersion and serious viscosity increase of hexagonal boron nitride. Yu Kangkang [6] prepared a new spherical boron nitride silicone rubber thermal conductive composite with the advantages of uniform dispersion, dense microstructure and high thermal conductivity by high gravity deposition. Jiang Wenzheng et al. [7] and Jing et al. [8] introduced a polymer composite with three-dimensional boron nitride network structure. However, this polymer composite constructed by them still has some problems, such as easy agglomeration and difficult dispersion of boron nitride in the polymer matrix.

 https://doi.org/10.1051/matecconf/202235801065

Graphene is composed of single or few layers of carbon atoms, with high strength and modulus, and its in-plane thermal conductivity is as high as 3500-5300 W / (m · K). It is the material with the highest thermal conductivity known at present. Its mechanical and thermal conductivity are far higher than ordinary thermal conductivity materials, and has been widely used in various fields [9]. Wang et al. [10] found that adding graphene oxide to epoxy resin can improve the thermal conductivity of composites. In terms of the material characteristics of graphene, some scholars have found that the random dispersion of graphene in the matrix will disrupt the unidirectional heat conduction mechanism in the composite and reduce the effective thermal conductivity of graphene [3] [11]. At present, some scholars use three-dimensional graphene preform as thermal conductive filler, and then composite it with polymer by vacuum impregnation to obtain composite thermal interface material, which effectively improves the thermal conductivity of polymer [12]. Qi et al. [13] prepared the phase change thermal interface material through CVD method, which increased the thermal conductivity from 0.27 W/ (m. K) to 1.22 W/ (m. K).

Inspired by the above, building a three-dimensional filler network in the polymer matrix can efficiently build a heat conduction path. Compared with the randomly dispersed filler system, the composites with three-dimensional heat conduction network structure show higher heat conduction properties. Therefore, the composites with three-dimensional graphene thermal conductive structure are simulated by finite element analysis.

2. Methods

In order to verify whether the existence of graphene and 3D network structure can promote the heat transfer ability of the material, we analyzed it through COMSOL finite element simulation, and revealed the total heat flux and temperature distribution of the material during heating. The boundary conditions are shown in the Figure 1. The outer walls on the left and right sides are thermally insulated, the temperature at one end of the upper and lower ends is 273K, the other end is 323K, and the inward heat flux from the hot end is 0.1W.

3. Results and discussion

The finite element model established in the Figure 2 shows three cases: one is that graphene is randomly dispersed in the PDMS matrix, the other is that graphene forms a 3D network structure in the PDMS matrix, and the other is that there is only PDMS matrix. Because graphene and PDMS are solid, solid heat transfer interface is selected. After the boundary setting is completed, the established model is meshed, and the physical field is selected to control the super refinement of the mesh, so as to improve the calculation accuracy and obtain more accurate results. We can see from the temperature distribution image that the top temperature of the composites with graphene 3D network structure is significantly higher than that of the composites with graphene random dispersion in Figure 2(a) compared with Figure 2(c), which indicates that the 3D network structure has better heat transfer ability, which is consistent with the simulation conclusion of Xu Xingxing [3] and the experimental conclusion of Wenzheng Jiang et al. [7]. Comparing Figure 2(a) with Figure 2(b), in the case of the same 3D network structure, it is obvious that the top temperature of the thermal conductive material with graphene is higher, which shows that the addition of graphene can well improve the thermal conductivity of the thermal conductive material. Moreover, in the heat flux distribution diagram, it is obvious from Figure 2(d) and Figure 2(f) that most of the heat flux is obviously located in graphene, indicating the super thermal conductivity of graphene. The comparison between Figure 2(d) and Figure 2(e) shows that the heat flow forms a connecting path along the heat transfer direction in the 3D network composite, while in the random composite, the random dispersion of graphene will hinder the heat flow conduction, which indirectly proves the super heat transfer ability of the 3D composite. This simulation result is consistent with the simulation conclusion of Jingchao Li et al. [8].

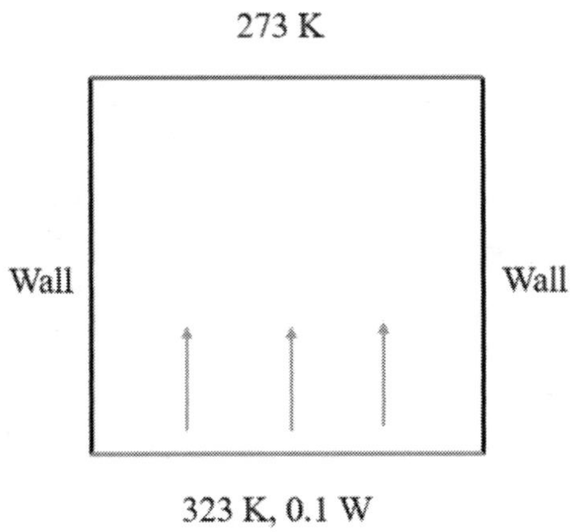

Figure 1. Boundary conditions.

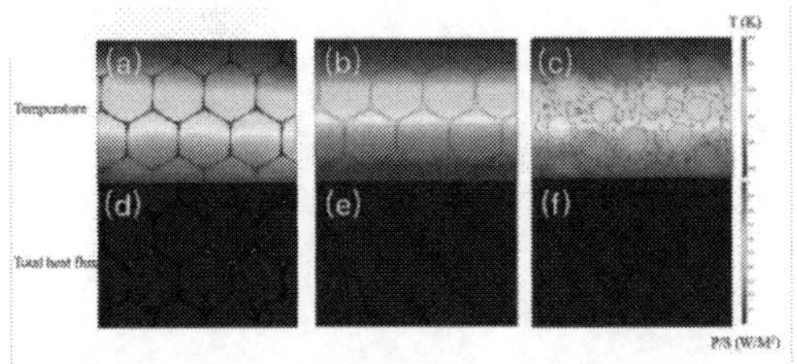

Fifure 2. Simulate temperature distribution (a)(b)(c) and total heat flux distribution (d)(e)(f) as well as the thermal conductivity values of the 3D graphe-PDMS composites, the 3D PDMS and the random graphe-PDMS composites based on a heat source acquired by finite element analysis.

4. Conclusion

To sum up, adding graphene as filler into PDMS matrix can improve the thermal conductivity of graphene PDMS composites, and the 3D thermal conductivity network structure formed by graphene as filler also greatly improves the thermal conductivity of composites. This shows that 3D graphene PDMS has broad application prospects in the field of heat dissipation in the future.

References

1. Alexander A, Balandin. Thermal properties of graphene and nanostructured carbon materials. Nature materials, 2011, 10(8): 569-81.DOI: 10.1038/nmat3064.
2. Dong, Y. Research progress of high thermal conductivity materials. Contemporary Chemical Research, 2019(10):179-180. (in Chinese)
3. Xu, X. Study on heat conduction model and mechanism of graphene / PDMS composites. Chongqing University of Posts and Telecommunications, 2018. (in Chinese)
4. Zhang, Z., Qu, J., Feng, Y. et al. Assembly of graphene aligned polymer composites for thermal conductive applications. Composites Common , 2018, 9, 33−41.
5. Li, B., Liu, Y., Sun, B. et al. Properties and thermal conduction mechanism of polymer matrix thermal conductive composites. Journal of Chemical Engineering, 2009,60(10):2650-2655. (in Chinese)
6. Yu, K. Preparation and properties of new boron nitride / silicone rubber thermal conductive composites. Tianjin Polytechnic University, 2021. (in Chinese)
7. Jiang, W., Lin, Y., Jiang, P. et al. Three dimensional boron nitride structure and its thermal conductive insulating polymer nanocomposites. Journal of Electrical Engineering, 2021,16(02):12-24. (in Chinese)
8. Li, J., Li, F., Zhao, X. et al. Jelly-Inspired Construction of the Three-Dimensional Interconnected BN Network for Lightweight, Thermally Conductive, and Electrically Insulating Rubber Composites. *ACS Applied Electronic Materials* **2020** *2* (6), 1661-1669
9. Liu, D. Preparation and properties of copper based high thermal conductivity composites. Guizhou University, 2020. (in Chinese)
10. Wang, S., Tambraparni, M., Qiu, J. et al. Thermal expansion of graphene composites. Macromolecules, 2009,42(14):5251-5255.
11. Chu, K., Wu, Q., Jia, C. et al. Fabrication and effective thermal conductivity of multi-walled carbon nanotubes reinforced Cu matrix composites for heat sink applications. Composites Science & Technology, 2010,70(2): 298-304.
12. Huang, Y. Preparation and thermal conductivity of three-dimensional graphene / Polystyrene Composites. Hunan University, 2019. (in Chinese)
13. Qi, G., Yang, J., Bao, R. et al. Hierarchical graphene foam-based phase change materials with enhanced thermal conductivity and shape stability for efficient solar-to-thermal energy conversion and storage. Nano Research, 2016,10(3): 802-813.
14. Mao, Y., Fan, H., Liu, Y. Research and application of heat conducting polymer materials. Scientific Chinese, 2016(18):36. (in Chinese)
15. Chen, H., Shi, Z., Qi, Y. et al. Heat transfer simulation analysis of medium frequency furnace pipeline based on COMSOL waste heat recovery system. Journal of Shenyang Architecture University (NATURAL SCIENCE EDITION), 2021, 37(06):1107-1114. (in Chinese)
16. Lan, L. Numerical simulation of heat transfer characteristics of corrugated fins based on COMSOL. Journal of Jiujiang University (NATURAL SCIENCE EDITION), 2020,35(01):32-38. DOI:10.19717/j.cnki.jjun.2020.01.008. (in Chinese)
17. Shen, Z. Research on three-dimensional heat transfer and energy saving utilization in thermal bridge based

on COMSOL simulation technology. Wuhan Polytechnic University, 2017. (in Chinese)

18. Lv, J., Ma, G., Guo, J. et al. Temperature field analysis of autoclave forming die based on COMSOL. Journal of North University of China (NATURAL SCIENCE EDITION) , 2021,42(05):412-418. (in Chinese)
19. Huang, Z. Preparation and properties of high thermal conductivity composites based on graphene. Anhui University, 2019. (in Chinese)
20. Lin, Y., Jia, Y., Alva, G. et al. Review on thermal conductivity enhancement, thermal properties and applications of phase change materials in thermal energy storage. Renewable & Sustainable Energy Reviews, 2018,82(3): 2730-2742

Molecular Dynamics Study on the Effect of Doping Defects on the Properties of CL-20/HMX Co-crystal Explosive

Jiahao Yang [1,*], Yuling Wang [1] and Chengzhou Wang [2]

[1] Rocket Force University of Engineering, School of Nuclear Engineering, 710025 Xi'an, China
[2] Troop No. 96037 of PLA, 721012Baoji, China

Abstract. In order to explore the influence of doping defects on the properties of CL-20 / HMX eutectic explosive, the models with different doping rates were calculated by molecular dynamics method. The results show that compared with the non-doped crystal model, the binding energy in the doped model is reduced by 0.332%~11.805%; The initiation bond length increased by 0.9%~4.6%; The binding diatomic interaction energy is reduced by 2.31%~15.46%; The cohesive energy density decreased by 19.75%~50.39%; The decreasing range of density is 0.697%~4.450%; The reduction range of detonation velocity is 0.744%~5.449%; The reduction range of explosion pressure is 1.864%~12.841%; The bulk modulus decreased by 0.4844~6.7461GPa, the shear modulus decreased by 0.9442~5.2329GPa, and the tensile modulus decreased by 2.0431~12.5374GPa. It shows that doping defects will reduce the stability, safety and detonation performance of explosives, and make the hardness, fracture resistance and rigid strength of explosives worse.

1 Introduction

CL-20, the full name of Hexanitrohexaazaisowurzane, is one of the single substance explosives with the highest energy density at present. It has good application prospects, but it also has the disadvantages of high sensitivity and high cost [1, 2]. As one of the best comprehensive single substance explosives at present, HMX has the advantages of good detonation performance, high energy and good thermal stability. Therefore, CL-20 / HMX eutectic explosive composed of CL-20 and HMX has attracted extensive attention of scholars at home and abroad in recent years.

CL-20 be doped with Ethyl Acetate impurity ($C_4H_8O_2$) [3] in the preparation process, resulting in Ethyl Acetate molecules in the crystal of CL-20 / HMX explosive, resulting in doping defects, which will affect the performance of the explosive. As the main factor affecting the properties of explosives, crystal defects have been widely concerned by scholars, and the study of their influence on the properties of explosives has become a subject of great significance. Hang Gui-yun et al. [4] founded in the study of HMX / NQ eutectic explosive defects that the defects reduce the stability and detonation performance of the explosive and increase the sensitivity. Sun Ting et al. [5] studied the eutectic structure and properties of CL-20 / HMX by MD method, they found that the sensitivity can be reduced by introducing HMX into CL-20 system. Hu Jing-wei et al. [6] studied the effect of vacancy defects on CL-20 explosive and found that vacancy defects improved the thermal decomposition rate of CL-20 molecules and thus increased the sensitivity of the explosive.

In this paper, molecular dynamics method was used to calculate different doping rate models. Through the analysis of calculation data, it provides a reference for predicting the performance of CL-20 / HMX explosive.

2 Model and calculation method

2.1. Original model establishment

CL-20 / HMX eutectic explosive is composed of CL-20 and HMX in a molar ratio of 2:1. It belongs to monoclinic system. The space group is P21 / C, and the lattice parameters are a=1.635nm, b=0.994nm, c=1.214nm, β= 99.233° [7]. The CL-20 / HMX single crystal model obtained from reference [7] is shown in Figure 1 (a), which is extended to 3×2×2 The supercell model is shown in Figure 1 (b), which is marked as model-1.

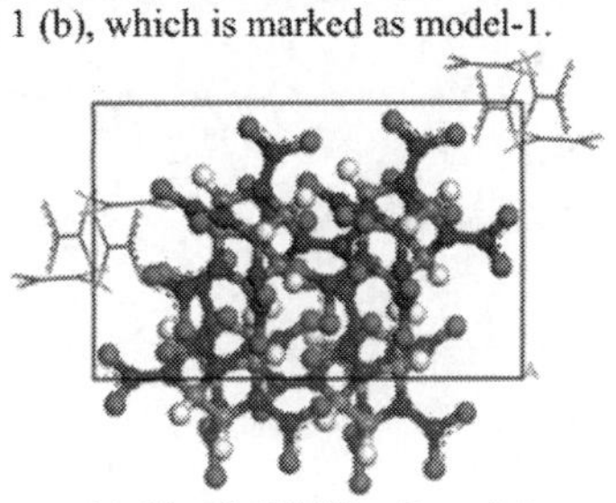

(a) CL-20 / HMX cell model

* Corresponding author: dashuai199708@163.com

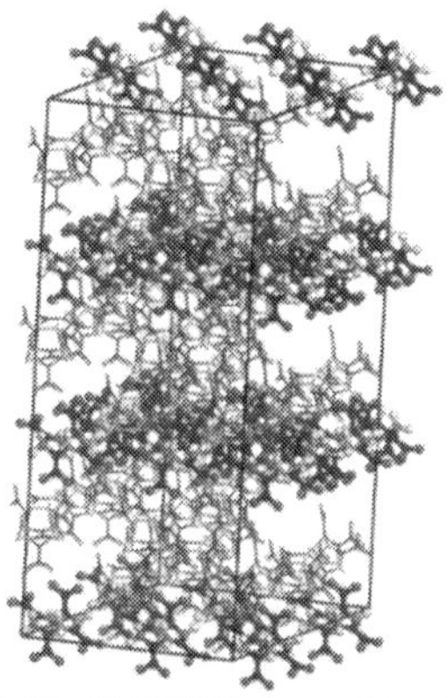

(b) CL-20 / HMX supercell model

Fig. 1. Single cell model and supercell model

2.2 Establishment of doping model

By replacing CL-20 molecule in the original model-1 with Ethyl Acetate impurity, five doping models with doping rates of 2.78%, 5.56%, 8.33%, 11.11% and 13.89% were formed. Figure 2 shows the doping model-5.

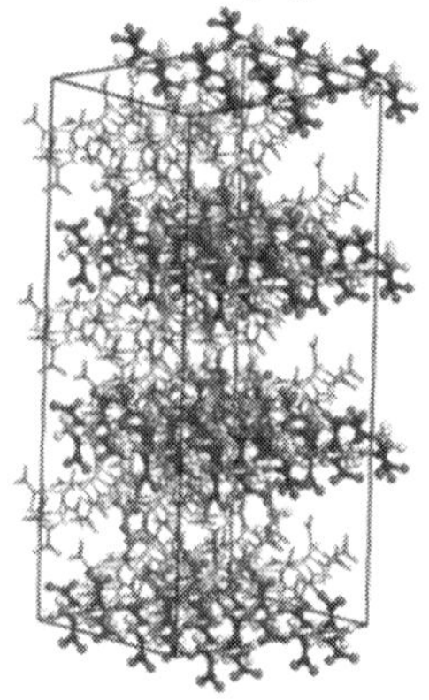

Fig. 2. Doping model-5

2.3 Calculation condition setting

Firstly, the geometric optimization of the model structure is carried out to make its energy converge. Then carry out molecular dynamics calculation, adopt NVT ensemble and compass force field, set the temperature as 295K and the pressure as 0.1MPa. Set the step size as 1fs, save the results every 100 steps, and calculate 10000 steps in total.

3 Results and discussion

3.1 Balance discrimination

Take model-2 as an example, in the process of molecular dynamics calculation, the temperature reaches equilibrium after 4000fs; The energy gradually stabilized after 2000fs. Therefore, the calculation results can be statistically analysed

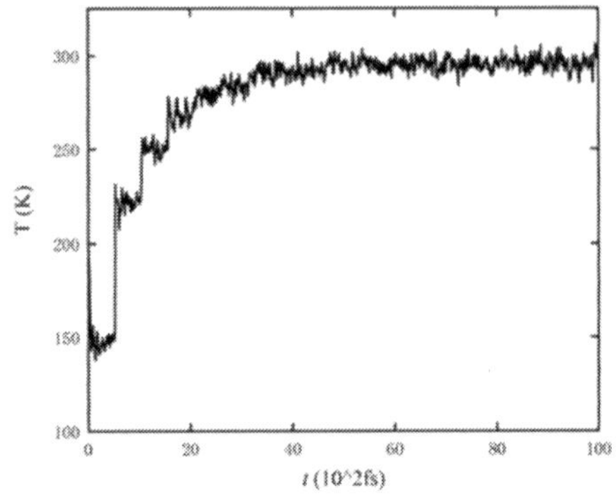

(a)T-t curve

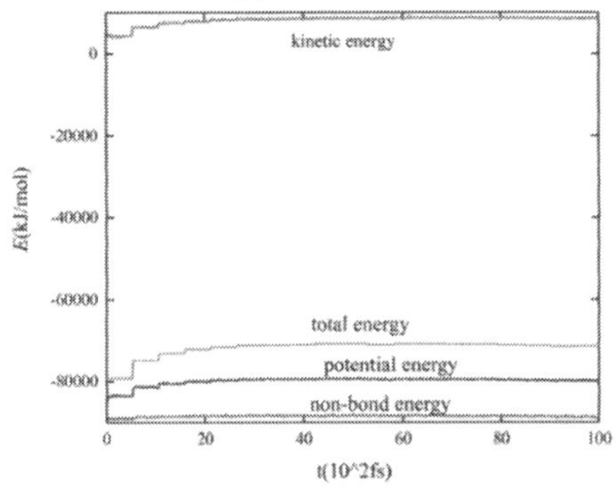

(b)E-t curve

Fig. 3. Variation curve of temperature and energy

3.2 Binding energy

The calculation formula of binding energy is

$$E_b = -E_{inter} = (E_{CL\text{-}20} + E_{HMX}) - E_{total} \qquad (1)$$

In the above formula, E_b is the binding energy; E_{inter} is intermolecular force; $E_{CL\text{-}20}$ is the total energy of CL-20 molecule; E_{HMX} is the total energy of HMX molecule; E_{total} is the total energy when the system reaches equilibrium.

The binding energies of six models are 468.16、466.60、458.91、446.43、424.55、412.90kJ/mol. The above data show that the binding energy of model-1 is the largest. With the increase of doping rate, the binding energy decreases gradually, and the reduction range of doping model is 0.332% ~ 11.805%. The binding energy reflects the tightness of explosive molecules, and the decrease of binding energy indicates that the increase of doping rate has an impact on the stability of explosive structure.

3.3 Sensitivity

It is usually used to reduce the explosive susceptibility of high explosives. According to the "hot spot theory" [8] and "initiation bond thought" [9], combined with the research of Xiao Ji-jun et al. [10-13], the change of sensitivity in this paper is obtained by analysing the initiation bond length, initiation bond diatomic interaction energy and cohesive energy density of explosive crystal.

3.3.1 Initiation bond length

CL-20 is the main explosive, due to its high sensitivity and sensitivity to external stimuli, it is more likely to react than HMX. Therefore, the $N\text{-}NO_2$ bond in CL-20 is selected as the initiation bond [14,15]. Take model-3 as an example, the bond length distribution of its initiation bond is shown in Figure 4, which is approximately Gaussian distribution.

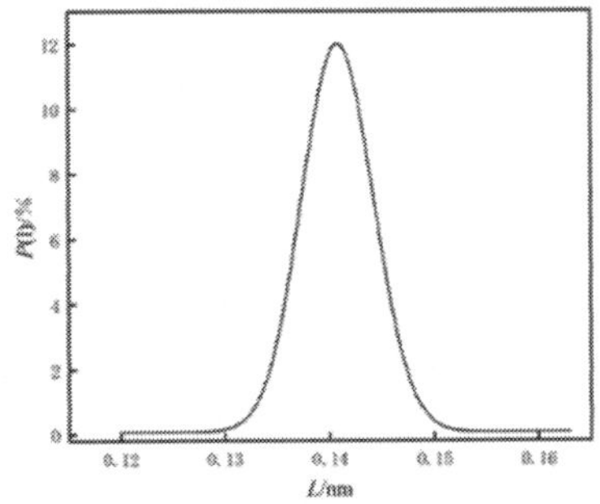

Fig. 4. Initiation bond length distribution

Table. 1. Average bond length and maximum bond length

model	L_{ave}/nm	L_{max}/nm
1	0.1755	0.2162
2	0.1740	0.2184
3	0.1770	0.2210
4	0.1765	0.2213
5	0.1770	0.2224
6	0.1785	0.2257

It can be seen from table 1 that the average bond length has little correlation with the doping rate, and the maximum bond length increases with the increase of the doping rate, but the overall range is small. The maximum bond length of non-doped model is 0.2162 nm; The increase range of the five doping models is 0.9% ~ 4.6%. Due to the high energy of molecules with the maximum bond length, like activated molecules, they are easy to cause explosion, so they have high sensitivity and low safety.

3.3.2 Initiation bonding diatomic interaction energy

Calculation formula of bonding diatomic interaction energy ($E_{N\text{-}N}$) of initiating bond

$$E_{N\text{-}N} = \frac{E_1 - E_2}{n} \tag{2}$$

In the above formula: E_1 is the total energy corresponding to the equilibrium of the system; E_2 is the total energy when the system reaches equilibrium after fixing all nitrogen atoms in CL-20; n is the number of $N\text{-}NO_2$ bonds contained in CL-20 molecule in the system.

After calculation, the initiation bond bonding diatomic interaction energies of the six models are 118.29、115.47、111.62、109.13、108.62、100kJ/mol. It can be seen from the above data that the diatomic interaction energy of initiation bond decreases with the increase of doping rate, and the decrease range is 2.31% ~ 15.46%. The diatomic interaction energy of initiation bond is a reflection of the stability of initiation bond. The higher the energy, the less likely the initiation bond is to break. Therefore, the increase of doping rate will reduce the stability of initiation bond and increase the sensitivity.

3.3.3 Cohesive energy density

Cohesive energy density (*CED*) refers to the energy required for 1mol condensate in unit volume to overcome the intensification of intermolecular interaction, with the unit of kJ/cm^3 and the calculation formula of [16]

$$CED = \frac{H_V - RT}{V_m} \tag{3}$$

In the above formula: H_V is molar evaporation heat, kJ/mol; RT is the expansion work done during material gasification, kJ/mol; V_m is the molar volume, cm^3/mol. The cohesive energy density is equal to the sum of van der Waals force and electrostatic force. Its essence is a non-bond force, which can reflect the strength of intermolecular interaction in the mixed system. Through calculation, the cohesive energy density, van der Waals force and electrostatic force are obtained.

Table. 2. Cohesive energy density, van der Waals force and electrostatic force(kJ/cm^3)

model	*CED*	vdW	Electrostatic
1	1.554	1.372	0.182
2	1.247	1.129	0.145
3	1.218	1.016	0.112
4	0.976	0.904	0.072
5	0.895	0.837	0.058
6	0.771	0.749	0.022

It can be seen from the data in Table 2 that the three values of the non-doped model are the highest, and the cohesive energy density of the five doped models decreases by 19.75% ~ 50.39%. Cohesive energy density reflects the difficulty of explosive changing from condensed state to gaseous state. Therefore, the increase of doping rate will reduce the cohesive energy density, make the explosive easier to change into gaseous state, and increase the sensitivity of the explosive.

3.4 Detonation performance

Detonation performance reflects the power and energy density of explosives. In this paper, the modified nitrogen equivalent method is used to calculate the detonation parameters of explosives. Detonation parameter calculation formula of the modified nitrogen equivalent method is as follows

$$V_D = (690 + 1160\rho)\sum N_{ch} \tag{4}$$

https://doi.org/10.1051/matecconf/202235801066

$$p = 1.106(\rho \sum N_{ch})^2 - 0.84 \tag{5}$$

$$\sum N_{ch} = \frac{100}{M_r}(p_i N_{p_i} + \sum B_K N_{B_K} + \sum G_j N_{G_j}) \tag{6}$$

In the above formula: V_D is explosion speed, m/s; p is explosion pressure, GPa; ρ is the explosive density, g/cm^3; $\sum N_{ch}$ is the corrected nitrogen equivalent; p_i is the mole number of the i-th detonation product produced by 1mol explosive; N_{pi} is the nitrogen equivalent coefficient of the i-th detonation product; B_K and N_{BK} is the number of occurrence of the K-th chemical bond and the nitrogen equivalent coefficient; G_j and N_{Gj} is the occurrence times and nitrogen equivalent coefficient of the j-th group.
The calculated results are shown in Table 3.

Table 错误!文档中没有指定样式的文字。. 3. Detonation parameters

model	ρ/g·cm^{-3}	υ_D/m·s^{-1}	p/GPa
1	2.001	9202.2	39.17
2	1.951	9133.7	38.44
3	1.901	8948.6	36.42
4	1.851	8832.2	35.27
5	1.801	8771.3	34.75
6	1.751	8700.8	34.14

It can be seen from the above table that the density, detonation velocity and detonation pressure of the non-doped model are the highest, and the density, detonation velocity and detonation pressure of the five doped models decrease with the increase of doping rate, with the reduction ranges of 0.697% ~ 4.450%, 0.744% ~ 5.449% and 1.864% ~ 12.841% respectively. The reason may be that the energy of ethyl acetate is lower than CL-20 and destroys the stability of the explosive.

3.5 Mechanical property

The mechanical properties of the model after dynamic calculation are analysed. According to Hooke's law, the following calculation formula is obtained.

$$E = \frac{9KG}{3K + G} \tag{7}$$

$$\mu = \frac{3K - 2G}{2(K + G)} \tag{8}$$

In the above formula: K is the bulk modulus, indicating the hardness of the material; G is the shear modulus, indicating the fracture resistance of the material; E is the tensile modulus, which represents the rigidity measurement of the material; μ is Poisson's ratio.

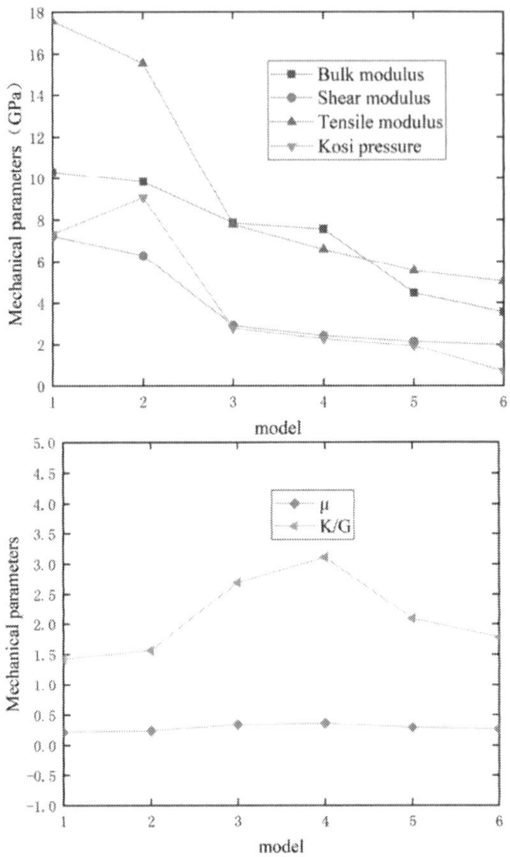

Fig. 5. Mechanical properties

In the figure above: Poisson's ratio and K / G are elastic constants without unit; Cauchy pressure indicates the ductility of the material; K / g indicates the toughness of the material.
It can be seen from Fig. 5 that with the increase of doping rate, the bulk modulus, shear modulus and tensile modulus of the explosive decrease, indicating that with the increase of doping rate, the hardness, fracture resistance and rigid strength of the explosive decrease,; When the doping rate is 2.78%, the maximum Cauchy pressure is 9.062GPa, which indicates that the doping defects improve the ductility of the crystal to a certain extent, but the overall trend reflects that the higher the doping rate is, the better the ductility is; When the doping rate is 8.33%, the maximum K/G is 3.11. The overall trend is to increase first and then decrease with the increase of doping rate, indicating that the doping defects can increase the toughness of the crystal to a certain extent, but the more the better; The Poisson's ratio is between 0.216 and 0.355, which is relatively stable, indicating that the doping defects have little effect on the change of Poisson's ratio.

https://doi.org/10.1051/matecconf/202235801066

4 Conclusion

(1) Compared with the non-doped model, the binding energy, van der Waals force and electrostatic force of the doped model are reduced, which leads to poor stability. The higher the doping rate, the lower the binding energy and the worse the stability of the explosive.
(2) The maximum bond length of initiation bond of explosives increases with the increase of doping rate, and the diatomic energy and cohesive energy density of initiation bond decrease with the increase of doping rate. These two changes will increase the sensitivity of explosives.
(3) The density of explosive decreases with the increase of doping rate, which reduces its detonation velocity and detonation pressure, which has an adverse impact on the detonation performance of explosive.
(4) Doping defects will affect the mechanical properties of explosives. The higher the doping rate is, the smaller the bulk modulus, shear modulus and tensile modulus of the explosive are, and the hardness, fracture resistance and rigid strength of the explosive are reduced.
Based on the above conclusions, with the increase of doping rate, the stability, safety, detonation performance and mechanical properties of explosives decrease, which is contrary to the current goal of pursuing high energy density and low sensitivity explosives, and the hardness, fracture resistance and rigid strength decrease, which makes the reliability of materials not strong. Therefore, doping defects should be minimized in the manufacture of explosives.

References

1. Yuxiang, O. U. , Z. Meng , and J. Liu . "Advance in high energy density compound CL-20—— developments of synthesis route and production technologies of CL-20." *Chemical Industry and Engineering Progress* (2007).
2. Nair, U. R. , et al. "Hexanitrohexaazaisowurtzitane (CL-20) and CL-20-based formulations (review)." *Combustion, Explosion and Shock Waves* 41.2(2005):121-132.
3. Liu, K. , et al. "Determination of Organic Impurity in Reference Materials CL-20 by Quantitative Nuclear Magnetic Resonance(qNMR)." *Journal of Instrumental Analysis* (2017).
4. Hang, G. Y. , et al. "Investigating the Effect of Adulteration Defect on the Properties of HMX/NQ Cocrystal Explosive by Molecular Dynamics Method." *Huozhayao Xuebao/Chinese Journal of Explosives and Propellants* 42.2(2019):145-151.
5. Sun, T. , et al. "Molecular Dynamics Simulation of Interface Interactions and Mechanical Properties of CL-20/HMX Cocrystal and Its Based PBXs." *Acta Chimica Sinica* 72.9(2014):1036.
6. Hu Jingwei et al "Molecular dynamics simulation of initial chemical reaction of CL-20 containing defects at high temperature." *Energetic materials* 29.6 (2021): 11
7. Bolton, O. , et al. "High Power Explosive with Good Sensitivity: A 2:1 Cocrystal of CL-20:HMX." *Crystal Growth & Design* 12.9(2012):4311.
8. Bowden, F. P. . "Initiation and Growth of Explosion in Liquids and Solids." *American Journal of Physics* 20.4(1952):250.
9. Kamlet, M. J. , and H. G. Adolph . "The relationship of Impact Sensitivity with Structure of Organic High Explosives. II. Polynitroaromatic explosives." *Propellants, Explosives, Pyrotechnics* (1979).
10. Liu, Q. , et al. "Molecular Dynamics Simulation on Sensitivity and Mechanical Properties of ε-CL-20 Crystal at Different Temperatures." *Chinese Journal of Explosives & Propellants* 37.2(2014):7-12+17.
11. Sun, T. , et al. "Molecular Dynamics Simulation of Interface Interactions and Mechanical Properties of CL-20/HMX Cocrystal and Its Based PBXs." *Acta Chimica Sinica* 72.9(2014):1036.
12. XU, et al. "Molecular dynamics simulations on the structures and properties of ε-CL-20-based PBXs —— Primary theoretical studies on HEDM formulation design." *Science in China* (2007).
13. Sun, T. , et al. "Molecular Dynamics Simulation of Compatibility,Interface Interactions and Mechanical Properties of CL-20/DNB Cocrystal Based PBXs." *Chinese Journal of Energetic Materials* (2015).
14. Xu, X. J. , et al. "Theoretical Study on Pyrolysis Mechanism for-Hexanitrohexaazaisowurtzitane." *Chinese Journal of Organic Chemistry* 25.5(2005):536-539.
15. Geetha, M. , et al. "Studies on CL-20: The most powerful high energy material." *Journal of Thermal Analysis & Calorimetry* 73.3(2003):913-922.
16. Zhao, L. , et al. "Molecular dynamics study on the relationships of modeling, structural structure and energy properties with sensitivity for RDX-based PBXs." *Scientia Sinica(Chimica)* (2013).

Research on the influence of the geometric characteristics of surrounding rock fractures on the macroscopic mechanical properties of rock-like materials based on PFC

Xiaoye Zhu

Shandong Normal University, Jinan, China

Abstract. The surrounding rock of coal mining face often contains numerous micro-cracks, and these micro-cracks greatly restrict the mechanical strength of the rock. Mechanical properties of rock degraded by joints and fissures in rock,, taking the Mesozoic three-stream pluton granite in Dandong City, Liaodong uplift area as an example,, establish a particle flow model and use the "trial and error method" to repeatedly calibrate the microscopic parameters until they are consistent with the actual situation,, uniaxial compression tests were performed on rock masses with different fissures, and 16 orthogonal discrete element numerical simulation experiments with 4 factors and 4 levels were carried out. Based on multiple linear regression analysis, the influence of four factors on the mechanical properties of granite was analysed, and the influence of single factor on the macro-mechanical properties of granite was studied through numerical experiments. The results show that the effect of the crack geometry on the peak strength of granite is greater than the effect on the elastic modulus of granite; The order of the influence degree of the crack geometric characteristic parameters on the peak strength is the crack Y-axis position component > the crack X-axis position component > the crack length > the crack inclination angle; The order of the influence degree of the crack geometric characteristic parameters on the elastic modulus is the crack X-axis position component > the crack Y-axis position component > the crack length > the crack inclination angle; The crack initiation position gradually moves to both ends of the crack as the angle increases. The specimen is mainly dominated by tension cracks, and the crack tip is accompanied by local shear failure; The longer the crack length, the lower the peak strength of the granite, and the worse the stability; The positional change of primary microcracks in granite mainly changes the peak strength and post-peak stress-strain state of the rock sample, but has no significant effect on the elastic modulus. The research results have certain reference value for predicting the mechanical properties of fractured granite.

Keyword: Rock mechanics; Particle flow; Microfractures; Peak strength

1. Introduction

In addition to pores, the surrounding rock of the coal mining face usually contains natural cracks or fissures of different sizes (such as faults, joints, micro-cracks, etc.). The existence of these discontinuous surfaces causes the mechanical properties of the rock to deteriorate to varying degrees. Many geotechnical buckling failures are caused by crack/crack propagation inside the rock, and the fracture characteristics and mechanical properties of the cracked rock mass usually determine the peak strength and failure mode of the rock mass. Therefore, it is of great engineering significance to study the relationship between the geometric characteristics of cracks in rock mass, the law of expansion after loading and the macroscopic strength of rock mass.

Many scholars at home and abroad have carried out a large number of uniaxial and triaxial compression experimental studies on natural rock masses with different lithologies or similar materials containing fractures, and accumulated many rich results in fracture propagation and rock failure in fractured rock masses. Zhang Guokai et al. [1] used high-strength gypsum-like materials to make specimens, used MTS rock mechanics testing machine to conduct uniaxial compression tests on rocks with single fractures, and used acoustic emission, sound waves and camera tests to obtain fracture initiation pressure, peak value. The strength is consistent with the change of the fissure inclination, showing a trend of first decreasing and then increasing. Xiao Taoli et al. [2] used high-strength silica fume mortar to make specimens, and carried out triaxial loading tests under different confining pressures, showing that the peak strength of rock mass will gradually decrease with the increase of single fracture length. Zhou Hui et al. [3] conducted direct shear tests using cement mortar rock materials with prefabricated fractures, and

 https://doi.org/10.1051/matecconf/202235801067

found that the longer the joint fracture, the lower the shear strength and cohesion of the rock, and for the joint fracture of the same length, its different positions in the rock have very different effects on the strength of the rock. When it is in the middle of the specimen, the shear strength and cohesion of the rock are the smallest, and the peak shear strength and cohesion of the rock are the largest when the joint crack is located at the rear end of the specimen. Liu Hongyan [4] conducted uniaxial compression tests on prefabricated jointed rock masses using similar materials, and found that with the increase of the number of parallel joint groups, the peak strength of the specimen gradually decreased. Ke Zhiqiang et al. [5] used cement mortar and other materials to make columnar jointed rock mass models with different column inclination angles and distribution of lateral joints. And through uniaxial compression test, the influence of column inclination and lateral joints on the anisotropic mechanical properties and failure mechanism of rock mass is studied. It is of scientific value to select materials similar to natural rock mechanics and properties to study rock crack propagation and failure modes. However, due to the proportion of various ingredients and various errors in the experiment, the results of experimental research using similar materials have great uncertainty. Therefore, many scholars use the method of prefabricated fractures in natural rock mass to carry out experimental research on rock mechanical characteristics. Guo Qifeng et al. [6] used GAW2000 rigidity testing machine to conduct uniaxial compression test on prefabricated single-fracture granite samples. It is concluded that the uniaxial compressive strength of the rock sample is the largest when the crack orientation is parallel to the loading direction, and the rock strength decreases with the increasing angle between the crack orientation and the loading direction. Yin Qian et al. [7] used YNS2000 universal servo press to conduct uniaxial compression test on sandstone samples with longitudinal fractures, and studied the effect of strength deterioration of sandstone. It is found that with the increasing of the prefabricated fissure offset, the bearing capacity of the specimen first increases, then decreases, and then increases again, indicating that the position of the fissure has an important influence on the rock failure mode.

With the development of computer numerical simulation technology, the combination of numerical simulation software and laboratory test is an effective method to analyze the failure mode of fractured rock mass. In the aspect of rock mass failure caused by crack propagation in fractured rock mass, the previous studies on cracks mostly focus on a single variable, while the comprehensive variable analysis of cracks shows less impact. Wang et al.[8] used RFPA numerical simulation software for rock fracture process to study the effect of original fracture length on fracture propagation and evolution under specific stress conditions.

2. Numerical model establishment of crack propagation and buckling failure in granite with fissures

The test rock samples were taken from the Sanguliu rock mass granite quarry in Sanguliu Village, Dandong City, Liaoning Province.The formation age of granite is Mesozoic, and its main mineral components are quartz, alkaline feldspar, plagioclase, and a small amount of biotite.The rock is unweathered, gray-white on the outside, dense, and has no obvious cracks on the surface, but the permeability test shows that it contains many cracks.

Fig.1 The granite samples obtained from the Sanguliu rock mass of Dandong city

The granite samples used in the test were cut into the standard size of rock mechanics, that is, Φ25 mm × 50 mm, as shown in Figure 1.Using the MTS815.03 rock triaxial testing machine, the triaxial compression experiments of three rock samples under several different confining pressures (5 MPa, 10 MPa and 20 MPa) were carried out.When the axial loading pressure is continuously increased to cause the rock to become unstable and fail, record the peak stress (that is, the compressive strength of the rock 1).The elastic modulus, cohesion and internal friction angle of the rock were obtained according to relevant calculations (Table 1).

Table 1 Mechanical properties of granite samples obtained from Dandong city

Sample serial number	E/(GPa)	1/(MPa)	3/(MPa)	Cohesion (MPa)	Internal friction angle (o)
1	23.465	242.5473	5		
2	23.5692	339.0023	10	19.65	62.68
3	26.06	498.9676	20		

2.1 Numerical Model Construction

Under the condition of external force,in order to intuitively observe the propagation and evolution of internal cracks,the failure process of rock instability, and the correlation between the geometric characteristics of cracks and the macro-mechanical properties of rock samples,in this paper, relevant numerical experiments are carried out using the particle flow numerical simulation software PFC2D.PFC2D is a numerical simulation software developed by ITASCA based on the principle of

 https://doi.org/10.1051/matecconf/202235801067

discrete element stress analysis. It is often used to simulate the instability and failure process of rocks, and can be applied to both microscopic damage and macroscopic damage of rock materials.Based on the physical and mechanical properties of granite, domestic and foreign scholars generally use the bonding model to study the instability and failure process of granite.The bonding model is divided into a contact bonding model and a parallel bonding model. The parallel bonding model can transmit forces and moments [9], and is suitable for materials with cement inside the rock, and the cemented part can also simulate the initiation and propagation of cracks.Therefore, the parallel bonding model is suitable for the study of granite cracks.In this paper, the parallel bond model is used as the constitutive model to study the expansion process of the newly-created cracks and the instability failure of the rock samples after the granite with primary cracks is loaded.The two-dimensional geometric model of the standard rock sample in this paper is set to be 100 mm long and 50 mm wide, and the uniaxial compression model is determined by deleting the "walls" on both sides.

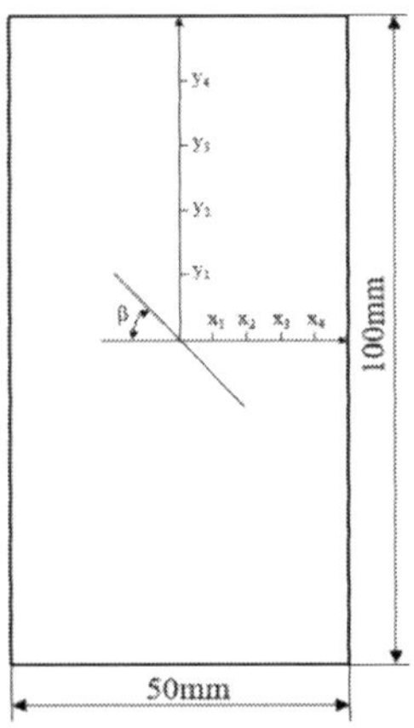

Fig.2 Geometric parameters of single cracked specimen

Fig.3 PFC2D simulation of granite particle arrangement

2.2 Microscopic parameters of the parallel bonding model

In the parallel bond model, the macro-mechanical properties of rock (peak strength, Poisson's ratio, elastic modulus) are mainly affected by microscopic parameters such as particle contact modulus, particle stiffness ratio, particle size, particle normal strength, tangential strength, parallel bond contact modulus, parallel bond stiffness ratio, etc[10]. The parallel bonding model has many meso-parameters. In order to reduce the difficulty of calibration, some assumptions are generally made to reduce the number of meso-parameters.The calibration of the meso-parameters is based on multiple laboratory tests, so as to obtain results that are more consistent with the macro-mechanical properties.

When determining the meso-parameters of the parallel bonding model, the following assumptions are followed [11-16]:

1.The uniaxial compressive strength is less than the shear strength, that is, $\sigma<\tau=c+\sigma\tan\varphi$, in order to obtain a suitable uniaxial tension-compression ratio.

2.$\gamma=1.0$

3. $R_{max}/R_{min}=1.66$

4.$\rho=2830$ kg/m3

5.$\mu=0.5$

6. $E_p=E_c$, $k_n/k_s=\bar{k}_n/\ \bar{k}_s$

The uniaxial compression test is carried out by the particle flow numerical simulation software PFC2D. The microscopic parameters are repeatedly modified by the "trial and error method" by comparing the macroscopic parameters of the indoor test.Finally, the macroscopic mechanical parameters obtained by the uniaxial compression numerical simulation test and the macroscopic mechanical parameters of the laboratory test have an error within 5%.The meso-parameters of the final parallel bonding model are shown in Table 2.

Table 2 PFC2D simulation meso parameters

Model height L/mm	Model width W/mm	Minimum particle radius Rmin/mm	Particle radius ratio Rmax/Rmin	Particle density ρ/(kg/m3)	Particle Contact Modulus Ec/GPa	Particle stiffness ratio kn/ks	Particle friction factor μ	Bond radius factor λ	Parallel Bond Modulus Ep/GPa	Parallel bond stiffness ratio $\bar{k}_n/\bar{k}_s$	Parallel bond normal stress σp/MPa	Parallel bond tangential stress τp/MPa
100	50	3	1.66	2830	15	2.5	0.5	1	15	2.5	15	15

3. Orthogonal model tests for particle flow simulation of fractured granite

3.1 Design of orthogonal simulation test scheme

As a rule of thumb, the orthogonal test can select uniform discrete levels and representative factors to design the experiment, which can greatly reduce the experimental workload.In addition, variance analysis can also be used to analyze the influence of different factors on the dependent variable. Therefore, this paper adopts the orthogonal test to study the influence of single-crack geometric characteristics on the macro-mechanical properties of granite.

When carrying out the orthogonal test of the geometric characteristic parameters of the crack, we adopted a design scheme with 4 levels and 4 factors (shown as Table 3). The 4 factors are the geometric characteristic parameters of the crack, and the 4 levels are the values of the average discrete distribution of each geometric parameter. The primary crack geometric feature parameters include: ① Crack inclination angle x1, ② Crack length x2, ③Crack X-axis position component x3, ④ Crack Y-axis position component x4;The macro-mechanical parameters of rock samples considered include:①peak strength y1, ②elastic modulus y2.

When the particle flow numerical simulation test is carried out, the correlation between the four factors is not considered, and the orthogonal table is used to obtain a specific test plan with four factors and four levels (show as Table 4).The inclination angle of the primary crack (that is, the angle between the crack direction and the X-axis direction) is set to four levels of 0, 30, 60 and 90 respectively; Considering the influence of specimen length, width and boundary mechanical effects, the crack lengths are set to four grades of 5, 10, 15 and 20 mm, respectively; Considering the axial symmetry of the specimen under uniaxial compression, a Cartesian coordinate system is established with the center of the specimen, which is divided into five equal parts from the origin of the X and Y axes to the boundary of the specimen, respectively. Take the middle 4 coordinate points as the values of the 4 levels of the X and Y axes, as shown in Table 4.Table 4 also lists the simulation results of macroscopic mechanical parameters (peak strength, elastic modulus) for each experiment.

Table 3 Orthogonal test level design table

factor / level	Angle x_1 (°)	Crack length x_2 (mm)	X-axis components x_3 (mm)	Y-axis component x_4 (mm)
1	0	5	5	10
2	30	10	10	20
3	60	15	15	30
4	90	20	20	40

Table 4 Granite particle flow orthogonal simulation test plan

Crack geometric feature parameters				Macro-mechanical parameters	
Angle x_1 (°)	Crack length x_2 (mm)	x-axis component x_3 (mm)	y-axis component x_4 (mm)	peak intensity y_1 (MPa)	Elastic Modulus y_2 (MPa)
0	5	5	10	48.5	30.1
0	10	20	30	37	28
0	15	10	40	30.7	26.7
0	20	15	20	29.7	22.2
30	5	15	30	41.1	29.1
30	10	10	10	40.2	27.2
30	15	20	20	38.8	27.4
30	20	5	40	29.8	23.5
60	5	20	40	39.6	28.1
60	10	5	20	47.9	29.8
60	15	15	10	37.2	26.4
60	20	10	30	32.6	28.1
90	5	10	20	51.2	31.2
90	10	15	40	50.9	31.1
90	15	5	30	50.8	30.8
90	20	20	10	50.3	30.9

3.2 Analysis of Numerical Simulation Test Results

Through the particle flow numerical simulation test, the macroscopic mechanical parameters of the rock are obtained, including①peak intensity y_1, ②elastic modulus y_2,as shown in Table 5.

Table 5. Results of macro parameters

number	peak intensity y_1 (MPa)	Elastic Modulus y_2 (MPa)
1	48.5	30.1
2	37	28
3	30.7	26.7
4	29.7	22.2
5	41.1	29.1
6	40.2	27.2
7	38.8	27.4
8	29.8	23.5
9	39.6	28.1
10	47.9	29.8
11	37.2	26.4
12	32.6	28.1
13	51.2	31.2
14	50.9	31.1
15	50.8	30.8
16	50.3	30.9

3.2.1 Analysis of variance

In this experiment, the effects of four mesoscopic parameters: ① crack inclination angle x1, ② crack length x2, ③ crack X-axis position component x3, and ④ crack Y-axis position component x4, on macro-mechanical parameters of rock are discussed.Therefore, a multivariate analysis of variance is required.

SPSS software was used for multivariate analysis of variance.First, the model mode is selected to be customized, the interaction between factors is ignored, and the main effect of each factor is analyzed.Secondly, the significance level of the orthogonal test in this paper is selected as $\alpha = 0.05$, and the analysis results are shown in Table 6-Table 7.

Table 6 Analysis of variance of peak intensity

source of variance	deviation sum of squares	degrees of freedom	sum of mean squares	F value	Sig.
Angle x_1 (°)	526.982	3	175.661	12.160	0.035
Crack length x_2 (mm)	230.437	3	76.812	5.317	0.102
X-axis components x_3 (mm)	71.092	3	23.697	1.640	0.347
Y-axis components x_4 (mm)	84.257	3	28.086	1.944	0.299
e	43.337	3	14.446		

Table 7. Analysis of variance of the elastic modulus

source of variance	deviation sum of squares	degrees of freedom	sum of mean squares	F value	Sig.
Angle x_1 (°)	47.627	3	15.876	3.036	0.193
Crack length x_2 (mm)	27.788	3	9.263	1.771	0.325
X-axis components x_3 (mm)	5.148	3	1.716	0.328	0.808
Y-axis components x_4 (mm)	7.448	3	2.483	0.475	0.722
e	15.688	3	5.229		

According to the magnitude of the F statistic, the primary and secondary order of the influence of each factor on the macro-mechanical parameters of rock can be judged.It can be seen from Table 5 that the order of the influence of the crack geometric characteristic parameters on the peak strength is: crack inclination angle > crack length > Y-axis position component > crack X-axis position component.Among them, the crack inclination angle is the most significant influencing factor, followed by the crack length, the Y-axis position component and the X-axis position component of the grain.Due to the crack length, the Y-axis position component and the X-axis position component of the pattern, the deviation sum of the above three parameters is not lower than the experimental error.It shows that each parameter has a certain effect on the peak intensity, but has no significant effect.It can be seen from Table 6 that the order of influence of the crack geometric characteristic parameters on the elastic modulus is: crack inclination angle > crack length > Y-axis position component > crack X-axis position component.The crack inclination angle and crack length have a certain influence on the elastic modulus, but the influence is not significant.

From the above variance analysis results, it can be seen that the influence of the four meso-parameters on the macro-parameters is not the same.The crack inclination angle is a significant factor affecting the strength characteristics of rock, and the main influencing factor is the crack length.

3.2.2 Regression analysis

According to the calculation results of the above numerical simulation experiments, the relationship between the macro parameters and the main significant factors can be established.Using the stepwise regression method, the probability threshold value of variable entry is 0.05, and the probability threshold value of deletion is 0.10.A significance level of $\alpha = 0.05$ was taken as the standard for the regression coefficient test.The significant influencing factor of the peak strength is the crack inclination angle, and the linear regression analysis is carried out to obtain the formula (1). Its fitting coefficient is R^2=0.648.

$$y_1=0.169x_1+34.295$$

It can be seen from the above formula that the fitting coefficient of linear regression is small, and the fitting effect is relatively poor.It shows that the relationship between the meso-parameters and the crack inclination is complex, and there may be a nonlinear relationship.What kind of nonlinear relationship exists will be univariate analysis in the follow-up research.

4. Mechanical characteristics of fractured granite under the influence of single factor

4.1 Experimental Design

From a single factor (such as length, dip angle and location) to study the effect of crack geometry on the

https://doi.org/10.1051/matecconf/202235801067

macroscopic mechanical properties of granite, the tests are divided into three groups:①The crack is located in the center of the specimen, the length of the crack is 20 mm, and the angle between the crack direction and the X axis is set to 0, 15, 30, 45, 60, 75 and 90 respectively;②The crack is located in the center of the specimen, the angle between the crack direction and the X axis is 0, 15, 30, 45, 60, 75, 90, and the crack lengths are changed to 10 mm, 15 mm, 20 mm, 25 mm, 30 mm, and 35mm;③ The angle between the crack direction and the X axis is 45 mm, the crack length is 15 mm, and 4 positions are set as the crack center points. In order to avoid the adverse effect of the boundary effect on the rock fracture, the cracks are arranged within a reasonable size range [17].

4.2 Influence of crack inclination angle on rock strength

The crack inclination angle has a significant effect on the initiation position and failure mode of the new crack.Figure 3 shows the generation and expansion of new cracks caused by uniaxial compression in the simulated granite under the condition of different crack inclination angles at a certain number of calculation steps.It can be seen from the figure that when the crack inclination angle is 0, the crack initiation position is in the middle of the prefabricated crack. With the progress of the axial loading,the new cracks start from the middle and continue to expand to both ends of the specimen along the direction of external load (that is, the direction of maximum principal stress).The specimen is mainly dominated by tension cracks, and the crack tip is accompanied by local shear crack failure.

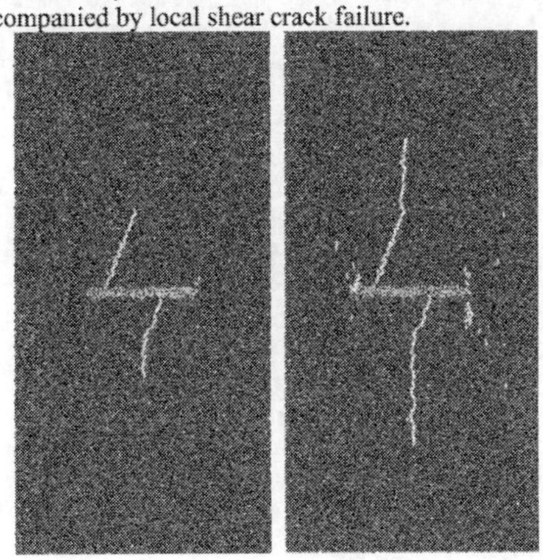

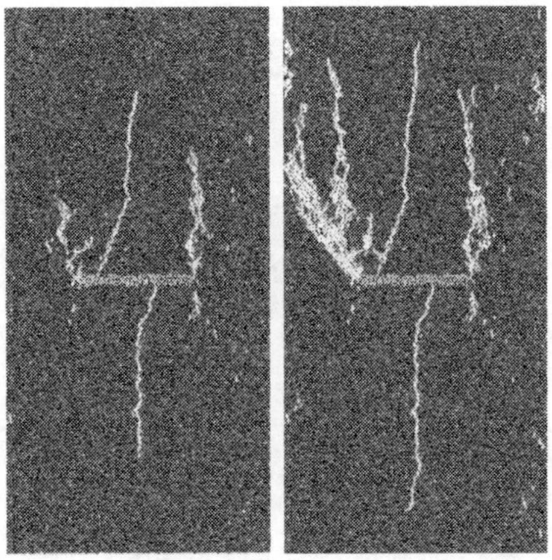

Fig.4 Uniaxial compression failure diagram of horizontal fracture

With the increase of the crack inclination angle, the crack initiation position of the new crack gradually shifts from the middle of the original crack to the crack tip.When the inclination angle of the primary crack is 15°, the crack initiation position is about 1/2 from the middle of the crack to the two ends. The new crack is dominated by tension and extends to both ends of the specimen along the external load direction;When the inclination angle of the primary crack is 30°, the crack initiation position is located at 2/3 of the middle of the primary crack, and the shear damage at the tip also increases gradually;When the inclination angle of the primary crack is 45°, the initiation position of the new crack is transferred to the tip of the primary crack, and the tensile failure and the local shear failure intersect each other;When the inclination angle of the primary crack is 60°, the specimen starts to be dominated by shear failure cracks, and accompanied by local tensile failure, the new cracks gradually expand along the tangential direction of the primary crack;When the inclination angle of the primary crack is 75°, an inclined shear crack penetrates the specimen along the direction of the primary crack and finally forms a fracture surface, which is a typical shear failure mode;When the inclination angle of the primary crack is 90°, a series of shear cracks are formed at the tip of the primary crack under the influence of shear stress, and the cracks gradually expand until they penetrate the boundary of the rock sample (Fig. 3a).

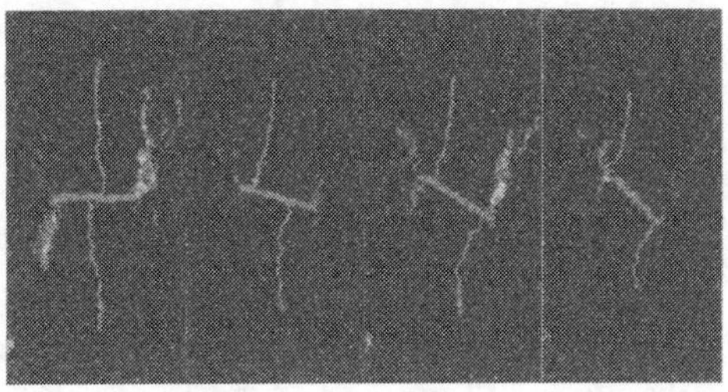

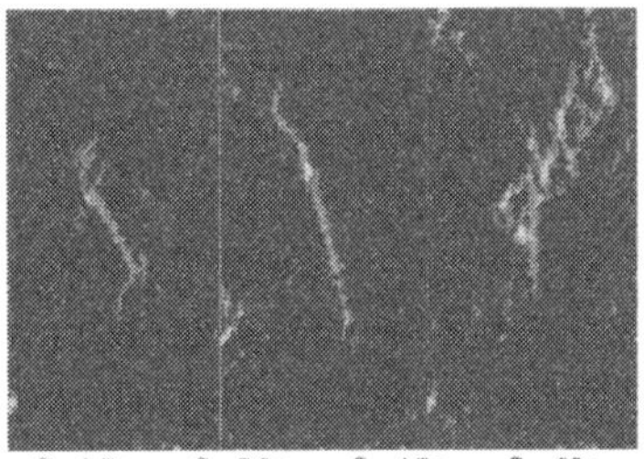

β=0° β=15° β=30° β=45° β=60° β=75° β=90°

Fig.5 Distribution of new cracks in rock samples with different dip angles of primary microcracks under uniaxial compression

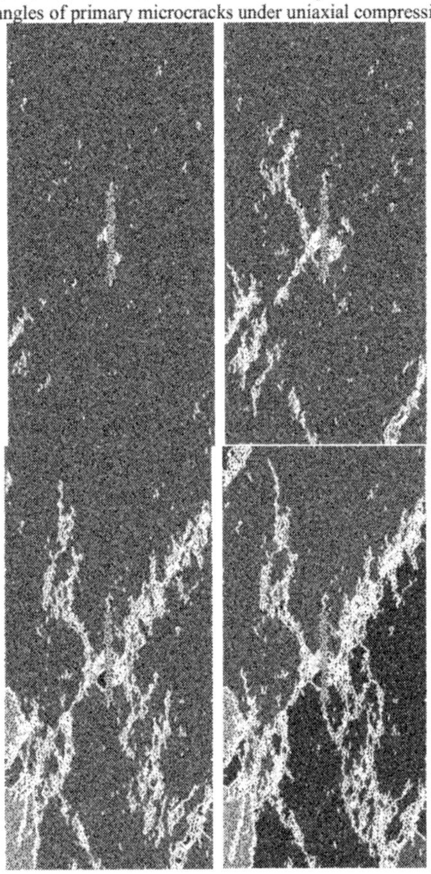

Fig.6 Uniaxial compression failure diagram of vertical fracture

Figure 4 shows the stress-strain relationship curve of rock caused by uniaxial compression under the condition of different inclination angles of primary cracks.It can be found that with the increase of the inclination angle of the primary crack in the rock, the uniaxial compressive strength of the rock increases continuously.When the inclination angle of the primary crack is 0°, the uniaxial compressive strength is the smallest, which is 29.6 MPa;Before the inclination angle of the primary crack increases to 30°, the upward trend of the peak value is relatively gentle;When the inclination angle increases to 45°, the peak value increases to 39.5 MPa, and the peak intensity is 15.8% higher than that of 30°;When the inclination angle of the primary crack increases from 45° to 90°, the peak strength increases to 53.4 MPa.Compared with the original crack inclination angle of 45°, the peak strength is increased by 35.2%.Compared with the obvious increase from 0 to 45°, it can be seen that45 is the dividing point of the peak intensity increase rate.When the inclination angle of the primary crack is less than 45°, the peak strength of the rock rises gently. After the inclination angle of the primary crack reaches 45°, the peak strength of the rock increases rapidly, which is consistent with the research of other scholars.

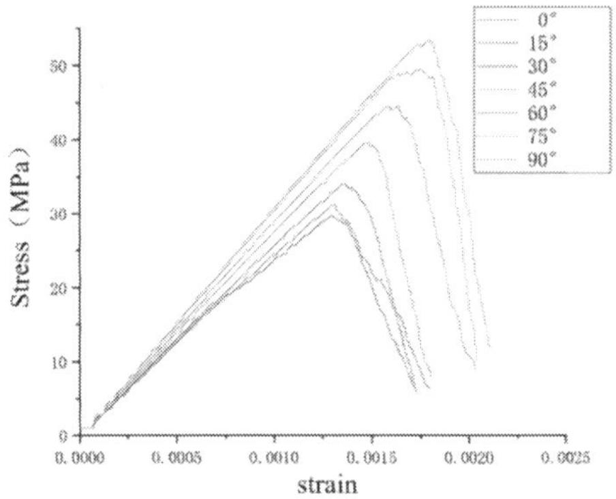

Fig.7 Stress-strain curves of specimens with cracks at different inclination angles

4.3 Influence of primary crack length and dip angle on rock strength

Fig. 5 shows the uniaxial compressive strength curves of rock samples with different dip angles and lengths of primary cracks.It can be seen from the figure that under the condition that the inclination angle of the primary crack remains unchanged, except for the case where the inclination angle is 90°, the peak strength of the rock sample basically shows a decreasing trend with the increase of the primary crack length.For example, when the inclination angle of the primary crack is 0°, the peak strength of the rock sample is reduced by 81.2% when the primary crack length is 35 mm and the crack length is 10 mm.However, when the inclination angle of the primary crack is 90°, that is, when the direction of the crack is consistent with the direction of force loading, the peak strength of the rock sample basically does not change with the increase of the crack length.The inclination angle of the primary crack is positively correlated with the peak strength of the rock sample, which means that the larger the crack inclination angle, the higher the peak strength of the rock sample.When the primary crack length is 10 mm, the peak strength of the rock sample when the crack inclination angle is 0 is 38 MPa,however, under the condition of the same crack length, when the crack inclination angle is 0°, the peak strength of the rock sample is 55 MPa.It can be seen that the inclination angle of the primary crack and the length of the crack in the rock sample have a greater influence on the peak strength of the rock sample. The smaller the crack inclination angle and the longer the length, the lower the peak intensity of the rock sample.

 https://doi.org/10.1051/matecconf/202235801067

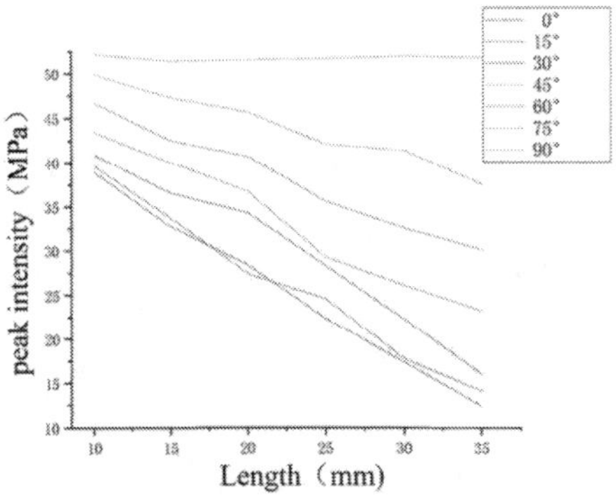

Fig. 8 Variation curve of peak strength of specimen under cracks with different inclination angles and lengths

4.4 Influence of crack position on rock peak strength

Since the uniaxial compression test is axisymmetric, a numerical simulation experiment of the failure evolution of rock samples under uniaxial compression conditions with primary cracks at four different positions is designed in this paper (Fig. 9). That is, the midpoint coordinates of the primary crack are located at point 1 (x_4, y_2), point 2 (x_2,y_2), point 3 (x_4,y_4), and point 4 (x_2,y_4).

Figure 7 shows the peak intensity of the rock samples when the primary cracks are in 4 different positions, and the difference between them is relatively small, ranging from 2 to 4 MPa.Compared with the peak intensity of relatively complete rocks, the intensity decreases between 23% and 31%, and the peak intensity generally shows a trend of high in the middle and low in the surrounding area.From the point of view of the failure form of the rock sample, when the primary crack is close to the edge of the rock sample (point 3), the crack is more likely to penetrate to the boundary, resulting in a significant decrease in the peak strength of the rock sample.

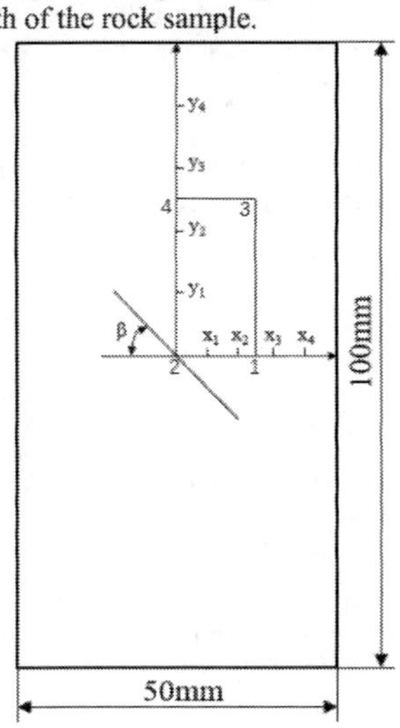

Fig.9 Schematic diagram of crack center position distribution

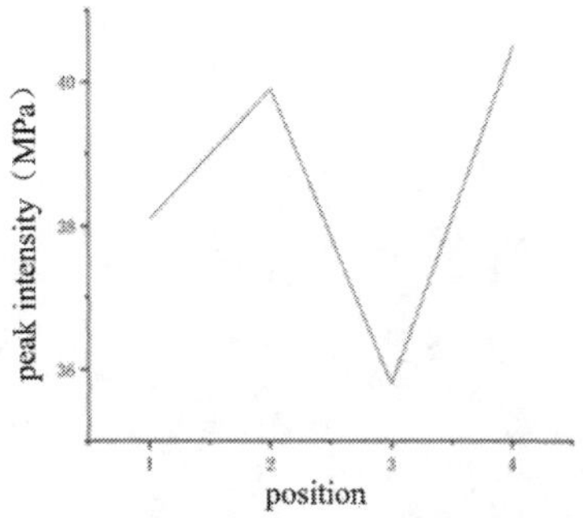

Fig.10 Peak intensity change curves at different positions

Figure 11 shows the effect of primary crack location on the stress-strain relationship of rock samples in uniaxial compression experiments.It can be seen that the location of the primary crack mainly changes the peak intensity of the rock sample and the stress-strain state after the peak, but has no significant effect on the elastic modulus of the rock sample (that is, the slope of the straight line in the stress-strain relationship curve).

The characteristics of the stress-strain relationship curve are mainly as follows:1) When the primary crack is located in the middle of the rock sample (point 2 and point 4), there is only one peak strength, and the stress of the rock sample falls back rapidly after the peak, which is consistent with the stress-strain relationship curve of the classical general rock;2) When the primary crack is located at the edge of the rock sample (such as point 3), the post-peak stress of the rock sample falls back slowly.And after the stress reaches the first peak strength (that is, the uniaxial compressive strength), there is a small increase after the post-peak stress has a large fall back; When the primary crack is located at the lateral boundary in the middle of the rock sample (such as point 1).After the stress first reached its peak, it did not fall back rapidly, but with the continuous fluctuation of the stress, a larger peak strength appeared than that in the linear elastic stage.where the second peak strength is the uniaxial compressive strength.

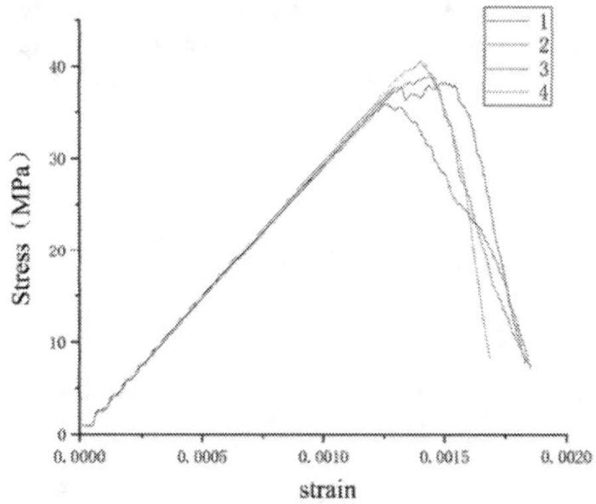

Fig.11 Stress-strain curves of specimens at different positions

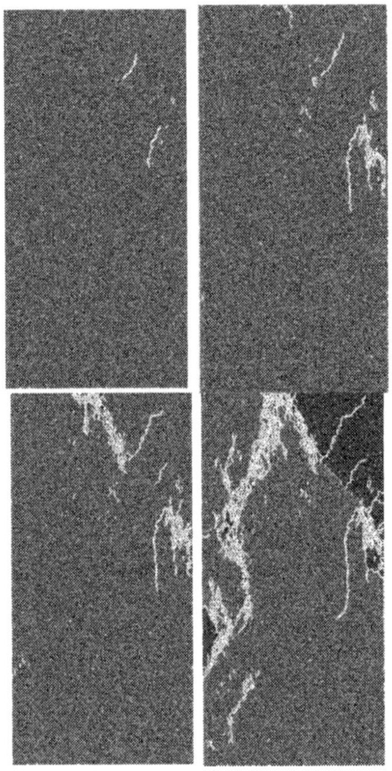

Fig.12 The fracture process of fractured rock at the same location

5. Conclusion

When the crack inclination angle is 0°, the crack initiation position is in the middle of the prefabricated crack, and gradually moves to both ends of the crack as the angle increases.With the continuous loading, the crack starts from the middle and expands to both ends of the specimen along the direction of external load. The specimen is mainly composed of tension cracks, and the crack tip is damaged by local shear cracks;The longer the crack length, the lower the peak strength and the worse the stability.The change of crack position mainly changed the peak strength and post-peak stress-strain state of the rock sample, but had no obvious effect on the elastic modulus.
2) The influence of the crack geometric characteristic parameters on the peak strength is as follows: the position component of the Y axis of the crack > the position component of the X axis of the crack > the length of the crack > the inclination angle of the crack. The influence of the position offset in the normal direction of the loading axis, and the influence of the position offset along the crack (X, Y axis) on the peak strength are also much greater than the effects of the crack length and inclination angle.
3) The influence of the crack geometric characteristic parameters on the elastic modulus is as follows: the position component of the X axis of the crack > the position component of the Y axis of the crack > the length of the crack > the inclination angle of the crack, and the position offset of the crack along the normal direction of the loading axis affects the elastic modulus It is slightly larger than the influence of the position offset of the crack along the loading axis, and the influence of the position offset of the crack (X, Y axis) on the elastic modulus is also much greater than that of the crack length and inclination angle.

References

1. Zhang Guokai, Li Haibo, Wang Mingyang, Li Xiaofeng. Study on crack propagation characteristics of single-fracture rock based on acoustic test and camera technology [J]. Geotechnical Mechanics, 2019, 40(S1): 63-72+81.
2. Xiao Taoli, Li Xinping, Jia Shanpo. Triaxial test study and mechanical analysis of structural plane effect of deep single-fracture rock mass [J]. Chinese Journal of Rock Mechanics and Engineering, 2012, 31(08): 1666-1673.
3. Zhou Hui, Meng Fanzhen, Zhang Chuanqing, Lu Jingjing, Xu Rongchao. Experimental study on the effect of fractures of different positions and sizes on rock mass failure [J]. Chinese Journal of Rock Mechanics and Engineering, 2015, 34(S1): 3018-3028.
4. Liu Hongyan, Huang Yushi, Li Kaibing, Zhang Jihong. Experimental study on strength and failure mode of prefabricated jointed rock mass specimens [J]. Rock and Soil Mechanics, 2013,34(05):1235-1241+1246.
5. Ke Zhiqiang, Wang Huanling, Xu Weiya, Lin Zhinan, Ji Hua. Experimental study on mechanical properties of columnar jointed rock mass with lateral joints [J]. Rock and Soil Mechanics, 2019, 40(02): 660-667.
6. Guo Qifeng, Wu Xu, Cai Meifeng, Xi Xun, Ren Fenhua, Miao Shengjun. Strength characteristics and failure mode test of prefabricated fractured granite [J]. Journal of Engineering Science, 2019, 41(01): 43-52.
7. Yin Qian, Jing Hongwen, Su Haijian, Zhao Honghui. Strength degradation and loading rate effect of sandstone with longitudinal fractures [J]. Chinese Journal of Mining and Safety Engineering, 2016, 33(01): 128-133.
8. Wang Guoyan, Yu Guangming, Song Chuanwang. Influence of geometric elements of initial fissures on the evolution of rock fissures [J]. Journal of Liaoning University of Engineering and Technology (Natural Science Edition), 2011, 30(05): 681-684.
9. Shi Chong, Xu Weiya. Numerical simulation skills and practice of particle flow [M]. Beijing: China Construction Industry Press, 2015
10. Huang Yanhua, Yang Shengqi. Particle flow simulation of macro-mechanical behavior of non-coplanar double-fractured red sandstone [J]. Chinese Journal of Rock Mechanics and Engineering, 2014, 33(08): 1644-1653.
11. Chen Pengyu. Research status of PFC~(2D) simulation of crack propagation characteristics of

fractured rocks [J]. Chinese Journal of Engineering Geology, 2018, 26(02): 528-539.

12. Yuan Guangxiang, Zhang Luqing, Zeng Qingli, Li Jianyong, Huang Zhiquan, Wang Hongjian, Deng Xubiao. Study on the correlation between mineral composition of granite and its uniaxial compressive strength [J]. Hydrogeology and Engineering Geology, 2018, 45(05): 93-100.
13. Budong Yang.Yue Jiao, Shuting Lei. A study on the effects of microparameterson macroproperties for specimens created by bonded particles[J]. EngineeringComputations: International Journal for Computer-Aided Engineering and Software,2006,23(6).
14. D.O. Potyondy, P.A. Cundall. A bondedparticle model for rock[J]. International Journal of Rock Mechanics and Mining Sciences, 2004, 41(8).
15. Zhao Guoyan, Dai Bing, Ma Chi. Study on the influence of mesoscopic parameters on macroscopic properties in the parallel bonding model.
16. Jiang Mingjing. Fang Wei, Sima Jun. Microparameter calibration of parallel bonding model for simulated rock [Xiao.Journal of Shandong University (Engineering Edition). 2015,45(04):50-56.
17. Fu Zhen, Cai Yongen. Numerical analysis of the effect of rock specimen size on crack stress field [J]. Chinese Journal of Seismology, 2005(03):301-308+355.

Microwave-assisted Synthesis of Greigite Fe_3S_4 nanosheets wrapped in an rGO Matrix as anode material for Sodium-ion batteries

Ikhwan Choi, Bolin Liu, Youqi Zhu and Chuanbao Cao *

School of Materials Science and Engineering, Beijing Institute of Technology, Beijing, China

*Corresponding author e-mail: cbcao@bit.edu.cn

Abstract. Greigite (Fe_3S_4), which have ferromagnetic in inverse thiospinel (AB_2S_4), is widely researched to use an adsorbent and biomedical field because non-toxicity and abundant in nature. Iron-based materials are known to have a high theoretical capacity because of their multivalent state including redox pairs, but still suffer from collapse and aggregate during the charge/discharge process. Here, the synthesized Fe_3S_4 nanosheet structure materials wrapped with reduced graphene oxide (Fe_3S_4NSs@rGO) were used as an anode electrode material for sodium-ion batteries (SIBs). The nano-sheet structure facilitates ion diffusion through expanded surface area, and rGO can effectively improve electrochemical conductivity and structure stability. As-prepared Fe_3S_4NSs@rGO were used as a host material to insert Na-ion via a conversion process, and the stabilized structure maintains the high capacity and long cycle performance. Thus, the Fe_3S_4NSs@rGO deliver a reversible capacity of 950 mAh g^{-1} after 200 cycles at a current density of 1A g^{-1} and 524 mAh g^{-1} after 400 cycles at a current density of 2A g^{-1}, which is much higher than reported materials.

1. Introduction

Research on renewable energy technologies has been continually developed in response to worries about fossil fuels and environmental problems, which are restricted in their availability. Many new alternative energy production technologies, such as, nuclear power, hydroelectricity, and solar power, are being introduced into the marketplace. Additionally, research on energy conversion devices for effectively storing power generated as a result of technical advancements is ongoing. [1-5] The battery industry, which was founded years ago on the findings of Professor John Goodenough and his team of lithium-ion battery researchers, has been able to get one step closer to achieving a clean energy future that is reliable and efficient by collaborating with smart grids. [6]

Sodium metal, the world's fourth-largest resource under the surface of the earth, has the potential to be used in a next-generation battery. Li^+ and Na^+ have similar electrochemical and chemical properties. In terms of the Standard Hydrogen Electrode (SHE), there is only a small gap of about 300 mV; Na (-2.71V vs SHE), Li (-3.04V vs SHE). [7] High-power performance, low-cost, ease of demand and supply, adaptability to operate in a variety of climatic conditions, and the possibility of free recycling in the face of environmental pollution and ecological influences are all factors that contribute to the compatibility of both sodium-ion batteries and energy storage system (ESS) industries. [8-10] The kinetics of Na^+ is related to the rate performance of the SIBs, which is defined by ion diffusion and electron conductivity. [11, 12] In light of the critical role the anode material plays in sodium energy storage performance, much effort must be expended to achieve outstanding reversibility, sodium dynamics, and a large intrinsic capacity.[13]

Iron sulfide is unique among Transition metal chalcogens (TMCs) in terms of its cheap price, abundant reserves, and lack of toxicity. Numerous investigations have been conducted to determine the electrochemical characteristics and performance of FeS and FeS_2. [14-17] However, electrochemical studies into Fe_3S_4 are inadequate. In recent years, the nanostructure design has been widely investigated as an effective strategy to solve poor capacity retention and cycling problems by shortening the diffusion path and allowing access to electrolytes with a larger surface area. [18-20] Qingjie Zhang et al., obtained nanocrystal Fe_3S_4 anode materials which have a specific capacity of 548 mAh g^{-1} at 0.2 A g^{-1} via solvothermal synthesis method. [21] However, the capacity performance of nanocrystal Fe_3S_4 needs further develop.

Herein, A low-cost, high-performance anode material is obtained through the synthesis of an Fe_3S_4 composite. Reduced graphene oxide (rGO) is widely used in many studies as a promising compound in SIBs to reinforce electrochemical and cyclic performances. [22-24] When the rGO is doped onto the anode material, the original structure of the material is still maintained and the rGO matrix prevents volume expansion during charge and discharge process, resulting in superior storage dynamics. [25] There are no report on the electrochemical

* Corresponding author: cbcao@bit.edu.cn

performance studies of the sodium-ion battery using the Fe_3S_4 nanosheets structure with rGO composite as the anode material. A Fe_3S_4NSs@rGO compound was obtained by microwave-assisted synthesis. Through simple microwave-assisted synthesize, a long reaction time of 10 hours or more, which requires a conventional hydrothermal method, was drastically reduced to within 10 minutes. The Fe_3S_4NSs@rGO composites deliver a capacity of 950 mAh g^{-1} even after 1A g^{-1}, 200 cycles at a wide operating voltage of 0–3V and were able to reach a higher capacity of 524 mAh g^{-1} compared to the existing discharge capacity of Fe_3S_4NSs of 365.8 mAh g^{-1}.

2. Experimental

2.1 Preparation of Fe_3S_4NSs and Fe_3S_4NSs@rGO composites

The reagent used in the experiment is analytical grade, and no further purification was performed during the experiment. Graphene oxide (GO) powders were synthesized by a modified Hummers process by coating natural graphite flake. [26] 0.05g of GO powder was dissolved in 10ml ethylene glycol solvent and stirred for 30 minutes under ultrasonic conditions. Combine 16mmol of $FeCl_3$ and 3mmol of thiourea in an ethylene glycol solvent and agitate until the solution turns transparent yellow. After mixing GO with the solution, agitate for another 10 minutes and heat at 155 °C for 9 minutes in a microwave assisted heating system (MARS 6CEM Corporation, USA). The microwave has been set to 500W. The sediment formed during the heating process was washed several times with deionized water and ethanol through a centrifuge (7500 rpm, 3 min), and then obtained The Fe_3S_4@NSs composites through evaporating residual ethanol. The Fe_3S_4@NSs compound is synthesized in the same method without adding GO in the process.

2.2 Characterization.

The crystal structure of Fe_3S_4NSs@rGO was characterized by the X-ray diffractometer (XRD) of Bruker D8 and proceeded at a scan rate of 5° min^{-1}. To examine the surface elemental and valence state distribution of Fe_3S_4NSs@rGO, X-ray photoelectron spectroscopy (XPS) was conducted using the PHI Quantera model. The structures and sizes of samples were examined using a scanning electron microscope (SEM, Hitachi S-4800). EDAX PW9900 energy dispersive spectrometry (EDS) was examined to analyze the distributions of Fe, S, and C elements in Fe_3S_4NSs@rGO. The morphologies of prepared Fe_3S_4NSs@rGO were examined using JEM-2100F TEM equipment.

2.3 Electrochemical Measurements.

The electrochemical properties of the prepared sample were irradiated through coin-type CR2016 half-cells. As-prepared product was mixed in an 8:1:1 weight ratio with Fe_3S_4NSs@rGO, super p conductive addictive and poly(vinylidene fluoride) binder, and N-methyl-2-pyrrolidone (NMP) was used as a solvent. Electrode was manufactured through a slurry coating process, Cu-foil was used as the current collector. It was dried in an oven at 80°C for 24h to eliminate the residual solvent. The amount of active material coated on Cu-foil was ~1.3 mg. The dried electrodes were sliced into disks measuring 1.1 cm. Sodium metal was used as a counter electrode, and the electrolyte was about 0.3mL of prepared $NaPF_6$ (Aladdin≥ 99.99%) in DEC/EC with a molar ratio of 1:1. The separator was made of glass fiber (Whatman, Grade GF/D), and the coin cell was assembled in an Ar-filled glove box with less than 1 ppm of water and oxygen content. Galvanostatic measurements were performed using a CT2001A Land battery measuring instrument within a voltage window of 0–3 V and at various current densities. On a CHI 660E electrochemical workstation, cyclic voltammetry (CV) was performed at room temperature. Through the same process, Fe_3S_4NSs also tested electrochemical measurements.

3. Results and Discussion

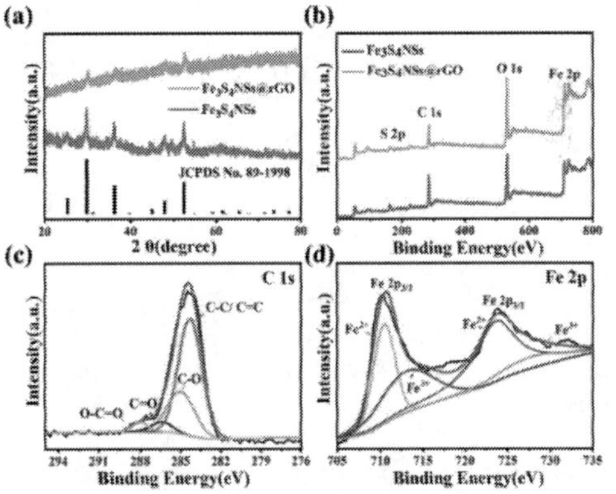

Figure 1. (a) XRD patterns of the Fe_3S_4NSs and Fe_3S_4NSs@rGO composites. (b) Total XPS spectra of the Fe_3S_4NSs and Fe_3S_4NSs@rGO composites. (c-d) High-resolution spectrum of C 1s and Fe 2p.

In this study, thiourea interacts stable with Fe^{2+} to provide sulfur and regulate its structure. The Fe_3S_4 nanosheet structure was prepared through microwave-assisted method. The crystal structure of Fe_3S_4NSs and Fe_3S_4NSs@rGO were characterized as shown Figure 1a. The strong peaks at 29.94°, 36.34° and 52.3° can be attributed Fe_3S_4 cubic phase (JCPDS No. 89-1998) with the face-centered cubic unit composed of 24 iron atoms and 32 sulfur atoms and small crystal sizes can be identified through the presence of broad diffraction peaks. There is no difference between Fe_3S_4NSs and Fe_3S_4NSs@rGO peaks, indicating that the insertion of rGO has no effect on the structure of Fe_3S_4.

Figure 1b-d illustrate the XPS analysis to validate the valence state and element composition of the Fe_3S_4NSs@rGO composites. In Figure 1b, C, O, Fe, and S elements are detected in both spectrums of the Fe_3S_4NSs and Fe_3S_4NSs@rGO. At presented in Figure 1c, C 1s spectrum can be deconvolved into four peaks of 288, 286.4, 285.1, and 284.3 eV, which correspond to O=C–O,

C=O, C–O, and C–C/C=C bonds, respectively.[27] In Figure 1d, two pairs major peaks, Fe $2p_{3/2}$ and Fe $2p_{1/2}$, are deconvolved to correspond the elements of Fe^{2+} at 710.5, 723.9eV peaks and Fe^{3+} at 713.9 and 727.4eV peaks. [28-30]

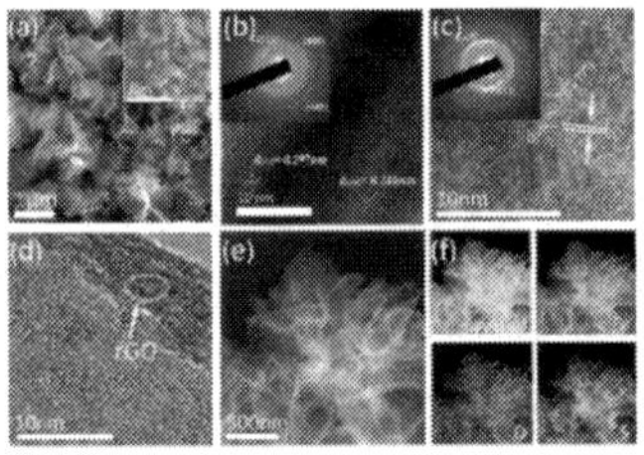

Figure 2 Structural characterization of the pure Fe_3S_4NSs and Fe_3S_4NSs@rGO composites. (a) SEM images of the Fe_3S_4NSs and inside Fe_3S_4NSs@rGO composites. (b) HRTEM images of the pure Fe_3S_4NSs with inside SAED. (c) HRTEM image of the Fe_3S_4NSs@rGO. (Inset: SAED pattern) (d-e) TEM images of Fe_3S_4NSs@rGO. (f) EDS mapping of the Fe_3S_4NSs@rGO.

The size and morphology of Fe_3S_4NSs and Fe_3S_4NSs@rGO were analyzed by SEM, TEM, HRTEM, and mapping images (Figure 2). As shown in Figure 2a, both the Fe_3S_4NSs and Fe_3S_4NSs@rGO exhibit the same 2D ultra-thin nanosheet morphology. The nanosheet structures with a large surface area provide sufficient contact area with electrolyte and reaction sites, which can shorten the diffusion path of sodium ions. [31] SEM images show that Fe_3S_4NSs@rGO has smaller particle size compared to pure Fe_3S_4NSs, confirming that the rGO controls the size of the nanosheet by suppressing excessive growth of the structure due to the interaction between the matrix of Fe_3S_4 which can be contributed to the reduction of GO. [28, 32] Figure 2b indexes the (311), (400), and (440) planes of the Fe_3S_4 via the selected area electron diffraction (SAED) pattern and provides the same diffraction pattern in HRTEM by measuring the distance d=0.297 nm of (311) plane and the d=0.246 nm of (400) plane. Figure 2c also illustrates the same plane of the greigite Fe_3S_4 in SAED pattern, but the distance (d=0.261nm) in the (400) plane of the Fe_3S_4NSs@rGO was extended compared with the existing 0.246 nm which can boost the sodium insertion kinetics. The presence of rGO could be confirmed through Figure 2d. COOH, COH, etc., which are the main functional groups of rGO, are easily coupled with Fe^{2+} ions. In addition, rGO has a strong surface interaction and buffers the volume expansion of Fe_3S_4 during the sodiation/desodiation processes.[33] Figure 2e–f prove that Fe and S elements are uniformly distributed.

In Figure 3a, the Fe_3S_4NSs@rGO is scanned at a wide voltage range of 0–3 V and 0.5mV s^{-1} rate to confirm the CV plot. The same peaks are found in both the first and second cycles. Peaks were found at charge curve in 1.57V and 1.95V. In the discharge curves, 1.52, 0.71, and 0.44V were measured, respectively. The two cycle curves overlapped each other showing that the charging/discharging process maintains a stable curve after the initial cycle. In Figure 3b, the charge/discharge curve of Fe_3S_4NSs@rGO can be confirmed. The Fe_3S_4NSs@rGO exhibits the initial discharge capacity of 1418 mAh g^{-1} and the charge capacity of 914.4 mAh g^{-1} with 64.83 % coulombic efficiency at 1 A g^{-1}. A charge capacity lower than the discharge capacity resulted from Na^{+} ions with other irreversible decomposition reactions in the anode in the solid electrolyte interphase (SEI) formation process. [34, 35]

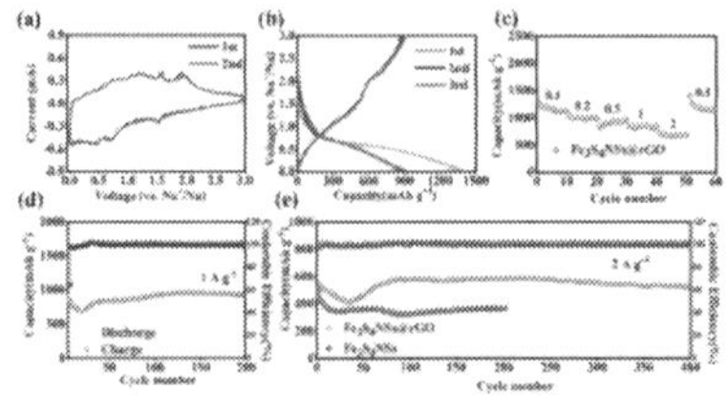

Figure 3. Electrochemical performance of Fe_3S_4NSs@rGO composites compared with pure Fe_3S_4NSs. (a) The CV curves of Fe_3S_4NSs@rGO at scanning rate of 0.5mVs^{-1}. (b) Galvanostatic charge-discharge curve. (c)Rate performance in the voltage range of 0-3V (d) Cycling performance with columbic efficiency at 1A g^{-1}. (e)Cycling data between pure Fe_3S_4NSs and Fe_3S_4NSs@rGO.

In the second and third cycles, discharge capacities of 919 mAh g^{-1} and 899 mAh g^{-1} could be maintained. The rate performance of Fe_3S_4NSs@rGO can be confirmed in Figure 3c. The Fe_3S_4NSs@rGO delivers specific capacities of 1258.7, 1020.9, 821.4, 847.2, and 727.1 mAh g^{-1} at 0.1, 0.2, 0.5, and 1 and 2 A g^{-1} current densities. After that, when it returns to 0.1 A g^{-1} current density, the capacity returns to a high capacity of 1405.8 mAh g^{-1} for excellent rate and reversible performance. To further measure the cycle stability, an attractive capacity of 950 mAh g^{-1} with a stable coulombic efficiency of 100% is obtained at 1 A g^{-1} current density after 200 cycles as shown in Figure 3d. Due to the strong electrical conductivity and mobility of rGO, the electrode's contact resistance decreases, facilitating the rapid removal of electrons. [28] Additionally, the incremental specific capacities in following cycles occur from the formation of a polymeric gel-like coating in the electrode, which is attributed to kinetic activation. [36] Figure 3e shows the long-term cycle performance of Fe_3S_4NSs and Fe_3S_4NSs@rGO at a relatively high current of 2 A g^{-1}. Fe_3S_4NSs shows a capacity of 365.8 mAh g^{-1} after 200 cycles, but Fe_3S_4NSs@rGO shows a capacity twice as high as 524 mAh g^{-1} even after 400 cycles. The increased discharge capacity of Fe_3S_4NSs@rGO is mostly due to the linked rGO network delaying the aggregation of of Fe_3S_4 and reduces the diffusion path for Na-ion shuttling.[23]

4. Conclusion

In summary, Fe_3S_4NSs@rGO was obtained using simple microwave-assist synthesis. The effect of Fe_3S_4NSs@rGO on the structural and electrochemical properties was evaluated compared with the pure Fe_3S_4NSs. Morphological and structural results showed that rGO is successfully composited with Fe_3S_4NSs while maintaining the structure integrity, and the rGO complex is evenly dispersed and doped into the prepared material.

The rGO complex did not cause additional structural deformation and demonstrated that the interaction between the matrixes between Fe_3S_4NSs improves electrical conductivity and the cycle performance of the electrodes. It shows excellent capacities of 1258.7, 1020.9, 821.4, 847.2, and 727.1 mAh g^{-1} at 0.1, 0.2, 0.5, 1, and 2 A g^{-1}. Especially, it maintaines a capacity of 950 mAh g^{-1} even after 200 cycles at 1 A g^{-1} and shows competitive electrochemical performance compared to pure Fe_3S_4NSs.

5. Conflicts of interest

There are no conflicts to declare.

References

1. FENG M, LU Y, YANG Y, et al. Bioinspired greigite magnetic nanocrystals: chemical synthesis and biomedicine applications [J]. Scientific Reports, 2013, 3(1).
2. ROBERTS A P, CHANG L, ROWAN C J, et al. Magnetic properties of sedimentary greigite (Fe3S4): An update [J]. Reviews of Geophysics, 2011, 49(1).
3. WANG X, CAI W, WANG G, et al. One-step fabrication of high performance micro/nanostructured Fe3S4–C magnetic adsorbent with easy recovery and regeneration properties [J]. CrystEngComm, 2013, 15(15).
4. ZHENG J, CAO Y, FU J-R, et al. Facile synthesis of magnetic Fe3S4 nanosheets and their application in lithium-ion storage [J]. Journal of Alloys and Compounds, 2016, 668: 27-32.
5. ARMAND M, AXMANN P, BRESSER D, et al. Lithium-ion batteries – Current state of the art and anticipated developments [J]. Journal of Power Sources, 2020, 479.
6. K. Mizushima, P. C. Jones, P. J. Wiseman, J. B. Goodenough, LixCoO2 A new cathode material for batteries of high energy density [J] Mater. Res. Bull. 1980, 15, 783. [J].
7. Claude Delmas, Jean-Jacques Braconnier, et al. Electrochemical intercalation of sodium in NaxCoO2 bronzes[J]. Solid State Ionics,1981, 3(4): 165-169. [J].
8. LIU T, ZHANG Y, JIANG Z, et al. Exploring competitive features of stationary sodium ion batteries for electrochemical energy storage [J]. Energy & Environmental Science, 2019, 12(5): 1512-33.
9. TAPIA-RUIZ N, ARMSTRONG A R, ALPTEKIN H, et al. 2021 roadmap for sodium-ion batteries [J]. Journal of Physics: Energy, 2021, 3(3).
10. PALOMARES V, CASAS-CABANAS M, CASTILLO-MARTÍNEZ E, et al. Update on Na-based battery materials. A growing research path [J]. Energy & Environmental Science, 2013, 6(8).
11. LAN T, MA Q, TSAI C L, et al. Ionic Conductivity of Na3V2P3O12 as a Function of Electrochemical Potential and its Impact on Battery Performance [J]. Batteries & Supercaps, 2020, 4(3): 479-84.
12. ZHOU P, LIU X, WENG J, et al. Synthesis, structure, and electrochemical properties of O′3-type monoclinic NaNi0.8Co0.15Al0.05O2 cathode materials for sodium-ion batteries [J]. Journal of Materials Chemistry A, 2019, 7(2): 657-63.
13. YE L, WANG W, ZHANG B, et al. Regeneration of well-performed anode material for sodium ion battery from waste lithium cobalt oxide via a facile sulfuration process [J]. Materials Today Energy, 2022, 25.
14. MWIZERWA J P, ZHANG Q, HAN F, et al. Sulfur-Embedded FeS2 as a High-Performance Cathode for Room Temperature All-Solid-State Lithium-Sulfur Batteries [J]. ACS Appl Mater Interfaces, 2020, 12(16): 18519-25.
15. SHEN Y, ZHANG Q, WANG Y, et al. A Pyrite Iron Disulfide Cathode with a Copper Current Collector for High-Energy Reversible Magnesium-Ion Storage [J]. Adv Mater, 2021: e2103881.
16. WU J, LIANG Y, BAI P, et al. Microwave-assisted synthesis of pyrite FeS2 microspheres with strong absorption performance [J]. RSC Advances, 2015, 5(80): 65575-82.
17. ZHANG S S. The redox mechanism of FeS2 in non-aqueous electrolytes for lithium and sodium batteries [J]. Journal of Materials Chemistry A, 2015, 3(15): 7689-94.
18. WANG W, LI W, WANG S, et al. Structural design of anode materials for sodium-ion batteries [J]. Journal of Materials Chemistry A, 2018, 6(15): 6183-205.
19. LIU Q, HU Z, ZOU C, et al. Structural engineering of electrode materials to boost high-performance sodium-ion batteries [J]. Cell Reports Physical Science, 2021, 2(9).
20. WANG L, LIU B, ZHU Y, et al. General metal-organic framework-derived strategy to synthesize yolk-shell carbon-encapsulated nickelic spheres for sodium-ion batteries [J]. J Colloid Interface Sci, 2022, 613: 23-34.
21. LI Q, WEI Q, ZUO W, et al. Greigite Fe3S4 as a new anode material for high-performance sodium-ion batteries [J]. Chem Sci, 2017, 8(1): 160-4.
22. WANG Z, ZHANG Y, PENG H, et al. Engineering kinetics-favorable 2D graphene@CuS with long-term cycling stability for rechargeable magnesium batteries [J]. Electrochimica Acta, 2022, 407.
23. KANDULA S, BAE J, CHO J, et al. Gram-scale synthesis of rGO wrapped porous α-Fe2O3 as an advanced anode material for Na-ion batteries with superior cyclic stability [J]. Composites Part B: Engineering, 2021, 220.
24. WU C, ZHENG J, LI J, et al. Fe3S4@reduced graphene oxide composites as novel anode materials for high performance alkaline secondary batteries [J]. Journal of Alloys and Compounds, 2022, 895.

25. FENG C, ZHANG L, YANG M, et al. One-Pot Synthesis of Copper Sulfide Nanowires/Reduced Graphene Oxide Nanocomposites with Excellent Lithium-Storage Properties as Anode Materials for Lithium-Ion Batteries [J]. ACS Appl Mater Interfaces, 2015, 7(29): 15726-34.
26. Hummers, W. S.; Offeman, R. E. Preparation of Graphitic Oxide. J. Am. Chem. Soc. 1958, 80, 1339−1339. [J].
27. LUO J, HU Y, XIAO L, et al. Synthesis of 3D flower-like Fe3S4 microspheres and quasi-sphere Fe3S4-RGO hybrid-architectures with enhanced electromagnetic wave absorption [J]. Nanotechnology, 2019, 31(8): 085708.
28. XU Q-T, LI J-C, XUE H-G, et al. Effective combination of FeS2 microspheres and Fe3S4 microcubes with rGO as anode material for high-capacity and long-cycle lithium-ion batteries [J]. Journal of Power Sources, 2018, 396: 675-82.
29. FU Z, JIANG T, LIU Z, et al. Highly photoactive Ti-doped α-Fe2O3 nanorod arrays photoanode prepared by a hydrothermal method for photoelectrochemical water splitting [J]. Electrochimica Acta, 2014, 129: 358-63.
30. XIA J, JIAO J, DAI B, et al. Facile synthesis of FeS2 nanocrystals and their magnetic and electrochemical properties [J]. RSC Advances, 2013, 3(17).
31. XIE D, CAI S, SUN X, et al. FeS/ZnS nanoflower composites as high performance anode materials for sodium ion batteries [J]. Inorganic Chemistry Communications, 2020, 111.
32. HEIFT D. Iron Sulfide Materials: Catalysts for Electrochemical Hydrogen Evolution [J]. Inorganics, 2019, 7(6).
33. GUO S P, LI J C, XIAO J R, et al. Fe3S4 Nanoparticles Wrapped in an rGO Matrix for Promising Energy Storage: Outstanding Cyclic and Rate Performance [J]. ACS Appl Mater Interfaces, 2017, 9 (43): 37694-701.
34. ZHANG S, LI H, WANG Z, et al. A strongly coupled Au/Fe3O4/GO hybrid material with enhanced nanozyme activity for highly sensitive colorimetric detection, and rapid and efficient removal of Hg(2+) in aqueous solutions [J]. Nanoscale, 2015, 7(18): 8495-502.
35. ZHANG W-M, WU X-L, HU J-S, et al. Carbon Coated Fe3O4Nanospindles as a Superior Anode Material for Lithium-Ion Batteries [J]. Advanced Functional Materials, 2008, 18(24): 3941-6.
36. CABANA J, MONCONDUIT L, LARCHER D, et al. Beyond intercalation-based Li-ion batteries: the state of the art and challenges of electrode materials reacting through conversion reactions [J]. Adv Mater, 2010, 22(35): E170-92.

AUTHOR INDEX

AUTHOR INDEX

AUTHOR INDEX